Scientific Protocols for Fire Investigation

As a 30-plus year fire investigator and lecturer, I have had the opportunity to read several books advertised as being resources for fire investigators. Of those, *Scientific Protocols for Fire Investigation*, is one of the best. John Lentini has blended his extensive experience as a scientist, practical fire investigator and a central figure in the improvement of our profession in creating a must-have resource for all fire investigators. *Scientific Protocols*, now in its third edition, continues to excel at covering a broad variety of topics that are scientifically sound, well-presented, legally compelling and interesting to read. While NFPA 921 offers sound guidance to fire and explosion investigators, *Scientific Protocols* weaves many of 921's principles together with actual examples from a variety of cases and investigations. It is unsurpassed in coupling sound scientific principles with real-life fire examples and delivers common-sense suggestions of how to apply them in fire investigations. This book should be required reading for all professional fire investigators and those seeking to broaden their knowledge of the field.

—Steve Carman, Carman Fire Investigations, Grass Valley, California

Scientific Protocols for Fire Investigation stands out as the crème de la crème on the topic of fire investigation. The book is easy to read and understand and serves as the most credible in the world-wide fire investigation community.

—Dennis J. Merkley CEO, Fire Facts Incorporated, Toronto, Ontario, Canada

Scientific Protocols is a must-have reference for any serious fire investigator or fire litigator. Lentini is at the forefront of fire science and that is reflected in his book.

—Stuart A. Sklar, Trial Attorney, Farmington Hills, Michigan

Lentini's brilliant monograph gives us a giant leg up in approaching the challenges of fire investigation.

—Dr. Bernard Cuzzillo, Fire Protection Engineer, Berkeley, California

Scientific Protocols for Fire Investigation is a text that all professional fire investigators must read.

—Wayne Chapdelaine, Fire Dynamics Analysts, Halifax, Nova Scotia, Canada

Battle lines have been drawn around methodology, qualifications, and the reliable application of the scientific method. John Lentini has been a leader in these battles. Lentini's *Scientific Protocols* promotes the use of the scientific method, identifies the required science for qualification, and provides examples of relevant applications. The enhanced third edition must be found on the bookshelves of any educated fire investigator.

—Douglas J. Carpenter, Vice President/Principal Engineer, Combustion Science & Engineering, Inc., Columbia MD

CRC SERIES IN PROTOCOLS IN FORENSIC SCIENCE

Norah Rudin and Keith Inman

Scientific Protocols for Fire Investigation, Third Edition
John J. Lentini

Scientific Protocols for Forensic Examination of Clothing
Jane Moira Taupin and Chesterene Cwiklik

Ethics in Forensic Science: Professional Standards for the Practice of Criminalistics
Peter D. Barnett

Principles and Practice of Criminalistics: The Profession of Forensic Science
Keith Inman and Norah Rudin

Scientific Protocols for Fire Investigation

Third Edition

John J. Lentini

CRC Press is an imprint of the
Taylor & Francis Group, an **informa** business

CRC Press
Taylor & Francis Group
6000 Broken Sound Parkway NW, Suite 300
Boca Raton, FL 33487-2742

Printed on acid-free paper

International Standard Book Number-13: 978-1-138-03701-4 (Hardback)
International Standard Book Number-13: 978-1-138-03702-1 (Paperback)

Library of Congress Cataloging-in-Publication Data

Names: Lentini, John J., author.
Title: Scientific protocols for fire investigation / John J. Lentini.
Description: Third edition. | Boca Raton, FL : CRC Press, Taylor & Francis Group, [2018] | Includes bibliographical references and index.
Identifiers: LCCN 2018016141 | ISBN 9781138037014 (hardback : alk. paper) | ISBN 9781138037021 (pbk. : alk. paper) | ISBN 9781315178097 (ebook)
Subjects: LCSH: Fire investigation.
Classification: LCC TH9180 .L46 2018 | DDC 363.37/65--dc23
LC record available at https://lccn.loc.gov/2018016141

Visit the Taylor & Francis Web site at
http://www.taylorandfrancis.com

and the CRC Press Web site at
http://www.crcpress.com

Dedication

This book is dedicated to the following citizens, for whom a second look at their fire made all the difference (indicates conviction vacated)*

Bruce Aslanian
Zeiden Salem Ammar
Joseph Awe*
Michele Owen Black
Nelson Brown
Kristine Bunch*
Lawrence Butcher
Barbara Bylenga
Sonia Cacy*
Paul Camiolo
Weldon Wayne Carr*
Rachel Casey
Donte Casson
Irma Castro
Maynard Clark
Melissa Clark
Andrew Currie
Kenneth and Ricky Daniels
Valerie and Scott Dahlman
Louis DiNicola*
Michael Dutko
William Fortner
Harold Fowler
James Frascatore
A. Stanley Freeman
David Lee Gavitt*
Ray Girdler*
Robert Gibson
Adam Gray*
Judith Slappey Gray
Robert Hancock
Jean and Stephen Hanley
Rebecca and Stephen Haun
James Hebshie*
Cecelia Hernandez
David and Linda Herndon
Robert Howell
James Hugney*
Terry Jackson
Danny and Christine Jewell
Eve and Manson Johnson
Thomas Lance
Han Tak Lee*
Gerald Lewis
Beverly Jean Long
Tonya Lucas*
Charles Martin
Amanda Maynard
Arturo Mesta
John Metcalf
Nafiz Muzleh
Joan Nellenbach
Pedro Oliva
Shawn Porter
Kimberly Post
Kazem Pourghafari
Ray Price
Bryan Purdie
Davey James Reedy*
Sasheena Reynolds
Daryl Rice
Karina and Juan Rojo
Victor Rosario*
Bruce Rothschild
Barbara Scott
Charles Schuttloffell
Lauren Shaw
Larry Sipes
Jermaine Smith
Terry Lynn Souders
George Souliotes*
Paul and Karen Stanley
Terri Strickland
Kevin Sunde
Louis Taylor
Kum Sun Tucker
Amaury Villalobos*
Jerome and Karen Vinciarelli
Michael Weber
Ernest Ray Willis*

Contents

Abbreviations and acronyms

American Association for the Advancement of Science (AAAS): publisher of *Science* magazine.
Academy of Forensic Sciences (AAFS): a professional organization of over 6,600 members that serves as an umbrella for forensic science certification and accreditation bodies. Publisher of the *Journal of Forensic Sciences*.
American Bar Association (ABA).
American Board of Criminalistics (ABC): an FSAB accredited certifying body that offers certification in various forensic science disciplines, including fire debris analysis.
ANSI-ASQ National Accreditation Board (ANAB): an organization offering third party accreditation to forensic science organizations to ISO 17020 and ISO 17025. Merged with ASCLD-LAB in 2016.
American National Standards Institute (ANSI): a standards development and accreditation organization, and the US representative to ISO.
Academy Standards Board (ASB): a standards development organization that is part of AAFS. ASB produces standards in areas of forensic science not covered by other standards development organizations (SDOs).
American Society of Crime Laboratory Directors (ASCLD).
American Society of Crime Laboratory Directors-Laboratory Accreditation Board (ASCLD-LAB).
American Society for Quality (ASQ): provides the quality community with training and professional certifications.
American Society for Testing and Materials (ASTM): now known simply as ASTM International. An ANSI accredited SDO with over 30,000 members that produces more than 12,000 voluntary consensus standards, including standards for forensic science.
Applied Technical Services, Inc (ATS): a multidiscipline testing and consulting firm where the author was employed from 1978 to 2006.
Bureau of Alcohol Tobacco Firearms and Explosives (ATF): an agency of the US Department of Justice (DOJ).
Building Officials and Code Administrators International (BOCA): an association of professionals employed in the establishment and enforcement of Building Codes. Publisher of the *International Building Code*.
British thermal unit (Btu): the amount of heat needed to raise one pound of water at maximum density through one degree Fahrenheit, equivalent to 1,055 joules.
Consolidated Fire and Smoke Transfer (CFAST): a zone model available from NIST.
Certified Fire and Explosion Investigator (CFEI): the principal certification offered by NAFI.
Certified Fire Investigator (CFI): the principal certification offered by IAAI.
Consumer Products Safety Commission (CPSC): an independent federal (US) regulatory agency.
Forensic Specialties Accreditation Board (FSAB): a corporation developed by AAFS that accredits certifying bodies in forensic science, including the IAAI certification programs and the ABC.
Fire Dynamics Simulator (FDS): a field model available from NIST.
GC-FID: gas chromatography with a flame ionization detector.
GC-MS: gas chromatography with a mass spectral detector.
Heating Ventilating and Air Conditioning (HVAC).
Heavy Petroleum Distillate (HPD).
International Association of Arson Investigators (IAAI): a professional organization of more than 9,000 individuals. IAAI offers training certification for fire investigators.
International Electrotechnical Commission (IEC): an SDO that publishes international standards for electrical, electronic and related technologies.
International Organization for Standardization (ISO): an international standards development organization.
Ignitable Liquid Residue (ILR).
Lower Explosive Limit (LEL).
Lower Flammable Limit (LFL). (The same as LEL).
Light Petroleum Distillate (LPD).
Medium Petroleum Distillate (MPD).

Material Safety Data Sheet (MSDS): OSHA changed the term in 2012 to SDS (Safety Data Sheet). Safety Data Sheets have more specific content and formatting requirements.
National Association of Fire Investigators (NAFI): a professional organization that offers training and certification for fire investigators.
National Fire Incident Reporting System (NFIRS): a standard reporting system overseen by the US Fire Administration.
National Fire Protection Association (NFPA): an ANSI accredited SDO that produces standards, recommended practices, codes, and guides for the fire service. Publisher of NFPA 921, *Guide for Fire and Explosion Investigations* and NFPA 1033, *Standard for Professional Qualifications for Fire Investigator.*
National Institute of Justice (NIJ): the research arm of the United States Department of Justice (DOJ)
National Institute of Standards and Technology (NIST): formerly the National Bureau of standards. An agency of the US Department of Commerce, and home of OSAC.
Overfilling Protection Device (OPD).
Organization of Scientific Area Committees (OSAC): an organization sponsored by NIJ and NIST devoted to identifying and promulgating voluntary consensus standards in the forensic sciences, including fire debris analysis and fire and explosion scene investigation.
Occupational Safety and Health Administration (OSHA): agency of the US Department of Labor.
Propane Education and Research Council (PERC).
Polyvinylchloride (PVC): a thermoplastic used in plumbing and electrical insulation.
Scientific Area Committee (SAC): five such committees exist in OSAC. Each committee has between four and seven subcommittees, one for each forensic science discipline.
Science Advisory Workgroup (SAW)
Standards Development Organization (SDO)
Scientific Fire Analysis, LLC (SFA): the author's consulting firm.
Society of Fire Protection Engineers (SFPE): publisher of the *SFPE Handbook*. Publisher, with NFPA, of *Fire Technology.*
Thermal Cutoff (TCO)
Upper Explosive Limit (UEL).
Upper Flammable Limit (UFL) (The same as UEL).
Underwriters Laboratories (UL): a private SDO that "lists" products that meet the requirements of its standards. UL also conducts fire research.
Volts Alternating Current (VAC).
Volts Direct Current (VDC).

Acknowledgments

This book is a product of 45 years of experience in forensic science and fire investigation. During that time, I have been privileged to work with colleagues and clients who have shaped my thinking, guided my career in interesting directions, and made it possible for me to make a living while continuing along a path of lifelong learning. It is appropriate to offer my thanks here for the wisdom, guidance, and time these people have provided.

My original mentors in the forensic science profession, Larry Howard, Byron Dawson, Roger Parian, and Kelly Fite, made me acutely aware of the importance of doing good work and the consequences of error. My colleagues from the various standardization bodies, the IAAI Forensic Science Committee, the American Board of Criminalistics, ASTM Committee E30 on Forensic Sciences, the National Fire Protection Association (NFPA) Technical Committee on Fire Investigations, the Organization of Scientific Area Committees (OSAC) Subcommittee on Fire and Explosion Investigations, and the Texas Fire Marshal's Science Advisory Workgroup, have all been instrumental in specifically improving my work product and, more generally, that of our profession. Attending meetings with these dedicated professionals has provided me with some of my most valuable learning experiences. These leaders of the forensic science and fire investigation professions who have motivated and inspired me include Andy Armstrong, Susan Ballou, Peter Barnett, Craig Beyler, Doug Carpenter, Dan Churchward, John Comery, Chris Connealy, Dick Custer, Peter De Forest, Julia Dolan, Jim Doyle, Itiel Dror, Mary Lou Fultz, Kathy Higgins, Max Houck, Ron Kelly, Thom Kubic, Dale Mann, Reta Newman, Bud Nelson, Larry Presley, Tony Putorti, Carlos Rabren, Rick Roby, Marie Samples, Carl Selavka, Ron Singer, Denny Smith, Rick Strobel, Rick and Kary Tontarski, Peter Tytell, Chris Wood, and David Smith, one of the finest investigators and human beings it has been my privilege to know.

I owe much to my collaborators on various research and teaching projects conducted over the years: Andy Armstrong introduced me to the benefits of diethyl ether as a solvent; Richard Henderson joined me in the Oakland study; David Smith also came to Oakland and coordinated the Tucson mobile home experiment described in Chapter 3.

Laurel Mason, Mary Lou Fultz, Julia Dolan, Eric Stauffer, and Cheryl Cherry all participated in projects related to obtaining a deeper understanding of fire debris analysis.

I am also indebted to the fire investigators and engineers whose work I have admired and who have shared their experiences and insights with me, including Steve Avato, George Barnes, Vyto Babrauskas, Steve Carman, Andrew Cox, Mick Gardiner, Mark Goodson, Dan Gottuk, Dan Hebert, Dan Heenan, Ron Hopkins, Gerald Hurst, David Icove, Rick Jones, Pat Kennedy, Daniel Madrzykowski, Jack Malooley, Michael Marquardt, Jamie Novak, Robert Schaal, Richard Vicars, Lee West, and the members of the NFPA Technical Committee on Fire Investigations.

I would never have been able to collaborate with these esteemed colleagues were it not for the support of my clients, who paid the bills, referred me to their friends, and provided me with the constructive criticism so necessary if one is to succeed in this profession. Among this group are Linda Ambrose, Charlie Arnold, Dave Bessho, Larry Bowman, Ken Burian, Archie Carpenter, Al Dugan, Mike Dutko, David Eliassen, Buck Fannin, Bob Gallantucci, Peter Goldberger, Doug Grose, Clark Hamilton, Peter Hart, Grant Law, Bob Lemons, Arnie Levinson, Jeff McConnaughey, Suzanne Michael, Dan Mullin, Al Nalibotsky, Lloyd Parker, Ed Pihl, Stuart Sklar, Kevin Smith, Kevin Sweeney, Wayne Taylor, Ty Tyler, Karl Vanzo, Peter Vogt, Don Waltz, Joe Wheeler, Ken White, Pamela Wilk, and the man who put my children through college and has become a good friend as well as a client, Mike McKenzie. His ethics and his expertise as a fire litigator are a credit to the legal profession.

My work has attracted the attention of several outstanding journalists and documentarians, who have helped bring public attention to the problems in fire investigation and allowed me to address a wider audience. These include Joe Bailey, Radley Balko, Tom Berman, Jessie Deeter, David Grann, Wendy Halloran, Randi Kaye, Dave Mann, Steve Mills, Steve Mims, Maurice Possley, Liliana Segura, and Scott Simon.

I also owe a debt of gratitude to my colleagues and the staff at Applied Technical Services (ATS), where I practiced for 28 years prior to "retiring" to my one-man consultancy. The late James F. Hills thought he saw some potential and gave me a job. He supported my professional development as well as the development of my fire investigation practice. His son, Jim J. Hills, continued that support. I have had the tremendous advantage of working under the same roof with expert chemists and metallurgists such as Phil Rogers, Semih Genculu, James Lane, and Bob Wiebe, who have helped extend the range of issues that I could address. Jeff Morrill has been my friend and colleague since 1990, and I thank him for relieving me of the necessity to investigate vehicle fires. Dick Underwood was my electrical engineering consultant for the entire time I worked at ATS and continues to provide guidance today. I learned about electricity from him, and he learned about fire (and grammar) from me. He graciously agreed to provide an expert review of Chapter 6. Leslie Macumber and Daniel Schuh provided invaluable assistance in preparing the artwork for this text.

I also want to acknowledge Doug Carpenter for his insights and review of Chapters 2 and 3, Steve Carman for his review of Chapter 3, and Julia Dolan for her meticulous review of Chapter 5. This level of review is exceedingly time consuming, especially for people with other demands on their time. These reviews were critical to my ability to cover these highly technical subjects. Mark Goodson and Rich Vicars contributed significantly to Chapter 6. I also want to thank my editors Norah Rudin and Keith Inman for their continuous encouragement, for the many hours of work they put into the production of this text, and for helping me write the book I wanted to write. My special thanks go to Becky McEldowney, my first editor at Taylor & Francis Group, for taking the bold step of allowing the text to be printed in full color, making it the first full-color text on fire investigation. Publishing this text in color has prompted other fire textbook authors to follow that example.

Introduction

Fire investigation is a forensic science that is a world unto itself. The investigator who ventures here risks exposure not only to a unique and possibly dangerous physical environment but also to scientific, professional, and personal challenges not found in any other field of forensic science. Scientists from traditional forensic science laboratories may feel overwhelmed and unprepared for the analytical challenges, most of which makes mastery of a laboratory analysis seem like child's play. At first blush, it seems like nothing we learned in our study of chemistry or physics can explain the chaos that presents itself after nearly every fire, but with patience, practice, and a careful application of the scientific method, the truth can usually be teased out of the ashes.

Unlike a homicide, a robbery, or almost any other incident that requires investigation, a fire is unique in that the first major task, and often the most daunting one, is to determine whether a crime has been committed. Few other fields of investigation exist in which this is true. Unexplained death comes to mind, but in that case, a clearly defined set of protocols, the forensic autopsy, exists that will usually resolve the question unequivocally. The medical examiner analogy is a useful one because the fire investigator is called upon to perform a forensic "autopsy" of a structure or vehicle to determine the cause of the fire. The deviations from the analogy, however, are what make it interesting.

The medical examiner performing the autopsy has an undergraduate degree, usually in a natural science, a four-year medical education, and several more years of internship and residency in pathology or forensic medicine. The fire investigator, on the other hand, may have no education beyond high school. His training typically consists of a 40-hour "basic arson" school, followed by an 80-hour "advanced arson" school, and perhaps some continuing education taught by people with the same training and more experience. Certainly, many skilled fire investigators can and do perform careful, science-based investigations, even without the benefit of formal scientific training, but that is unfortunately not the rule. While there exists no standardized curriculum for the education of criminalists, there is at least an expectation of an undergraduate degree in a physical science.

The lack of any established curriculum for fire investigation is problematic. An investigator may possess superb critical thinking and problem-solving skills, but those skills are of no use if the basic facts of chemistry and physics are absent from the investigator's toolbox. In modern practice, such gaps can be exploited to the point where the investigator's work is no longer presentable. The requirements of NFPA 1033 explicitly disqualify an investigator who lacks scientific knowledge.

Because society asks fire investigators to perform both scientific functions and law enforcement functions, the scope of work of fire investigators tends to be much broader than that of the medical examiner or other forensic scientist. While determining the origin and cause of the fire is the investigator's primary task, he[1] is frequently charged with "putting it all together." In assuming the role of *principal investigator,* the fire investigator is responsible for making sure that all of the data make sense in the context of the hypothetical fire scenario. Medical examiners and many other forensic scientists properly focus on narrow issues, but in fire investigation, a narrow focus can lead to important data being overlooked. It is important, however, that the investigator "put it all together" in the correct order. Prejudgments based on "investigative" findings that are not relevant to the fire's behavior must not cloud the investigator's scientific work.

Methodology is another area where the analogy to a medical examiner breaks down. The methodology of the medical examiner is likely to be very predictable in that he or she will follow a written, peer-reviewed protocol. The methodology of the fire investigator, on the other hand, depends almost entirely on the individual investigator and by whom he or she is employed. Until 2000, a constant, and sometimes fierce, debate took place on what standards, if any, fire investigators should be held to; on whether fire investigation is an art, a science, or a mixture of both; and on the level

[1] Throughout this text, the masculine pronoun applies with equal effect to the feminine (but it is a fact that the fire investigation profession is overwhelmingly male).

of training and certification required to perform this difficult job. The Supreme Court's decision in *Kumho*, followed by the International Association of Arson Investigators' (IAAI's) and the U.S. Department of Justice's (USDOJ's) endorsement of NFPA 921, basically settled the debate about standards, and the courts have followed suit.

The professional environment in which the medical examiner practices tends to be collegial. Doctors tend to treat other doctors with respect. If there is a difference of opinion about the cause of death, they will make an effort to resolve the issue. Medical associations and medical boards promulgate standards and regulate practice, and those organizations are generally held in high regard. If two fire investigators disagree about the cause of a fire, however, the issue is usually left for the court to resolve (as if judges and juries are better at recognizing good science than the investigators). Failing to recognize the difference between a personal and a professional opinion, two investigators may just agree to disagree, saying that "everyone is entitled to their own opinion," as if it is acceptable that one of them is wrong (and in most cases, at least one of them is). The oft-heard refrain is, "I have to call it like I see it."

Fire investigators may (or may not) belong to one or more professional organizations, and those organizations have historically been reluctant to admit that any standards even existed. In 1997, the largest of these organizations, the IAAI, filed an *amicus curiae* brief in the case of *Michigan Millers v. Benfield* (and later in *Kumho v. Carmichael*), which argued, in effect, that fire investigators should not be held to *Daubert* standards because fire investigation is a "less scientific" discipline. Of course, the Supreme Court ruled unanimously against that misguided proposition and, turning the IAAI's argument on its head, stated further that the "less scientific" disciplines required even *more* scrutiny under *Daubert*. Fortunately, the recent leaders of the IAAI have embraced the scientific approach to fire investigation, and the organization is now at the forefront of bringing the profession up to speed. The fire investigation profession continues to advance a scientific approach, despite the dwindling rear-guard's opposition. As the great scientist Max Planck quipped, "Science advances one funeral at a time." How is it possible that an individual with no formal scientific training and no certification, filled with misconceptions about the phenomenon in which he professes to have expertise, gets to opine before a jury on issues of life and death? More important, what can be done about it?

This situation arose, quite simply, by default. Somebody has to do the work. For a variety of reasons,[2] forensic scientists, with few exceptions, have left the field of fire scene investigation to the nonscientists. They have been content to participate in the modest task of determining whether a sample of debris contains ignitable liquid residue. While reliable chemical analysis is important, the bulk of hypothesis formation and testing (when the investigator chooses to follow the scientific method) takes place in the field, in the dark, dirty, smelly, burned-out hulks of former residences, offices, and factories. A main purpose of this volume is to encourage interested scientists and engineers to overcome their natural aversion to disorder and to bring their scientific talents and knowledge to a field sorely in need of these resources. Of equal importance, this book encourages fire investigators to recognize that they participate in our justice system as forensic scientists.

This text begins, as all good science books do, with a retrospective of the discipline. It demonstrates that the history of science and the history of fire investigation are inextricably intertwined and have been since the time when people believed in only four elements: earth, air, fire, and water. The development of modern chemistry, physics, and fluid dynamics is intimately related to the understanding of the phenomenon of fire. There is also a discussion on the evolution of a standard of care for fire investigation and the necessity for changes in our training infrastructure.

The second chapter discusses the basic chemistry and physics of fire. The concepts of work, energy, power, flux, and temperature are presented in a manner that I hope will be comprehensible even to individuals without much training (or memory of their training) in science. This chapter is basically a "refresher" of high school-level chemistry and physics. Since 2009, changes in the minimum requirements of NFPA 1033, *Standard for Professional Qualifications for Fire Investigator*, have brought an end to the days when a fire investigator could claim that such basic knowledge is unnecessary.

Chapter 3 lays out a short description of the science of fire dynamics and explores how things burn and how they interact with their surroundings while doing so. The concepts of flammability and fire pattern development and the

[2] One of the primary factors that prevent properly educated scientists and engineers from becoming fire investigation is the low pay. State and local governments, strapped for cash, would sooner promote a firefighter or police officer into fire investigation than pay a college graduate with a science degree a reasonable salary.

development of fire modeling as a tool for fire investigation are presented. The oversimplification presented to many juries, "heat rises," is supplemented with an explanation of what happens when that heat encounters an obstruction, such as a ceiling, and how that might affect the artifacts left after a fire. Common misconceptions about fire dynamics are covered, as are the application and misapplication of computerized fire modeling to the understanding of fire behavior. The development and refinement of this valuable engineering tool have resulted in fire protection engineers developing an interest in fire investigation and reconstruction. Unfortunately, like any tool, modeling is subject to misuse and abuse, so cautions are provided.

Chapter 4 proposes a practical procedure for conducting fire scene inspections. This starts with understanding the purpose of the investigation and includes an understanding of the mindset required to optimize the chance for a correct determination. Evidence documentation and collection, damage assessment, hypothesis formation and testing, and methods of reporting and record keeping are discussed. Recent work that highlights what may be a startlingly high error rate in origin determination is discussed.

Chapter 5 discusses the laboratory examination of fire debris to test for the presence of ignitable liquid residues. This chapter is written for fire debris analysts, and the portions beyond the introduction will likely be difficult to comprehend for anyone without at least an undergraduate understanding of organic chemistry. Once the development of the field has been absorbed, nonchemists should feel free to skip over the balance of this chapter.

Chapter 6 describes in some detail the laboratory examination of potential ignition sources and the testing of ignition scenarios. Examples of the common systems and appliances that are regularly examined in a fire laboratory are provided. Common failure modes for each of the appliances and systems are discussed so that hypotheses regarding this evidence can be formulated and tested. This chapter is where the photographs begin in earnest. Specific indicators of common failure scenarios are provided, and there is a new section provided by Richard Vicars describing printed circuit board fires and low voltage ignition sources. Fires involving lithium ion batteries and metal oxide varistors are also discussed for the first time in this edition.

Chapter 7 recounts 30 fires where the principles set forth in the preceding chapters are applied to actual cases. These are all cases either investigated or reviewed by me. Most of these cases provide a lesson in how to (or how not to) conduct an investigation.

Chapter 8 traces the evolution of the mythology of arson investigation and the recent acceptance by many conscientious investigators that what they learned, taught, and testified to was just plain wrong. I hope that readers will come away with an understanding of how the myths originated and why they have persisted. Some examples of the impacts of these myths on real cases are also provided.

Chapter 9 describes seven common root causes of errors in fire investigation as well as the steps one can take to avoid those errors. It provides a method for evaluating the work product of fire investigators. Finally, there are a few horror stories about investigations gone terribly wrong, with descriptions of how those faulty investigations affected the lives of real people, and analyses of the errors that led to false conclusions.

In the final chapter, the professional practice of fire investigation is discussed. This includes a description of a quality assurance program, business practices, and the fundamentals of being an expert witness. This chapter includes practical advice for giving testimony in depositions and at trial.

Whenever possible, I attempt to use real cases involving real structures and real people to illustrate points. This book is not just about catching arsonists. It is about finding answers. The most interesting fires, and often the most challenging ones, are frequently not incendiary fires but accidents. Accidental fires are far more likely to result in litigation than incendiary fires. This is not only because the financial stakes tend to be much higher than in arson cases but also because of the trend in the insurance industry, which supplies most of the dollars to hire the investigators in the private sector, toward an attempt to recover insured losses from third-party manufacturers or service providers. The issues in civil cases surrounding fires are frequently more complicated than those in the criminal cases, although many civil trials are merely arson trials with a less demanding burden of proof.

To the extent possible, the focus of this text is on the practical application of scientific principles to the practice of fire investigation. A certain amount of theoretical exposition is required, but if the reader is looking for a deep understanding of the theoretical basis of ignition or fire spread or modeling, that will be found by reading the references.

The goal here is to acquaint the reader with the existence and the significance of the central concepts necessary to accurately determine the origin and cause of a fire or to critically review the work of another investigator. I focus on how to approach the fire scene, the steps to determine the fire's cause, and how to present that determination in a way that is meaningful to clients.

A word about the investigator's mindset is appropriate here. Many fire investigators come from a firefighting background. The experience of observing the behavior of real fires can be quite valuable, but the skill set and the mindset of firefighting are quite different from that required for fire investigating. When fighting a fire, there is not a lot of time for critical thinking. Training is often the key to successfully extinguishing the fire and surviving the experience.

Firefighters are successful almost all of the time. By the time firefighters leave the scene, the fire is extinguished. On the other hand, when a fire investigator leaves the scene, there is a significant probability that the cause of the fire will not have been determined. Investigators must learn to accept this possibility. Sometimes, "undetermined" is the only answer that the evidence will permit. This failure to accomplish the assigned task is something to which firefighters are unaccustomed, but if they are to make the transition to becoming successful investigators, their goal should be adjusted from "always successful" to "usually successful."

Certain categories of fire are beyond the scope of this text quite simply because they are beyond the scope of the author's expertise. Vehicle fire investigations were handled by a trusted associate. Wildland fire investigations have only been a minor part of our practice. The focus of this text is entirely on fires in residential, commercial, and industrial structures.

This text relies heavily on documents that this author believes represent the *standard of care* in fire investigation. The perspective presented was gained by participating in the development of these documents as well as by observing first-hand the results of investigations where the principles they espouse were ignored. The practice of fire investigation, and the standards that guide that practice, have evolved considerably since the first edition of this book. Throughout the 1990s, there was a focus on persuading investigators that fire does not always burn upward and outward. More recently, the focus has shifted to understanding the effects on ventilation on fire behavior and fire patterns. There are times when it seems that the more we learn about fire, the less we know.

Fire investigations have many possible outcomes, and a common outcome is the filing of civil or criminal litigation. All fire investigations, therefore, should proceed according to the best possible science. Good science should necessarily mean that the investigation is of "litigation quality." Whether the investigation leads to the settlement of an insurance claim or to a civil or criminal courtroom, the stakes are high. The accurate determination of fire cause is a prerequisite if the courts are to avoid miscarriages of justice, which have been disturbingly numerous. (See the dedication of this text for a list of examples.) Life, liberty, and serious dollars are at stake in any large fire loss. The intent of this book is to promote greater use of the scientific method in fire cause determination and the entry of more forensic scientists and engineers into the field, with the ultimate purpose of providing the accurate determinations that our stakeholders—our clients, our employers, the courts, and the public—have every right to expect.

John J. Lentini, CFI, D-ABC
Islamorada, Florida

Author

John J. Lentini has been at the center of most of the important developments in fire investigation for the past 35 years. He began his career at the Georgia Bureau of Investigation Crime Laboratory in 1974. There he learned criminalistics in general and fire debris analysis in particular and received an introduction to fire scene investigation. He went into private practice in 1977 and spent the next 10 years working 100 to 150 fire scenes per year, mostly for insurance companies that had doubts about the legitimacy of their insureds' fire losses. At the same time, he managed a fire debris analysis laboratory with a nationwide clientele.

He has been at the center of the standardization of both laboratory and field investigations of fires. As co-chair of the International Association of Arson Investigators' (IAAI's) Forensic Science Committee, he was the principal author of the first laboratory standards published by the IAAI in 1988. He oversaw the acceptance of those standards by ASTM Committee E30 on Forensic Sciences, where he served for two terms as chair of the Criminalistics Subcommittee and for three terms as main committee chair.

When the American Board of Criminalistics (ABC) was incorporated in 1993, he was the first civilian elected to the ABC board of directors. He served two terms on the board; was the principal organizer of the ABC *Operations Manual*; and served as the first editor of the ABC newsletter, *Certification News*. He was a co-author of the examination for the certification of fire debris analysts.

Lentini has been a certified fire investigator since certification first became available and was among the first group of individuals certified by the ABC as fellows in fire debris analysis. He is one of the few individuals in the world who holds certifications for both laboratory and field work.

He has been a contributor to the development of National Fire Protection Association (NFPA) 921, *Guide for Fire and Explosion Investigations,* and has been a member of the NFPA's Technical Committee on Fire Investigations since 1996. On that committee, he has been instrumental in the development of committee positions on accelerant detection canines, the concept of the "negative corpus," and the acceptance of the scientific method.

When the National Institute of Standards and Technology (NIST) and the National Institute of Justice (NIJ) set up the Organization of Scientific Area Committees (OSAC) in 2013, he was one of the first fire investigators invited to serve. In 2016, he joined the Texas Fire Marshal's Science Advisory Workgroup, which is the first agency to conduct retrospective case reviews, a practice endorsed by the IAAI but, so far, adopted only in Texas. In 2015, he was selected as the Society of Fire Protection Engineers *Person of the Year* for his work in advancing the fire investigation profession and helping to prevent or reverse many wrongful arson convictions.

Lentini is the author of more than 30 publications, both in the peer-reviewed literature and in the trade journals of fire investigation, the insurance industry, and the legal profession. His study of the Oakland Hills fire in 1991 resulted in a rethinking of much of the conventional wisdom in fire investigation, and his laboratory work has resulted in research papers that are standard works in the field. Lentini has personally conducted more than 2,000 fire scene inspections and has been accepted as an expert witness on more than 200 occasions. He is a frequent invited speaker on the subjects of the standard of care in fire investigation, laboratory analysis of fire debris, and the progress of standardization in the forensic sciences.

After managing the fire investigation department at Applied Technical Services in Marietta, Georgia, for 28 years, he moved to the Florida Keys in 2006 and now provides training and fire investigation consultation, doing business as Scientific Fire Analysis. He offers a three-day course based on the contents of this book.

Mr. Lentini can be contacted via e-mail at scientific.fire@yahoo.com. His website is www.firescientist.com.

CHAPTER 1

Fire and science

There is no more open door by which you can enter into the study of natural philosophy than by considering the physical phenomena of a candle. There is not a law under which any part of this universe is governed which does not come into play, and is not touched upon, in these phenomena.

—Michael Faraday
The Chemical History of a Candle, 1848 and 1860

LEARNING OBJECTIVES

After reviewing this chapter, the reader should be able to:

- Appreciate the relationship between science and the study of fire
- Understand the history of various hypotheses that were used to explain fire
- Identify the major scientists who advanced the study of fire
- Understand the development and importance of NFPA 921 *Guide for Fire and Explosion Investigations* and NFPA 1033 *Standard for Professional Qualifications for Fire Investigator*
- Understand the roles of agencies and individuals involved in fire investigation

1.1 INTRODUCTION

For the forensic scientist trying to determine whether the investigation of fires is compatible with his or her scientific training, or for the fire investigator trying to determine whether science has an appropriate place in the investigation of fire scenes, the Christmas lectures of Michael Faraday to the Royal Institution in London provide the answer. Fires, as chaotic and irreproducible as they may appear at first glance, can be understood through careful observation (Figure 1.1) and the application of fundamental scientific principles. Fire investigations, conducted according to the established rules set forth in the scientific method, can be expected to yield determinations that not only are logically defensible but also, at least to a first approximation, are true. While holding out the promise that science can explain fire, Faraday's observation also means that fire is an extremely complicated phenomenon and therefore difficult to understand and explain.

1.2 ARGUMENT VERSUS EXPERIMENT

For the first few millennia of civilization, there may have been scientific processes taking place, but this method of finding the truth was not articulated as such. Argument, the method of Aristotle, was the chief method of finding what then passed for the truth. The methodology of the ancient Greeks espoused rigorous logic in making arguments, but unfortunately, Aristotle did not have much use for experiment. He was the

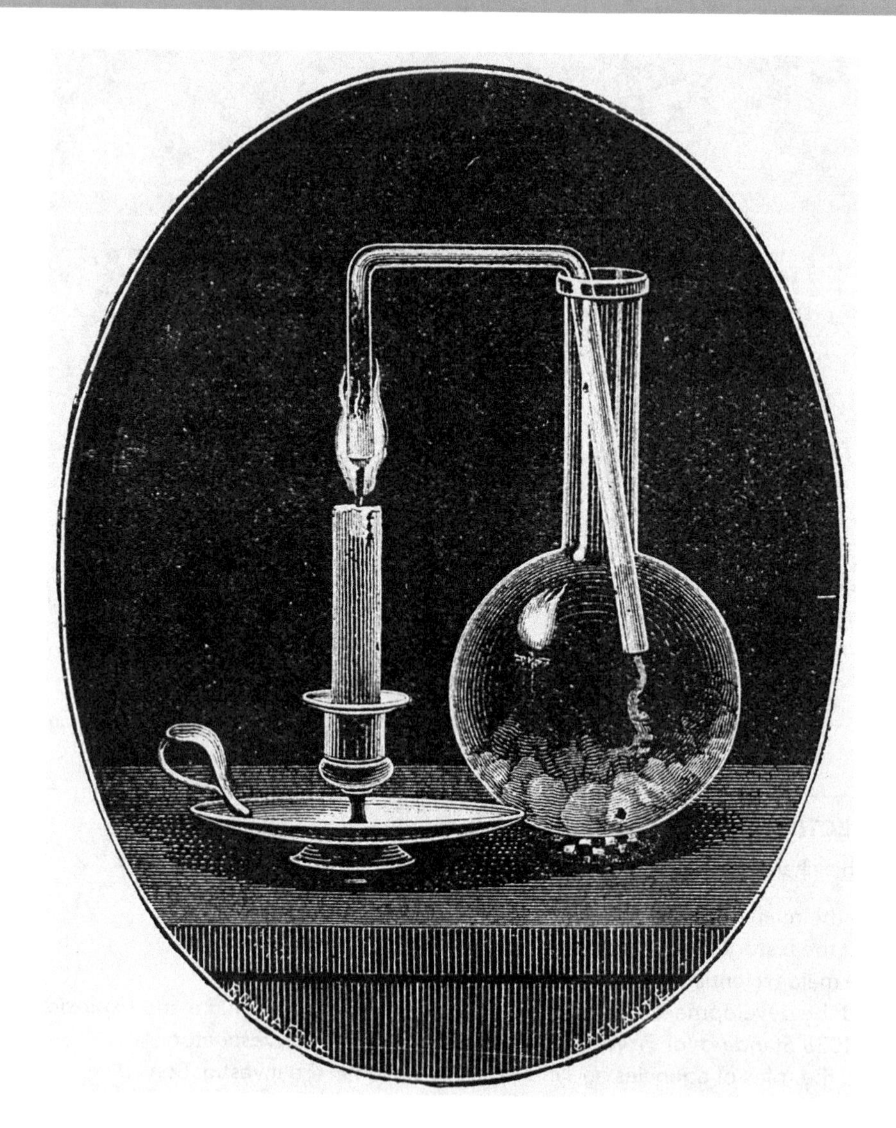

Figure 1.1 Faraday's apparatus for investigating the different parts of a candle flame. An ignited candle in a holder is next to a glass flask; a glass tube extends from the flame of the candle through the neck and into the globe of the flask. (Printed in Mendeleev's *History of Chemistry*, 1897. Courtesy of the National Library of Medicine, Bethesda, MD.)

original proponent of the *ipse dixit*[1] school of expertise, which until only recently has held sway not only in fire investigation but in many of the other forensic sciences as well.

Scientific methodology and the utility of experiments were not widely accepted until the Renaissance. One of the first "scientists" of the Renaissance was Albertus Magnus, a thirteenth-century Dominican friar who understood that the first steps toward knowing nature are observation, description, and classification. Influenced by Arab philosophers who had rediscovered classical Greek knowledge, he traveled through Europe asking questions of fishermen, hunters, beekeepers, bird catchers, and craftspeople. He produced two books, one on zoology and one on botany.

The universality of the scientific method was recognized by Roger Bacon in his *Opus Maius*, describing the activities of Peter of Maracourt, who was famous for his work on magnetism.

[1] *Ipse dixit* is Latin for "he himself said it." It is used in the legal world to refer to something that is asserted but unproved. Loosely translated, it means "it is so because I say it is so." The US Supreme Court, in *GE v. Joiner* (1997), disparaged the use of this logic. The Court stated that "nothing in either *Daubert* or the Federal Rules of Evidence requires a district court to admit opinion evidence that is connected to existing data only by the *ipse dixit* of the expert. A court may conclude that there is simply too great an analytical gap between the data and the opinion proffered."

What others strive to see dimly and blindly, like bats in twilight, he gazes at in the full light of day because he is a master of experiments. Through experiment he gains knowledge of natural things medical, chemical, and indeed of everything in the heavens or earth [1].

By the end of the sixteenth century, the scientific method gained more acceptance as a result of the work of Copernicus, Galileo, and Kepler. The old scholastic method of judging a theory by how well it could be argued failed miserably when faced with new scientific data. In 1620, Francis Bacon, in a major work called *The New System* (because it replaced Aristotle's old system), wrote

> Argumentation cannot suffice for the discovery of new work, since the subtlety of nature is greater many times than the subtlety of argument [2].

Bacon proposed a data management process with four essential components that would bring into being our modern view of knowledge: finding, judging, recording, and communicating. Bacon's new system was followed shortly by Rene Descartes's *Discourse on Method,* in which he set out rules for seeking certainty. The secret lay in what he called "methodical doubt," by which all things except "self-evident truths" were to be questioned until they had proved themselves true. Bacon's empiricism and Descartes' methodical doubt produced a new investigative technique, the adherents of which formed themselves into academies. The first, the Accademia del Cimento, opened in Rome in 1657 with the motto, "Test and test again." Similar societies sprang up all over Europe. On November 28, 1660, a group of 12 men, including Christopher Wren, Robert Boyle, and Robert Hooke, met at Gresham College, London, to set up "a Colledge for the promoting of Physico-Mathematicall Experimentall Learning." This "colledge" became the Royal Society of London for Improving Natural Knowledge. The Royal Society's purpose was to learn about natural philosophy by having an experiment repeated before a number of society members in an act called "multiple witnessing." The Society's motto *Nullius in verba,* loosely translated, means "take nobody's word for it"—the opposite of *ipse dixit.* In his 1667 *History of the Royal Society,* Thomas Spratt explained the value of the new experiments:

> Transgression of the law is idolatry:[2] The reason of men's condemning all jurisdiction and power proceeds from their idolizing their own wit. They make their own prudence omnipotence; they suppose themselves infallible; they set up their own opinions and worship them. But this vain idolatry will inevitably fail before experimental knowledge, which is an enemy to all manner of false superstitions, especially compared to that of men's adoring themselves and their own fancies [3].

1.3 FIRE AND THE ENLIGHTENMENT

Fire was one of the earliest phenomena subjected to and at least partly understood as a result of the experimental method. Until the eighteenth century, fire was thought to be a substance unto itself, one of the four "elements" described by the ancient Greeks. Greek philosophers believed in the four elements of earth, air, water, and fire. Aristotle proposed that elements also contained two of the following qualities: heat, cold, moisture, and dryness. For example, fire was hot and dry, water was cold and moist, air was hot and moist, and earth was cold and dry. Aristotle was the teacher of the alchemists. Through them, his ideas dominated chemistry for nearly 2,000 years.

The next explanation of fire to come along was the *phlogiston* theory, proposed by two German physicians, university professors Johann Becher and Georg Stahl. Phlogiston explained how air initially supports combustion and then does not. It also addressed some of the shortcomings of Aristotle's theory, particularly its vague notions of chemical change. Becher proposed that *terra pinguis* (fatty earth) was present in all flammable materials; this substance was given off during burning, and the resulting ash was the true material. Stahl proposed the term *phlogiston* from the Greek word *phlogistos* ("burning"). Stahl believed that living matter contained a soul differing in composition from nonliving matter (the vitalism theory held sway through the mid-nineteenth century). Stahl outlined his medical theories in *The True Theory of Medicine* (1708), and the book had great influence throughout Europe. The Becher/Stahl theory explained burning, oxidation, calcination (metal residue after combustion), and breathing in the following way:

- Flames extinguish because air becomes saturated with phlogiston.
- Charcoal leaves little residue upon burning because it is nearly pure phlogiston.
- Mice die in airtight spaces because the air is saturated with phlogiston.
- When calx (metal oxide) is heated with charcoal, metals are restored because phlogiston is transferred from charcoal to calx.

---

[2] *Idols* are false beliefs that today would be described as "cognitive biases."

Becher and Stahl derived these conclusions outside the laboratory, but other scientists in laboratories were finding that metals such as magnesium *gained weight* during combustion. If phlogiston is given off when a metal forms a calx, why does the calx weigh more than the metal? Stahl attributed the weight increase to air entering the metal to fill the vacuum left after the phlogiston escaped.

Joseph Priestley, a true believer in the phlogiston theory (Sidebar 1.1), discovered oxygen in 1774[3] and named it "dephlogistocated air." A copy of the cover of his treatise on "Different Kinds of Air" is shown in Figure 1.2. He subjected mercuric oxide to sunlight in a closed container and obtained "air" that he said was five or six times as good as common air. It allowed candles to burn longer and mice to breathe longer when placed in a bell jar.

Using the scientific method, Antoine Lavoisier (Figure 1.3), a French "natural philosopher" and a correspondent of Priestley, began to study combustion in 1772. Lavoisier's interest in combustion was fueled by his natural curiosity, but it was also job related. He proposed a method to improve street lighting in Paris and was appointed to the National Gunpowder Commission. It was in the latter role that he was able to equip a state-of-the-art laboratory at the Paris Arsenal. One of the most impressive pieces of equipment in Lavoisier's laboratory was a balance he invented, a balance capable of measuring changes as small as half a milligram. This level of precision was required to measure the change of weight in small volumes of gases.

Upon observing the increase in weight of some substances, particularly metals, when burned, Lavoisier was required to postulate a "negative weight" for some phlogiston and a "positive weight" for other phlogiston. Although perfectly capable of holding two competing thoughts in his head at once, this was a position he could not justify. As a result of his having to discard the phlogiston theory, Lavoisier went on to discover (or at least come very close to understanding)

SIDEBAR 1.1 The phlogiston theory of combustion

Phlogiston was described as a weightless or nearly weightless elastic (compressible) substance that was released into the air on burning. Phlogiston was thought to be a major constituent of good fuels. Candles and coal were thus considered nearly pure phlogiston. Sand and rock contained no phlogiston and therefore would not burn. Water, which extinguished fires, had negative phlogiston.

Substances were described in terms of their phlogiston content. Metals were considered rich in phlogiston. Earth was considered poor in phlogiston because it would not burn. Gases were described in terms of their phlogiston content or their burning characteristics. The following observations were thought to support the theory of phlogiston:

- Combustibles lose weight when they burn because of the loss of phlogiston.
- Fire burns out in an enclosed container because of the saturation of the air with phlogiston.
- Metal oxides (called calxes) are reduced to native metals when placed in contact with burning charcoal because the charcoal imparts its phlogiston to the metal.
- Charcoal leaves very little residue when it burns because it is made up almost entirely of phlogiston.

There was a fly in the ointment, however; some metals gained weight when they were burned, so if they lost phlogiston to the air, that phlogiston must have had a "negative weight." Also, mercuric oxide (mercury calx) would turn back into mercury with simple heating, even in the absence of charcoal.

Joseph Priestley fell out of favor with the scientific community because of his lifelong defense of the phlogiston theory.[4] His 1775 treatise describing the discovery of oxygen was entitled "Experiments and Observations on Different Kinds of Air." A copy of the cover of the text is reproduced as Figure 1.2 on the next page. At the time, all gases were referred to as "airs." Oxygen was called dephlogistocated air, while nitrogen was called phlogistocated air, mephitic air, or azote. Carbon dioxide was called fixed air and hydrogen was called inflammable air.

Priestley published a lengthy defense of the phlogiston theory entitled "Considerations on the Doctrine of Phlogiston, and the Decomposition of Water," in 1796, more than 20 years after his original text. In it, he found fault with the finer details of Lavoisier's experiments, such as his inability to account for two colors of iron oxide, and he questioned the accuracy of the scales used in the measurement of hydrogen, oxygen, and water in an experiment demonstrating that water is the combination of "inflammable air and dephlogistocated air." He stated, "But the apparatus employed does not appear to me to admit of so much accuracy as the conclusion requires; and there is too much of correction, allowance, and computation in deducing the result."

Some of Priestley's arguments in defense of the old theory seem to echo today in the fire investigation community as some investigators schooled in "old fireman's tales" defend the validity of these tales and attack new knowledge.

[3] Actually, Carl Scheele, a Swede, discovered oxygen 2 years before Priestley, but Priestley gets the credit because he published first.

[4] Priestley was out of favor with many groups, including the mob that burned his house and chapel in 1791, because they did not care for his religious beliefs—he was a Dissenter from Calvinism—or for his support for the French Revolution. After being forced to resign from the Royal Society and being burned in effigy, he fled to Pennsylvania, where he founded the first Unitarian Church in America. His parishioners included Thomas Jefferson and Benjamin Franklin.

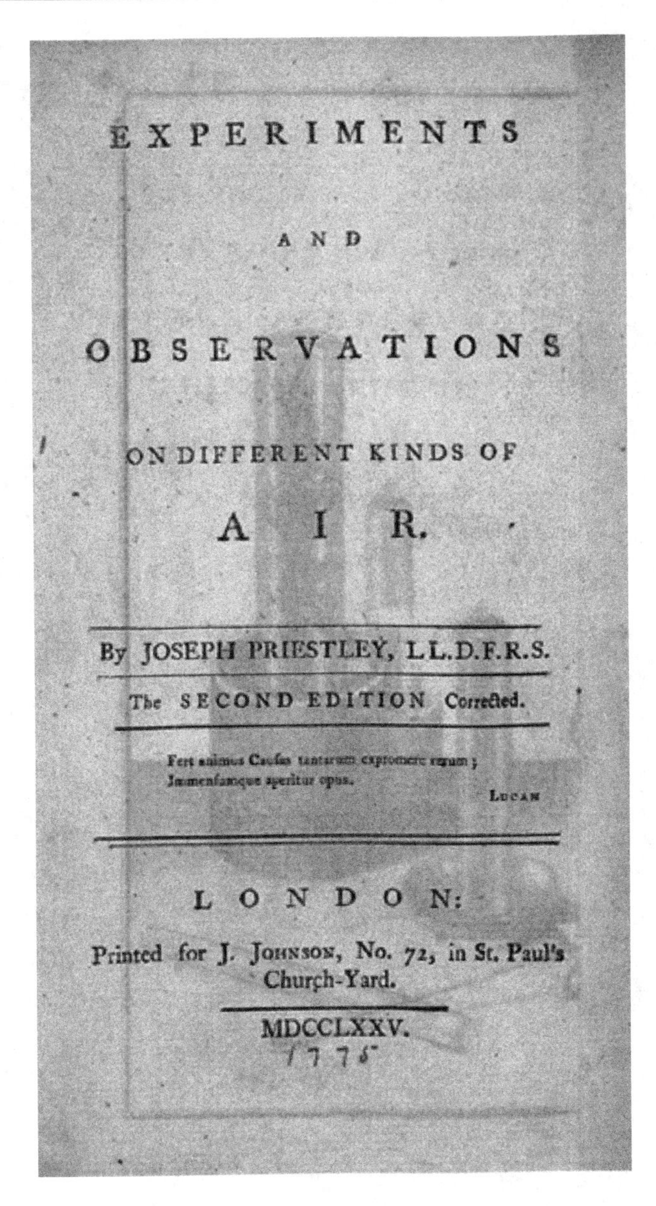

EXPERIMENTS

AND

OBSERVATIONS

ON DIFFERENT KINDS OF

AIR.

By JOSEPH PRIESTLEY, LL.D.F.R.S.

The SECOND EDITION Corrected.

Fert animus Causas tantarum expromere rerum;
Immensumque aperitur opus.

LUCAN

LONDON:

Printed for J. Johnson, No. 72, in St. Paul's Church-Yard.

MDCCLXXV.

1775

Figure 1.2 Joseph Priestley's 1775 Treatise on "different kinds of air." (Courtesy of the National Library of Medicine, Bethesda, MD.)

the nature of combustion and the role of oxygen in supporting both fire and life (Figure 1.4). In so doing, Lavoisier also discovered the law of conservation of mass, one of the fundamental tenets of modern chemistry, which states that the sum of the weight of the products of a reaction equals the sum of the weight of the reactants. The concept of "balancing" a chemical equation was born. In 1789, Lavoisier published the first modern text of chemistry: *Traité Élémentaire de Chimie*.[5] In this book, Lavoisier elaborated a new system of chemical nomenclature that is still in use today. This system is based on the names of the elements, and it designates to the compounds a name describing the elements with which they are formed, replacing the fanciful names used by the alchemists or the phlogiston adherents. Table salt, a compound of sodium and chlorine, is called sodium chloride. The gas formed by hydrogen and sulfur is hydrogen sulfide. An acid that contains sulfur is sulfuric acid. In addition to the new nomenclature, the treatise discussed chemistry in the light of Lavoisier's new theory of combustion.

[5] Lavoisier has been called the father of modern chemistry. This title is sometimes awarded to John Dalton, who proposed that the atoms of each element have a characteristic atomic weight and that atoms were the combining units in chemical reactions.

Figure 1.3 Lavoisier in his laboratory. (Courtesy of the National Library of Medicine, Bethesda, MD.)

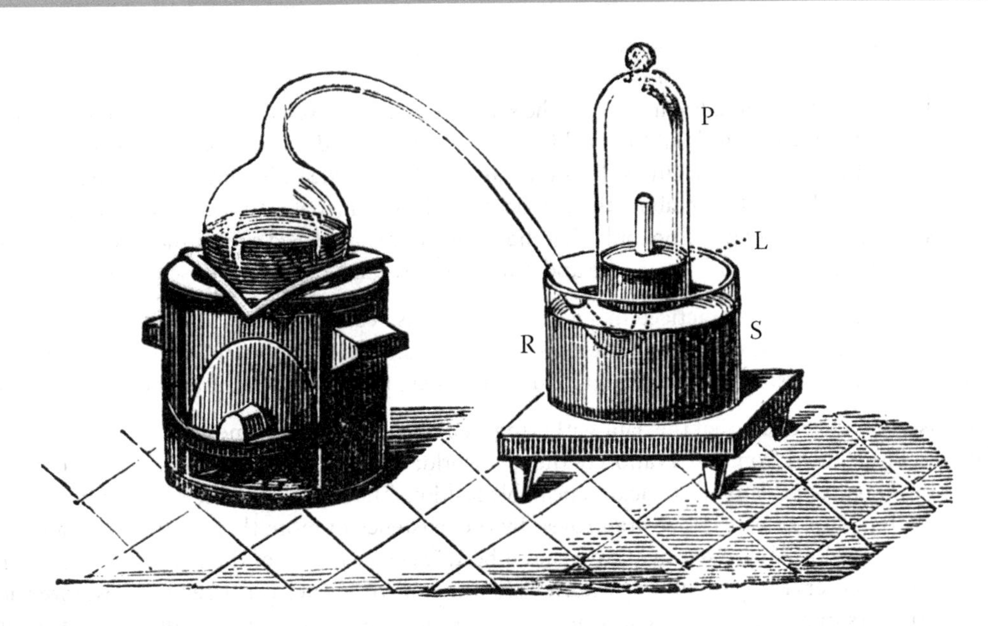

Figure 1.4 Lavoisier's apparatus for determining the composition of air and the increase in the weight of metals upon calcining (oxidizing). (Printed in Mendeleev's *History of Chemistry*, 1897. Courtesy of the National Library of Medicine, Bethesda, MD.)

Lavoisier disproved the existence of one improbable substance but immediately postulated the existence of another. "Caloric" was proposed to be a weightless elastic fluid that could be transferred from one body to another but not "created" or "destroyed." The theory was refuted by Benjamin Thompson, Count Rumford, who proved by experiments on a cannon-boring machine that heat was a form of energy, not a form of matter.[6] Rumford's work formed the basis of the science of thermodynamics.[7]

Also building on the work of Lavoisier, in 1802, Joseph Louis Gay-Lussac discovered the law of combining volumes, also known as the law of constant proportions. The law states the following: in reactions between gases, the volumes of gases always combine in simple ratios. For example, 2 liters (L) of hydrogen plus 1 L of oxygen produces 2 L of water vapor. Any excess gas does not react.

In 1811, Amedeo Avogadro published an article in *Journal de Physique* that clearly drew the distinction between the molecule and the atom. He pointed out that Dalton had confused the concepts of atoms and molecules. The "atoms" of hydrogen and oxygen are, in reality, "molecules" containing two atoms each. Thus, two molecules of hydrogen can combine with one molecule of oxygen to produce two molecules of water. Avogadro's hypothesis—**equal volumes of gas contain equal numbers of molecules**—was not widely accepted until after his death. The number of molecules in a "mole," the weight of a compound equal to the sum of all the atomic weights of the atoms in the compound, has been called Avogadro's number in his honor. The number is almost incomprehensible—6.022×10^{23}.

By the time of Faraday's second Christmas lecture in 1860, most of the fundamentals of the behavior of fire had been documented. Adolf Fick had derived the laws of diffusion. Joseph Fourier discovered, among other things, the laws of heat transfer. In 1866, Ludwig Boltzmann and James Maxwell used statistical mechanics to describe the microscopic behavior of atoms and molecules responsible for the macroscopic phenomena associated with heat and pressure. The collective knowledge of the behavior of fire allowed the industrial revolution to move forward as new and better ways were found to harness the energy contained in fossil fuels. In the late nineteenth and early twentieth centuries, Max Planck and Albert Einstein filled in most of the remaining gaps in our understanding of light, heat, and radiation.

---

6 Not only did Rumford disprove Lavoisier's Caloric theory, he married Lavoisier's widow after the great scientist was beheaded in the French Reign of Terror.

7 Rumford also invented a recipe for baking soda, thermal underwear, a drip coffee maker and a more efficient fireplace, versions of which are still sold and touted as models of energy efficiency and safety. Rumford fireplaces are tall and shallow to reflect more heat and have streamlined throats to eliminate turbulence and carry away the smoke with little loss of heated room air.

1.4 THE SCIENTIFIC APPROACH TO FIRE INVESTIGATION

The "natural philosophers" of the enlightenment and the scientists who followed them not only described the world but also formulated a system for learning about the world. The scientific method is a way of thinking that provides a set of techniques for testing the validity of the almost limitless range of human ideas. It allows us to separate the valid from the merely plausible. Virtually any beliefs about the empirical world are amenable to such testing. Indeed, the historical evolution of the scientific method reveals not only that the study of fire has played a prominent role in the development of science but also that the beliefs about the nature and causes of fires can be subjected to scientific testing.

This means that science not only is practiced in a laboratory but also can be applied in the field. A scientific opinion is one based on a specific process of defining the problem; data collection and analysis; hypothesis formation; and, most important, hypothesis testing. This is true regardless of where that process is performed or who performs it.

Some nonscientists seem to believe that the data-gathering step does not exist if some analytical machine is not used. Actually, data can be gathered from observation of the real world; from previous experience; or, as is most common, by simply thinking about a problem in a logical, repeatable fashion. The utility of the scientific method and its applicability to almost all of human experience are proven by the existence of more than 20,000 separate scientific and technological disciplines. Rachel Carson may have said it best: "Science is part of the reality of living; it is the what, the how and the why of everything in our experience" [4]. The type of science practiced by most forensic scientists, and by fire investigators in particular, is usually at odds with the general public's view of what constitutes science. Science is frequently thought of as experiments done in laboratories by people in white laboratory coats or as predicting future events through the replication of standardized events. Both of these activities can be characterized as science, but neither forensic scientists nor fire investigators can be expected to replicate a homicide, a burglary, a plane crash, or a fire—at least not on a regular basis. Our inability to reproduce historical events is not a result of a lack of will but of the asymmetry of time, which moves in only one direction. It is not possible to re-create or to undo the past.[8] There exist, in fact, many branches of science that are unable to reproduce past events or accurately predict future ones. Sciences in this category include geology; astronomy; archaeology; paleontology; and, in many cases, forensic science and fire investigation. In geology, for example, the theory of continental drift is widely accepted, but certainly no one has reproduced the movement of the continents. In astronomy, although it is possible to predict with great accuracy the positions of heavenly bodies, nobody has found a way to reproduce the birth of a star. For these types of inquiries, the goal is to understand the cause of a particular event, as opposed to predicting future events. This type of science, the goal of which is to understand causation, is categorized as *historical science*.

One difficulty in the practice of historical science is that more information is present than is required to infer the cause of an event. For example, in the narrow sense, a fire requires a combination of fuel, an ignition source, and an oxidizer, yet many factors are discovered during an investigation that might contribute to the cause of a particular fire. Using the scientific method, an investigator identifies and isolates each of these factors, poses hypotheses, and proposes tests (or relies on previous controlled experiments) that will either refute or fail to refute each hypothesis. Thus, historical science combines information from the past with empirical testing (either from validation studies or testing for a specific circumstance) to draw inferences about the most likely cause of the event.

1.5 MODERN FIRE ANALYSIS

Quantitative treatment of fire dynamics began during World War II when Sir Geoffrey Taylor described air entrainment in plumes above ditches filled with burning gasoline, which he used to clear fog from runways. Further development of the quantitative description of fires was funded by the US Office of Civil Defense in the 1950s when the National Science Foundation Committee on Fire Research, chaired by Hoyt Hottel of the Massachusetts Institute of Technology (MIT), and later by Howard Emmons of Harvard, began conducting and collecting research. During the next 20 years, fire research became the province of the fire protection engineering community. In the mid-1970s, James Quintiere developed a simple mathematical model of fire behavior, and his approach to the subject influenced the thinking of other researchers in the field. Practical mathematical models of fire, however, did not become possible until the advent of sufficient computing power to simultaneously solve

[8] Even if we could or did, we would leave evidence of having done it.

for the variables involved in fluid dynamics, heat transfer, radiation, and chemical reactions. Engineers at the National Bureau of Standards (now the National Institute of Standards and Technology [NIST]) and elsewhere quickly developed and merged fire-modeling programs. This was accomplished under the leadership of individuals such as Harold Nelson, who not only could do the math but also could explain it to the less gifted. The first textbook that rigorously treated the subject of fire dynamics was not published until 1985, when Wiley published Dougal Drysdale's *Introduction to Fire Dynamics.* Quintiere published a more user-friendly treatment of the subject, *Principles of Fire Behavior,* in 1997. Unfortunately, until the mid-1980s, very little of the newly discovered knowledge about the behavior of fire was passed on to fire investigators.

Fire investigation developed on a parallel track. The science of fire investigation resulted from the need to know what caused fires and how to prevent fires. These investigations were delegated largely to police and fire departments. Paul L. Kirk wrote the first science-based text on fire investigation in 1969. Kirk, one of the greatest forensic scientists of the twentieth century, authored a book entitled (simply enough) *Fire Investigation.* While Kirk realized then that there were many fire investigators making improper inferences about fire artifacts, he believed in the concept of "normal" fire behavior and did not address the ramifications of the compartment fire or flashover (covered in this text in Chapter 3). Kirk realized that fire investigations were carried out largely by individuals who could not access the highly technical publications that dealt with the behavior of fire, and he tailored his work to help both scientists and nonscientists comprehend the basic principles of fire behavior, as they were understood at the time.

John DeHaan addressed many of the shortcomings in Kirk's original work when he wrote the second edition in 1983, and he has continued to improve the book with each subsequent edition (the eighth edition, now authored by Icove and Haines, was published by Brady in 2018). Other texts on fire investigation published in the last three decades of the twentieth century, with few exceptions, were authored by well-meaning but poorly informed individuals who perpetuated misconceptions, thereby contributing to many erroneous determinations of fire cause. These texts and the misconceptions are described in detail in Chapter 8.

The early 1980s were years of decisive change for fire investigators. For the first time, computers were able to come to grips with the complexities of simultaneously modeling fluid dynamics, heat transfer, mass loss, and chemistry. The computational ability was resident in only a few large facilities, notably NIST. As cheaper and faster computers became generally available, more individuals had the ability to learn how to model fires.

In addition to the computational advances, there were technological advances in the study and measurement of real fires. In 1982, Vytenis (Vyto) Babrauskas, working at the National Bureau of Standards, invented the cone calorimeter, which allowed for a comprehensive understanding of ignition requirements, burning rate, heat release rate, and smoke production in real burning materials. The advent of the furniture calorimeter and the room calorimeter, larger versions of the bench-top cone calorimeter, allowed for comparisons and calibration of computer models with real fires.

In 1984, the NFPA released *Countdown to Disaster,* a 16-minute (min) video of a test fire that mimicked the behavior of an ordinary residential fire involving an ordinary upholstered chair. The spectacular flashover that occurred is still used to illustrate how compartment fires behave in a manner entirely beyond the comprehension of anyone who has not personally witnessed the phenomenon. *Countdown to Disaster* destroyed the simple "heat rises" concept that many fire investigators used (and some still use unfortunately) to determine the origin of fires.

In the late 1980s, the knowledge that flashover could account for what had previously been called "suspicious" fire patterns and "abnormally rapid" fire spread led to some highly publicized reversals of arson convictions. Two criminal cases from Arizona, *State v. Knapp* [5] and *State v. Girdler* [6], continue to reverberate even today. Arguments raised by the defendants in those cases were derided by the large majority of fire investigators as "the flashover defense." As time went by, the true effects of fires that progressed beyond flashover became more widely known. Flashover, as a plausible explanation for compartment fire artifacts, became more widely accepted, and most fire investigators today are aware of the implications for the investigation of a fire that has progressed to full room involvement. Nonetheless, because many fire investigators are trained by mentors, who were trained by mentors, the belief structure of investigators who learned their craft 30 or 40 years ago still influences many current practitioners.

These incorrect beliefs were passed on through weekend or weeklong "seminars" in which newcomers to the field were taught to "recognize arson." Fires were set using flammable liquids and then extinguished before the compartment reached flashover, and the easily recognized patterns were shown as an example of what to look for. Usually, there was little input from scientists or fire protection engineers in the design of these "burn exercises." The scientific community left the fire investigation community pretty much alone, and the mutual isolation was unhealthy for both. The misconceptions were passed along. To compound the problem, the myths were collected and published in numerous fire investigation textbooks, giving them added credibility. Clearly, the training infrastructure was badly in need of repair.

Just how badly the infrastructure was in need of repair became evident in 2005, when a group of certified fire investigators from the Bureau of Alcohol, Tobacco, Firearms and Explosives (ATF) conducted a test in Las Vegas. They built two test rooms, set them on fire, and allowed them to burn for just a few minutes beyond flashover. They then asked 53 fire investigators to walk through the rooms and choose the **quadrant** where the fire had originated. Three investigators correctly identified the quadrant of origin in the first test room, and a different three correctly identified the quadrant in the second test room. The investigators had been misled by patterns generated as a result of ventilation-controlled, post-flashover burning [7].

When word of this experiment got out, there was widespread concern in the fire investigation community, although the results should not have surprised anyone. For years, an informal version of the Las Vegas experiment had been conducted at the Federal Law Enforcement Training Center in Glynnco, Georgia, at the opening of an advanced arson school. Although the results were intentionally not documented, participants in this informal survey reported that the success rate was on the order of 8%–10%. The Las Vegas experiment was repeated in 2007 in Oklahoma City, but this time there were three rooms. One room was allowed to burn for 30 seconds beyond flashover, a second room was allowed to burn for 70 seconds beyond flashover, and a third room was allowed to burn for 3 min beyond flashover. The results confirmed the error rate that the investigators obtained in Las Vegas.

Of the 70 participants, 84% correctly identified the quadrant of origin of the 30-second fire. This fell to 69% for the 70-second fire and to 25% for the 3-minute fire. Thus, for a fire that burned 3 min beyond flashover, trained fire investigators did no better at selecting the quadrant of origin than any untrained group of people choosing at random.

The last slide of the ATF presentation on the 2005 Las Vegas experiments states, "The 'old-days' of locating the point of origin of a post-flashover fire by relying on the 'lowest burn and deepest char' are OVER!" While this is certainly true, it is also late in coming.

Back in 1997, a report was issued based on a series of 10 laboratory test fires and 10 acquired structure test fires that concluded:

> When flashover conditions have been produced in a room, patterns which are located at low levels on the walls (as low as the floor) may be produced in areas not related to the origin. These patterns may be produced by the burning of furniture items or ventilation effects. Accurate origin determination cannot be made based solely on the presence of areas of low burning when flashover conditions existed [8].

Apparently, the fire investigation community was unable to accept the contents of that report, even though it was funded by the US Fire Administration. Such denial is becoming increasingly rare.

While the experimental design of the ATF origin tests does not control for all of the variables involved, the low success rate, no better than random chance, calls into question the very validity of any origin determination in a fully involved compartment based on investigators "reading" fire patterns. Certainly, it is still possible to accurately interpret fire patterns, but it is absolutely necessary that ventilation be one of the first factors considered when trying to interpret patterns in a fully involved (ventilation-controlled) fire. In such a fire, the dominant factor that controls burn pattern development is the availability of oxygen. Fire patterns that are generated after flashover offer little insight to assist the investigator in finding the origin. Such considerations will be discussed in detail in Chapter 3.

1.6 NFPA 921

The perception of a profession plagued by misconceptions caused the Standards Council of the NFPA to form a Technical Committee on Fire Investigations in 1985. Thirty members, representing a balance of the law enforcement, engineering, and insurance communities, were recruited. The Technical Committee was one of the first venues where fire investigators and fire protection engineers worked together on the same project. After 7 years, the first edition of

NFPA 921, *Guide for Fire and Explosion Investigations*, was released. The intent of the document was (and remains) "to assist in improving the fire investigation process and the quality of information on fires resulting from the fire investigation process [9]." The document is currently updated every 3 or 4 years, with the ninth cycle having been completed with the approval of the 2017 edition, a text more than twice the length of the 1992 edition. NFPA 921 is the single most important treatise ever published in the field of fire investigation. Anyone who holds him- or herself out as an expert in the field will be measured against this text, which the US Justice Department has described as a "benchmark [10]." With each new edition, NFPA 921 incorporates new data from research into fire behavior, and input from the fire investigation, legal, insurance, engineering, and scientific communities.

The most significant contribution of NFPA 921 is the chapter on basic methodology, which recommends the systematic approach of the scientific method that is used in the physical sciences. It is no coincidence that the NFPA, which advocates the production of scientifically based codes and standards, urges fire investigators to use the methodology that helped people understand not only fire but all of chemistry and physics as well. NFPA 1033, discussed in the following, requires: "The fire investigator shall employ all elements of the scientific method as the operating analytical process throughout the investigation and for the drawing of conclusions."

To say that the early editions of NFPA 921 were not universally embraced by the fire investigation community would be a serious understatement. As with any new standard, NFPA 921 aroused the ire of those accustomed to working according to their own subjective criteria. In 1999, these individuals organized a write-in campaign that urged the NFPA to delete any reference to science in the document. Part of the driving force behind this campaign was the 11th Circuit Court decision called *Michigan Millers Mutual v. Benfield*. In that case, the appeals court upheld the exclusion of an investigator's testimony and held that, because the investigator stated that he was knowledgeable in "fire science," the admissibility of his testimony was subject to a *Daubert* challenge [11]. This ruling sent shudders through the world of fire investigation. Certain leaders of the fire investigation community began urging fire investigators to avoid any mention of the word *science* and thus avoid the *Daubert* challenge that it might bring. An *amicus curiae* brief was filed on behalf of the International Association of Arson Investigators (IAAI) that argued, "Cause and origin investigations, by their very nature, are 'less scientific' than envisioned by *Daubert*" [12], essentially asking that the Court "grandfather" the folly that "traditional" methods represented. The Supreme Court, in the case of *Kumho Tire v. Carmichael*, its next interpretation of *Daubert* and Rule 702, turned this argument on its head when it held unanimously that *Daubert* applied to all expert testimony, and Rule 702 made no relevant distinction between "scientific knowledge" and "technical" and "other specialized" knowledge. In *Daubert,* the Court specified that it is the noun *knowledge*, not the adjectives *scientific* or *technical*, that establishes a standard of evidentiary reliability [13]. To the extent that parties wanted to advance a "less scientific" methodology, the Court ruled in *Kumho* that the methodology required more scrutiny.

Although the Supreme Court had already made all fire investigator testimony subject to a reliability inquiry, more than a hundred comments were submitted to the NFPA for the 2001 edition of NFPA 921, protesting the Technical Committee's decision to continue to recommend the use of the scientific method. Many arguments were made in support of a "systematic approach" instead of the scientific method. These arguments revealed a misunderstanding on the part of many submitters as to the elements of a scientific investigation. Some of the arguments included the following:

- Scientific research is important; however, fire investigation is not an exact science.
- Fire scenes are not laboratories. The scientific method is to be used when there can be "scientific testing."
- Fire investigators should not be subjected to extraneous questions concerning the scientific method if they are truly conducting a quality fire investigation and have the proper conclusion about the origin and cause of a fire.
- "Systematic" is a clearer definition of what we do. "Scientific" implies an exact science, which is impossible in true fire investigation.
- The use of the wording "scientific method" is often confusing.
- Scientific method implies that only scientists are qualified to investigate fires and offer opinions as to the origin and cause of fires.

Perhaps the most telling argument stated, "This proposal is a confusing attempt to mitigate the disagreement between the fire investigation and the scientific worlds competing for control of this document [14]."

The Technical Committee and the NFPA saw no such competition, nor did they acknowledge that there are two different "worlds," but the argument brightly illuminates a major problem in fire investigation today.

One minor change was made in the description of the scientific method, which the Committee hoped would eliminate the contention that in order to "scientifically" test a hypothesis, one needed to rebuild the building and then burn it down again. In the paragraph on hypothesis testing, the following sentence was added: "This testing of the hypothesis may be either cognitive or experimental." This explained the proposition that as long as deductive reasoning was used to test the hypothesis, an actual experiment was not a necessity. Some fire investigators took this as a license to just think about the fire scene and rely on their experience, as they had in the past, but the subsequent edition, published in 2004, provided more details about the use of inductive and deductive reasoning.

The 2008 and subsequent editions of NFPA 921 were completed with far less rancor than the previous editions. Fire investigators and their clients were growing accustomed to the document, and people who objected to it were retiring. As the courts became more familiar with NFPA 921, a body of law developed holding that even though the NFPA called it a "guide" and not a "standard," the document provided the courts with guidance on what should be the standard of care in fire investigation. After the *Kumho* decision, and the embrace of NFPA 921 by the US Department of Justice, the IAAI stopped resisting and actually embraced the document. New resistance was exhibited, however, to the change in 2011 in which the document disparaged the methodology known as "negative corpus." The Committee made some changes in the 2014 edition as a result of these comments, but the disparagement of negative corpus methodology remains in the document to this day.

It might be said that 2000 marked the point where NFPA 921 became "generally accepted" by the fire investigation community. That year saw endorsement of the document by both the IAAI and the US Department of Justice (DOJ). Most of the court rulings recognizing NFPA 921 as the standard of care came after that date.

1.7 NFPA 1033

The evolution of another document certified the importance of NFPA 921 to the fire investigation community. NFPA 1033, *Standard for Professional Qualifications for Fire Investigator,* started as a subunit of NFPA 1031, *Professional Qualifications of Fire Inspector, Fire Investigator, and Fire Prevention Education Officer,* which was adopted by NFPA in 1977. The first edition of NFPA 1033, a stand-alone *Standard for Professional Qualifications for Fire Investigator*, was adopted by the NFPA in 1987. The 2003 edition of the document was updated, and while it mentioned NFPA 921 (once, in the appendix),[9] most of the changes were made to bring it into conformance with a new manual of style.

The 2009 edition of NFPA 1033 included significant changes, the most important of which was the addition of paragraph 1.3.8, which lists 13 subjects that fire investigators were supposed to have mastered. Here is the new paragraph:

1.3.8* The investigator shall have and maintain at a minimum an up-to-date basic knowledge of the following topics beyond the high school level at a post-secondary education level:

1. Fire science
2. Fire chemistry
3. Thermodynamics
4. Thermometry
5. Fire dynamics
6. Explosion dynamics
7. Computer fire modeling
8. Fire investigation

[9] Here is the entirety of the reference to NFPA 921 in the 2003 edition: A.1.3.7 Fire investigation technology and practices are changing rapidly. It is essential for an investigator's performance and knowledge to remain current. It is recommended that investigators be familiar with the technical information and procedural guidance presented in materials such as NFPA 921 and *Fire Protection Handbook*. Three additional references to NFPA 921 occur in the Appendix of the 2009 edition. In the 2014 edition, NFPA 921 takes a far more prominent role, as the source of much of the knowledge base that fire investigators are required to have.

9. Fire analysis
10. Fire investigation methodology
11. Fire investigation technology
12. Hazardous materials
13. Failure analysis and analytical tools

In 2014, NFPA 1033 was updated to add three more topics to the list, which is now at 1.3.7:

14. Fire protection systems
15. Evidence documentation, collection, and preservation
16. Electricity and electrical systems [15]

These changes in the 2009 and 2014 editions of NFPA 1033 have made it a much more useful tool for challenging fire investigators. It is a short document with **mandatory requirements**, and the addition in 2009 of a list of subjects on which the fire investigator is required to have knowledge beyond the high school level has made challenging fire investigators a straightforward exercise by asking questions such as, "What are the basic units of energy?," "What are the basic units of power?," "What is the difference between energy and power?"

Investigators who are unable to answer these types of questions may have a difficult time persuading the court that they are qualified according to NFPA 1033, but failing to answer these questions correctly tends to short-circuit the process. Instead of trying to convince *a judge* that one is qualified despite the lack of mandatory knowledge, the task becomes trying to convince *a sponsoring attorney* that one is qualified. Once an investigator's shortcomings are revealed, counsel may decide that he or she no longer wishes to sponsor this "expert's" opinion testimony. The treatment of the subjects in Chapters 2, 3, and 4 will provide fire investigators with the knowledge base now required to meet this minimum standard.

1.8 CFITrainer.net

One major improvement in the infrastructure for training fire investigators made its debut in 2005, with the introduction of CFITrainer.net. Sponsored by the IAAI, this Web-based tool brings current knowledge to anyone who wants it, and, as of this writing, it does so at no cost to the user. CFITrainer.net has incorporated the lessons of the Las Vegas and Oklahoma City experiments as well as others and is an invaluable tool for investigators to obtain training not available elsewhere. This online training is acceptable for use by fire investigators seeking certification or recertification.

There are dozens of training modules that take anywhere from one to four hours to complete on CFITrainer.net. A partial list of available courses follows:

- Accreditation, Certification and Certificates for Fire Investigators
- An Analysis of the Station Nightclub Fire
- Arc Mapping Basics
- Basic Electricity
- Critical Thinking Solves Cases
- The Deposition: Format Content and Preparation
- The Deposition: Questioning Tactics and Effective Responses
- Digital Photography and the Fire Investigator
- Documenting the Event
- DNA
- Effective Investigation and Testimony
- Electrical Safety
- Ethical Duties Beyond the Fire Scene
- Ethics and Social Media

- Ethics and the Fire Investigator
- Evidence Examination: What Happens at the Lab?
- Explosion Dynamics
- Fire and Explosion Investigations: Utilizing NFPA 1033 and 921
- Fire Dynamics Calculations
- Fire Investigator Scene Safety
- Fire Protection Systems
- Fundamentals of Interviewing
- Fundamentals of Residential Building Construction
- The HAZWOPER Standard
- How First Responders Impact the Fire Investigation
- Insurance and the Fire Investigation
- Introduction to Evidence
- Introduction to Fire Dynamics and Modeling
- Investigating Fatal Fires
- Investigating Motor Vehicle Fires
- Investigating Natural Gas Systems
- Investigation of Fire and Explosion Incidents Involving a Line of Duty Death
- Legal Aspects of Investigating Youth-Set Fires
- MagneTek: A Case Study in the Daubert Challenge
- Managing Complex Fire Scene Investigations
- Motive, Means, and Opportunity: Determining Responsibility in an Arson Case
- Motor Vehicles: The Engine and the Ignition, Electrical and Fuel Systems
- NFPA 1033 and Your Career
- Physical Evidence at the Fire Scene
- Postflashover Fires
- The Potential Value of Electronic Evidence in Fire Investigations
- Preparation for the Marine Fire Scene
- Process of Elimination
- Residential Electrical Systems
- Residential Natural Gas Systems
- The Scientific Method for Fire and Explosion Investigation
- Search and Seizure
- Understanding Fire through the Candle Experiments
- Vacant and Abandoned Buildings: Hazards and Solutions
- A Ventilation-Focused Approach to the Impact of Building Structures and Systems on Fire Development
- Wildland Fires Investigation

Like the "list of 16" from NFPA 1033, the diverse array of course modules illustrates the multidisciplinary nature of fire investigations.

All of the modules in the CFITrainer.net curriculum are thoroughly vetted prior to being made available and are compliant with the principles and practices espoused by NFPA 921 and NFPA 1033. As this book went to press, the IAAI was experimenting with on-site, hands-on courses designed to allow fire investigators to demonstrate skills they have learned through their studies on CFITrainer.net.

1.9 SCIENCE, LAW, AND LAW ENFORCEMENT: OVERCOMING POTENTIAL BIASES

The power of science in the search for truth lies in its **independence**. When an expert is presenting an interpretation of a postfire artifact to a jury, that jury should have confidence that the interpretation is based only on what the science allows and not on extraneous contextual information that may have been developed elsewhere in the investigation.

As an example, consider the defendant charged with arson based on the fact that she had solicited several individuals to set the fire, and failing to find a willing accomplice, stated that she was going to set it herself on a particular date. After the fire, she admitted to a friend that she had in fact set the fire. The fire was so destructive, however, that when the scene was examined, the investigators correctly concluded that neither the origin nor the cause of the fire could be determined. This did not prevent the woman from being convicted, as the other evidence in the case made her guilt plain for all to see. The jury did not need and would have been ill served by an expert who concluded that the fire was intentionally set based on the witness information. That was the prosecutor's function.

Forensic scientists (**all fire investigators function as forensic scientists**) have duties that are distinct yet complementary to the duties of law enforcement officers or lawyers. Different duties imply different ethical obligations. If a fire investigator wishes to have the ability to tell a court that his conclusions are based on science, then the conclusions of the investigator must be based solely on the demonstrable scientific evidence and not on collateral information developed elsewhere in the investigation.

NFPA 921 recognizes the potential for the introduction of information that is not relevant to science in its discussion of expectation bias and context bias. There is a distinction between a witness statement regarding the placement of furniture or the observation of flames and a witness statement regarding a subject's demeanor or character. The first kind of witness statement can assist in the location of the origin or the understanding of fire patterns, while the second introduces unwanted bias into the investigation.

Fire investigation is not unique among the forensic sciences in terms of its susceptibility to contextual bias. A fingerprint analyst could be improperly influenced by being told that DNA from the subject was matched to a crime scene. During an international terrorism investigation, an FBI analyst "determined" that a particular type of explosive, pentaerythritol tetranitrate (PETN, the explosive found in "det. cord"), had been detected, but he had insufficient chemical analysis data to confirm his analysis. He admitted to the court that part of his "confirmation" came from a field agent who found detonator cord, a PETN-based explosive, in the defendant's garbage. The chemist underwent a withering cross-examination and the defendant was acquitted. A later investigation concluded that the chemist's approach reflected "a fundamental misunderstanding of the role of a forensic scientist" that he "should have known not to state his scientific conclusions more strongly than could be supported by the underlying analytical results," and that "his performance was wholly inadequate and unprofessional [16]."

Fire investigators should attempt to avoid learning about motive, means, and opportunity until the fire scene inspection has been completed. This approach to evidence examination is known as sequential unmasking. The investigator first needs to learn what the physical evidence at the fire scene means, without being influenced by other investigative data. This is not to say that the investigation can be conducted in a vacuum or without any context, but to the extent possible, the determination of the origin and cause of the fire should be made based on empirical evidence.

Ideally, the determination of the origin and cause of the fire would be conducted independently of the investigation into responsibility. Unfortunately, the resources for such an approach are frequently unavailable, leaving the fire investigator with the unenviable task of trying to block out the potentially contaminating effects of information that is not relevant to fire behavior from the origin and cause determination. As will be discussed in Chapter 9, it is contamination by context effects that has led to many seriously flawed fire cause determinations.

A 2016 court decision from Kentucky spells out some of the problems that result from having the "scientist" fire investigator also serve as the "detective" fire investigator. The order from the court vacated a 2006 arson conviction. The court wrote:

> There may be something about the state of fire science and the nature of arson investigation that tends to permit errors. Placing the duties of investigation and prosecution upon the same persons expected to give an objective scientific analysis may be common practice in arson cases and this may be part of the problem. The more reliable expert testimony in criminal cases is presented by persons not directly involved in the general investigation and prosecution of the case.

> In the interest of justice, the Commonwealth is urged to re-examine the case utilizing trained fire investigators not previously involved. Scientific evidence should be evaluated separately from the other evidence in the case. This may help avoid the implication of "confirmation bias" and the inadvertent mixing of fire folklore, personal opinion and other non-scientific factors [17].

This author has previously urged bifurcation of duties at the scene and bifurcation of issues in trials, as a means of minimizing expectation bias in fire cases [18]. A similar approach called linear sequential unmasking has been applied in other forensic sciences [19]. Until that becomes routine, however, fire investigators need to develop strategies to mitigate the effects of bias.

One means of combating context bias, as well as documenting use of the scientific method, is contemporaneous documentation of observations and data. (When you learned a fact can be as important as the fact itself.) Then, write down all the competing hypotheses that you are able to develop. Under each hypothesis, document the data that supports that hypothesis and data that refutes it. The hypothesis with the lowest number of refuting data points is the one likely to prevail. This method of scientific deduction is useful for testing hypotheses in cases where the data are incomplete, ambiguous, or conflicting. Intelligence analysts are tasked with predicting future events using data that may also be incomplete, ambiguous, or conflicting. The United States Central Intelligence Agency has compiled a list of strategies to overcome biases in a book called *The Psychology of Intelligence Analysis* [20].

1.10 CONCLUSION

The history of science and the study of fire are inextricably intertwined. The scientific method is particularly well suited to the testing of hypotheses about the origin and cause of fires.

In the past 40 years, major advances have taken place in our understanding of the behavior of fire, in our ability to mathematically describe fire phenomena, and in our approach to solving problems in fire investigation. The most important of these advances is the general (if sometimes reluctant) agreement to seek a science-based consensus approach to understanding the cause of fire.

Review questions

1. Why should fire investigators follow the guidance provided by NFPA 921?
   a. Because many courts have recognized NFPA 921 as the standard of care for fire investigations.
   b. Because the US Department of Justice recommends NFPA 921 for investigating large fires and incendiary fires.
   c. Because NFPA 921 is recognized as the authoritative guide by the IAAI, NAFI, and many insurance companies.
   d. Because failing to follow NFPA 921 may result in exclusion of your testimony.
   e. All of the above.
2. In the NFPA standards development system, what is a guide?
   a. A document that is advisory or informative in nature but is not suitable for adoption into law.
   b. A standard that is an extensive compilation of provisions covering broad subject matter that is suitable for adoption into law.
   c. An organized or established procedure intended to form a network arranged to achieve specific goals.
   d. A written organizational directive that establishes or prescribes specific operational or administrative methods to be followed routinely, which can be varied due to operational need in the performance of designated operations or actions.
   e. A guide is the same thing as a recommended practice.
3. Which of the following statements about NFPA 1033 are true?
   a. If it is not adopted into law in my jurisdiction, following NFPA 1033 is optional.
   b. It is a standard applicable to public sector investigators.
   c. It is a standard applicable to private sector investigators.
   d. It requires that all fire investigators be certified.

i. I and III only
ii. I and II only
iii. II and III only
iv. I, II, and III
v. II, III, and IV

4. "Knowing where to look it up" is sufficient knowledge to satisfy the requirement for maintaining "an up-to-date basic knowledge" of the topics included in the "list of 16."
   a. True
   b. False
5. *Ipse dixit* is Latin for "he himself said it." This type of testimony was disparaged by which Supreme Court ruling?
   a. *Daubert v. Merrill Dow Pharmaceuticals*
   b. *General Electric v. Joiner*
   c. *Kumho Tire v. Patrick Carmichael*
   d. *Frye v. United States*
   e. *Weisgram v. Marley*

Questions for discussion

1. Research the history of the phlogiston theory of combustion and describe the reasons that it was eventually disproved.
2. Discuss the advantages and disadvantages of having more than one investigator working on an investigation.
3. Why is it important to distinguish fact testimony from opinion testimony?
4. How have the study of fire and the study of chemistry affected each other?
5. How have NFPA 921 and NFPA 1033 influenced the way fires are investigated?

References

1. Burke, J. L., and Ornstein, R. (1995) *The Axemaker's Gift*, Tarcher Putnam, New York, p. 117.
2. Sagan, C. (1995) *The Demon-Haunted World*, Random House, New York, p. 211.
3. Burke, J. L., and Ornstein, R. (1995) *The Axemaker's Gift*, Tarcher Putnam, New York, p. 159.
4. Carson, R. (1952) Acceptance speech of the National Book Award for Nonfiction for The Sea Around Us. National Book Foundation. Available at http://www.nationalbook.org/nbaacceptspeech_rcarson.html#.WlUr3ZM-egR (last visited on January 9, 2018).
5. *State of Arizona, Appellee, v. John Henry Knapp, Appellant*, No. 3106, Supreme Court of Arizona, 127 Ariz. 65; 618 P.2d 235; 1980 Ariz. LEXIS 273, 1980.
6. *State of Arizona, Plaintiff, v. Ray Girdler, Jr., Defendant*, in the Superior Court of the State of Arizona in and for the County of Yavapai, No. 9809.
7. Carman, S. (2008) Improving the understanding of post-flashover fire behavior, in *Proceedings of the 3rd International Symposium on Fire Investigations Science and Technology (ISFI)*. Available at http://www.carmanfireinvestigations.com (last visited on January 9, 2018).
8. Shanley, J. H., Jr., and Kennedy, P. M. (1996) Program for the study of fire patterns, in *National Institute of Standards and Technology Annual Conference on Fire Research: Book of Abstracts 149*, 150. Available at http://fire.nist.gov/bfrlpubs/fire96/PDF/f96156.pdf (last visited on January 9, 2018).

9. NFPA 921 (1992–2017) *Guide for Fire and Explosion Investigation*, NFPA, Quincy, MA, p. 1.

10. Technical Working Group on Fire and Arson Scene Investigation (2000) *Fire and Arson Scene Evidence: A Guide for Public Safety Personnel*, NJC181584, USDOJ, OJP, NIJ, p. 5. Available at https://www.ncjrs.gov/pdffiles1/nij/181584.pdf (last visited on January 9, 2018).

11. *Michigan Millers Mutual Insurance Company v. Janelle R. Benfield*, 140 F.3d 915 (11th Cir. 1998).

12. Burke, P. W. (1997) Amicus curiae brief filed on behalf of the International Association of Arson Investigators, in *Michigan Millers Mutual Insurance Company v. Janelle R. Benfield.*

13. *Kumho Tire Co. v. Carmichael*, (97-1709) 526 U.S. 137 (1999) 131 F. 3d 1433.

14. NFPA (2000) Report on Comments, NFPA 921, Fall 2000 Meeting, NFPA, Quincy, MA, p. 437.

15. NFPA (2009, 2014) NFPA 1033, *Standard for Professional Qualifications for Fire Investigator*, NFPA, Quincy, MA, p. 6.

16. USDOJ/OIG (1997) Special Report, The FBI Laboratory: An investigation into laboratory practices and alleged misconduct in explosives-related and other cases. Available at http://www.justice.gov/oig/special/9704a/ (last visited on January 9, 2018).

17. Gill, T. L. (2016) Circuit Court Judge, *Commonwealth v. Robert Yell*, Logan Circuit Court, Case No. 04-CR-00232, December 28.

18. Lentini, J. J. (2008) Toward a more scientific determination: Minimizing expectation bias in fire investigations, *Proceedings of the 3rd International Symposium on Fire Investigations Science and Technology (ISFI)*, NAFI, Sarasota, FL.

19. Dror, I. E. (2013) Practical solutions to cognitive and human factor challenges in forensic science, *Forensic Science Policy & Management*, 4(3–4):1–9.

20. Heuer, R. J. (1999) *The Psychology of Intelligence Analysis*, Center for the Study of Intelligence, U.S. Central Intelligence Agency. Available at https://www.cia.gov/library/center-for-the-study-of-intelligence/csi-publications/books-and-monographs/psychology-of-intelligence-analysis/PsychofIntelNew.pdf.

CHAPTER 2

The chemistry and physics of combustion

I'm sorry. If you don't know H_2O, you will not be rendering opinion testimony in my courtroom.

—Hon. J. Michael Ryan
Washington, DC Courts

LEARNING OBJECTIVES

After reviewing this chapter, the reader should be able to:

Demonstrate knowledge of fire chemistry and physics beyond the high school level as required by NFPA 1033 including:

- Knowing the definitions of atoms, molecules, elements, compounds and mixtures
- Knowing how to set up and balance a simple combustion equation
- Understanding the states of matter, and the behavior of solids, liquids, and gases when exposed to fire
- Appreciating the relationship between fire and energy
- Understanding the concepts of energy, power, and flux, and knowing the units used in the measurement of each of these properties

2.1 BASIC CHEMISTRY

Fire is defined as a rapid oxidation process with the evolution of light and heat in varying intensities. To understand fire, it is first necessary to understand the nature of fuels and the processes that allow fuels to burn. Without a basic knowledge of chemical principles, the fire investigator will be at a loss to understand the artifacts left behind.

All matter is composed of atoms. Atoms, with rare exceptions (called rare gases), are bonded together in molecules or crystals containing two or more atoms. Those atoms can be all the same, in which case the substance is called an *element.* If the atoms are not all the same, the substance is called a *compound.* Pure elements or pure compounds are relatively rare, except in a laboratory. Most elements and compounds are found in *mixtures.* A mixture contains more than one type of substance, but the components of a mixture are not chemically bound to each other. Air, for example, is a mixture of gases.

The closest thing to a pure element that a fire investigator is likely to find is the copper used to manufacture electrical conductors or pipes for water or gas. Copper in this form is usually more than 99% pure. Pure compounds are easier to find. Water is a pure compound, as are salt, sugar, baking soda, and dry ice. A solution is a special case of a mixture wherein one or more substances are dissolved in another substance. Most flammable liquids are solutions containing many substances that are soluble in each other.

Carbon-based molecules are called organic compounds because their ultimate source is living things. Most of the compounds found in petroleum can be traced back to a plant or animal source. The long-chain hydrocarbons found in kerosene, for example, started out as long-chain fatty acids [1].

Although atoms and molecules actually exist in three dimensions, for simplicity, they are presented in two dimensions as lines or as balls on sticks. Atoms and molecules are actually three-dimensional. The positively charged proton in the nucleus of the atom is about 2,000 times the size of the negatively charged electron that "orbits" the nucleus, and the interior of an atom consists almost entirely of empty space. The electron buzzing around the nucleus has been likened to a fly in a cathedral. Thus, while sketches and models can help us understand the relative orientation of atoms in a molecule, they do not "look" anything like the drawings. In fact, no one has actually seen most of the molecules in which we are interested, but their structure has been inferred from the way the molecules interact with energy in the form of electromagnetic radiation such as infrared, ultraviolet, or X-rays.

The atomic weight of an element depends on the number of protons and neutrons in the nucleus of the atom. Hydrogen, which has only one proton and no neutrons, has an atomic weight of 1. Carbon, which has six protons and six neutrons, has an atomic weight of 12 and is the standard by which all other atomic weights are determined. A "mole" of carbon weighs 12 grams (g). A mole contains an incomprehensibly large number of molecules, 6.022×10^{23} (Avogadro's number). The *molecular weight* of a compound is the weight in grams of a "mole" of that substance. The molecular weight is simply the sum of all the atomic weights of the atoms in the molecule. A mole of hydrogen gas (H_2) weighs 2 g. A mole of methane gas (CH_4) weighs 16 g. A mole of toluene (C_7H_8) weighs 92 g. Carbon dioxide (CO_2) and propane (C_3H_8) both have the same molecular weight—44 g.

2.2 FIRE AND ENERGY

Energy: A property of matter manifested as an ability to perform work, either by moving an object against a force or by transferring heat. Measured in joules (J).

Energy is defined as a *property of matter* manifested as an ability to perform work, either by moving an object against a force or by transferring heat. Chemical processes either absorb or give off energy. Reactions or phase changes that absorb energy are called endothermic. Reactions or phase changes that give off energy are called exothermic. Reaction rates increase with increasing temperature, so the energy given off in a reaction can increase the reaction rate, resulting in the release of even more energy.[1] This process can result in a phenomenon called thermal runaway. Fire is an exothermic chemical reaction that gives off energy in the form of heat and light. This is the energy that makes fire useful or destructive. *The understanding of fire requires a firm grasp of the basic concepts of energy.*

Such a grasp may seem more elusive as we examine the concept more closely. The first concept that must be addressed is the distinction between energy and temperature. When matter absorbs energy, its temperature increases. The molecules that make up a substance are constantly in motion. Increased temperature is manifested by an increase in molecular motion or molecular vibration. If the air in a car's tires is hotter, the molecules of oxygen and nitrogen are moving faster, colliding with the walls of the tire more frequently and increasing the pressure. (The typical speed of an oxygen molecule at room temperature is about 480 meters per second (m/s) or 1,080 miles per hour.) The kinetic theory of gases, worked out by Maxwell and Boltzmann, states that the typical speed of a gas molecule is proportional to the square root of the absolute temperature. An increase in temperature also results in solids melting, liquids vaporizing, and molecular bond vibrations increasing. Given sufficient energy, these bonds will break, resulting in the formation of new, smaller molecules.

Temperature: The measureable effect of the absorption of energy by matter. Measured in Kelvins (K) or in degrees C or degrees F.

Most of what is known about energy involves the transformation of energy from one form to another. For example, gasoline contains energy, and when put in a car, it can be burned in the engine and used to move the vehicle. This process converts chemical energy to heat energy, and heat energy into mechanical energy. The power company burns fuel to boil water to move a turbine that spins a magnet inside a coil of wires to produce electrical energy.

[1] An approximate rule of thumb is that reaction rates double with every increase of 10°C. This is known as the Arrhenius equation.

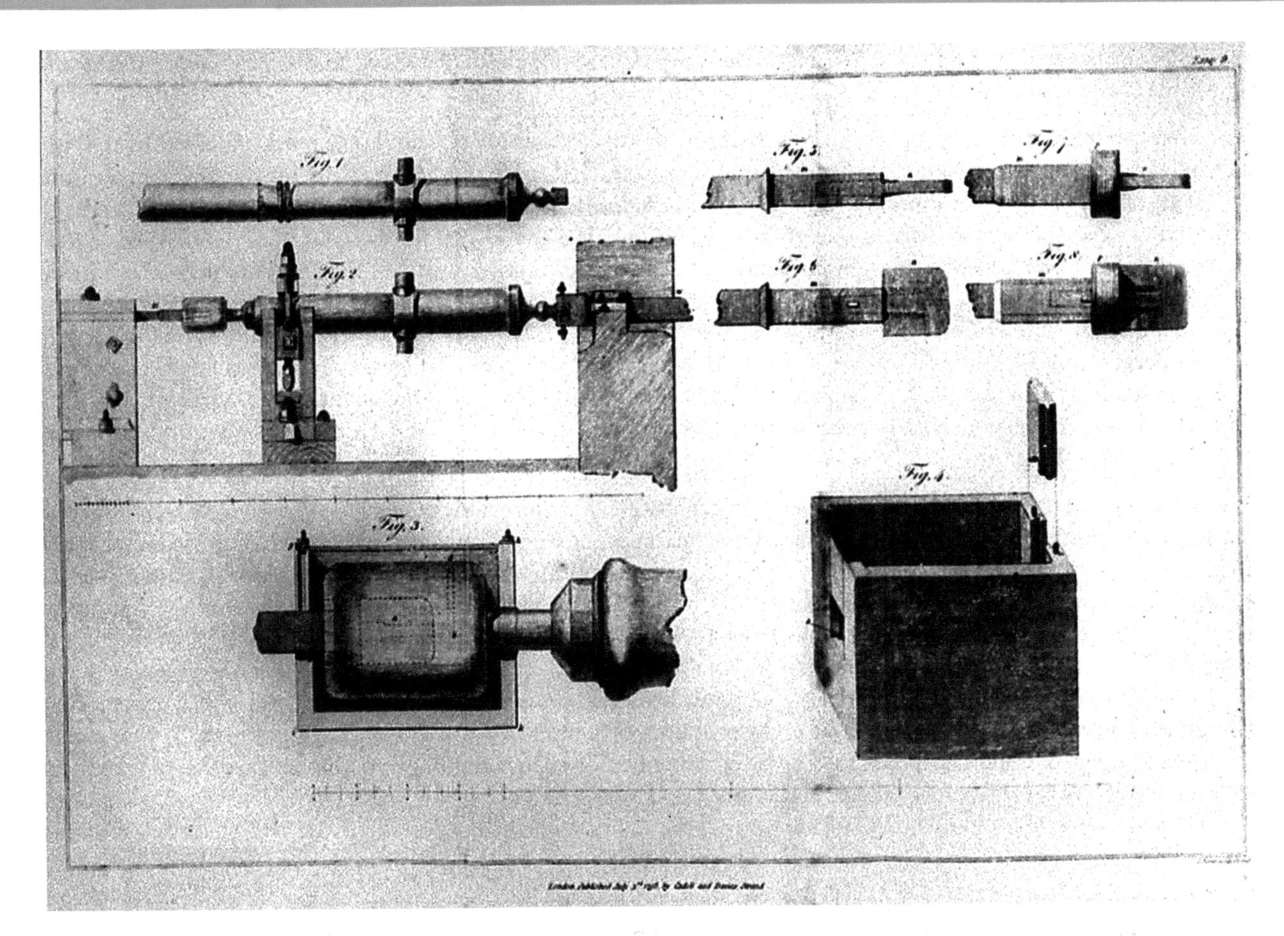

Figure 2.1 Count Rumford's cannon boring experiment, in which the friction involved was sufficient to boil water. (Courtesy of the National Library of Medicine, Bethesda, MD.)

When that electrical energy is passed through a filament to make light, or through a resistance element to heat water or air, it is turned into heat energy. When supplied to a motor, the electrical energy is converted into mechanical energy.

It is useful to think of energy as *the ability to do work*. Count Rumford learned from his experiments with friction that the work of boring a cannon barrel produced heat energy, which he used to boil water (Figure 2.1). It was his insight that heat is actually a form of work that provided the understanding of the concepts of energy transfer.

If a glass of ice water is placed in a room, heat will flow from the room into the glass until the ice melts. Eventually, the water will be the same temperature as the room and heat transfer will cease. Energy transfer that takes place exclusively by virtue of a temperature difference is called a heat flow. This concept of flow led early chemists to the caloric theory—something was flowing. While Lavoisier and others thought it was a substance that was flowing, Rumford and Joule proved that it was energy.

Heat flow or heat transfer is the transport of heat energy from one point to another caused by a difference of temperature. Heat transfer is responsible for much of the physical evidence (fire patterns) used by investigators when they attempt to establish a fire's origin and cause.

In order to understand fire, it is necessary to understand the units that are used to measure a fire's properties. The unit of work in any system of measurement is the unit of force multiplied by the unit of distance. In the metric system, the unit of force is the *newton* and the unit of distance is the *meter*. A newton is that force that gives one kilogram (kg) an acceleration of one meter per second per second. One newton-meter equals one *joule*, which is the basic unit of energy. The English equivalent of the newton-meter is the foot-pound (ft-lb) (its equivalence to the newton-meter would be more apparent if it were called the "pound-foot"). Although the pound is used in everyday life as a unit of quantity of matter, properly speaking, it is a unit of force or weight. Thus, a pound of butter is that quantity that has a weight of one pound. The ft-lb is a unit of work equal to the work done by lifting a mass of one pound (0.454 kg) vertically against gravity (9.8 meters per second per second) through a distance of one foot (0.305 meter). Doing just a little math or consulting any good conversion program reveals that 1 ft-lb = 1.356 J (0.454 × 9.8 × 0.305).

But what can movement of a weight over a distance tell us about heat transfer? The work involved in moving a standard weight over a standard distance against a standard force is equivalent to raising the temperature of a body of water by a fixed number of degrees. The "calorie" was originally defined as the amount of energy required to raise the temperature of one gram of water one degree Celsius. As the quantitative measurement of heat transfer became more precise, it was discovered that it takes more energy to raise the temperature of a gram of water from 90°C to 91°C than it does to raise it from 30°C to 31°C. This variability required that the definition be refined, and the calorie is now known as the "15° calorie," that is, the quantity of heat required to change the temperature of one gram of water from 14.5°C to 15.5°C. A corresponding unit, defined in terms of degrees Fahrenheit and pounds of water, is the British thermal unit (Btu).[2] One Btu is the quantity of heat required to raise the temperature of one pound of water from 63°F to 64°F (17.222°C to 17.777°C). Because the quantity of water is greater (454 g) and the temperature increase is less (0.555°C), 1 Btu = 252 calories (454 × 0.555 = 252).

We have discussed the equivalence of work and heat, but relating this concept to our everyday experience requires that we consider an important dimension that has been left out thus far: time. The amount of work required to raise a given weight to a given height is the same whether it takes a second, a minute, or an hour. Likewise, the amount of heat necessary to raise the temperature of a gram of water from 14.5°C to 15.5°C is the same, regardless of how long it takes (assuming a perfectly insulated system). The amount of work done *per unit time* is the quantity of interest. The rate at which work is done is called power. Power is defined as a *property of a process* (such as fire) that describes the amount of energy that is emitted, transferred, or received per unit time and is measured in joules per second (J/s) or watts (W). Appliances that use energy are defined by the power they produce, either in Btu per hour or in watts. A watt (W) is one joule (J) per second. The time component is built in. When the energy consumption or energy output is reported in Btus, the time element must be added. The size of a fire (or an ignition source) can be described in terms of watts or, more commonly, kilowatts (kW) or megawatts (MW). The fire investigator who has the ability to relate the energy output of a fire to everyday heating and cooking appliances is well on the way to understanding and being able to explain the phenomenon of fire.

Power: A property of a process (such as fire) that describes the amount of energy that is emitted, transferred, or received per unit time and is measured in joules per second or watts (W).

A fire investigator needs to understand and be able to describe ignition sources and fires in terms of their size in watts. For example, consider a controlled fire, the natural gas burner in a 40,000 Btu/hour water heater. Most people have a rough idea of the size of that flame. What is its output in watts? The watt, as a unit of power, already contains a time factor, as one watt equals one joule per second. The energy output of gas appliances is usually expressed in "Btus," but that is shorthand for "Btu *per hour*."

$$1 \text{ Btu} = 1054.8 \text{ J}$$

$$1 \text{ W} = 1 \text{ J/s}$$

Example 1: A 40,000 Btu/hour water heater burner delivers about 11.72 kW

$$40{,}000 \text{ Btu/hr} = 11.11 \text{ Btu/s}$$

$$11.11 \text{ Btu/s} \times 1054.8 \text{ J/Btu} = 11{,}720 \text{ J/s}, 11{,}720 \text{ W or } 11.72 \text{ kW}$$

Example 2: A 12,000 Btu stove top burner delivers about 3,500 W

$$12{,}000 \text{ Btu/hr} = 3.33 \text{ Btu/s}$$

$$3.33 \text{ Btu/s} \times 1054.8 \text{ J/Btu} = 3516 \text{ W}$$

[2] To honor great scientists, we name units of measure after them but do not capitalize the *name* of the unit, for example, the watt, named after James Watt. The *abbreviation* for a unit named after a person is capitalized, as is the B for "British" in British thermal unit. The temperature scales are always capitalized, whether written out or abbreviated, except for kelvins.

Example 3: A 125,000 Btu gas furnace delivers 36,625 W

What the power company sells is actually not power (watts), but energy (joules). The meter measures energy consumption in kilowatt-hours (kWh). A kilowatt is 1,000 J/s. A kilowatt-hour equals 1,000 J/s times 3,600 seconds per hour, or 3.6 million joules, or 3,413 Btu. Table 2.1 provides some useful energy and power conversion factors.

The size of a fire in kilowatts is known as its *heat release rate* (HRR). The HRR is the single most important property of a fire because it allows us to predict how that fire will behave and to relate the fire to our everyday experience. The heat release rate affects the temperature of the fire, its ability to entrain air (draw fresh air into the fire plume), and the identity of the chemical species produced in the fire.

Table 2.1 Energy conversion factors

Energy, Work, or Quantity of Heat	
1 joule (J)	1 newton · meter (N · m)
	0.7376 foot-pounds (ft-lb)
	9.48×10^{-4} Btu
	2.778×10^{-4} watt-hour
	10^7 erg
1 kilojoule (kJ)	1000 N · m
	737.6 ft-lb
	0.948 Btu
	0.2778 watt-hr
	10^{10} erg
1 kilowatt-hour (kWh)	3.6×10^6 J
	3,600 kJ
	3,413 Btu
	2.655×10^6 ft-lb
1 British thermal unit (Btu)	1,054.8 J
	2.928×10^{-4} kilowatt-hr
	252 gram-calories
Power or Heat Release Rate	
1 watt (W)	1 Joule/s
	10^7 ergs/s
	3.4129 Btu/hr
	0.05692 Btu/min
	1.341×10^{-3} horsepower (hp)
1 kilowatt (kW)	1 kJ/s
	10^{10} ergs/s
	3,413 Btu/hr
	56.92 Btu/min
	0.9523 Btu/s
	1341 hp
1 Btu/hr	0.2931 W
	0.2162 ft-lb/s
1000 Btu/hr	2,931 W
	2.931 kW
1000 Btu/s	1,050 kW
1 horsepower (hp)	745.7 W
	0.7457 kW
	42.44 Btu/min
	2546.4 Btu/hr

Note: Converting values from one system to another can introduce additional significant figures and false precision. See Chapter 3 for a discussion of the appropriate use and reporting of numerical values.

The size of a fire, or any energy source, is important to know, but it is equally important to know *how that energy is distributed.* Thirty-six kilowatts spread evenly throughout a structure by a furnace's circulation fan will keep it comfortable on a cold winter day. Confining or focusing that energy can result in dramatically different consequences. The concept of *heat flux* is therefore an important consideration. Heat flux is a measure of the rate of energy falling on or flowing through a surface. The radiant heat flux from a fire is a measure of the heat release rate of a fire in kilowatts, multiplied by the radiant fraction (that portion of energy transmitted as radiation, as opposed to convection), divided by the area over which the energy is spread in square meters. Heat flux is measured in units of *power per unit area*, or kilowatts per square meter (kW/m^2). (Some texts have used the CGS system[3] and reported radiant heat flux in watts per square centimeter (W/cm^2). There are 10,000 square centimeters (cm^2) [100 × 100] in a square meter and 1,000 W in a kilowatt, so 20 kW/m^2 = 2 W/cm^2.) Using the smaller units is sometimes helpful when considering the competency of an ignition source.

Flux: A measure of the rate of energy falling on or flowing through a surface, measured in kW/m^2 or W/cm^2.

One final parameter with which a fire investigator should be familiar is total energy impact. This is calculated by multiplying the heat flux by the time of exposure in seconds. A surface that receives a heat flux of 100 kW/m^2 for 2 minutes has received 100 $kJ/s/m^2$ for 120 s, or 12 MJ/m^2. It is the total energy impact that determines the degree of damage or injury that a fire may cause. Just as it is possible to pass your hand through a candle flame without injury if you keep your hand moving, a brief exposure to a short-lived fire is unlikely to cause much damage to dimensional timber.

The noonday sun bathes the earth with a radiant heat flux of approximately 1.4 kW/m^2, and about 0.7–1 kW/m^2 makes it to the earth's surface, depending on the season, the time of day, and the location. This is enough energy to cause a sunburn in 30 min or less. We cannot increase the heat release rate of the sun, but we can increase the radiant heat flux by focusing the energy that falls on a large area onto a smaller area. If we use a magnifying glass or a concave mirror to decrease the area by 96%, the radiant heat flux shoots up to 25 kW/m^2, enough to ignite most combustibles, as shown in Figure 2.2. Figure 2.3a shows how sunlight focused by a concave makeup mirror burned one stripe per day into the underside of the soffit outside the window where the mirror was located. Figure 2.3b shows a similar fire, caused by sunlight being focused through a "bubble window," a popular architectural feature in the United Kingdom.

Thirty seconds of exposure to a 4.5 kW/m^2 radiant heat source can cause a second-degree burn. Twenty kilowatts per square meter is generally accepted as the radiant heat flux required to bring an average residential compartment to flashover. (Flashover will be discussed in detail in Chapter 3.) Therefore, if flashover has been reached in a compartment, one can calculate the minimum heat release rate of the fire in that compartment by multiplying the area in square meters by 20 kW. A square room 12 ft on a side will flash over if the fire inside releases 267 kW ($144\ ft^2 = 13.37\ m^2 \times 20\ kW/m^2 = 267\ kW$). Keep in mind, however, that fires typically release their energy as conduction and convection as well as radiation. The radiation might account for only 20%–60% of the energy, and less than half of that reaches the floor. In addition, energy is lost to the walls and ceilings, and there are convective losses out of any openings in the enclosure. Therefore, to get 267 kW on the floor, the fire must be approximately 800 kW or more. Likewise, if the heat release rate of the fuel in the room is known, one can predict whether a fire on a particular fuel package is sufficient to bring the room to flashover, or failing that, ignite a second fuel package, which may provide sufficient additional heat to achieve flashover. Table 2.2 describes the effects of some typical radiant heat fluxes.

HUMANS ARE EXOTHERMIC

The average human consumes fuel (food) and inhales oxygen (air) and releases carbon dioxide and heat—about 100 W—when not doing anything strenuous. Releasing 100 W continuously for 24 hours results in a net energy release of 2,400 Wh or 2.4 kWh, or just over 2,000 kilocalories. Kilocalories are also called Calories (with a capital C). This is why a 2,000 Calorie diet is considered average.

[3] CGS means "centimeter/gram/second," as opposed to MKS, which means "meter/kilogram/second."

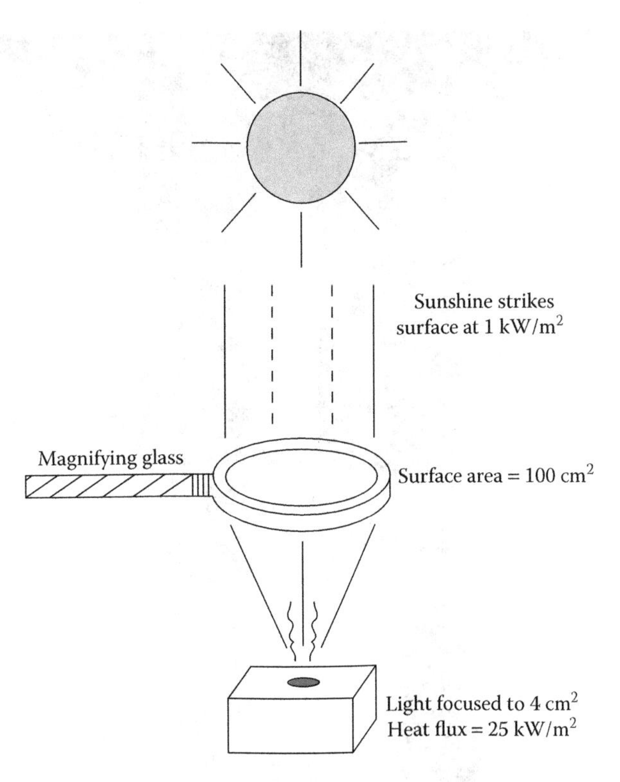

Figure 2.2 Using a magnifying glass to increase the radiant heat flux from the sun.

Figure 2.3 (a) Burn lines scorched into the underside of the eaves by a makeup mirror. (Courtesy of the David M. Smith, Associated Fire Consultants, Bisbee, AZ.)

Figure 2.3 (b) Origin of a fire that started on a curtain, caused by sunlight focused through the round lens in this "bubble window." (Courtesy of the Mick Gardiner, Gardiner Associates Training and Research [GATR], Cambridge, England.)

Table 2.2 Typical radiant heat fluxes

Approximate radiant heat flux (kW/m^2)	Comment or observed effect
170	Maximum heat flux as currently measured in a post-flashover fire compartment.
80	Heat flux for protective clothing Thermal Protective Performance (TPP) Test.[a]
52	Fiberboard ignites spontaneously after 5 seconds.[b]
29	Wood ignites spontaneously after prolonged exposure.[b]
20	Heat flux on a residential family room floor at the beginning of flashover.[c]
20	Human skin experiences pain with a 2-second exposure and blisters in 4 seconds with second-degree burn injury.[d]
15	Human skin experiences pain with a 3-second exposure and blisters in 6 seconds with second-degree burn injury.[d]
12.5	Wood volatiles ignite with extended exposure[e] and piloted ignition.
10	Human skin experiences pain with a 5-second exposure and blisters in 10 seconds with second-degree burn injury.[d]
5	Human skin experiences pain with a 13-second exposure and blisters in 29 seconds with second-degree burn injury.[d]
2.5	Human skin experiences pain with a 33-second exposure and blisters in 79 seconds with second-degree burn injury.[d]
2.5	Common thermal radiation exposure while fire fighting.[f] This energy level may cause burn injuries with prolonged exposure.
1.0	Nominal solar constant on a clear summer day.[g]

Source: NFPA 921, *Guide for Fire and Explosion Investigations*, NFPA, Quincy, MA, 2017. With permission.
Note: The unit kW/m^2 defines the amount of heat energy or flux that strikes a known surface area of an object. The unit kW represents 1000 watts of energy and the unit m^2 represents the surface area of a square measuring 1 m long and 1 m wide. For example, 1.4 kW/m^2 represents 1.4 multiplied by 1000 and equals 1400 watts of energy. This surface area may be that of the human skin or any other material.

[a] From NFPA 1971.
[b] From Lawson, "Fire and Atomic Bomb."
[c] From Fang and Breese, "Fire Development in Residential Basement Rooms."
[d] From Society of Fire Protection Engineering Guide: "Predicting 1st and 2nd Degree Skin Burns from Thermal Radiation." March 2000.
[e] From Lawson and Simms, "The Ignition of Wood by Radiation." pp. 288–292.
[f] From U.S. Fire administration, "Minimum Standards on Structural Fire Fighting Protective Clothing and Equipment." 1997.
[g] *SFPE Handbook of Fire Protection Engineering*, 2nd edition.

2.3 STATES OF MATTER

Ordinary matter, whether combustible or not, is found in one of three states or phases: (1) solid, (2) liquid, or (3) gas. Figure 2.4 is a schematic representation of the three states of matter and the changes that accompany a change of state. Melting and vaporization are endothermic changes. Freezing and condensation are exothermic. Liquids and solids are referred to as *condensed phases* because they take up much less space than a gas. To a first approximation, a liquid will take up 10% more space than a solid of the same material. A gas (or vapor) of that liquid will take up 100–1,200 times as much space as the liquid, depending on its molecular weight and the relative attraction that exists between its molecules. Because equal volumes of gases contain equal numbers of molecules, the volume of vapor that a given volume of liquid will produce can be estimated if its molecular weight and its density are known. A change of state does not involve a change in chemical composition.

Because fire is a gaseous phenomenon, we start with the gases and work our way up the density scale to the liquids and then to the solids.

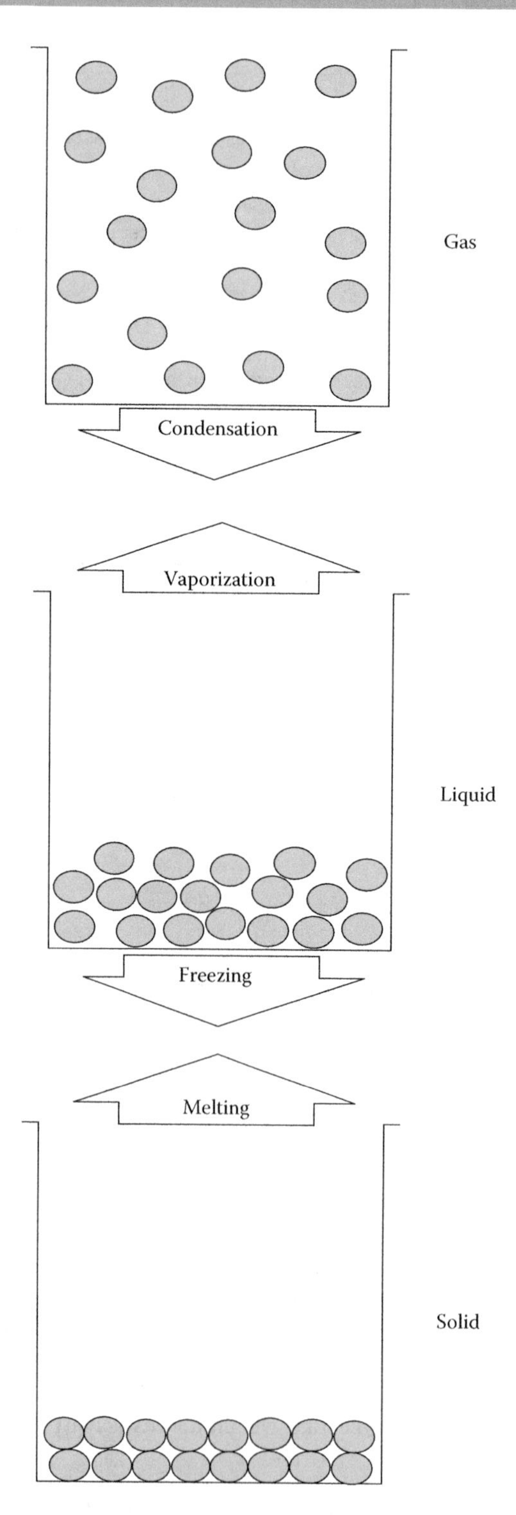

Figure 2.4 Schematic representation of the three states of matter.

2.4 THE BEHAVIOR OF GASES

Gases are the simplest form of matter in that the individual components, atoms, or more commonly, molecules, are not tightly bound to one another. A gas can be an elemental gas or it can be a compound. Elemental gases, with the exception of the inert or noble gases (i.e., helium, neon, argon, krypton, xenon, and radon), are usually found as diatomic gases; that is, there are two atoms of the same element bound to each other. Thus, hydrogen, oxygen, and nitrogen occur as H_2, O_2, and N_2, respectively. Gases of compounds, by definition, have at least two different atoms bound together. Carbon monoxide is represented as CO, and methane is represented as CH_4. (See Sidebar 2.1.)

SIDEBAR 2.1 Chemical nomenclature

Learning any discipline, scientific or otherwise, first requires a mastery of the nomenclature, or terminology, of the discipline. Chemistry is no different. There are more than 2 million organic compounds, so there must be some method for keeping the names straight. The naming of compounds is the purview of the International Union of Pure and Applied Chemistry (IUPAC). Before IUPAC, compounds were named by their discoverers or had ancient Arabic, Greek, or Latin names. Such names are called "trivial" names, but trivial or not, many substances are more well known by these names than by the official IUPAC names. The petrochemical industry in particular uses many non-IUPAC names for common compounds.

Many compounds have a Greek prefix denoting the number of carbon atoms contained in the molecule and a suffix denoting the type of functional groups present. The simplest series of molecules is called the *alkanes*. Alkanes are also known as paraffins (an obsolete term), and as saturated hydrocarbons, because they contain only carbon and hydrogen, and because they contain only single bonds. (They are saturated with hydrogen.) Alkanes all have the formula C_nH_{2n+2}, where *n* is a whole number. The simplest alkane is methane, CH_4. Methane is the major component of natural gas. It is shaped like a tetrahedron, with a carbon atom in the center and hydrogen atoms at each corner. For simplicity, two-dimensional representations will be used in most drawings in this text. Methane looks like this:

```
    H
    |
H — C — H
    |
    H
```

Methane

A two-carbon alkane is called ethane (C_2H_6), which looks like this:

```
    H   H
    |   |
H — C — C — H
    |   |
    H   H
```

Ethane

Inserting a (–CH_2–) group between the two carbons makes propane (C_3H_8):

```
    H   H   H
    |   |   |
H — C — C — C — H
    |   |   |
    H   H   H
```

Propane

Butane (C_4H_{10}), pentane (C_5H_{12}), and hexane (C_6H_{14}) are shown in the following:

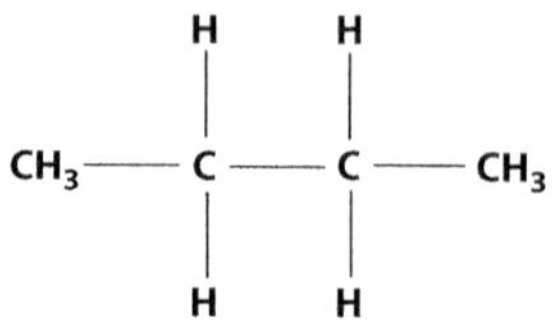

Butane

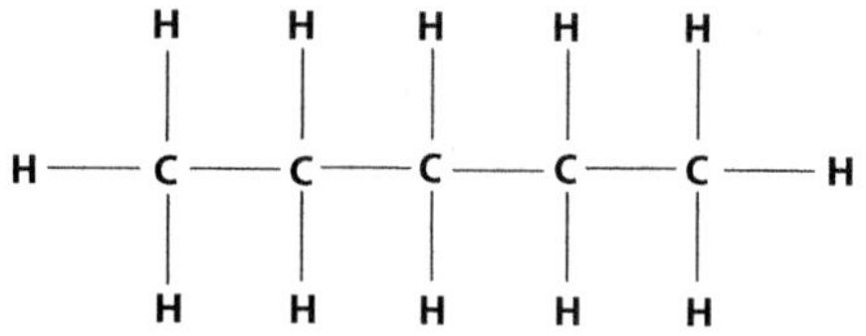

Pentane

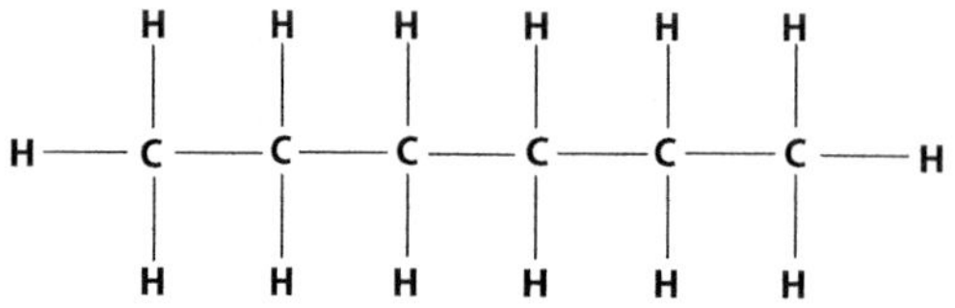

Hexane

Molecules having the same chemical formula but a different arrangement of atoms are called *isomers*. One can see that there are two ways to arrange 4 carbon atoms and 10 hydrogen atoms. The carbons can be in a straight line as shown earlier, or they can have a "branched" structure, such as

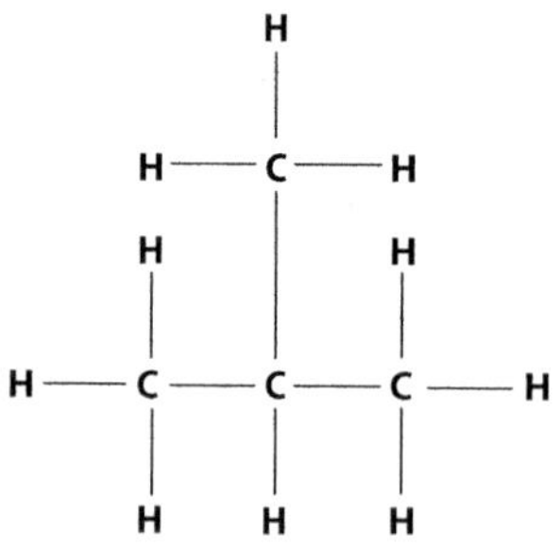

Isobutane

This molecule is called isobutane, or to use the IUPAC name, 2-methylpropane—there is a propane "backbone" with a methyl ($-CH_3$) group attached at the middle carbon, in position 2. For pentane, a similar rearrangement can be performed to make 2-methylbutane (or isopentane):

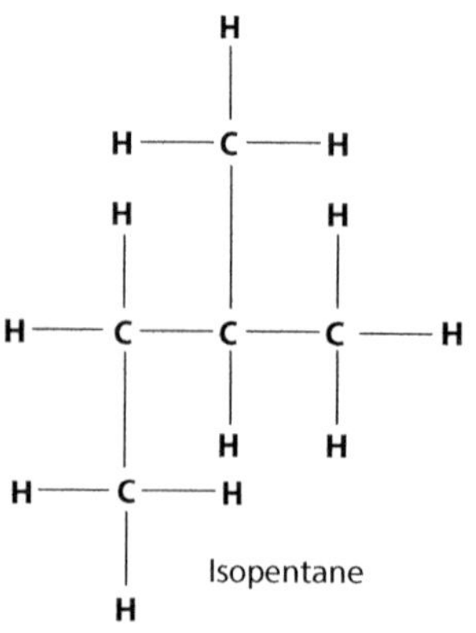

Isopentane

But there is yet another way to arrange 5 carbon atoms and 12 hydrogen atoms. Two methyl groups can be attached to a propane backbone to make 2,2-dimethylpropane, also known as neopentane.

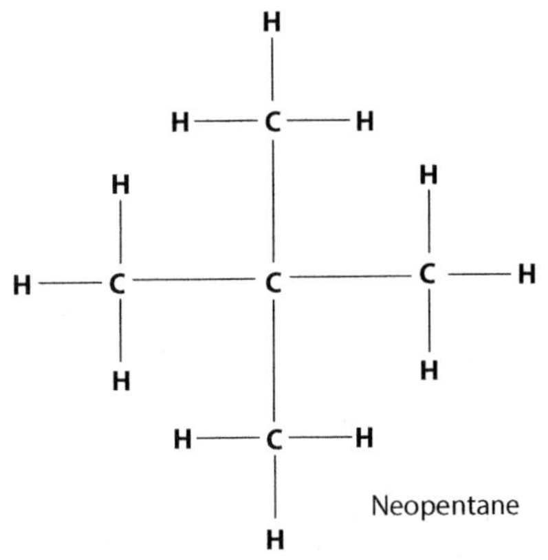

Neopentane

As the number of carbon atoms increases, the number of possible isomers also increases. There are 2 isomers of butane, 3 of pentane, 5 of hexane, 9 of heptane, and 15 of octane.

If one of the bonds in an alkane is changed to a double bond, it becomes an *alkene*. It will have two fewer hydrogen atoms and has the formula C_nH_{2n}. Alkenes are named similarly to alkanes but end with the suffix *-ene* instead of *-ane*. Their names are ethene (or ethylene), propene or propylene, butene, pentene, etc. Alkenes are also called olefins. They are the monomers used to form polyolefin plastics (polyethylene, polypropylene, and polybutylene).

Another major class of organic compounds with which fire investigators should be familiar is *aromatic* compounds. These contain a six-membered benzene ring. A large percentage of the compounds in gasoline are aromatics. Benzene is the simplest aromatic compound. It looks like a hexagon. The ring can be "substituted" at one or more of the six corners. Adding one methyl group yields toluene or methylbenzene. Benzene and toluene are shown in the following.

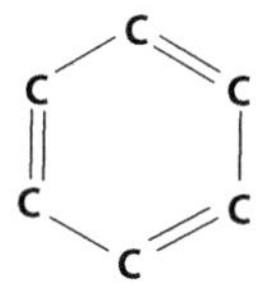

Benzene 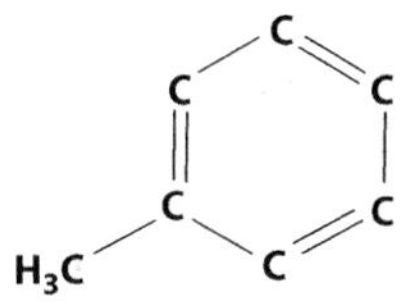 Toluene

Adding two methyl groups yields xylene or dimethylbenzene. There are three different isomers of xylene, known as ortho-, meta-, and para-xylene. The three isomers of xylene are shown in the following.

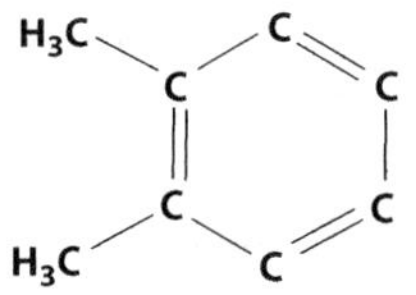

O-xylene 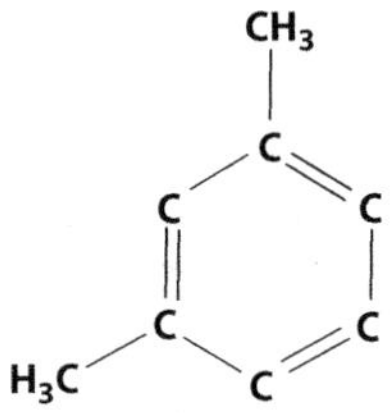 M-xylene 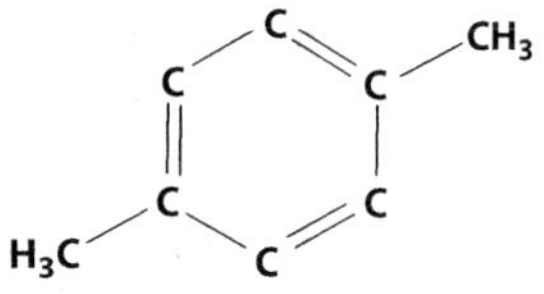 P-xylene

Organic compounds can also form rings that are saturated (i.e., not benzene rings). The most commonly encountered such ring is cyclohexane. Like benzene, cyclohexane can be substituted. Cyclohexane, methylcyclohexane, and butylcyclohexane are shown here.

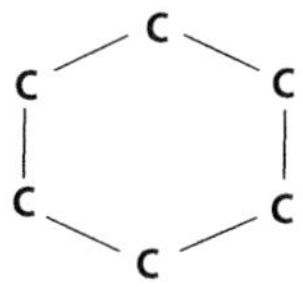

Cyclohexane 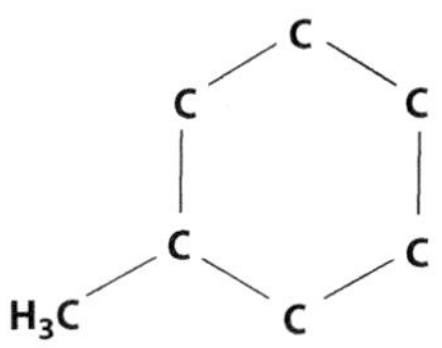 Methylcyclohexane 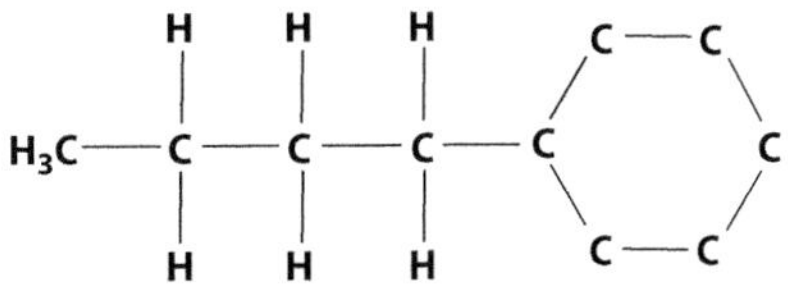 Butylcyclohexane

Some organic compounds contain multiple benzene rings fused together. These are called polynuclear aromatics (PNAs) or polynuclear aromatic hydrocarbons (PAHs). Naphthalene is the simplest such compound. Naphthalenes can be substituted. Naphthalene and 1- and 2-methylnaphthalene are shown here.

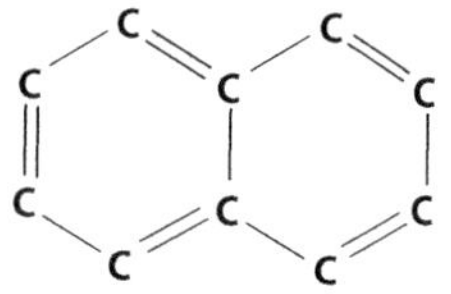

Naphthalene 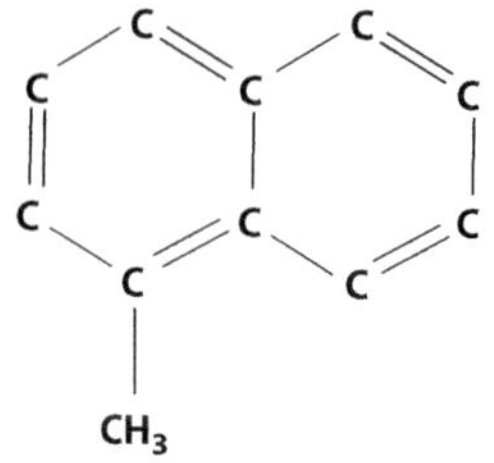 1-methylnaphthalene 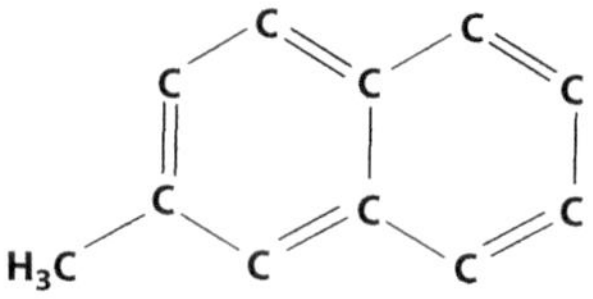 2-methylnaphthalene

Other compounds that might be of interest to fire investigators are alcohols, which substitute a hydroxyl (-OH) group for a hydrogen. Methanol, ethanol, and isopropanol are found in consumer products. These three light alcohols are shown here.

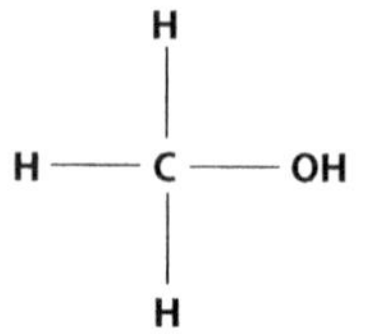

Methanol 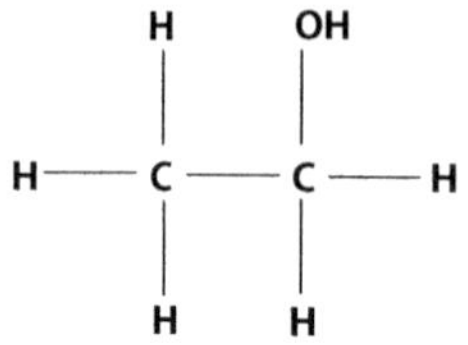 Ethanol 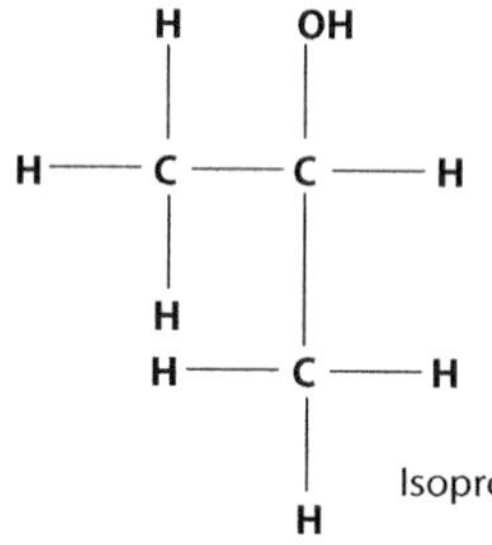 Isopropanol

Ketones contain a carboxyl group (C = O). Commonly encountered ketones include acetone (dimethyl ketone) and methyl ethyl ketone (MEK) shown here.

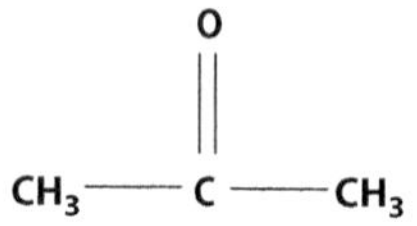

Acetone 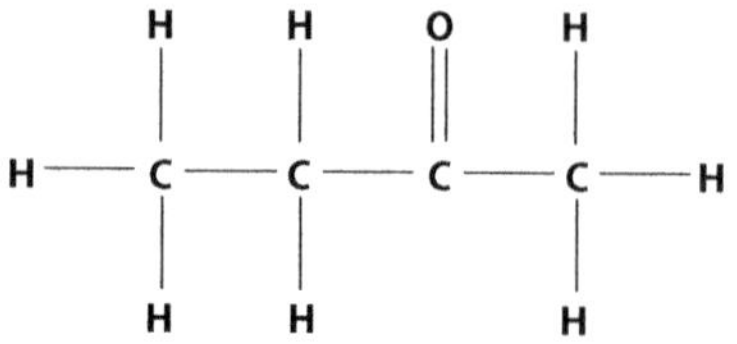 Methyl ethyl ketone

Other classes of compounds are contained in petroleum products, are produced in fires, or are found in the background of our everyday environment. Information about any kind of compound known to science can be found at the IUPAC Web site.

A gas exists in a state characterized by very low density and viscosity relative to liquids and solids. Gases expand and contract with changes in pressure and temperature and are uniformly distributed throughout the internal volume of any container.

For most purposes, it can be assumed that gases obey the *ideal gas law:*

$$PV = nRT$$

where:

P is the pressure
V is the volume
n is the number of moles
R is the universal gas constant
T is the absolute temperature

- Reducing the size of a closed container of gas by half compresses the gas and doubles the pressure (e.g., a piston or a bicycle pump).
- Doubling the number of molecules of gas, by adding more, and allowing the container to expand doubles the volume (e.g., a rubber balloon).
- Doubling the number of molecules and holding the volume constant doubles the pressure (e.g., an automobile tire).
- Raising the temperature from 298 K (25°C, 77°F) to 596 K (324°C, 613°F) doubles the volume if the pressure stays constant (e.g., a compartment fire) or doubles the pressure if the volume stays constant (e.g., a propane cylinder, or any closed container in a fire). The math only works when one uses the absolute temperature—200°C is not "twice as hot" as 100°C. See Sidebar 2.2 for a discussion of temperature scales.

True gases are different from vapors. Gases exist in the gaseous state at standard temperature and pressure (STP; 1 atm, 760 torr, at 0°C, or 273 K, or 32°F), while vapors are the gas phase of substances that are liquids or solids at standard temperature and pressure.

SIDEBAR 2.2 Temperature scales

There are four temperature scales currently in common use. Two of them—the Fahrenheit[4] and Celsius[5] scales—are based on the behavior of water and are known as empirical temperature scales. Water freezes at 32°F or 0°C. Water boils at 212°F or 100°C. Thus, 180 (212 – 32) Fahrenheit degrees equals 100 Celsius degrees. To convert Fahrenheit to Celsius, first subtract 32 and then multiply by 5/9. To convert Celsius to Fahrenheit, multiply by 9/5 and then add 32. (The Fahrenheit and Celsius scales converge at –40°)

The other two scales are based on only one point, absolute zero, where all molecular motion stops. The Rankine[6] scale measures from absolute zero in degrees equivalent to Fahrenheit degrees. Absolute zero is minus 459°F. Water freezes at 491 degrees Rankine and boils at 671 degrees R. The Kelvin[7] scale measures from absolute zero in the equivalent of degrees Celsius, but the term *degrees* is not used. Water freezes at 273.15 K and boils at 373.15 K.

When considering differences in temperature, a difference in kelvins is the same as a difference in degrees C. A difference in degrees R is the same as a difference in degrees F. However, when performing calculations involving thermodynamics or fire dynamics, the temperature is usually the absolute temperature. Discussing something that is "twice as hot" requires the use of an absolute temperature scale.

To consider the expansion of a gas based on temperature alone, start with a volume of one liter at room temperature (298 K). The gas will double in volume if the temperature is raised to 596 K, and it will triple if the temperature is raised to 894 K. Because most people do not think in kelvins, the temperature can be converted to Fahrenheit or Celsius (e.g., 298 K = 25°C = 77°F; 596 K = 323°C = 613°F; and 894 K = 621°C = 1150°F).

This text uses both Fahrenheit and Celsius scales to describe temperatures, but absolute temperature is reported only in kelvins.

The most familiar gas is air, a *mixture* containing 78% nitrogen; 20.95% oxygen; 0.93% argon; 0.04% carbon dioxide (and rising); and trace amounts of neon, helium, methane, krypton, nitrous oxide, hydrogen, xenon, and ozone (O_3) [2]. Hereafter, for simplicity's sake, air is considered a gaseous substance containing approximately 80% nitrogen and approximately 20% oxygen. The average molecular weight of air can be calculated by knowing the molecular weight of nitrogen (N_2, 28), the molecular weight of oxygen (O_2, 32), and the relative proportions of the two. This gives rise to the following equation:

$$0.8(28) + 0.2(32) \approx 29$$

A mole (Avogadro's number, 6.022×10^{23} molecules) of any ideal gas at STP occupies a volume of 22.4 liters (approximately 24 L at room temperature). Dividing the weight of a mole of air by its volume yields a density of approximately 1.3 g/L.

---

[4] Named for Daniel Fahrenheit (1686–1736), a German physicist and instrument maker who invented the alcohol thermometer in 1709.

[5] Named for Anders Celsius (1701–1744), a Swedish astronomer. His first scale had water freezing at 100 and boiling at 0. The order was reversed after his death. Celsius's research proved that the freezing temperature is independent of atmospheric pressure, while boiling temperature is dependent on atmospheric pressure. The Celsius scale is also known as the centigrade scale, but Celsius is the preferred term.

[6] Named for William Rankine (1820–1872), a Scottish civil engineer who studied steam engines and contributed to the understanding of thermodynamics.

[7] Named for William Thompson, Baron Kelvin (1824–1907), another Scot who worked with Rankine on thermodynamics problems and explored the concept of irreversible processes.

It is important to know the density of a gas, particularly a fuel gas, to predict how it might behave. But the density of gases and vapors is usually not expressed in weight per unit volume but rather as a ratio of the density of the gas to the density of air. This ratio is exactly the same as the ratio of the molecular weight of the gas divided by the molecular weight of air (calculated earlier as 29 g/mole). Gases or vapors having a molecular weight greater than 29 g/mole are "heavier than air," and those with a molecular weight less than 29 g/mole are "lighter than air." To make density calculations simpler, a value of 1 is assigned to air, and a gas or vapor is described by its *specific gravity* (its density relative to air) rather than by its absolute density.

Thirteen gases are lighter than air: hydrogen, helium, hydrogen cyanide, hydrogen fluoride, methane, ethylene, diborane, illuminating gas, carbon monoxide, acetylene, neon, nitrogen and ammonia (4H MEDIC ANNA). Illuminating gas is an obsolete form of natural gas made from coal. *All other gases and vapors (except for water vapor) are heavier than air* when at the same temperature as the air. Gases heavier than air at room temperature can be lighter than air when heated.

The combustion of hydrogen with oxygen to produce water is the simplest combustion reaction and was one of the first gaseous reactions studied. Two volumes of hydrogen mixed with one volume of oxygen produce two volumes of water vapor. The chemical equation (see Sidebar 2.3 on balancing equations) looks like this:

$$2H_2 + O_2 \rightarrow 2H_2O$$

SIDEBAR 2.3 Balancing chemical equations: a quick refresher

NFPA 1033 requires that a fire investigator have knowledge beyond the high school level of "fire chemistry." Faced with a qualifications challenge, it would be difficult to persuade an adversary that one need not know how simple fuels combine with oxygen. The judge quoted at the beginning of this chapter was certainly not persuaded of that. The following is high school chemistry.

The law of conservation of matter states that matter can be neither created nor destroyed; it only can be changed in form. When describing chemical processes, therefore, the number and kinds of atoms in the starting materials (reactants) must equal the number and kinds of atoms in the products. The law of constant proportions (also known as the law of multiple proportions) makes the process of balancing equations no more complex than first-year algebra. There will always be an equation where the amounts of each reactant can be expressed as a whole number.

Consider the combustion of hydrogen in air to make water. Hydrogen and oxygen are diatomic gases. H_2 and O_2 combine to make H_2O. Expressed this way, it can be seen that there are two atoms of oxygen to start with but only one oxygen atom in the water. Because the water contains two hydrogen atoms for each oxygen atom, it stands to reason that one balances the equation by doubling the amount of hydrogen. The correct expression is

$$2H_2 + O_2 \rightarrow 2H_2O$$

When carbon-based fuels burn, the products of complete combustion are carbon dioxide (CO_2) and water. The secret to balancing hydrocarbon combustion equations is to first balance the carbons. For the burning of methane (the major component of natural gas), this is simple because there is one carbon atom on each side of the equation. However, $CH_4 + O_2 \rightarrow CO_2 + H_2O$ is not balanced with respect to hydrogen (four reactant atoms but only two product atoms), nor is it balanced with respect to oxygen (two reactant atoms but three product atoms). The hydrogen can be balanced by doubling the water, as in $CH_4 + O_2 \rightarrow CO_2 + 2H_2O$, but this still leaves an oxygen deficiency (two reactant atoms but four product atoms). Doubling the oxygen in the reactants balances the equation:

$$CH_4 + 2O_2 \rightarrow CO_2 + 2H_2O$$

Because air is only one-fifth oxygen (actually, it is 20.95% by volume, but for our purposes, 20% is close enough and easier to use), it can be seen that 10 volumes of air are required for each volume of methane burned.

Now, consider the burning of a larger molecule with 10 carbon atoms, decane, $C_{10}H_{22}$. The products are still carbon dioxide and water. Start by balancing the carbons: $C_{10}H_{22} + O_2 \rightarrow 10CO_2 + H_2O$. It is immediately clear that there will be more than one water molecule. In fact, there are 11 water molecules:

$$C_{10}H_{22} + O_2 \rightarrow 10CO_2 + 11H_2O$$

Now, there are 31 oxygen atoms in the products, so there must be 15.5 oxygen molecules in the reactants:

$$C_{10}H_{22} + 15.5O_2 \rightarrow 10CO_2 + 11H_2O$$

To use whole numbers, simply double each term as follows:

$$2C_{10}H_{22} + 31O_2 \rightarrow 20CO_2 + 22H_2O$$

One can see that burning a volume of decane vapor requires 77.5 volumes of air (15.5 × 5). If there is a deficiency of air, say, 65 volumes instead of 77.5, the reaction will produce carbon monoxide (CO):

$$2C_{10}H_{22} + 26O_2 \rightarrow 10CO_2 + 22H_2O + 10CO$$

Because these are all even numbers, the equation can be simplified to

$$C_{10}H_{22} + 13O_2 \rightarrow 5CO_2 + 11H_2O + 5CO$$

Simple algebra!

The chemical equation describing the combustion of the simplest organic fuel, methane, is only slightly more complicated. In this case, one volume of methane combines with two volumes of oxygen to produce two volumes of water vapor and one volume of carbon dioxide:

$$CH_4 + 2O_2 \rightarrow 2H_2O + CO_2$$

As simple as this reaction seems, it has been studied in some detail to understand all the intermediate steps that can, and usually do, take place. For example, if there is insufficient oxygen, burning methane, usually encountered as burning natural gas, will produce carbon monoxide:

$$CH_4 + 1.5O_2 \rightarrow 2H_2O + CO \qquad \text{or} \qquad 2CH_4 + 3O_2 \rightarrow 4H_2O + 2CO$$

The previous *stoichiometric* equations only list the starting materials and the end products. A great many steps exist between reactants and products, involving the transitory existence and reaction of atoms and free radicals. In fact, it is known that the combustion of methane, the simplest of all the fossil fuels, involves no fewer than 40 reactions and 48 different species (molecules, atoms, or radicals) that can be detected at some point during combustion. Table 2.3 lists some of these reactions. The finer details of these processes are beyond the scope of this book.

It is worthwhile to note that most combustion occurs in air, not in pure oxygen, and to have the required 2 volumes of oxygen, it is actually necessary to have almost 10 volumes of air because each volume of oxygen is accompanied by 4 volumes of nitrogen.

It is also worthwhile to note that these equations for the combustion of methane are silent on the subject of heat produced, but clearly, the products of the reaction are at a significantly higher temperature than the starting materials. Thus, despite the fact that three volumes of reactants (one volume of CH_4 and two volumes of O_2) yield three volumes

Table 2.3 Intermediate reactions that occur in combustion of methane

	CH_4	+	M	=	$\cdot CH_3$	+	H•	M	a
	CH_4	+	•OH	=	$\cdot CH_3$	+	H_2O		b
	CH_4	+	H•	=	$\cdot CH_3$	+	H_2		c
	CH_4	+	•O•	=	$\cdot CH_3$	+	•OH		d
	O_2	+	H•		•O•	+	•OH		e
	CH_3	+	O_2	=	CH_2O	+	•OH		f
	CH_2O	+	•O•	=	•CHO	+	•OH		g
	CH_2O	+	•OH	=	•CHO	+	H_2O		h
	CH_2O	+	H•	=	•CHO	+	H_2		i
	H_2	+	•O•	=	H•	+	•OH		j
	H_2	+	•OH	=	H•	+	H_2O		k
	CHO	+	O•	=	CO	+	•OH		1
	CHO	+	•OH	=	CO	+	H_2O		m
	CHO	+	H•	=	CO	+	H_2		n
	CO	+	•OH	=	CO_2	+	H•		o
H•	+	•OH	=	M	=	H_2O	+	M	p
H•	+	H•	=	M	=	H_2	+	M	q
H•	+	O_2	=	M	=	$HO_2\cdot$	+	M	r

Source: Drysdale, D., *An Introduction to Fire Dynamics*, Wiley-Interscience, New York, 1985. With permission.
Note: This reaction scheme is by no means complete. Many radical-radical reactions, including those of the $HO^{\bullet}_2$ radical, have been omitted. M is any "third body" participating in radical recombination reactions (p–r) and dissociation reactions such as a.

of product (two volumes of water vapor and one volume of CO_2), the products will occupy a greater volume than the reactants simply because they are hotter.

The amount of heat released when a specific amount of substance is completely burned is its *heat of combustion*. Heat of combustion is measured as energy per unit mass. This is usually reported as kilojoules per kilogram (kJ/kg) but can be reported as Btu/lb or calories/gram, or, if the mass is converted to volume, kJ/m^3 or Btu/ft^3. The heat release rate is then the heat of combustion multiplied by the *mass loss rate* (the rate at which the fuel is consumed, in grams/second).

A cubic foot of natural gas (the kind that comes out of the pipe from the gas company) has an energy content of approximately 1,000 Btu or over a million joules. If all this energy is released in one second, the explosion will have a heat release rate of 1 MW. If burned over an hour, as is typical in a pilot light, the heat release rate is less than 300 W. The energy content is the same in both cases—what makes the difference is how rapidly that energy is released. When one considers the energy available if a 40,000 Btu/hr burner is left open without a pilot light for an hour, one begins to understand the necessity for safety devices on gas-burning appliances.

Methane is the most common fuel gas that fire investigators are likely to encounter. Methane is recovered from gas wells as a fossil fuel and is the cleanest burning of all the fossil fuels. Although exact percentages differ with geographic areas and there are no standards that specify its composition, natural gas is mostly methane; with lesser amounts of nitrogen, ethane, propane; and with traces of butane, pentane, hexane, carbon dioxide, and oxygen [4]. Methane recovered as a fossil fuel is known as *thermogenic gas*.

Methane is also produced by the decomposition of organic substances in oxygen-poor environments. This gas is known as landfill gas, swamp gas, or sewer gas. Gas produced by the recent decomposition of organic materials is called biogenic gas and is characterized by the absence of the higher-molecular-weight species (ethane, propane, etc.). Distinguishing the source of "fugitive" methane (a sometimes critical question for gas utility companies) is accomplished using gas chromatography. A comparison of thermogenic gas and biogenic gas is shown in Figure 2.5. Note that, because of the higher economic value of the larger molecules, gas refiners have found better ways of removing them from commercial natural gas, so it is now possible to find natural gas with little or no larger alkane gases.

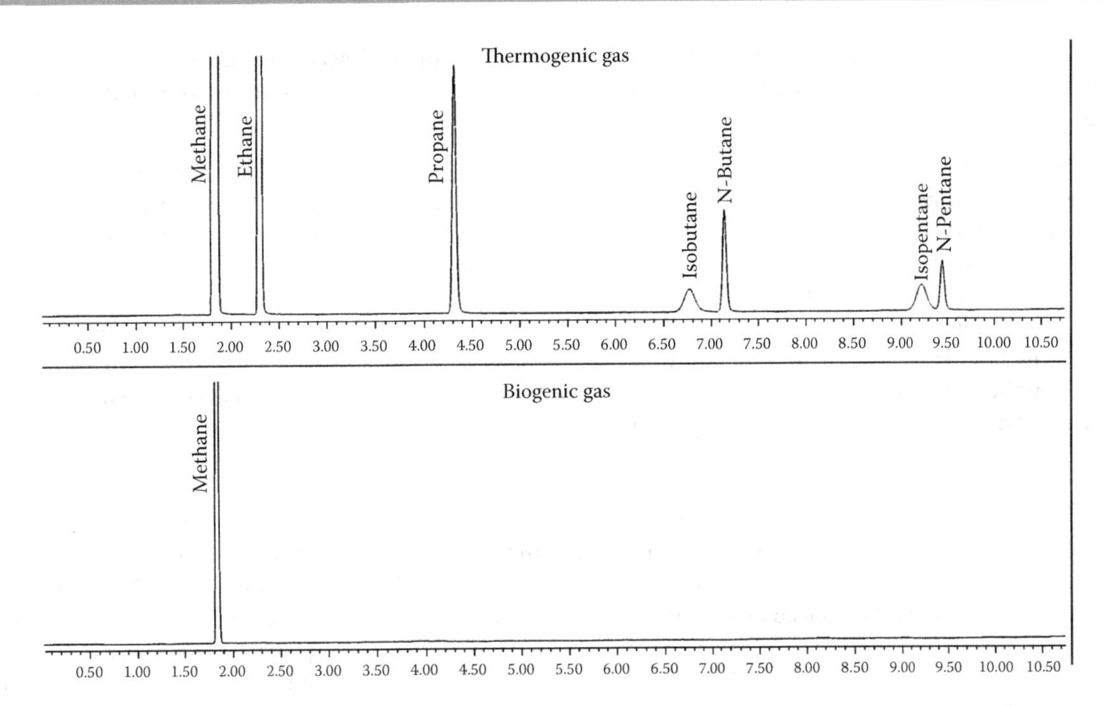

Figure 2.5 Comparison of thermogenic gas (top) with biogenic gas (bottom). Chromatograms have been normalized so that the propane peak in the natural gas is at 100% of scale. Methane and ethane peaks were plotted off scale so that the C_4 and C_5 alkanes can be seen.

Propane is the next most common fuel gas that an investigator is likely to encounter. Unlike methane, which is lighter than air (its molecular weight is 16, so its specific gravity is 0.55 [16 ÷ 29 = 0.55]), propane has a molecular weight of 44 and a specific gravity of 44 ÷ 29 = 1.51. Propane leaking from an open gas line or appliance tends to "puddle" on the floor of the compartment in which it is released. People who have survived ignitions of fugitive propane have described the event as standing in a sea of fire. An interesting demonstration of the density of propane and its spreading characteristics can be made using a cup of water and a piece of dry ice. (Dry ice is solid carbon dioxide and has the same specific gravity as propane.) When the dry ice is dropped into the water, gaseous carbon dioxide vapor bubbles out, flows downward around the cup, and spreads out on whatever surface is beneath the cup. This behavior is exactly the same as the behavior of propane, but the demonstration is safe because the carbon dioxide cannot ignite.

Propane has the molecular formula C_3H_8. The equation for the complete combustion of propane is

$$C_3H_8 + 5O_2 \rightarrow 3CO_2 + 4H_2O$$

In this case, there are six volumes of reactants and seven volumes of product, but it will still be the increase in temperature rather than the increase in the number of molecules that is mostly responsible for the increased volume of the products. Note that, while one volume of methane requires 10 volumes of air for complete combustion, a volume of propane requires 25 volumes of air. The burning of a cubic foot of propane produces approximately 2,500 Btu (see Sidebar 2.3).

This indicates that, volume for volume, there is more energy in propane than in natural gas. It also suggests a relationship between the amount of oxygen consumed in a reaction and the amount of energy produced. This relationship forms the foundation of the energy measurement used in oxygen consumption calorimetry. An oxygen consumption calorimeter, used extensively in fire research, measures the depletion of oxygen as a result of flaming combustion processes. Small calorimeters might measure the heat release rate of a block of foam, while large furniture calorimeters might measure the output of a burning chair or sofa. These results are achieved by measuring how much oxygen is consumed. Hydrocarbon fuels are remarkably consistent. Within a range of ±5%, completely burned hydrocarbon fuels release 13.1 kJ/g of oxygen consumed.

Other fuel gases that an investigator is likely to encounter are butane and acetylene. Butane is the largest of the organic fuel gases because molecules larger than butane are liquids at room temperature. Even so, thermogenic gas (natural gas) does contain very small but detectable quantities of isopentane and pentane in the vapor phase.

Acetylene is a highly energetic molecule containing two carbon atoms and two hydrogen atoms in a straight-line array. Carbon atoms have four bonds available to them, but in the acetylene molecule, three of those bonds are used to bind the two carbons together. This triple bond releases far more energy than a single bond when it is broken. Acetylene is also highly sensitive to shock and heat and cannot safely be stored at pressures above 15 psig (or about twice atmospheric pressure—1 atmosphere = 14.7 psi). Pressure can be measured in absolute terms (psi) or in relationship to atmospheric pressure (psig or gauge). Commercial acetylene is sold in tanks with both a porous rocklike substance and liquid acetone. The acetylene is dissolved in the acetone, much like carbon dioxide is dissolved in soda water. When the valve on the acetylene tank is opened, the acetylene bubbles out as a result of the reduction in pressure. (The same thing happens when opening a bottle of soda, beer, or champagne.)

2.5 STOICHIOMETRY AND FLAMMABLE LIMITS

Stoichiometry deals with the relative proportions of substances—reactants and products—in a chemical reaction. In the earlier description of the reaction of methane with oxygen, we referred to volumes of methane and volumes of oxygen or air. Because the molecular weights of these substances are known, we could have also stated that one molar volume (16 g) of methane requires two molar volumes (64 g) of oxygen and produces two molar volumes (36 g) of water vapor and one molar volume (44 g) of carbon dioxide. A cubic foot of methane may shrink when cooled or expand when heated, but a gram is always a gram, regardless of volume. Every chemical reaction has characteristic proportions, and in the gas phase at standard temperature, the use of volumes is appropriate, but remember that it is actually the mass of material that is reacting.

A *stoichiometric mixture* exists when there is exactly the minimum amount of air required to react with the fuel completely. In these proportions, the reaction will be most efficient and the resulting flame temperatures will be highest. At the stoichiometric proportion, the energy required to ignite the mixture will also be lowest. This ignition energy is measured in millijoules. The minimum ignition temperature is likewise lowest at the stoichiometric concentration. For methane in air, Babrauskas reports that the auto ignition temperature (AIT) is 640°C, and the minimum ignition energy is 0.3 mJ [5].

A *flammable gas mixture* is a mixture of gases through which flames can propagate. The flame is initiated in the mixture by means of an external source, and the *limits of flammability* can be defined as the limiting composition of a combustible gas and air mixture beyond which the mixture will not ignite and continue to burn. If there is insufficient fuel, the mixture is said to be too *lean,* and if there is insufficient air, the mixture is said to be too *rich.* As can be seen from the equation describing the production of carbon monoxide (see Sidebar 2.3), it is possible for combustion to occur over a range of concentrations. This range is known as *the flammable range.* Figure 2.6 is a graphic illustration of the concept of flammable range.

Flammable limits (the term is interchangeable with *explosive limits*) define the concentrations below which the fuel/air mixture is too lean (lower flammable limit, lower explosive limit [LEL]) and above which the fuel/air mixture is too rich to burn (upper flammable limit, upper explosive limit [UEL]).

The limits of flammability are determined by laboratory analysis, and various environmental factors such as temperature and pressure; the geometry and size of the confining vessel can have an impact on the outcome of these measurements. Flammable ranges tend to be broader at higher temperatures; that is, the lower limit decreases and the upper limit increases. At higher pressures, the lower limit tends to remain nearly the same, but the upper limit increases. In narrow vessels, where there is substantial heat loss to the walls, the flammability limits are narrowed [6].

Knowing the volumes of air required for combustion based on stoichiometry, one should be able to predict where the middle of the flammable range will be. Not surprisingly, for methane, this is 10% (remember that 1 volume of methane requires 10 volumes of air to burn), and the flammable limits of methane in air are 5%–15%. These considerations regarding the flammable limits of gases also apply to vapors of flammable or combustible liquids, usually referred to as *ignitable liquids.*

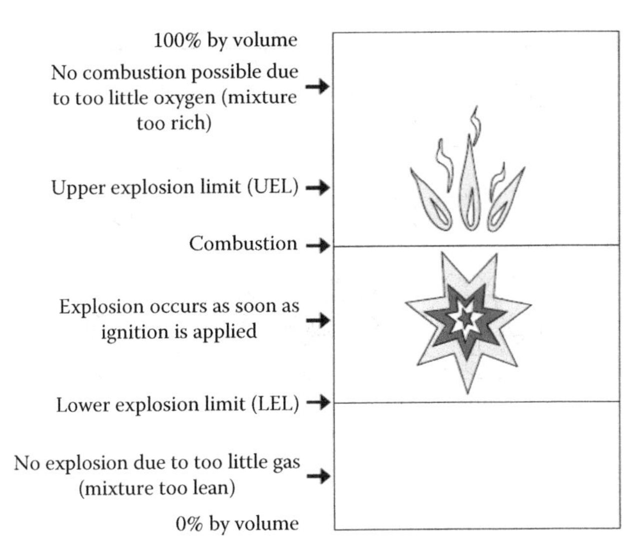

Figure 2.6 Schematic representation of the concept of flammable limits (also known as explosive limits).

2.6 THE BEHAVIOR OF LIQUIDS

Any liquid, ignitable or not, when placed in a closed container, begins to vaporize. If the temperature is held constant, the liquid eventually appears to stop vaporizing. Appearances can be deceiving. The liquid continues to vaporize, but when the liquid and the vapor above it are in a state known as *equilibrium,* the rate of vaporization equals the rate of condensation. *Relative humidity* is a term applied to water in air. At any given temperature, the concentration of water vapor above a still pool in a closed container reaches a constant, and the air is said to be *saturated.* Such air has a relative humidity of 100%. The percentage of water in the air versus the amount of water that the air could potentially hold if it were saturated is called the relative humidity.

For the vapors of liquids other than water, the amount of liquid in the vapor phase can be described as its concentration (percentage by volume in air), but the amount of vapor is usually determined by measuring its pressure. The concentration of a vapor above its corresponding liquid is referred to as the *vapor pressure* or partial pressure.

Vapor pressure is generally expressed in millimeters of mercury (mmHg), or torr. This measurement describes the height of a column of mercury in an evacuated tube (a barometer) that can be supported by the surrounding pressure. The air around us exerts a pressure of approximately 760 mmHg, or about 29.92 inches of mercury. This is the combined pressure of all the oxygen, nitrogen, and other gases in the atmosphere. Under most circumstances, a substance present at a concentration of 1% in air has a vapor pressure of 7.6 mmHg. If the vapor pressure is 76 mmHg, the substance has a concentration of 10%. If the vapor pressure is 760 mmHg, then the vapor concentration is 100%. This only occurs when the liquid that is producing the vapor is at its boiling point.

As long as there is equilibrium, the vapor pressure, or concentration, of a substance in the vapor phase above its liquid is constant, or fixed, for a given temperature. This is an important property of any liquid and has been measured and reported for thousands of liquids. It is important to note that equilibrium can only be achieved in a closed container, as shown in Figure 2.7.

Liquids with a high vapor pressure have a relatively high tendency to evaporate and are said to be *volatile* liquids. Examples of volatile liquids include ether, acetone, gasoline, and water. The term *volatile* applies equally to ignitable and nonignitable liquids. Nonvolatile liquids such as heavy oils and waxes, which have very low vapor pressures, can be made to burn if a means can be found to vaporize a sufficient number of molecules. Generally, the vapor pressure of a liquid (or a volatile solid) is inversely proportional to its molecular weight. Small molecules have high vapor pressures, and large molecules have low vapor pressures.

Combustion of ignitable liquids occurs only in the vapor phase. The bulk liquid temperature at which the concentration of vapors above the liquid in a still pool enclosed in a specified apparatus will ignite when exposed to a flame is known as the *flash point.* This is not the same as the ignition point, which describes the temperature of the ignition source, as opposed to the temperature of the liquid.

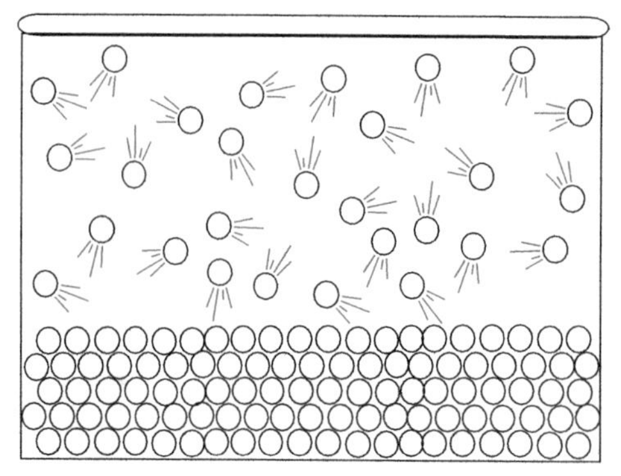

Closed container: evaporation rate equals
Condensation rate;
Equilibrium achieved

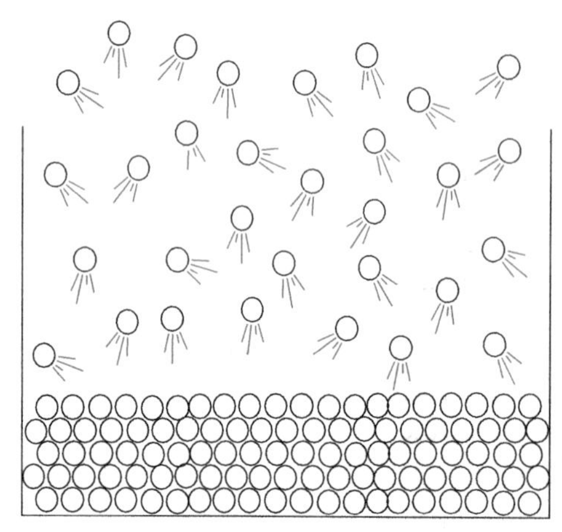

Open container: evaporation rate exceeds
Condensation rate;
Equilibrium never achieved

Figure 2.7 Schematic representation of the concept of vapor phase equilibrium.

It is entirely possible for a liquid to produce sufficient vapors to burn when the liquid is below its flash point. Flash point describes a very particular set of circumstances, that is, the laboratory apparatus shown in Figure 2.8, where the flame is held very close to the surface of the liquid and the liquid is gradually heated. At a certain temperature in this particular apparatus, the liquid can be seen to flash when the flame is applied to the vapor phase.

Warming the ignitable liquid is not the only way to achieve a concentration of vapors within the flammable range. Consider a kerosene lamp. Kerosene has a flash point in excess of 100°F, but it is easy to light a kerosene lamp at temperatures well below that. The reason for this is that, in the immediate vicinity of the wick, the kerosene vapors are present above their lower flammable limit. The wick acts to increase the surface area to such an extent that even a very low rate of evaporation results in a sufficiently high local concentration of kerosene molecules in the air to support combustion. Another example of combustion when a substance is below its flash point is a wax candle.

Caution is required when thinking about ways to increase surface area. If all that were required was increasing the total area, it should be possible to light a swimming pool full of diesel fuel on fire, simply because of the large surface area of the swimming pool. What should be considered is an increase in the surface area *as compared to* a flat pool of liquid. It is no easier to light a swimming pool full of diesel fuel than it is to ignite a small cup, but both can be successfully ignited if a wick is inserted.

To illustrate this point, consider a cylinder of air above a liquid pool. It is easy to see that adding a wick to the pool increases the vaporization of liquid into that cylinder because it increases the number of places

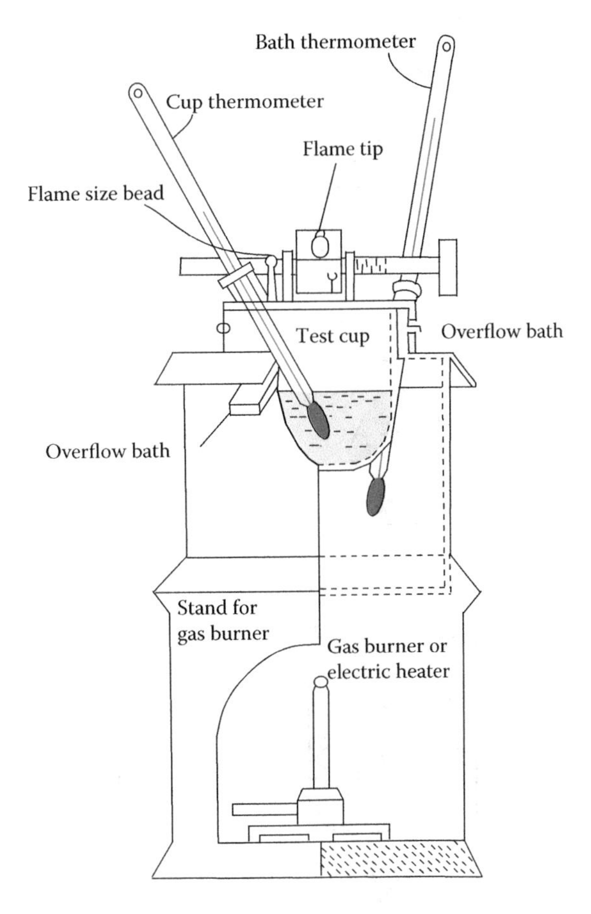

Figure 2.8 Tag closed cup flash point apparatus.

where evaporation can occur by increasing the surface area. Using the swimming pool analogy merely increases the diameter of the cylinder.

Another way to increase the concentration of vapor in air without heating the liquid is to atomize it mechanically. If the droplets can be made small enough, a relatively cool liquid (i.e., one below its flash point) can be made to vaporize sufficiently for combustion. Atomization such as this takes place in diesel engines when first started and in certain types of weed burners and flamethrowers. Figure 2.9 shows the various ways of producing the lower explosive limit.

Although liquids can form flammable vapor/air mixtures below their flash point, this is not to say that the flash point is insignificant. Flash points determine the temperature required to ignite a liquid present in a container, or as a puddle or pool. Flash points are useful principally in determining the *relative flammability* of one liquid compared to another.

Flash point can be measured in several different ways. While the Tag closed cup flash point apparatus shown in Figure 2.8 is the most widely used flash point tester, other standard methods might be more appropriate for a particular sample. The Tag tester is useful for most ignitable liquids, provided they are not viscous and do not contain nonflammable compounds having a higher vapor pressure than the flammable compounds. For viscous liquids, or for liquids with a flash point in excess of 93°C (200°F), the *Pensky–Martens* method, which involves stirring the sample, is more appropriate than the Tag tester. Most flash point testers require sample volumes of 50 mL. If there is a limited supply of liquid, the *Setaflash* tester, which requires only 2 mL, is more appropriate. If the liquid in question contains volatile nonflammable compounds such as methylene chloride, an erroneously high flash point may be determined. Such liquids should be tested at various stages of evaporation to correctly assess their flammability. This author has tested a mixture of methylene chloride and hydrocarbon solvent with a flash point of just over 100°F when poured straight from the container, but after being allowed to evaporate for a few minutes, the loss of methylene chloride caused the flash point to drop below 100°F. Thus, a mixture that was properly labeled (according to NFPA 30) [7] "combustible" quickly became a "flammable" liquid under conditions of expected use. Many paint stripping liquids

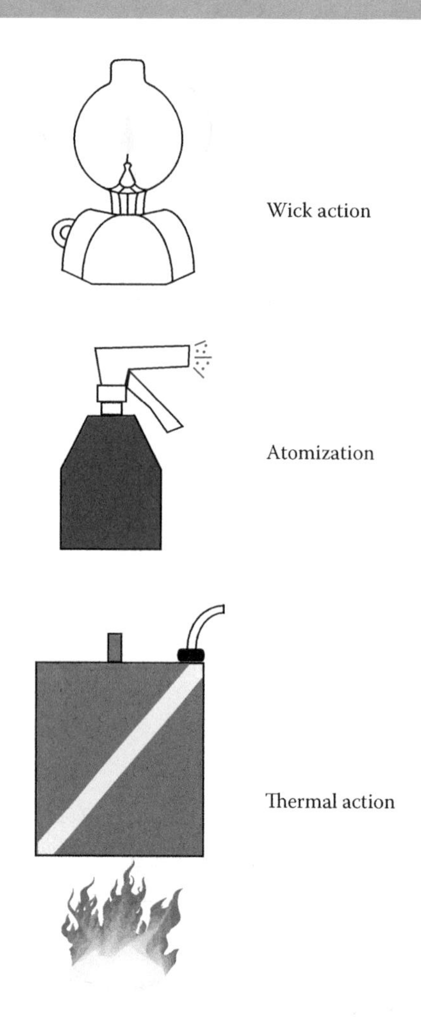

Figure 2.9 Different means of achieving the LEL of an ignitable liquid.

and pastes contain methylene chloride and are likely to present difficulties when assessing their potential hazards. ASTM[8] E502, *Standard Test Method for Selection and Use of ASTM Standards for the Determination of Flash Point of Chemicals by Closed Cup Methods*, provides useful guidance on the issue of flash point determination [8].

It is unusual for an investigator to encounter a "pure" ignitable liquid. Almost all ignitable liquids are actually mixtures of many different compounds. Any of the common fuels derived from petroleum are likely to have 50–500 compounds present in various proportions. What is interesting about mixtures is that the compounds with lower molecular weights tend to dominate the situation in the vapor phase. Conversely, after exposure to a fire, the lighter components are present at very reduced concentrations, and the relative concentrations of the heavier components are greater.

It is well known that the addition of even a small quantity of gasoline to a large quantity of kerosene can cause a dramatic reduction in the flash point of the mixture. Figure 2.10 shows the results of an experiment in which increasing quantities of gasoline were added to kerosene, which had an initial flash point of 120°F. The flash point dropped rapidly, to well below room temperature at 5% and to 0°F at 20%. Table 2.4 shows the vapor pressures of representative compounds found in gasoline and kerosene and explains why gasoline is so much more volatile.

The behavior of solutions of liquids is dictated by *Raoult's law*, which states that the vapor pressure of a compound above a mixture of liquids is equal to the vapor pressure of the pure compound times the mole fraction of the

[8] ASTM is the abbreviation for the American Society for Testing and Materials, a voluntary standards development organization. Now known as ASTM International, ASTM promulgates more than 12,000 standards (test methods, practices, guides, and specifications) for all types of goods and services, including forensic sciences.

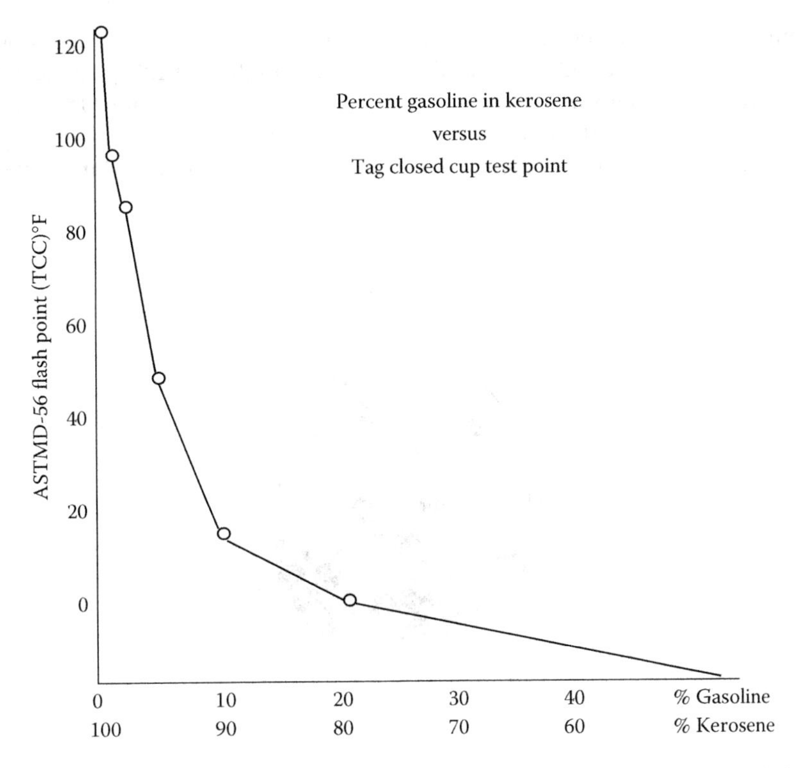

Figure 2.10 Raoult's law in action: a decrease in flash point with increase in concentration of gasoline in kerosene.

Table 2.4 Comparison or boiling points and vapor pressures of gasoline and kerosene components

				Vapor pressure in mmHg at:				
		Boiling point		0°C	20°C	40°C		60°C
Compound	Formula	°C	°F	(32°F)	(68°F)	(104°F)		(140°F)
Gasoline Components								
iso-Pentane[a]	C_5H_{12}	27.8	(82)	297	598	>760	>760	(b.p. 27.8°C)
n-Pentane	C_5H_{12}	36.1	(97)	222	430	>760	>760	(b.p. 36.1°C)
iso-Hexane[b]	C_6H_{14}	60.3	(140)	72.7	207	385	754	
n-Hexane	C_6H_{14}	68.7	(155)	47.6	137	315	596	
n-Heptane	C_7H_{16}	98.4	(209)	12.6	37.2	94.5	251	
Toluene	C_7H_8	110.6	(231)	8.3	21.5	64.5	165	
Xylenes[c]	C_8H_{10}	144.0	(292)	2.6	7.7	21.9	50	
Kerosene Components								
n-Undecane	$C_{11}H_{24}$	195.8	(385)	<1	<1	2.6	7.0	
n-Dodecane	$C_{12}H_{26}$	216.2	(421)	<1	<1	<1	2.6	
n-Tridecane	$C_{13}H_{28}$	234.0	(453)	<1	<1	<1	1.0	
n-Tetradecane	$C_{14}H_{30}$	252.5	(487)	<1	<1	<1	<1	(1 mm 76.4°C)
n-Pentadecane	$C_{15}H_{32}$	270.5	(519)	<1	<1	<1	<1	(1 mm 91.6°C)
n-Hexadecane	$C_{16}H_{34}$	287.5	(550)	<1	<1	<1	<1	(1 mm 105.3°C)
n-Heptadecane	$C_{17}H_{36}$	303.0	(577)	<1	<1	<1	<1	(1 mm 115°C)

Source: *Handbook of Chemistry and Physics*, 98th Edition, CRC Press, Boca Raton, FL, 2017–2018.
[a] 2-Methylbutane.
[b] 2-Methylpentane.
[c] Mixture of three isomers.

compound in the liquid. This phenomenon is illustrated more easily by examining the behavior of a mixture of two pure compounds, rather than gasoline and kerosene. For example, a mixture of 10% pentane (an extremely flammable component of cigarette lighter fluids whose vapor pressure equals 430 mmHg) and 90% toluene (a flammable aromatic solvent whose vapor pressure is 21.7 mmHg) produces vapors above the liquid consisting of 67% pentane and 33% toluene. As shown in Figure 2.11, the addition of only 10% of the more volatile pentane causes the vapor pressure,

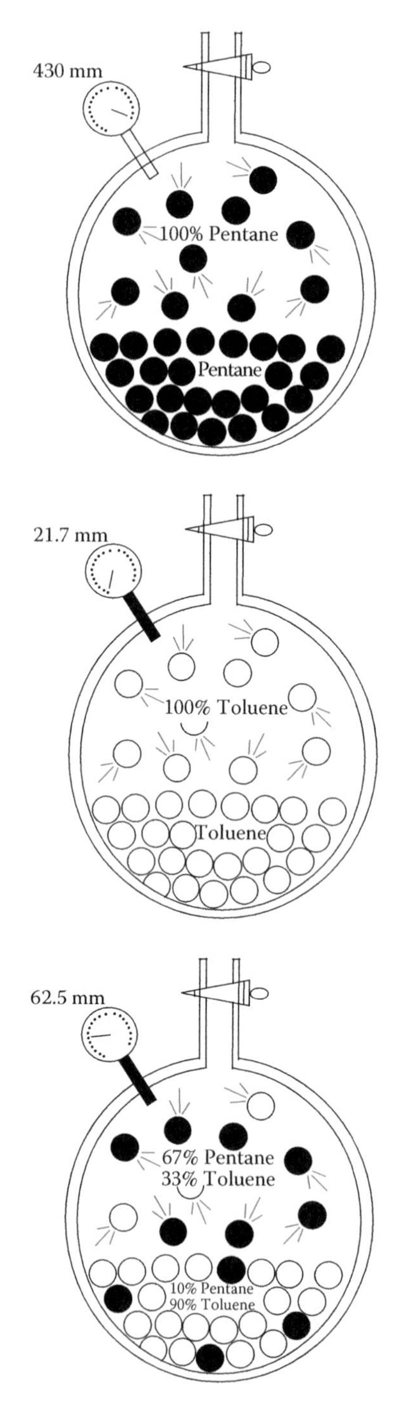

Figure 2.11 Raoult's law in action. The top vessel contains 100% pentane and the vapor pressure is 430 mmHg. The center vessel contains 100% toluene, and the vapor pressure is 21.7 mmHg. The bottom vessel contains 10% pentane and 90% toluene. The addition of 10% pentane has caused the vapor pressure to triple, to 62.5 mmHg. Two-thirds of the molecules in the vapor space are pentane, but only 10% of the molecules in the liquid phase are pentane.

and thus the concentration of combustible molecules above the liquid, to triple. Because of its high vapor pressure, pentane has a much lower flash point than does toluene: 40°F (4°C) for toluene versus −40°F or C[9] for pentane.

If a mixture of toluene and pentane is exposed to a fire, the residue may consist almost entirely of toluene. For more complex mixtures, such as gasoline, the residue after exposure to a fire will be substantially different from "fresh" unevaporated gasoline. Figure 2.12 shows a series of gas chromatograms for gasoline in different stages of evaporation. (Gas chromatography is explained in great detail in Chapter 5.) The peaks at the left side of the chromatogram represent lighter components, and those at the right represent heavier components. Notice how the height of the peaks changes with evaporation. If a quart of water is allowed to evaporate until only a pint is left, the chemical makeup of the "evaporated" water will be exactly the same as the unevaporated water. The same is true of all pure compounds, but not of

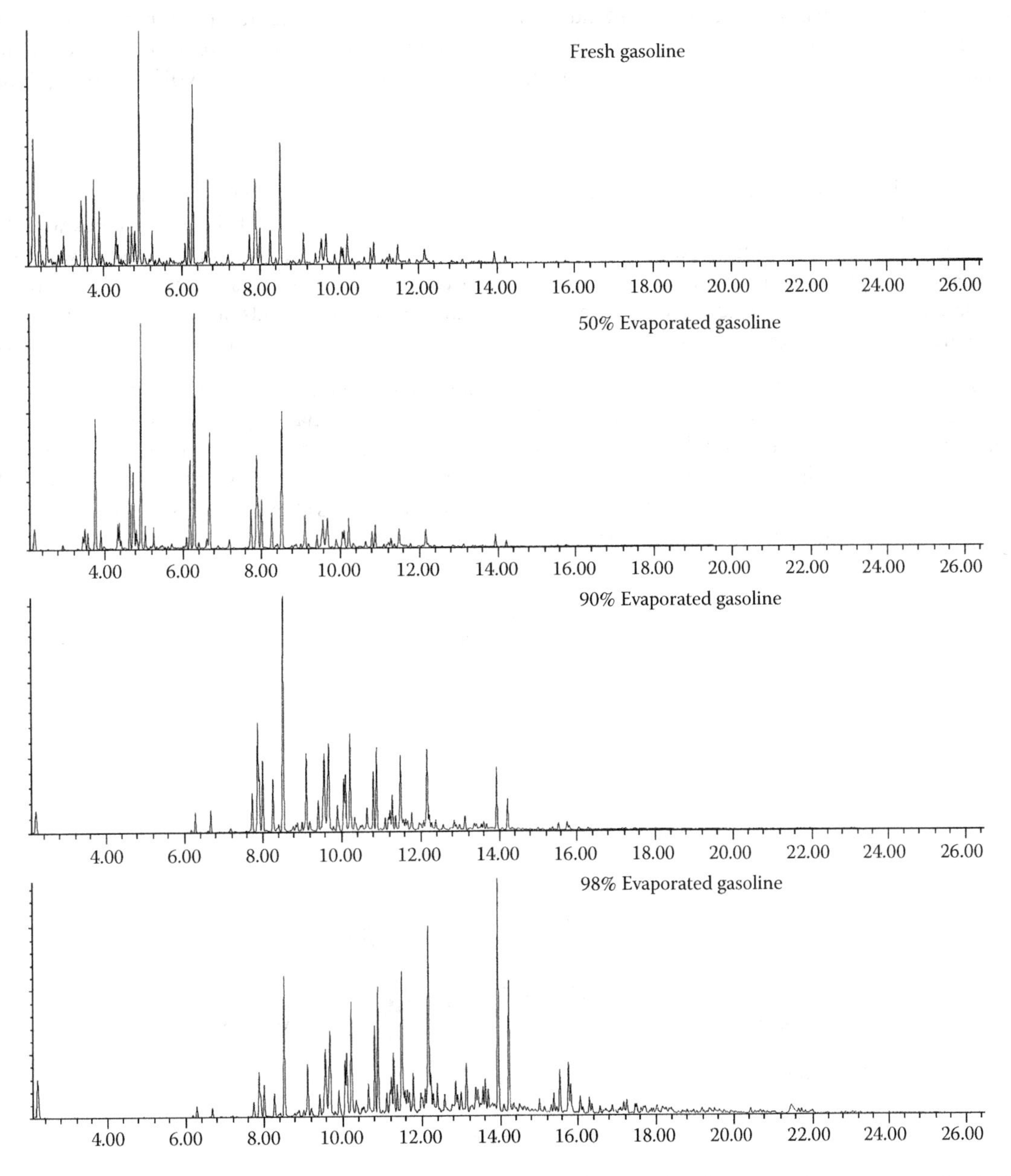

Figure 2.12 Four chromatograms showing gasoline in different stages of evaporation. As evaporation progresses, the lighter compounds represented by the peaks at the left side of the chart are lost, making the residue relatively richer in the heavier compounds on the right side of the chart.

[9] The Fahrenheit and Celsius scales converge at −40°.

mixtures or solutions. If the same test is conducted with gasoline, the chemical makeup of the evaporated gasoline will be substantially different from the unevaporated gasoline because gasoline consists of a mixture of different compounds having a wide variation in molecular weight and vapor pressure. Simply put, upon exposure to heat, the composition of pure liquids will not change, and the composition of liquids that are mixtures will change.

Flash point is sometimes confused with *ignition point*. The flash point of a liquid is the temperature at which the vapors above a still pool of the liquid can be ignited when exposed to a competent ignition source. The temperature of that ignition source is always significantly higher than the flash point. The flash point of gasoline, for example, is –40°F, but its ignition point is anywhere from 550°F to 850°F, depending on its octane rating. The vapors over a liquid heated above its flash point *can* burn, but those vapors *will* burn only when exposed to a source of heat that is hotter than the ignition point. Note that flash point applies to the temperature of the bulk liquid, while ignition point applies to the temperature of the vapor at the place where the ignition source is located. This can be an almost infinitesimally small space, so a tiny spark that has a temperature at or above the ignition point of a vapor can ignite a huge volume of vapor that is well below the ignition point.

2.7 THE BEHAVIOR OF SOLIDS

Compared to solids, describing the behavior of gases and liquids is relatively easy. If solids were simply liquids at a temperature below their freezing point, also known as the melting point, describing the behavior of solids would also be relatively easy. A candle is an example of this simple kind of solid. The heat from the flame melts the solid wax to produce liquid wax, and the liquid is then vaporized so it can combine with oxygen in the air. Most solids, however, and almost all combustible solids, do not exist as liquids or gases and can be transformed into liquids or gases only by breaking down large molecules into new and smaller molecules. There is no gaseous molecule of cellulose, or wool, or many polymers.

The principles of explosive limits, discussed earlier, apply equally well in a formal sense to the volatiles produced by burning solids, but the vapors and gases result from decomposition rather than vaporization. Whether an ignition source is capable of igniting a particular target fuel depends on whether it can cause the production of sufficient vapors to burn. And whether the fuel will keep burning depends on whether the resulting flame can transfer sufficient energy to the surface of the solid so that it continues to produce vapors above their lower flammable limit. The concept of a burning solid fuel is shown in Figure 2.13.

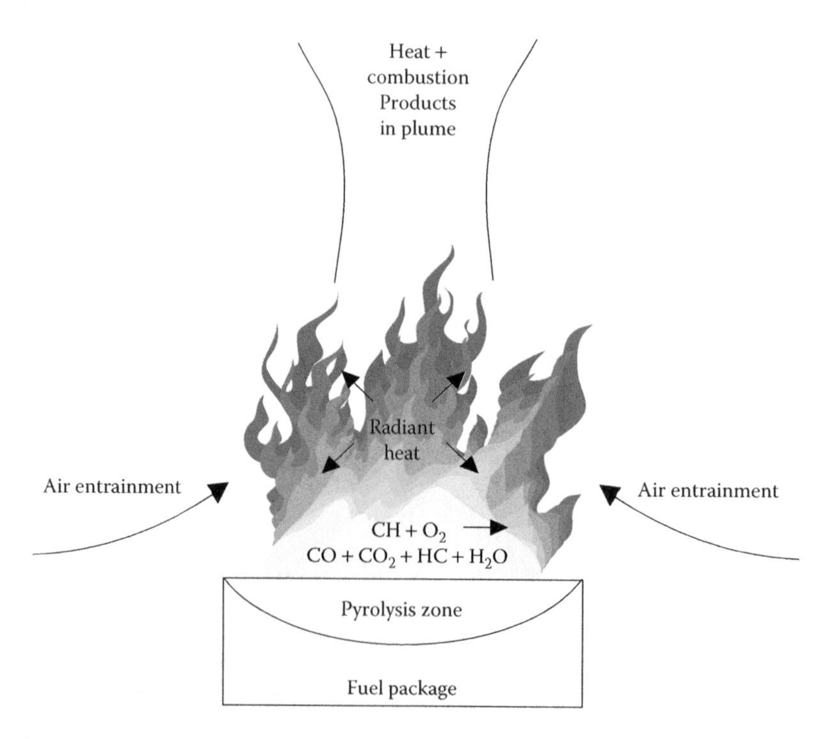

Figure 2.13 Schematic representation of some of the processes taking place in the combustion of a solid fuel. Combustion continues as long as the flame radiates enough heat onto the surface of the solid to produce a sufficient quantity of vapors and gases.

Solids generally fall into three categories: (1) metals and metal compounds, (2) refractory materials (such as concrete, glasses, and ceramics), and (3) polymers. With very few exceptions, all the combustible solids likely to be encountered by a fire investigator fall into the third category. Polymers are macromolecules that can have molecular weights in the thousands or even millions. The basic unit of a polymer is called a monomer. Simple molecules that are bonded together in a line, a sheet, or a three-dimensional matrix make polymers.

Polymers can be either natural or synthetic. They can be in the form of solids, foams, fibers, or films. There are very few "pure" polymers. Most substances called polymers have significant fractions of other ingredients, such as fillers or plasticizers, added to them to give them desirable properties. Some polymers can contain up to 50% plasticizer. A pure polymer is generally both hard and brittle; plasticizers make polymers soft and pliable.

Wood is a mixture of three natural polymers, the most widely known of which is cellulose. Cellulose, which is the most abundant organic molecule on earth, is a *polysaccharide,* which means "many sugars." It is a polymer of the molecule glucose, and its chemical structure is shown in Figure 2.14. Hemicellulose refers to any one of several polysaccharides that are more complex than sugar but less complex than cellulose. Hemicellulose occurs chiefly in plant cell walls. Lignin is the most abundant natural aromatic (benzene ring–containing) polymer and is found in all vascular plants. Lignin together with cellulose and hemicellulose are the major cell wall components of the fibers of all wood and grass species [2]. *Proteins* are another kind of natural polymer and include silk and wool.

The six most common synthetic polymers are polyethylene, polypropylene, polyvinyl chloride, polystyrene, polyester, and polyurethane. The chemical structures of some common polymers are shown in Figure 2.15.

Polymers respond to heat by undergoing *pyrolysis,* which is an irreversible decomposition caused by the breaking of chemical bonds. Note that pyrolysis is an endothermic reaction; that is, it requires energy. The energy causes the atoms within the solid to vibrate more and more rapidly, until the chemical bonds between the atoms break. Pyrolysis occurs in several different ways, the simplest of which is known as monomer reversion. The large molecules simply break down into their constituent parts. Polystyrene, polytetrafluoroethylene (PTFE, also known as Teflon), and polymethylmethacrylate (PMMA) are three of the few polymers that essentially "depolymerize" when exposed to heat. Their behavior is not much different from the behavior of the wax in a candle. Once the polymer has melted, these monomers can then be vaporized and burned in the gas phase. Other polymers, particularly the polyolefins (polyethylene, polypropylene, polybutylene), undergo a process known as random scission, resulting in the production of organic molecules that are larger than the starting monomers but small enough to vaporize, and behave similarly to combustible liquids. A third type of pyrolysis involves the breaking off of side groups on the polymer chain. This is known as side group scission. Polymers that undergo side group scission as their major pyrolysis route include polyvinyl chloride and polyvinyl acetate. More complex reactions are likely to occur after a polymer has its side groups stripped off. Most polymers, when exposed to heat, undergo a combination of all three types of pyrolysis. Generally, pyrolysis results in charring or melting of the polymer. Solids that are not polymers can behave differently.

Solids exposed to heat can be expected to do one of three things: (1) melt, (2) dehydrate, or (3) char. Substances that char include wood and many of the thermosetting plastics. A thermosetting plastic is a polymer that does not melt or soften on exposure to heat and maintains (more or less) its original shape. Polymers that soften and deform on exposure to heat are known as *thermoplastics.* Because only crystalline solids have sharp melting points, and because most polymers are not crystalline solids but mixtures of crystalline and are amorphous (lacking a clearly defined

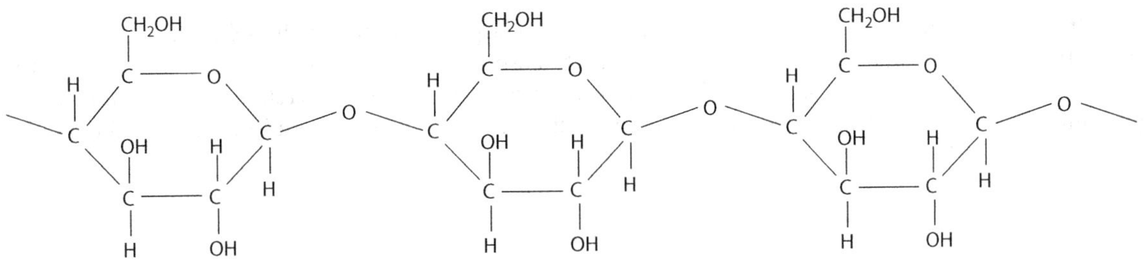

Figure 2.14 Chemical structure of cellulose.

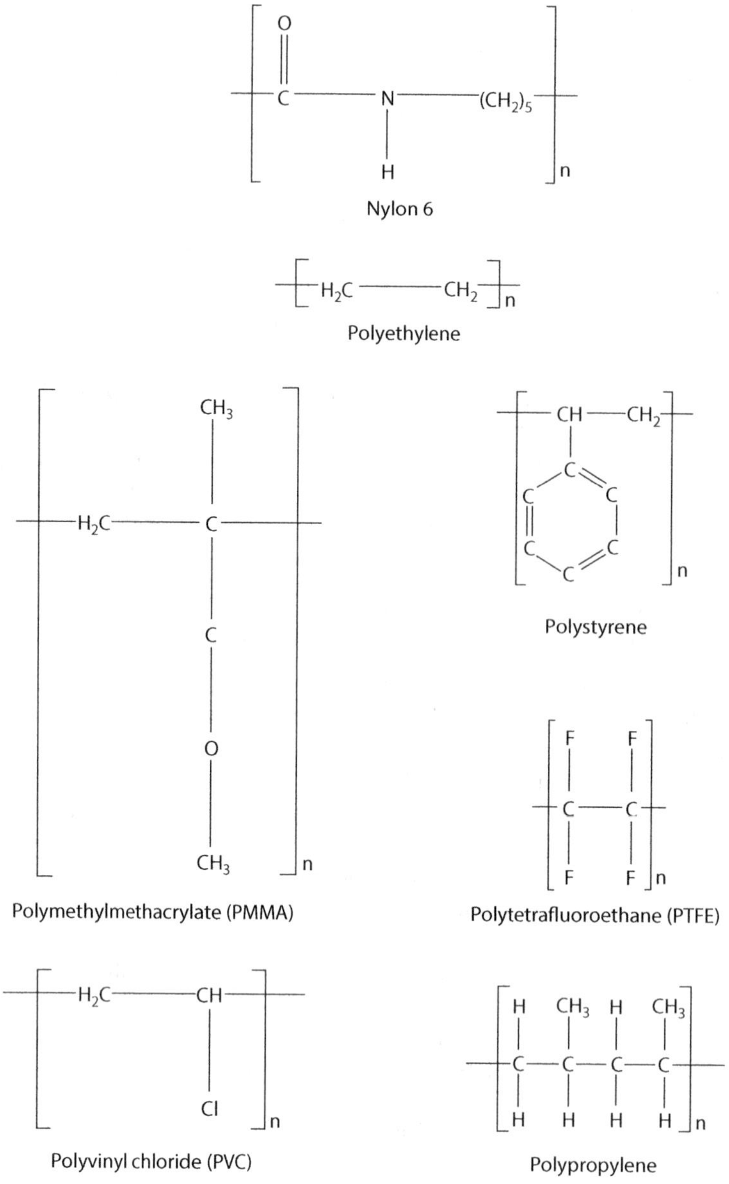

Nylon 6

Polyethylene

Polymethylmethacrylate (PMMA)

Polystyrene

Polytetrafluoroethane (PTFE)

Polyvinyl chloride (PVC)

Polypropylene

Figure 2.15 Chemical structures of some common polymers.

form) arrangements of molecules, polymers that melt typically have what is called a "glass transition temperature." (Glass also has a glass transition temperature and not a melting point.) Pure metals have a very sharp melting point, as do most substances that consist of crystals bound together.

In dehydration, water molecules that are chemically bound to larger molecules in the solid matrix are released by breaking the weak chemical bonds. This process involves the absorption of energy, which in turn provides for the fire resistance of substances such as gypsum wallboard and concrete. When the water is released, *calcination* takes place in gypsum wallboard. The released water vaporizes under fire conditions, and the resulting steam is thought to play a role in the *spalling*[10] of concrete.

[10] Spalling may be caused by steam pressure overcoming the tensile strength of concrete, by differential expansion of the cement and aggregate, or by rapid heating of concrete. For a detailed discussion of the causes of spalling, see Chapter 8.

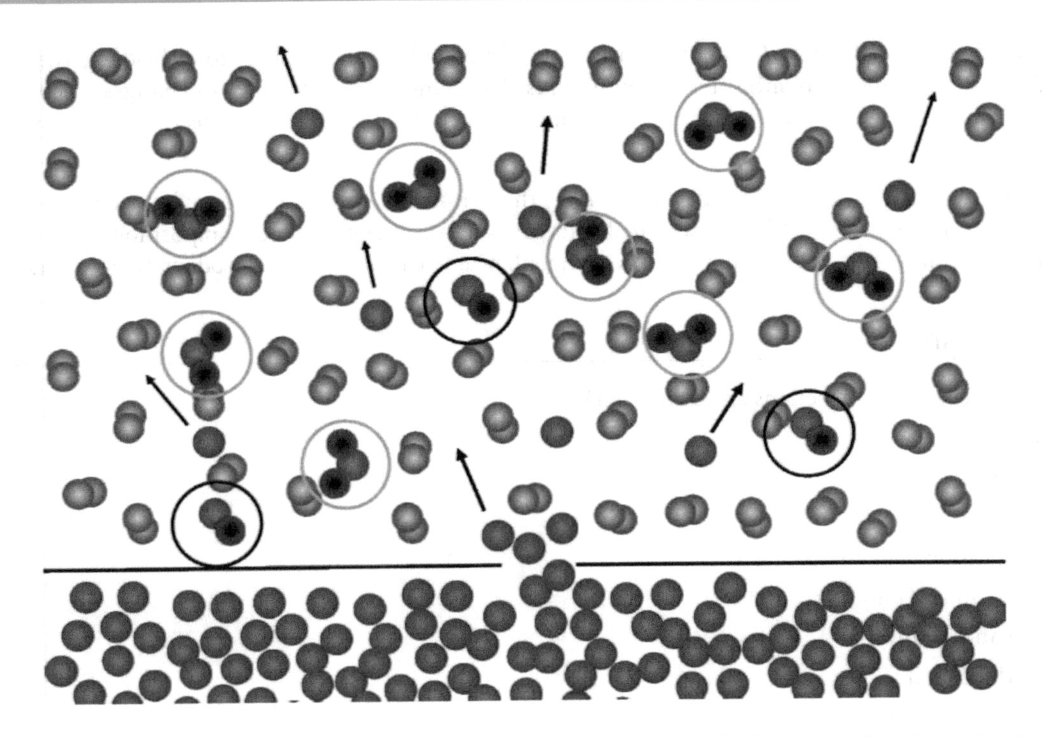

Figure 2.16 Schematic representation of the behavior of the molecules involved in the combustion of a carbon-based solid. (Drawing courtesy of the Steve Carman, Carman & Associates Fire Investigation, Grass Valley, CA.)

The charring behavior is of most interest to fire investigators. For most solids, ignition occurs when volatiles (small molecules) are created and then driven off by an external heat source, and these volatiles ignite and burn in air. This flaming combustion above the solid's surface provides more heat for additional pyrolysis and volatilization. The solid left behind is known as *char*. Char can insulate the material below it from the heat source because char is generally less conductive than the solid from which it originates. The insulating properties of char make determination of a "charring rate" difficult because this rate changes over time. It is possible to determine an average charring rate for substances exposed to a calibrated energy source such as a furnace, but real fires are not calibrated. In the fire resistance furnace described in ASTM E119, *Standard Test Methods for Fire Tests of Building Construction and Materials* [9], the charring rate of most common woods has been found to range between 0.5 and 0.8 mm/min [10].

The susceptibility of a solid material to ignition depends on many factors, and the subject of ignition, except for a discussion of potential sources of ignition, is far too large to be covered in this text. *The Ignition Handbook* by Vytenis Babrauskas provides a detailed examination of all factors affecting ignition and is highly recommended to the reader who wishes to study the subject in greater depth. For the purposes of this text, the focus is on the measurable flammability properties of solids related to ignitibility, flame spread rate, burning rate, and heat release rate.

The important point to remember about solids is that *in order to burn, they must be volatilized.* Figure 2.16 shows a schematic representation of the volatilization of a solid carbonaceous fuel exposed to heat. The fuel molecules (red) are vaporized and react with the O_2 molecules (blue) to produce CO_2 (circled in gray) and CO (circled in black). As will be shown in the next chapter, the gas phase can become saturated with combustion products and depleted of oxygen. The behavior of solids in fires depends greatly on their form, geometry, and surroundings, and is covered in Chapter 3 on fire dynamics.

2.8 CONCLUSION

Fire is an exothermic chemical reaction, a reaction that gives off energy. The energy is measured in joules (J). The rate at which the energy is released is known as power and is measured as joules per second or watts (W). The amount of power transmitted to a given area is known as heat flux and is measured in kilowatts per square meter or watts per square centimeter.

All matter is composed of atoms. Compounds are made of molecules, two or more atoms held together by chemical bonds. Most substances are mixtures of compounds. The composition of mixtures changes when exposed to heat. Most combustible solids are polymers, macromolecules consisting of long chains of repeating units called monomers.

Gases will burn when present at concentrations within the flammable limits and exposed to a competent ignition source. Liquids and solids must be transformed into gases by vaporization or decomposition in order to burn. Combustion continues as long as the flames radiate enough heat to cause continued production of vapors and gases.

Review questions

1. Which of the following reactions are exothermic?
   a. The melting of ice
   b. The boiling of water
   c. Setting of concrete
   d. Combustion of methane
      i. III and IV only
      ii. I and II only
      iii. II, III, and IV
      iv. I, II, III, and IV
2. How many volumes of air are required to burn one volume of natural gas (methane)?
   a. 25
   b. 10
   c. 5
   d. 2
3. What is the approximate energy content of a cubic foot of propane gas?
   a. 2,500 Btu
   b. 1,000 Btu
   c. 5,000 Btu
   d. 25,000 Btu
4. In the metric system, what is the basic unit for measuring energy?
   a. Watt (W)
   b. British thermal unit (Btu)
   c. Calorie (cal)
   d. Joule (J)
5. Which of the following ignitable liquids have vapors lighter than air?
   a. Gasoline
   b. Diethyl ether
   c. Cigarette lighter fluid
   d. None of the above

Questions for discussion

1. Describe the difference between an atom and a molecule.
2. Why is it important to be able to define energy, power, and flux, and what is the difference between energy and power?
3. Describe how flammable limits are related to balancing a chemical equation.

4. Why is it important that a fire investigator understands chemical reactions?
5. What is the difference between heat and temperature?

References

1. Stout, S., Uhler, A., McCarthy, K., and Emsbo-Mattingly, S. (2002) Chemical fingerprinting of hydrocarbons, in Murphy, B. and Morrison, R. (Eds.), *Introduction to Environmental Forensics*, Academic Press, Burlington, MA, p. 168.
2. *Merck Index* (2001) 13th ed., Merck Publishing, Rahway, NJ, p. 983.
3. Drysdale, D. (1985) *An Introduction to Fire Dynamics*, Wiley-Interscience, New York.
4. NFPA 921 (2017) *Guide for Fire and Explosion Investigations*, NFPA, Quincy, MA.
5. Babrauskas, V. (2003) *Ignition Handbook*, Fire Science Publishers, Issaquah, WA, p. 1043.
6. Segeler, C. (1965) *Gas Engineers Handbook*, 1st ed., Industrial Press, New York, 2/73.
7. NFPA 30 (2018) *Flammable and Combustible Liquids Code*, National Fire Protection Association, Quincy, MA.
8. ASTM E502 (2013) *Standard Test Method for Selection and Use of ASTM Standards for the Determination of Flash Point of Chemicals by Closed Cup Methods*, Annual Book of Standards, Volume 14.05, ASTM International, West Conshohocken, PA.
9. ASTM E119 (2016) *Standard Test Methods for Fire Tests of Building Construction and Materials*, Annual Book of Standards, Volume 4.07, ASTM International, West Conshohocken, PA.
10. Babrauskas, V. (2004) Wood char depth: Interpretation in fire investigations, in *Proceedings of the 10th International Fire Science and Engineering (Interflam) Conference*, Interscience Communications, London, UK.

CHAPTER 3

Fire dynamics and fire pattern development

A case can be made for fire being, next to the life processes, the most complex of phenomena to understand.

—Hoyt Hottel

LEARNING OBJECTIVES

After reviewing this chapter, the reader should be able to:

- List the different kinds of ignition
- Describe the various kinds of flames
- Understand the concept of flammability, and describe how different aspects of flammability are measured
- Discuss the means by which fire patterns are generated, and know which patterns are significant in origin determination
- Understand the critical role of oxygen in fully involved compartment fires
- Describe the three different kinds of fire models and their usefulness in fire investigation

3.1 INTRODUCTION

This chapter is intended as an introduction to the concepts of fire dynamics, but one chapter cannot come close to being a comprehensive treatment of this very large subject. The basic concepts of fire dynamics were not even set down in a single volume until 1985, when Dougal Drysdale compiled a series of lectures into his textbook entitled *An Introduction to Fire Dynamics*. It is an excellent book but one that was designed to be read and understood by fire protection engineers.

"Fire dynamics without tears" (i.e., with most differential equations excised) was not available to the average practitioner in fire or to firefighters, code officials, and investigators until James Quintiere's text, *Principles of Fire Behavior*, was published in 1998. Prior to that, the only text that came close to the vernacular was found in the "Basic Fire Science" chapter of NFPA 921. Some examples of simple fire dynamics equations are shown in Sidebar 3.1.

Chapter 2 covered some of the basic chemistry and physics of combustion. This chapter deals with the concepts of flames, heat transfer, ignition, and flame spread; the interaction of the fire with its surroundings in a compartment; and the use (or misuse) of computer models.

3.2 IGNITION

For ignition to occur, the substance under consideration first must be capable of propagating self-sustained combustion or, in the case of explosive materials, an exothermic decomposition wave. Ignition is defined as the process by which this propagation begins.

Table 3.1 Thermal properties of selected materials[a]

Material	Thermal conductivity (*k*) (W/mK)	Density (ρ) (kg/m³)	Heat capacity (*c*) (J/kg-K)	Thermal inertia (kρc) ($W^2 \times s/k^2\ m^4$)
Copper	387	8940	380	1301×10^6
Concrete	0.8–1.4	1900–2300	880	$1.34–2.83 \times 10^6$
Gypsum plaster	0.48	1440	840	0.581×10^6
Oak	0.17	800	2380	0.324×10^6
Pine (yellow)	0.14	640	2850	0.255×10^6
Polyethylene	0.35	940	1900	0.625×10^6
Polystyrene (rigid)	0.11	1100	1200	0.145×10^6
Polyvinyl chloride	0.16	1400	1050	0.235×10^6
Polyurethane	0.034	20	1400	0.000952×10^6

Source: Drysdale, D., (1999) An Introduction to Fire Dynamics, 2nd Edition, John Wiley & Sons.
[a] Typical values. Properties vary with temperature.

Ignition of solids occurs when the heat generation rate in a given volume of material exceeds the heat dissipation rate, and that process continues (and the rate of the reaction accelerates) as the temperature rises further. The heat causes chemical bonds to break, and the material decomposes into volatile substances. These volatiles either ignite in the presence of a pilot or autoignite. Piloted ignition temperatures for solids are generally in the range of 250°C to 450°C, while autoignition temperatures are usually more than 500°C. Either the pilot (a preexisting flame) or the hot environment provides the ignition energy for the gases and vapors evolved by heating. At some point, the rate of heat generation is controlled by the rate of production of the volatiles, and a more or less stable flame is created. For thin materials (paper or fabrics less than approximately 2 mm thick), whether or when a target fuel will ignite depends on its *thermal inertia*, which is the square root of the product of the thickness of the material, its density, and specific heat capacity, which is the amount of heat in joules, required to change the temperature of one gram of a substance by one Kelvin, expressed as $J\ m^{-1}K^{-1}$. For thick materials, rather than thickness, a fuel's thermal conductivity (k), the ability to conduct heat away from the surface, becomes important, in addition to density (ρ) and specific heat capacity (c). Thermal conductivity is measured in watts per meter-kelvin expressed as $Wm^{-1}K^{-1}$, and density measured in kilogram per cubic meter is expressed as $Kg\ m^{-3}$. The units of thermal inertia are expressed as the typographically cumbersome $J\ m^{-2}\ K^{-1}\ s^{-1/2}$. The unit "tiu" for thermal inertia unit has been proposed. Table 3.1 shows typical values for all of the thermal properties discussed earlier.

When thinking about thermal inertia and its effect on ignition temperature or ignition time, it may be best to think of the three factors, conductivity, density, and heat capacity, individually. Ignition temperatures are lower for poor conductors, for low-density solids, and for solids with a low heat capacity. Thus, polyurethane foam ignites sooner than a polyethylene trashcan, which will ignite sooner than a particleboard desk. When or whether a fuel ignites also depends heavily on the heat flux incident on its surface. The critical heat flux, expressed as kilowatt/square meter (kW/m^2), is a threshold value below which ignition does not occur.

Ignition occurs more easily, that is, with the addition of less energy, in mixtures of vapors or dusts in air than on solids. Fuel/air mixtures just slightly richer than stoichiometric are the most easily ignited, with minimum ignition energies (MIEs) on the order of 0.2 millijoules (mJ) for hydrocarbons and 0.01 mJ for hydrogen [1]. Depending on the particle size distribution, MIEs for dust clouds are on the order of tens to hundreds of millijoules [2].

3.3 SELF-HEATING AND SPONTANEOUS IGNITION

NFPA 921 defines self-heating as the result of exothermic reactions, occurring spontaneously in some materials under certain conditions, whereby heat is liberated at a rate sufficient to raise the temperature of the material. Self-ignition is defined as resulting from self-heating.

Spontaneous ignition follows the same rules as piloted ignition, or ignition by radiant heat. Instead of the heat source being outside the fuel, however, the heat source is a chemical reaction within the fuel pile. When the rate of heat generation exceeds the rate of heat dissipation, the temperature increases, and if it increases to the ignition point of the fuel, a fire can result. A delicate balance is required to meet these conditions. First and foremost, the material must be prone to oxidation at the ambient temperature. As the ambient temperature is raised, more and more substances meet this

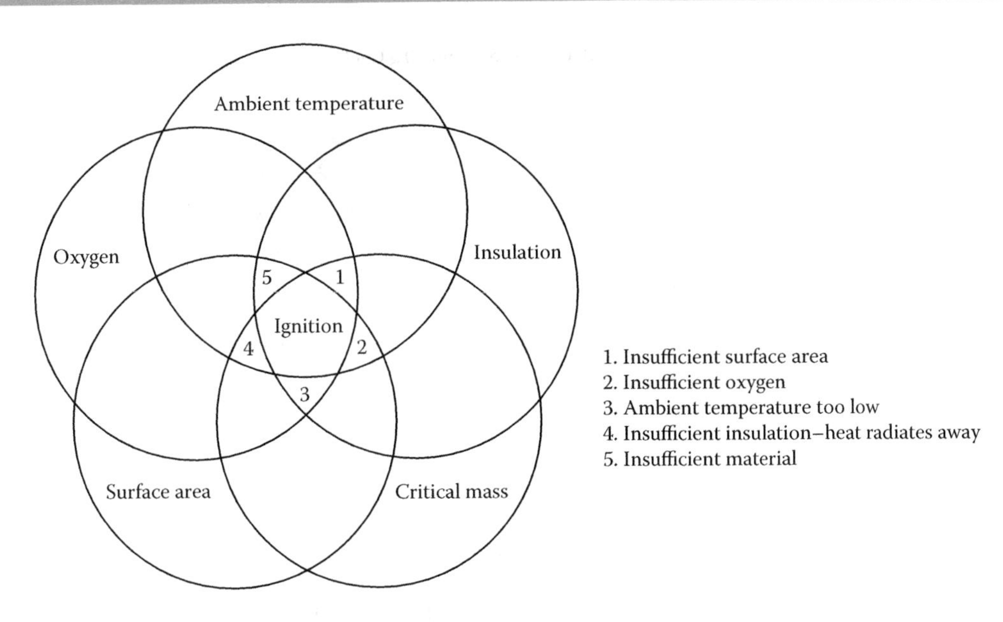

Figure 3.1 Interactions required for spontaneous ignition.

criterion. At a certain point, it becomes academic whether the actual ignition was spontaneous or the result of absorption of energy from a hot environment. This discussion centers on ambient temperatures below 200°F. At these temperatures, the materials most commonly encountered that undergo spontaneous heating are unsaturated animal or vegetable oils, or the terpenes found in most plants. Certain finely divided dusts containing drying oils or unreacted polymers are also subject to spontaneous heating. Auto paint overspray is an example of such a fuel. The double bonds in these substances oxidize, resulting in the release of heat. If the pile is sufficiently large, it can provide enough insulation to keep the heat within the pile and increase the rate of reaction. There is a balance between sufficient insulation and sufficient exposure to oxygen. Without both, the process stops. A special case of spontaneous heating is that of the haystack. This involves a two-step process. First, biological activity in the stack increases the ambient temperature to the point where, although the organisms die, there is sufficient energy to increase the rate of oxidation of the oils in the plant material.

Although it is spontaneous, self-heating is usually a very slow process and might eventually lead to smoldering combustion. This is because the heat is generated at the center of the fuel, but there is usually insufficient oxygen for flaming until the surface of the fuel becomes sufficiently hot. The timing of spontaneous heating is not always predictable, however. Although it usually requires several hours, times under 15 minutes have been recorded [3], as well as times greater than three days [4]. A simple diagram showing the balance required for spontaneous ignition is presented in Figure 3.1. Most often, one or more of the required conditions are not met, and while oxidation and even heating occur, no ignition results. Babrauskas provides a rigorous treatment of self-heating [5], and much has been written and argued in the literature, particularly with regard to the low-temperature ignition of solid fuels [6]. (Low temperature ignition of solid fuels has not gained sufficient support to be accepted as evidence [7].) A discussion of means of testing hypotheses about spontaneous ignition can be found in Chapter 6.

3.4 CHEMICAL IGNITION

Certain substances burst into flame immediately on exposure to air or moisture. Most of these substances are either *pyrophoric* metals or exotic organic peroxides. Alkali metals such as lithium, sodium, and potassium must be stored under organic liquids such as kerosene or mineral oil in order to exclude oxygen. Other metals become pyrophoric when finely powdered. NFPA 400, the *Hazardous Materials Code,* defines "pyrophoric material" as a chemical with an autoignition temperature below 130°F (54.4°C) [8].

Fire investigators are more likely to encounter chemical ignitions resulting from the contamination of oxidizers used as swimming pool sanitizers. These include calcium hypochlorite and "stabilized" chlorine compounds such as sodium dichloroisocyanuric acid. When contaminated with an oxidizable material, these oxidizers produce intense flames and highly toxic fumes, and may undergo decomposition reactions that are almost impossible to stop until

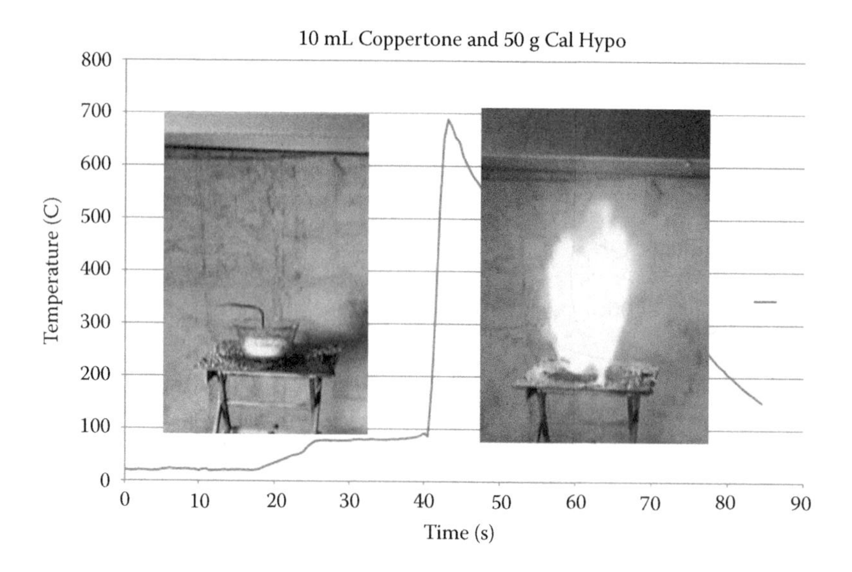

Figure 3.2 Reaction of 10 mL of suntan oil with pool sanitizer (50 g of 68% calcium hypochlorite). Forty seconds after the suntan oil was applied, a violent reaction ensued.

the supply of chemical is exhausted. The speed and intensity of the chemical reaction of pool sanitizers are far greater than the combustion of ordinary fuels. Figure 3.2 shows one such reaction.

Regulations for the labeling, storage, and handling of oxidizers are based on NFPA 400, the Hazardous Materials Code. Oxidizers were previously regulated under NFPA 430, but in 2010, NFPA 430, NFPA 432, NFPA 434, and NFPA 490 were withdrawn as separate documents and included in their entirety in NFPA 400. These codes covered hazardous material categories found in building and fire codes such as corrosives, flammable solids, pyrophoric substances, toxic and highly toxic materials, unstable materials, and water-reactive materials, as well as compressed gases and cryogenic fluids.

3.5 SMOLDERING IGNITION

Smoldering is a slow combustion process that occurs between the oxygen in the air and a solid fuel on the surface of the fuel. Smoldering involves pyrolysis, but because of limited ventilation, insufficient fuel vapor is produced to create or sustain a flame. Limited ventilation also keeps the oxygen concentration below the stoichiometric balance, so smoldering combustion tends to produce far more carbon monoxide than it does carbon dioxide. Although the rate of CO production per unit *mass* in smoldering combustion exceeds that of flaming combustion, the rate of CO combustion per unit *time* may be far less than that of flaming combustion, depending on the speed of the smolder reaction.

Smoldering rates depend heavily on ventilation and typically range from 1 to 5 mm/min [9]. It is often a change of ventilation that results in the smoldering combustion making the transition to flaming combustion. Because of the large number of variables involved, there is no way to predict how long a particular fuel will smolder before it makes the transition to flaming combustion, or even if that transition will ever be made.

Smoldering can occur either before or after flaming combustion. A smoldering ignition source such as a cigarette can induce smoldering on cellulosic materials as well as on polyurethane foam. In order to be a candidate for smoldering, the fuel must have the ability to form a *porous char* and resist melting to the extent that the molten fuel blocks the pores.

Most materials do not smolder. Common materials that can smolder if conditions are favorable include the following:

- Sawdust and other finely divided cellulosic materials
- Paper
- Leather
- Latex foam
- Polyurethane foam
- Charcoal

- Cigarettes
- Dusts
- Forest duff

Smoldering can take place when flaming combustion consumes much of the oxygen in a compartment. Some disagreement exists about the distinction between glowing combustion and smoldering combustion. Babrauskas states that smoldering combustion that exhibits a visible glow can be referred to as glowing combustion, but glowing combustion can also be caused by imposing an external heat flux to a fuel in a situation where the pyrolysates are driven off quickly, yet the surface continues to glow.

The pyrolysis and vaporization of fuel that precedes flaming combustion is sometimes erroneously referred to as smoldering. Pyrolysis and vaporization are endothermic reactions, whereas smoldering is an exothermic oxidation reaction. It is not smoldering if the fuel only produces vapors when the heat source is applied but does not oxidize.

3.6 FLAMES

A flame is a luminous zone of burning gases or vapors and fine, suspended matter where combustion is taking place. The kind of flame we are usually interested in is called a *diffusion* flame. In a diffusion flame, fuel gases or vapors and oxygen combine in a reaction zone because of *differences in concentration*. Substances move, or diffuse, from areas of higher concentration to areas of lower concentration. Thus, the wax vapors in a candle move away from the wick, where they are most concentrated and too rich to burn, into the reaction zone where the wax vapors are broken into smaller molecules that combine with oxygen from the air moving toward the flame. The oxygen moves toward the flame because the local convection causes a partial vacuum near the reaction zone. Figure 3.3 shows a candle flame in cross section. All diffusion flames, including large turbulent flames, behave in a similar manner, but candle flames are easier to study and illustrate (Figure 3.4).

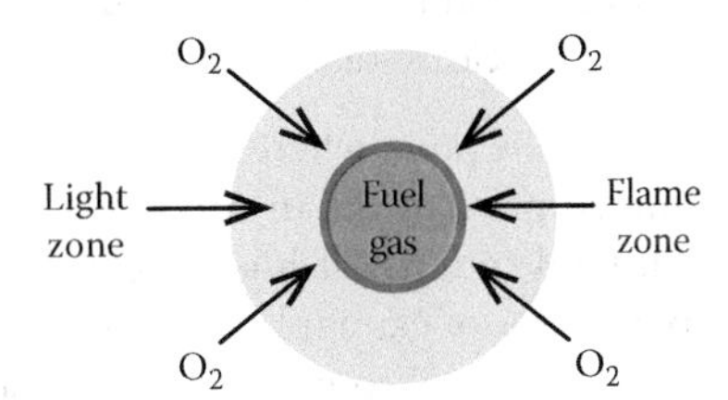

Figure 3.3 Diagram showing the reaction of oxygen with the fuel gas produced by the vaporization of the wax.

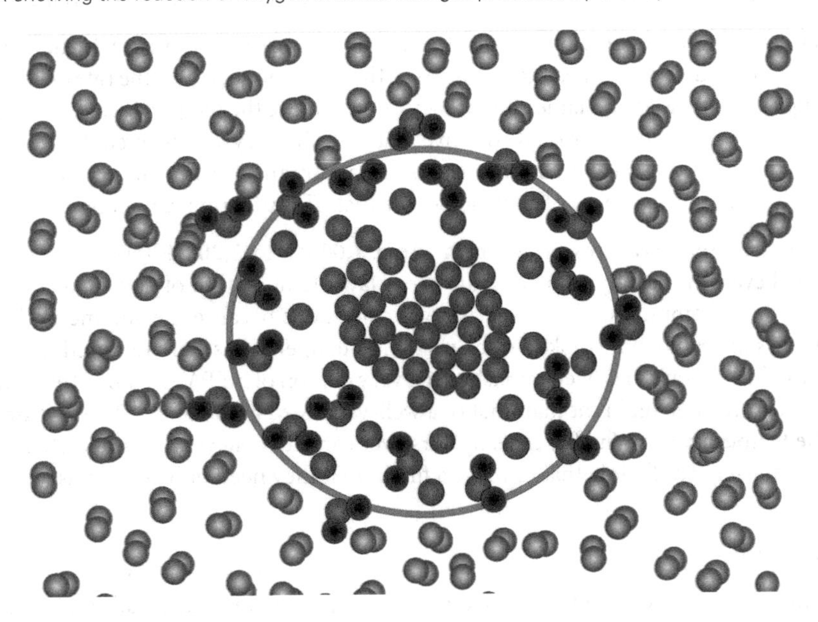

Figure 3.4 Schematic representation of the combustion of the fuel molecules (red) reacting with oxygen molecules (blue) to produce carbon dioxide. (Drawing courtesy of Steve Carman.)

Diffusion flames can be either *laminar* (orderly) or *turbulent*. Any flame taller than a foot is likely to contain sufficient disorder to be considered turbulent [10,11]. A candle flame is the classic example of a laminar diffusion flame. Because of their orderly nature, laminar flames have been studied extensively to aid in the understanding of what is happening in more complex flames.

In fact, the study of simple systems forms the basis for much of what is known about fire. Fire protection engineers have sometimes been criticized for their reliance on controlled and reproducible laboratory experiments, but those experiments allow us to understand the fundamental principles as well as to draw conclusions about fires that happen outside the laboratory. For this reason, fire protection engineers have spent much time studying the behavior of circular or square pools of burning flammable liquids, or blocks of pure polymethyl methacrylate (PMMA), also known as acrylic as well as by the trade names Plexiglas and Lucite. These scientists understand not only the differences between the field and the laboratory but also that it is necessary to start somewhere.

Flames have been studied extensively using tiny temperature-measuring devices called thermocouples to measure temperature variations in different parts of the flame and using optical devices such as spectrophotometers to measure the concentration of molecular species within the flame. In 1861, Michael Faraday, in his Christmas lectures on *The Chemical History of a Candle*, was able to sample gases from different parts of the flame and demonstrate the composition of those gases by observing their behavior. In 1985, Smyth et al. [12] produced a fascinating series of diagrams showing the symmetrical variations in temperature, fluid velocity, and chemical species (water, hydrogen, CO, CO_2, as well as various hydrocarbons) in a laminar natural gas flame. See Quintiere's *Principles of Fire Behavior* for a more complete discussion of Smyth's work.

While laminar flames allow us to make inferences about the reactions occurring in all flames, most of the time, fire investigators look at the aftermath of turbulent diffusion flames. The orderliness of the laminar flame is lost because of the tremendous buoyancy of the products of combustion, the large size of the flame, and the irregular shape of the fuel packages. The reaction zone in a turbulent diffusion flame has far more surface area than the reaction zone in a laminar flame, so turbulent flames tend to have higher heat release rates as a result, but the processes taking place are the same.

Unlike diffusion flames, where the flame itself is the point where the fuel and oxygen come together, *premixed* flames are those where the fuel and oxygen are mixed prior to the introduction of the ignition source. Most gas-burning appliances employ premixed flames, which tend to burn at higher temperatures because the fuel and oxygen are close to the stoichiometric mixture. Except in cases of gas appliance malfunctions, fire investigators are likely to see premixed flames as ignition sources rather than as fuel sources. It should be noted here that, although a premixed natural or propane gas flame burns with a blue color, seeing a blue flame in a fire does not necessarily mean that gas is burning. In fact, both natural gas and LP gas burn with a yellow flame if the gas is not premixed.

Flame temperatures vary widely depending on whether the flame is premixed or diffuse, whether it is laminar or turbulent, the amount of fuel involved, and its *net* energy release (i.e., the difference between the rate of heat release and the rate of heat loss). One thing we do know about flames is that they never achieve the temperatures reported in the literature as the "adiabatic flame temperature." *Adiabatic* is an adjective that describes a system wherein there are no heat losses. It is a theoretical maximum temperature that can be achieved only by a substance burning in a perfectly insulated space with more than enough oxygen to cause complete combustion of the fuel. This situation cannot be attained in the real world.

In everyday structure fires, temperatures range from about 1,000°F to 2,200°F (600°C to 1,200°C). The start of flashover (a transitional event) is usually defined as an upper layer temperature of 500°C to 600°C. Temperatures in the 1,300°C range are not uncommon under post-flashover conditions because the volume of the compartment can become filled with flames instead of hot smoke. In the past, high temperatures were reported to result from the use of accelerants, but it is now known (and has been published in every edition of NFPA 921 since 1992) that the temperature of a well-ventilated wood fire is no different than that of a well-ventilated gasoline fire. **Ventilation, not the nature of the fuel, controls the temperature.** (Why fire investigators ever came to believe that accelerated fires burn hotter is a mystery. People have known for millennia that to make a fire hotter, they need only to increase the airflow.)

3.7 FLAMMABILITY

The term *flammability* encompasses a number of different properties of a material, and many tests have been designed to measure these properties. The properties of a fuel that might be of interest to a fire investigator include the following:

- Energy required for ignition, measured in millijoules
- Flash point, measured in degrees
- Burning rate or mass loss rate, measured in grams per second
- Energy content, measured in joules per gram (or kilowatt hours per gram or British thermal units [Btu] per pound)
- Heat release rate, measured in kilowatts or megawatts (Heat release rate is the same as energy release rate.)
- Flame spread rating, measured in arbitrary units compared to the rate of flame spread on red oak (flame spread can also be expressed as distance per unit time [cm/s])
- Thermal properties: density, conductivity, heat capacity, thermal inertia

One or all of these measures of flammability might be of interest to the fire investigator, depending on the issues involved in a particular case.

For gases, vapors, and dusts, two relevant properties are flammability limits and minimum ignition energy. Both properties are difficult to measure and require specialized and expensive apparatus. Fortunately, the values for most common fuels can be found in the literature.

Flash Point of a Liquid: The lowest temperature of a liquid, as determined by specific laboratory tests, at which the liquid gives off vapors at a sufficient rate to support a momentary flame across its surface.

For ignitable liquids, the most important property is the flash point, the temperature at which a still pool of liquid produces a vapor/air mixture equal to the lower flammable limit. The fire point, the temperature at which the flash will be sustained, is generally a few degrees higher than the flash point, although this is not the case for the lower alcohols [13]. Flash points are relatively easy, though tedious, to determine and are well known for most common liquid fuels. Other tests of liquid flammability, which are less likely to have searchable literature values, include the flame projection test and the closed drum test described in ASTM D3065, *Standard Test Methods for Flammability of Aerosol Products.*

Flammability is one of the four hazards required by NFPA 704 to be listed in the "diamond." The other hazards are "health," "instability," and "other." For ignitable liquids, five degrees of hazard are listed in the red square at the top of the diamond:

0—Materials that will not burn under typical fire conditions, including intrinsically noncombustible materials such as concrete, stone, and sand.

1—Materials that must be preheated before ignition can occur.

2—Materials that must be moderately heated or exposed to relatively high ambient temperatures before ignition can occur.

3—Liquids and solids (including finely divided suspended solids) that can be ignited under almost all ambient temperature conditions.

4—Materials that rapidly or completely vaporize at atmospheric pressure and normal ambient temperature or that are readily dispersed in air and burn readily.

The Occupational Safety and Health Administration (OSHA), on the other hand, has almost the exact opposite ranking in its Hazard Communication System. OSHA has adopted a globally harmonized standard (GHS) in which 1 is the highest hazard and 4 is the lowest. NFPA and OSHA are aware of the potential for confusion, and this subject is discussed in the Annex of NFPA 704 [14].

Autoignition Temperature: The lowest temperature at which a combustible material ignites in air without a spark or flame.

The autoignition temperature is important for all phases of fuel, although it is difficult to determine accurately for most practical situations; but again, there are generally accepted values for many fuels in the literature.

Flame Spread: The movement of the flame front across the surface of a material that is burning (or exposed to an ignition flame), where the exposed surface is not yet fully involved.

NFPA 92, *Standard for Smoke Control Systems,* defines *flame spread* as the movement of the flame front across the surface of a material that is burning (or exposed to an ignition flame), where the exposed surface is not yet fully involved. Physically, flame spread can be treated as a succession of ignitions resulting from the heat energy produced by the burning portion of a material, its flame, and any other incident heat energy imposed on the unburned surface [15]. Flame spread is the result of heating of the material ahead of the flame front to its ignition point. The source of the heating can be the flame and/or an external heat source (e.g., the hot upper layer in a compartment). Flame spread in solids generally occurs in one of two modes: either (1) on the surface of the burning material or (2) by radiation. Surface flame spread depends on the ability of the existing flame to cause pyrolysis and the production of sufficient vapors to maintain combustion. Flame spread by radiation requires that the existing flame produce a sufficient radiant heat flux to ignite the target fuel. In a given situation, the amount of radiation that will be "sufficient" radiation is a function of the following:

- Distance between the flame and the target
- Angle of incidence (the target's "view" of the flame)
- Presence or absence of a "pilot" (i.e., an existing flame in the vicinity of the pyrolyzing target fuel)
- Presence or absence of other burning fuel packages or a hot gas layer
- Inherent properties of the target fuel
- Orientation of the fuel (horizontal or vertical—this can make a huge difference)
- Location of the flame on the fuel package (top or bottom)

All these factors influence the rate of heat transfer to the surface of the fuel, which ultimately determines how fast the flames spread across the surface. The critical heat flux for piloted ignition is always lower than that for nonpiloted ignition. For most materials, the critical piloted heat flux is in the range of 10 to 15 kW/m^2, with some materials exhibiting higher values. This is in contrast with minimum heat fluxes for smoldering ignition of 7 to 8 kW/m^2 [16].

Interior finishes are required to meet certain flame spread specifications or critical heat flux specifications. The use of inappropriate materials has resulted in many tragic fire losses and may be of interest to the fire investigator. One such loss was the Station Nightclub fire of February 2003, in which polyurethane foam, used illegally as an interior finish, caused the fire to spread violently [17]. Flame spread rating is determined using a Steiner tunnel test, and the method of testing is described in ASTM E84 (also known as NFPA 255), *Standard Test Method for Surface Burning Characteristics of Building Materials.* In this curious test, the material being tested is suspended on the ceiling of a tunnel 25 ft long and 18 in wide. A small draft (3.8 mm of water column) is provided. A natural gas flame that provides 5,000 Btu (5.3 MJ) per minute (88 kW) is applied 1 ft from the end, and temperatures, smoke density, and the movement of the flame front are recorded. Noncombustible cement board is used to determine the zero point, and select-grade red oak flooring is used to determine the values that define a flame spread rating of 100. Typically, it takes 5 to 6 min for the flame to travel 25 ft over the surface of the oak. Materials can have a flame spread higher or lower than the oak. The tunnel test apparatus is shown in Figure 3.5.

The *Life Safety Code*, NFPA 101, defines three "classes" of materials used for wall or ceiling finish according to their flame spread rating. Class A has a rating of 25 or less; that is, the flame spreads only 25% as quickly as it spreads on red oak, so it requires 20 to 25 min for the flames to reach the end of the tunnel, if they ever do. Class B materials have a flame spread rating of 26 to 75. Class C materials have a flame spread rating of 76 to 200. The Building Officials and Code Administrators International (BOCA) *National Building Code* and the *Uniform Building Code* use classes designated I, II, and III instead of A, B, and C.

Some confusion is possible when reading Roman numeral ratings because there is a second classification used by the NFPA for floor coverings that depends on the critical radiant flux required to spread a flame over a flooring sample. A Class I floor finish requires a critical radiant heat flux of 45 kW/m^2, and Class II requires a critical radiant heat flux of 22 kW/m^2. The test method for determining critical radiant heat flux is set forth in ASTM E648 (also known as NFPA 253), *Standard Test Method for Critical Radiant Flux of Floor-Covering Systems Using a Radiant Heat Energy Source.* In this test, a sample measuring about 8 × 40 in. of the flooring is mounted horizontally on the bottom of the test chamber facing upward as actually installed (unlike the "tunnel test," where it is installed on the "ceiling" of the test apparatus) and exposed to a radiant gas-fired panel set up on a 30° angle, 14 cm above the sample. The test runs for as long as it takes the sample to stop burning after a 10-min exposure. The distance that the flooring material burns is then measured. The radiant heat energy source apparatus is shown in Figure 3.6.

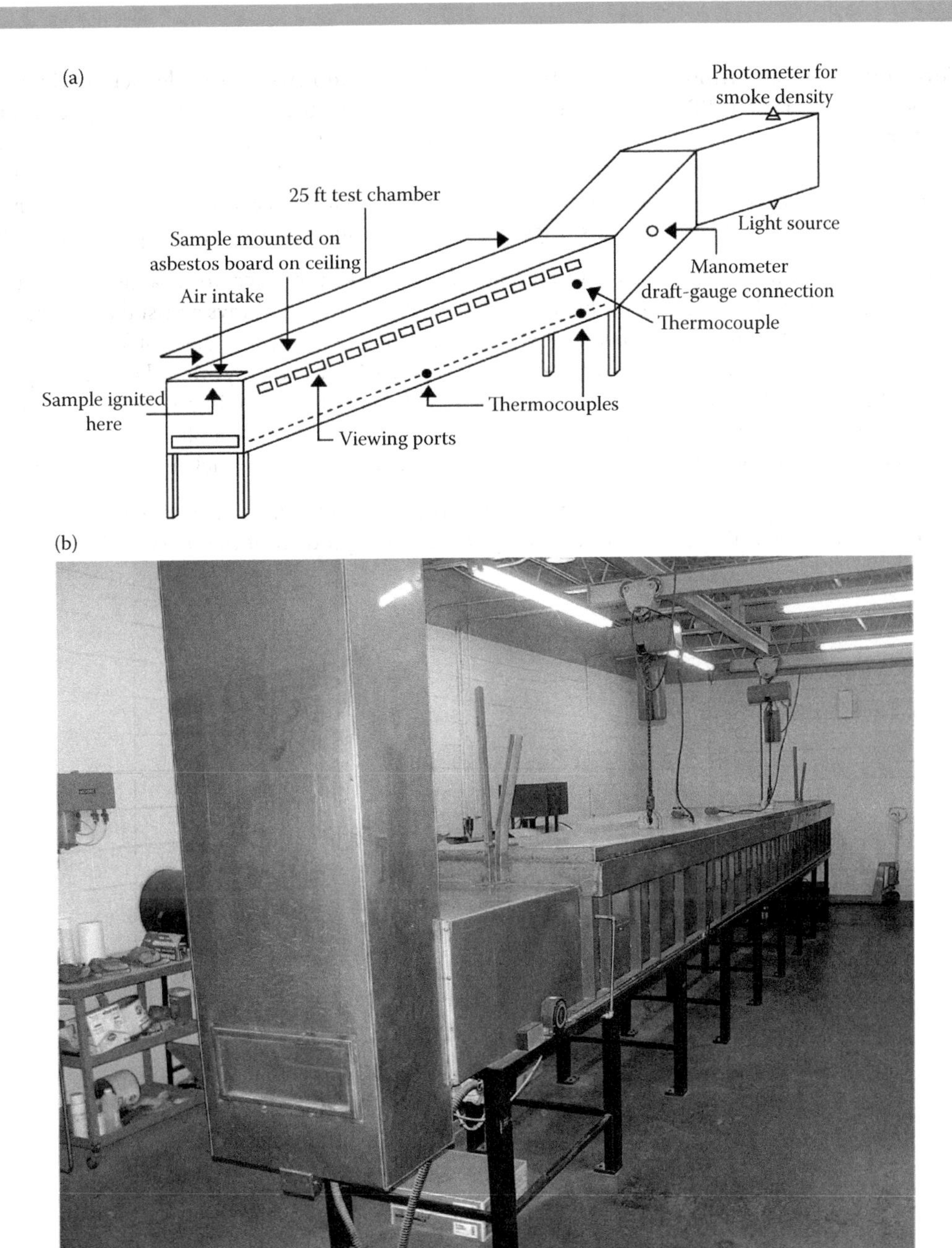

Figure 3.5 (a) Schematic view of a Steiner tunnel test apparatus. (b) Steiner tunnel test apparatus. Hoists are used to open the top of the chamber so that the 25 ft long × 2 ft wide test sample, mounted on cement board, can be loaded. The room is darkened for the test so that the operator can view the progress of the flame front. (Courtesy of Commercial Testing, Dalton, GA.)

Certainly there are materials that ignite more easily than Class II floor coverings or whose flame spread rating is faster than Class C wall or ceiling finishes. However, the use of such materials is not permitted in exits or in certain kinds of occupancies. Many furnishings and other fuel packages in a structure may not meet the flammability requirements for interior finishes.

Until recently, furnishings were required to be resistant to cigarette ignition and were classified (again using Roman numerals) according to their ability to resist ignition when tested using NFPA 260, *Standard Methods of Test and Classification System for Cigarette Ignition Resistance of Components of Upholstered Furniture.* Class I materials exhibit a char length of less than 45 mm (1.75 in.), and Class II materials exhibit a char length of 45 mm or greater.

Cigarette ignition resistance tests for furnishings are rapidly becoming a thing of the past. Modern furniture regulations require that the furniture be tested according to more stringent tests developed at the California Bureau of Home Furnishings. (Many furniture manufacturers regard the California tests as de facto national standards.) Technical Bulletin 117, revised in 2013, requires a series of flame ignition tests in addition to cigarette ignition tests. More comprehensive testing of mattresses is specified in Technical Bulletin 603, which requires a room calorimeter to run the test. The variety of tests that might be relevant to a particular item of furniture in a particular occupancy is too large to be discussed in more detail here, but the information is available at the California Bureau of Electronic and Appliance Repair, Home Furnishings and Thermal Insulation website (http://www.bearhfti.ca.gov).

Ease of ignition is one characteristic of flammability that should be taken into account, particularly when the investigator is proposing a specific fuel package as the first fuel ignited. It is almost always a mistake, for example, to posit a piece of dimensional lumber as the first fuel ignited. The flame spread rating of a piece of kiln-dried spruce might be twice that of red oak, but it is exceedingly difficult to ignite, say, a 2 × 4 with anything short of a large fire. Just try lighting a piece of timber with a blowtorch; it goes out when the flame is removed. In any house with copper plumbing, there will be numerous places where the structural timbers are scorched by a plumber's torch, but unless the torch ignites insulation board or some other easily ignited fuel, the result is local blackening, not a fire.

The most important property of any fuel that contributes to the spread of a fire is its *heat release rate* (HRR). In the past, fuels were classified according to their total energy content (the product of the mass of the fuel times the heat

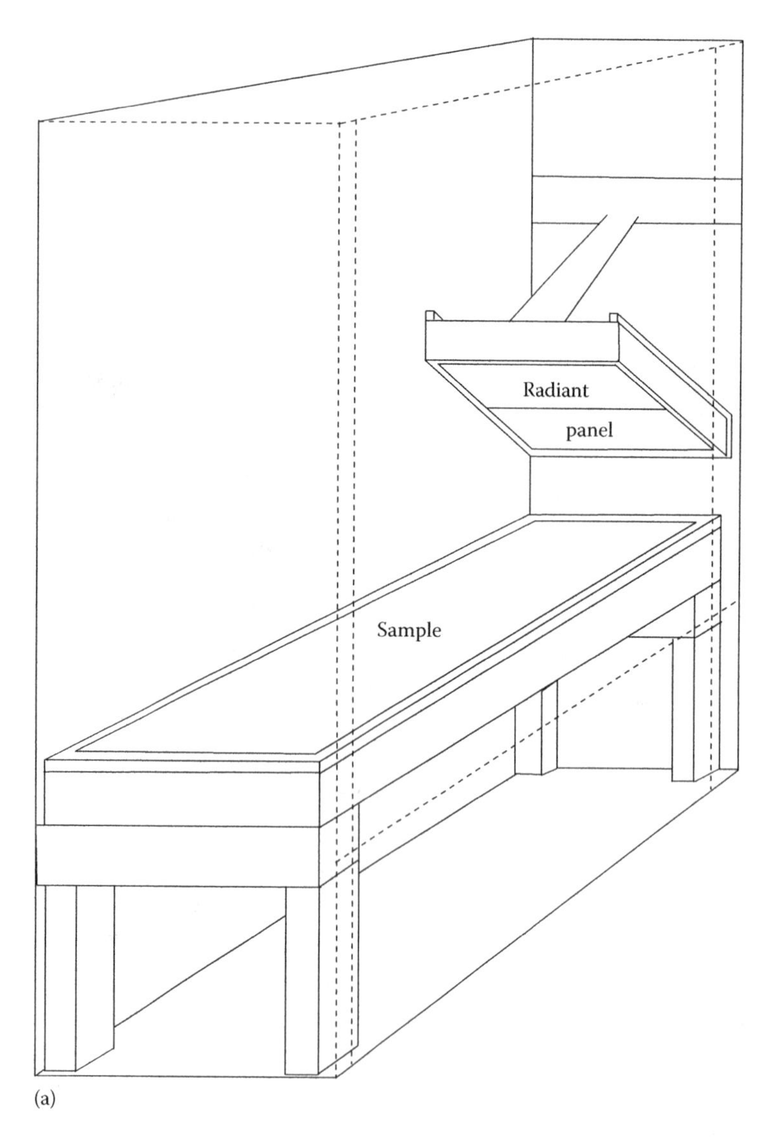

Figure 3.6 (a) Schematic view of a radiant panel test apparatus.

Figure 3.6 (b) Radiant panel test apparatus. The test is run for 5 min with exposure to the radiant panel only. Next, a T-burner is applied to one end of the test sample and held there for 5 min. Finally, the burner is shut down, but the radiant panel stays ignited. The test is over when the sample stops burning. (Courtesy of Commercial Testing, Dalton, GA.)

of combustion), but the total fuel load has no bearing on the growth of a fire in its pre-flashover phase [18]. While knowing the fuel load is sometimes useful for determining how much energy a given fuel package will *ultimately* contribute to a fire, it is the *rate* at which that energy is released that determines the course of the fire. The frequently used example is that of a pound of sawdust and a pound of wood. Both have the same energy content, about 8,000 Btu (8.4 MJ), but the sawdust's energy can be released in a much shorter time (and it is much easier to ignite). Thus, two materials with the same energy content can have dramatically different energy release rates.

The energy content of a fuel, reported as energy per unit mass (joules per gram, kilojoules per kilogram, or Btus per pound), is determined using an oxygen bomb calorimeter. In the apparatus shown in Figure 3.7, a small amount of the fuel is carefully weighed, mixed with an excess of oxygen, and ignited in a closed container immersed in a water bath. The increase in temperature of the water bath is converted to joules, Btu, or calories. The rate of energy release is not measured, just the total energy.

The HRR is the product of the burning rate measured in g/s and the heat of combustion, measured in J/g.[1] For substances with a constant heat release rate, such as ignitable liquids in configurations such as circular pools, the heat release rate can be calculated, and much has been learned about the behavior of fire by observing burning pools of liquid. For most solid fuels, however, the process of burning causes changes in the HRR. Where the HRR is not a constant value but changing with time, the HRR curve must be characterized by more than the peak HRR. Such characterizations include the growth rate, the decay rate, the peak HRR, and the total area under the HRR curve. (The HRR curve is a plot of the heat release rate over time.) To date, the heat release rate of a specific fuel cannot be calculated from first principles or fundamental properties—it can only be measured. Estimates of HRRs of fuel packages that have not been measured should be viewed with skepticism, and even those that have been measured have probably not been measured inside a compartment.

The HRR of a fuel package can be measured using an oxygen consumption calorimeter, which measures, among other things, the amount of energy in watts required to ignite the fuel, the mass loss during combustion, and the amount of oxygen consumed by the combustion. This wealth of information is one reason that regulatory authorities are requiring oxygen consumption calorimetry to replace the simpler form of fuel package testing, like cigarette ignition resistance tests. Oxygen consumption calorimeters can be small (cone calorimeters), medium (furniture calorimeters), or large (room calorimeters). Oxygen consumption calorimetry is illustrated in Figure 3.8.

---

[1] In the expression J/g × g/s, the grams cancel each other out, leaving J/s, or W.

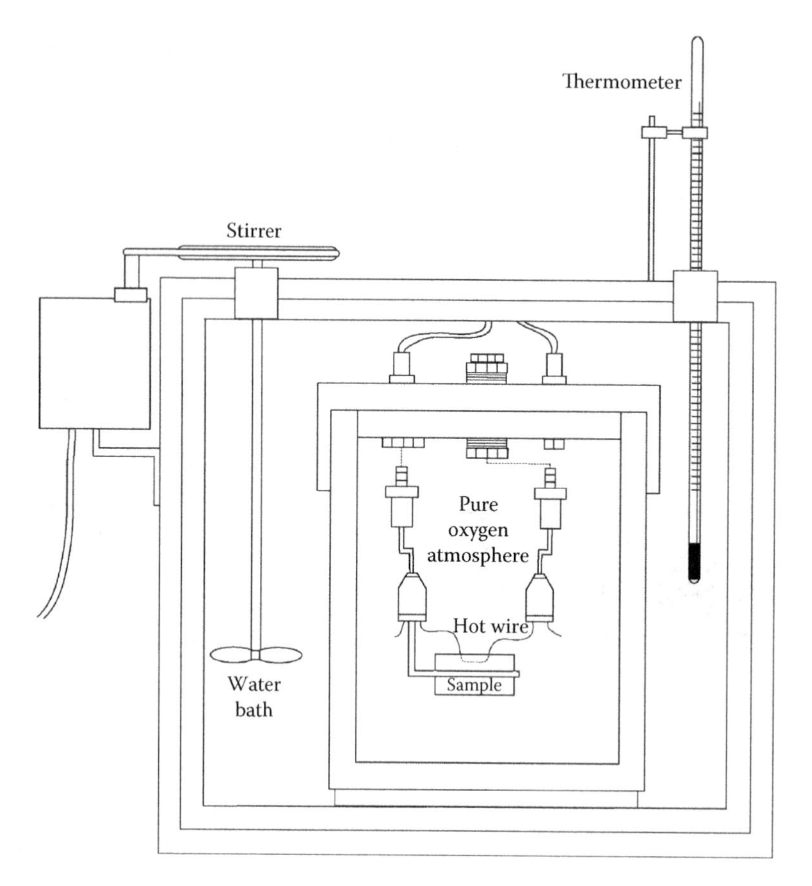

Figure 3.7 Schematic view of an oxygen bomb calorimeter.

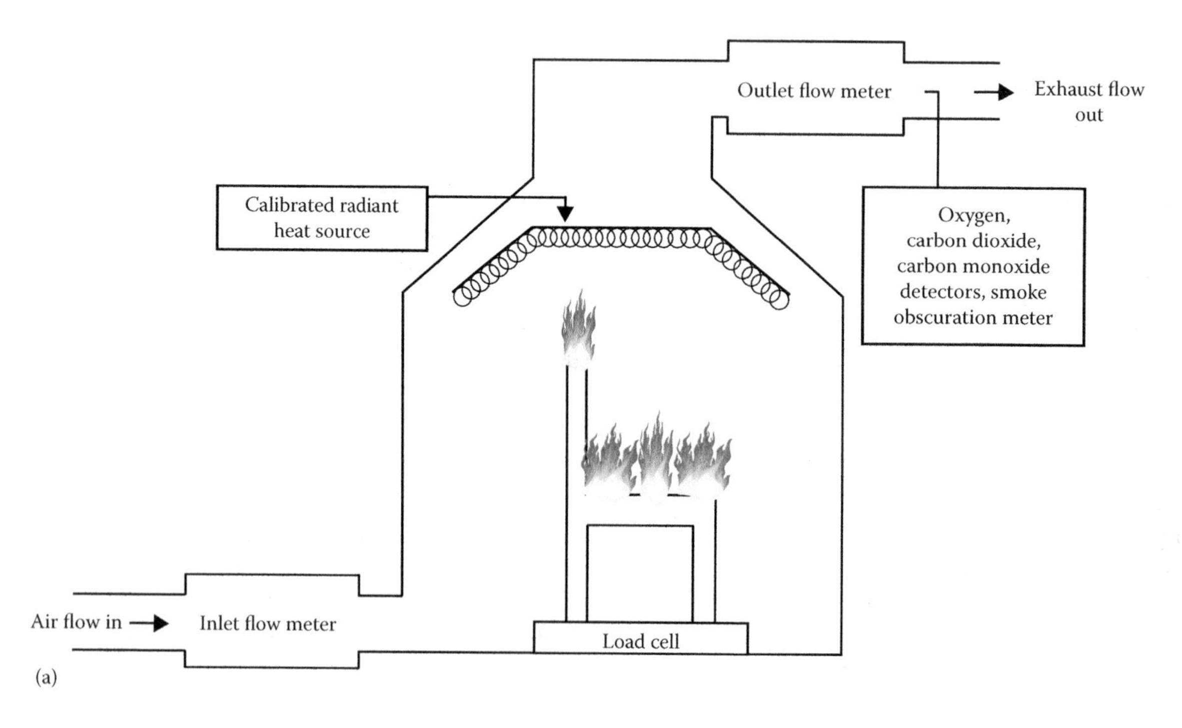

Figure 3.8 (a) Schematic view of an oxygen consumption calorimeter.

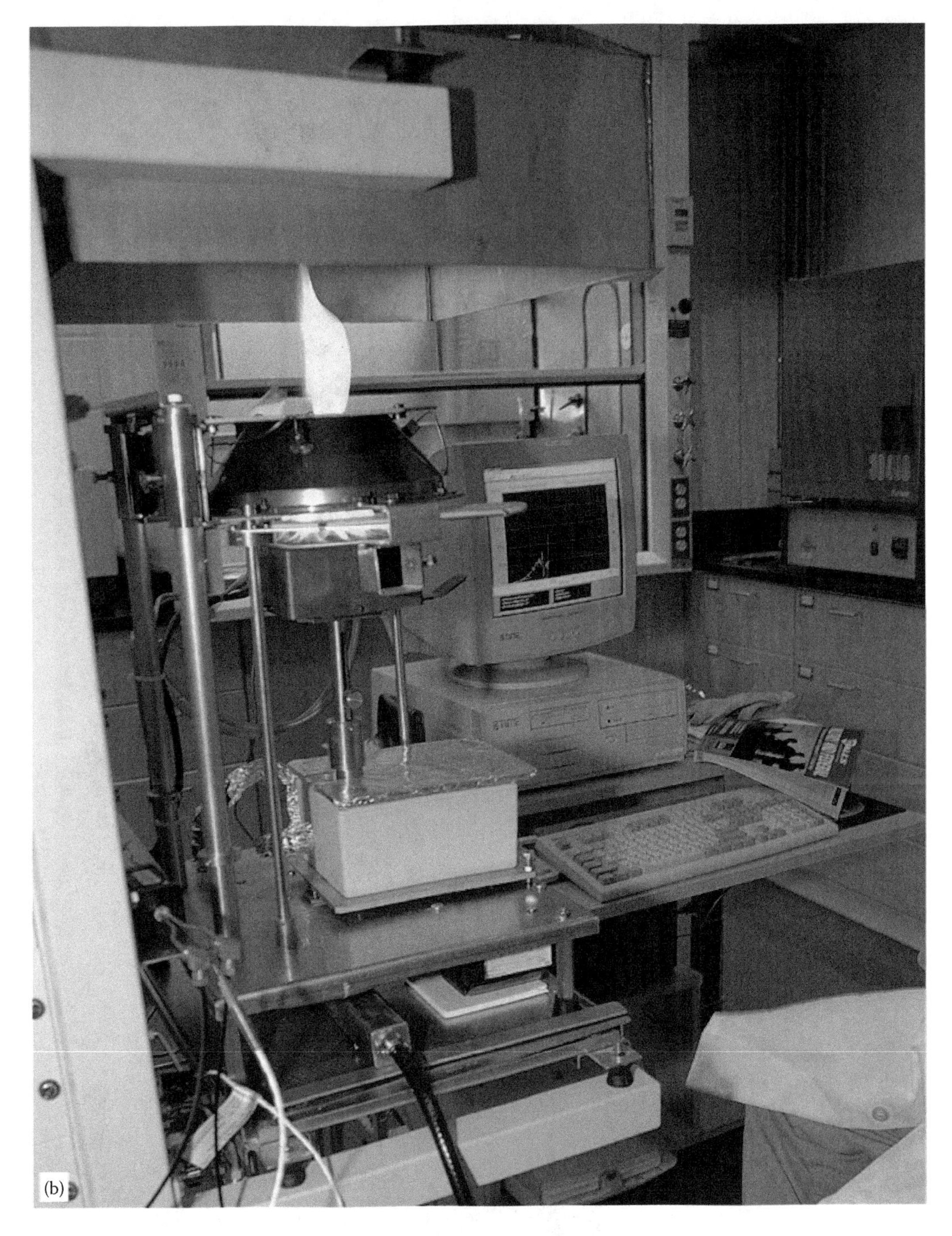

Figure 3.8 (b) Cone calorimeter in operation. (Courtesy of Bureau of Alcohol, Tobacco, Firearms and Explosives Fire Research Laboratory, Beltsville, MD.)

Figure 3.8 (c) Furniture calorimeter in operation. Note that the furniture is burning "in the open." Different results would be obtained if the furniture were in a compartment, where radiation feedback from the compartment and oxygen concentrations might differ. (Courtesy of Bureau of Alcohol, Tobacco, Firearms and Explosives Fire Research Laboratory, Beltsville, MD.)

Some typical values for the peak HRRs of common fuels are given in Table 3.2. Note that fuel configuration and flame location are critical to the outcome of any test of HRR. Folded newspaper has a much lower HRR than crumpled newspaper. If one ignites a crumpled sheet of newspaper on the top, it burns downward slowly; but if the bottom is ignited, it burns upward rapidly. There is a 10 kW difference, depending on which end is ignited. Mitler and Tu reported that the location of the ignition source on a larger fuel package, such as a chair, affects the time at which the peak HRR is reached, but it will not significantly affect the shape of the HRR curve or the peak heat release rate [19]. Stroup et al. found that placement of a chair in a room likewise affected the time required to reach the peak HRR and the shape of the curve, but not the peak HRR [20]. Test results showing the effect of an enclosure are presented in Figure 3.9. Kranzy and

Table 3.2 Peak heat release rates of some common fuels

Fuel	Peak HRR
Cigarette, not puffed	5 W
Match, wooden	80 W
Newspaper, folded double sheet, 22 g (bottom ignition)	4 kW
Newspaper, crumpled double sheet, 22 g (top ignition)	7 kW
Newspaper, crumpled double sheet, 22 g (bottom ignition)	17 kW
Polyethylene waste basket (285 g) filled with 12 milk cartons (390 g)	50 kW
Plastic trash bags filled with paper trash	120–350 kW
1 ft^2 pool of gasoline or kerosene	300 kW
Cotton upholstered chair	300–400 kW
Polyurethane upholstered chair	1350–2000 kW
Polyurethane mattress	800–2600 kW
Polyurethane upholstered sofa	3000 kW
Furnished living room	4000–8000 kW

Source: NFPA 921, Table 5.4.2.1. With permission.

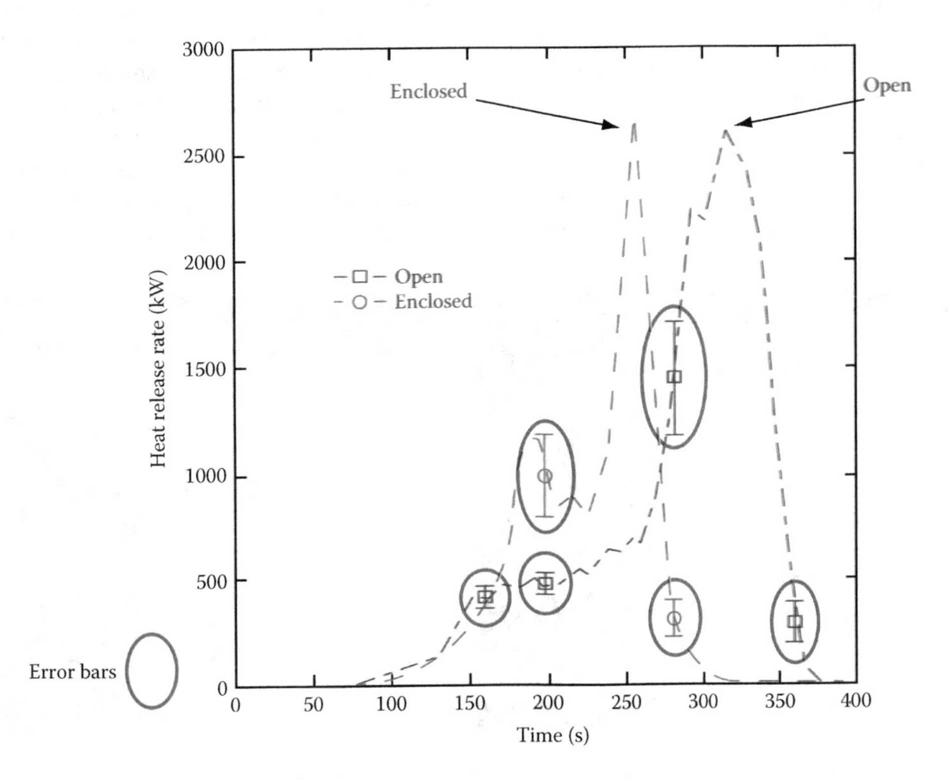

Figure 3.9 Results of two tests performed on two identical armchairs. (Data courtesy of NIST, Gaithersburg, MD.)

Babrauskas found the effect of an enclosure on HRR measurements to be minimal at HRRs below 600 kW. At 1,200 kW, a value far exceeded in most fully involved compartments, enclosure increased the measured HRR by 50% [21].

3.8 COMPARTMENT FIRES

While knowledge of the relevant properties of a fuel is important, it is even more important for the fire investigator to understand the relationship that exists between a fuel on fire and its surroundings. **Fuel packages behave differently inside structures** (compartments) than they do when they are unconfined.

Most people have some familiarity with fire behavior because of exposure to campfires, brush fires, and trash fires, and everyone knows that heat rises. This all makes sense outdoors but represents the simplest case of fire, and fire investigators who fail to appreciate the significant differences between unconfined (free-burning) fires and compartment fires run the risk of making gross misinterpretations of what they observe on the fire scene.

Ceiling Jet: A relatively thin layer of flowing hot gases that develops under a horizontal surface (e.g., ceiling) as a result of plume impingement and the flowing gas being forced to move horizontally.

Consider the simplest case of a fuel package interacting with its surroundings: a typical wastebasket fire with a heat release rate of 150 kW. A plume forms above the burning fuel package, just as it would outdoors, but the situation changes when the plume meets the ceiling. At that point, a ceiling jet forms as the fire is directed sideways. Figure 3.10a shows the structure of the plume. As the air is entrained, the rising smoke is diluted and cooled. Typical temperatures are shown in Figure 3.10b.

If the wastebasket is in the center of a large room, unaffected by walls, the flame height will be about 1.3 m (just over 4 ft). If we move the wastebasket next to a wall, cool air can no longer be entrained in the plume from the direction of the wall. The length of the mixing region of the flame is extended, and the flame will be taller, but not twice as tall. If we move the wastebasket into the corner, the flames will be taller still because only 25% of the circumference of the plume is open. Thus, it is possible for a fuel package in the corner of a room to appear much more heavily damaged than one might otherwise expect. This is called the "corner effect." It can lead to erroneous determinations as to the origin of the fire. This effect is demonstrated in Figure 3.11. In the first photo, a couch cushion is ignited in the center of a room. Figure 3.11a shows the cushion 17 s after ignition. Placing an identical cushion against a wall results in a slightly taller flame, as shown in Figure 3.11b. The effect is not dramatic because the cushion was ignited on the side away from the wall, so air entrainment from that direction was not significant at the outset. When the cushion is placed in the corner, as shown in Figure 3.11c, there is a dramatic increase in the height of the flame. The presence of a ceiling to confine the products of combustion makes the picture even more interesting. While the position of the

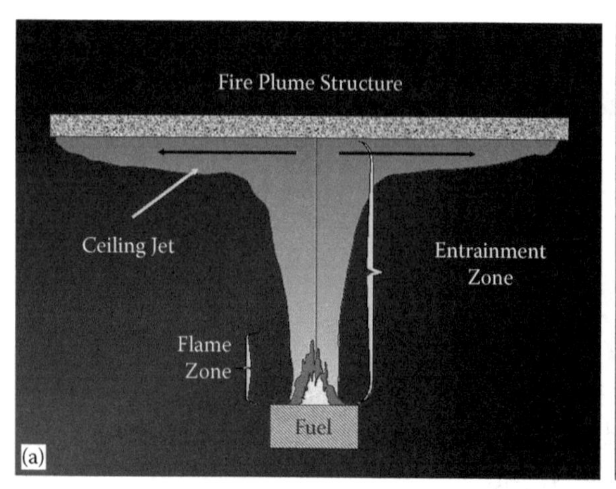

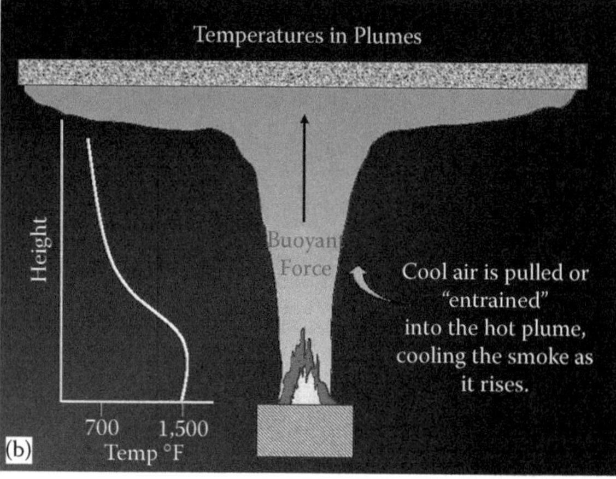

Figure 3.10 (a) Structure of a fire plume confined by a ceiling. (b) Typical fire plume temperatures in a developing fire. (Drawings courtesy of Steve Carman.)

Figure 3.11 The effect of location in a compartment on flame height. (a) Couch cushion burning in the center of a room. (b) identical cushion burning against a wall; and (c) identical cushion burning in a corner. All photos were taken at 17 ± 1 s after ignition. (Courtesy of Jamie Novak, Novak Investigations, Lindstrom, MN.)

fuel package in the compartment is still important, what happens when the plume is blocked by the ceiling is outside the realm of most people's experience.

The burning cushions in Figure 3.11 demonstrate the consequences of moving fuel packages around on the floor, in the X-Y plane. It is also important to consider vertical movement, in the Z plane. Raising the height of the flame increases the probability that the descending hot gas layer will starve the flames at the origin of oxygen. This in turn can result in the production of confusing fire patterns. It is possible for a fire with an elevated origin to oscillate between an oxygen-starved, or underventilated, fire and one with sufficient oxygen to form a plume. If the hot gas layer descends and deprives the flame of oxygen, the fire becomes smaller, air entrainment is reduced, and the rate of temperature rise in the hot gas layer decreases. The hot gas layer may then rise in this instance, allowing air to reach the flame once again. At this point, the temperature in the hot gas layer rises and the cycle can repeat itself. When one considers that cooking is one of the leading fire causes, it becomes apparent that many fires have elevated origins [22].

Flashover: A transition phase in the development of a compartment fire in which surfaces exposed to thermal radiation reach ignition temperature more or less simultaneously and fire spreads rapidly throughout the space, resulting in full room involvement or total involvement of the compartment or enclosed space.

Figure 3.12 shows the typical development of a compartment fire. Once the ceiling jet reaches the walls, a hot gas layer begins to develop. The particles, aerosols, and gases (smoke) in that layer absorb the energy released by the burning fuel package. Simultaneously, these particles also reemit some of the energy they have absorbed. The layer grows thicker and hotter as the burning fuel package produces more smoke and the combustion process releases more energy. The hot gas layer becomes a heat source itself, emitting some of its energy in the form of radiation, which moves in all directions (the word *radiation* comes from the same root as *radius*, as in a sphere), including down.

Thus, even without any "help," the hot gas layer has the capacity to ignite objects below it. This ignition by radiation, known as *flashover*, typically occurs when the temperature of the hot gas layer reaches 500°C to 600°C. The radiant heat flux at the floor of the compartment is about 20 kW/m^2 (or 2 W/cm^2). The transition from pre-flashover burning to post-flashover burning has been described as changing from "a fire in a room" to "a room on fire." The room will continue to burn long after flashover occurs, but the fundamental nature of the fire changes.

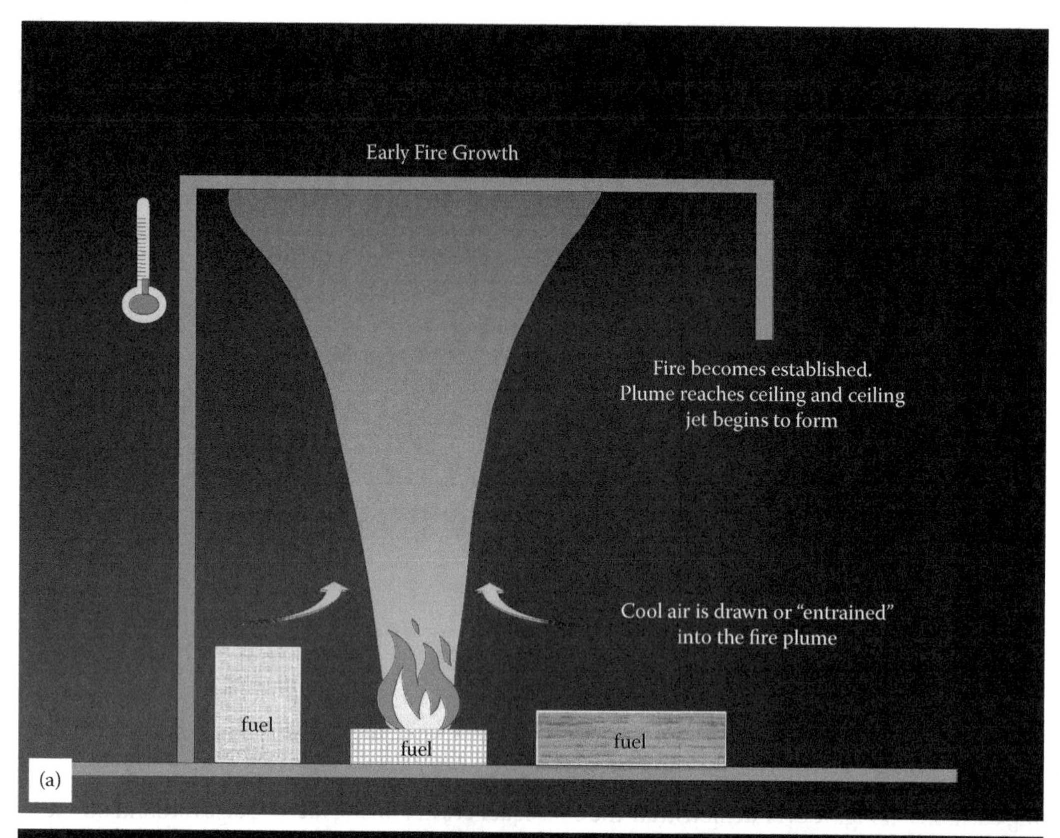

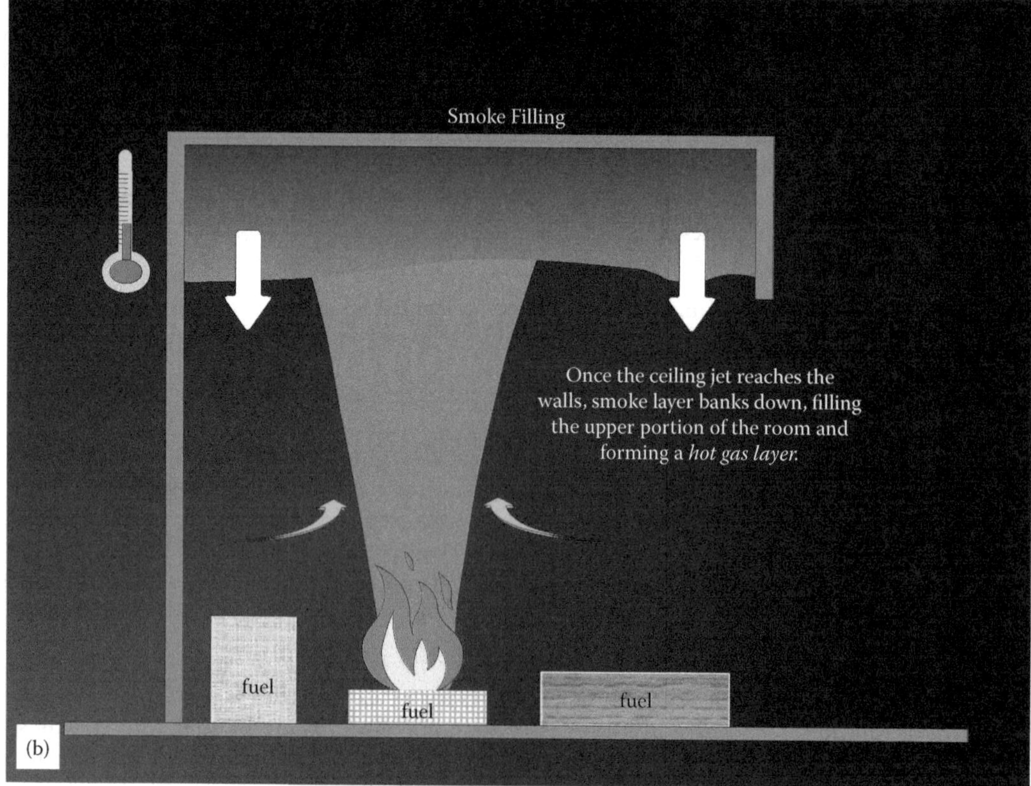

Figure 3.12 Sequence of events [(a) through (d)] in the development of a typical compartment fire. (Drawings courtesy of Steve Carman.)

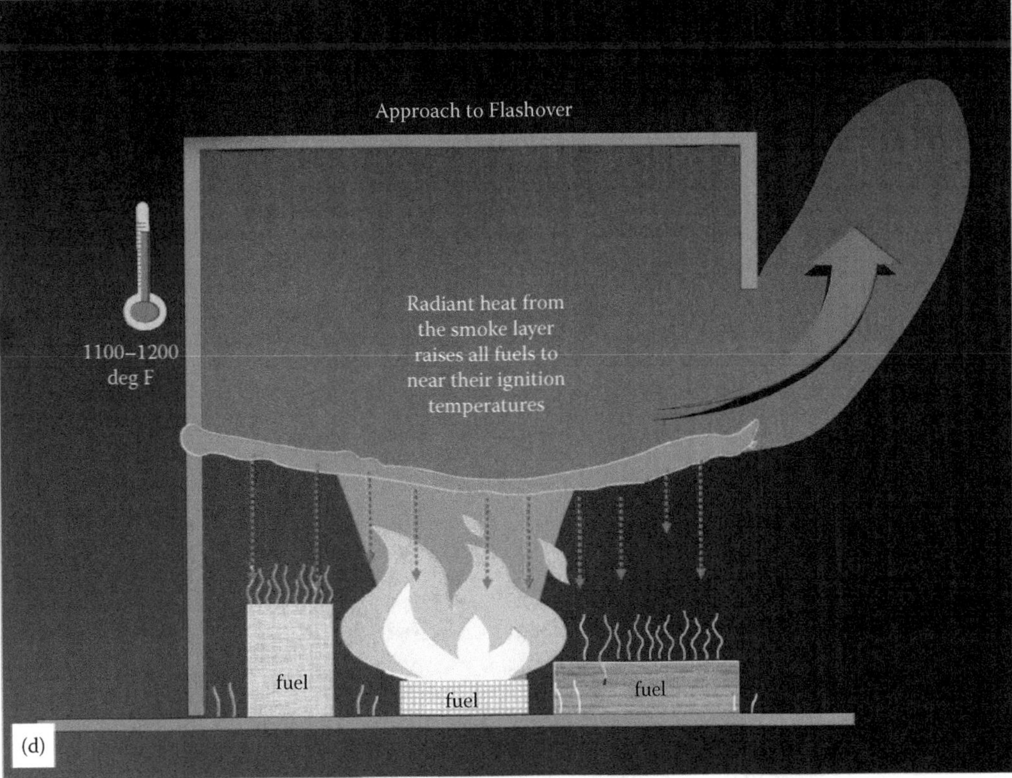

Figure 3.12 Sequence of events [(a) through (d)] in the development of a typical compartment fire. (Drawings courtesy of Steve Carman.)

In the free-burning phase, the fire grows as additional fuel becomes ignited. The fire is a "fuel-controlled fire." There is plenty of oxygen, so the size of the fire is controlled by the surface area of the fuel involved. The size of the fire in watts (and area) is roughly a function of the square of the time [23]. If the fire is extinguished prior to the involvement of other significant fuel packages, determination of the origin of the fire is a straightforward exercise.

In a post-flashover fire, all of the exposed fuel is already ignited so the growth of the fire is limited by the amount of air available to combine with the burning fuel. Oxygen becomes scarce because most of it was consumed in the flashover event. This is a "ventilation-controlled fire." Figure 3.13a shows what happens on the molecular level. Unburned fuel molecules and the products of combustion vastly outnumber the oxygen molecules, and **the fire at the origin may actually stop flaming**. Flaming fire can take place only near the green-shaded space. Figure 3.13b shows measurements that demonstrate dependence of temperature (also HRR) on oxygen concentration. In a series of experiments conducted at Underwriters Laboratories (UL), the oxygen in the room became depleted at the moment of flashover, and this resulted in an immediate temperature drop. This experiment was repeated in several full-scale fires, all showing similar results.

This phenomenon was recognized by a change in the 2017 edition of NFPA 921. Figure 3.14 is a comparison of the full room involvement diagrams from the 2014 and the 2017 editions.

Fire does not "seek" oxygen—it just cannot burn where there is none. In a fully involved compartment, the most intense parts of the fire will necessarily be in those locations where a good air supply exists, such as around windows and doors; where there is a preexisting penetration of a wall, floor, or ceiling; or where the fire itself creates a vent by burning through an exterior surface or breaking a window. The determination of the first fuel package ignited in this situation is much more difficult. The investigator who bases an origin determination on the area of lowest burning or deepest char in a fully involved compartment may be easily misled by a ventilation-generated fire pattern.

Ventilation-controlled fires frequently produce V-shaped patterns around doors, and for years, such patterns were misinterpreted as evidence of "accelerant" poured through the doorway. After all, when an arsonist is pouring liquid accelerant throughout a structure, he necessarily has to walk through a doorway to communicate the fire from one room to the next. The finding of a V-shaped pattern around the living room doorway was one of the factors that led investigators to conclude, erroneously, that the Lime Street fire was the result of the ignition of a trail of gasoline [24].

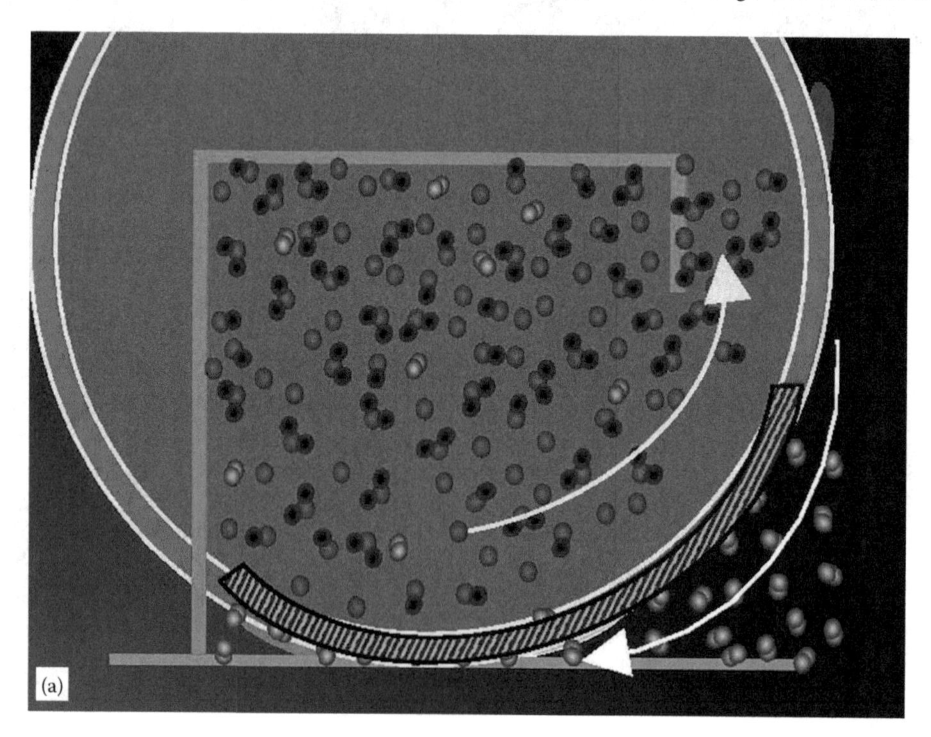

Figure 3.13 (a) Flashover depletes the available oxygen. Consequently, flaming combustion can only take place where there is an air supply.

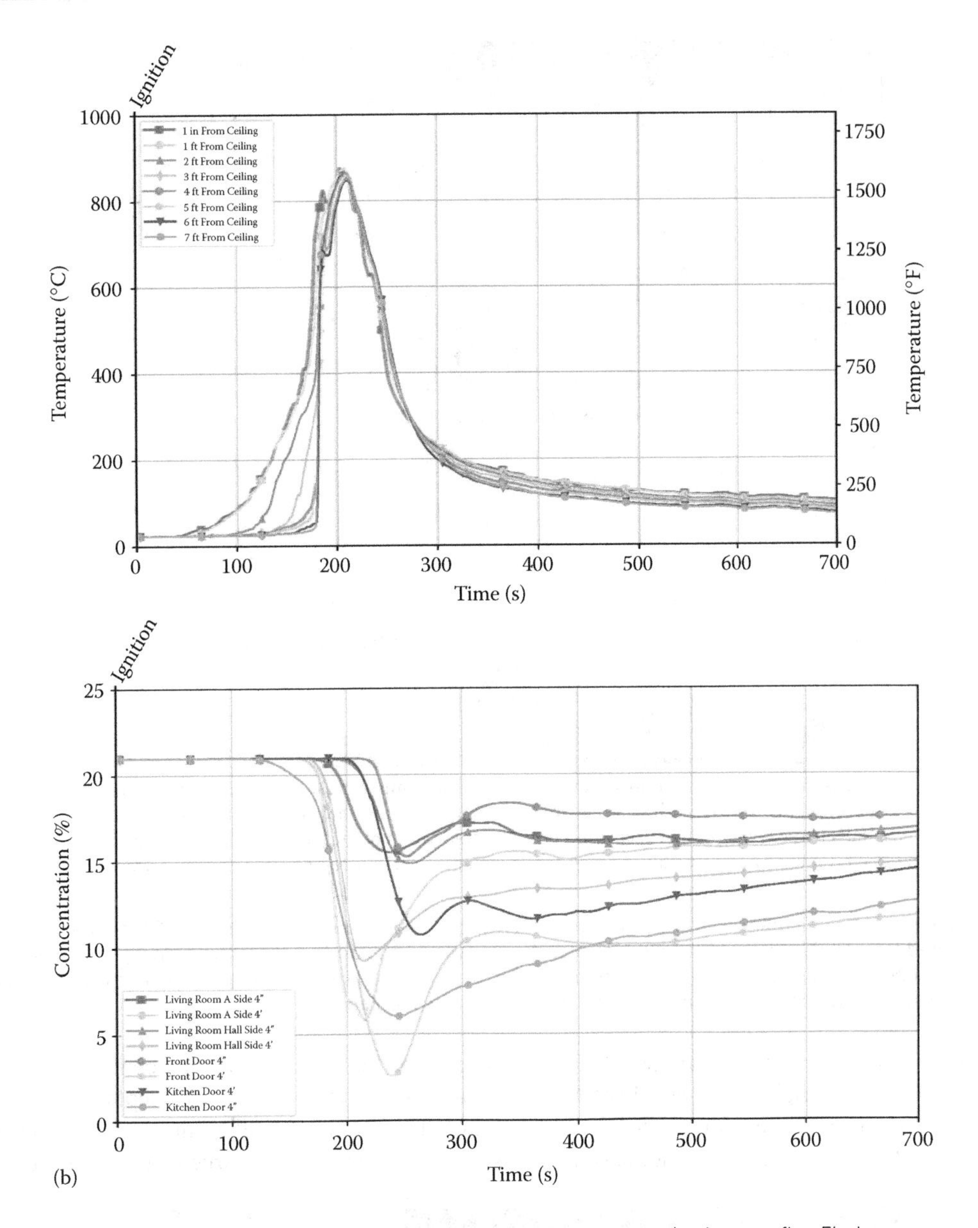

Figure 3.13 (b) Graph showing the relationship of temperature to oxygen concentration in a test fire. Flashover occurred at just over 200 s. The oxygen concentration dropped at the same time, and the temperature dropped precipitously as well.

The fire pattern on the hallway floor just outside the door also exhibited features that were erroneously characterized as having been made by burning gasoline. The investigator could hardly be blamed, however, as the laboratory analyst incorrectly identified gasoline on a sample of this flooring.

Another feature of ventilation-controlled fires is that they can produce V-shaped patterns that can be mistaken for plume patterns from a burning fuel source. If there is no naturally occurring fuel package or ignition source at the base of the V, rather than inferring that there must have been a different origin, some fire investigators infer instead that fuel must have been added in the form of flammable liquid, and the ignition source must have been an open flame that was removed. For this reason, NFPA 921 states that if there is no competent ignition source at a hypothetical origin, that origin hypothesis should be treated with "increased scrutiny [25]."

The importance of oxygen availability cannot be overstated. Where there is no oxygen, there is no flame. Uskitul et al. demonstrated this nicely on a small scale. A video showing the experiment can be downloaded at https://www.App.

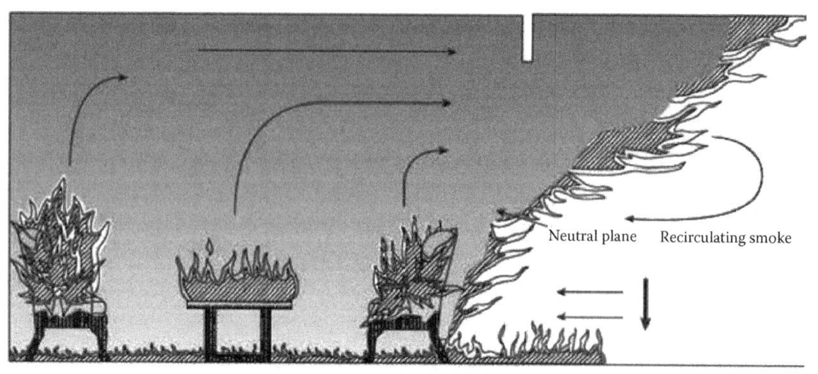

Postflashover or Full Room Involvement in Compartment Fire.

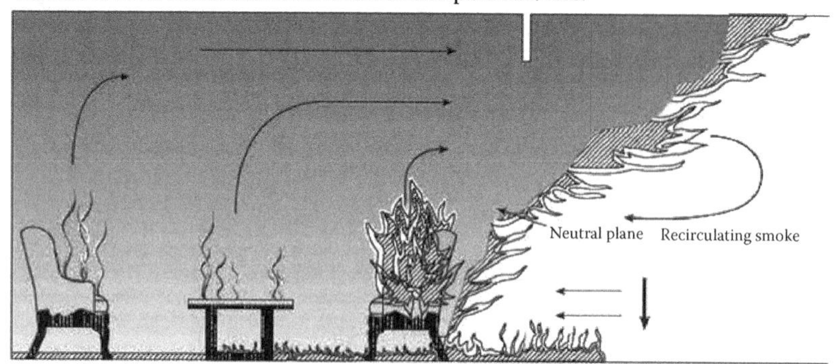

Postflashover or Full Room Involvement in a Typical Compartment Fire. Although pyrolysis can continue throughout the compartment, flaming combustion will only occur where there is sufficient oxygen present. Depending on the momentum of the entraining air, flaming combustion may occur within the ventilation stream at various depths into the compartment.

Figure 3.14 Comparison of diagrams of full room involvement from the 2014 edition of NFPA 921 (top) and the 2017 edition (bottom). (Reprinted from NFPA 921-2014, and 2017 *Guide for Fire and Explosion Investigations*, Copyright © 2014, 2017 National Fire Protection Association, Quincy, MA, with permission. This reprinted material is not the complete and official position of NFPA on the referenced subject, which is represented only by the standard in its entirety.)

box.com/s/jw4l3ovzs653qvcvr3nwf2bear9ayam8. Figure 3.15 shows a burning pool of heptane inside a small enclosure with a vent off to the lower right. After burning for two minutes, the enclosure filled with heptane molecules and the products of combustion, and the flame shifted to the vent [26]. The heptane is still boiling in the photo on the right in Figure 3.15, but the fire is taking place at the vent. This same thing happens on a large scale in a compartment fire.

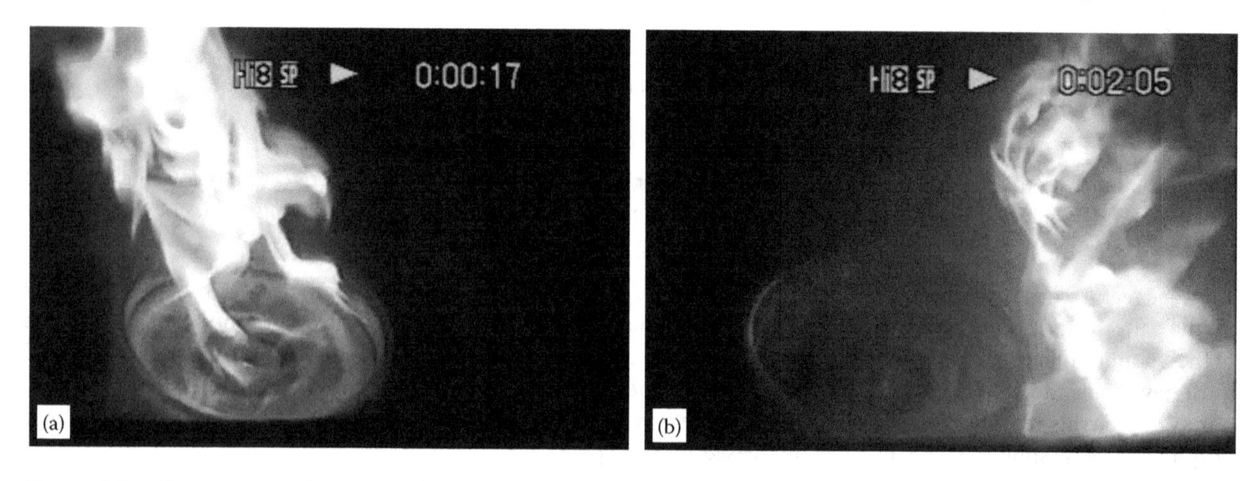

Figure 3.15 Flame migrates from the surface of a pool of heptane (a) to the location of the air supply (b) in this elegant experiment by Uskitul et al.

Just prior to flashover, as radiation becomes the predominant means of heat transfer, patterns are produced when surfaces are heated or ignited by thermal radiation. The edges of such patterns are often coincident with the edges of objects that cast a shadow, or that lay directly on a surface, preventing the radiation from reaching it. Thermal radiation is a line-of-sight phenomenon similar to light. It does not bend around corners. This can produce differences in exposure that can create patterns consistent with blockages to the line of sight. Such patterns are known as *protection* patterns. When the object that blocks the radiation is a regular shape, the pattern will be regular. When the object is an irregularly shaped object such as a piece of fallen drywall or an article of bedding or clothing, the pattern takes on an irregular shape. If the object is located directly on the protected surface, protection patterns can be quite sharp. Protection patterns and the confusion that they can cause are described at length in Chapter 8 under "Lines of Demarcation." Another type of protection pattern is one where the intervening object blocks the deposition of smoke. These patterns tend to be found in fires of lesser intensity than the fires that produce radiant heat protection patterns.

Because of the major differences in fire patterns produced by compartment fires compared to unconfined fires, one of the fire investigator's first tasks is to identify those compartments that became fully involved and those that did not. When the artifacts produced by full room involvement were first publicized in the late 1980s and early 1990s, many traditional fire investigators saw this information as nothing more than an effort by defense attorneys and experts to "explain away" what the original arson investigator believed was evidence of the use of an accelerant. A large cadre of investigators derided this new information as "the flashover defense." The debate raged in newsletters and in seminars, but eventually, especially after the general acceptance of NFPA 921, the reality of full room involvement and its effect on fire patterns was accepted [27].

Even today, however, investigators hope to avoid questions about the complexities of fire pattern interpretation in a fully involved compartment and simply declare (and often erroneously) that flashover did not occur. Fortunately, the determination as to whether flashover has occurred is not a difficult one to make, but there are some publications that erroneously describe what evidence to observe. The U.S. Army's Field Manual, *Law Enforcement Investigations,* states that "The big tipoff to a flashover fire is that burning is only on the top surface of items in the room [28]." Carroll's 1979 book, *Physical and Technical Aspects of Fire and Arson Investigations,* shows a photograph (Figure 3.11) of a room that has definitely not gone to flashover, but the caption states that "This room was ignited by flashover [29]." This misinformation notwithstanding, definitive clues are available from eyewitnesses as well as external and internal fire damage.

When flashover occurs in a compartment, one thing that usually happens is that the windows break. In fact, the breakage of the windows is often considered the event that defines when flashover has occurred. The breakage of glass is not due to increased pressure but to thermal stresses caused by the vast difference in temperature between the exposed glass and the unexposed glass under the edge of the frame. If the firefighters report fire blowing out the window, it is likely that flashover has occurred in that compartment. If there are large areas of burning above the window, as shown in Figure 3.16, it is likely that flashover has occurred.

Inside the compartment, the finding of floor-to-ceiling charring is the artifact that tells us that flashover has occurred. Observe the condition of the baseboards in particular. If there is nearly uniform charring around the room, burning by radiant heat is the only likely explanation. Uniform charring on the trim around doorways and windows is another indicator, as is widespread charring on the floor.

Burning on table legs, once thought an indicator of accelerants burning on the floor, is a common effect of flashover. After flashover, the flames fill the compartment and impinge on all exposed surfaces, including the undersides of tables and shelves.

Flashover initially causes a more or less uniform level of fire damage, and if a fire is extinguished not long after flashover, the uniformity zmight still be present. The uniformity begins to deteriorate almost immediately, however, as the effects of ventilation take over. In addition, there will almost always be some protected areas where some object prevents the radiant heat from reaching a surface. Minor variations from uniformity are not cause to state that flashover has not occurred.

Figure 3.16 Fire pattern above a window. Patterns such as these indicate that flashover, followed by full room involvement, has almost certainly occurred. (Courtesy of Mick Gardiner, Gardiner Associates Training and Research [GATR] Cambridge, England.)

The real question we should be asking is: Has the room became *fully involved* in fire? Full room involvement is the natural sequel to flashover, but it is possible for a compartment to become fully involved without going through flashover. This is particularly true in large compartments such as warehouses. The interpretation of the fire patterns is no different than if a clearly defined moment of flashover was observed. Unless something or someone interferes, a compartment fire is likely to progress through flashover to full room involvement, and the fire investigator must be prepared to interpret all postfire artifacts in that context.

While it is true that full room involvement frequently results in the production of patterns of damage that are subject to misinterpretation, it is not true that flashover destroys all patterns that preexisted the flashover event. Patterns that existed prior to flashover certainly change and might be obscured by subsequently produced patterns, but the outlines of any given pattern, including patterns produced by burning ignitable liquids, might well persist beyond flashover [30]. The challenge for fire investigators is to recognize these patterns for what they are. Meeting this challenge becomes more difficult as the fire burns for longer periods, and it becomes almost impossible if the gypsum wallboard or other interior finish falls off the walls and ceiling.

3.9 PLUME PATTERN DEVELOPMENT

Fire patterns are defined as the visible or measurable effects that remain after a fire [31]. Most patterns are recorded on two-dimensional surfaces, at the place where those surfaces intersect with the three-dimensional fire. Patterns can be caused by radiation, contact with the hot gas layer, contact with hot coals, or contact with the fire plume (direct flame impingement). After flashover, patterns are produced by local air supplies providing the oxygen to ignite clouds of unburned fuels. It is the goal of the fire investigator to locate the *first* plume pattern generated by the fire. This pattern will be at or very near to the origin of the fire.

Plume patterns produced prior to full room involvement include triangles, columns, and cones. NFPA 921 calls these patterns "truncated cone patterns." If one considers flaming fire at its incipient stage, one can visualize an inverted cone[2] that generally reflects the shape of the flame. If this cone is intersected, or truncated, by a wall, the fire produces a triangle pattern on the wall, as shown in Figure 3.17. Triangular patterns, sometimes referred to as inverted cone patterns, are the result of incompletely developed fires but do not necessarily indicate a separate point of origin. They are frequently caused by "drop down." Such patterns can be the result of a late involvement and early extinguishment (or simply burning out) of a secondary fuel, such as window curtains.

As the fire progresses, the products of combustion are sent upward in a buoyant plume that assumes the shape of a column. The intersection of a wall with the column produces a vertical pattern with roughly parallel sides, as shown in Figure 3.18. Columnar patterns, while produced by fires more developed than the fires that produce triangular patterns, usually evolve further if the fire is not extinguished. The fire on the sofa shown in Figure 3.18b was extinguished with a garden hose just a few minutes after ignition. The columnar pattern would have developed into a cone-shaped pattern had the fire not been extinguished.

When the fire continues to grow and becomes obstructed by a ceiling, a conical pattern results. This cone can be described as V-shaped if the base of the plume is close to the wall, as shown in Figure 3.19. Simple geometry tells us that as the base of the plume is moved farther away from the wall, the intersection of the cone with the wall will be higher and the bottom of the cone will be rounder, resulting in a U-shaped pattern, as shown in Figure 3.20. Where the cone intersects the ceiling, a radial burn pattern is produced. This can be a full circle if the plume is not intersected by a wall, or it can be a partial circle like the one shown in Figure 3.21.

Because the production of inverted cones, columns, and cones follows a sequence, these patterns usually overlay one another, and the first pattern created is frequently obscured by the successor pattern.

The astute reader will have noticed that all of the plume patterns presented thus far have occurred in compartments that did not achieve full room involvement. Such patterns are easily recognized and assist in illustrating the concept

[2] Strictly speaking, a geometric cone is wide at the bottom and comes to a point at its top. Fire investigators have adopted an inverted definition.

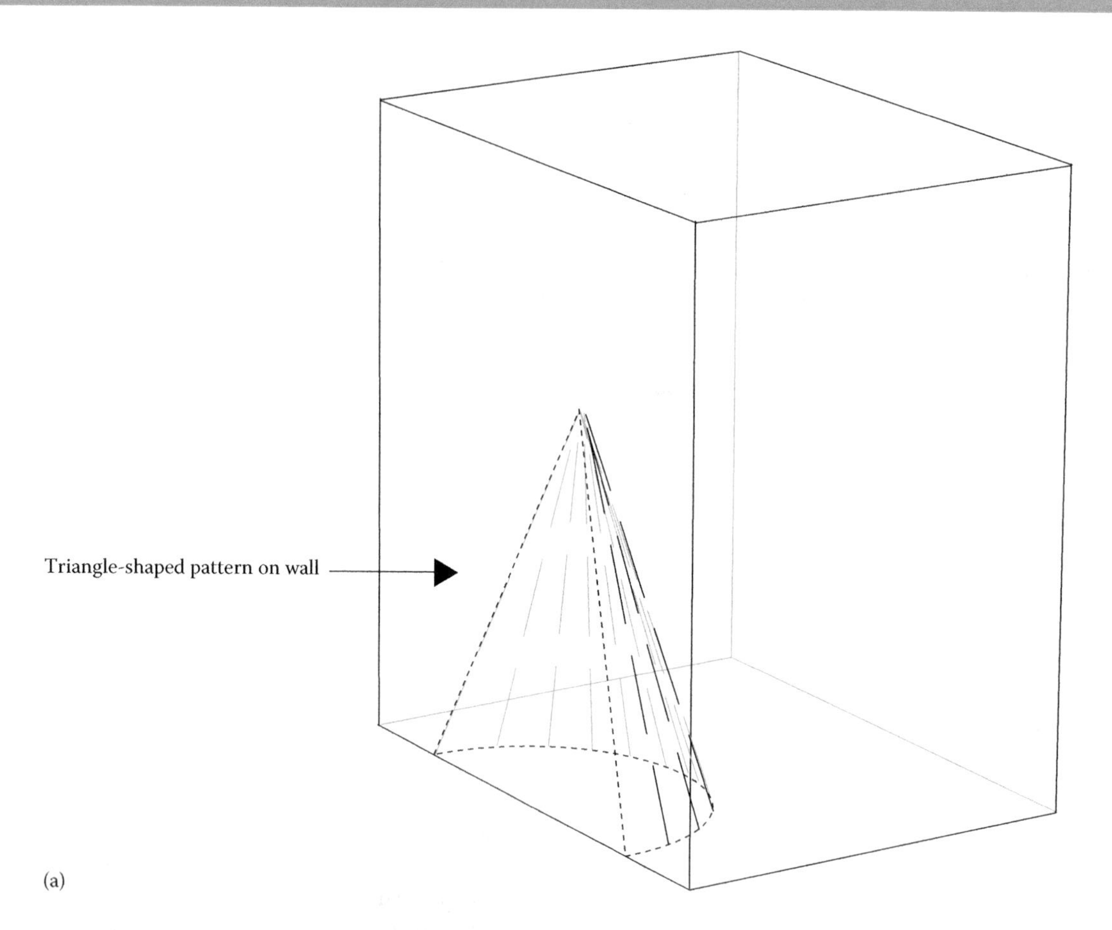

Figure 3.17 (a) Production of an inverted cone pattern.

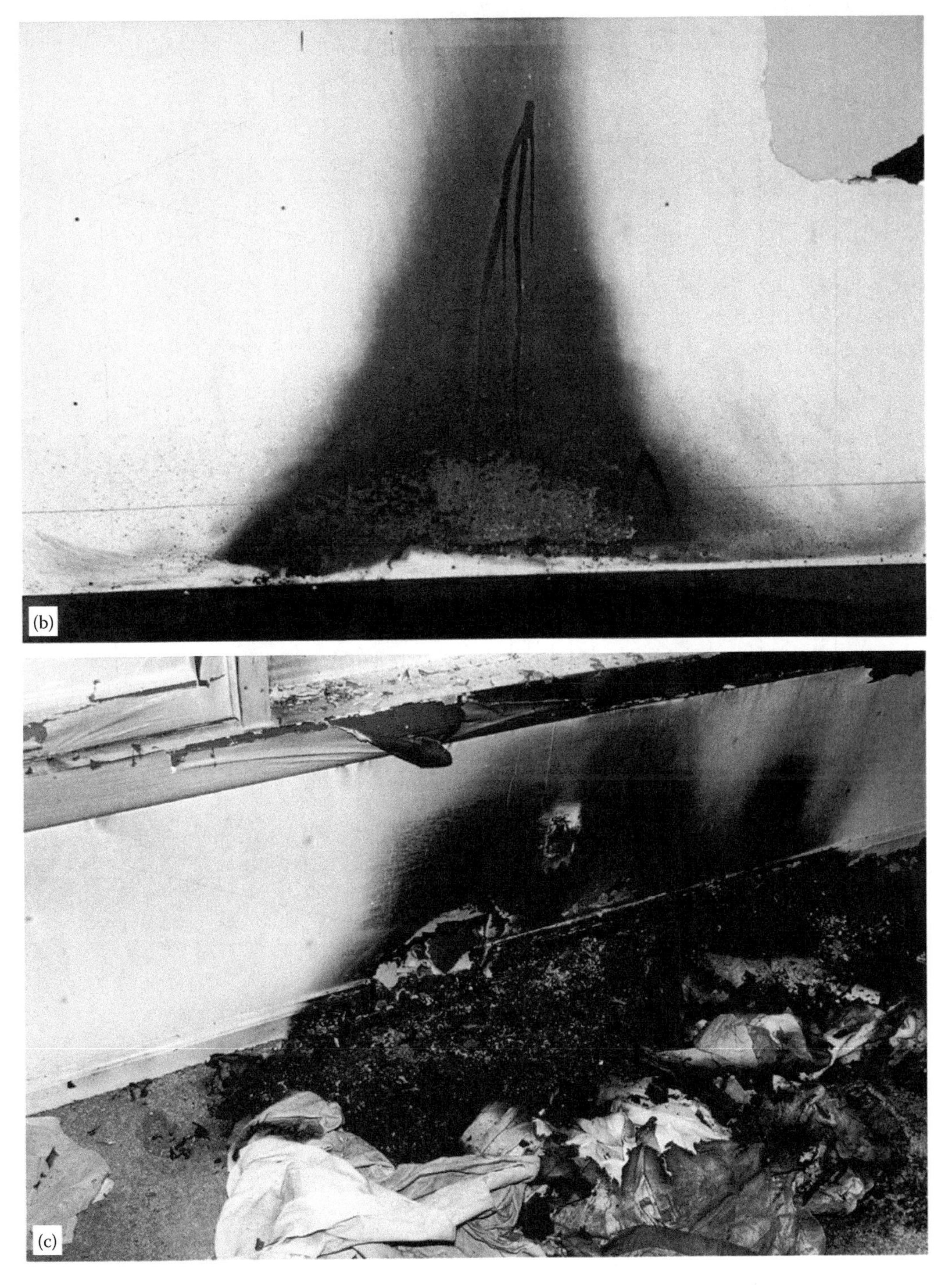

Figure 3.17 (b) Example of an inverted cone pattern. This kind of pattern is frequently caused by ignitable liquids burning on the floor. These patterns are usually seen when the fire does not develop further. This pattern was generated in a test fire by a small pile of burning newspaper. (c) Triangular pattern more typical of those seen at fire scenes. This was a separate set fire, involving kerosene.

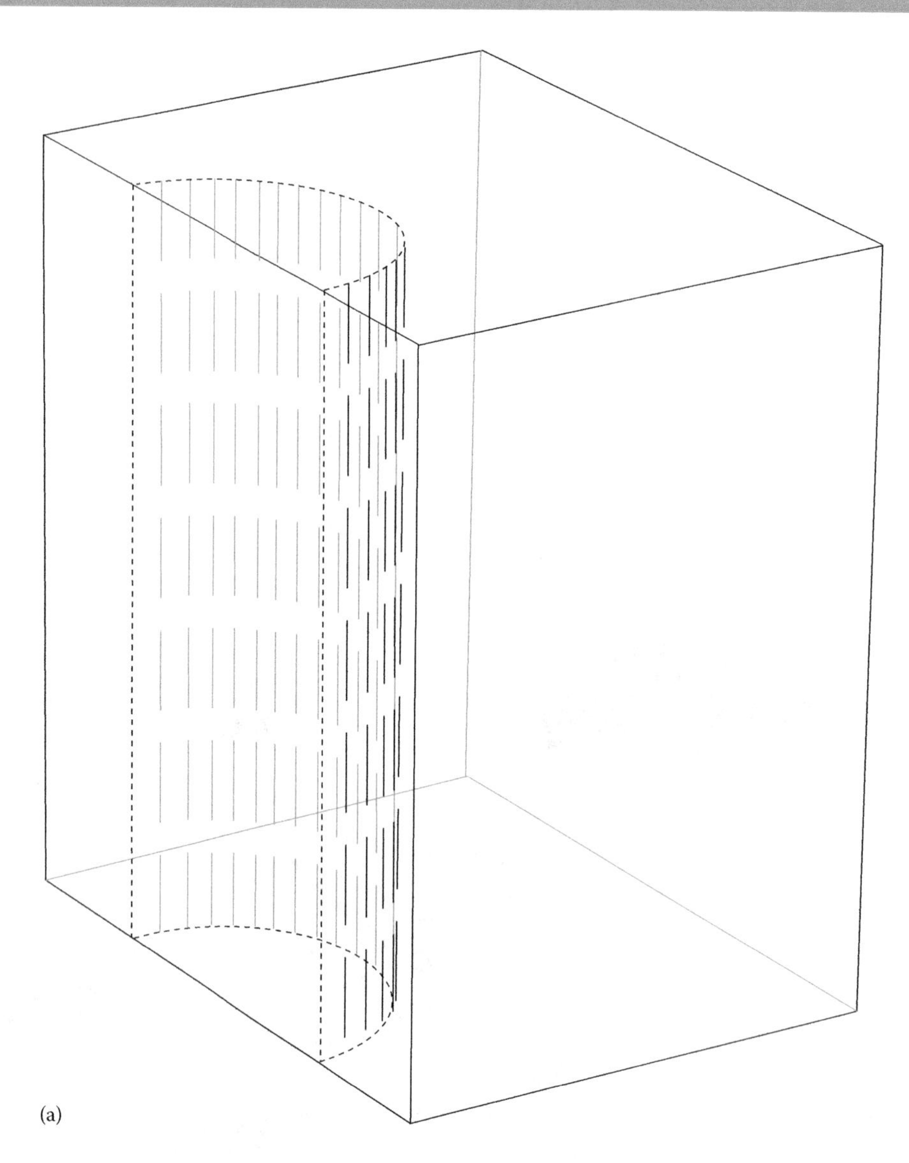

Figure 3.18 (a) Production of a columnar pattern.

Figure 3.18 (b) Columnar pattern. Because the pattern shape changes when the plume interacts with a ceiling, columnar patterns are short-lived and uncommon indoors. The fire that caused this pattern was extinguished before it had much time to interact with the ceiling.

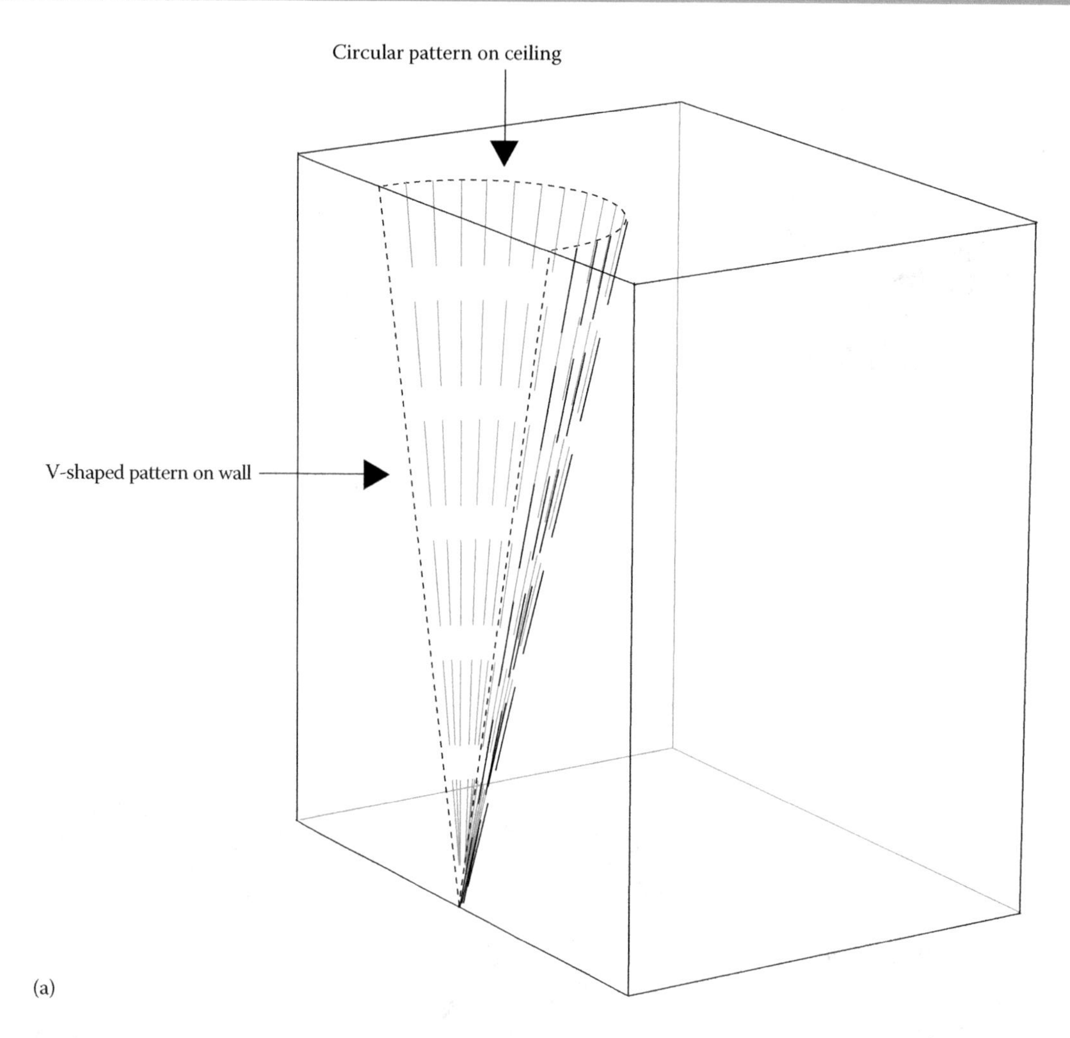

Figure 3.19 (a) V-shaped burn pattern produced at the intersection with a wall, and semicircular pattern produced on the ceiling by a cone-shaped fire plume.

Figure 3.19 (b) Typical V-pattern at the origin of a fire. There was a trashcan at the base of the V. Note the small protection pattern on the floor caused by the presence of the melted polyethylene trashcan.

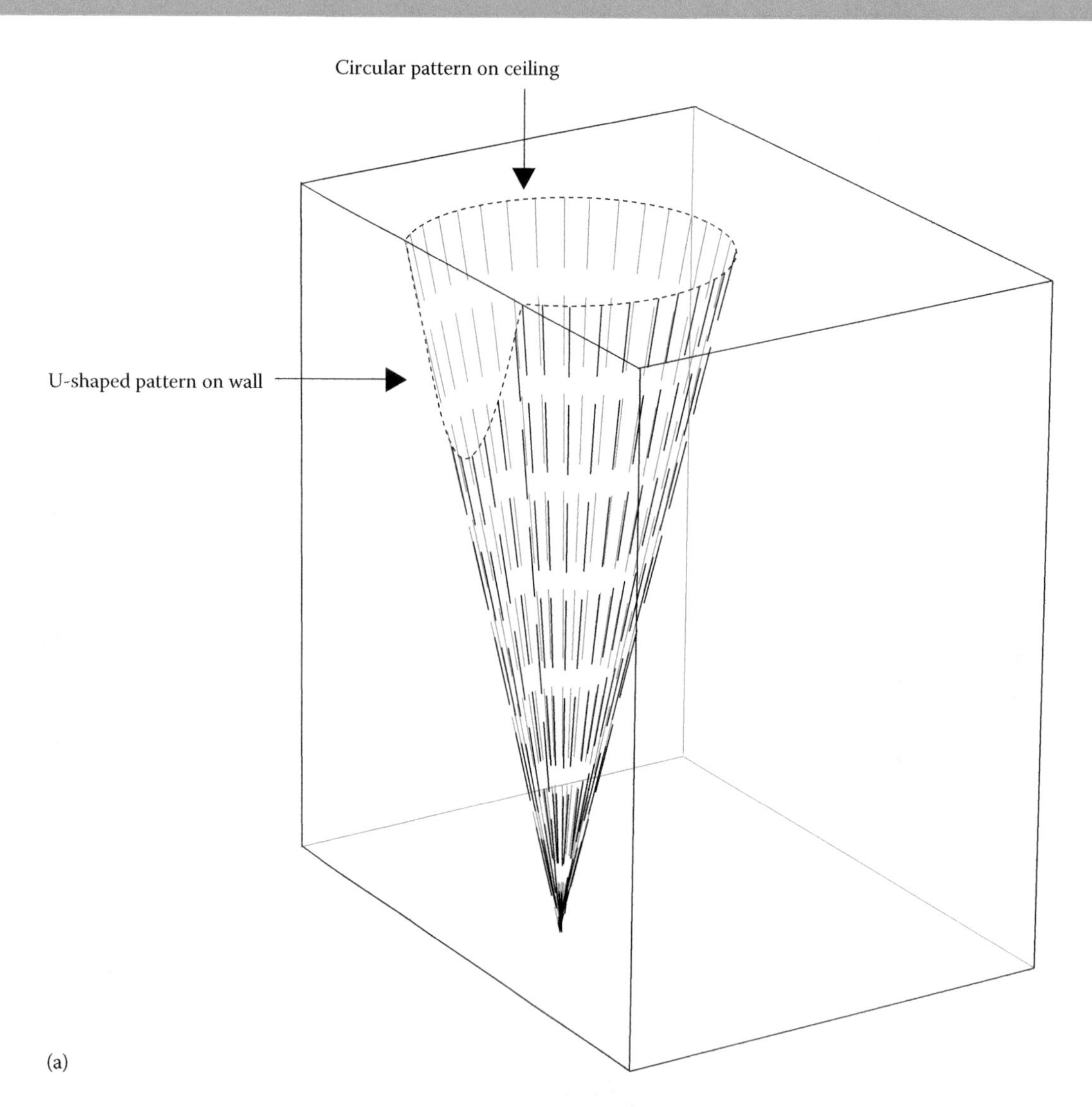

Figure 3.20 (a) U-shaped burn pattern produced at the intersection with a wall, and circular pattern produced on the ceiling by a cone-shaped fire plume.

Figure 3.20 (b) U-shaped fire pattern on the wall in the background. There was a V-shaped pattern with its base behind the door. The U was produced on the wall farther from the origin.

Figure 3.21 Semicircular pattern produced on a ceiling by a fire plume. The plume in this case was confined between two wall studs and then broke out of the enclosure at the top of the wall. The switch box and wires were added as part of a reconstruction.

of plume-generated patterns, but fire investigation is seldom as simple as this. One of the more difficult tasks facing the investigator is recognizing patterns generated *after* the compartment became fully involved. The next section deals with the kind of patterns that make fire investigation more complicated—those generated after the fire becomes ventilation controlled.

3.10 VENTILATION-GENERATED PATTERNS

In ventilation-controlled fires, patterns are generated around doors and windows as well as around any places where air leakage existed before the fire, or where such ventilation was created as a result of the fire. When flashover occurs, and every exposed combustible surface ignites almost simultaneously, the fire can consume almost all the oxygen in the room. After that happens, diminished burning occurs where the oxygen is depleted. The fuel at the origin of the fire will not necessarily continue to burn, and much larger patterns independent of the origin can be generated. These new patterns can obscure older, plume-generated patterns, or even if the older patterns persist, the new patterns can be confusing.

Figure 3.22 shows the original, plume-generated pattern produced in the 2005 Las Vegas experiment conducted by the Bureau of Alcohol, Tobacco, Firearms and Explosives (ATF) and described by Carman in 2008. Figure 3.23 shows the ventilation-generated pattern on the wall opposite the doorway opening. This was the pattern that led many investigators to mistakenly identify this quadrant of the room as the quadrant of origin. In order to understand what was happening, Carman modeled the fire using Fire Dynamics Simulator (FDS). Figure 3.24 is a Smokeview output showing heat flux 60 s after the fire was ignited next to the bed. What happens to the oxygen concentration is instructive. Figure 3.25 is a "slice" showing the oxygen concentration at the center of the doorway. There is clearly depletion of the oxygen in the upper level, but plenty of oxygen, the original 21%, at the floor level. At 290 s (just after flashover), most of the oxygen in the room has been depleted (the blue color showing near zero), and only the immediate vicinity of the door exhibits a substantial oxygen concentration. This is shown in Figure 3.26a. A plan view of the oxygen concentration 8 in above the floor, also at 290 s, is shown in Figure 3.26b. This oxygen concentration suggests the existence of a "floor jet." This inflowing oxygen collided with the wall opposite the doorway and moved upward, mixing with the

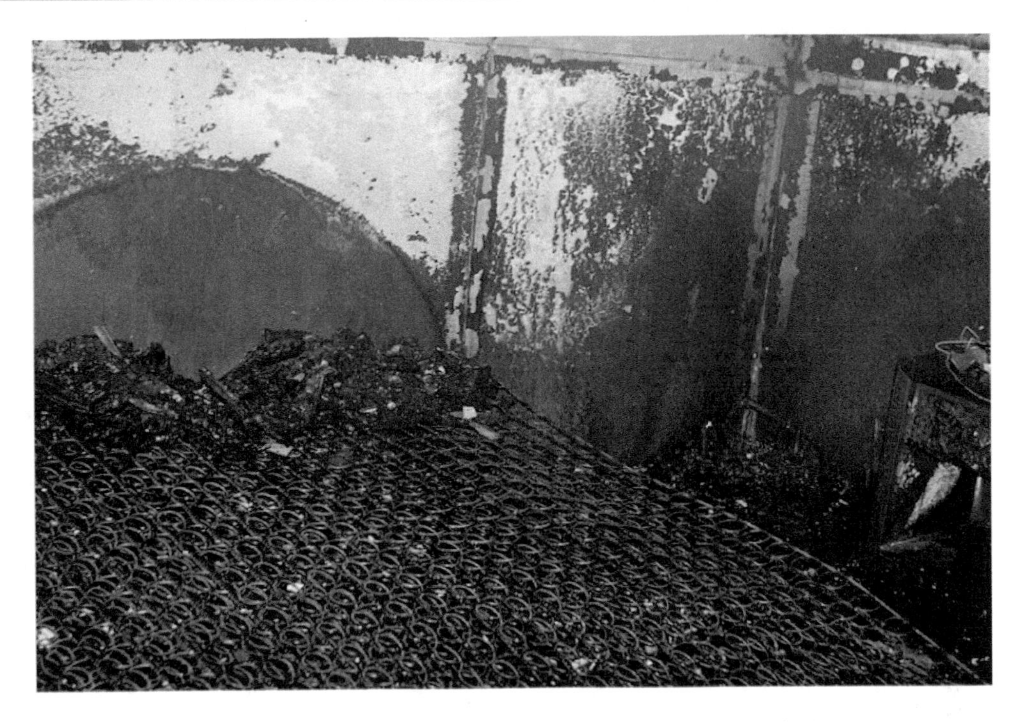

Figure 3.22 Roughly V-shaped pattern on the wall next to the bed, where a test fire was ignited.

Figure 3.23 Large, conspicuous ventilation-produced pattern on the wall opposite the door, incorrectly identified as the origin of the fire by many investigators in the 2005 Las Vegas test. There was no competent ignition source or fuel source at the base of the pattern. The wide base on the pattern may have led some investigators to believe that an ignitable liquid was spread along the wall.

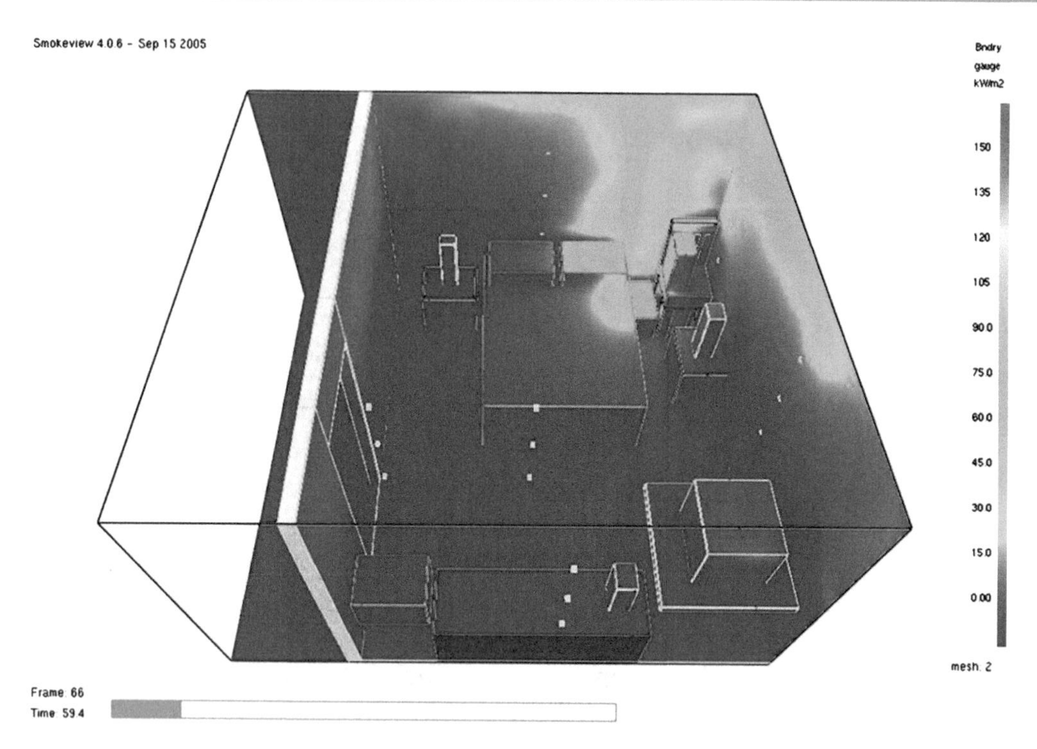

Figure 3.24 Smokeview output, from a test fire set to mimic the Las Vegas fire, showing estimated heat flux as it existed 60 s after ignition.

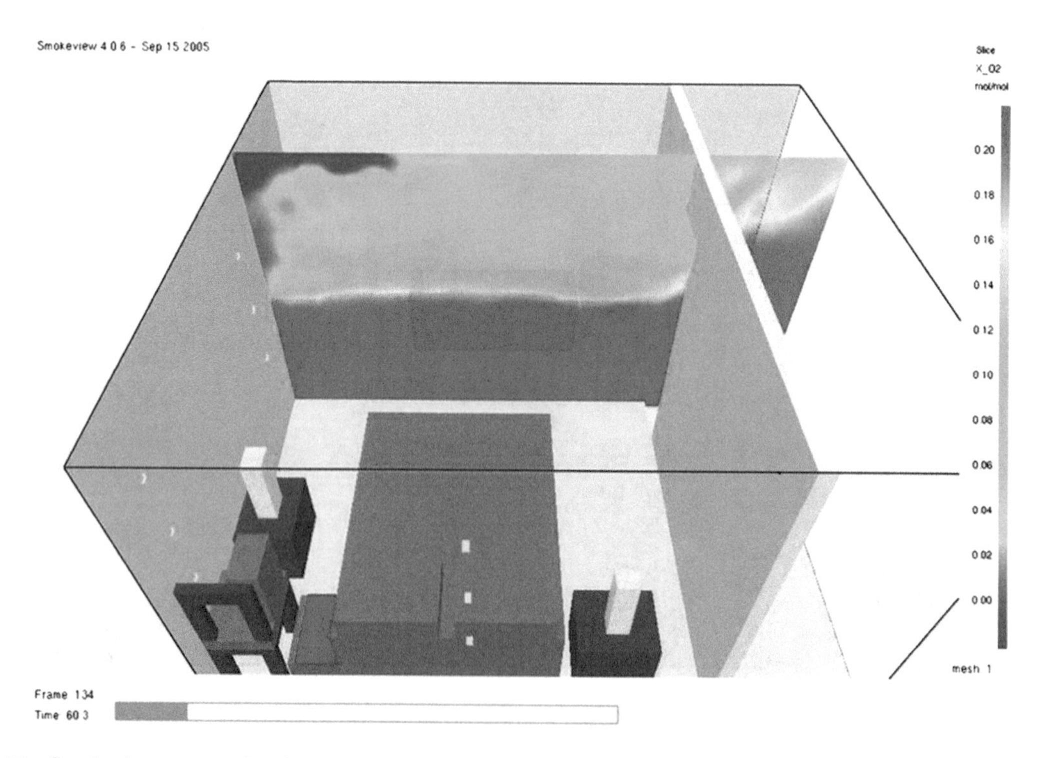

Figure 3.25 Smokeview output, showing a slice predicting oxygen concentration in the plane of the doorway, 60 s after ignition.

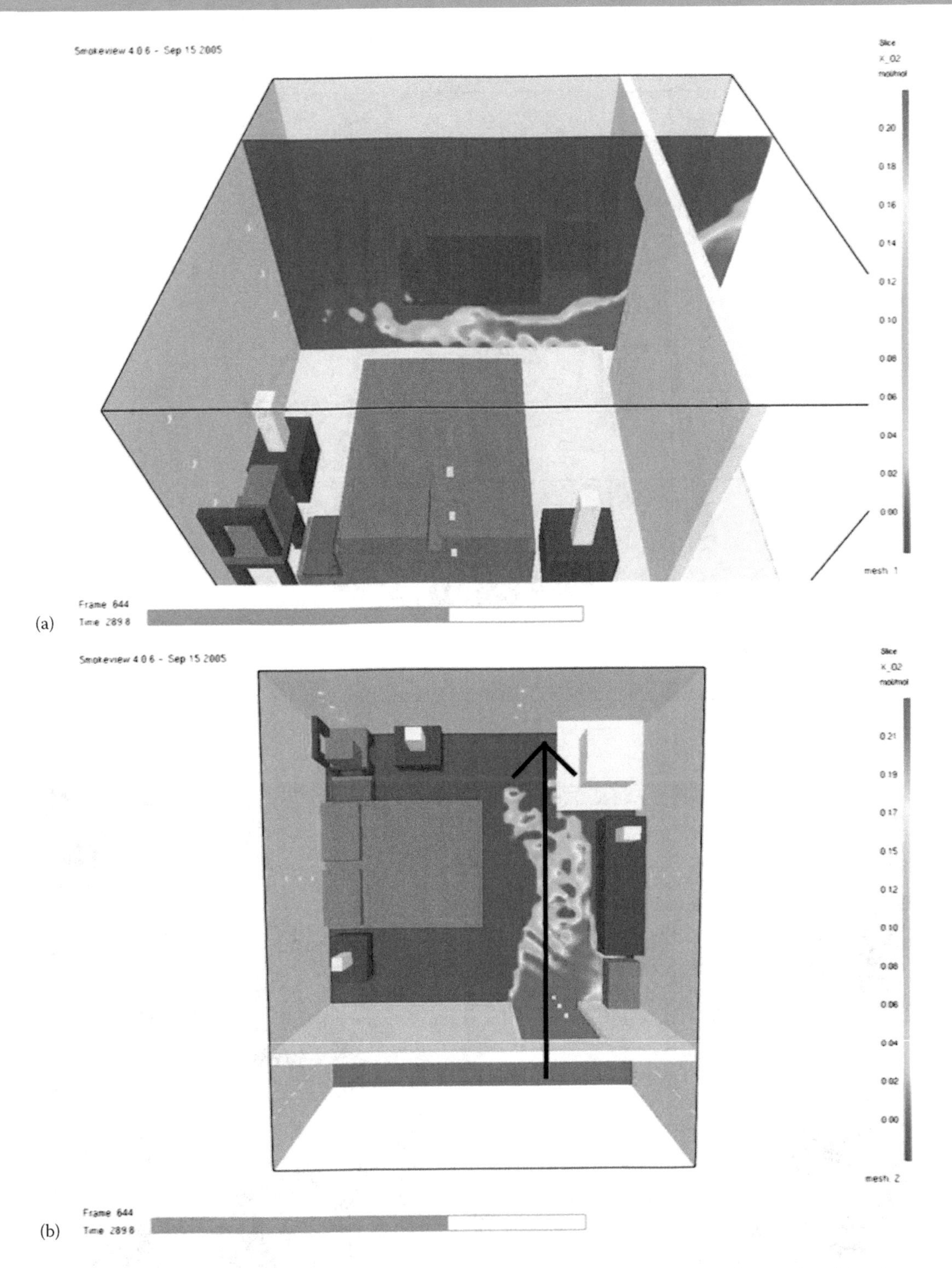

Figure 3.26 (a) Smokeview output, showing the same slice in the plane of the doorway, 290 s after ignition (12 s after the model calculated that upper layer reached 600°C). (b) Smokeview output, showing a plan view of the oxygen concentration 8 inches above the floor, 290 s after ignition. The arrow indicates the path of the "floor jet."

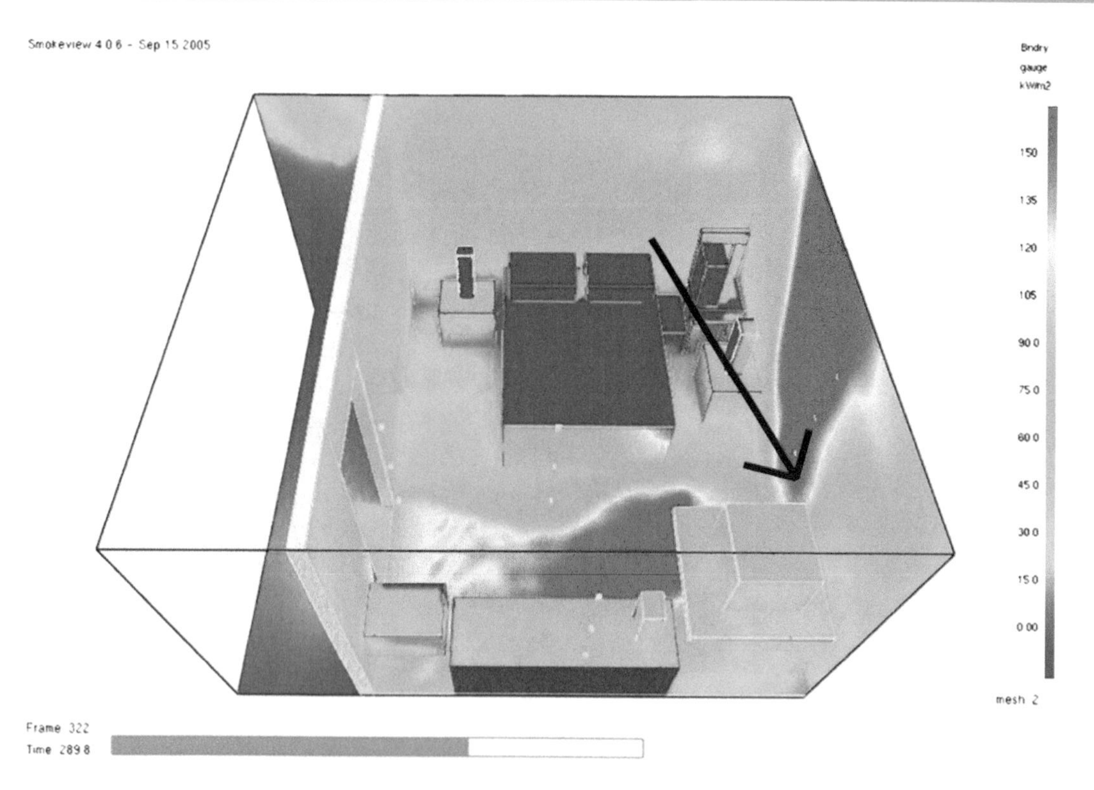

Figure 3.27 Smokeview output, showing the predicted heat flux on the wall opposite the doorway 290 s after ignition.

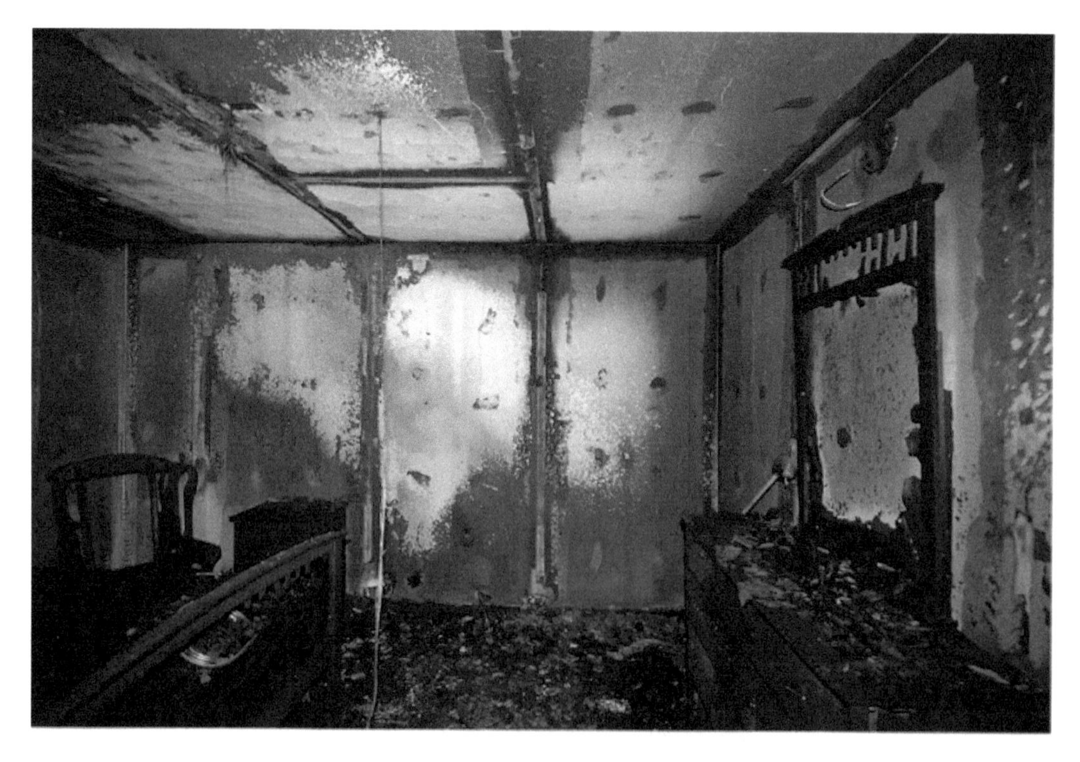

Figure 3.28 V-shaped pattern caused by post-flashover ventilation in a test fire. Such patterns can be confused with plume patterns.

Figure 3.29 (a) V-shaped pattern caused by post-flashover ventilation in a real fire. The fire had burned through the exterior wall opposite this interior wall. (b) Flooring damaged by the same floor jet that caused the V-shaped pattern in Figure 3.29a. The floor level burning was interpreted erroneously as evidence of an ignitable liquid burning there.

Figure 3.30 Anomalous fire pattern that was produced on the wall and above the nightstand before the nightstand ignited.

unburned fuel, and resulted in a heat flux of more than 150 kW/m^2, as shown by the heat flux gauge in Figure 3.27. Clearly, the total energy impact (heat flux multiplied by time of exposure, expressed as kJ/m^2) on this surface far exceeded the total energy impact on the wall next to the origin.

The shape of patterns produced by ventilation-controlled fires is not always predictable, but V-shaped patterns are not uncommon. Figure 3.28 shows a sharp V-pattern produced in a test fire that was due entirely to ventilation. Despite its V-shape, suggesting a plume, video of the pattern development showed a progression from the top down. These patterns can even occur above a fuel package and lead to further confusion. Such a V-pattern from a real-world fire is shown in Figure 3.29a. The absence of a fuel or ignition source at the base of the V-pattern, and the damage caused to the floor covering by the floor jet, shown in Figure 3.29b, led one investigator to mistakenly conclude that he was looking at the origin of an incendiary fire. He apparently believed that the ignitable liquid fire lasted long enough to completely penetrate the drywall and cause *calcination*[3] on the back surface of the drywall from the next room. Even if it took only 5 min to completely remove the first layer of drywall, any ignitable liquid would have been long gone.

The pattern shown in Figure 3.30, located directly above a nightstand, was not located near or even opposite an opening. It occurred simply in a place where a cloud of fuel-rich smoke managed to mix with sufficient oxygen to burn. The pattern was produced before the nightstand ignited.

Prior to flashover, the plumes that exist above fuel packages tend to remain more or less stationary. Areas of ventilation tend to move around and change, however, as conditions in the compartment change. Unburned fuels can travel long distances from the fuel package where they originated prior to mixing with an oxygen supply. Such movements are not necessarily predictable, and the fire investigator must accept that he is likely to observe fire patterns for which no explanation is available.

[3] Calcination of gypsum wallboard involves driving the free and chemically bound water out of the gypsum ($CaSO_4 \cdot 2H_2O$), thus changing the gypsum to another substance, anhydrite ($CaSO_4$). Deeper calcination results from greater total heat exposure, measured in kilojoules per square meter.

3.11 PENETRATIONS THROUGH FLOORS

A hole burned through a floor is the ultimate "low burn." Much has been written about how to tell whether a hole burned from above or below, but in most cases, the fire's direction is obvious. There is usually far more damage on the side from which the fire emanates. Figure 3.31 shows the classic means of determining the fire's direction, that is, examination of the direction of beveling. While such an examination can be useful, the investigator needs to remember that the edges of the hole only record the direction of the fire's *last* movement through the hole. A fire can burn down through a floor and then ignite combustible materials below the floor and burn back up.

One factor that is very unlikely to cause a fire to burn downward through a floor is the presence of a burning ignitable liquid. Such downward burning may be possible on carpeted floors, where the melted carpet fibers allow the fire to burn for an extended period of time. Most volatile liquids burn off very quickly, however, and **do not have the time** to cause penetration of the floor. In some cases, burning on a smooth floor does not even cause scorching. Recent work by Mealy et al. shows that gasoline spills tend to burn out within 60 s on smooth surfaces but carpeted surfaces tended to stay ignited for up to 8 min (Figure 3.32) [32]. Melted polymers and burning debris, which burn for much longer, are far more likely to cause penetrations. Mealy's results mirror those obtained by Sanderson in 2001, where he was unable to burn a hole through a floor using accelerants and was likewise unable to cause aluminum to melt [33]. As long ago as 1969, Kirk warned that investigators who attribute holes in the floor to a flammable liquid are likely to be incorrect [34].

Air moving over hot coals can cause the HRR to increase, resulting in holes in floors. Once the hole is created, additional ventilation is provided, and the fire temperatures can increase further still. Ventilation can also account for damage to the underside of doors due to burning of the surface or the flow of hot gases across the surface.

The structure of the flooring system can significantly influence the resulting fire patterns. Fires that burn up through a floor are likely to be influenced by the presence of floor joists, and the holes in the floor tend to be rectangular, reflecting the shape of the support structure. A fire that burns from the top down, on the other hand, is unlikely to be influenced by the underlying floor joists, although the flooring directly on top of the joists may remain because the joists keep it from falling down.

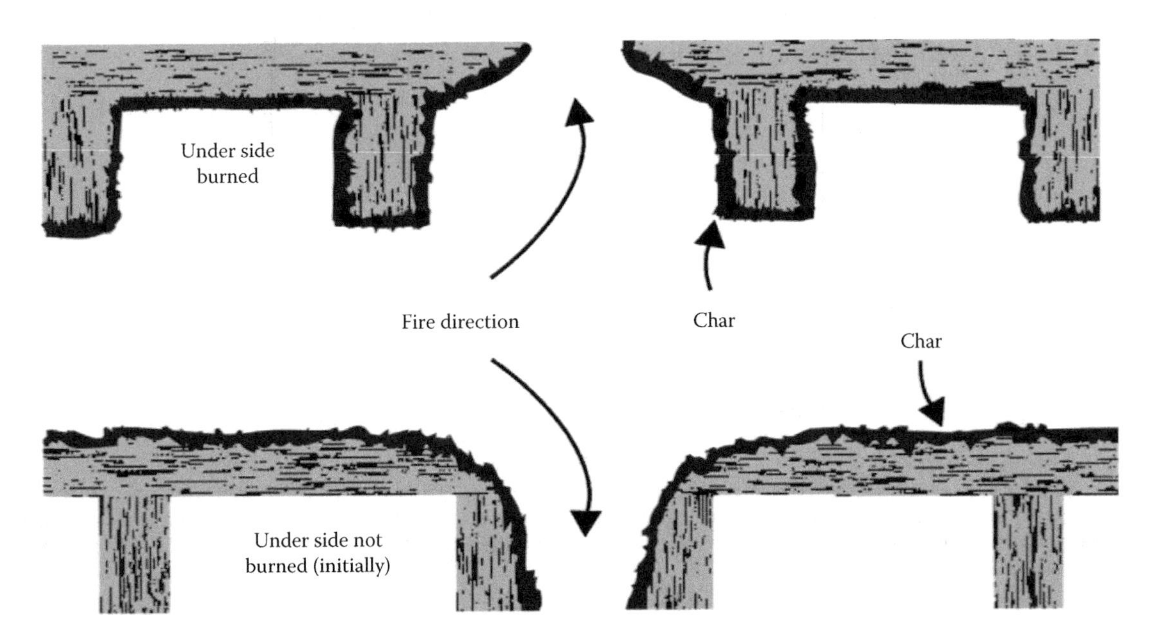

Figure 3.31 Fire patterns resulting from burning from above and below.

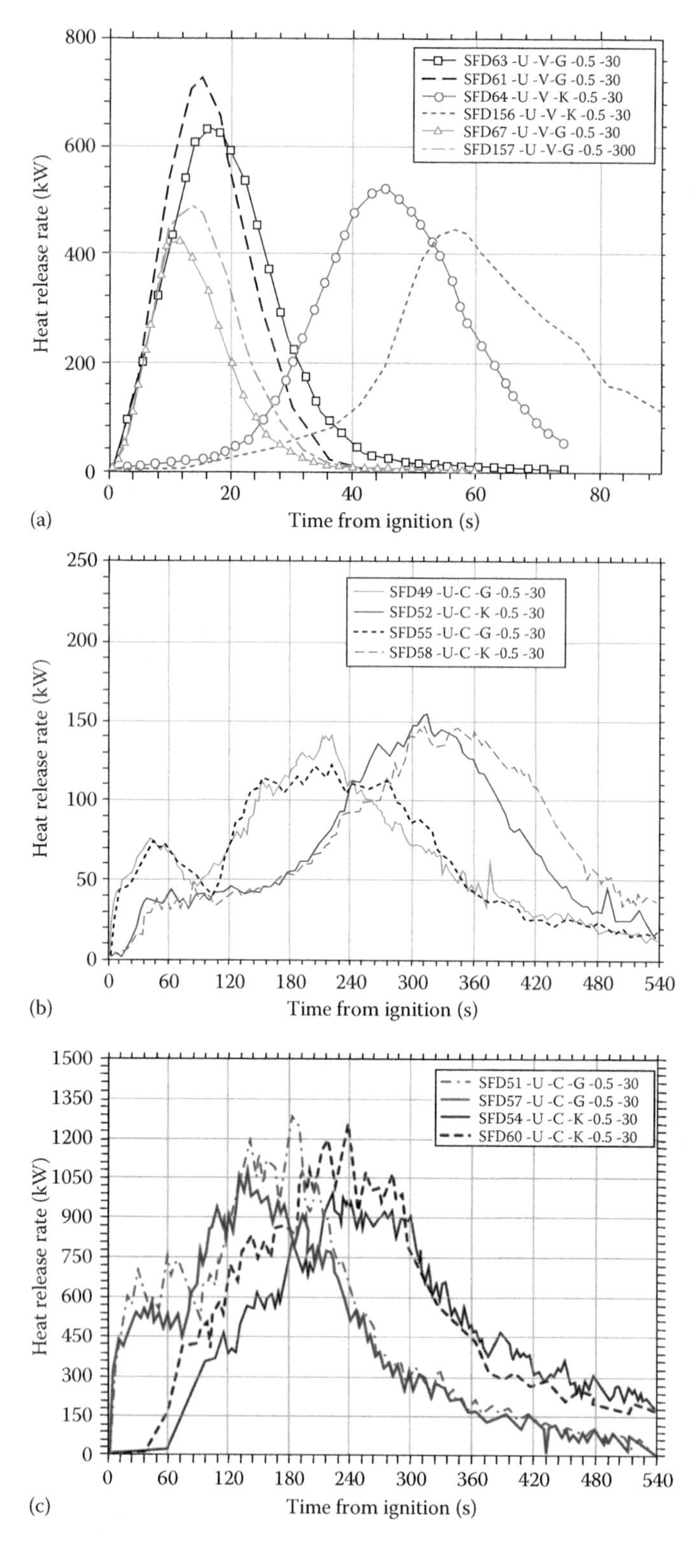

Figure 3.32 (a) Heat release rate of 0.5 L of gasoline and kerosene spills on vinyl flooring. (b) Pouring the liquids on carpeted surfaces led to lower HRRs over a longer period of time. (c) Increasing the amount of fuel led to higher HRRs, but no significant increase in the time of burning.

SIDEBAR 3.1 Some simple fire dynamics equations

The relationship between heat release rate and the average flame height (the point where the top of the flame exists 50% of the time) is given by the equation

$$HRR = \frac{79.18H_f^{5/2}}{k}$$

where:

HRR is the heat release rate in watts
H_f is the flame height in meters
k is the wall effect factor; k = 1 when there are no nearby walls, k = 2 when the fuel package is at a wall, and k = 4 when the fuel package is in a corner

The reverse equation, which allows calculation of flame height from the heat release rate Q, is $H_f = 0.174\ (kQ)^{0.4}$ (It helps to have a calculator with a y^x key to perform these operations.)

For a small trash fire at 100 kW:
In the center of a room: $H_f = 0.174\ (1 \times 100)^{0.4} = 1.1$ m
The same fire against the wall: $H_f = 0.174\ (2 \times 100)^{0.4} = 1.45$ m
The same fire in the corner: $H_f = 0.174\ (4 \times 100)^{0.4} = 1.9$ m
For a large upholstered chair fire at 700 kW:
Against the wall: $H_f = 0.174\ (2 \times 700)^{0.4} = 3.15$ m
In the corner: $H_f = 0.174\ (4 \times 700)^{0.4} = 4.16$ m

It is intuitively obvious that there should be some relationship between flame height and heat release rate, but why this particular set of numbers?

The fire engineering literature is replete with such equations. These equations are largely based on measurements of test fires and the preparation of best-fit curves applied to observed phenomena. The data are plotted on a linear, logarithmic, or semilogarithmic graph (whichever works best), and the equation that best describes the line is the one selected.

Another example of such an equation is one that describes the relationship between the amount of energy required for flashover and the size of the ventilation openings in a compartment. It has been determined by experiment that the minimum size of a fire that can cause flashover is a function of the ventilation (both air in and smoke out) provided through an opening and the amount of heat lost to the surfaces of the compartment. The ventilation factor is the product of the area of the opening (A_o) and the square root of the height of the opening (h_o). The effect of cooling by heat transfer to the bounding surface (ceiling, walls, and floor) increases the amount of energy required and is accounted for by a second area term, A_w.

The equation for the minimum heat release rate for flashover in kilowatts is

$$HRR = (378A_0)(h_0)^{0.5} + 7.8A_w$$

In the case of multiple openings, it is necessary to treat them as separate entries A_{o1}, h_{o1}, A_{o2}, h_{o2}, and add them together. (This assumes that the tops of each opening are at the same height—if one opening is higher than the others, the equation is less accurate.)

Example: A 10 × 13 ft room (3 × 4 m) with an 8 ft (2.44 m) ceiling, with an open 4 × 6 ft window, 3 ft above the floor, and two 3 × 7 ft doors.

Door 1:	$A_{o1} = 1.95\ m^2$	$h_{o1} = 2.13$ m
Door 2:	$A_{o2} = 1.95\ m^2$	$h_{o2} = 2.13$ m
Window:	$A_{o3} = 2.23\ m^2$	$h_{o3} = 2.13$ m
	Total: $6.13\ m^2$	

If the window is closed, A_o is reduced from 6.13 to 3.9 m^2, while A_w increases by the area of the window (now treated as a wall), and the energy requirement for flashover is reduced. The equation becomes

$$HRR = (378 \times 3.9)\ (2.13)^{0.5} + (378 \times 2.23)^{0.5} + 7.8 \times 44.78 = 2500 \text{ kW or } 2.5 \text{ MW}$$

Closing the window dramatically reduces the HRR required for flashover. The plot in Figure 3.33 shows that there is an increase in the required power as the amount of ventilation increases. Increasing the size of the room also increases the power requirement. Conversely, less power is required to flash over a small room with the windows closed, but there is a lower limit to this, as some ventilation is required to sustain flaming combustion. The equations in these examples are two simple cases of fire modeling. We can use the equations here to get an idea of the effect on fire growth of changing certain parameters, such as the location of a fuel package or the amount of ventilation. The numerical factors in these equations are reported to three or four significant figures. Keep in mind that such precision is impossible in fire testing, where variability on the order of 30% is common.

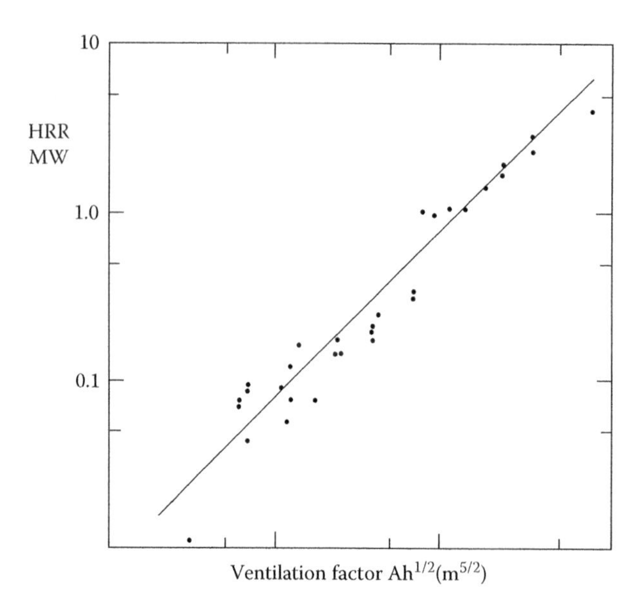

Figure 3.33 Heat release rate required for flashover as a function of the ventilation factor. Note that even when plotted on a double logarithmic grid, the data do not line up perfectly. (Reprinted from NFPA 921-2004, *Guide for Fire and Explosion Investigations*, Copyright © 2004, National Fire Protection Association, Quincy, MA, 2017. With permission. This reprinted material is not the complete and official position of NFPA on the referenced subject, which is represented only by the standard in its entirety.)

3.12 HORIZONS, MOVEMENT, AND INTENSITY PATTERNS

Still another kind of pattern is created by the interaction of the hot gas layer with walls. These patterns appear as "heat horizons" or "smoke horizons" on walls. Such patterns are typically straight lines parallel to the floor and ceiling (if the ceiling is not slanted), as shown in Figures 3.34 and 3.35. These patterns often do not survive flashover because the hot gas layer descends to floor level. Finding an intact smoke or heat horizon in a room is usually sufficient grounds to conclude that flashover did not take place in that compartment. (If the origin is elevated, this may still be the compartment of origin.)

Movement of the fire from one compartment to another creates patterns that are recorded on the walls and ceilings, usually at doorways, such as the pattern shown in Figure 3.36. Such patterns typically start out at the top of the doorway and gradually move downward if the fire is allowed to grow. Such patterns may actually be recording the top of a fire plume.

Fire patterns record what happened during the course of a fire but often provide no information of the *sequence* of pattern development. What we see after a fire is the total record of what happened, without necessarily being able to see the sequential order of the creation of each pattern. For example, Figure 3.37 a shows the pattern on the outside of a mobile home burned in a test fire. The large V-shaped pattern shows that the lowest and most intense fire was located in the living room. Further examination revealed a hole in the floor in front of the living room sofa. Any fire investigator coming upon this scene with no witness statements would certainly be drawn by the patterns to the living room and would probably conclude that the fire started there. The fire actually started in the kitchen on the counter; during the 28 min that this fire burned, it flashed over the kitchen and then extended to the living room, where there was more fuel and (eventually) better ventilation. The fire in progress is shown in Figure 3.37b. "Reading" the fire patterns without considering witness statements would have resulted in the common error of confusing evidence of intensity with evidence of duration.

Figure 3.34 Smoke horizon produced by the hot gas layer.

Figure 3.35 Heat horizon produced by the hot gas layer. The pattern on the walls shows evidence of confinement of the hot gas layer, so one can conclude that it was "recorded" on the walls before the roof burned off. (Courtesy of Mick Gardiner, Gardiner Associates Training and Research [GATR] Cambridge, England.)

Figure 3.36 Movement pattern at a doorway. (Courtesy of Mick Gardiner, Gardiner Associates Training and Research [GATR] Cambridge, England.)

Figure 3.37 (a) V-pattern produced late in a 28-min test fire. The fire began in the kitchen at the left of the photograph, but the most intense burning occurred in the living room. (b) Test fire that produced the pattern shown in (a). Two flashover events were recorded: one in the kitchen at the left and a second one in the living room at the right. The living room contained a much larger fuel load and eventually had a better air supply.

3.13 CLEAN BURN

Clean burn has historically been described as "a phenomenon that appears on noncombustible surfaces when the soot and smoke condensate that would normally be found adhering to the surface is burned off." This phenomenon was believed to produce a clean area adjacent to areas darkened by the products of combustion and was attributed to direct flame contact or intense radiated heat [35].

Research conducted at the ATF Fire Research Laboratory in 2008 indicates that there are other ways that clean burning can be produced, and in fact testing indicates that a different mechanism is at work. Three identical burn cells were ignited and allowed to burn for 10 s beyond flashover in the first cell and for two minutes beyond flashover in the second and third cells. A clean burn at the origin appeared in test number 1, much as it did in the 2005 Las Vegas test. In test numbers 2 and 3, however, no clean burning was evident at the origin, but when the depth of calcination was measured in tests 2 and 3, it was comparable to the calcination depth above the origin in test number 1. Conspicuous, V-shaped clean burn patterns, one of which is shown in Figure 3.28, were observed on the wall of the rooms opposite the doorway openings for the second and third tests, but a review of the progress of the fire, which was captured on video, showed that no soot was present on the wall before the clean burn patterns were generated.

What appears to have happened in the case of the origin pattern is that it became "re-sooted" in the two tests that burned for two minutes beyond flashover. The researchers speculated that this resulted from a "flattening" of the temperature gradients, as well as a very sooty environment created by an upholstered chair near the origin. With respect to the large V-shaped patterns, these were not caused by plumes but by a cloud of fuel-rich smoke in the hot gas layer finding a source of oxygen. The walls exposed to this flame impingement were too hot for any subsequent deposition of soot particles to occur [36].

Investigators interpreting these patterns according to the historical definition of clean burn, which implies a sequence of deposition and removal by flame contact, may use the clean burn to infer the direction of fire progression, duration of fire exposure, and possibly the location of burning fuel packages. Some may tend to attribute severe damage to a longer period of burning and thus associate the patterns with fire origin. Such interpretations are likely to be incorrect. While more research needs to be done on the dynamics of clean burn production, investigators need to be aware that, as with many fire investigation notions, historical methodology may lead to incorrect interpretations.

3.14 ELECTRICAL PATTERNS

One type of pattern that frequently does provide sequence information is electrical arc damage. By a process of "arc mapping" (also known as arc tracing or arc survey), the investigator can determine which parts of the electrical system were compromised by the fire while the system was still energized, and sometimes the sequence of events can be inferred.

On a given circuit, arc severing cuts off the power supply downstream of the severed point; so if there is more than one point of arc severing on a circuit, one can conclude that the arc downstream of the severed point occurred first. In the case of a "black hole," the electrical system might provide the only evidence of the fire's progression. Figure 3.38 shows the remains of a sofa burned in a test fire and illustrates the power of arc mapping. The sofa was ignited with a

Figure 3.38 (a) This test fire was started on the right side of the sofa, but heavier damage, including a clean burn on the ceiling, occurred on the left. A grid of Romex cable attached to the ceiling showed arc severing above the true origin. (Courtesy of Jack Olsen, Olsen Engineering, Penn Hills, PA.)

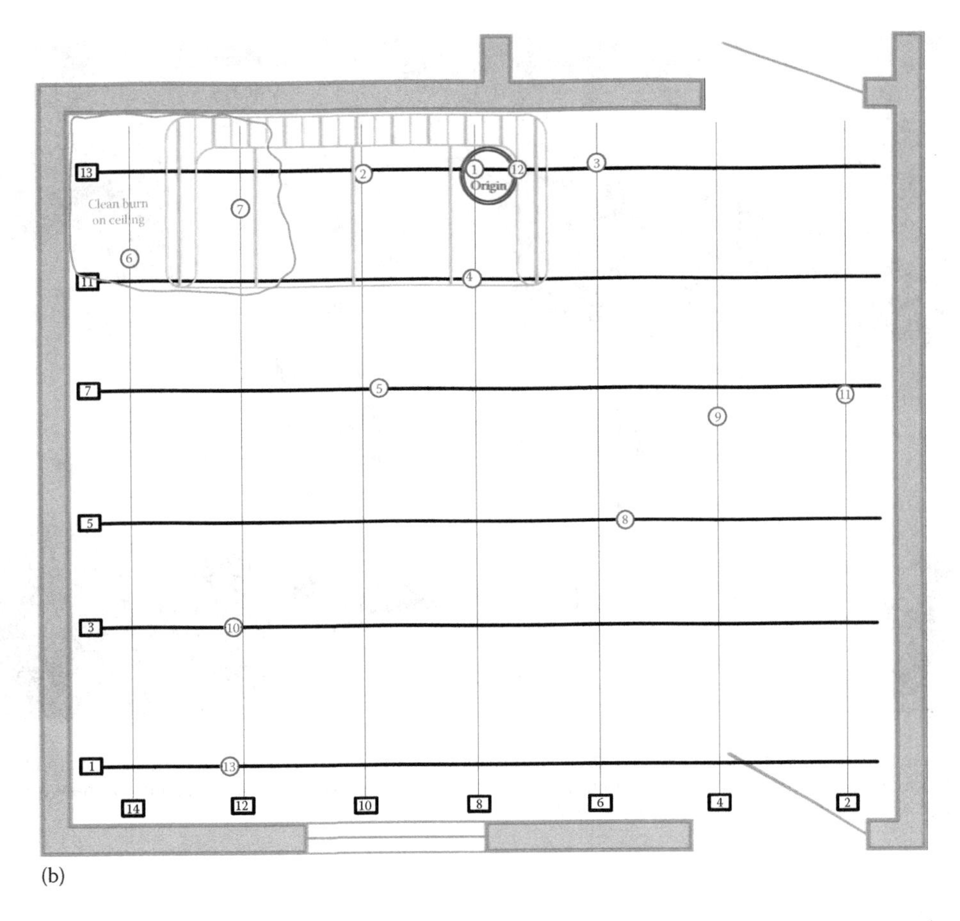

(b)

Figure 3.38 (b) Sketch of the test fire and Romex grid on the ceiling of the room shown in (a). Numbers on the grid show the location and order of the production of electrical arc damage. This test demonstrates the validity of arc mapping to assist in the determination of the origin of a fire, but the limitations of this technique must be considered.

small ball of gasoline-soaked cotton applied to the right-hand cushion. The pattern of damage would have led many investigators to believe that the fire began on the left side of the sofa. Finding an ignition source in the corner, such as a halogen lamp, might even lead to an erroneous cause determination. In this experimental burn, however, a grid of nonmetallic sheathed cable (Romex) circuits was attached to the ceiling over the sofa. The arcing that occurred in that grid showed a sequence that allowed for a correct determination of the origin at the right side.

Despite the experimental validation of arc mapping and the theoretical simplicity of the concept, arc mapping does have serious limitations. Research has shown that a heat flux in excess of 24 kW/m^2 is required to cause arcing through char in a typical nonmetallic sheathed cable. Such heat fluxes are unlikely to occur in branch circuit cables behind a wall or ceiling, unless that membrane is breached. Almost all branch circuit wiring is protected by wall covering or ceiling covering, so it is not necessarily going to be exposed early in the fire. In fact, most of the early arcing that occurs as a result of a fire is going to take place on exposed appliance power cords rather than on branch circuits.

Arc mapping is also limited by the investigator's proficiency at recognizing arc sites. As the wire diameter decreases, distinguishing arc melting from environmental melting becomes more difficult, particularly with the stranded wires that make up most power cords. The following is a list of characteristics generally associated with electrical as opposed to environmental melting:

- Sharp demarcation between damaged and undamaged area
- Round, smooth shape of artifact
- Localized point of contact
- Identifiable corresponding area of damage on the opposing conductor

Figure 3.39 Mass transfer from the grounding conductor (top) to the ungrounded hot conductor. This photograph shows a typical arc bead. The conductors exhibit high porosity in the melted area and striations from the manufacturing of the wire in the unmelted area. (Courtesy of ATF Fire Research Laboratory, Beltsville, MD.)

- Resolidification waves
- Copper drawing lines visible outside the damaged area
- Localized, round depressions
- Small beads and divots over a limited area

In most cases in which mass transfer between conductors occurs, the mass is transferred from the grounded or grounding conductor to the ungrounded (hot) conductor. This mass transfer may allow for the identification of the hot leg. Figure 3.39 shows this mass transfer.

Metallurgical characteristics can also help distinguish between arcing and environmental melting, but it is unlikely that the fire investigator will have the opportunity to examine these on the scene. Some characteristics of environmental melting that disconfirm arcing include the following:

- Visible effects of gravity in the artifact
- Extended area of damage without a sharp demarcation from undamaged material
- Gradual necking of the conductor (assuming this is not due to mechanical break)
- Blisters on the surface (assuming gross overload was ruled out)

Examples of both kinds of melting are shown in Figure 3.40.

Arcing need not always take place between conductors in a cable. A potential difference between a conductor and a grounded surface can induce arcing at that site. One common grounded surface where arcing can be seen is in structures that employ rigid metal conduit rather than nonmetallic sheathed cable. Arcing between an energized conductor and its conduit can result in "blowholes"; finding a blowhole almost certainly means that an arc site has been detected. The absence of a blowhole, however, does not necessarily indicate the absence of arcing inside the conduit. An arc survey that consists of walking through a building looking for blowholes in metal conduit is inadequate, as it is entirely possible for arcing to take place without the production of a blowhole, as shown in Figure 3.41.

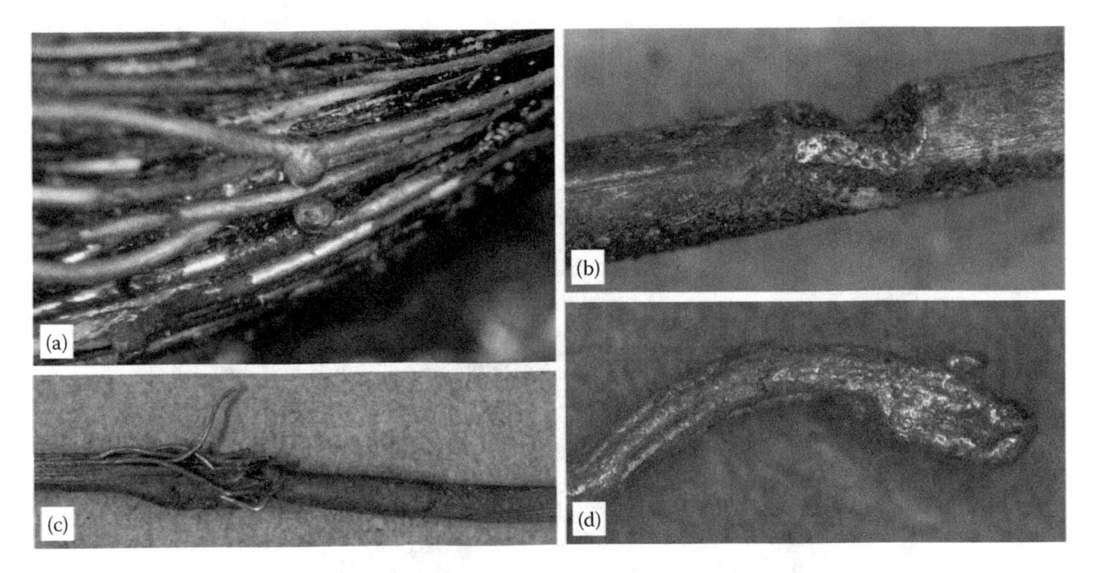

Figure 3.40 Examples [(a) through (d)] of conductors damaged by electrical melting (top) and fire melting (bottom). (Courtesy of Mark Goodson, Goodson Engineering, Denton, TX.)

Figure 3.41 This conductor arced to the interior wall of its conduit and did *not* create a blowhole. (Courtesy of ATF Fire Research Laboratory, Beltsville, MD.)

It is necessary to remove the conductors from the conduit (a difficult and time-consuming process) in order to make certain that no arcing has occurred inside.

Arc mapping as a tool for origin determination has not been without controversy. The fact that most conductors in an electrical system are not compromised early in the fire has led one author to conclude that it is useful in less than 1% of cases. The author stated, "It is essential to identify explicitly the exact hypothesis being invoked, and how the conclusions follow from that hypothesis [37]." This position provoked a response from researchers at the ATF Fire

Figure 3.42 (a) The homeowner had constructed a heat exchanger behind the stone façade surrounding this fireplace. The heat exchanger was in use at the time of the fire. (b) The location of this arced wire behind the stone façade allowed a determination that the fire started in the hidden space.

Research Laboratory, who argued against removing this useful tool from the investigator's toolbox [38]. This author has found arc mapping to be a useful technique on several occasions, and recommends always considering its use to both develop and test hypotheses, just as one would use any other pattern.

As an example, the fatal fire that damaged the den shown in Figure 3.42a caused severe damage throughout the room and caused the floor of the room above to be consumed. The homeowner had built a clever (but not safe) heat exchanger next to the fireplace to supplement the central furnace. The wiring in the room exhibited no arcing, except in the area behind the stone façade. Because of how the heat exchanger fan was powered, the arcing damage shown in Figure 3.42b could only have occurred if the origin of the fire was behind the stone façade, where no human access was possible. Thus, arc mapping allowed a fire that had been deemed "undetermined but suspicious" could be correctly classified as an accidental fire.

3.15 VIRTUAL FIRE PATTERNS

Events that occur during a fire can also provide a pattern of sorts if the structure is equipped with devices that respond to the fire and those devices are connected to any kind of recording device. Fire alarms and sprinklers are an obvious source of such data, but motion detectors are known to respond as well. Smoke alarms and motion detectors connected to a monitoring station that is remote from the fire can indicate when a fire first impinged on a particular part of the structure.

With the proliferation of Internet-connected devices in our homes and offices, a new analysis should now be considered during the origin determination phase of a fire investigation. This analysis will be called information technology (IT) mapping.

In an office, individual computers or terminals are tied to the network by either network cable or WiFi, and the network server or cloud routinely polls these terminals to assess their status. As any one of these devices succumbs to high heat or fire, they naturally go offline and no longer communicate with the network. These are discrete data points as to when the fire impinged on the device—indicating fire movement direction. The network logs this inactivity by station, offering the investigator a matrix of data points as to how the fire progressed.

The average household has seven Internet-connected devices, with the most complex homes having more than 15 devices on the network at one time [39]. These devices include not only our computers and printers but also smartphones, tablets, televisions, streaming TV devices, video cameras, Internet-enabled appliances, and devices like

Amazon Echo or Google Home. Additional devices can include security systems, thermostats, and even garage door openers. Given the types of devices that can be talking to the network or the cloud, the possibilities are numerous.

Many of these devices are regularly reporting into the network regarding their status, and some even communicate via the cloud. So these Internet-of-things (IOT) devices have unintentionally become monitors with respect to fire development. As a fire spreads through a structure that has 7 to 15 Internet-connected devices in operation, each device will stop reporting as the fire moves throughout, failing or physically destroying each device. This reporting log can be analyzed and interrogated to develop a 3D model depicting a visual timeline showing where the fire first interacted with a device and how it progressed. A fire investigator should at least be aware that such information may exist for any fire scene that included such devices on the premises. IT mapping may become yet another powerful tool for the investigator to deploy when determining the area of origin.

3.16 FIRE MODELING

Many processes take place simultaneously in a compartment fire. Energy is being released by the burning fuel and transferred to the surrounding fluids and solids in the environment. The temperature of the room is increasing. A fire plume transports the products of combustion upward, and a hot gas layer forms and then grows deeper and hotter. The gas layer radiates energy onto other fuel packages in the room and conducts energy into the boundary surfaces. Chemical bonds are being broken and new ones are being formed. The concentrations of gaseous species in the room are changing as oxygen is consumed, and carbon dioxide, carbon monoxide, water, and other combustion products are generated. Figure 3.43 shows some of these processes pictorially, and Figure 3.44 shows, on an idealized, dimensionless graph, the general trends of what is happening with respect to these various processes during the course of a compartment fire [40]. Most of these phenomena are capable of being measured, but how reproducible those measurements may be is a subject that has not yet been settled.

A model is an attempt to use quantitative information to mathematically describe how some or all of these processes will change over time under specific conditions. The algebraic equations used to compute flame height and ventilation factor are simple examples of fire modeling, which is based on the idea that fire might be studied numerically. The algebraic models are known as hand calculations or correlations. The more complex models use multiple differential equations (calculus), which must all be solved simultaneously but using numerical methods. The models actually use algebra, not calculus, to solve for unknown values as discrete points in the center of each cell in a three-dimensional matrix that describes the compartment. This requires a computer, as well as the ability to describe the structure and its contents on a three-dimensional grid.

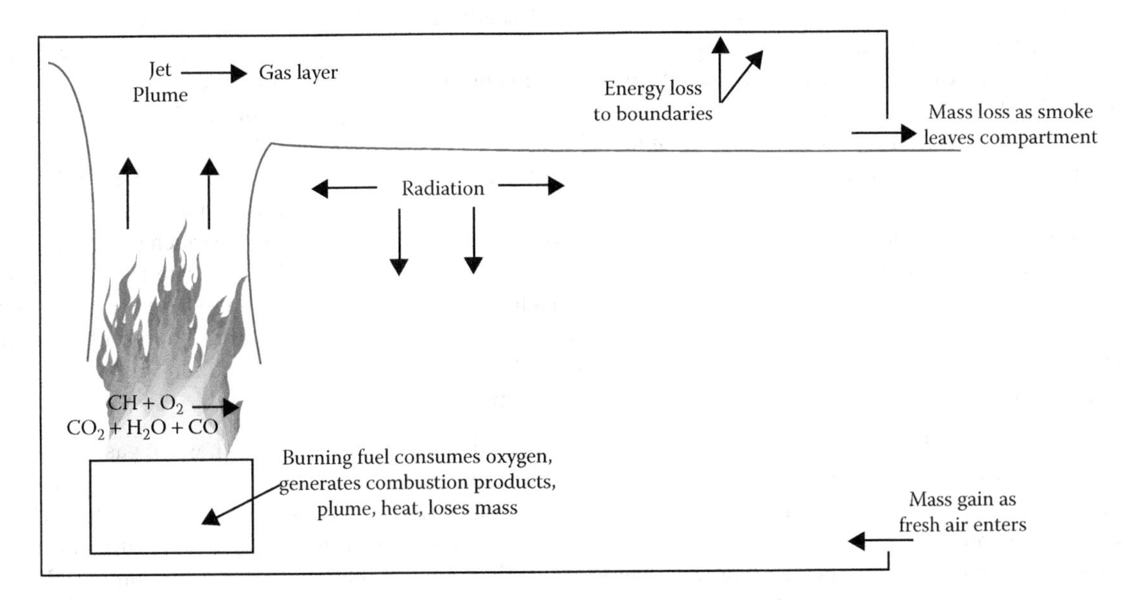

Figure 3.43 Schematic representation of some of the measurable processes taking place in a compartment fire.

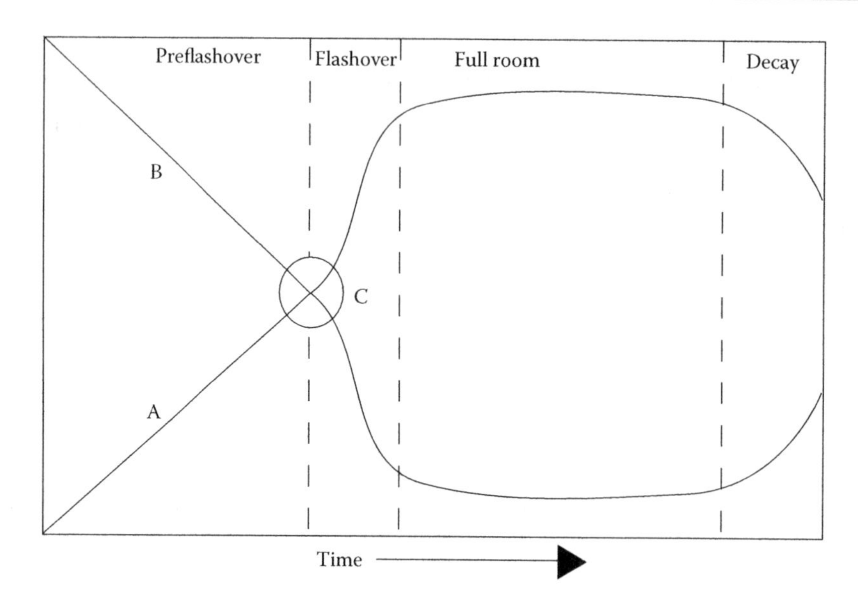

Figure 3.44 Graphical representation of processes taking place during a compartment fire. Curve A represents the following properties: temperature, radiant heat flux, CO and CO_2 concentrations, and upper layer depth. Curve B represents the following properties: ratio of convected heat transfer to radiant heat transfer, oxygen concentration, height of lower layer, and survivability. (From Kennedy, P., and Kennedy, K., Flashover and fire analysis: a discussion of the practical use of flashover analysis in fire investigations, in *Proceedings of the 10th International Fire Science and Engineering (Interflam) Conference*, Interscience Communications, London, UK, July 5–7, 2004, pp. 1101. With permission.)

Fire models were not initially designed to be used in fire investigations. They have been developed by fire protection engineers, largely as a means to avoid actual testing or to leverage data collected in fire tests to avoid additional testing. Some fire protection engineers will state (not entirely in jest) that in the twenty-first century, their whole reason for existence is to eliminate the fire resistance test. Fire models are the means to that end. Of course, live fire tests are necessary to validate any fire model.

The fundamental equations describing heat transfer, fluid dynamics, and combustion have been known for over a century, but the complexity involved in simultaneously solving for all the important variables, especially in an uncontrolled fire, was too daunting for even the most mathematically savvy engineers. There were simply too many things happening at once. The number of equations is large not only due to multiple phenomena occurring simultaneously but also because the associated variables are highly coupled. A change in one parameter, such as the concentration of CO_2, causes several other parameters to change. Other impediments to mathematical modeling include the facts that fire scenarios can be varied almost infinitely and the fuels under consideration were not designed to be fuels. They were designed to be chairs, beds, and building materials. It was not until the beginning of the information age that there was sufficient computing power for the development of models that could be applied to fires, but only simple models, zone models that divided a compartment into two layers, could be handled. However, it was a start. Researchers knew that the best understanding of a fire would come when the compartment could be divided into smaller and smaller cells, ranging from a cubic foot down to a cubic inch. These models, known as field models, provide much more information but require more inputs and take much more computing power to run. Even as recently as the late 1980s, the amount of computing power required was beyond the reach of all but a few researchers lucky enough to have very fast computers with very big memories. In the mid-1990s, it was not uncommon for a computer to require two months of number crunching to test a single scenario using computational fluid dynamics (CFD). In fact, such long runs are common today because the larger, faster computers are being asked to do more calculations.

Some of the first quantitative analyses of fire plumes were undertaken during World War II, when gasoline burning in ditches was used to clear fog from runways. Sir Geoffrey Taylor prepared a paper for the U.S. Government entitled "Dynamics of a Hot Gas Rising in Air." Taylor's work, which described the properties of a heat-driven plume, formed the basis of zone modeling. During the 1940s and 1950s, government offices in the United States, United Kingdom, and Japan were set up to study the quantitative aspects of fire. The First International Symposium on Fire Research, sponsored by the National Academy of Sciences, was held in Washington, DC, in November of 1959. Initially, burning liquid pool fires were studied because of their simplicity; but by the mid-1960s, there was a consensus in the scientific community that it might be possible to model more complex fire phenomena. In 1976, James Quintiere presented a paper entitled "Growth of Fires in Building Compartments" at an ASTM symposium. The first published fire model, by the Illinois Institute of Technology Research Institute (IITRI) in the mid-1970s, was influenced by Quintiere's approach and demonstrated the potential of this tool for use in fire protection engineering design. Other models soon followed, and as each was published, the next generation merged the equations and algorithms from previous models, so that more and more aspects of fire development could be taken into account. In 1986, Harold "Bud" Nelson of the National Bureau of Standards (NBS) now the National Institute of Standards and Technology (NIST) merged several hand equations with an egress time model and a sprinkler response model to produce FIREFORM, which he later expanded into FIRE SIMULATOR and FPETOOL, which was released in 1990. British and Australian teams produced similar program suites of fire engineering calculations. Multi-room models were first seen in the early 1980s. A model called Fire and Smoke Transport (FAST) was released by the Center for Fire Research at NBS in 1985. It was merged with a faster numerical problem solver and became the Consolidated Compartment Fire and Smoke Transport (CFAST) model, which is still in use today [41]. CFAST is a *zone model* that divides each compartment into only two zones—an upper and a lower zone—connected by the fire plume.

Field models, which were developed both in the United States and elsewhere, utilized CFD to model the behavior of a fire in many individual cells. These models solve multiple simultaneous differential equations to balance mass, energy, and momentum in all of these thousands or millions of cells. NIST's current model is the Fire Dynamics Simulator (FDS). Perhaps one of NIST's most useful achievements in the field of modeling is its development of Smokeview, a program that transforms the output of FDS into a three-dimensional view of the fire in progress, which can be tuned to examine smoke, particles, gas temperatures, boundary temperatures, and chemical species [42]. Numerous examples of the output of the Smokeview program can be viewed at the NIST website, www.fire.nist.gov.

Although CFD models such as FDS examine a fire in much finer detail than zone models, there are trade-offs. Once the data describing the compartment or structure have been input, zone models can be run in a few minutes of processor time, while a CFD model may take days, weeks, or months. There is a place for both kinds of models. Multiple scenarios can be run using a zone model to select one or two scenarios for CFD modeling.

Because the NIST models have been largely developed at taxpayer expense, the NIST website allows anyone to download any of their models and user manuals at no charge. Because proprietary models can cost thousands or tens of thousands of dollars, the NIST models are likely to see the most widespread use. The current version of CFAST, a zone model, includes a graphical user interface (GUI). The fire dynamics simulator (FDS) field model does not yet incorporate a GUI for input, so describing a building can be incredibly tedious. David Sheppard of the ATF Fire Research Laboratory has developed a program that can take the output of a computer-aided drafting (CAD) program and convert it to the many lines of code that must be supplied to a model. Many of the commercial field models do include a GUI.

NIST mathematicians and fire protection engineers have used their models to assist fire investigators studying most of the major events that have occurred in the past few years. These include the Station Nightclub fire, the Cook County Administration Building fire, and the World Trade Center attacks. Models can be useful in developing or testing hypotheses, but care must be used in their interpretation. As with any computer simulation, the garbage in, garbage out (GIGO) rule applies. Models require the use of assumptions and approximations. More complex models make fewer simplifications but require more data input. If an incorrect assumption is used or a parameter is incorrect, an incorrect answer is the likely result.

A model does not take the postfire artifacts and run the fire in reverse to find the origin. The proper use of the model is to propose an ignition scenario and then run the model forward in time to see if the model accurately predicts the outcome. One of the best uses of a fire model is to test the effects of changing a significant parameter by asking "what if" questions. What if we had sprinklers in place? What would have happened if the stairwell door had not been propped open, or the smoke detector had batteries in it, or the interior finish had been rated Class A instead of Class D?

Answers that a fire protection engineer might consider to be "in relatively good agreement" may be too imprecise to address certain questions in the context of a forensic investigation. For example, a study of the response time of a smoke detector revealed that three different models predict three different activation times, and three different maximum escape times, ranging from 2.7 to 8.2 min. (For a discussion of precision in fire testing and modeling, see Sidebar 3.2.) In one scenario, the CFAST model predicted a longer smoke detector activation time than the FDS model, while in a

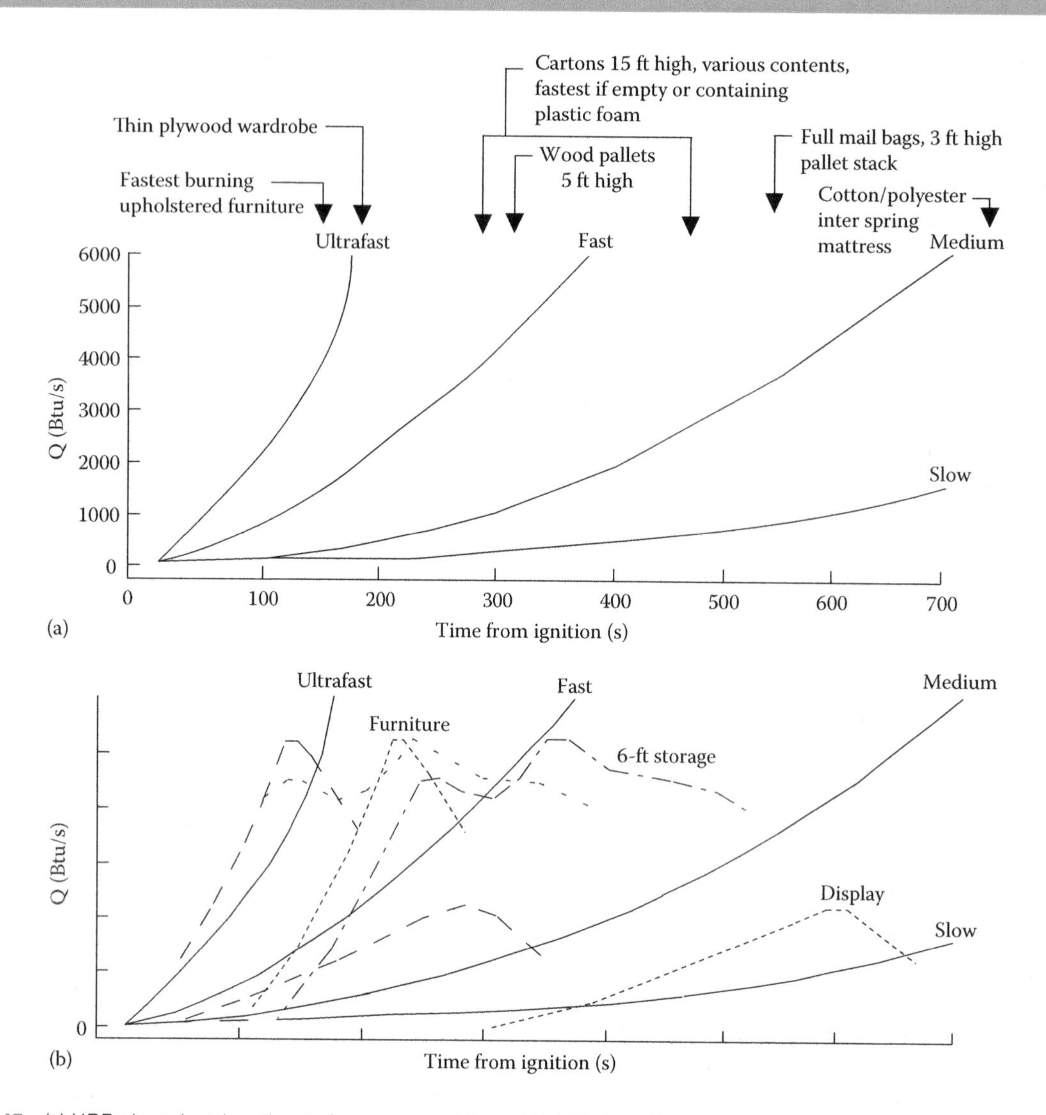

Figure 3.45 (a) HRR plotted against time in four T-squared fires. (b) HRR plotted against time in four T-squared fires, with data from real test fires plotted on the same graph. (From Nelson, H. E., An engineering analysis of the early stages of fire development—The fire at the Dupont Plaza Hotel and Casino, December 31, 1986. U.S. Dept of Commerce, NBSIR 87-3560, 1987. With permission [55].)

second scenario, the results were reversed [43]. These response times are based on different fire models and assumptions (zone versus CFD) as well as different activation criteria (rate of temperature rise versus smoke concentration). Zone models do not model the smoke transport very well in the early stages of a fire because they assume the layer is uniform across the entire ceiling at the start of the calculation. The temperature rise criterion for activation does not take into account velocity and the lag time associated with the increase in smoke concentration inside the detector.

Earlier models generally required the user to input one of four rates of fire growth: slow, medium, fast, or ultrafast. These are "t-squared" fires: theoretical constructs that assume that fires grow in proportion to the square of the time, a generally close assumption but one that does not work well for the "incubation" period while the fire establishes itself. Figure 3.45a shows the plot of time versus HRR for these four ideal fires, with some representative fuel packages that might produce such fires. Figure 3.45b shows those same four graphs overlaid with real fire test data. The time to reach an intensity (i.e., HRR) of 1,000 Btu/s (1,055 kW) is 600 s for a slow fire, 300 s for a medium fire, 150 s for a fast fire, and 75 s for an ultrafast fire. Inputting a different rate of fire growth changes the output of the model. NFPA 72, *The National Fire Alarm Code,* contains an annex entitled "Engineering Guide for Automatic Fire Detector Spacing" that provides data from several dozen furniture calorimeter tests that indicate that slow, medium, and fast fires can all be created when typical furnishings are set on fire. Ultrafast is a description usually reserved for only the fastest-burning "normal fuel" packages and for fires involving ignitable liquids. The newer models now allow for nearly endless customizations of the fire growth rate curves, but some still have difficulty modeling flame spread. With this increased input flexibility comes the necessity to justify those inputs with actual data.

The uncertainty associated with the predictive abilities of models is their principal drawback. While the measurements taken in actual fire tests can have uncertainties of up to 30%, real tests involving real fires still have more credibility than computer models in some quarters [44]. Confronted with a computer model that predicts a fire resistance of 2 h for an architectural assembly, a fire official might demand proof that the model is valid. Confronted with a hypothesis that a fire began or spread in a particular way based on a model, a party to fire litigation might ask for similar proof.

If an investigator were to conduct five identical fire experiments, the value for any given variable (temperature, CO concentration, smoke density, etc.) at a particular point in space and time would vary from test to test, and if enough tests were run (a very expensive proposition), the "error bars" for each value could be determined, assuming accurate measurement capabilities. If the investigator puts the same data into a computer model, however, only one value comes out. Both CFAST and FDS come with the following disclaimer in their user manuals:

The software package is a computer model that may or may not have predictive capability when applied to a specific set of factual circumstances. Lack of accurate predictions by the model could lead to erroneous conclusions with regard to fire safety. All results should be evaluated by an informed user [45,46].

ASTM Committee E05 on Fire Standards has promulgated two standards[4] related to the evaluation of computer models for fires: ASTM E1355, *Standard Guide for Evaluating the Predictive Capability of Deterministic Fire Models* [47], and ASTM E1591, *Standard Guide for Obtaining Data for Deterministic Fire Models* [48]. Although many of the provisions in the guides are applicable to other types of fire models, the focus is on zone models.

The definitive guidance for selecting and using models to answer questions about a fire can be found in the Society of Fire Protection Engineers (SFPE) *Engineering Guide*, "Guidelines for Substantiating a Fire Model for a Given Application [49]."

What does the availability of models mean for the fire investigator? That depends entirely on the nature of the question that the fire investigator asks. A model will not locate the origin of the fire, nor will it determine the cause. There has been a disturbing trend for fire investigators to use *hand models*[5] or spreadsheet calculators such as "CFI Calculator" in inappropriate ways. Models simply lack the ability to resolve many issues that concern the fire investigator.

When fire protection engineers are designing a sprinkler system, they have the option of using a model to help them, but they do not base their fire safety engineering decisions solely on the output of the model. It is a relatively simple matter to appropriately overengineer the system, for example, using 15 sprinkler heads in the final design even if the model predicts that 10 sprinkler heads will do the job.

Some fire investigators estimate the HRR required to bring a room to flash over, or cause a fire to spread between two fuel packages (Sidebar 3.3), then they estimate the HRR of a proposed single fuel package, and if that package is "insufficient," these investigators declare that there must have been two or more points of origin. If insufficient physical evidence from the fire scene exists to reach a conclusion as to the origin and cause of the fire independent of the model, relying on the model to answer these questions is invalid and irresponsible. The model was simply not designed for that application. Examples of "successful" modeling often include a comparison of the output of the model with a videotape of the actual fire. The Station Nightclub case is a good example of such a success story. **The only reason that the model can so successfully mimic the videotape is that the videotape existed.** The first time the model was run, it predicted flashover in less than six seconds. Dozens of iterations of data entry were required to get the model to agree with the videotape. As Salley et al. have demonstrated, models (even the most highly sophisticated ones) are tools that must be used very carefully to test hypotheses (See sidebar 3.2). If the evidence at the fire scene is insufficient to even formulate a testable hypothesis, the model output amounts to nothing more than computerized speculation. People are impressed with numbers, but the mere existence of data that can be quantified and manipulated is no guarantee of valid results.

Given that warning, and the disclaimers that accompany all modeling tools, there may be issues that models can address. For example, it may be a useful thing to have a Smokeview video to illustrate an investigator's opinion. Heat

[4] At one time, there were four standards, but ASTM Committee E05 decided to defer to the expertise of the SFPE and withdrew two of them, E1472 and E1895, which are now "historical" standards.

[5] Hand models are equations that can be solved using a calculator and do not require the use of a computer. The Babrauskas and Thomas equations for determining the HRR required for flashover are examples of hand models.

flux analyses can sometimes help in the understanding of fire pattern production. If a model is going to be used to achieve these ends, more data must be collected from the fire scene. The first step in setting up a model is the accurate description of the relevant compartment(s), including the composition and location of all significant fuel packages, the interior finish, and the size and location of all ventilation openings. The more accurate the description of the scene, the more likely it is that the model will accurately predict the outcome of a particular scenario.

Certainly, it is possible to run models without all of the necessary data, but like all computing, the quality of what comes out can be no better than the quality of what goes in. Multiple scenarios can be run to determine what the effect might be of a range of possibilities for wall covering, for example, but getting the accurate data while on the scene eliminates the need for modeling alternate, or "bracketing," scenarios. Even with a perfect description of the structure, a model might not accurately reproduce a fire because the description of the fire itself is input by the user. The user tells the program the initial HRR and specifies the rate of growth of the fire. As has been shown by the NIST chair tests, estimating the HRR of a particular fuel package, even with multiple tests, is an uncertain proposition.

Despite the uncertainties involved, modeling as a tool to test hypotheses is becoming more common in fire investigation. Carvel reported using a CFD model named JASMINE (a BRE/FRS program) to test five different origin scenarios in a building where the damage prevented any determination of even the room of origin. The model was run to see which scenario best matched witness observations, and of the five scenarios proposed, the model clearly favored one. Such "investigations" represent an interesting trend but one that requires constant vigilance. In the case of Carvel's fire, the model and the witness observations were all he had to work with [50].

An interesting comparison of model predictions versus real-world fire behavior was conducted in 2006 by Rein et al. In this exercise, conducted at Dalmarnock (Glasgow) Scotland. 10 teams of modelers, 8 using FDS4 and 2 using the 2000 edition of CFAST, were asked to predict fire behavior in a typical apartment in a high-rise building. The modeling teams were provided with far more information than is typically available to a modeler investigating a real-world (non-experimental) fire, but unlike many other comparisons of model "predictions" versus actual fires, the modeling teams were not given much of the experimental data.[6] They were asked to predict time to flashover and upper layer gas temperature, as well as other parameters. The predictions varied widely from each other, in part due to the wide variation in literature values for thermal properties of many of the fuels, and they varied widely from the experimental results. The authors of the study reported "the accuracy to predict fire growth (i.e., evolution of the heat release rate) is, in general, poor [51]." They further reported similar results obtained by other investigators as well.

While modeling is an interesting tool, it is, in this author's view, "not ready for prime time" use in fire investigation. One can use models to make conservative engineering decisions, but using them to "predict" the behavior of a particular fire is likely to lead to error. Until models can be shown to accurately describe what is going to happen without the modeler being provided with a videotape of the fire from its ignition until its extinguishment, the output of any model should be viewed with extreme skepticism. If the classification of the fire cannot stand on its own without the use of a model, then the classification should remain undetermined.

SIDEBAR 3.2 What do the numbers mean?

The equations and models used by fire protection engineers have the ability to spit out results of incredible apparent precision when predicting fire behavior. Unfortunately, this precision is not justified by the data. Fire protection engineers know that despite the apparent precision, the results are only estimates, and ballpark estimates at that. Fire investigators need to appreciate this as well. When a calculation produces a result for the minimum HRR needed for flashover of 498.75 MW, that number does not mean that the minimum HRR is known that precisely. Such a number actually means "somewhere in the range of 400 to 600 kW, or in that ballpark."

Consider the graph shown in Sidebar 3.1 plotting HRR versus ventilation factor. Note that the axes are labeled in a nonlinear fashion. The graph paper is known as three-decade log paper and is intended to provide an "order of magnitude" view of the data. Despite the fact that a "best fit" line can be drawn through the data points, and the slope of that line can be calculated to four significant figures, it must be noted that very few of the data points actually touch the line.

Some fire investigators have taken to using models inappropriately. It is possible to "estimate" the release rate of a fuel package such as a chair or sofa, then "estimate" the amount of energy needed to drive a room to flashover, and conduct a comparison. If the HRR estimated for the fuel package is less than the HRR estimated to be required for flashover, some fire investigators think that they can credibly state that the fuel package was incapable of driving the room to flashover by itself, and some investigators

[6] The following quote is attributed to Sir Winston Churchill: "I always avoid prophesying beforehand because it is much better to prophesy after the event has already taken place."

have even gone so far as to opine that a second fuel package must have been necessary, and therefore the fire must have had two points of origin. Such thinking indicates a lack of appreciation for the value of models.

Research conducted by Daniel Madrzykowski at the National Institute of Standards and Technology [52] attempted to quantify some of the uncertainties inherent in the equations and models. The first factor that he considered was that the oxygen consumption calorimeter, which measures the HRR and other important properties of the fire, has a measurement uncertainty of ±11%. The calibration chart is shown below in Figure 3.46.

Madrzykowski burned three known fuels—natural gas, gasoline, and polyurethane foam—and measured flame height, flame width, and the area of the patterns produced by these flames. Even for the simplest of fuels, natural gas, there was a 25% uncertainty in flame width and a 33% uncertainty in the fire pattern area. For polyurethane foam, there was a 50% uncertainty in flame height and a 57% uncertainty in fire pattern area.

The takeaway message from these experiments is that equations and models should be used with an appropriate appreciation for their uncertainty when attempting to apply them to a fire investigation. Madrzykowski's work supports work done at the Nuclear Regulatory Commission (NRC) by Salley et al., from 2007 to 2010 [53], which compared measurements from test fires with predictions provided by different kinds of models. The NRC researchers found deviations of up to 60% between the predictions and the measured results. Their data are shown in Figure 3.47.

The output of models can be used to help gain an understanding of the relationships that exist between the building and the fire, and they can certainly be useful in designing sprinkler systems and locating exits. For such purposes, error rates of ±30% or more are perfectly acceptable. Such uncertainty is not acceptable for reaching conclusions about the cause of a fire.

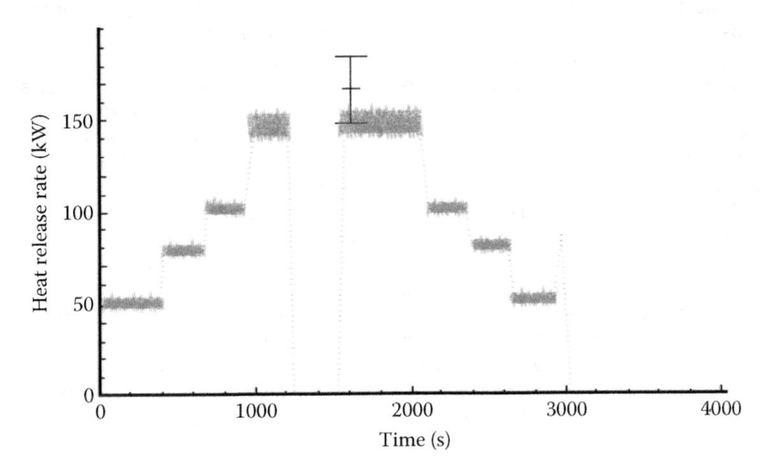

Figure 3.46 Calibration data for the NIST oxygen consumption calorimeter used in Madrzykowski's tests on repeatability of fire patterns. The uncertainty is ±11%.

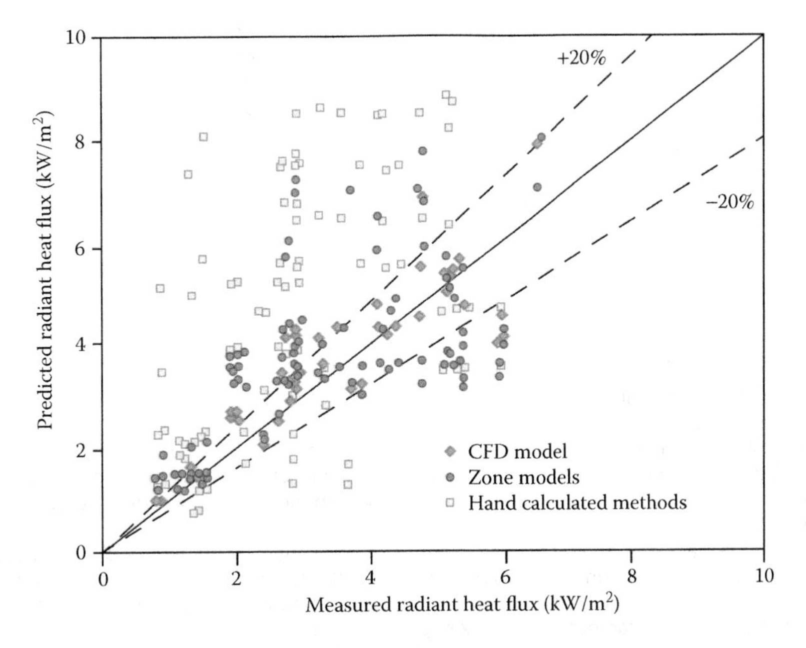

Figure 3.47 Comparison of the heat flux predicted by various models plotted against measured heat flux in fire tests.

In 2014, NFPA 921 recognized the problem of false precision in fire investigations. A note was added to Table 1.4 in that document showing units of measure, which stated that converting from one system of measurement to another usually introduces additional significant figures to a value. The converted values should be rounded off so that they include no more significant figures than the original measured or reported values.

More important, paragraph 1.5 on measurement uncertainty was added. It states,

> The reproducibility of measurements reported in this guide may be very high, such as density measurements of pure substances, or more variable, such as gas temperatures, heat release rates, or event times in test fires. Therefore, all reported measurements, or factors in equations should be evaluated to assess whether the level of precision expressed is appropriate or broadly applicable.

Numbers can be powerful, but fire investigators using them should appreciate what they mean.

SIDEBAR 3.3 Will the next item ignite?

In the event that a fire investigator is able to identify the fire's origin as being on a particular piece of furniture, the question arises as to whether it is possible to predict whether a fire starting on that piece of furniture could spread to an adjacent piece of furniture or other fuel package. Clearly, the distance between the burning fuel package and the target fuel package is critical.

Fire spread does not readily lend itself to quantitative analysis, and a practical application of formulas can be difficult, but it may be useful to perform the estimation simply to see if a particular fire spread scenario is feasible.

Fire is likely to spread from one fuel package to another by radiation, so an important consideration is how much of the total energy released is released as radiation as opposed to convection. Fuels exhibit an inherent radiant heat fraction, depending on how sooty their flames are. Sootier fuels release more heat via radiation than clean-burning fuels. Natural gas and methanol have radiant energy fractions of 15% to 20%, while gasoline and polyurethane have radiant energy fractions of 40% to 60%. Simply changing the estimated radiant fraction can have a significant impact on calculated heat fluxes.

The shape of a burning object also affects how much energy it radiates and in what direction. For purposes of obtaining an estimate, however, the simplest course is to consider the burning fuel as a point source (an approximation) and consider also that the radiant energy is distributed evenly around the surface of a sphere (another approximation) at any given distance. The area of a sphere is given by $4\pi r^2$, so a sphere having a radius of 1 m has a surface area of 12.5 m^2. A sphere with a radius of 1.5 m has an area of 28.3 m^2. Generally, the energy distribution at a distance from a point source exhibits an inverse square relationship. The energy flux from a nonpoint source does not fall off as quickly with distance, so the values given in the following table are underestimates. This table gives approximate heat flux values at 1, 1.5, and 2 m for fuel packages such as chairs having different HRRs and radiant fractions of 0.5 and 0.4.

Radiant fraction = 0.5 chair HRR (kW)	Heat flux at 1 m kW/m^2	Heat flux at 1.5 m kW/m^2	Heat flux at 2 m kW/m^2
800	32	14	8
1,000	40	18	10
1,500	60	27	15
2,000	80	35	20
2,500	100	44	25
Radiant fraction = 0.4 chair HRR (kW)			
800	26	11	6
1,000	32	14	8
1,500	48	21	12
2,000	64	28	16
2,500	80	35	20

Yellow indicates likely ignition in under 1 min.

The table shows approximate heat flux at fixed distances from a burning chair. Whether the target ignites depends on distance, radiant fraction, "view," and the HRR of the burning chair.

If the chair is releasing a fixed amount of energy (another approximation) and only ignited on one side, one would expect that the radiation imposed on the target would be increased to the extent that all of the energy is directed to a hemisphere rather than to a sphere. If the target fuel is positioned at an angle with respect to the burning fuel, more radiant energy will be required because of the difference in "view."

Experimental testing of fuel packages suggests that 1.5 m is a limit beyond which the involvement of the second item is unlikely prior to flashover [54]. (Of course, once flashover occurs, all of the fuel packages in the room will ignite.) Given that neither the HRR nor the radiant fraction can be known with any certainty, a cautious approach to deciding whether the target fuel will ignite would be to assume that any target fuel within 2 m probably will ignite.

3.17 CONCLUSION

Ignition is the process by which self-sustained combustion begins. Vapors or dust clouds are more easily ignited than solids. A flame is a luminous volume where combustion occurs. Flames can be either premixed or diffused, laminar or turbulent.

Flammability can describe the ease of ignition, the rate of fire spread, the heat release rate (HRR), the mass loss rate, or other properties related to the fire behavior of a fuel.

The energy released by a burning fuel depends on the chemical makeup, form, orientation with respect to the fire, and location within an enclosure.

Compartment fires behave differently than the free-burning fires with which most people are familiar. As fire investigators learn more about post-flashover burning, it becomes clear that the understanding of the mechanisms of post-flashover pattern development is still in its infancy.

Equations and models can enhance the understanding of a fire in a building, but these tools designed for fire protection engineering do not yet have sufficient resolution to determine the origin or cause of a fire.

Review questions

1. What is the name of the process by which self-sustaining combustion begins?
   a. Pyrolysis
   b. Decomposition
   c. Thermal runaway
   d. Ignition
2. What kind of flame is typically found on a candle?
   a. A premixed laminar flame
   b. A turbulent diffusion flame
   c. A laminar diffusion flame
   d. A turbulent luminous flame
3. What units are used to describe a fire's heat release rate?
   a. Joules or kilojoules
   b. Kilowatts per square meter
   c. Kilocalories
   d. Watts or kilowatts
4. What units are used to describe radiant heat flux?
   a. Joules or kilojoules
   b. Kilowatts per square meter
   c. Kilocalories
   d. Watts or kilowatts
5. Using the Heskastad equation, you estimate that a flame six feet tall has an HRR of 536.75 kW. Given that the uncertainty of this calculation is ±30%, what is the most accurate way to describe the HRR of the flame?
   a. 536.75 kW
   b. Between 530 and 540 kW
   c. Between 500 and 550 kW
   d. Between 400 and 650 kW

Questions for discussion

1. What is the difference between pyrolysis and smoldering combustion?
2. Why is an ignitable liquid not a likely explanation for a hole burned in a hardwood floor?
3. Why do ventilation-generated patterns provide less insight into the fire's origin than plume-generated patterns?
4. If a compartment is fully involved in fire for five minutes after flashover, what should be the approach to evaluate ignition sources in that compartment? Why?
5. Discuss the appropriate use of computer fire modeling in fire investigation.

References

1. Drysdale, D. (1985) *An Introduction to Fire Dynamics*, Wiley-Interscience, NY, p. 80.
2. Eckhoff, R. K. (1991) *Dust Explosions in the Process Industries*, Butterworth-Heinemann, Oxford, UK, p. 562.
3. Minick, C. (2005) Pros and cons of oil finishes, *Fine Woodworking*, 177, May/June.
4. Lentini, J. (2009) Watching paint dry: Testing spontaneous ignition hypotheses, *Presentation to the AAFS Criminalistics Section*, Denver, CO, 40. Available at https://app.box.com/s/4owqh32oxt1fjjl5r3zu9pgv34osvtqd (last visited on January 14, 2018).
5. Babrauskas, V. (2003) *Ignition Handbook*, Fire Science Publishers, Issaquah, WA, pp. 1043–1044.
6. Cuzzillo, B. (1997) Pyrophoria, Doctoral dissertation, University of California at Berkeley, CA.
7. 10th Circuit Court of Appeals (2004) *Truck Insurance Exchange v. Magnetek* 360 F. 3d 1206.
8. NFPA 400 (2016) *Hazardous Materials Code*, National Fire Protection Association, Quincy, MA, 3.3.60.8.
9. Quintiere, J. (1998) *Principles of Fire Behavior*, Delmar Publishers, Albany, NY, p. 38.
10. Babrauskas, V. (2003) *Ignition Handbook*, Fire Science Publishers, Issaquah, WA, p. 315.
11. Quintiere, J. (1998) *Principles of Fire Behavior*, Delmar Publishers, Albany, NY, p. 26.
12. Smyth, K. et al. (1985) Soot inception in a methane/air diffusion flame as characterized by detailed species profiles, *Combustion and Flame*, 62:157.
13. Drysdale, D. (1985) *An Introduction to Fire Dynamics*, Wiley-Interscience, New York, p. 197.
14. NFPA 704 (2017) *Standard System for the Identification of the Hazards of Materials for Emergency Response*, National Fire Protection Association, Quincy, MA, Annex G.
15. NFPA 92 (2015) *Standard for Smoke Control Systems*, National Fire Protection Association, Quincy, MA, p. 52.
16. NFPA 921 (2017) *Guide for Fire and Explosion Investigations*, National Fire Protection Association, Quincy, MA, p. 41.
17. NIST (2005) *Key Findings and Recommendations for Improvement: NIST Investigation of the Station Nightclub Fire*, Available at NIST.gov.
18. NFPA 921 (2017) *Guide for Fire and Explosion Investigations*, National Fire Protection Association, Quincy, MA, p. 31.
19. Mitler, H. E., and Tu, K. M. (1994) Effect of ignition location on heat release rate of burning upholstered furniture, in *Proceedings of the Annual Conference on Fire Research*, NIST, Gaithersburg, MD.
20. Stroup, D. W., et al. (2001) Upholstered chair fire tests using a California technical bulletin 133 burner ignition source, Report of Test FR 4012. Available at fire.nist.gov
21. Krasny, J., et al. (2001) *Fire Behavior of Upholstered Furniture and Mattresses*, Andrew Publishing, Norwich, NY, p. 62. Available at fire.nist.gov/bfrlpubs/fire85/PDF/f85003.pdf.

22. Carman, S. (2011) Investigation of an elevated fire—Perspectives on the "Z-Factor," *Proceedings of Fire and Materials*, San Francisco, CA, January 2011, Interscience Communications, London, UK.

23. Hurley, M., and Bukowski, R. (2008) *Fire Protection Handbook*, 20th ed., NFPA Quincy, MA, pp. 3–127.

24. Powell, R., DeHaan, J., and Lentini, J., (1992) The Lime Street Fire, three perspectives, *The Fire and Arson Investigator*, 43(1), pp. 41-59.

25. NFPA 921 (2017) *Guide for Fire and Explosion Investigations*, National Fire Protection Association, Quincy, MA, p. 215.

26. Utiskul, Y., Quintiere, J. G., Rangwala, A. S., Ringwelski, B. A., Wakatsuki, K., Naruse, T. (2005) Compartment fire phenomena under limited ventilation. *Fire Safety Journal*, 40(4):367–390.

27. Lentini, J. J. (2012), The evolution of fire investigation and its impact on Arson Cases, *Criminal Justice*, 12(1), pp. 12-17.

28. U.S. Army (1985) *Law Enforcement Investigations*, Field Manual 19–20, p. 219.

29. Carroll, J. (1979) *Physical and Technical Aspects of Fire and Arson Investigation*, Charles C Thomas, Springfield, IL, p.73.

30. Shanley, J. (1997) *USFA Fire Burn Pattern Tests*, United States Fire Administration, Emmitsburg, MD, FA 178.

31. NFPA 921 (2017) *Guide for Fire and Explosion Investigations*, National Fire Protection Association, Quincy, MA, p. 15.

32. Mealy, C., Benfer, E., and Gottuk D. (2011) Fire dynamics and forensic analysis of liquid fuel fires, NCJRS, Final Report, Grant No. 2008-DN-BX-K168, Available at: https://www.ncjrs.gov/pdffiles1/nij/grants/238704.pdf (last visited on January 14, 2018).

33. Sanderson, J. (2001) Floor-level burning: What kinds of damage do burning accelerants cause? *Fire Findings*, 9(1): 1.

34. Kirk, P. (1969) *Fire Investigation*, John Wiley & Sons, New York, p. 74.

35. NFPA 921 (2017) *Guide for Fire and Explosion Investigations*, National Fire Protection Association, Quincy, MA, p. 14.

36. Carman, S. (2010) "Clean burn" fire patterns—A new perspective for interpretation, in *4th International Symposium on Fire Investigations Science and Technology* (*ISFI*), Available at http://www.carmanfire.com.

37. Babrauskas, V. (2017) Arc mapping: New science or new myth? *Presentation to the 2017 Fire and Materials Conference*, Available at https://www.doctorfire.com/ArcMappingFM.pdf (last visited on January 14, 2018).

38. ATF Fire Research Laboratory (2017) Technical Bulletin 002, Available at https://www.atf.gov/file/114497/download (last visited on January 14, 2018).

39. Maclean, J. (2016). Households use an average of seven connected devices every day. *cantech letter.* https://www.cantech-letter.com/2016/08/households-now-use-average-seven-connected-devices-every-day-report/ (last visited on January 13, 2018).

40. Kennedy, P., and Kennedy, K. (2004) Flashover and fire analysis: A discussion of the practical use of flashover analysis in fire investigations, in *Proceedings of the 10th International Fire Science and Engineering* (*Interflam*) *Conference*, Interscience Communications, London, UK, July 5–7, 2004, 1101.

41. Nelson, H. E. (2002) From phlogiston to computational fluid dynamics, *Fire Protection Engineering*, 13: 9–17.

42. Forney, G. (2013) *User's Guide for Smokeview (Version 6)—A Tool for Visualizing Fire Dynamics Simulation Data*, NIST Special Publication 1017-1, National Institute of Standards and Technology, Gaithersburg, MD, Available at http://fire.nist.gov/bfrlpubs/fire07/PDF/f07050.pdf (last visited on January 14, 2018).

43. Rein, G., Bar-Ilkan, A., Alvares, N., and Fernandez-Pello, A., (2004) Estimating the performance of enclosure fire models by correlating forensic evidence of accidental fires, in *Proceedings of the 10th International Fire Science and Engineering (Interflam) Conference*, Interscience Communications, London, UK, July 5–7, 2004, 1183.

44. Janssens, M. L., (2002) Evaluating computer fire models, *Fire Protection Engineering*, 13: 19.

45. McGrattan, K., et al., (2010) *Fire Dynamics Simulator (Version 5) User's Guide*, NIST Special Publication 1019-5, National Institute of Standards and Technology, Gaithersburg, MD.

46. Peacock, R. D., et al., (2000) *A User's Guide for FAST: Engineering Tools for Estimating Fire Growth and Smoke Transport*, NIST Special Publication 921, National Institute of Standards and Technology, Gaithersburg, MD, iv.

47. ASTM E1355-12, *Standard Guide for Evaluating the Predictive Capability of Deterministic Fire Models*, ASTM International, West Conshohocken, PA.

48. E1591-13, *Standard Guide for Obtaining Data for Deterministic Fire Models*, ASTM International, West Conshohocken, PA.

49. SFPE (2011) Engineering Guide, *Guidelines for Substantiating a Fire Model for a Given Application*, SFPE G.06 2011, SFPE Bethesda, MD, 2011.

50. Carvel, R.O. (2004) Fire investigation using CFD simulations of a fire in a discotheque, in *Proceedings of the 10th International Fire Science and Engineering (Interflam) Conference*, Interscience Communications, London, UK, July 5–7, 2004, 1207.

51. Rein, G., et al. (2009) Round-robin study of a priori modelling predictions of the Dalmarnock Fire Test One, *Fire Safety Journal* 44(4):590–602.

52. Madrzykowski, D. (2010) Fire pattern repeatability: A laboratory study on gypsum wallboard, in *Proceedings of the 4th International Symposium on Fire Investigation Science and Technology*, Columbia, MD, September 27–29.

53. Salley, M., et al. (2010) *Verifying and Validating Current Fire Models for Use in Nuclear Power Plant Applications*, NUREG-1824, U.S. Nuclear Regulatory Commission, Rockville, MD.

54. Krasny, J., et al. (1985) Fire behavior of upholstered furniture, Available at https://www.nist.gov/publications/fire-behavior-upholstered-furniture-nbs-monograph-173, 52 (Last visited on January 14, 2018).

55. Nelson, H. E. (1987). An engineering analysis of the early stages of fire development–The fire at the Dupont Plaza Hotel and Casino, December 31, 1986 U.S. Dept of Commerce, NBSIR 87-3560.

CHAPTER 4

Fire investigation procedures

It is easier to do a job right than to explain why you didn't.

—Martin Van Buren

LEARNING OBJECTIVES

After reviewing this chapter, the reader should be able to:

- List the steps taken in the scientific investigation of a fire
- Plan and conduct an investigation
- Discuss the concept of negative corpus fire cause determination
- Avoid spoliation or destruction of evidence
- Write a report of the investigation that will satisfy the requirements of a court of law

4.1 INTRODUCTION

This chapter proposes the basic steps to follow in carrying out a systematic fire investigation. As any guideline will tell you, each fire investigation is different from every other, if for no other reason than that fires happen at different places and at different times. There are sufficient similarities among fires, however, that it is essential to develop a standard approach to the process. Not every step will be necessary or practical in every investigation, and there are certainly situations that will not be addressed here that require some initiative and creativity on the part of the investigator. Many of the routine steps are possible only if the investigator is the "first responder," or the one who is on the scene before it has been significantly altered. Many investigations do not even start until other investigators have had a 6- or 12-month head start and have reached a conclusion that, for example, a particular product failure caused the fire. An investigator retained by the product manufacturer will thus be relying on the documentation produced by the earlier responders.

Procedures also vary according to the problem being addressed. In many cases, the cause of the fire is well known, but the investigator may be tasked with addressing a particular question about a particular system. Regardless of the reason why a fire investigator was called or when he or she was called, it is important to have a written procedure. The purpose of written procedure protocols and policies is to ensure reproducibility of results and consistency of quality over an extended period of time. Lack of written procedure protocols or policies can result in inconsistent and/or poor quality. An example of a written general procedure for the investigation of fires appears in Sidebar 4.1 at the end of this chapter, but what follows now is a detailed look at how one should conduct a fire investigation.

4.2 RECOGNIZE THE NEED

Before we get into the subject of *how* to conduct a fire investigation, we first need to understand *why* one conducts a fire investigation. This is the first step of the scientific method, described in NFPA 921 as "Recognize the Need." What does the entity calling for the investigation need? The main task of most fire investigators is to determine *where* the fire started and *how* it started; the word *cause*, however, often has many deeper meanings, depending on the context. In addition to the cause of a fire, one might want to understand the cause of a fire death. One might have a really simple *fire* cause determination, such as an obvious cooking fire, but the next question becomes, What was the cause of the victim's failure to escape? Was the victim incapacitated? Was there something wrong with the building? Were the exits blocked? Did the smoke alarm function properly, and if not, why not? Did the fire spread too quickly, and if so, why? Fire investigators are frequently asked to determine why a fire spread the way it did. This could be due to the nature of the first fuel ignited; the interior finish; the nature of the building ventilation system; or the failure of a fire protection system, such as a sprinkler or vent damper or automatic door closure; or because someone propped open the door to the exit stairway.

In many fires, the cause of the fire (the circumstances that brought the fuel and the ignition source together) is reasonably well known because there are eyewitnesses. For example, this author has examined many restaurant fires where the cook stated that the deep fat fryer overheated and set the building on fire. An inspection of the scene may well verify that, but people are going to want to understand *what caused* that deep fat fryer to malfunction. They are going to want to know whether there was a defect in the design or manufacture of the fryer, or whether there was a broken thermostat or a safety device that had been defeated. Even after the cause of the ignition is known, however, people are going to want to understand what caused the fire suppression system to fail to extinguish the fire.

Even if no criminal case is brought, the person who wired around the safety device (or more likely, his employer) and caused the loss may be civilly liable to the owner of the building. The owner of the restaurant may be civilly liable to the tenant next door if he failed to maintain the fire suppression system. All of the various parties can retain their own fire investigators, who will be required to work together so that the wheels of justice can move forward.[1]

In the public sector, the first goal is usually to determine whether the fire was accidental or incendiary. That is sometimes the limited assignment of the first investigator sent out by an insurance carrier, but usually an insurance carrier is looking for someone, anyone, who might be held responsible for the fire other than its named insured.

4.3 THE NULL HYPOTHESIS: ACCIDENTAL CAUSE

The most important tool a fire investigator can bring to a scene is an open mind. Although fires are common,[2] they can be very difficult to investigate, as evidenced by the large number of incorrect arson determinations that have come to light over the years. (See the Dedication of this text for a list of examples.) Given the uncertainty associated with fire origin and cause determination, as discussed in Chapter 3, it is necessary to put in place some minimal provision to prevent serious errors. One way to prevent incorrect determinations of arson is to start with the premise that the fire is accidental until evidence proves otherwise. In 18 states,[3] it is the law that *all* fires are presumed accidental or providential until proven otherwise. But fire investigators are advised by NFPA 921 to approach a fire without presumption [1].

In an ideal world, starting with no presumption might be acceptable, but the track record of fire investigations does not justify such a position. There have been too many accidental fires erroneously called arson. In this author's view, it is both a scientific error as well as an ethical error to approach a fire investigation with a presumption of any cause other than accidental. It violates the spirit of many state laws, and it deprives potential suspects of the

[1] It has been said that these wheels turn very slowly, but they grind exceedingly fine.

[2] According to NFPA, in 2015, there were 1,345,500 fires reported in the United States. These fires caused 3,280 civilian deaths, 15,700 civilian injuries, and $14.3 billion in property damage.

- 501,500 were structure fires, causing 2,685 civilian deaths, 13,000 civilian injuries, and $10.3 billion in property damage.
- 204,500 were vehicle fires, causing 500 civilian fire deaths, 1,875 civilian fire injuries, and $1.8 billion in property damage.
- 639,500 were outside and other fires, causing 95 civilian fire deaths, 825 civilian fire injuries, and $252 million in property damage.

Each day, there were an average of 1,000 home structure fires.

[3] Arkansas, Georgia, Hawaii, Indiana, Maryland, Michigan, Missouri, Montana, Nebraska, North Carolina, Oregon, Pennsylvania, Tennessee, Texas, Vermont, Virginia, Washington, and West Virginia.

presumption of innocence. In many cases, if a fire is incendiary, there is only one logical suspect. If a fire is erroneously called incendiary because an investigator failed to keep an open mind or failed to start with a presumption of accidental cause, the unwitting survivor will be denied his or her right to the presumption of innocence. Even if the falsely accused citizen is later exonerated, he or she will have been severely traumatized, and the individual and his or her family will be financially ruined. It is easy enough to approach a fire with the idea that it was an accident and then demand that the fire prove itself to be incendiary. According to NFPA 921, an accidental fire is one for which the proven cause does not involve an intentional human act to ignite or spread fire into an area or under circumstances when and where there should not be a fire [2]. Prior to a fire scene being processed, what is known is that a competent ignition source came together with a competent fuel source and an ignition sequence followed. Unless the investigator finds evidence of human involvement, the data and the evidence support an accidental hypothesis for the initial classification of the fire cause.

Classical science is defined by the notion of *hypothesis testing*. In very simple terms, the scientist proposes a hypothesis, performs experiments or makes observations to test the hypothesis, and obtains results that either tend to confirm or invalidate the hypothesis [3]. A hypothesis is a supposition or proposed explanation made on the basis of limited evidence as a starting point for further investigation [4]. A valid hypothesis is capable of being disproved by evidence. Carrying a presumption of accidental cause into a fire investigation simply means that the first hypothesis proposed and tested should be "this fire was an accident." This kind of hypothesis is known as the null hypothesis. It is accepted as being true until it is disproved, and as with any hypothesis, an investigator should try very hard to disprove it.

Such an approach will not allow a guilty party to escape detection if the investigator is diligent, and it will almost certainly ensure that innocent citizens are not falsely accused of arson or murder. Incendiary fire classifications made using this approach are more reliable.

Because one should approach a fire with the presumption of innocence (accidental cause), that is *not* to say that the scene should not be physically treated as a crime scene. As long as the potential exists for the fire to be determined as incendiary, site security should be in place, along with mandatory procedures designed to protect the integrity of any criminal prosecution.

The controlling legal authority of a fire investigator's right of entry is the Supreme Court's interpretation of the Fourth Amendment in two Michigan cases, *Michigan v. Tyler* and *Michigan v. Clifford* [5,6]. In *Tyler*, the Court held that an "expectation of privacy" exists even after a building burns, so a warrant is required except when there are "exigent circumstances." This means that firefighters have a right to make a forceful, unannounced, nonconsensual, warrantless entry to extinguish the fire, make sure it will not rekindle, and make a determination of the cause. Once they leave the building, however, the Fourth Amendment goes into effect. In *Clifford*, the Court held that the need for a warrant existed even in cases of "administrative" searches, that is, those searches that start out being conducted for purposes other than the collection of evidence in a criminal investigation. However, most entries are made under exigent circumstances or with the consent of the owner. Once the exigent circumstances have ceased to exist, and if the owner is unavailable to give permission to enter, it is easier to obtain a warrant than to later explain why no warrant was obtained. The law on a fire investigator's right to enter is well settled. Mann (1984) provides a detailed analysis of the role of search warrants in fire investigations [7]. In civil investigations, a policyholder who refuses permission for an investigator to enter and conduct an investigation risks voiding the contract of insurance, but the investigator still needs permission to enter and risks being charged with trespass if he or she enters without it.

FOURTH AMENDMENT

The right of the people to be secure in their persons, houses, papers, and effects, against unreasonable searches and seizures, shall not be violated, and no Warrants shall issue, but upon probable cause, supported by Oath or affirmation, and particularly describing the place to be searched, and the persons or things to be seized.

4.4 NEGATIVE CORPUS METHODOLOGY

In contrast to the presumption (or null hypothesis) that a fire is accidental, some investigators carry with them the presumption that the fire was intentionally set. These investigators will not admit that they carry such a presumption, but their use of negative corpus methodology demonstrates that they do in fact carry such a presumption into every fire scene. Negative corpus is shorthand for negative *corpus delicti*—no "body of the crime." The fire investigator believes that he is sufficiently talented to be aware of and to detect all possible accidental ignition sources and is able to find evidence of such a fire cause in every case. Thus, finding no accidental cause, even in the complete absence of evidence supporting an incendiary cause, the investigator feels comfortable in declaring, "All accidental causes have been eliminated; therefore, the fire must have been intentionally set."

Such declarations are sometimes possible, but only in rare cases. For example, if there is a small fire confined to a bedroom closet and there are no lights or wires or other devices in the closet, it may be possible to state with certainty that the ignition source was an open flame, even if no match or lighter is found. Until the 2011 edition, NFPA 921 allowed for such a determination in cases "where the area of origin is clearly defined and all other potential heat sources at the origin can be examined and credibly eliminated [8]." What was not clearly defined was the phrase "clearly defined."[4] This "ambiguity" was exploited by some investigators, who misconstrued the language to mean that it allowed them to state that "clearly defined" means whatever they want it to mean (*ipse dixit*). The intent of the passage was to allow a reasonable determination to be made when the ignition source was absent, but not to give license simply to declare a fire incendiary because no accidental cause could be found.[5] The Technical Committee of Fire Investigation's discussion of the issue began in the mid-1990s, and the first description of the proper use of the phrase "process of elimination" appeared in the 1998 edition. The committee attempted to tighten up the language through three subsequent editions, but by the 2011 edition, the consensus was that, in order to avoid misuse of this reasonable methodology, the concept of the negative corpus should be directly addressed and condemned.

"Clearly defined," in this investigator's view, means that anyone, even someone completely untrained in the investigation of fires, could look at the damage and unhesitatingly, without fear of contradiction, point at the location and state, "That's where the fire started." Figure 4.1 shows what this author had in mind when signing off on the "clearly defined" language.[6]

So, the "Cause" chapter of NFPA 921 no longer contains the language about "clearly defined" origins. Instead, the Technical Committee inserted language about inappropriate use of the process of elimination and disparaged the negative corpus methodology. Here is the language from the 2011 edition:

> **18.6.5 Inappropriate use of the process of elimination** The process of determining the ignition source for a fire by eliminating all ignition sources found, known, or believed to have been present in the area of origin, and then claiming such methodology is proof of an ignition source for which there is no evidence of its existence, is referred to by some investigators as "negative corpus." Negative corpus has typically been used in classifying fires as incendiary, although the process has also been used to characterize fires classified as accidental. This process is not consistent with the scientific method, is inappropriate, should not be used because it generates untestable hypotheses, and may result in incorrect determinations of the ignition source and first fuel ignited. Any hypothesis formulated for the causal factors (e.g., first fuel, ignition source, and ignition sequence), must be based on facts. Those facts are derived from evidence, observations, calculations, experiments, and the laws of science. Speculative information cannot be included in the analysis.
>
> **18.6.5.1 Cause undetermined** In the circumstance where all hypothesized fire causes have been eliminated and the investigator is left with no hypothesis that is evidenced by the facts of the investigation, the only choice for the investigator is to opine that the fire cause, or specific causal factors, remain undetermined. It is improper to base hypotheses on the absence of any supportive evidence. That is, it is improper to opine a specific ignition source that has no evidence to support it even though all other hypothesized sources were eliminated [9].

---

[4] In NFPA documents, if no definition is provided, readers are advised to consult *Merriam-Webster's Collegiate Dictionary*, 11th edition, for the ordinarily accepted meaning.

[5] **Disclaimer:** Please note that any "interpretations" of the language of NFPA 921 in this text are the author's and not those of the NFPA or the Technical Committee on Fire Investigations—although it is hoped that at least some committee members would support these views. Although the author was involved in the drafting and "wordsmithing" of many of the passages cited herein, these interpretations do not constitute a "formal interpretation" of any part of the document.

[6] There is a disturbing tendency among some fire investigators to overuse the words *clear* and *obvious*. Such misuse of the language is an argumentative trick designed to make readers or listeners feel inferior if they are unable to see the "obvious pour pattern," or the "clear and distinct trail of burning," or the "obvious origin of the fire." If the information is not obvious or clear *to the untrained eye*, then it is neither obvious nor clear. This is "The Emperor's New Clothes" argument. Such arguments render the entire investigation suspect.

Figure 4.1 A "clearly defined" origin. The fire began in the closet in the corner. There were no lights or other potential ignition sources in the closet so a determination that the ignition source was an open flame can be justified. This photo also shows the development of a heat horizon, a smoke horizon, and a well-formed V-pattern caused by the intersection of the fire plume with the wardrobe door. (Courtesy of Mick Gardiner, Gardiner Associates Training and Research [GATR] Cambridge, England.)

After publication of the 2011 edition of NFPA 921, significant concern was expressed by users of negative corpus methodology, and there were several proposals to go back to the 2008 language. For the 2014 edition, the language was slightly modified and in 2017 it was modified slightly again. The 2017 edition says:

> **19.6.5* Appropriate Use** The process of elimination is an integral part of the scientific method. All potential ignition sources present, or believed to be present in the area of origin should be identified and alternative hypotheses should be considered and challenged against the facts. Elimination of a testable hypothesis by disproving the hypothesis with reliable evidence is a fundamental part of the scientific method. However, the process of elimination can be used inappropriately. Identifying the ignition source for a fire by believing to have eliminated all ignition sources found, known, or suspected to have been present in the area of origin, and for which no supporting evidence exists, is referred to by some investigators as *negative corpus*. Determination of the ignition source must be based on data or logical inferences drawn from that data. Negative corpus has typically been used in classifying fires as incendiary, although the process has also been used to characterize fires classified as accidental. The negative corpus process is not consistent with the scientific method, is inappropriate, and should not be used because it generates untestable hypotheses, and may result in incorrect determinations of the ignition source and first fuel ignited. Any hypotheses formulated for the causal factors (e.g., first fuel, ignition source, and ignition sequence), must be based on the analysis of facts and logical inferences that flow from those facts. Those facts and logical inferences are derived from evidence, observations, calculations, experiments, and the laws of science. Speculative information cannot be included in the analysis [10].

The operative statement disparaging negative corpus methodology did not change from the 2011 edition to the 2017 edition. Negative corpus methodology as a basis for finding a fire to be incendiary is now generally frowned upon. At least one arson conviction has been overturned based on the change in NFPA 921 on the subject of negative corpus [11]. This case is discussed in detail in Chapter 9.

NFPA 921-17 AT 19.6.5

The negative corpus process is not consistent with the scientific method, is inappropriate, and should not be used because it generates untestable hypotheses, and may result in incorrect determinations of the ignition source and first fuel ignited.

Failure to disprove the null hypothesis can be turned around by stating, "No evidence of an intentionally set fire was found; therefore the fire must have been accidental." This is at least some scientific justification for this statement if no evidence of human involvement can be proven.

4.5 PLANNING THE INVESTIGATION

When the fire investigation assignment first comes in, getting sufficient information allows the formulation of a reasonable plan. Learn the basics about when the fire occurred, which fire department responded, who owns and leases the building, what the building's occupancy (use) was, and what other parties may have been affected by the fire. Private sector investigators may need to coordinate their efforts with both public sector investigators and private fire investigators retained by other parties. Learn to what extent suppression and overhaul may have altered the scene. At this stage, the investigator may be able to decide whether it is necessary to recruit additional talent or obtain heavy equipment to assist with the investigation. If there is a credible suspicion of an electrical failure, it is a good practice to bring a forensic electrical engineer to the site. If there is a collapse of a steel truss roof over a 50,000 ft^2 retail store, it will probably be necessary to rent a crane and hire a cutting and welding contractor.

It is useful to learn as early as possible whether there were any eyewitnesses to the fire.[7] This can save an immense amount of work. Witnesses might have photographed or videotaped the fire in progress. If a building has a security system, there may be records of alarm activations or possibly videotapes from security cameras (though the cameras always seem to aim just to the left or right of where we would like). Generally, alarm-monitoring companies respond only to inquiries coming directly from their customer, and even then, many are nervous because they understand that they may be held liable if their system failed to respond to a fire.

For most occupancies, the use of the structure is apparent, and the influence of activities of the occupants or the kind of fuels present on the spread of the fire will be simple enough to determine. For industrial occupancies, it is usually necessary for the investigator to become familiar with the peculiarities of the processes taking place by consulting with the plant manager or someone knowledgeable about the systems. The fire history of the facility should also be determined.

Prior to visiting the scene, it is necessary to ensure that one has permission to do so. For public sector responders, authority to enter and investigate most often comes in the form of exigent circumstances or permission from the owner, but it can also take the form of a search warrant or an order of the court. Private sector responders usually require permission from the owner. Having made as many preparations as practical, the site visit is the next step.

4.6 THE INITIAL SURVEY: SAFETY FIRST

The investigator should make his or her presence known prior to conducting any investigation, assuming anyone else is present. Investigating fire scenes alone is not a good idea for a number of reasons. First and foremost, it is not wise to enter a dangerous place alone, and fire scenes are dangerous places. Should the investigator become injured or trapped, he is in a world of hurt if he is on the scene by himself. If it is necessary for the investigator to enter the scene alone, he should have a cell phone, and someone on the outside should know where he is and when he is expected to return. It is always useful, and sometimes essential, to have a second brain and a second pair of eyes on the scene. Review the exterior of the scene for safety hazards prior to going inside. Safety hazards include the obvious ones, such as a lack of visibility, structural instability, energized electrical circuits, and leaking fuel gases, but the investigator should also consider the less obvious hazards, such as fire gases (CO and HCN), toxic hazards such as asbestos, and

[7] Eyewitnesses are not always reliable, but their reported observations, even the reported observations of persons who might be suspects, constitute data that must be evaluated. Particularly in cases where full room involvement has occurred, eyewitness accounts may tell the investigator more than the fire artifacts can.

biohazards. For an excellent lesson in safety issues at fire scenes, investigators are urged to sign on to CFITrainer. net and complete the Fire Scene Safety module [12]. The Safety chapter of NFPA 921 (Chapter 13 in the 2017 edition) should be required reading for anyone who enters a fire scene.

If conditions that prevent the inspection from being safely conducted exist, steps must be taken to alleviate those conditions before the investigation can proceed. Waiting for daylight is always a good idea. Not only will the investigator be safer, but natural light also frequently allows investigators to see more than they can under artificial light. The emergency is over. The fire is out and there is nobody who needs rescuing. Fire investigators will be injured often enough even if they try to be safe. There exists no reason for fire investigators to intentionally take risks with their own or their colleagues' safety. As with any activity involving danger, it is important to remember that conditions change during the investigation, so investigators should continually reassess conditions. For example, moving debris from one place to another affects the loading of the structure, and changing weather conditions can cause previously stable elements to collapse.

Changes can also occur to the investigator, who may become overheated, fatigued, or dehydrated doing the physically demanding work of clearing debris and reconstructing the scene. These hazards can be just as deadly as the physical hazards associated with the building. Taking scheduled breaks and staying hydrated are essential.

Deviations from best safety practices must be avoided. Failing to use personal protective equipment (PPE) seldom has immediate consequences, so some investigators continue to deviate. Eventually, such deviations can result in serious bodily injury or illness caused by chronic exposure to hazardous materials.

A brief walk-through is considered useful by some investigators and is indispensable in some scenes. For a single residence, however, the best plan starts with the exterior and then moves to the interior, starting with the area of least damage and then moving to the area of greatest damage, but any system that results in a complete inspection of all rooms is adequate. The area of greatest damage should not be automatically construed as the origin of the fire. Except in the most obvious of cases, it is best not to use the walk-through to formulate hypotheses. The best approach is to keep an open mind for as long as possible. Do not hurry the process. Although the investigator is obviously making observations at this point, one should avoid drawing inferences from those observations. One determination that must be made during the walk-through, however, is whether any compartments became fully involved. The rules for interpreting fire damage change in such compartments because the fires in those compartments have become ventilation-controlled.

Ideally, the walk-through will help the investigator formulate a general plan for conducting the site inspection—nothing more. A good rule at this point in the investigation is EOMSHIP: eyes open, mouth shut, hands in pockets.

4.7 DOCUMENTATION

Regardless of when one arrives at the site, its condition should be documented before the investigator makes any changes (or any further changes). Documentation can take many forms, but the best form is still photography. If there is an observation made, it should be documented. With the advent of digital photography, there no longer exists any reason not to take a photograph of every pertinent artifact.[8] Two years after an inspection, the important photos will not be the ones taken but the ones not taken.

Images should be collected at the highest resolution available on the camera, either in RAW or FINE JPG format. RAW files are more versatile, but are quite large and can quickly fill up a memory card. For most purposes, JPEG files provide sufficient resolution. Keep in mind that most modes of photography now document the *order* in which photographs were taken and, if the camera's clock is set properly, the time the image was collected. (The camera's date and time should be verified every day the camera is used.)

A logical investigation is reflected in the order of the images. Digital cameras also serve to document the time spent at a scene, or at least the time between the first and last photographs. Such information can be used to assess a fire investigator's approach to a scene or, conversely, to counter assertions that an investigation was conducted hastily or unsystematically.

[8] There never was a good reason for not taking enough photographs, but some agencies have cited budgetary constraints. In this author's view, if an agency cannot afford to do the job properly, they should find someone else to do the job.

Other forms of documentation are also essential. Written notes document those observations not well suited to photography, and video documentation frequently helps orient photographs.

Another critical piece of documentation, required on all but the simplest investigations, is the sketch. For most buildings, the minimum sketch should be a floor plan. Sometimes, particularly in commercial or industrial buildings, an "as built" drawing of the building can be obtained from the building owner/operator (assuming the fire did not destroy such documents). Many commercial occupancies have fire escape plans framed on the wall in several places. These are usually sufficient, at least as the starting point for a sketch. A floor plan is necessary because the people reading an investigator's report (or listening to his testimony) will not be familiar with the building. A floor plan is indispensable for getting other people oriented so that they can understand what the photographs show and what they might mean. A floor plan is also essential for documenting ventilation openings. For every room that has become fully involved, the floor plan should indicate whether a particular opening served as an intake or an exhaust. The intakes are likely to produce ventilation-generated patterns that offer little or no insight to the origin of the fire, because such patterns are created only after full room involvement. (Unless, the origin was located in the ventilation path.)

The documentation prior to disturbance of any evidence takes time, and during that time the investigator can make the observations necessary to help him or her decide how to reconstruct the fire scene. This time should also be used to become familiar with the building. Document the utility supplies. Are the electricity and gas supplies still turned on? Were they disconnected before or during the fire? What were the interior and exterior finishes? Sometimes these details are very important, and other times they only become important if the investigator fails to document them. After working 50 additional fire scenes, it is unlikely that the investigator will be able to remember whether the furnace was gas or electric unless it is documented during the initial inspection. Fire scenes are rich in data. It can be difficult to remember everything that needs to be documented. A preprinted form for documenting common building characteristics, such as the one shown in Figure 4.2, helps the investigator avoid missing important details.

Graph paper (10 × 10, 8½ × 11 inch paper ruled with bold lines in 1-inch squares and fine lines in 1/10-inch squares) makes the task of sketching much easier. If an error is made while drawing the sketch, it is likely to be caught while the sketch is being drawn. This is not the case with freehand sketches. For most middle-class residences, a rough scale of 1 in. equals 10 ft allows the building to fit on one sheet and also allows for sufficient details to be noted. If more precision is required, a scale of 1 in. equals 5 ft will allow most rooms to fit on a single sheet. Used properly, graph paper allows for drawing the sketch without actually writing down each measurement (because the lines are drawn to scale).

The level of detail on the sketch is left to the investigator and is dictated by the needs of the investigation. In many cases, a simple plan view showing rooms and doorways will suffice. Because ventilation is almost always an issue, the location of the windows should be included. If fire modeling is contemplated, then not only is it necessary to draw the doors and windows but also the height of each opening, the ceiling height, the location of each fuel package, and the composition and thickness of all interior finishes. The best time to gather this information is while the building is being sketched. Getting this information later may not be possible.

The sketch is as good a place as any to orient the building to the earth. A compass is an inexpensive device that should be in every investigator's camera case (or present as an app on his or her cell phone). When a report refers to the left or right side of the building, it is apparent that the investigator forgot to determine which way was north. Left and right can be very confusing descriptions, as it is then necessary to say where one is standing and which direction one is facing. Compass directions are much more consistent and easier to understand. North and south are also easier for laypeople to understand than "A side" or "C side."

One curious feature found on many sketches is the notation "not to scale," even when the investigator has made careful measurements. An approximate scale on the sketch allows the reader to understand that this is not a blueprint, and that the investigator does not care if a wall is 14 ft or 14 ft 3 in. long, but still lets the reader know that this is approximately what the building looked like. Such a scale might look like this: |~10 ft|. Writing out "1 in. equals 10 ft" is only useful if the sketch is not enlarged or reduced.

FIRE SCENE FIELD NOTES

JOB # ________ Policy or Claim # ________________________

Date of Loss ________________ Time__________
Assignment Rcvd ______________________
Date of Investigation ______________________
Building Last Occupied ______________________

Insured: ________________________________
Address: ________________________________
____

CONSTRUCTION Approx. Dimensions ______________ Square Feet ______ # Floors _____ # Bedrooms ___ # Baths _______

Orientation Front door faces [ ]N [ ]E [ ]S [ ]W

Exterior Finish	**Interior Finish**			
	Floors	**Subfloor**	**Ceilings**	**Walls**
[ ] Frame	[ ] Carpet	[ ] Plywood	[ ] Sheetrock	[ ] Sheetrock
[ ] Metal/Plastic Siding	[ ] Tile/Linoleum	[ ] Plank	[ ] Plaster/Lath	[ ] Plaster/Lath
[ ] Brick Veneer	[ ] Hardwood	[ ] Particle Board	[ ] Panel	[ ] Panel
[ ] Stone Veneer	[ ] Plywood	[ ] Tile	[ ] Tile	[ ] Ply Panel
[ ] Brick	[ ] Particle Board	[ ] Slab	[ ] __________	[ ] _____
[ ] Mobile Home	[ ] __________	[ ] __________		
[ ] Other__________				

Other Construction Data: ________________________________

HVAC: Make of Heater: ________________________

[ ] Central	[ ] Natural Gas	[ ] LP Gas	[ ] Other
[ ] Electric	[ ] Forced Air	[ ] Forced Air	# Units
[ ] Forced Air	[ ] Floor Furnace	[ ] Floor Furnace	[ ] Wood Stove
[ ] Baseboard	[ ] Space Heaters	[ ] Space Heaters	[ ] Fireplace
[ ] Ceiling	[ ] Vented	[ ] Aboveground Tank	[ ] Coal
[ ] Wall Mounted	[ ] Unvented	[ ] Submerged Tank	[ ] Kerosene
[ ] Space Heaters	[ ] Open Line Flow ______	% Full ____ ____PSI	[ ] __________
[ ] Heat Pump		[ ] Open Line Flow _____	

Other Heating System Data: ________________________________

Air Conditioning: [ ] Central [ ] Room Units Make: ________________________

DETECTION/SUPPRESSION Smoke Detectors: _______ Fire Extinguishers: ______________ Sprinklers: __________

SECURITY	**Appliances: (type, brand)**	**Furniture (#)**
Days Since Fire __________	Refrigerator __________	______ Beds
# Doors Open __________	Range __________	______ Dressers
# Windows Open __________	Oven __________	______ Couches
Insureds Home __________	Dishwasher __________	______ Chairs
# of Occupants __________	Microwave __________	______ Dining
Smokers? __________	Freezer __________	__________
Melted Metals [ ] Fe [ ] Cu [ ] Al	Washer __________	__________
__________	Dryer __________	**Clothing:**
__________	Water Heater __________	[ ] Hangers
__________	Compactor __________	[ ] Cloth

INVENTORY

[ ] Pots & Pans	[ ] Food	[ ] Televisions ________	Other electronics ______________
[ ] Dishes	[ ] Guns	[ ] Computer	______________
[ ] Silverware	[ ] Photographs	[ ] Pets	______________

Explosions: [ ] None in Evidence
[ ] Before Fire
[ ] During Fire
[ ] Reported [ ] Before [ ] During

Weather: ______________
Lightning ______________
Wind ______________

Fire Suppression: [ ] Effective [ ] Ineffective [ ] Not attempted Response time ______________ Control time __________

Department(s) ________________________ Dept. Contact: ________________________
____

ATS 851A, 08/04

Figure 4.2 Checklist describing building conditions.

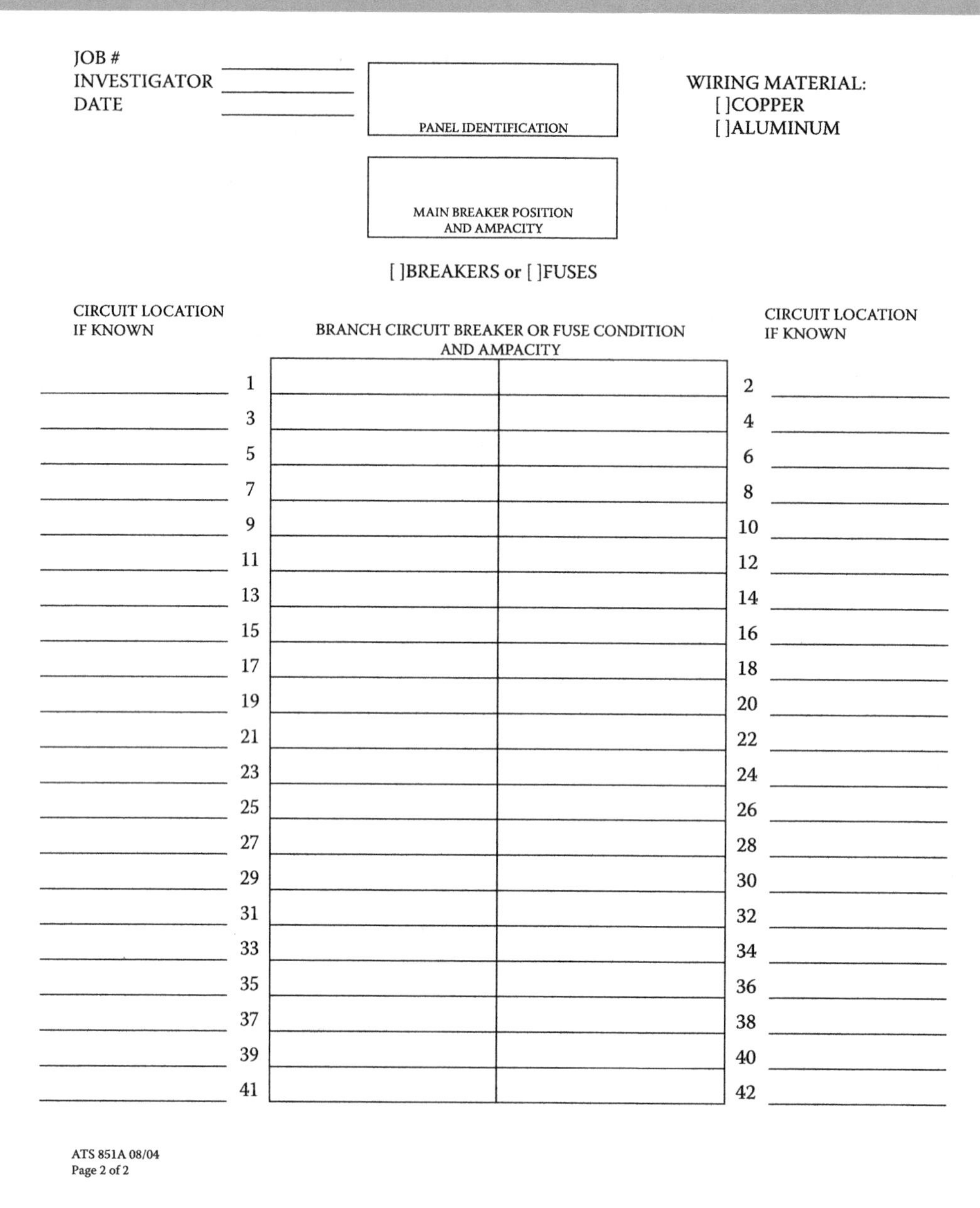

JOB # ____________
INVESTIGATOR ____________
DATE ____________

PANEL IDENTIFICATION

WIRING MATERIAL:
[ ]COPPER
[ ]ALUMINUM

MAIN BREAKER POSITION AND AMPACITY

[ ]BREAKERS or [ ]FUSES

CIRCUIT LOCATION IF KNOWN		BRANCH CIRCUIT BREAKER OR FUSE CONDITION AND AMPACITY			CIRCUIT LOCATION IF KNOWN
	1			2	
	3			4	
	5			6	
	7			8	
	9			10	
	11			12	
	13			14	
	15			16	
	17			18	
	19			20	
	21			22	
	23			24	
	25			26	
	27			28	
	29			30	
	31			32	
	33			34	
	35			36	
	37			38	
	39			40	
	41			42	

ATS 851A 08/04
Page 2 of 2

Figure 4.2 Checklist describing building conditions.

4.8 RECONSTRUCTION

The purpose of fire scene reconstruction is to re-create as nearly as practicable the prefire positions of contents and structural components [13]. During the process of sketching, documenting, and becoming thoroughly familiar with the fire scene, the investigator has a chance to make observations about the fire and to formulate a plan for reconstruction. This almost always involves debris removal and may require several hours or days of strenuous physical activity to remove the debris that covers the artifacts to allow a determination of origin and cause. Although there is heavy equipment that can accomplish this debris removal, starting with heavy equipment will almost always destroy more evidence than it uncovers. Too many investigations have been compromised by bobcats and backhoes used by investigators looking for "pour patterns" or spalling. The causative agent frequently lands in a pile of debris in the parking lot, with no possible way to determine where it was located before it was run over by the heavy equipment. The problem with heavy equipment, efficient though it can be, is that instead of clearing out a doorway or a hallway, it takes down the walls on either side.

As unglamorous and time-consuming as it may be, if the investigator is looking for the origin and cause, the best tool is usually the flat-headed shovel. It may take 10 such shovels, a few wheelbarrows, and 10 off-duty firefighters to operate them, but that is the way to find the cause of the fire. Sometime after the burned roofing material is carried outside, interesting artifacts will begin to emerge. It is necessary to stop periodically and document this potential evidence as it is found. It is frequently necessary to move things out of the way so that debris clearing can continue, and then move those things back to their original location to understand their role, if any, in the progress of the fire.

In planning debris removal operations, try to plan as far ahead as possible. Nothing is more discouraging than to realize that the debris that was just moved from the hallway into a bedroom needs to be moved again. In a well-executed debris removal operation, the debris is moved only once.

Once the debris has been moved out of the way, items should be placed back into their original locations, if possible. Tables and chairs frequently leave "footprints," protected areas on floors that allow for exact repositioning of the furniture. Use caution with freestanding items such as tables, however, as it is possible to reposition them exactly backward.

Once the reconstruction is as complete as it is going to get, any portions of the scene that have been changed should be rephotographed, paying particular attention to patterns that indicate movement or intensity of the fire, fuel packages, and ignition sources.

A common mistake in the documentation of the scene is the failure to take overall shots of the area where close-ups are going to be taken. Photographs of fire patterns or of small objects that might be ignition sources are just not as useful if there is no photograph that places the subject of the close-up into its context. The documentation of a fire scene, and indeed the entire investigative process, should entail moving from the general to the specific, "zooming in" on the relevant evidence and ultimately the cause of the fire. With the right equipment, photographs can be enlarged using the camera's electronics. Figure 4.3a shows the power supply wire to a kitchen exhaust fan located over a pan

Figure 4.3 Digital photographs [(a) and (b)] enlarged on the scene using the camera's electronics.

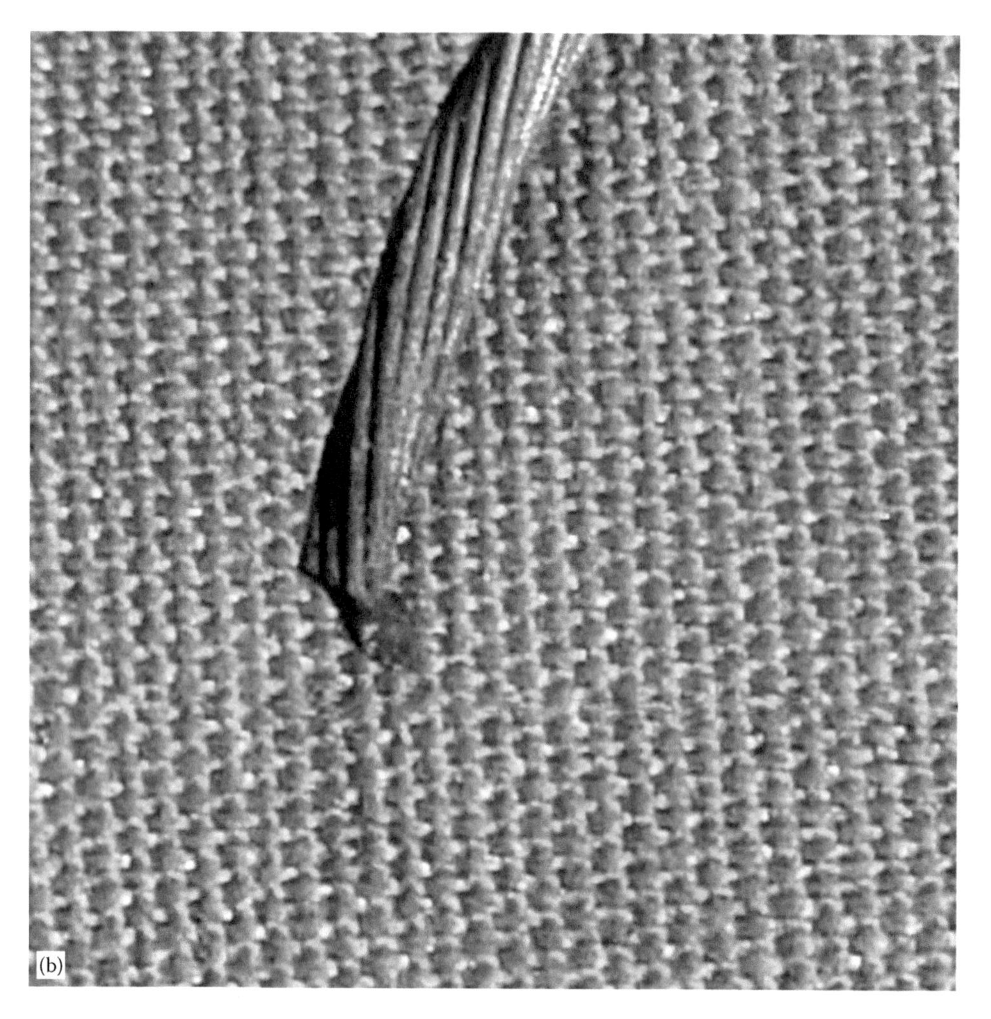

Figure 4.3 Digital photographs [(a) and (b)] enlarged on the scene using the camera's electronics.

of burning solvent in a clandestine methamphetamine lab. The digital camera can be used as an on-site magnifier, as shown in Figure 4.3b. The resolution of the image on the viewfinder did not degrade until a 25× magnification was used. A blue cloth is used for the background. This reduces glare and serves as a check on the level of detail that can be captured. Light blue or gray file folders can be used if a blue cloth is unavailable.

Some of the most important artifacts[9] are those that allow the collection *of sequential data*. The fire scene "records" each movement of the fire from one place to another, but often these patterns do not help us learn *when* a particular event occurred, or in which order two or more events occurred. Patterns that give sequential data are the most instructive. Sometimes, the orientation of fire patterns on an object allows one to draw inferences about the object's position when the pattern was created. Fallen objects such as ceiling pieces might exhibit smoke or heat damage on the top or bottom surface. It is important to examine this "debris" as it is being removed. Some surfaces exhibit streaking, caused by extinguishment water or condensation flowing downward. Such streaks can be used to orient formerly vertical objects. Sometimes, the task of moving debris is so large that the focus on such details is lost.

The evidence is now exposed and has been documented. The fire investigator is presented with what is likely all the physical evidence there is to see. The work that has been done should now allow for the development of a credible hypothesis as to where the fire started and what may have caused it.

---

[9] This word has several meanings. When used in this text, *artifact* means a structure or feature left behind by the action of the fire.

4.9 INVENTORY

Inventory is physical evidence that frequently requires documentation. The investigator should, at the very least, determine that the inventory is consistent with the reported type and level of occupancy. Are there sufficient beds, linens, toiletries, and food for the number of persons reported to be living in a residence?[10] Private fire investigators are often asked to include an inventory section in their reports to verify the accuracy of an insurance claim. Assuming it is the kind of item not likely to have been consumed in the fire, the absence of some items may indicate a fraudulent claim. In fact, it may indicate that those items were removed before the fire based on the homeowner's knowledge that the fire was going to happen. Prior removal of contents, or substitution of contents, can be used as evidence of prior knowledge and may overcome a presumption that a fire was accidental. An accident is, by definition, unexpected.

This author has investigated many fires where the homeowner-arsonist could not part with his beloved guns or his televisions, so he removed these items before setting the fire. Then, in an attempt to "double dip," the missing items were claimed on the inventory. The homeowner may have even been able to document the purchase of items removed prior to the fire. Sentimental or irreplaceable items, such as family bibles,[11] are another class of items frequently removed prior to intentionally setting a fire. Because these items have little or no economic value, the claim that some burglar stole them and then set fire to the house does not exactly inspire credibility.

Substituted inventory is an even better indicator of what happened and who is responsible. In one fire investigated by the author, the homeowner first put his valuables in storage, then went to the town dump and obtained what he hoped would be sufficient junk furniture and appliances to support a claim after those materials had been burned to powder. He made the mistake of using too much gasoline, however, and the fire went out for lack of air. There was a trail of broken picture tube glass leading from the driveway into the living room, where a television set with a broken picture tube sat, with not a scorch mark on it.

Another investigation involved a homeowner who succeeded in burning his house to powder but something did not look right about the six television sets found in the debris. A careful examination revealed that they all employed vacuum tube electronics, yet the homeowner was able to produce receipts proving that he had, in fact, purchased the six new solid-state televisions that he wanted the insurance company to replace. The only logical inference to draw from this evidence was that the old televisions had been substituted for new ones prior to the fire. While there was insufficient physical evidence to determine how the fire started, we could determine that somebody knew it was going to happen—and planned accordingly.

In one of the author's first fraud investigations, the house was reduced to ashes but the homeowner had stashed his coin collection in the trunk of his car, where it was later spotted by an alert insurance adjuster. Valuable property is sometimes stored in detached structures. If there are outbuildings on the property, the investigator should document their contents to make sure that undamaged property is not later claimed as having been destroyed in the fire.

Simply finding inventory where it apparently does not belong is insufficient to prove prior knowledge. There may well be a legitimate excuse for the photos and other mementoes being found in the chicken house. In such a case, the homeowners must deny removing anything prior to the fire.

Caution is required in determining that inventory is removed or even absent. One insurance company was held to be dealing in bad faith when its retained investigator reported that he could not find clothing and other items of combustible inventory in a house that burned to powder. The court stated, "To accept the position … that failure to recover and identify every item in a residence is conclusive of anything … shocks the conscience of this Court. Such a position would render every homeowner in Alabama without insurance coverage [14]."

4.10 AVOIDING SPOLIATION[12]

It is now necessary to interrupt the flow of this discussion to discuss an issue that frequently interrupts the flow of fire investigations. *Spoliation* is defined as the "loss, destruction, or material alteration of an object or document that is evidence or potential evidence in a legal proceeding by one who has the responsibility for its

[10] Do *not* open the refrigerator to check the food supply until just before you are ready to leave, especially in the summer.

[11] Photographs do not necessarily have the sentimental value that they once possessed. Because digital photos can be stored in the cloud, they are easily replaced.

[12] The pronunciation of this word is frequently mangled. The correct pronunciation is spo-lee-AY-shun.

preservation." There often comes a time when an investigator realizes that a particular device may have malfunctioned and caused the fire. This might be a light switch or a ceiling fan, a cooking or heating appliance, a computer, or any one of thousands of manufactured products. Maybe a contractor performing a service, such as refinishing the floor or reroofing a commercial building, was on site shortly before the fire, and the evidence seems to point to some careless act on the part of one of the contractor's employees. When this occurs, it is the investigator's job to *stop any further activity* that might prejudice the rights of an entity that may soon become a party to litigation. The investigation can resume at a later time. If the property is insured, the insurance carrier will look to whoever caused the fire for compensation. That party, in turn, will likely insist on the opportunity to view the evidence in place. Failure to accommodate potential defendants can result in sanctions against the plaintiff seeking damages, up to and including dismissal of the lawsuit. Spoliation is not often an issue in criminal cases, but it is likely to come up in the future, so public sector investigators should not dismiss the concept. In one Minnesota case, *Tollefson,* the court dismissed an arson charge because the fire scene was demolished shortly before the defendant was indicted. The Minnesota Supreme Court held, in an unpublished opinion, "Destruction of the house before defendant was charged precluded defendant from retaining his own professional investigators to aid with his defense. The delay in charging defendant until after the house had been destroyed was prejudicial and contrary to the notion of fundamental fairness [15]." In a Pennsylvania case, where an electrical outlet at the origin of the fire was not preserved, the trial judge explicitly told jurors, "You may infer, if you choose to do so, that it would have been evidence unfavorable to the Commonwealth [16]." To this author's knowledge, sanctions on the public sector for spoliation have been limited to judicial remedies rather than monetary sanctions. To avoid spoliation, potential defendants must be put "on notice," be told when the investigation will proceed, and be allowed to send a representative to participate in the continuation of the investigation. The investigator should communicate with the client about the preliminary determination, and either the client (usually an insurance company or a lawyer representing an insurance company) or the investigator contacts the manufacturer, contractor, or other potential defendant to give them the bad news. A typical letter of notification is shown in Figure 4.4.

The time required for the potentially responsible party to engage its own investigator, and for that investigator to make arrangements to come to the scene, makes it necessary to cease operations and come back another day. Because bringing in other parties may involve considerable expense to those entities, once the investigator determines that a joint inspection of the scene is required, careful consideration should be given to finding *all* of the potentially responsible parties. This is necessary because, if it turns out that the first party notified is able to demonstrate, for example, that it was not their coffeemaker that started the fire, but the microwave oven next to it, it then becomes necessary to put the manufacturer of the microwave oven on notice and reconvene the investigation at a later date. Such complications can result in a fire scene inspection stretching out for several days over a period of months. Sometimes, it just makes sense to notify any party that may be responsible and let them decide whether to participate in the investigation. Some manufacturers have dollar criteria below which they will not send an investigator. Others will respond to notification by asking that everything be documented and their product be preserved for later laboratory examination. If a potentially responsible party is notified but chooses not to participate in the investigation, that party will not later be able to make a successful claim of spoliation.

Spoliation is a relatively new concept for many investigators. In the past, an investigator would go to the site and determine the cause, and if that cause was an appliance, it would be collected and taken to an electrical engineer for further evaluation. If the engineer found evidence that the appliance was responsible for the fire, the insurance carrier would make a claim against the manufacturer and the case would proceed. Such behavior by a fire investigator in the twenty-first century, however, not only prevents the investigator's client from collecting from the manufacturer of the defective product, but also it is likely to get the investigator sued, even if he or she has correctly identified the cause of the fire and the appliance is, in fact, defective.

Spoliation has become the second line of defense[13] in product liability litigation. This is possibly a result of the legal community becoming aware of a relatively high rate of error (see Chapter 9) in the determination of the origin and cause of fires. Defendants no longer simply accept the first determination, and the courts have supported their position.

[13] After "paternity," that is, "This is someone else's product."

<Date>

<__________>Corporation
Legal Department
Street Address
City, State 00000
via facsimile 555-555-5555

Dear <__________>Corporation:

<____________________> has been retained to investigate a fire loss that may have been caused by a <model> <type of appliance> manufactured by <________>Corporation. The unit is less than <age> years old.

The fire occurred on <date of loss> at <loss location>. The <____________> Fire Dept extinguished the fire and did an investigation.

Damage to the structure is estimated to be $<__________>. <________> Insurance Co, which is our client, affords coverage.

The <appliance> is still in place. We will soon be examining the fire scene and collecting evidence.

Please have your legal department contact me at <__________> to arrange a mutually convenient time for your designated representative to view the evidence prior to any destructive examination.

Sincerely

AGENCY

Investigator

Figure 4.4 A typical notification letter. The letter should contain sufficient information to allow the recipient to decide how to proceed. The event that triggered the notification should be described in sufficient detail that existing public records can be obtained. The dollar value helps the reader decide whether to retain an independent investigator or an attorney.

Not only is it necessary to let the manufacturer see its defective product, but also it is often necessary to *let them see it in place*: to let them have their own investigator confirm that the product was, in fact, at the origin of the fire and to allow that investigator to independently rule out other potential sources of ignition. A manufacturer of clothes dryers, defending itself in a product liability action, will likely be successful if the washing machine was not at least preserved for inspection by the manufacturer's chosen investigator. A coffeemaker manufacturer might claim spoliation if its investigator was not allowed to see the kitchen range before it was sent to the landfill, even if there is clear evidence of overheating on the element.

Claims of spoliation can be taken to ridiculous extremes, and realizing this, the NFPA's Technical Committee on Fire Investigations has undertaken to offer guidance not only on what activities constitute spoliation but also on those activities that should *not* be considered spoliation. It is usually necessary to conduct significant debris removal prior to uncovering the cause of the fire.[14] Debris removal to uncover the area of origin should not be considered spoliation.

[14] It is almost axiomatic that the debris is always deepest at the origin. Also, if it is raining on the day of the inspection, the roof will be almost completely destroyed directly over the origin.

Removal of the offending device from the scene should not be considered spoliation if such removal is necessary either to protect the device or to identify its manufacturer. Further, it may be possible to eliminate an appliance by a simple nondestructive examination off-site. Note that any device once suspected but later eliminated must be preserved so that it can be examined by other potentially responsible parties. The destruction of an "innocent" coffeemaker (or failure to preserve it) can result in the failure of a claim against, for example, a tenant who left a pan of food cooking on the range.

Scenarios such as this one can influence criminal cases as well. It is no longer acceptable for a public sector investigator to "eliminate" a reasonable potential source of accidental ignition and then allow it to be destroyed. A defendant may then claim that he has been prejudiced by the destruction of potentially exculpatory evidence. Such a claim will be especially effective if the only "evidence" of an incendiary cause is the investigator's inability to find any source of accidental ignition in the area identified as the origin.

Claims of spoliation seeking monetary damages from private-sector investigators for losing evidence are common. Similar claims seeking damages from public agencies for destroying evidence are not common, but public sector investigators should not dismiss the idea of preserving evidence for the civil investigation. Running over a fire scene with a bobcat or backhoe to look for "pour patterns" or spalling, particularly if no such "evidence" is uncovered, makes it next to impossible to discover the real cause of the fire.

4.11 ORIGIN DETERMINATION

It is simple enough to say, "Develop a hypothesis regarding the origin of the fire." Actually doing so is more problematic. The determination of the origin of a fire is a fire investigator's most important and difficult task. If one does not find the correct origin, it is practically impossible to find the correct cause.[15] Conversely, if one is able to correctly determine the origin, it is frequently not difficult to find the cause. NFPA 921 provides a long list of factors that should be "considered" when trying to locate the origin, but the guidance is necessarily incomplete. What do you do after you consider the effects of ventilation? Other texts offer advice on fire pattern interpretation, for example, to examine the beveled edge of a hole to determine which way a fire came through a wall or a floor. What those sketches show is the direction the fire was moving when it *last* penetrated the floor or wall, which is sometimes all we can hope for.

Rooms that have become fully involved require special attention. The fire patterns in these rooms are more likely to have been caused by ventilation than by interaction of fuel packages and compartment boundaries with the fire plume. It is likely that the origin is located in the first room to have flashed over, and this is why witness statements about the progress of the fire can be so important. Having determined that a room flashed over (or became fully involved), the investigator needs to start identifying patterns that were caused by ventilation. It is only those patterns *not* caused by ventilation that may lead the investigator to the origin (Figure 4.5). Because of the sometimes random nature of ventilation patterns, identifying patterns caused by ventilation may not be simple. It is also possible that a ventilation pattern has been laid down directly on top of the fire pattern produced by the first fuel package that was burning.

What this means is that, in the absence of definitive evidence, the investigator should not narrow the area of origin in a fully involved compartment to an area smaller than the entire compartment. The experiments conducted by Carman and his colleagues demonstrate that relying on the "lowest and deepest char" results in erroneous origin determinations. It is no longer valid (and apparently never was valid) to opine that a potential ignition source in a fully involved compartment could be ruled out because it was "outside the area of origin."

One clue that a hypothesized origin is really a ventilation-generated pattern is the absence of any evidence of arcing on exposed electrical wires. Exposed energized wires, particularly power cords, near an origin *should* exhibit arcing damage. The branch circuit wiring directly below the ventilation-generated V-pattern shown in the previous chapter (Figure 3.29) exhibited no evidence of electrical activity.

[15] Except when the point of origin is academic. If a structure is thoroughly doused with gasoline, nobody usually cares where the bad guy was standing when he dropped the match. In cases of fuel gas explosions, if one finds the source of the leaking fuel, the cause is determined. The ignition source may not be important. Certainly there was one, but whether it was a refrigerator compressor switch or a light switch is usually academic.

Figure 4.5 (a) Rectangular base of a fire pattern caused by a "drop-down" fire behind the wall.

Figure 4.5 (b) The cross-member stopped the hot coals and the coals produced the pattern.

In 2013, Andrew Cox published a method of studying fire patterns designed to weed out those patterns generated late in the fire's progress. This technique, called origin matrix analysis, involves dividing a fully involved compartment into segments, and considering where in the compartment the best air supply existed. A pattern adjacent to an opening or directly across the room from the opening at floor level is likely the result of the enhanced air supply. This pattern may be the lowest and deepest pattern in the room, but it may have absolutely nothing to do with the area of origin. Figure 4.6 shows the concept graphically. No matter where the fire started, if it burns for more than a few minutes beyond flashover, the fire patterns will be the same. If the wall covering falls down, there will likely be no meaningful fire patterns at all.

What this means for the fire investigator is that, under circumstances of prolonged, fully involved burning, it is not valid to narrow the origin to any volume smaller than the entire compartment. Therefore, **every potential ignition source and every potential first fuel ignited in that compartment must be considered and documented**. This daunting task must be accomplished if the investigator hopes to avoid calling the cause undetermined (though doing the analysis may not yield definitive results). One approach to documenting the search for ignition sources and first fuels has been proposed by Bilancia. Using the Bilancia Ignition Matrix™, potential ignition sources are listed as column headings, and potential first fuels are listed in rows under each column. Preparing this matrix is actually a straightforward exercise using the "Table" feature of MSWord. This produces an exhaustive pair-wise comparison designed to systematically evaluate numerous ignition sources and document how each was or was not competent to ignite a particular first fuel.

At each cell, the investigator notes:

1. Is this ignition source competent to ignite this fuel?
2. Is this ignition source close enough to this fuel to be capable of igniting it?
3. Is there evidence of ignition?
4. Is there a means for a fire ignited in this fuel to ignite the main fuel?

Actually, if the answer to any of the four questions is "no," there is no need to answer the other questions. Most pairs can be eliminated because the proposed ignition source and the proposed first fuel are too far apart. An example of a matrix table is shown in Figure 4.7.

This approach encourages the investigator to consider a comprehensive range of alternative hypotheses and consider each one on the basis of factors such as: heat release rate, heat flux, separation distances, thermal inertia, and routes of fire spread. The completed matrix also documents the thoroughness of the investigation and shows compliance with ASTM E678, *Standard Practice for the Evaluation of Scientific or Technical Data.*

Some additional factors to consider when looking for the origin are the artifacts produced by the confinement of a hot gas layer. If a smoke horizon or a heat horizon appears on a wall, in all likelihood, that artifact was recorded on the wall *before* the ceiling in that room failed. This can be very useful, at least when it comes to placing the origin on the correct level of a structure. The absence of evidence of confinement can be almost as useful. Fires that originate in the attic frequently cause ventilation of the roof before the ceiling fails, so there is no confinement. Lightning fires, like those described in Chapter 7, frequently ignite roofs and attics, resulting in characteristic damage.

Structural framing also causes fire to penetrate walls and floors in predictable ways, and the artifacts of a surface penetration can help identify which side of the surface was involved first. If a fire attacks a floor from above, the fire is not going to be influenced by the joists under the floor. If it attacks the floor from below, then frequently the holes in the floor will be rectangular, with the joists directing the flow of hot gases. Beware, however, of the unburned fuel

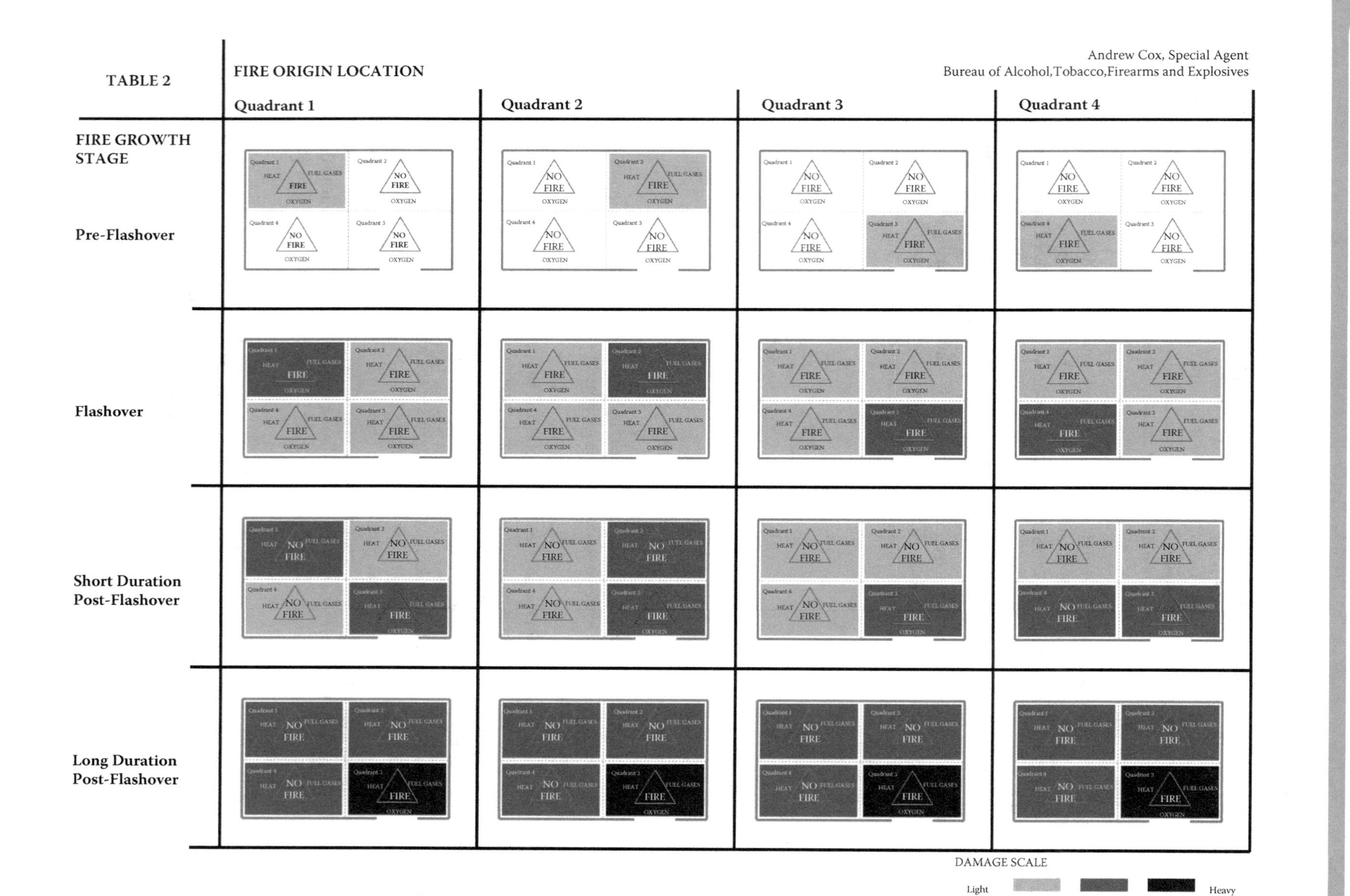

Figure 4.6 This table shows the expected fire patterns in four different fires, with origins in each of the four quadrants. For long-duration, post-flashover fires (fires that burn fully involved for more than 3 minutes) the expected fire patterns are all alike.

	Ignition Source #1	Ignition Source #2	Ignition Source #3	Ignition Source #4
First Fuel #1	1. competent? 2. close? 3. ignition? 4. spread?	1. competent? 2. close? 3. ignition? 4. spread?	1. competent? 2. close? 3. ignition? 4. spread?	1. competent? 2. close? 3. ignition? 4. spread?
First Fuel #2	1. competent? 2. close? 3. ignition? 4. spread?	1. competent? 2. close? 3. ignition? 4. spread?	1. competent? 2. close? 3. ignition? 4. spread?	1. competent? 2. close? 3. ignition? 4. spread?
First Fuel #3	1. competent? 2. close? 3. ignition? 4. spread?	1. competent? 2. close? 3. ignition? 4. spread?	1. competent? 2. close? 3. ignition? 4. spread?	1. competent? 2. close? 3. ignition? 4. spread?
First Fuel #4	1. competent? 2. close? 3. ignition? 4. spread?	1. competent? 2. close? 3. ignition? 4. spread?	1. competent? 2. close? 3. ignition? 4. spread?	1. competent? 2. close? 3. ignition? 4. spread?

Figure 4.7 Table used to document pairwise consideration of ignition sources and first fuels.

that falls through a hole from above, then continues to burn upward from under the floor. Wall studs can also direct or contain the fire. Straight up and down holes or patterns in the drywall, with edges coincident with the wall studs, indicate the fire came through from the other side. Figure 4.5 shows such a pattern. Frequently, such determinations are no-brainers, but it is easy to be misled by other factors, such as the lowest burn or heaviest burn. That is when one needs to consider the effects of ventilation.

Other times, the smoke and heat horizons might be the only evidence remaining. In the case where a structure collapses and burns in the basement, one does not usually expect to see a smoke or a heat horizon on the foundation walls unless the fire started in the basement.[16] If the fire started upstairs and burned its way into the basement, there would have been a hole where the fire penetrated downward, thus preventing the formation of a hot gas layer there.

If there are humans, particularly dead humans, in the structure, their activities must be considered. The individual in Figure 4.8 caught his shirt on fire in the kitchen but did not remain there. His presence in the hallway in the area of deepest and heaviest burning, and in an area where there were no potential sources of accidental ignition, could have triggered a major criminal investigation. Fortunately, an astute investigator discovered the pan of burned food in the kitchen.

Examination of damage to the electrical system is frequently useful in determining the origin or in testing a hypothesis about the origin. An electrical circuit has the capacity to act as a fire detector. When the insulation on an electrical circuit is compromised by a fire, several things can happen. The carbonized insulation becomes conductive, and arcing through char can occur. This may or may not sever the circuit. Frequently, the current meets sufficient resistance to flow that it does not trip the circuit breaker, and the circuit can be attacked in another area, either upstream (toward the service panel) or downstream (away from the service panel). When the arc causes the wire to sever, current flow downstream ceases and the point of arc severing furthest downstream can be said to have preceded any other arc severing on that circuit. Usually, the fire department or power company disconnects power to the structure sometime during the fire, and any electrical activity observed necessarily occurred before that happened. Sometimes the fire itself cuts off the power when it burns the service drop or causes the service entrance cable to arc to the case of the distribution panel. In a fire scene that exhibits electrical damage in only one

[16] What one "expects" to find is based on one's experience and hopefully on some science and logic. It is the proper "calibration" of one's expectations that allows one to formulate credible hypotheses.

Figure 4.8 The only significant fire damage in this residence was in the immediate vicinity of the body. The victim's clothing was ignited by a small cooking fire that self-extinguished. (Courtesy of Mick Gardiner, Gardiner Associates, Stanmore, Middlesex, UK.)

part of the structure, the search for the origin should focus on that area. Electrical arc damage is sometimes the only kind of artifact that survives a fire.

4.12 EVIDENCE COLLECTION AND PRESERVATION

Fire investigators collect several types of evidence. The most common are samples of flooring, furniture, and other materials suspected of containing ignitable liquid residues. Ignition sources comprise another type of evidence collected to determine their possible involvement in causing the fire. Evidence such as fuel packages or interior finishes may also be collected to test for flammability. Components of fire protection systems such as smoke alarms and sprinkler heads may require laboratory examination. As with other facets of the investigation, evidence collection is more successful if the fire investigator has a plan. If there is more than one investigator on the scene, all should agree to the plan for evidence collection.

With respect to samples collected for ignitable liquid residue analysis, the samples least likely to contain residues should, if possible, be collected first. This minimizes the risk of cross-contamination. It is strongly recommended that comparison samples be collected and that they be collected prior to collection of evidence samples. Select materials that are identical, or nearly identical, to the materials suspected of containing ignitable liquid residue. If the compartment has become flooded, collection of a comparison sample is absolutely imperative. A recent study has shown that extinguishment water has the ability to spread ignitable liquids across a room, potentially from an area where the presence is usual and incidental to an area that has no innocent explanation [17]. When sampling an irregular pattern suspected of being caused by a flammable liquid, research has shown that the center of the pattern is more likely to test positive for ignitable liquid residue than the edges. The liquid itself actually keeps the center of the pattern cooler, and fire damage tends to spread outward from the area where the liquid is located [18].

Label the evidence containers with the following information:

- Unique case number or file number identifying the fire scene
- Sample's serial number
- Description of the substrate material
- Location of the sample
- Date collected
- Initials of the investigator collecting the sample

The labeled containers should be placed at the location where the sample will be collected, and they should be photographed in place before the sample is placed in the can. A second photograph should be taken showing the sample in the can next to the former location of the sample. The location of the sample is the single most important attribute of the sample, so it is important that this information is thoroughly documented. When collecting samples for ignitable liquid residue analysis, the investigator should wear disposable gloves and change gloves between each sample. The gloves should be left behind at the sample location and not placed in the sample container.[17]

Some investigators prefer metal cans and some prefer glass for the collection of samples for ignitable liquid residue analysis. Glass has the advantage of transparency and resistance to corrosion, but it breaks. Metal cans do not break, but in the time between the site inspection and any trial, they are likely to corrode. Corrosion can be avoided or delayed by the use of polyester-lined paint cans. The coating is opaque gray or opaque tan and is the kind used for water-based paints. Such a coating will not influence the laboratory analysis, but it is good practice to save one can from each batch purchased, in case it becomes necessary to prove this.

Polyester or nylon evidence bags made for the purpose are suitable fire debris containers, but their only real advantages are their light weight and small volume, which makes it possible for the investigator to carry a large number of them. Major disadvantages include difficulty in sealing, poor containment of wet samples, and susceptibility to punctures or tears. Most laboratories find it necessary to repackage evidence delivered in bags.

Evidence tape is not necessary unless one's agency has a policy that calls for its use. It is important that the container be sealed tightly, but tamper-evident tape is an extravagance (and in this author's view, a nuisance) in the context of a fire investigation. If someone is going to intentionally contaminate a sample, tamper-evident tape will not stop them. In any case, by the time such containers are shown to a jury, they will have been opened at least twice.

The location of any sample collected should be noted on the investigator's sketch. An evidence transmittal form (Figure 4.9) should be completed as soon as practical after collecting the samples. Samples should not be held longer than absolutely necessary prior to submission to the laboratory. While hand delivery of samples is thought to be

[17] It was once suggested that the investigator place the gloves inside the sample container as a means of documenting their use, but some gloves release extraneous compounds that interfere with laboratory analysis. To document the use of gloves, photograph each pair of gloves next to the sample.

APPLIED TECHNICAL SERVICES, INCORPORATED

1190 Atlanta Industrial Drive Marietta, GA 30066 (770) 423-1400 Fax (770) 424-6415 e-mail fire@atslab.com

EVIDENCE TRANSMITTAL

Date:________________ Verbal Report To: ________________

Phone #: ________________

Submitted By:________________ Written Report To: ________________

File #:________________ Invoice To: ________________

Insured:________________

Claim Number:________________

Date of Loss:________________

ATS Reference #________________

Description of Evidence: Container Size, Type of Material, Condition of Material (burned or unburned) — **Location Collected**

1.________________

2.________________

3.________________

4.________________

5.________________

6.________________

Special Instructions:________________

Chain of Evidence
(Signature Required)

From:	To:	Date:	Time:
From:	To:	Date:	Time:
From:	To:	Date:	Time:
From:	To:	Date:	Time:

ATS 800 (10/99)

Professional Engineers
Design • Consulting • Testing and Inspection
Members in AAFS, ACS, ASM, ASME, ASNT, ASQC, ASTM, AWS, FSCT, IAAI, NACE, NCSL, NFPA, SAFS
GEORGIA SOCIETY OF PROFESSIONAL ENGINEERS, NATIONAL SOCIETY OF PROFESSIONAL ENGINEERS

Figure 4.9 Evidence transmittal used for submission of fire debris samples to a private laboratory.

useful in maintaining the chain of custody, the use of a common carrier such as UPS or FedEx is a perfectly acceptable way to deliver samples to the laboratory. If hand delivery is likely to create a significant delay, then the use of a common carrier is preferable. In 44 years of practice, this author has never had a problem establishing a chain of custody when a common carrier was used.

In the context of a joint inspection, all parties should have the opportunity to request that certain items of evidence be preserved for later examination. The person responsible for collection of potential ignition sources or other evidence (usually the investigator retained by the property owner or their insurance carrier) should photograph each item in place and then label the item with a label or tag that contains all of the information listed earlier for ignitable liquid residue samples. Once all interested parties have had a chance to view the items in place, they should be carefully transported to a testing facility. Wrapping the items with plastic cling wrap, available at low cost at trailer rental facilities and other places, will prevent the loss of loose parts of appliances.

4.13 FATAL FIRES

With one major exception, determining the origin and cause of the fatal fire should involve the exact same steps as determining the origin and cause of a fire involving no fatalities. That exception is the information provided by the victim or victims. When possible, the victims should be examined in place, but this is frequently not possible. In every case, both an autopsy and a toxicology screen should be conducted. The toxicology screen should, at a minimum, include a thorough drug and alcohol screening, as well as tests for carboxyhemoglobin (COHb) and cyanide.

The autopsy reveals whether a victim was breathing during the fire. Soot particles can usually be identified deep in the airway and lungs. The tissues of victims with high levels of carbon monoxide in their blood assume a bright cherry red coloration. Victims found some distance from the origin are normally expected to exhibit high levels of carbon monoxide, particularly in the case where the compartment of origin has become ventilation controlled. Because the fire becomes starved for oxygen, a fire's output of carbon monoxide increases dramatically in post-flashover burning. In well-ventilated fires, the level of CO produced may be as little as a few hundred parts per million (i.e., 0.02%). In underventilated, smoldering, or post-flashover fires, however, CO concentrations of 1% to 10% (10,000 to 100,000 ppm) can be produced. Elevated CO concentrations can also develop during fire suppression [19].

Interpreting carbon monoxide levels requires some knowledge of the victim's lifestyle and exposure to medical treatment. Heavy smokers may have COHb levels as high as 15%, while nonsmokers typically have levels on the order of 1%. Because of its stability, carbon monoxide can be measured in victims days or weeks after death. If victims were removed from the fire prior to death, they may have been provided with oxygen, which would tend to lower the CO level in the blood.

While victims remote from the origin typically exhibit high concentrations of COHb, it does not necessarily follow that victims close to the origin have lower levels of COHb. A recent study by Hill, involving 85 victims, revealed that COHb levels were not a reliable indicator of the victim's location [20]. Figure 4.10 shows a comparison of victim location versus COHb concentration.

Finding that a victim was not alive during the fire raises serious questions regarding the cause and manner of death. Toxicology findings should be double-checked, preferably retested using a different method of chemical analysis in the case of an "unexpected" result. Low COHb is a classic unexpected result in a victim found in a fire scene.

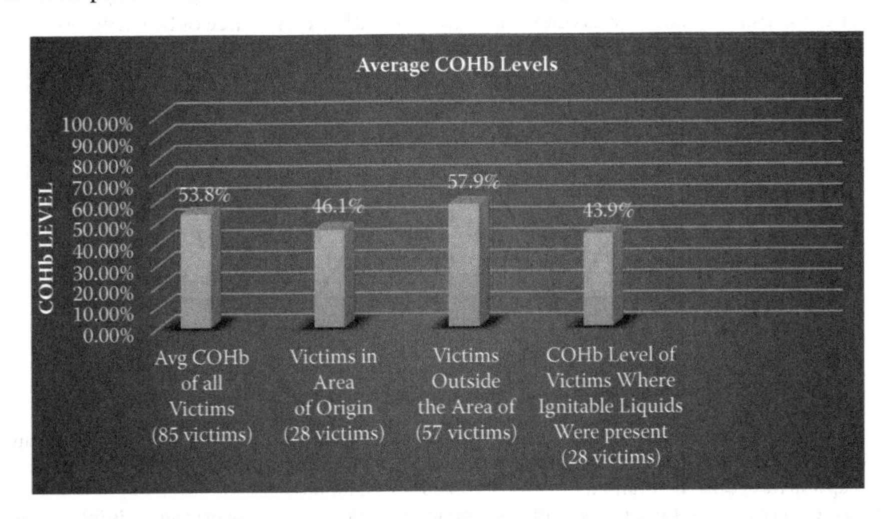

Figure 4.10 Comparison of the COHb concentration in the blood of 85 fire victims compared to their location.

Laryngospasm, the reflexive muscular contraction of the vocal folds, is one possible reason for low COHb and negative findings of soot below the larynx, but this is a difficult diagnosis. Laryngospasm also requires an explanation of a rapid exposure to extreme heat.

Fire investigators should bear in mind that the role of the medical examiner (ME) is to determine cause and manner of death. The ME uses his or her extensive education, training and experience to determine cause of death but frequently relies on the fire investigator for manner of death. If the cause of death is determined to be smoke inhalation, the manner of death may be accidental, suicide, or homicide, depending on the fire investigator's determination as to the fire's cause.

4.14 HYPOTHESIS DEVELOPMENT AND TESTING

This is the critical stage of the investigation. Using his or her education, training, and experience, and the facts as they have been developed thus far, the investigator is likely to have sufficient information to formulate a fire scenario that accounts for the condition of the scene. Because this step depends heavily on the individual investigator's experience, it is a subjective exercise, and because it is subjective, it requires further testing if it is to be believed. Investigators who are trying to follow the scientific method as outlined in NFPA 921 might object to this characterization because they will be basing their hypothesis "solely on the empirical data that the investigator has collected." But the development of the hypothesis is, nonetheless, a subjective exercise using inductive reasoning, which involves starting with a particular experience and making generalizations. For example, an investigator whose mother preferred mild cheddar cheese may never have been exposed to sharp cheddar and may think that all cheddars are mild. Inductive reasoning leads to probabilities, not certainties. An investigator who has never seen a cigarette thrown into a wastebasket cause a fire might believe it does not happen and thus is unlikely to formulate (much less test) such a hypothesis [21].[18] Likewise, an investigator who has only seen electric service panels with holes arced through the case as a result of fires is unlikely to formulate a hypothesis that such an arc was the cause of a fire.

It is not only hypothesis formation that is influenced by experience. Our ability to perceive data is constrained by our experience and our expectations. We have a tendency to perceive what we expect to perceive. What do you see when you look at Figure 4.11?

Now refer to the footnote[19] to learn what you actually saw. Just as we tend to see what we expect to perceive, we tend not to see what we do not expect to see. This phenomenon was illustrated by an experiment conducted by asking 24 radiologists to study 5 CT scans for lung nodules. In one of the scans, researchers inserted an image of a gorilla 48 times larger than the average nodule. Because they were looking for nodules, not the gorilla, 20 of the 24 participants failed to notice the gorilla. The study showed that even expert searchers, operating in their domain of expertise, are vulnerable to "inattentional blindness." [22]

The overall hypothesis about the cause of the fire may include numerous subhypotheses, the accuracy of which may or may not influence the overall picture. For example, it may be part of the investigator's hypothesis that a particular door was open during the fire. If the door was open, then there should be evidence of burning on the exposed surfaces

Figure 4.11 People see what they expect to see.

[18] The editors of *Fire Findings* conducted an experiment to determine whether a cigarette thrown into a trashcan with paper would start a fire, and it required 132 trials before the first ignition. Had they stopped after 100 trials, they could well have published a report stating that cigarettes will not ignite paper in trash cans. They ran a total of 300 tests and got 5 ignitions.

[19] The article is written twice in each of the three phrases. This is commonly overlooked because perception is influenced by our expectations about how these familiar phrases are normally written.

> Just as we tend to see what we expect to perceive, we tend not to see what we do not expect to see.

of the doorjamb and the hinges. If, on the other hand, the door was closed, then there should be mirror images, much like inkblots, on the mating surfaces, as shown in Figure 4.12.

Note that the hypothesis "test" described earlier involves only observation and logic. No laboratory test is required. For hypotheses that state, "That fire pattern was caused by a liquid accelerant," a laboratory test *is* required. Except in the most obvious cases (none of which occur in a fully involved compartment!), it is not possible to credibly state that an accelerant caused a particular fire pattern unless there is confirmation by a laboratory that follows ASTM methodology.

Hypothesis testing is more objective than hypothesis formulation. The deductive test of a hypothesis should allow the investigator to construct a sentence with the structure "*If* A is true, *then* B will be true." *If* my hypothesis is true, *then* I should observe the following data. *If* the door was closed, *then* there should be mirror image patterns on the mating surfaces. *If* the fire pattern was caused by a liquid accelerant, *then* the laboratory should be able to detect the residue thereof. Unless the hypothesis test can be cast into this structure, it is not a valid hypothesis test.

Figure 4.12 Mirror patterns on the mating surfaces of a door hinge, showing that the door was closed at the time of the fire. Part of the initial investigator's hypothesis was that the exterior door was left open intentionally to provide ventilation. All three hinges were found in the closed position, with screws still in place.

On the other hand, simply being able to formulate a test of a hypothesis by deductive reasoning does not necessarily mean that the hypothesis is true. A hypothesis based on a faulty premise will produce a valid conclusion, but not a true one. For example, an investigator who believes that spalling of concrete is an indication of the presence of liquid accelerants may test the hypothesis by stating, "If this was an accelerated fire, then I should find spalling on the floor." A subsequent finding of spalling proves nothing, but the investigator can claim he conducted a "scientific" test of his hypothesis. To be both valid and true, *the initial premise must be true.*

Hypothesis testing requires that the investigator compare his or her hypothesis with *all* the relevant credible data, as well as with known scientific facts. It also requires that the investigator be willing to abandon a hypothesis if it is disproved and to seriously question that hypothesis if the data are merely "consistent" with it. The data must be *uniquely* consistent.

The identification of the origin of the fire is a hypothesis. Beware of the trap of circular logic. Once an investigator has identified an origin, every potential ignition source outside that origin is automatically "eliminated" because only ignition sources within the area of origin could have started the fire.

In those cases where the origin of the fire is less "clearly defined," the absence of any potential heat sources in the proposed area of origin is *data.* The absence of a heat source is, in fact, contradictory to the hypothesis that this is the origin and should cause the investigator to carefully reexamine the hypothetical origin. Is there another area that could reasonably be considered the origin of the fire? Will an investigator retained by another party agree with the hypothetical origin? Certainly, in a fully involved compartment, every potential heat source within that compartment requires careful examination. Heavier damage on one side of the compartment or the other is almost certainly a result of more fuel or more ventilation rather than the longest burning time. If there is more than one fully involved compartment, there must be credible data to allow the investigator to decide which one burned first.

The concept of credibility of data is worth exploring. According to the scientific method, if there are credible data that contradict the hypothesis, it is necessary to discard the hypothesis and formulate a new one. In a fire scene, there will always be more data than can be assimilated into a given hypothesis, and many of these data are extraneous. Extraneous data need not be taken into account, but contradictory data cannot simply be ignored. The most critical type of data, ignored at the fire investigator's peril, is eyewitness accounts of the fire's progress. Regardless of the investigator's skill and experience at reading fire patterns, 10 eyewitnesses who first saw the fire in a room other than where the investigator thinks it originated are probably not the ones who are mistaken. "I've got to call it as I see it" is a common expression among fire investigators, but maybe they need to shift their own view just slightly so they can see things more clearly. Although Sherlock Holmes made many gossamer leaps of logic, he was right on the mark in *The Boscombe Valley Mystery* when he said, "Circumstantial evidence is a very tricky thing; it may seem to point very straight to one thing, but if you shift your own point of view a little, you may find it pointing in an equally uncompromising manner to something entirely different."

The circumstantial evidence that fire investigators use to draw inferences and formulate hypotheses is frequently subject to more than one interpretation. It is seldom as clear as the "footprints in the snow" analogy used to explain the concept of circumstantial evidence to jurors. Figure 4.13, which is taken from a window above a garage, shows how instructive some circumstantial evidence can be. The evidence in this photograph allows one to draw many inferences, and such determinations are at least as reliable as an eyewitness account would be. It snowed. There was a vehicle parked in the driveway before it started snowing. Two people (probably a man and a woman judging by the footprints) walked around the vehicle and one of them left. The smaller footprints go around the outline of the vehicle and return to the garage. After the vehicle parked in the driveway moved, a second vehicle backed out of the garage and drove away. Four turkeys walked up the driveway at some point, and close examination of the tire tracks might shed light upon when that was. Three deer crossed the driveway sometime after the first vehicle left. If fires left such easy-to-read artifacts, anyone could do this work.

The circumstantial evidence from which fire investigators draw inferences is seldom as clear as the footprints in the snow. Figure 4.14 shows a different kind of circumstantial evidence, which is more subject to drawing an incorrect

Figure 4.13 Footprints in the snow. This circumstantial evidence allows one to draw numerous inferences regarding what happened and in what sequence it happened.

inference. Figure 4.14a clearly shows a jet trail, but what about Figure 4.14b? Is that a jet trail or a cloud? What should one make of the fire pattern shown in Figure 4.15? Are those sharp, continuous, irregular lines of demarcation between the burned and unburned areas indicative of the burning of a puddle of liquid fuel? Or are they simply a result of protection provided by the shrinking of the carpet to cover certain places while exposing others?

Data that contradict an investigator's hypothesis about either the origin or the cause need to be taken seriously. They cannot be discarded merely because "they do not fit." Hypotheses—all of them—need to fit the **relevant** data.

At this point, the investigator should ask, "What data are relevant?" Because so many investigators are tasked with law enforcement as well as scientific duties, they are likely to be exposed to irrelevant data, or data that is not relevant to the task of determining origin and cause. Such *task-irrelevant data*, also known as domain-irrelevant data, has the potential to introduce expectation bias into an investigator's conclusions. Task-irrelevant bias is a problem with all forensic disciplines, not just fire investigation. It is generally not possible to overcome the cognitive bias introduced by task-irrelevant data by an act of will. Shielding the scene investigator from irrelevant data by using two investigators, one acting as a case manager, and one acting strictly as an interpreter of the scene evidence and eyewitness statements, may help to overcome any biases. This approach has worked well in several of the forensic disciplines. Table 4.1 shows examples of relevant and irrelevant data. As with any lists, there are neither all-inclusive nor true in every case.

Once an investigator has developed a testable hypothesis and then tested it by comparing it against all of the relevant data, a most difficult step remains. That is the examination of any other hypotheses that explain the data equally well. By this point, it is difficult to maintain an open mind, but be certain that anyone who has an interest in finding an alternate hypothesis will put one forward. Sagan described the problem with alternate hypotheses succinctly:

> "Try not to get overly attached to a hypothesis just because it's yours. It's only a way station in the pursuit of knowledge. Ask yourself why you like the idea. Compare it fairly with the alternatives. See if you can find reasons for rejecting it. If you don't others will." [23]

Figure 4.14 The top photo (a) contains circumstantial evidence. The jet trail in the sky allows one to infer that a jet has recently passed by. In (b), the evidence is less clear. This could be a jet trail but, then again, it might just be a cirrus cloud. (Courtesy of Norah Rudin, Mountain View, CA.)

Figure 4.15 Were these burn patterns caused by a burning liquid, or are they the result of alternate exposure and protection of the floor by the carpet?

Table 4.1 List of task-relevant versus task-irrelevant data

Relevant data	Irrelevant data
Firefighters' observations relevant to the fire, scene security, and suppression activities	Financial records
Witness observations relevant to the fire and building contents	History of fires
Occupancy	Criminal record
History of defects	Claim file
Weather data	Marital strife
Prefire activities on the scene	Social media commentary
Ignitable liquid location	Gossip
Physical condition of the fire scene	Motive issues
Utilities	Financial strife
Victim injuries	House for sale—real estate activity
Security, detection, and alarm systems	Indications of deception or emotional state of victim
Overpressure damage	Personal records

Note: Task = Determining the origin and cause; compartmentalize information.

The investigator should be able to state that such alternate hypotheses were considered and be able to articulate exactly why those alternate hypotheses were discarded. In many cases where an investigator's findings are discredited, it is not because such findings are incorrect; it is because the investigator is unable to credibly articulate why his or her hypothesis is true and the alternate is not. Eliminating credible alternate hypotheses on the fly 2 years down the road is difficult. The task is made much easier by going through the exercise before selecting the final hypothesis. Consideration of hypotheses should be contemporaneously documented. Making a list of hypotheses, then recording data supporting and data contradicting the hypotheses, may help the investigator organize his or her thoughts and come to a decision as to which hypothesis best fits the data. It is the contradictory data that is likely to be the most helpful in this exercise.

Sometimes more than one hypothesis fits the data. In such cases, the investigator has the obligation to report on all credible hypotheses. Failing to do so is asking for trouble.

4.15 REPORTING PROCEDURE

The report may well be the single most important document generated during the fire investigation. It will be the document that prosecutors, defense attorneys, insurance adjusters, and other fire investigators look to in an effort to understand a fire investigator's procedures and opinions. In some jurisdictions, if the report does not address a particular issue, that issue will be out of bounds when the investigator testifies. If there is an error in the report, it will become the focus for cross-examination. If there is insufficient detail in the report, readers will not understand how an investigator reached his or her conclusions. The quality of the report frequently determines the fate of the case.

Reports can take many forms, ranging from a few comments at the end of a fire run report to a very detailed description, running a hundred pages or more. The format of any given report is usually governed by the requirements of the fire investigator's client, whether that is an insurance company, a police agency, a defense attorney, or a prosecutor. Some report formats are dictated by court rulings.

A routine fire investigation report should follow the format of a scientific report, containing a background section; a synopsis of important witness observations; a description of the investigator's observations, with a sketch and photographs to illustrate those observations; and a conclusion.

A convenient technique for "outlining" a report is to rename those images that will be used in the report and place them in a separate directory. Then number them in the order in which they will be discussed. With the sketch laying on the table and the images in order, the investigator is ready to produce a logical, well-ordered document.

The sketch should be a cleaned-up version of the graph paper sketch made during the site inspection. There are also several programs that allow the sketch to be produced on a tablet such as an iPad. There are many computer-aided drafting (CAD) programs that allow sketches to be imported into reports or emailed to interested persons. These range in cost from about $20 to $2000. A new feature found in some of the high-end programs allows the sketch to be imported directly into a computer-modeling program like FDS, saving a huge amount of time and effort. If the CAD output can be saved as a .dxf file containing 3DFACE entries, it can likely be imported directly into FDS using a conversion program available for free from NIST.gov.

ASTM E620 is the *Standard Practice for Reporting Opinions of Technical Experts* and generally applies to fire investigators. ASTM E620 requires that the report contain all facts that are pertinent to the opinion rendered and that there be an identification of those facts that are based on the investigator's observations, as opposed to those facts that the investigator has learned from other sources. It also requires that the report contain the logic and reasoning of the investigator whereby each of the opinions and conclusions were reached. There are times when it is appropriate to include a Discussion section in a report, particularly if there are alternate hypotheses that explain the condition of the scene. The Discussion section is the appropriate place to describe the relative merits of each of these alternate hypotheses.

Rule 26 of the *Federal Rules of Civil Procedure* sets down a very specific set of items to be included in the report of an expert in civil litigation. A Rule 26 report should first identify what the investigator was asked to do, what activities were undertaken in the investigation, and a list of all the materials that were reviewed prior to formulating the opinions to be expressed in the report. The next section of the report states the investigator's expert opinions and the bases for each of those opinions. In this section, it is best to divide the opinions into recognizable categories, such as origin, cause, code violations, responsibility, and so on. Subdividing the opinions will be particularly useful should a court agree with an adverse party that certain opinions are not appropriate or admissible at the time of trial. A Rule 26 report may simply reference a previously prepared report of a fire investigation as the basis for some of the opinions. Rule 26 reports are required to be signed by the expert.

Rule 26 next requires that the investigator provide a current resume, a list of all cases in which he or she has testified either at trial or by deposition in the preceding 4 years, and a list of all publications that the investigator has authored within the last 10 years. Finally, Rule 26 requires that the investigator disclose the amount of compensation received for working on the investigation.

The report requirements in criminal cases are only slightly less detailed. Federal Rule 16 states:

(G) *Expert Witnesses*. At the defendant's request, the government must give to the defendant a written summary of any testimony that the government intends to use under Rules 702, 703, or 705 of the Federal Rules of Evidence during its case-in-chief at trial. If the government requests discovery under subdivision (b)(1)(C)(ii) and the defendant complies, the government must, at the defendant's request, give to the defendant a written summary of testimony that the government intends to use under Rules 702, 703, or 705 of the Federal Rules of Evidence as evidence at trial on the issue of the defendant's mental condition. The summary provided under this subparagraph must describe the witness's opinions, the bases and reasons for those opinions, and the witness's qualifications.

Many states have requirements similar to the Federal Rules, and even in criminal cases, investigators who expect to be witnesses, either for the state or for the defense, may be required to prepare some kind of disclosure statement.

Sometimes the investigator's client requests no report. Sometimes not preparing a report is a bad idea because it is difficult to re-create one's thought processes two years down the road when the client calls back and states that now she or he does want a report. Many times, a client requests no report simply because the investigation has led to the conclusion that there will be no litigation forthcoming, the claim will be paid, and that will be the end of it. If this is the case, then no report is really necessary. On the other hand, if the investigation has the potential for evolving into litigation, the investigator would be wise to document his or her findings and thought processes in the form of a "memo to the file." When a report is later requested, it can be produced without going through the entire process of reviewing the data and formulating and testing hypotheses all over again. We write things down for two reasons: (1) to communicate with other people and (2) to help us remember. This second function of writing makes a memo to the file indispensable, even when the client requests that no report be written.

Sometimes a report deadline arrives before an investigator has been able to obtain or review all the relevant data. Even in such cases, it is frequently possible to arrive at certain opinions. In most cases, even if an investigator believes that all the data have been reviewed, it does not hurt to add a disclaimer to the effect that the investigator reserves the right to amend or supplement the report should additional information come to light.

Some investigators possess better writing skills than others. One should not allow readers to become distracted by spelling or grammatical errors in a report. If grammar is not your strong suit, have your report reviewed by someone with good grammar skills. Use simple declarative sentences. Avoid pretentious words. Say exactly what you mean, and have the report reviewed by a knowledgeable colleague. Communication is a difficult art form, and it is often impossible to critically review one's own writing after more than one or two drafts. If you know what you meant to say, that is how you will interpret what you wrote. A colleague will likely spot grammatical and typographical errors and unclear sentences on a first reading of your report. Sending out an unreviewed report is a prescription for miscommunication.

4.16 RECORD KEEPING

Each agency should have a file retention policy, and individuals within that agency should follow that policy. The policy should apply to entire files, however, and not to individual pieces of the file. If one is keeping a file on a particular fire investigation, one should keep all the data in that file. Fire investigators who destroy their notes should look forward to some uncomfortable questioning if the case goes to trial. The excuse that "it's our policy" is an unsatisfying one. The next question that arises is, "Why is that your policy?" In this author's opinion, the only reason to destroy notes or data is to deny information to a litigation adversary, hardly a reasonable justification. If documents are destroyed, people, particularly adverse parties, will infer that there was something in those documents that the person who destroyed them did not want revealed, whether or not that is the case. Arguments that the documents take up too much space or are dirty or unclear simply give rise to the suspicion that somebody is hiding something. (Scientists do not destroy their lab notes!) A few pages of notes are not going to be too large to store. If they are soiled, they can be placed in an envelope or a sheet protector, or photographed with a digital camera, or electronically scanned. It is simply easier to explain a notation than it is to explain why the sheet on which that notation was made was destroyed. Fire investigators necessarily work in the adversarial system of justice, and destroying evidence or notes generally allows one's adversaries to draw a negative inference about the reason for that destruction. The best policy is simply to maintain all case materials until all litigation is finally adjudicated and the time for all appeals has passed.

SIDEBAR 4.1

What follows is a general procedure for the investigation of fires. Having a written procedure may be of use if an organization plans to pursue accreditation (see Chapter 10) and helps to ensure consistent work.

1. SCOPE
   This procedure covers the determination of the origin and cause of structural fires.
2. REFERENCED DOCUMENTS (current versions of referenced standards)
   ASTM E620 *Standard Practice for Reporting Opinions of Scientific or Technical Experts*
   ASTM E678 *Standard Practice for Evaluation of Scientific or Technical Data*
   ASTM E860 *Standard Practice for Examining and Preparing Items That Are or May Become Involved in Criminal or Civil Litigation*
   ASTM E1188 *Standard Practice for Collection and Preservation of Information and Physical Items by a Technical Investigator*
   ASTM E1459, *Guide for Physical Evidence Labeling and Related Documentation*
   ASTM E1492, *Practice for Receiving, Documenting, Storing, and Retrieving Evidence in a Forensic Science Laboratory*
   NFPA 921, *Guide for Fire and Explosion Investigations*
3. PURPOSE
   The purpose of this procedure is to provide a standard approach to the investigation of structural fires where issues of origin, cause, spread, and responsibility are to be determined.
4. PROCEDURE
   4.1 Receiving the assignment.
      4.1.1 When the assignment first comes in, record the following information:
         - Date of assignment
         - Client's name, address, phone number, and email
         - Client's file identifier
         - Identity of building owner and occupants
         - Building location
         - Date and time of fire
         - Name of fire department(s)
         - Identity of public agency and individuals who are investigating the fire
         - Identity of fire victims and eyewitnesses if any
         - Miscellaneous information relevant to the task of origin and cause determination
         4.1.1.1 In the event that any information listed earlier is unavailable, make a notation to obtain the information from the appropriate source.
   4.2 Planning the investigation. Attempt to ascertain from the client, fire officials, the insured, or others the extent of the damage to the structure, the size of the structure, and what will be required in terms of personnel and equipment to clear the debris and investigate the fire.
      4.2.1 Make contact, if possible, with fire officials and attempt to obtain a copy of any run reports or investigative reports generated by the fire department or other public authorities prior to visiting the fire scene.
         4.2.1.1 Advise the public officials when the fire scene inspection will be conducted, and request an opportunity to share information when appropriate.
      4.2.2 Attempt to learn whether there are any civilian eyewitnesses to the fire and make an attempt to meet with them or interview them via telephone.
      4.2.3 Attempt to learn whether any witnesses photographed or videotaped the fire in progress.
   4.3 Site documentation.
      4.3.1 Determine whether the scene can be investigated safely.
         4.3.1.1 Document any safety hazards, and formulate a plan for mitigating those hazards.
      4.3.2 Prior to disturbing any physical evidence, document the site photographically and on videotape.
         4.3.2.1 Verify that the date and time on the camera are accurate.
         4.3.2.2 Document the condition of the grounds as well as the condition and contents of any outbuildings on the site.
      4.3.3 Observe the condition of the neighborhood and attempt to evaluate any impediments to firefighting.
      4.3.4 Observe the utilities supplying the house, including gas, electricity, and water.
      4.3.5 Take exterior photographs of the structure from all directions, if possible.
         4.3.5.1 For a residential structure, this will usually require at least four photographs.
         4.3.5.2 Take sufficient photographs to cover the entire exterior of the structure.
         4.3.5.3 For large structures, attempt to obtain overhead views.
         4.3.5.4 If available, download images of prefire conditions using Google Earth, or similar service.
         4.3.5.5 For every ventilation opening, attempt to determine whether it functioned as an inlet, an exhaust, or both.
      4.3.6 Inside the structure, photograph every room.
         4.3.6.1 Start with the room where there is the least amount of damage and proceed toward the room where there is the greatest amount of damage.
            4.3.6.1.1 Each room should be documented from one angle, and two are preferred: one from the doorway and one looking back toward the doorway.
            4.3.6.1.2 Use a placard or other label to identify which room is being photographed.

4.3.6.2 In a room where a fire is believed to have originated, documentation showing each wall is required. Documentation of the floor and ceiling is also required.
4.3.6.3 In larger structures where the damage is limited to a small portion of the building, only the damaged part of the building needs to be photographed.
4.3.7 Document the electric service distribution panel, the furnace, the water heater, major kitchen appliances, any other heat-producing appliances. Document the positions of valves and switches, keeping in mind that with switches, the down position is not always off.
4.3.7.1 It may be necessary to remove debris prior to detailed documentation of the appliances. Document this process.
4.3.7.2 If the site has been previously investigated, document the evidence of such investigative activities (e.g., debris piles indicating that material has been moved, smoke and fire protection patterns showing the former location of objects, etc.).
4.3.8 Document the existence and condition of any fire alarms, smoke detectors, security alarms or cameras, and fire suppression systems.
4.3.8.1 Attempt to obtain records from any of the previous systems. This will usually require cooperation of the building owner.
4.4 Debris removal. In most cases, it is necessary to remove substantial quantities of debris to see where items of furniture and other contents were located, and to observe the condition of the floors.
4.4.1 Debris removal is generally better accomplished using flat-headed shovels than using mechanical equipment.
4.4.2 It may be necessary to enlist the aid of knowledgeable individuals (preferably off-duty firefighters) to effect debris removal.
4.4.3 Once debris has been removed, document all the cleared-off surfaces.
4.4.4 Attempt to place items of furniture or their remains back into their original locations, and document the reconstructed scene.
4.5 Sketching the scene. Prepare a rough scale drawing of the structure, including all exterior walls, and at least those interior walls surrounding the fire-damaged area.
4.5.1 A complete floor plan is preferable to a partial floor plan.
4.5.2 It may be necessary to draw a complete floor plan of the structure, then draw a more detailed floor plan of the room or rooms of origin, specifically indicating the position of furniture, fire patterns, or items of evidence collected.
4.5.3 Prepare the sketch using graph paper (10 × 10 squares on the graph paper are preferred). This may be accomplished using an electronic tablet if preferred.
4.5.4 Using a compass or a cell phone application, determine which direction is north and so indicate on the sketch.
4.5.5 Place an approximate scale on the sketch, sign it, and date it.
4.6 Avoiding spoliation.
4.6.1 If at any time during the inspection, it becomes apparent that a potentially defective product or an improperly performed service caused the fire, cease all activities that might alter the scene.
4.6.2 Take steps to ensure that potentially responsible parties are notified of their potential liability, and make arrangements to return to the scene with representatives of those parties.
4.6.3 Refer to the "Legal Considerations" chapter of the current edition of NFPA 921 for guidance on avoiding spoliation and notifying interested parties. See also ASTM E860.
4.7 Evidence collection.
4.7.1 Prior to collecting any evidence, carefully plan the items to be collected and indicate their locations on the sketch.
4.7.2 If samples for ignitable liquid residue analysis are to be collected into containers, place the appropriately labeled containers in the location and photograph them.
4.7.2.1 Collect comparison samples, if possible, so that the background contribution to sample residues can be evaluated, if necessary.
4.7.3 Other items of evidence to be collected should be photographed in place prior to collection.
4.7.3.1 It is understood that some of these items will have been moved during the debris removal process.
4.7.4 Refer to ASTM E1188, E1459, and E1492 for more detailed instructions regarding the collection of physical evidence from fire scenes.
4.7.4.1 It may be necessary to collect items eliminated as the cause of the fire.
4.8 Prior to leaving the scene, complete a version of the Fire Scene Field Notes form shown in Figure 4.2.
5. HYPOTHESIS FORMATION
5.1 Based on the observations made at the fire scene, as well as any pertinent background information made available, generate a hypothesis regarding the origin and cause of the fire.
5.1.1 Consider only that data that is relevant to determining origin and cause. Considerations of motive or opportunity should not be a part of hypothesis formulation at this stage of the investigation.
5.1.2 Refer to ASTM E678 for guidance on developing and documenting hypotheses. This standard requires that an investigator "Prepare and maintain a logical and traceable record of analysis and deduction."
5.1.2.1 Use of the Bilancia Ignition Matrix™ is a convenient and credible means of documenting hypotheses considered.

6. HYPOTHESIS TESTING
   - 6.1 Compare all the known data, including, if appropriate, the analysis of fire debris samples, the examination of physical evidence by other personnel, fire scene observations, and witness statements to the hypothesis.
   - 6.2 Determine whether the data support the hypothesis.
   - 6.3 In the event that there are data that contradict the hypothesis, examine the data for credibility.
     - 6.3.1 If the data that contradict the hypothesis appear credible, formulate a new hypothesis and begin the process again.
   - 6.4 If all the known credible data are supportive of the hypothesis, consider whether alternative hypotheses are also supported by these data.
     - 6.4.1 If more than one hypothesis is supported by the data, it will be necessary to report all credible hypotheses.
   - 6.5 Once all reasonable hypotheses have been formulated and tested, prepare a report of the investigation.
7. REPORT
   - 7.1 A report can take several forms, depending on customer requirements, the nature of the hypothesis to be reported, and the forum in which it will be presented.
     - 7.1.1 All reports should conform to ASTM E620, *Standard Practice for Reporting Opinions of Scientific or Technical Experts*
     - 7.1.2 Reports prepared for use in federal civil litigation must conform to the requirements of Rule 26.
   - 7.2 Unless there is no possibility of litigation, if the customer requests no report, prepare a "memo to the file," outlining the background, and all pertinent information needed to prepare a report at a later date, including the identification of necessary photographs.
   - 7.3 If the customer requests a letter report due to the fact that there will be no follow-up investigations or litigation, prepare a letter report briefly describing your opinions as to the origin, cause, and responsibility of the fire, and a description of the steps taken to reach that opinion.
     - 7.3.1 Generally, photographs will be provided along with the letter report.
   - 7.4 Most investigations require the preparation of a Test Report that follows a specific format.
     - 7.4.1 The subject of the report should include sufficient information for the reader to identify which fire the report is about. Typically, this includes the property owner's name, a claim or policy number, a loss location, and a date of loss.
     - 7.4.2 The background section of the report should describe the information supplied to you by others, including the client, the fire department, the property owner or tenant, and eyewitnesses to the fire. Other information that is suitable for inclusion in the background section includes the name of the individual reporting the fire, the time the fire was reported, the name of the fire department(s) responding, the date of your investigation, and the identities of any individuals who were present during the inspection.
     - 7.4.3 The summary of results section briefly describes your opinion as to the origin and cause of the fire, and may also include information regarding the responsibility for the fire.
     - 7.4.4 In the observations section of the report, describe the observations made during the investigation, starting with the exterior, and then moving to the interior from the area of least damage to the area of origin.
       - 7.4.4.1 Once the origin has been identified in the report, describe the potential sources of ignition, the potential first fuels ignited, and your opinion as to the cause of the fire.
     - 7.4.5 After the observations section, if necessary, include a section entitled "Laboratory Analysis of Fire Debris" or "Laboratory Examination of Electrical Equipment," and so on, and describe any tests, particularly destructive tests, performed at the laboratory or elsewhere. If there were any other individuals witnessing such tests, their identity belongs in this section of the report.
     - 7.4.6 A discussion section may be appropriate to add to a report when there is information obtained second- or third-hand, or if it is necessary to discuss alternate hypotheses for which there is some support. This section is only necessary if it can be somehow related to your direct observations
     - 7.4.7 A conclusion section should be a reiteration of the summary of results.
     - 7.4.8 Photographs should appear at the end of the report, along with a diagram of the scene.
       - 7.4.8.1 Typically, the diagram is shown as Figure 1, and exterior photographs are shown as Figures 2 through 6 or 8 or 10, depending on how many exterior photographs are necessary to show the entire exterior.
     - 7.4.9 Appendices can be added when it is necessary to provide a copy of relevant Code sections or transcripts of witness interviews, or fire department reports.
8. DOCUMENTATION

   Every completed fire investigation file should contain the Investigation Assignment; Fire Scene Field Notes; records of evidence transmittal or storage (if any); a copy of all data generated during any laboratory analyses; the original sketch drawn on the scene; digital images; and a copy of the final report, letter report, or memo to the file.

4.17 CONCLUSION

To be in a position to conduct a credible investigation, each agency should have a written procedure for investigations that references NFPA 921 and appropriate ASTM standards. Fire investigators need to understand the purpose of their investigation and should formulate a plan for each one. No investigation should be undertaken until the investigative team makes sure that it can be done safely and legally.

Documentation of observations is critical, both to help investigators communicate their findings and to help them remember. Investigations frequently need to be suspended to allow interested parties to be given the opportunity to see a fire scene before it is significantly disturbed. Reconstruction and additional documentation, possibly including an evaluation of the inventory, are required once the initial evaluation is completed. Hypothesis formulation and testing are the most important functions of the investigator. Origin determination requires the formulation of a hypothesis that is validated by finding the cause within the area of origin. Finding no cause tends to invalidate the hypothetical origin and, in such a case, the investigator needs to reevaluate the origin determination. Reports can take many forms. They are used to communicate results and memorialize findings. Most reports should follow the basic outline of a scientific report so that the reader understands both the investigator's observations and his or her reasoning. Record keeping should be consistent, and destruction of documents should be avoided.

Review questions

1. Assuming he or she has the authority to investigate, and once he or she arrives on the scene, the **first** question a fire investigator should ask about a fire scene is which of the following?
   a. Were there any eyewitnesses?
   b. What did the first firefighter see on arrival?
   c. How many compartments flashed over?
   d. Can I conduct this investigation safely?
2. Which of the following activities constitute spoliation of evidence??
   a. Debris removal
   b. Unnecessary disassembly of an appliance
   c. Failing to protect potential alternate ignition sources
   d. Evidence disassembly to identify the manufacturer
      i. I, II, III, and IV
      ii. II and III only
      iii. II only
      iv. II, III, and IV
3. Which of the following cases or statutes protects public agencies from being sued for spoliation of evidence?
   a. NFPA 921
   b. *Michigan v. Clifford*
   c. The state fire marshal enabling statute
   d. No such protection exists
4. When collecting evidence samples for ignitable liquid residue testing, which attribute of the sample is the most important to document?
   a. Sample composition
   b. Sample location
   c. Sample size
   d. Use of gloves to collect the sample
5. Notes from a fire scene investigation be kept for how long?
   a. Until the report is written
   b. Until the report is approved
   c. Until the case either goes to trial or is dismissed
   d. Until all litigation is over

Questions for discussion

1. Why is negative corpus reasoning inconsistent with the scientific method?
2. Describe the kinds of documentation necessary for a fire scene investigation.
3. In preparing an origin and cause report, which facts are relevant and which are not?
4. What are the major features of spoliation of evidence? Why do the courts impose sanctions for spoliation?
5. Why is hypothesis development (as opposed to hypothesis testing) a subjective exercise?

References

1. NFPA 921 (2017) *Guide for Fire and Explosion Investigations*, National Fire Protection Association, Quincy, MA, §4.3.8, 20.
2. NFPA 921 (2017) *Guide for Fire and Explosion Investigations*, National Fire Protection Association, Quincy, MA, §24.1, 256.
3. Inman, K., and Rudin, N. (2000) *Principles and Practice of Criminalistics: The Profession of Forensic Science*, CRC Press, Boca Raton, FL, p. 5.
4. Stevenson, A., and Lindberg, C. A. Eds. (2010) *The New Oxford American Dictionary*, 3rd ed., Oxford University Press, New York.
5. *Michigan v. Tyler*, 436 U.S. 499, 1978.
6. *Michigan v. Clifford*, 464 U.S. 287, 1984.
7. Samuel A. Mann (1984) Criminal Procedure—The Role of the Search Warrant in Fire Investigations—*Michigan v. Clifford*, *7 Campbell Law Review*, 269. Available at https://scholarship.law.campbell.edu/cgi/viewcontent.cgi?referer=&httpsredir=1&article=1107&context=clr (last visited January 15, 2018).
8. NFPA 921 (2008) *Guide for Fire and Explosion Investigations*, National Fire Protection Association, Quincy, MA, §18.2, 156.
9. NFPA 921 (2011) *Guide for Fire and Explosion Investigations*, National Fire Protection Association, Quincy, MA, 2011, §18.6.5, 174.
10. NFPA 921 (2017) *Guide for Fire and Explosion Investigations*, National Fire Protection Association, Quincy, MA, §19.6.5, 220.
11. See the court's ruling in *Wisconsin v. Joseph Awe*, in the Circuit Court in and for Marquette County, NO. 07-CF-54, March 21, 2013. Available at http://www.stephenmeyerlaw.com/img/StateofWisconsin_v_Joseph_Awe.pdf (last visited January 15, 2018).
12. IAAI, (2018) CFITrainer.net, Fire Investigator Scene Safety.
13. NFPA 921 (2017) *Guide for Fire and Explosion Investigations*, National Fire Protection Association, Quincy, MA, §18.3.2, 204.
14. *United Services Auto. Asso. v. Wade*, 544 So. 2d 906, 917, Ala., 1989. Available at https://www.courtlistener.com/opinion/1772158/united-services-auto-assn-v-wade/? (last visited January 15, 2018).
15. Court of Appeals of Minnesota (1991) *State of Minnesota, Appellant, v. Gary Bruce Tollefson, Respondent*, No. C9-91-283, Minn. App. LEXIS 737, July 17, 1991.
16. In the Court of Common Pleas in and for the County of Montgomery, Pennsylvania, Criminal Division, Commonwealth of Pennsylvania versus Michele Owen Black, (2014) No. 925-13, Trial transcript at Volume 2, page 193 (page 391 of 408).
17. Black, J., Gelman, J., and Kuk, R. (2016) The possibility of ignitable liquid contamination in flooded compartments, *Fire and Arson Investigator*, 67(1):26–31.

18. Putorti, A. (2000) Flammable and combustible liquid spill/burn patterns, NIJ Report 604-00, U.S. Department of Justice, Office of Justice Programs, National Institute of Justice. Available at: http://fire.nist.gov/bfrlpubs/fire01/PDF/f01023.pdf (last visited January 15, 2018).

19. NFPA 921 (2017) *Guide for Fire and Explosion Investigations*, National Fire Protection Association, Quincy, MA, §25.2.1, 263.

20. Hill, D. (2106) Fire victims and origin analysis, An ATF case study, *Fire and Arson Investigator*, 66(4):28–39.

21. Sanderson, J. (Ed.) (1998) Cigarette fires in paper trash, *Fire Findings*, 6 (1):1.

22. Drew, T., Vo, M., and Wolfe, M. (2013) The invisible gorilla strikes again: Sustained inattentional blindness in expert observers. *Psychological Science* 24(9):1848–1853.

23. Sagan, C. (1995) *The Demon Haunted World*, Random House, New York, 210. Available at http://www.metaphysicspirit.com/books/The%20Demon-Haunted%20World.pdf (last visited January 15, 2018).

Chapter 5

Analysis of ignitable liquid residues

The identification of an ignitable liquid residue in samples from a fire scene can support the field investigator's opinion regarding the origin, fuel load, and incendiary nature of the fire.

—ASTM E1618

LEARNING OBJECTIVES

After reviewing this chapter, the reader should be able to:

- Describe the steps taken in the laboratory analysis of fire debris
- Understand the development of fire debris separation and analysis techniques
- Recognize the output modes of a gas chromatograph-mass spectrometer
- Appreciate the necessity of comparison samples in many cases
- Be aware of the widespread distribution of petroleum products in our household environment

Readers with a college level understanding of organic chemistry should be able to:

- List the ASTM methods and practices related to fire debris analysis
- Select an appropriate separation technique depending on the form of the sample
- Understand the advantages and disadvantages of available elution solvents and internal standards
- Process, analyze, and determine the presence of ignitable liquid residues (ILRs) in fire debris samples using gas chromatography–mass spectrometry (GC-MS)
- Understand the limits of the analytical techniques, and know how to improve sensitivity if necessary
- Write a standard report describing the findings of the fire debris analysis

5.1 INTRODUCTION

The laboratory analysis of fire debris is one of the most important hypothesis tests that can be performed in an investigation, especially when the investigator forms a hypothesis that the fire was set using ignitable liquids. As a result of what has been learned in the past 20 years about fire pattern production, laboratory analysis is the only valid way to conclusively determine that an ignitable liquid was used to start a fire, at least in a compartment that has become fully involved. Even when the compartment is not fully involved and there is a fire pattern similar to the one shown in Figure 5.1, the investigator still needs to determine the identity of the ignitable liquid.

In the past, the laboratory analysis was referred to as "the icing on the cake" because by the time samples were collected, the fire investigator had already decided what caused the fire, and the stated purpose of the laboratory analysis was merely to help determine the identity of the flammable or combustible liquid used to start the fire. Investigators

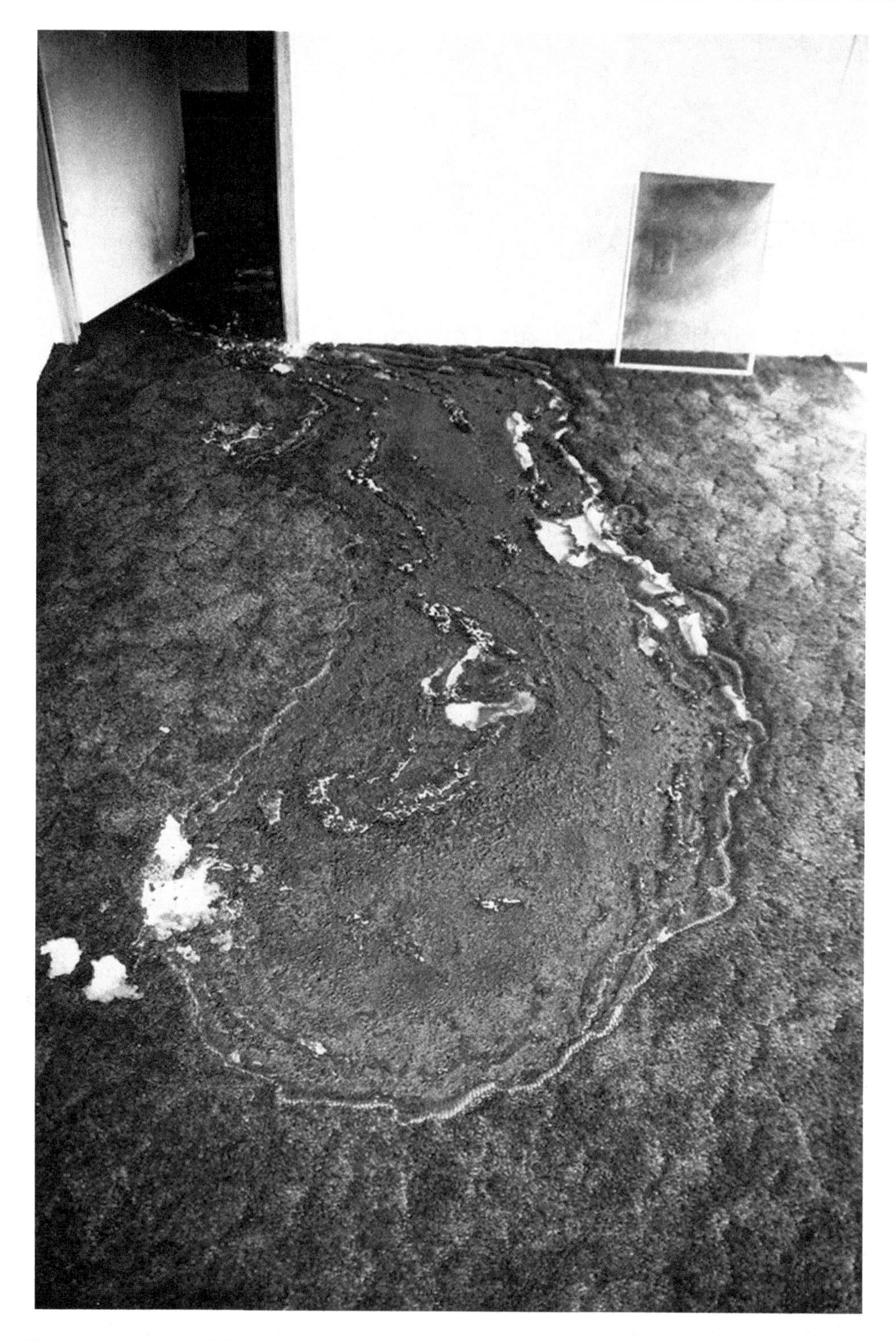

Figure 5.1 "Obvious pour pattern." The only surface that was burned in this mobile home was the floor. There were no furnishings in the house. The carpet was tested and found to be positive for the presence of a medium petroleum distillate such as mineral spirits or charcoal lighter fluid. This is one of the rare cases where visual observation alone can lead to valid conclusions about what caused the pattern.

were accustomed to receiving negative reports from their laboratory, even when they "knew" that a fire had been intentionally set with ignitable liquids.

In those days (prior to 1990), the term *ignitable liquid* had not yet been coined. "Flammable or combustible liquids" were generally referred to, even by people in the laboratory who had no clue how such liquids may have been used, as "accelerants." Many of the findings were, in fact, false negatives because laboratory methods were not sensitive enough to detect ILRs. The sensitivity has improved dramatically since this author began distilling fire debris in 1974, so much so that we can detect the petroleum products that are a natural part of our background. Many fire investigators still distrust negative reports from laboratories, however, based on this earlier experience.

False negatives were not the only problem with fire debris analysis. Some analysts, either because of a lack of skills or an eagerness to please their clients, found ILRs where they did not exist. Of the miscalled fires listed in the Dedication of this text, at least a third of them involved incorrect findings of the presence of an ILR, or an incorrect interpretation of the analytical findings. To this day, there are analysts who find an ILR, usually gasoline, which is not really there. It requires considerable skill and experience to differentiate between gasoline residues and polymer pyrolysis products, particularly when the concentration is low.

5.2 EVOLUTION OF SEPARATION TECHNIQUES

Paul Kirk reported in 1969 that the normal manner of isolating a liquid accelerant from other materials is "to distill the liquid from a solid residue in a current of steam." Kirk reported that the distillate could then be subjected to fractional distillation, flash point, refractive index, or density, but that a better procedure was to employ the gas–liquid chromatograph. Kirk stated, "From all of these laboratory procedures, the most important single piece of information that is made available is that a foreign flammable liquid was present at the fire scene. This alone is strong evidence for arson, at least after the possibility of accidental placing of the liquid is eliminated [1]."

Figure 5.2 shows a steam distillation apparatus. In this classical separation technique, the debris is covered with water and boiled, and the steam and other vapors are condensed in a trap that recycles the water and allows any nonmiscible oily liquids to float on top. There is actually a visible layer of liquid isolated from the sample. More often than not, this layer consisted of a drop or two, or simply a rainbow sheen on top of the column of water. This could be extracted with a solvent and analyzed, but even so, steam distillation was not a very sensitive technique. If the sample did not have a detectable petroleum odor, steam distillation was almost always ineffective at isolating any ignitable liquid.

In 1969, Kirk suggested that the debris could be heated in a closed container and the internal gaseous phase could be sampled and analyzed by gas chromatography, but he indicated that he was unaware of its use in routine analyses. By the mid-1970s, this technique, known as heated headspace and shown in Figure 5.3, was routinely used but had sensitivity limitations similar to steam distillation and was ineffective at isolating the heavier components of common combustible liquids such as diesel fuel. Heated headspace (Kirk called it a "shortcut") remains in use as a screening tool in some laboratories today.

The first quantum jump in sensitivity was reported in 1979 in the *Arson Analysis Newsletter*, when Joseph Chrostowski and Ronald Holmes reported on the collection and determination of accelerant vapors [2]. These two chemists, from the Bureau of Alcohol, Tobacco and Firearms' Philadelphia laboratory, employed a dry nitrogen purge and a vacuum pump to draw ignitable liquid vapors from a heated sample through a Pasteur pipette filled with activated coconut charcoal. The vapors were rinsed off the charcoal using carbon disulfide and analyzed by gas chromatography. Over the next decade, the apparatus for conducting this type of analysis, known as dynamic headspace concentration, resulted in the publication of many articles describing newer and more wonderful apparatuses. Dynamic headspace concentration remains a recognized analytical technique, but because it is both destructive and complicated, not many laboratories use it.

The case for *passive headspace concentration* (PHC) applied to fire debris, wherein an adsorbent package is placed in the sample container and heated, was made by John Juhala in 1982 [3]. Juhala used charcoal-coated copper wires

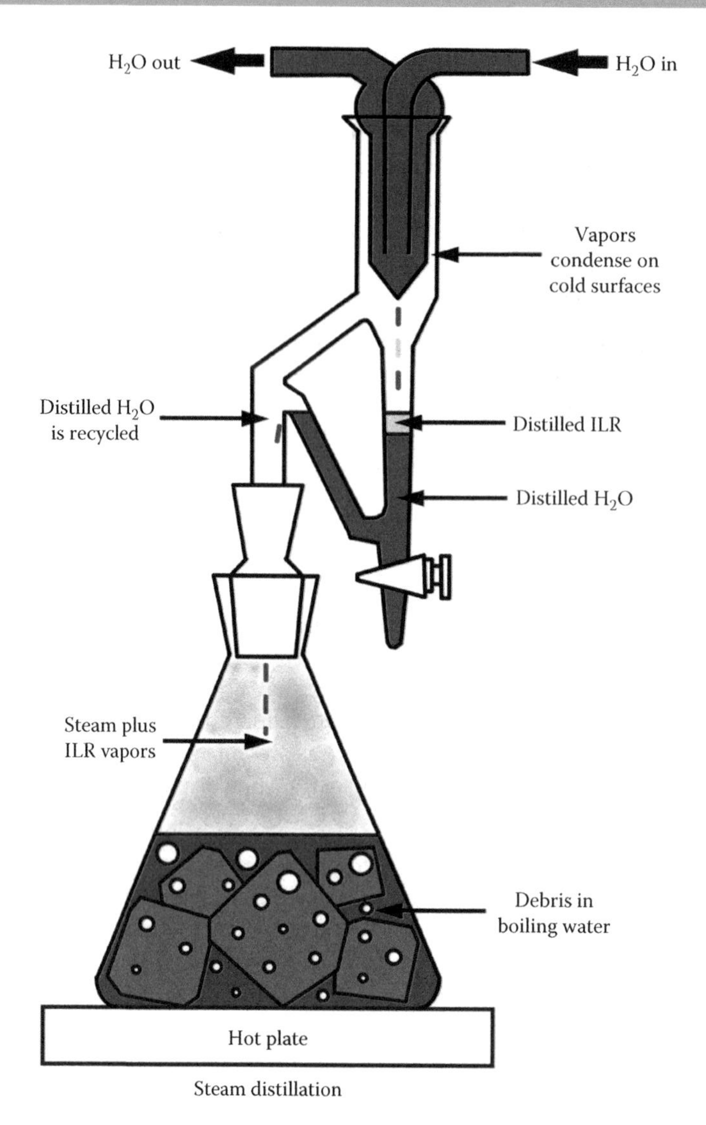

Figure 5.2 Steam distillation apparatus. The sample is boiled. Vapors condense on a "cold finger" and fall into the trap, which allows the water to recycle, while the immiscible oil layer builds up on top of the water column.

and Plexiglas beads. He reported an increase in sensitivity of two orders of magnitude over distillation and heated headspace analysis, but many laboratories had just completed setting up their dynamic systems. Consequently, adoption of passive techniques took some time, but gradually its advantages made it the dominant method of separation. Dietz [4] reported on an improved package for the adsorbent, called C-bags, in 1991, but these quickly gave way to activated carbon strips (ACSs), which required much less preparation. In 1993, Waters and Palmer [5] reported on the essentially nondestructive nature of ACS analysis, performing up to five consecutive analyses on the same sample with little discernible change, and no change in the ultimate classification of the residue. This separation technique is the method of choice in most laboratories today. Figure 5.4 is a conceptual drawing of the procedure for passive headspace concentration, and Figure 5.5 is a photograph of a typical adsorption device, which consists of a 10 × 10 mm square of finely divided activated charcoal, impregnated on a polytetrafluoroethylene (PTFE) strip. The technique was actually adapted from the industrial hygiene industry. Charcoal disk badges are worn by employees to determine their exposure to hazardous chemicals. The charcoal *adsorbs* a wide variety of organic compounds.

Figure 5.3 Using a gas-tight syringe to withdraw a sample of headspace from a fire debris sample. The headspace sample, about 500 μL, is injected directly into the GC-MS injection port.

Solid-phase microextraction (SPME) represents yet another kind of passive headspace concentration technique. The SPME fiber is a more active adsorber of most ignitable liquid residues than an activated carbon strip. Exposing an ACS to the headspace of a sample at elevated temperatures for 16 hours allows for the isolation of less than 0.1 μL of ignitable liquid residue if there is no competition from the substrate. An SPME fiber can accomplish the same feat in 20 minutes. The advantages and disadvantages of the various separation techniques are discussed later in this chapter.

Figure 5.4 Schematic drawing of passive headspace concentration using an activated carbon strip. Vapors are produced by heating the container with debris to 80°C. The ACS adsorbs the vapors for 16 h and then is rinsed with diethyl ether spiked with 100 ppm perchloroethylene, and the resulting solution is analyzed by GC-MS.

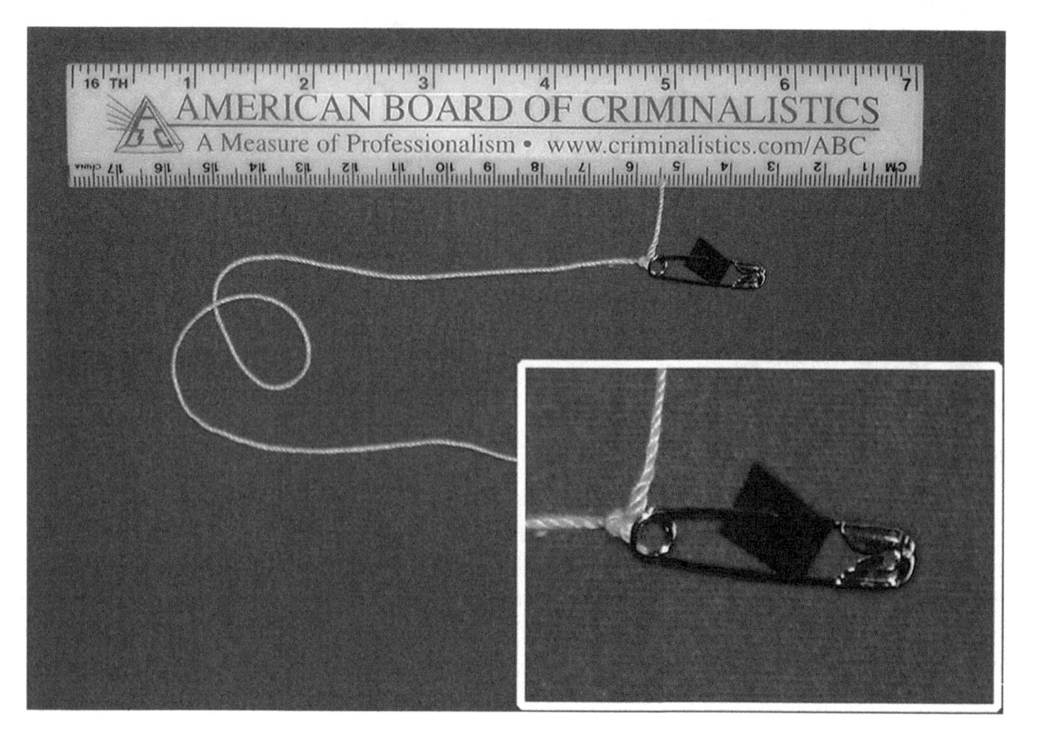

Figure 5.5 Close-up view of an activated carbon strip.

5.3 EVOLUTION OF ANALYTICAL TECHNIQUES

The development of analytical procedures has paralleled the development of standards for fire debris analysis. The first vague outline of a standard was not published until 1982 [6], and prior to that, analysts would report that a sample contained "an oily liquid" that had "sufficient similarities" to a known sample of gasoline or kerosene or diesel fuel.

As the ability of the separation step to isolate smaller and smaller quantities of ignitable liquid residue improved, the sensitivity of the analytical instruments also improved. In the 1950s and 1960s, extracts were analyzed by infrared (IR) or ultraviolet (UV) spectroscopy [7], but because most extracts were mixtures, these techniques were neither very sensitive nor very specific. The IR spectrum of gasoline looks very much like the IR spectrum of kerosene. Gas chromatography (GC) using pattern recognition techniques became the analytical method of choice beginning in the late 1960s. GC is actually a separation technique rather than an identification technique, but unlike separating the ignitable liquid residue from the sample matrix, GC works by separating similar compounds in an extract from each other.

In the 1970s, gas chromatography columns, the engine that makes the technique work, were glass or metal tubes, ¼ in. in diameter by 6 to 10 ft long. Chemists typically purchased empty columns and packed the columns themselves using a coated powdery substance (the stationary phase). It was known that ⅛-in. diameter columns provided better resolution than ¼-in. columns, but these had to be made from metal, and the chemist could not see inside the tube to check for gaps. These columns were usually purchased already packed. As the column manufacturers experimented with narrower and narrower columns, they went back to drawn glass tubes, coated on the inside with the oily stationary phase. There were problems with these early capillary columns, not the least of which was the forensic science community's resistance to change. As the bugs were worked out, capillary columns became the standard choice, but as late as 1990, packed columns were contemplated by ASTM E1387, *Standard Test Method for Flammable or Combustible Liquid Residues in Extracts from Fire Debris Samples by Gas Chromatography.* The standard "recommended" but did not require capillary columns in the 1995 edition. The 2001 edition requires "[a] capillary, bonded phase, methylsilicone or phenylmethylsilicone column or equivalent. Any column length or temperature program conditions can be used, provided that each component of the test mixture is adequately separated [8]." Very few laboratories use packed columns today.

GC detectors originally measured the change in thermal conductivity (TC) of the effluent from the column. Flame ionization detection (FID) improved the sensitivity by a couple of orders of magnitude over TC detectors. In a flame ionization detector, there is a hydrogen flame burning between two charged plates. The electrical conductivity, as indicated by the flow of current between the two plates, changes when a hydrocarbon compound comes through the hydrogen flame and is burned.

It is possible to look down the chimney of a flame ionization detector and actually see the invisible hydrogen flame turn yellow as the individual compounds are eluted from the column. This visible manifestation is the basis for the *flame photometric detector,* a device that measures the change in intensity of the visible light given off by the flame.

Even as early as 1976, some laboratories were using GC-MS [9]. At that time, mass spectrometers were expensive, were not terribly reliable, and required a computer (this was before the days of the PC, when mainframe computers took up half the room), and GC-MS was the exception rather than the rule. With the earlier instruments, the operator watched a strip chart recorder and pushed a button to take a mass spectrum when a peak indicated that a compound was coming off. This was a very labor-intensive process. There were people who argued that chemists had an obligation to use the best available technology, and the advantages of GC-MS over GC-FID required that MS be used. One of the leading proponents of GC-MS was Jack Nowicki, who also noted that GC-MS would make the previous accelerant classification system obsolete [10]. He was eventually proven correct. Most laboratories stayed with the FID methods because of the difficulties with implementing GC-MS and because they felt comfortable with their ability to read patterns using FID alone. By the early 1980s, mass spectrometry was still expensive, but its use had become more widespread in fire debris analysis, particularly in the better-funded laboratories. Public laboratories acquired GC-MS instruments for use in drug identification, and this was another reason that they became available for fire debris analysis. The instrumentation of the 1980s was more automated and would collect a mass spectrum several times per second, even if no peak was eluting. This resulted in a much more efficient process, but the data files were huge. Today's GC-MS, collecting data every tenth of a second, uses sophisticated software to keep the file size to around a megabyte, an amazing feat considering that 18,000 spectra can be collected during a single run.

In 1982, Martin Smith [11] published an article about a technique he had developed, called mass chromatography, which utilized a computer to separate the mass spectral signals according to the *functional groups* (e.g., alkanes, alkenes, aromatics, cycloalkanes, etc.) of the compounds that produced them. This technique allowed chemists to view many simple and easy-to-recognize patterns, as opposed to looking at one large complicated pattern. This powerful analytical tool is today known as extracted ion profiling or extracted ion chromatography and forms the basis of most identifications.

The development of the personal computer made it possible for average laboratories to control a mass spectrometer, and instrument manufacturers responded to the demand by producing bench-top models with increasing sensitivity and extraordinary robustness. The *quadrupole mass filter,* which is at the heart of the most popular mass selective detectors, has no moving parts and does not break.

5.4 EVOLUTION OF STANDARD METHODS

As the technology improved, so did the approach of the forensic science community to the problems of fire debris analysis. *Arson Analysis Newsletter* (*AAN*) continued publication through 1986, and much valuable information was exchanged. Through this informal journal, forensic scientists analyzing fire debris had a means of communicating with each other that was unavailable to scientists in many other disciplines. In 1982, based on work conducted at the Center for Fire Research at the National Bureau of Standards (NBS) and the Bureau of Alcohol, Tobacco and Firearms (ATF) National Laboratory, an "accelerant classification system" was published. Not only was there a description of five classes recognized as "usually identifiable by GC-FID patterning alone when recovered from fire debris," but also the authors (who were not named in the publication but are believed to include Dr. Philip Wineman and Dr. Mary Lou Fultz) published minimum requirements for class identification. This was the first time that anyone explained what "sufficient similarities" should mean and was a watershed moment in the history of fire debris analysis. Although the original publication of the classification system stated that the final report was not yet available and that the results of the evaluation would be printed in a future issue of *AAN,* that never happened. The classification system was used informally for the next 6 years.

The International Association of Arson Investigators (IAAI) has had, almost since its beginning, a standing committee of forensic scientists and engineers that it called upon to advise fire investigators about laboratory analysis issues. In 1987, IAAI president John Primrose approached the Forensic Science Committee and requested that it produce a position paper on what should appear in a laboratory report. It quickly became apparent to members of the committee that in order to prescribe the contents of a report, it would be necessary to first set down an acceptable method of analysis. Four sample preparation techniques (steam distillation, headspace analysis, solvent extraction, and dynamic headspace purge and trap) were the separation techniques described. Gas chromatography with flame ionization detection, photoionization, or mass spectral detection was required. Although the publication was called a "guideline," it contained the following sentence: "Unless a petroleum distillate has been identified by the pattern recognition techniques described in the following, it has not been sufficiently identified [12]." The guidelines then reproduced the NBS/ATF classification and identification scheme and described how some materials would not fall within the guidelines. Isoparaffinic hydrocarbons were cited as one example of an ignitable liquid not described in the guidelines. At that time, the only place that a fire debris chemist would likely encounter isoparaffinic hydrocarbons was in Gulf Lite® Charcoal Starter Fluid. With the introduction of the IAAI guidelines, fire debris chemists became accustomed to the idea that they needed to follow standard methods. In the private sector, laboratories advertised to the membership of the IAAI, their main source of clients, that they followed the guidelines. Laboratories that did not follow the guidelines lost business.

In 1990, ASTM Committee E30 on Forensic Sciences took the IAAI guidelines and formulated them into six different standards for the preparation, clean-up, and analysis of fire debris extracts. ASTM E1387 *Standard Test Method for Flammable or Combustible Liquid Residues in Extracts from Fire Debris Samples by Gas Chromatography* was originally published in 1990. In 1995, the cumbersome phrase "flammable or combustible liquid" was changed to "ignitable liquid," and the standard underwent several revisions, but it was allowed to expire in 2010 because it no longer represented the best practice. In 1993, Committee E30 enlisted the aid of Martin Smith, Jack Nowicki, and several other prominent chemists to draft E1618 *Guide for Fire Debris Analysis by Gas Chromatography–Mass Spectrometry.* The guide was revised in 1997, and in 2001, it was "promoted" to a standard test method. The most recent edition (as of January 2018) of E1618 is the 2014 edition. GC-MS now represents the best practice for ILR analysis.

The ignitable liquid classification scheme from 1982 was updated in an attempt to keep up with the ever-changing output of the petrochemical industry. New products were constantly being brought online, including "environmentally friendly" alternatives to solvents such as mineral spirits and fuels such as charcoal lighter fluid. When ASTM E1387 was first published, a "Class 0" was added to the original classification scheme to account for the liquids that did not fit into one of the five original classes. Further classification within Class 0 was possible, however, and so Classes 0.1 through 0.5 appeared in ASTM E1387-95. By the time the next revision was due, Class 0.6 had been created for de-aromatized distillates, but the committee realized it was time for a change, as the miscellaneous classes now outnumbered the original classes. In 2000, the system was completely revamped, resulting in nine different-named, but not numbered, classes, with subdivisions of light (C_4–C_9), medium (C_8–C_{13}), and heavy (C_8–C_{20+}) in eight of those nine classes [13].

The past 35 years have seen dramatic improvements in separation technology, in analytical technology, and in the scientific community's approach to fire debris analysis. In 1999, the U.S. Department of Justice Office of Law Enforcement Standards produced a report entitled "Forensic Sciences: Review of Status and Needs," compiled by more than 40 eminent forensic scientists. In reviewing the state of the art, fire debris analysis was described as a subdiscipline of trace analysis that is in good standing because there is sufficient published work on the analysis and interpretation of the material involved. Standard guides for the examination and interpretation of chemical residues in fire debris have been published through the consensus process of ASTM Committee E30 on Forensic Science. These standardization documents are often quoted in the scientific literature, helping to meet the requirements of the legal community [14].

Most of the other forensic disciplines discussed in this review were reported as still needing standardization and/or validation of standard methods. Fire debris analysts can point to a history of standardization that existed even before the *Daubert* court made it a necessity.

Further evolution of ASTM E1618 is likely, but the basic science is reasonably well settled. The OSAC Subcommittee on Fire Debris Analysis (as of 2018) was discussing proposing changes to E1618 by dividing it into four parts: instrumental quality assurance, a classification of ILRs, interpretation of data, and report writing [15]. Whether ASTM Committee E30 accepts these proposals remains to be seen. A working group within the American Association for the Advancement of Science (AAAS, the publisher of *Science* Magazine) stated in 2017, "[T]he Working Group believes the ASTM standard test methods for extraction, separation, and analysis of ILRs are sufficiently developed and mature and there is no reason for operational laboratories *not* to use these methods. All forensic practitioners should be made aware of these methods, should have access to them, and should be required to follow them if their analyses are to be admitted by courts [16]."

5.5 ISOLATING THE RESIDUE

5.5.1 Initial sample evaluation

Once the receipt of a sample has been documented and the chain of custody protected, the first critical step in the analysis of fire debris is the selection of a separation technique. Choosing an inappropriate technique could result in a false negative, a misidentification, or the destruction of evidence. The first step in this selection process (and the first step in any chemical analysis process) is *look at the sample.* One purpose of looking at the sample is to ensure that it is what it purports to be. Sample characteristics determine the most appropriate method for isolating any ignitable liquid residues that may be present. Once the visual examination has taken place, the next step is a "nasal appraisal." Occupational safety experts will no doubt frown on this recommendation, but it can be done carefully. There is no need to put one's nose in the can, although the analyst can be reasonably certain that the fire investigator who collected the sample has already done exactly that. Unless the sample is a liquid sample for comparison purposes, it can be safely appraised by removing the lid and waving the hand gently over the top of the sample to see if it exhibits any obvious odors. If there is an odor present, it becomes possible to do a rapid and accurate analysis by removing a small piece of the sample and extracting it with solvent. If the odor is very strong, it is advisable to remove a small piece of the sample and place it into a separate container for analysis.

5.5.2 Ignitable liquid residue isolation method selection

Solvent extraction, according to ASTM E1386, is an appropriate method for rinsing out empty containers, for extracting small aliquots of samples with a high concentration of ILR, and for isolating residues from very small samples. Not every investigator who comes to the laboratory has the experience and knowledge to know how to find the best samples, and it is not unusual for an inexperienced investigator to bring in a sample weighing only a few grams.

The vast majority of samples, however, are likely to be samples of burned building materials, floor coverings, and furnishings that do not exhibit a strong odor and are best analyzed by passive headspace concentration, as described in ASTM E1412. This technique is essentially nondestructive. If the analyst decides to use another technique later, such as looking for vegetable oils, running passive headspace concentration will not interfere with that.

Other methods of isolation have been studied thoroughly over the past 35 years, and while they have some utility, none match the advantages of passive headspace concentration using an activated carbon strip. Headspace sampling, which involves warming the container and sampling the vapors in the headspace directly and then injecting those samples into the gas chromatograph, is a good screening technique, but it does not result in the production of an archiveable extract, nor does it detect compounds much heavier than C_{15}. Dynamic headspace concentration, described in ASTM E1413, was useful in demonstrating the effectiveness of adsorption/elution as a valid approach to ILR isolation, but it is destructive and requires far more attention than passive headspace concentration, and the apparatus can be cumbersome and finicky. It is no more sensitive than passive headspace concentration, and it is possible for breakthrough (i.e., the loss of analyte out the effluent end of the tube) to occur. The only advantage that it offers is speed. Solid phase micro-extraction is another alternative, but it is very labor intensive and, like headspace sampling, does not have the potential to produce an archiveable sample that can be analyzed again. Using passive headspace concentration results in a solution that can be injected many times, and when the carbon strip is left in the solution, it will gradually readsorb ignitable liquid residues as the eluting solvent evaporates. (Juhala reported the readsorption by small portions of the activated charcoal that fell off his Plexiglas beads in 1982.) The solution can be reconstituted years later if a second look at the sample is desired. Because of the transient nature of many fire debris sample containers, the archived activated carbon strip is often the best evidence after a few years have passed.

Many European laboratories use a passive/dynamic technique using an adsorption medium known as *Tenax*, a kind of porous polymer. The sample is heated and an aliquot of the heated headspace (on the order of 100 mL) is removed and passed through a tube containing the Tenax. The tube is then subjected to rapid heating to thermally desorb the analyte, and inject it directly into the GC injection port. The method is reasonably rapid, and uses no solvent, but like SPME, the extract is not archiveable.

The only equipment required for passive headspace concentration is a convection oven, vials, ACSs, and a solvent dispenser. Caseload determines the required oven size. In our laboratory, the convection oven held up to 10 1-gallon cans and more than 20 1-quart cans.

Every laboratory should optimize the parameters in its ACS procedure to make sure it is getting the best results possible. "Good" results from ACS are those where the chromatogram of the concentrated headspace vapors of a standard closely matches the chromatogram of that same standard in the eluting solvent.

A 10×10 mm carbon strip is the minimum size recommended. This 100-mm^2 strip can easily accommodate the headspace vapors from 10 μL of any ignitable liquid placed on a Kimwipe in a quart can. It is possible to overload carbon strips, and this results in a preferential adsorption of heavier hydrocarbons over light hydrocarbons, and of aromatics over aliphatics, but this effect is generally not large enough to affect the identification. Usually, samples that are capable of overloading the carbon strip exhibit a strong odor, and the analyst can take an aliquot of the sample or reduce the analysis time.

For samples with a high water content, there exists a danger that the vapor pressure in the sample container will cause the lid to pop off. This has the potential for contaminating the sample oven. A "pressure relief device" is easy to construct for such samples. Puncture a small hole in the lid and cover with cellophane tape. Figure 5.6 shows a can so equipped.

The adsorption time of a typical ACS procedure is 16 h. The analyst is encouraged to experiment with different adsorption times and temperatures, with the goal of finding a balance between the maximum recovery and the

Figure 5.6 One-quart can equipped with a "pressure relief device," a short strip of cellophane tape over a small hole pierced in the lid.

minimum time necessary. Sixteen hours is convenient because the samples can be put in the oven at 4:00 p.m. and taken out at 8:00 a.m. the next day. One of the major advantages of ACS is that it requires very little attention from the analyst. Once the strip is in the can and the can is in the oven, nothing is going to happen until it is time to remove the strip from the can, put it in the vial, and add the eluting solvent. At this point in the analysis, however, the analyst must be extremely attentive to the procedure. Once the carbon strips are removed from the sample container, they have an identical appearance. At the point of placing the strips into their vials, an unrecoverable error can occur—a strip can be misidentified. The analyst should not allow him- or herself to be distracted by phone calls or other people in the lab. The operative word here is "focus."

Many laboratories use a preconcentration step, wherein they add approximately 500 μL of solvent to the strip in a vial; then, after the strip has had a chance to equilibrate, it is removed and the solvent is evaporated down to 100 μL or so. This results in a fivefold increase in concentration, but that increase can usually be achieved electronically with very little loss of signal or increase in noise. The first analysis, in this analyst's view, does not require the preconcentration step. That can be accomplished at a later time, if necessary, but preconcentration runs the risk of skewing the results if it is not done very carefully. The only safe way to evaporate the solvent is to pass a stream of dry nitrogen over it. Heating the solvent is a bad idea because lower molecular weight compounds are likely to evaporate with the solvent.

5.5.3 Solvent selection

The solvent used for the elution is another critical choice, not so much in terms of the quality of results but in terms of the analyst's quality of life. The most popular eluting solvent is carbon disulfide, a highly toxic, carcinogenic, teratogenic, smelly, nasty liquid that ignites upon exposure to boiling water. It does work very well to elute aromatics and aliphatics approximately equally from activated carbon strips, but so does diethyl ether. Studies indicate that carbon disulfide is superior to diethyl ether, or to pentane, the other solvent recommended by ASTM Committee E30 [17], but if standards are prepared using diethyl ether, the slight change in the chromatographic profile does not affect the identification. Lentini and Armstrong [18] found only marginal differences between diethyl ether and carbon disulfide when applied to carbon strips exposed to 10 μL samples of ignitable liquid residues. Comparing the relative health risks makes diethyl ether an obvious choice for this analysis.

Carbon disulfide was originally selected as a fire debris solvent because of its high desorption efficiency and its relatively quiet signal when passing through a flame ionization detector. When using a mass spectral detector, the advantage of its low signal disappears because the detector is turned off while the solvent is passing through.

Some concerns about the capability of diethyl ether to form explosive peroxides have been expressed, but that does not occur if the ether is kept in a refrigerator and used on a regular basis. Explosions of cans of ether have only been reported when those cans have been allowed to sit for years, unused, in the back of an unrefrigerated stockroom. Carbon disulfide, diethyl ether, and pentane are all highly flammable, but with respect to fire, carbon disulfide poses the greatest risk in that it has the lowest ignition temperature and the broadest flammable limits [19]. A comparison of the properties of the three solvents recommended by ASTM E1412 is shown in Table 5.1.

5.5.4 Internal standards

There are two places in the analysis of fire debris where the use of internal standards is appropriate. Addition of an internal standard to the sample itself allows the analyst to develop at least a qualitative feel for the "tenacity" of the sample and for the effectiveness of the isolation procedure. In our laboratory, this is accomplished by the addition of 0.5 μL of 3-phenyltoluene (actually, 20 μL of a 2.5% solution of 3-phenyltoluene in ether). In the eluting solvent, we use a second internal standard consisting of 0.01% (100 ppm) perchloroethylene. If the 3-phenyltoluene does not appear in the chromatogram, this means that we have an exceedingly tenacious sample and suggests that the tenacity of the sample might be the reason that the chromatogram appears so flat. If the perchloroethylene peak does not appear, it means something has gone wrong with the injection.

Comparison of the signal from the sample with the perchloroethylene signal allows for a semiquantitative determination of the amount of ILR present. For 10 μL standards of known ignitable liquids isolated according to ASTM E1412,

Table 5.1 Elution solvent comparison[a]

	Carbon disulfide	Diethyl ether	Pentane
Flash point	−22°F (−30°C)	−49°F (−45°C)	−40°F (−40°C)
LEL (% vol in air)	1.0	1.9	1.5
UEL (% vol in air)	50.0	48.0	7.8
Specific gravity	1.3	0.7	0.9
Boiling point	115°F (40°C)	95°F (35°C)	97°F (36°C)
Autoignition point	194°F (90°C)	356°F (180°C)	500°F (260°C)
Exposure limit, TWA[b]	4 ppm	400 ppm	600 ppm
Exposure limit, STEL[c]	12 ppm	500 ppm	750 ppm
Carcinogenic	Yes	No	No
Teratogenic	Yes	No	No
IDLH	500 ppm	19,000 ppm (LEL)	15,000 ppm (LEL)
FID signal	Small	Large	Very large
2017 cost per liter[d]	273.00	153.00	78.00

Note: IDLH = immediately dangerous to life and health; LEL = lower explosive limit.
[a] Refer to your laboratory's hazard communication literature or material safety data sheet (MSDS) for more complete information.
[b] Time weighted average for an 8-hour exposure.
[c] Short-term exposure limit (up to 15 min).
[d] Source VWR.com.

the two internal standard peaks are roughly the same order of magnitude as the sample peaks. Essential blanks include the 3-phenyltoluene applied to filter paper and the blank strip eluted with the spiked solvent. Some analysts may perceive the danger of being accused of "contaminating" a sample, but this is easily overcome by having a proper blank in the file. The advantages of using internal standards far outweigh this disadvantage.

5.5.5 Advantages and disadvantages of isolation methods

Two drawbacks have been cited for use of the ACS method: (1) the time required to perform the adsorption and (2) the relative lack of sensitivity compared to SPME. If laboratories were in the habit of providing same-day service, then the first argument might have some merit. Turnaround times generally range from 2 days to 2 months. In that context, a 16-hour versus a 15-minute analysis time is meaningless. With respect to sensitivity, the ACS method is capable of routinely detecting 0.1 μL of ignitable liquid from a nontenacious background, which should be low enough. Our goal is to help the fire investigator understand whether a *foreign* ignitable liquid was present at the fire scene. We now have the ability to detect the solvent in polyurethane finish 5 years or more after it has been applied to a hardwood floor. We have no need to be more sensitive than that. A technique that is capable of adsorbing significant quantities of ignitable liquid residue in 15 min is also capable of becoming contaminated much more easily than a carbon strip that might take 16 h to come to equilibrium with a dilute sample.

The use of a "screening" technique for fire debris samples is an issue that each laboratory needs to consider. There is usually no need to screen samples because an ACS separation is likely more sensitive. A sample that tests negative on a quick headspace analysis (per ASTM E1388) needs further testing anyway. If a request is made to look for light oxygenates (alcohols, acetone) or to get a ballpark estimate of analyte concentration (other than through a "nasal appraisal"), running a headspace analysis can be useful. Screening techniques also allow for a swift, if less than definitive, result. For routine analyses, however, passive headspace concentration, conducted according to ASTM E1412, should be the norm. Solvent extraction, as described in ASTM E1386, is appropriate for sampling aliquots of very strong samples or for extracting very small samples or empty containers.

Steam distillation might be selected in the odd case where one wants to produce a neat liquid extract of the fire debris. The benefit of this is that a vial of the liquid can be brought into a courtroom, shown to a jury, lit on fire on a Q-tip, and passed around. Because steam distillation is only appropriate on very concentrated samples, however, it is preferable to make sure that the sample is preserved, and the sample itself can be passed around for the jury to smell. For the most part, steam distillation is a technique whose time has come and gone. It should be noted that steam distillation has been relegated to "historical" status in the ASTM system.

A comparison of the advantages and disadvantages of the various isolation techniques is provided in Table 5.2.

Table 5.2 Comparison of ILR isolation techniques

Method	Advantages	Disadvantages
E1385 steam distillation (a historical standard, expired 2010)	Produces a visible liquid, simple to explain	Labor intensive, destructive, not sensitive, requires expensive glassware
E1386 solvent extraction	Useful for small samples and empty containers, does not cause significant fractionation, useful for distinguishing HPDs from each other	Labor intensive, expensive, coextracts nonvolatile substances, increased risk of fire, solvent exposure, destructive
E1388 headspace sampling	Rapid, more sensitive to lower alcohols, nondestructive	No archiveable sample, not sensitive to heavier compounds, poor reproducibility
E1412 passive headspace	Requires little analyst attention, sensitive, nondestructive, produces archiveable sample, inexpensive	Requires overnight sampling time
E1413 dynamic headspace, using activated charcoal	Rapid, sensitive, produces archiveable sample, inexpensive	Labor intensive, subject to breakthrough, destructive
E1413 dynamic headspace, using Tenax	Rapid, sensitive	Labor intensive, requires thermal desorption, no archiveable sample, destructive
E2154 SPME	Rapid, highly sensitive, useful for field sampling with portable GC-MS	Labor intensive, expensive, requires special injection port, reuse of fibers, no archiveable sample, more subject to contamination

5.6 ANALYZING THE ISOLATED IGNITABLE LIQUID RESIDUE

Despite the improvements in separation and detection technology, the overall approach to identification of ignitable liquid residues is the same as it was in the early 1970s. A chromatogram from the sample is compared with chromatograms from known standards, and the analyst determines whether there are "sufficient similarities" to make an identification. What has changed is the quantity of information available because of the increased resolution provided by capillary columns and the ability to obtain a mass spectrum up to 10 times per second, as well as the reaching of a consensus on the meaning of "sufficient." While there is more information to compare, the technique is still one *of pattern recognition and pattern matching.*

An argument can be made that when one looks at a mass spectrum, one is looking at structural details rather than simply matching patterns, but the patterns still have to match. The same argument has been made about structural elucidation in the use of Fourier transform infrared spectroscopy (FT-IR) for drug identification. While it would be nice to think that analysts routinely consider molecular structure, the day-to-day operation is one of pattern matching.

There exists a specific skill set required to compare chromatographic patterns. This skill set must be carefully learned and routinely used if it is to remain sharp. Pattern recognition is one "scientific" skill that has historically been problematic. Although fire debris analysts use the same equipment as drug analysts, the drug analyst typically looks for a single peak with a particular retention time. The fire debris analyst must compare an entire pattern of peaks produced by a sample to a pattern of peaks produced by a standard. This exercise includes a comparison of inter- and intragroup peak heights and can be quite complex. While it is true that many drug analysts also conduct ignitable liquid residue analyses, a different set of skills is involved.

Fire debris analysis also involves a different set of skills than those employed by environmental scientists, who are typically trying to quantify the components of an oil spill or contaminants at a Superfund site. Environmental methods typically assume the presence of gasoline or other petroleum products and then look for benzene, toluene, ethylbenzene, and xylene (BTEX) to quantify the amount present. Unless they are trying to identify the source of the spill, environmental analysts are usually not employing the same skill set used by fire debris analysts.

If we think about the nature of many petroleum products and the processes that are going on when we isolate those products from debris samples, we can begin to understand why chromatographic patterns look the way they do. Many petroleum products are straight-run distillates, particularly the medium and heavy petroleum distillates (MPD and HPD). The overall pattern of these products is a Gaussian (bell-shaped) distribution of peaks dominated by the normal alkanes. The patterns produced by kerosene and diesel fuel have been likened to a stegosaurus because the chromatograms bear a passing resemblance to the dinosaur's dorsal fin. A medium petroleum distillate looks like the same pattern, shrunken and coming off early.

A fractionation process similar to distillation occurs when we use PHC to isolate an ignitable liquid residue from a sample. This is caused by the very low vapor pressure of ILR components above C_{18} and by the selective adsorption of the heavier hydrocarbons on complex substrates. Hydrocarbons up to C_{23} can be captured from diesel fuel placed on a noncompeting substrate such as filter paper. That same fuel placed on charred wood, however, might exhibit a pattern that ends at C_{18}. If compounds do not get into the air in the headspace, they will not be adsorbed onto the carbon strip. The tenacity of the sample must be considered, particularly when trying to distinguish between kerosene and diesel fuel.

This author learned pattern recognition the old-fashioned way, that is, running dozens, then hundreds of standards, and learning what the patterns looked like. Today's analysts are fortunate in that the ASTM standards give examples of many of the patterns that an analyst is likely to see in positive samples, and there exists a detailed compilation of literally hundreds of patterns available in a standard text [20]. Both the ASTM standards and the *GC-MS Guide to Ignitable Liquids* provide sufficient information to allow the analyst to set up an instrument to provide patterns that look very much like the ones in the texts. While these texts are an important resource, it is imperative that every fire debris analysis laboratory have its own library of ignitable liquid residues. This provides for GC patterns with the exact same retention times and mass spectra with the exact same fragmentation patterns. The in-house library is also a quality assurance tool that lets the analyst know when there has been some drift in the instrument and when it is time to run a new set of standards or install a new column. Whatever approach is taken, it will take time to develop the ability to recognize ILR patterns.

The mass spectrometer provides the ability to simplify what can be very complex and confusing patterns. This is a result of the ability of the data analysis software to separate out only those peaks having particular ions present in the pattern. For example, if one wants to look at the alkanes, one only needs to obtain an extracted ion chromatogram for m/z (mass-to-charge ratio) 57, hereafter referred to as ion 57, and most of the balance of the components will disappear. It is thus possible to break down a total ion chromatogram (TIC) into its component parts. While there are more patterns to learn, they are simpler patterns and easier to remember. This approach to data analysis is known as *mass chromatography* and was first proposed by Smith in 1982 [11]. There are basically two ways to approach mass chromatography: (1) the single ion approach and (2) the multiple ion approach. Dolan [21] has proposed referring to the single ion chromatograms as extracted ion chromatograms (EICs) and multiple ion chromatograms as extracted ion profiles (EIPs). This text will adopt that terminology.

When working with software that scales the extracted ion chromatogram or profile to the tallest peak and when using multiple ions, some caution is advised. To the extent that the second, third, and fourth ions contribute to a pattern, they tend to make it more complicated, thus defeating some of the purpose of extracting the ions in the first place. To the extent that these additional ions do not change the pattern, they may convince the analyst that he or she is seeing more than is actually present. Finally, to the extent that these additional ions are present in substantially lower concentrations, they would be better observed on their own rather than in the profile. Figure 5.7 is a comparison of

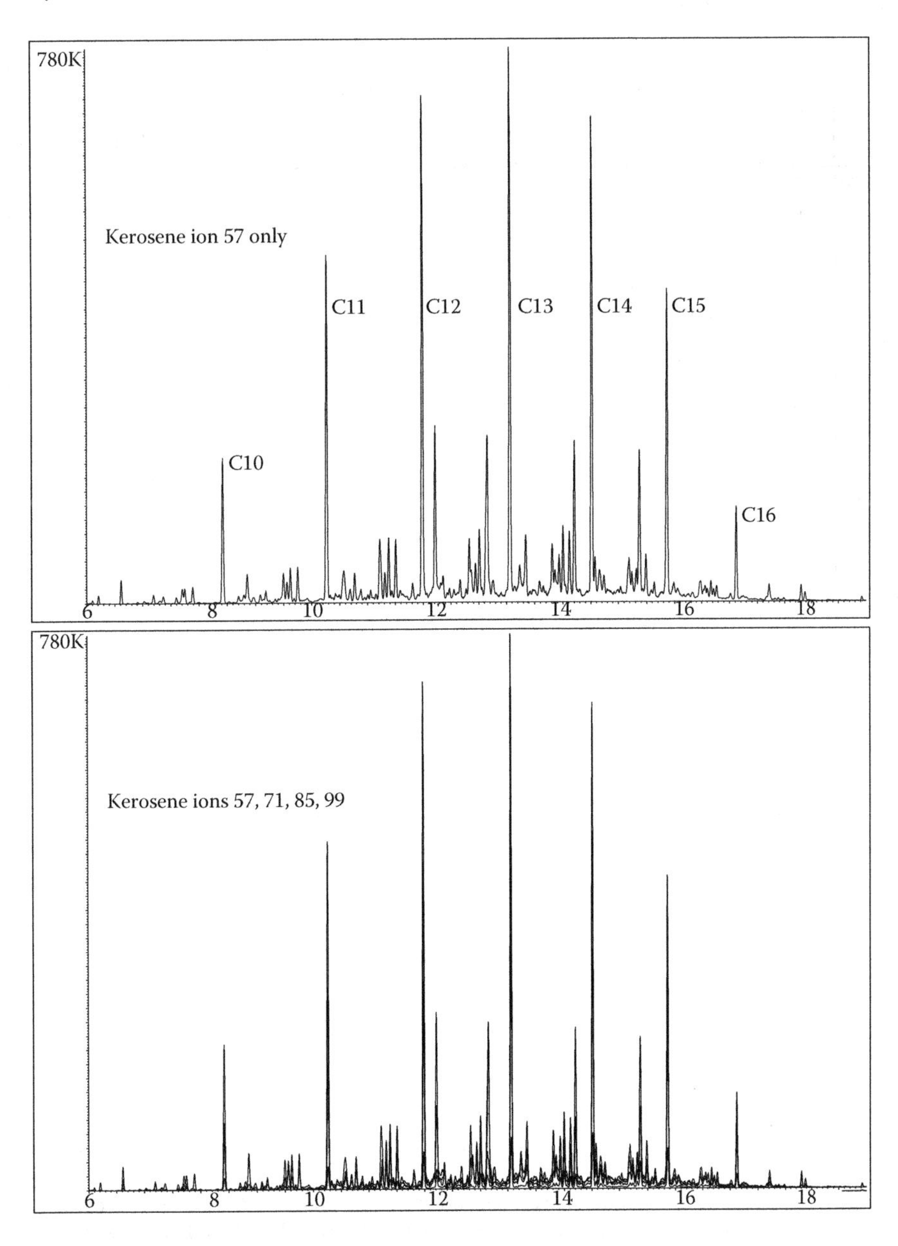

Figure 5.7 Ion profiling versus ion chromatography: comparison of the ion 57 chromatogram (top) with the ion profile combining ions 57, 71, 85, and 99 from a kerosene standard (bottom). The four ions in the profile are plotted in the merged format. Plotted individually, they are all very similar, with ion 57 presenting the tallest peaks.

ion 57 from a kerosene standard versus a profile based on ions 57, 71, 85, and 99. The profile is slightly more complicated, but because ion 57 is the *base peak*[1] for almost every component in the chromatogram, the ion 57 profile is the tallest.

Another example is shown in Figure 5.8, which presents ions 128, 142, and 156, the naphthalenes, from a gasoline sample. In the top chart, the ions are combined into a profile, which, although it gives the analyst an idea about the relative abundance of the three ions, shows very little detail for the dimethylnaphthalenes represented by ion 156.

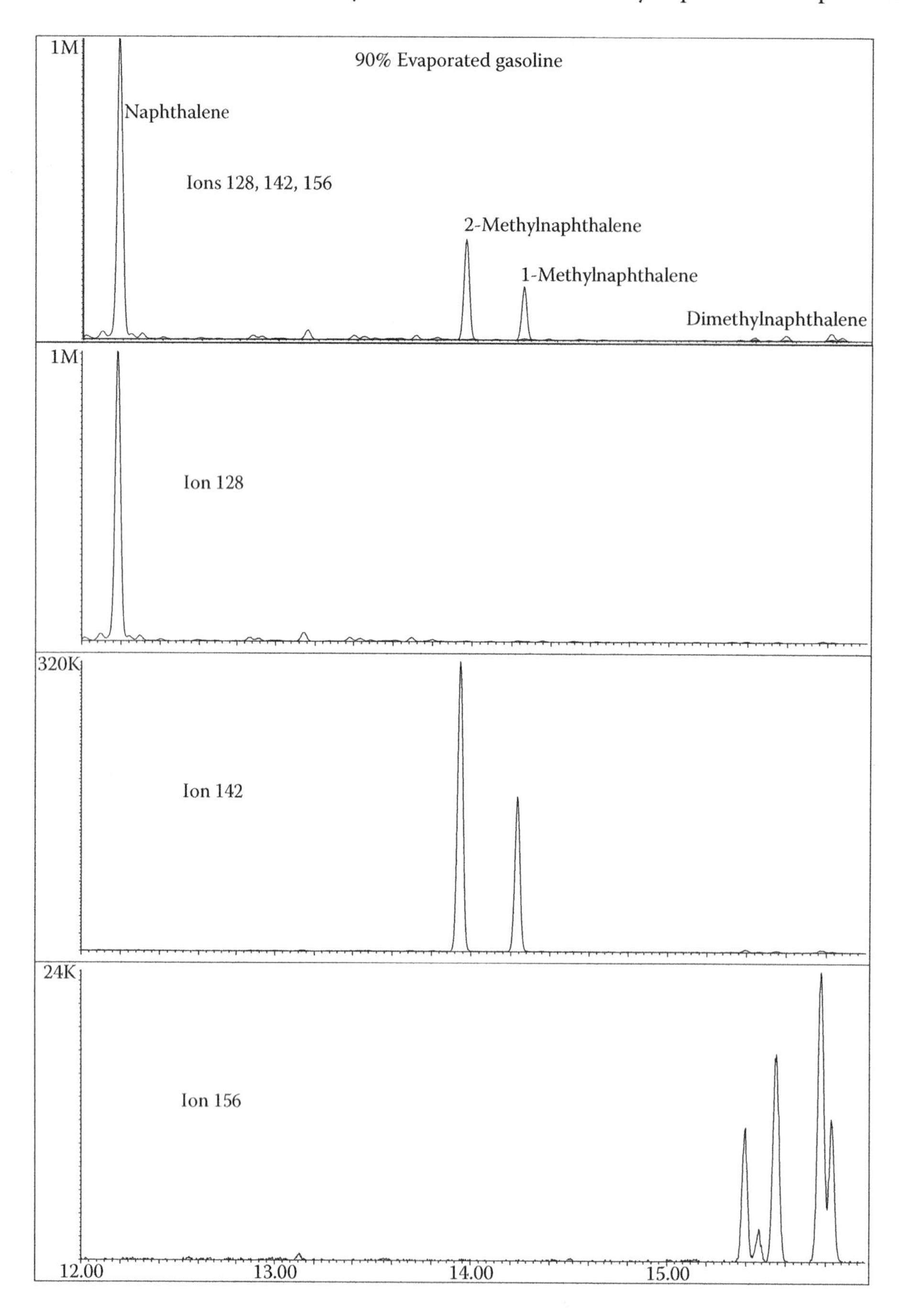

Figure 5.8 Ion profiling versus ion chromatography: comparison of the ion profile of the naphthalenes from a gasoline standard (top chart). Compare with the detail provided by presenting the three ions (128, 142, and 156) separately.

[1] Base peak: The tallest peak in a mass spectrum, due to the ion with the greatest relative abundance. Not to be confused with *molecular ion*. Base peaks are not always molecular ions, and molecular ions are not always base peaks.

Table 5.3 Ions used for extracted ion chromatography

Ion	Compounds
55	Alkenes, cycloalkanes
57, 71	Alkanes
83	Cycloalkanes
91	Toluene, xylenes
93	Terpenes
105	C_3 alkylbenzenes
117	Indan, methylindans
119	C_4 alkylbenzenes
131	Methyl, dimethylindans
142	Methylnaphthalenes
156	Dimethylnaphthalenes

When the extracted ions are presented separately, the analyst still gets the quantitative data by reading the abundance numbers next to the y-axis and also gets to see the fine details in the shape of the peaks at the right side of the chart. Table 5.3 lists the ions used in our laboratory to examine various classes of petroleum compounds.

A similar example is shown in Figure 5.9a. These charts show ions 91, 105, and 119 from a 75% evaporated gasoline standard. The effect is even more pronounced at 90% evaporation, shown in Figure 5.9b, when the ion 91 peaks are smaller. Whether an analyst chooses extracted ion chromatography or extracted ion profiling is largely a matter of

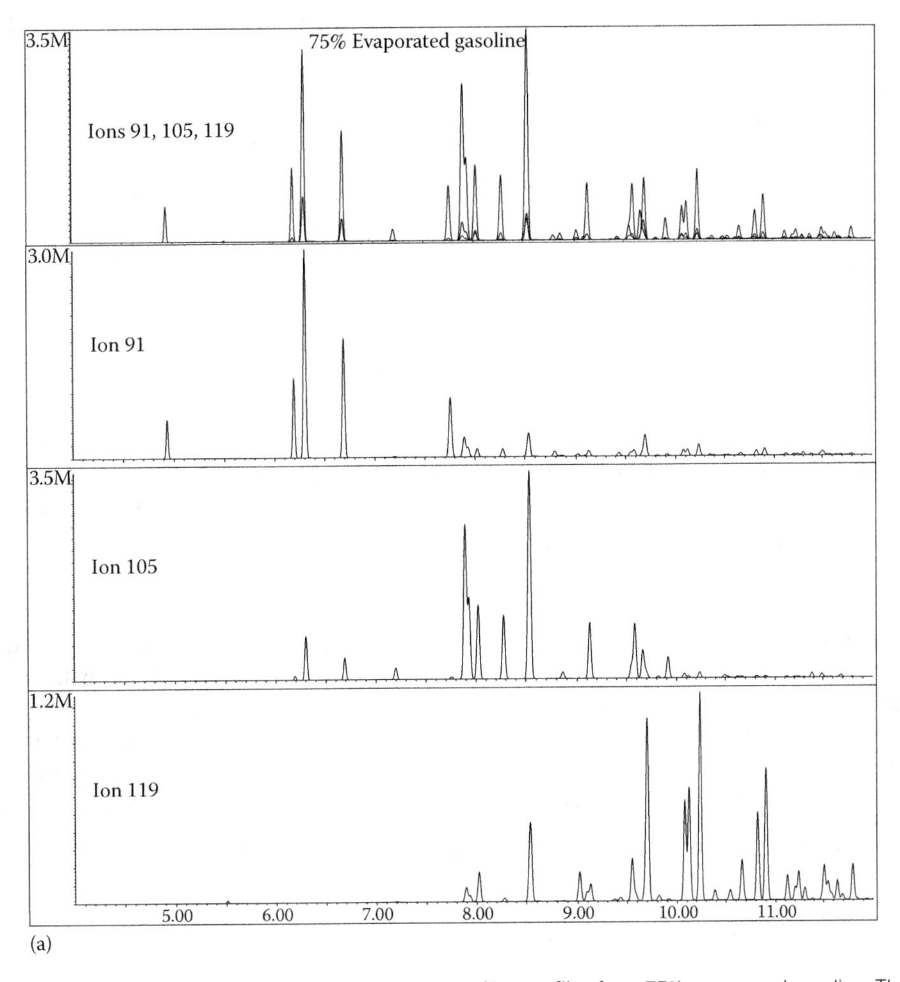

Figure 5.9 (a) Ion profiling versus ion chromatography: comparison of ion profiles from 75% evaporated gasoline. The top chart shows a combination of ions 91, 105, and 119. The next three charts show those ion chromatograms plotted independently.

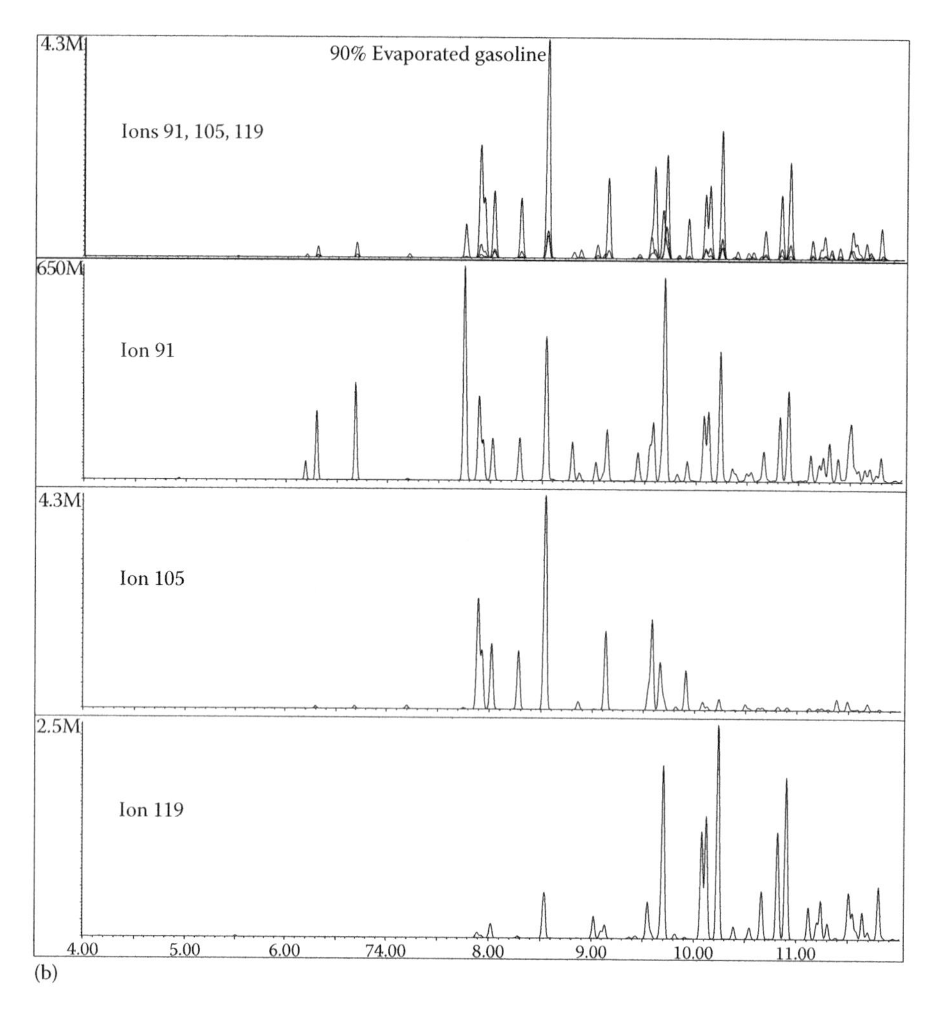

Figure 5.9 (b) Ion profiling versus ion chromatography: comparison of ion profiles from 90% evaporated gasoline. The top chart shows a combination of ions 91, 105, and 119. The next three charts show those ion chromatograms plotted independently. Note the improvement in the level of detail presented for the xylenes.

taste. It is easier for this author to use the simpler patterns and thereby avoid the eyestrain required to see the smaller peaks in the profiles. The typical set of six ion chromatograms used in our laboratory to document the presence of gasoline is shown in Figure 5.10, and the set of three ion chromatograms used for distillates is shown in Figure 5.11. The details of each are discussed in the following.

5.6.1 Criteria for identification

Most of the chromatograms that an analyst uses to make a positive identification of an ignitable liquid residue will match (i.e., exhibits "sufficient similarities" to the standard) at the level of the TIC. Samples with low concentrations of ILR, high backgrounds, or both can sometimes yield a positive identification for ILR *if the analyst is very careful.* Generally, but not always, if the TIC from the sample does not match the TIC from the standard, the sample is likely to be negative (i.e., determined to contain no detectable ignitable liquid residue). Examples of cases where this is not true are presented following this general discussion of criteria for identification of an ILR in routine cases.

ASTM E1618-14 identifies eight classes of ignitable liquids identifiable by GC-MS:

- Gasoline
- Petroleum distillates, including de-aromatized distillates
- Isoparaffinic products
- Aromatic products

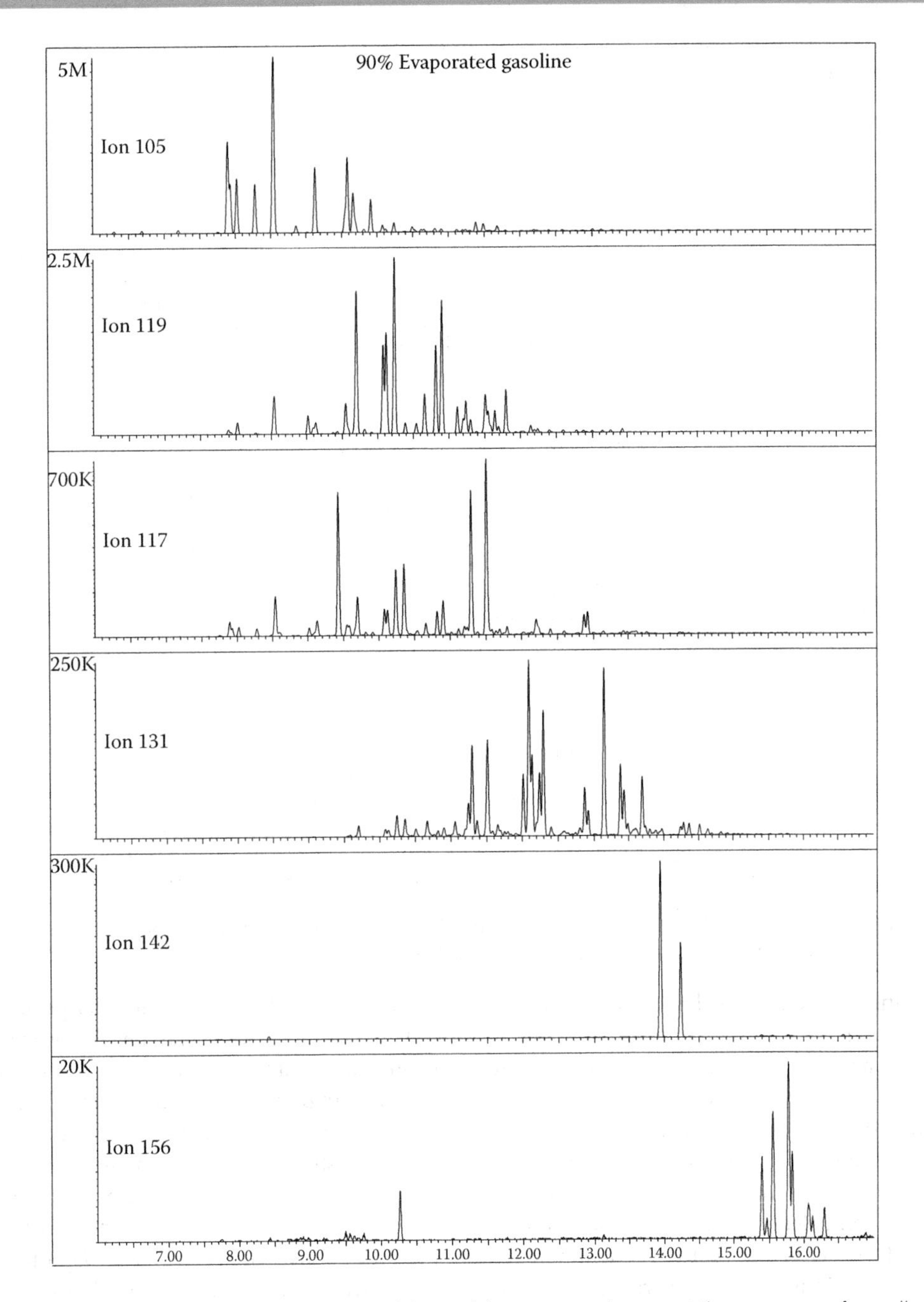

Figure 5.10 Typical set of six ion chromatograms used in our laboratory to document the presence of gasoline. Ions we use are 105 for C_3 alkylbenzenes, 119 for C_4 alkylbenzenes, 117 for indan and methylindans, 131 for dimethylindans, 142 for 2- and 1-methylnaphthalene, and 156 for dimethylnaphthalenes.

- Naphthenic–paraffinic products
- n-Alkane products
- Oxygenated solvents
- Miscellaneous

With the exception of gasoline and oxygenated solvents, each of the earlier classes can be placed into one of three ranges: light (C_4–C_9), medium (C_8–C_{13}), and heavy (C_8–C_{20+}).

This standard presents criteria for identifying the various kinds of compounds (alkanes, cycloalkanes, aromatics, and condensed ring aromatics) found in each one of these eight classes. The alkanes include both straight-chain and branched

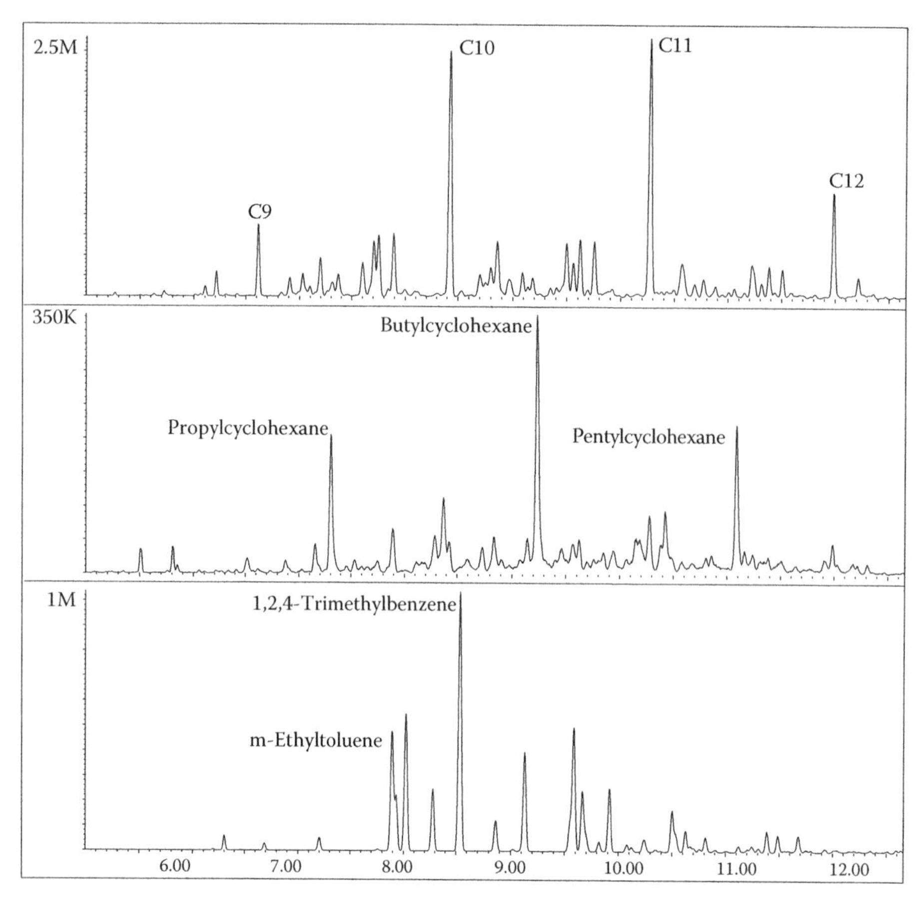

Figure 5.11 Typical set of three ion chromatograms used in our laboratory to document the presence of distillates. These are the ion 57, 83, and 105 chromatograms from Smokey Bear Charcoal Lighter, an MPD.

hydrocarbons, and can be extracted using ions 57 and 71. (Some isomers of alkanes exhibit taller peaks in ion 57, and some are taller in ion 71.) The cycloalkanes are mostly substituted cyclohexanes, which can be seen by extracting ion 83. Cycloalkanes also have a strong ion at 55, and as the length of the substituted alkane increases, ion 57 begins to dominate the mass spectra. *Aromatics* refers to alkyl-substituted benzenes with a single ring. These can be extracted using ions 91, 105, and 119. Ion 91 shows toluene and xylenes, ion 105 shows xylenes and C_3 alkylbenzenes, and ion 119 shows C_4 alkylbenzenes. There is some overlap in the extracted ion chromatograms, as shown previously in Figure 5.9. *Condensed ring aromatics* refers to indans and naphthalenes. Indans have a five-member ring attached to a benzene ring and can be substituted. The naphthalenes we usually see are naphthalene itself, 2- and 1-methylnaphthalene (written in that order because that is the order of elution from a nonpolar column), and the dimethylnaphthalenes. Naphthalene presents a single peak when ion 128 is extracted; there are two peaks for the methylnaphthalenes seen when ion 142 is extracted and eight peaks in the dimethylnaphthalene chromatogram seen when ion 156 is extracted. The indans can be visualized by extracting ions 117 and 134.

5.6.1.1 Identification of gasoline

Gasoline is the one class in the ASTM E1618 system in which the name of the class does not define its chemical composition. Some fuels sold as "gasoline" would be better described as isoparaffins. Aviation gasoline is an example of such a product. The composition of petroleum products as found in fire debris is influenced by three factors: (1) crude oil parentage, (2) the effects of petroleum refining processes, and (3) the effects of weathering. Gasoline is the way it is largely because of the second of these influences—petroleum refining. Although all crude stocks contain aromatics, aliphatic hydrocarbons comprise the bulk of most crude oils. Because the straight chain aliphatics cause knocking in gasoline engines, the value of the crude stock is enhanced by *cracking*, making small molecules out of larger ones, and by *reformation* through a process of dehydrogenation. Toluene and xylenes are the most abundant compounds produced in these processes. The identification of this product called reformate forms the basis for most gasoline identifications.

Alkylate is the product of a refinery process called alkylation, in which light unsaturated hydrocarbons such as propene or isobutene are combined with isobutane, resulting in a product that mainly consists of highly branched C_7–C_9 alkanes, with 2,2,4-trimethylpentane (iso-octane) as the main component. Most gasolines are a mixture of reformate and alkylate, though there are some pure alkylate gasolines, such as aviation gasoline. Identification of alkanes is one of the criteria for identifying gasoline per ASTM E1618, but the standard offers little guidance beyond the statement that alkanes must be present. A 2017 publication describes the use of alkylate components for classifying gasoline in fire debris samples [22]. This discussion will focus mainly on the identification of gasoline containing reformate as its major component.

Gasoline does contain numerous light aliphatics (as light as butane) when it is first pumped, but by the time it has been through a fire, most of the lighter ($< C_7$) aliphatics have evaporated. Consequently, toluene is usually one of the first tall peaks seen in a sample of gasoline, and once the gasoline has weathered to 50% or more, the toluene peak is much shorter than the C_3 alkylbenzene or xylene peaks. Gasoline changes considerably as it evaporates, which is what makes it one of the more difficult classes to identify. When a Gaussian distribution of normal alkanes changes because of evaporation, it is still a Gaussian distribution of normal alkanes, just a heavier one. Most of the samples of gasoline that this author has seen in samples of fire debris have been more than 75% evaporated. (Estimating the degree of evaporation is discussed later.)

In this author's experience, gasoline is the most frequently misidentified ignitable liquid residue. That is because many of the compounds present in gasoline as it comes from the pump are also produced when polymers, such as polyvinyl chloride (PVC, also known as vinyl) and polystyrene, degrade as a result of exposure to heat [23]. The key to avoiding misidentifications is making sure that the ratios *between* groups of compounds and *within* groups of compounds are consistent with the standard. Research conducted in the early 1990s showed that it is possible to confuse different classes of ILRs with each other using mass chromatograms unless the EIPs or EICs are compared with each other. Further, carpet pyrolyzate, a very common substance seen by fire debris analysts, was found to contain 37 of 43 gasoline target compounds [24].

Toluene is a very common pyrolysis product. It is an unusual fire debris sample that does not contain toluene at some level. Because it is one of the first compounds to evaporate, one does not expect to see a tall toluene peak in the absence of equally tall xylene peaks in a sample that is positive for gasoline. Figure 5.12 shows the chromatogram

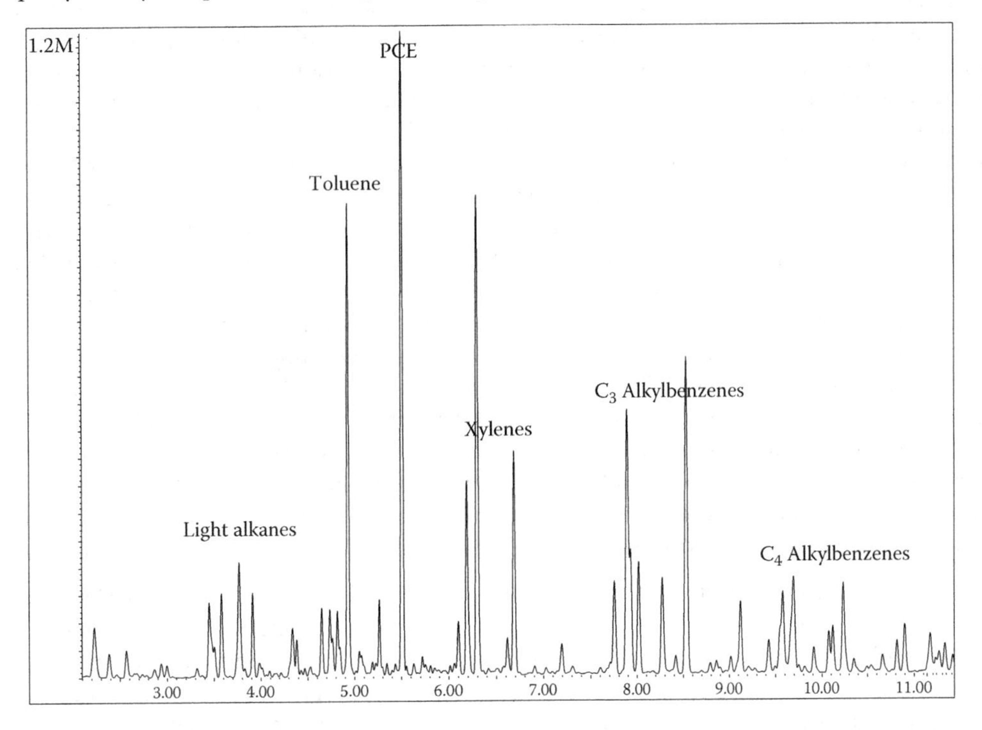

Figure 5.12 TIC of fresh gasoline. Ten microliters was spotted on a piece of filter paper, and the headspace was concentrated using ASTM E1412. The eluting solvent was diethyl ether spiked with 100 ppm perchloroethylene, shown here eluting between toluene and the xylenes.

of a 10 μL standard of gasoline adsorbed using an activated charcoal strip and eluted with diethyl ether spiked with 100 ppm perchloroethylene. The toluene and xylene peaks are almost equally tall. If a fire debris sample contains toluene from gasoline, it is accompanied by xylenes and the higher peak groupings of gasoline. Toluene that is not so accompanied comes from something other than gasoline.

Xylenes are also produced by the decomposition of plastics, but they are the first group of compounds that we can examine for correct *intergroup ratios*. Figure 5.13 shows ion 91, the base ion for xylenes, from gasoline in three different stages of evaporation; kerosene; and a medium petroleum distillate. The relative ratios for the three peaks are almost indistinguishable. Note that the ethylbenzene peak in the 50% evaporated gasoline is slightly lower. This trend continues as the degree of evaporation increases, but generally we find that if the xylenes in our sample are from an ignitable liquid, they exhibit this characteristic ratio. Note that there are three isomers of xylene *(ortho-*, *meta-*, and *para-)* but that the three peaks seen in our chromatograms actually represent four compounds (ethylbenzene followed by the three xylenes) because *meta-* and *para-*xylene cannot be resolved except in the longest columns. If xylenes are found in ratios other than the one shown in Figure 5.13, particularly if ethylbenzene is the tallest peak in the group, one can safely conclude that neither gasoline nor any other petroleum product was the sole source of the xylene. A word of caution is appropriate here. If the analyst wishes to hypothesize that the residue contains both gasoline *and* pyrolyzate, some proof (in the form of actual data) is necessary. *There is no published literature to date* that validates a finding of gasoline if the inter- and intragroup ratios of components do not match a known standard.

The next group to consider is the C_3 *alkylbenzenes*. This is by far the most important of the patterns in any sample of gasoline. Like the xylenes, it is also found in all petroleum products from which the aromatics have not been removed, and unless evaporation has decreased the concentration of the lighter compounds, the peak ratios are always the same. This is the group identified in the 1982 guidelines (and in every standard since) as "the m-ethyltoluene/pseudocumene five-peak group." This group appears in many samples that contain no gasoline, and it is the failure to exhibit the proper ratio of peaks that is sometimes the analyst's only clue that he or she is looking at data that can be easily misinterpreted. The group of peaks *must* be present to identify gasoline, but sometimes it is overemphasized and sometimes it is "seen" where it really is not present. Figure 5.14 shows one such sample where there was a misidentification of gasoline. The figure shows the suspect sample at the top, as well as four standards that exhibit the proper ratios of components in the C_3 alkylbenzene group. As with the xylenes, *meta-* and *para-*ethyltoluene cannot be resolved, although the resolution is somewhat better here than in the xylenes—we see a shoulder. In nearly all cases where the C_3 alkylbenzene group is, in fact, present as a component of a petroleum product, the pseudocumene peak will be the tallest. Because most of the samples found positive for gasoline contain highly evaporated residues, *the pseudocumene peak will be the tallest peak on the chart*—both the TIC and the EIC. Note that in the questioned sample at the top of Figure 5.14a, it is the second peak that is the tallest, not the pseudocumene peak. The first peak has the same retention time as m- and p-ethyltoluene, but the shape is wrong, and the mass spectrum indicates that the peak actually represents benzaldehyde. The mass spectrometer unequivocally identified the second peak in the questioned sample as 1,3,5-trimethylbenzene, but the third peak, which may have represented o-ethyltoluene, also contained a significant 118 ion, indicating the presence of α-methylstyrene, a common pyrolysis product. *Simply matching the components when the ratios are not right can lead to misidentifications.* The odds against three background components coeluting in exactly the right concentrations to skew the peaks in this group are pretty high. Three other samples from the same fire scene were similarly misidentified and exhibited similar peak ratios and mass spectral characteristics. **As with almost all misidentifications, there was not a good pattern match with the TIC**. Extracted ion chromatography or extracted ion profiling can be very useful; however, the analyst should remember that it is a spectral as well as a chromatographic technique. The spectra should be examined, especially when the peak ratios are "off."

A word about the use of the mass spectrometer beyond generating mass chromatograms is in order. Most analysts use the mass selective detector or mass spectrometer as a tool for generating extracted ion chromatograms and extracted ion profiles. Obviously, if one is looking at very simple mixtures or single components, the mass spectrum is necessary to make an identification. A sometimes overlooked function of the mass spectrometer is the evaluation of extracted ion chromatograms and profiles. As observed in Figure 5.14a, an extracted ion chromatogram with the intragroup

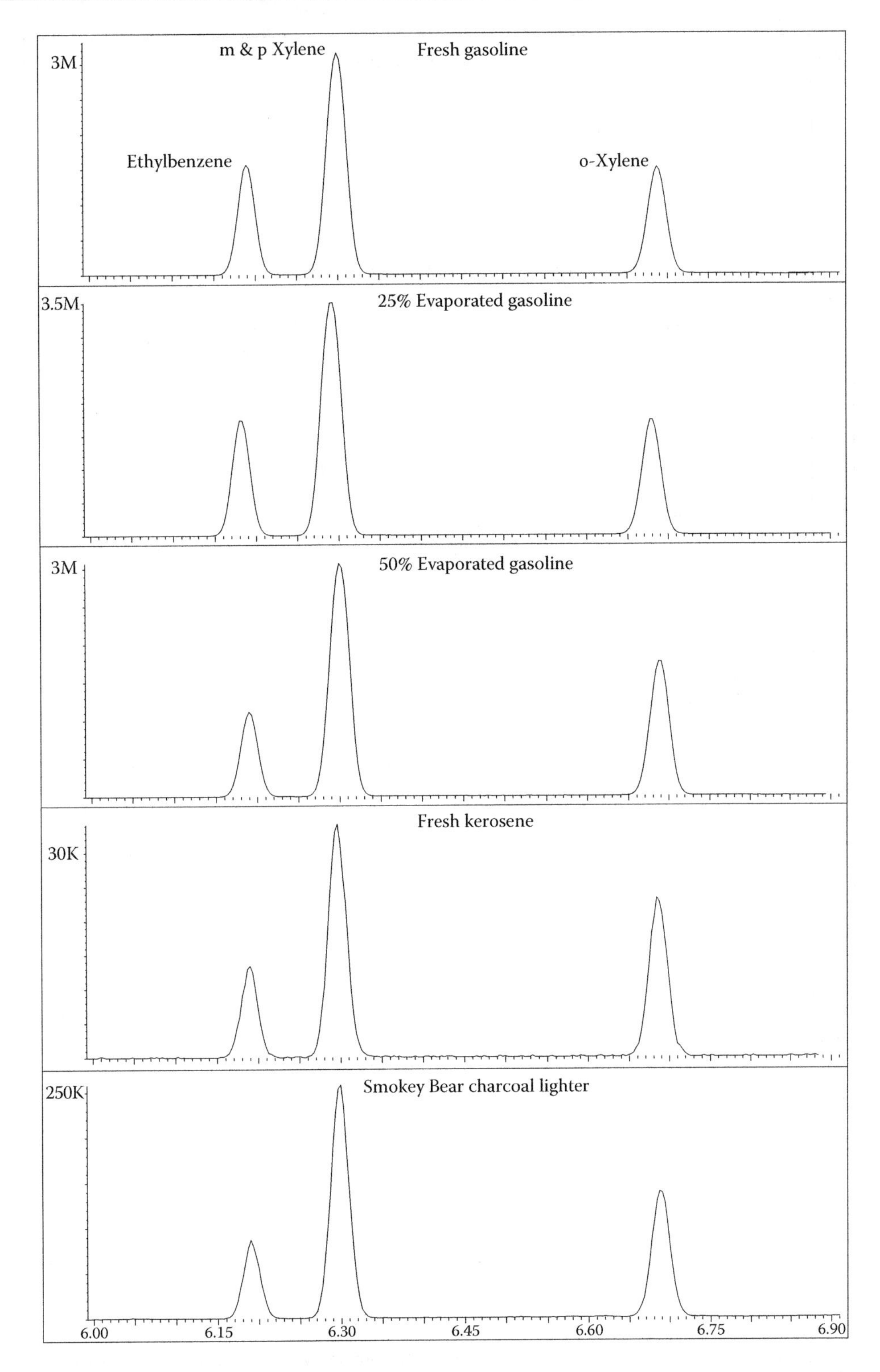

Figure 5.13 Ion chromatogram for ion 91 from gasoline in three different stages of evaporation, kerosene, and a medium petroleum distillate. Note that the relative ratios for the three peaks in the xylene group are almost indistinguishable. Evaporation causes the ethylbenzene peak in the 50% evaporated gasoline to be slightly shorter.

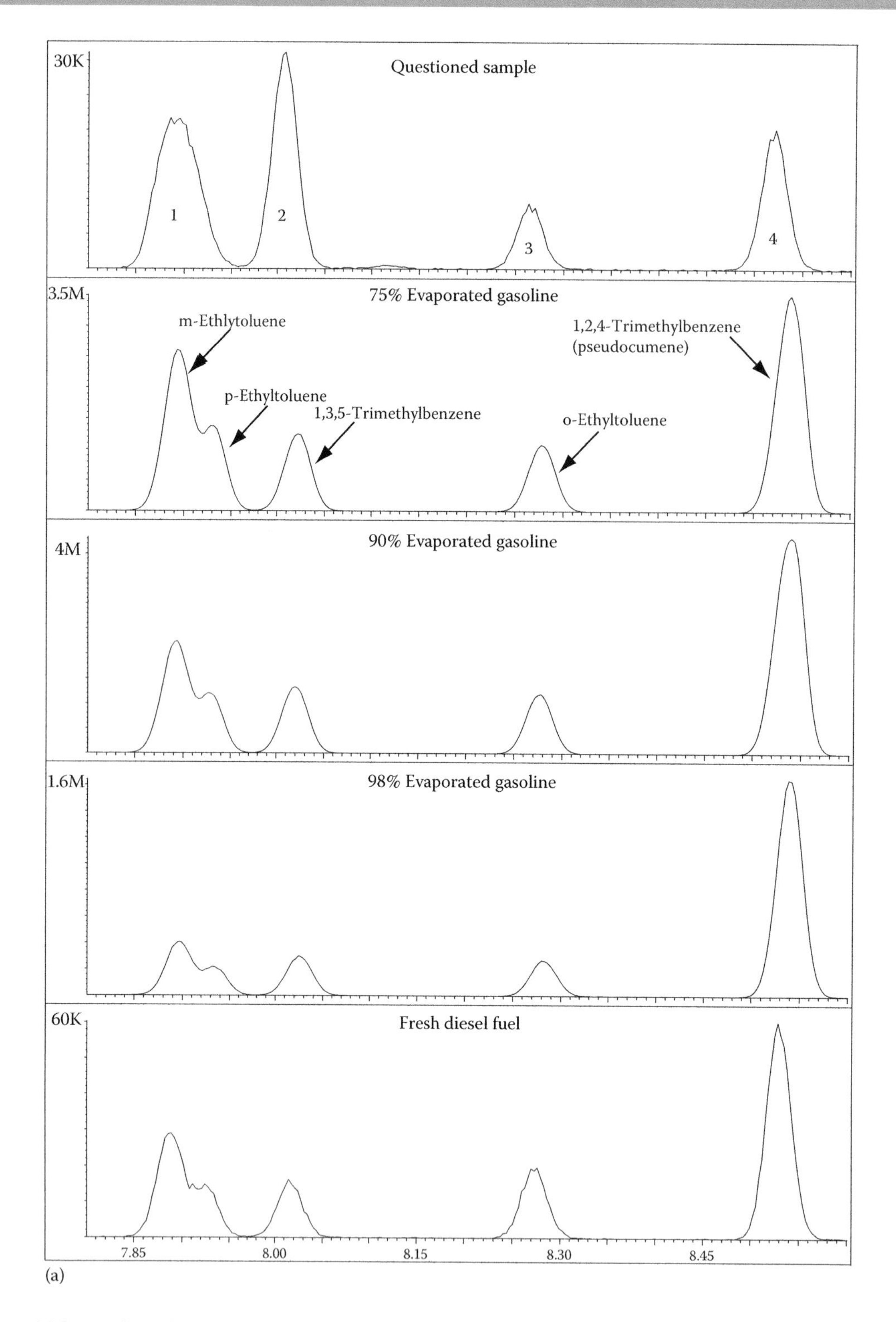

Figure 5.14 (a) Study of the C_3 alkylbenzene group. The top chromatogram shows an unknown sample. This is followed by gasoline in three stages of evaporation and fresh diesel fuel. The TIC of the unknown did not look like gasoline, but based on this EIC and a few other ion chromatography comparisons, an analyst called the sample positive for gasoline. The mass spectra shown in (b) through (e) showed that two of the peaks, #1 and #3, were not gasoline components.

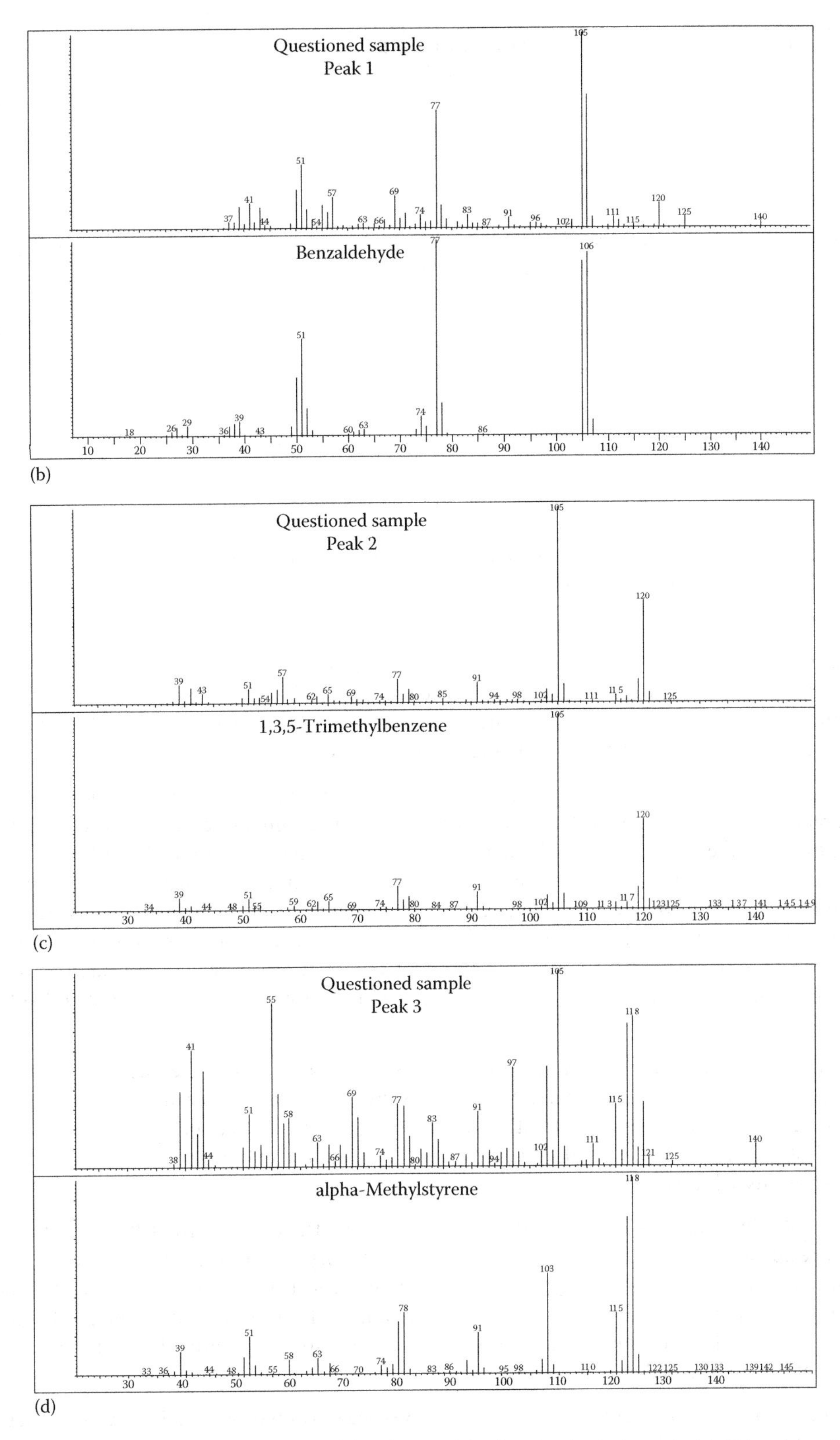

Figure 5.14 (b) Mass spectrum and library match of the first peak in the top chart shown in (a). This peak actually represents benzaldehyde and not m-ethyltoluene. (c) Mass spectrum of the second fully resolved peak in the top chart shown in (a). As with gasoline, this peak is identified as 1,3,5-trimethylbenzene. (d) Mass spectrum of the third major peak in the chart shown at the top of (a). This mass spectrum changed across the peak but clearly contained α-methylstyrene. 1-Decene, a common decomposition product, was also suggested by the spectrum. There may be some o-ethyltoluene coeluting as well, but that cannot be demonstrated.

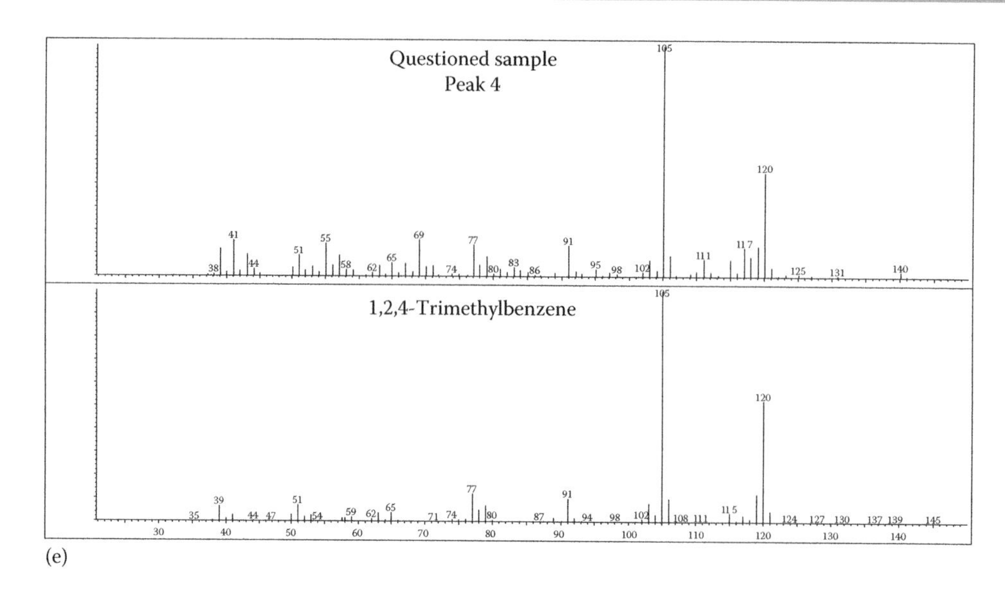

Figure 5.14 (e) Comparison of the mass spectrum fourth peak in the chart shown in (a) with 1,2,4-trimethylbenzene (pseudocumene). This is a reasonably straightforward match.

ratios just slightly "off" has the potential of being misleading. Figures 5.14b through e show the mass spectra of each of those five peaks, as well as the library's best match. The first peak, which is coincident with m-ethyltoluene, has a large peak at m/z 77, but no such ion is present in the spectrum of m-ethyltoluene. There may be some m-ethyltoluene hidden under this peak, but there is definitely also some benzaldehyde.

One way to determine whether there are coeluting compounds under a chromatographic peak is *to examine the mass spectra at different points across the peak.* If the peak represents a pure compound, the mass spectrum changes little, if at all. In Figure 5.14c, the second peak in the five-peak group is a near-perfect match for 1,3,5-trimethylbenzene, which is the third peak in the five-peak group required to identify gasoline. The third peak, whose mass spectrum is shown in Figure 5.14d, clearly contains more than one substance, but the strong peak at 118, as well as the retention time, indicates that α-methylstyrene is coeluting. Like the second peak, the fourth peak is a pure compound and a nearly perfect match for pseudocumene. Because the peak ratios are off, and especially because two of the four peaks in the ion 105 chromatogram represent compounds not found in gasoline, it must be concluded that the first identification was in error.

Background subtraction is a tool found in most data analysis software and can frequently resolve questions as to whether a particular compound is really there. A detailed evaluation of the quality of the spectra underlying a mass chromatogram, whether it is a single ion extraction or a multiple ion profile, should be carried out periodically, just to keep the analyst in practice, but it should routinely be carried out on any sample where either the peak ratios or retention times are "just a little off."

Note in Figure 5.14a that when a sample is sufficiently evaporated, the peaks at the left side of the chart begin to diminish. This result is expected. The initial guidelines for identifying gasoline stated that the m-ethyltoluene/pseudocumene five-peak group was still present in gasolines that had lost 90% of their fresh weight. In fact, this group does not disappear until the gasoline is more than 98% evaporated. If this group cannot be positively identified, an analyst is on very thin ice indeed when identifying a sample as containing gasoline. In such a case, there must be a peak-for-peak match of all the higher peak groupings.

As with many of the other components of gasoline, this five-peak group is present in almost all petroleum products from which the aromatics have not been removed. The bottom chart in Figure 5.14a is from unevaporated diesel fuel.

The C_4 *alkylbenzenes* are best viewed by extracting ion 119. Because there are more ways to build a C_4 alkylbenzene than a C_3, this is a more complex pattern, but it is present in almost all petroleum products. Figure 5.15 shows the

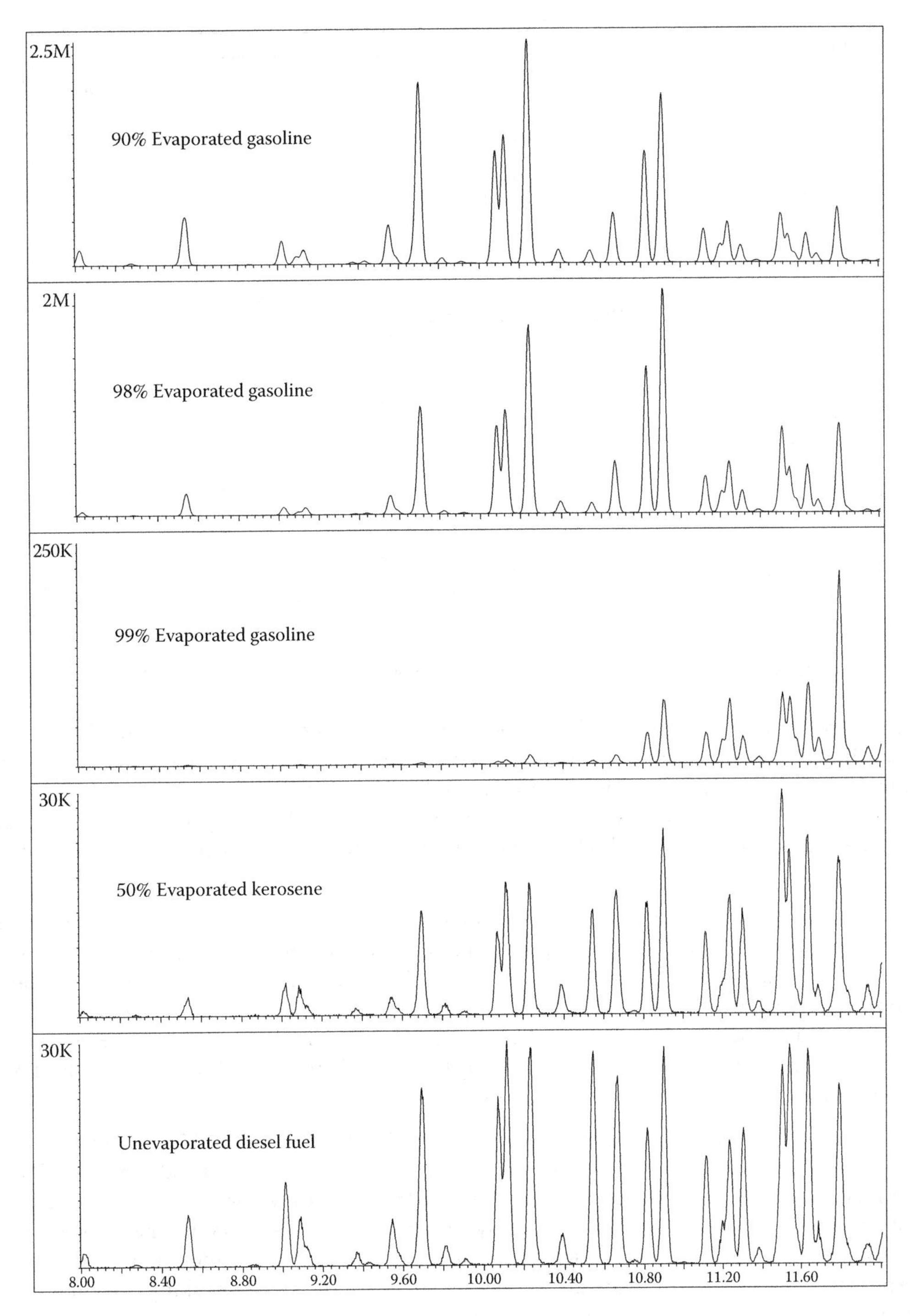

Figure 5.15 C_4 alkylbenzenes represented by ion 119 in gasoline in three different stages of evaporation, 50% evaporated kerosene, and unevaporated diesel fuel. All samples were prepared by spotting 10 μL on a piece of filter paper and processed using ACS adsorption/elution.

patterns found in highly evaporated gasoline, as well as in kerosene and diesel fuel, which are produced by the presence of the C_4 alkylbenzenes. Note that in kerosene and diesel fuel, this pattern is more complex than in gasoline.

The next group of compounds that must be present to make a solid identification of gasoline is the *indans*. These can be extracted using ions 117 and 131. The doublet at 11.3 and 11.5 min in the ion 117 chart is, in this analyst's experience, always present. It has traditionally not been an absolute requirement of the ASTM standards, but it probably

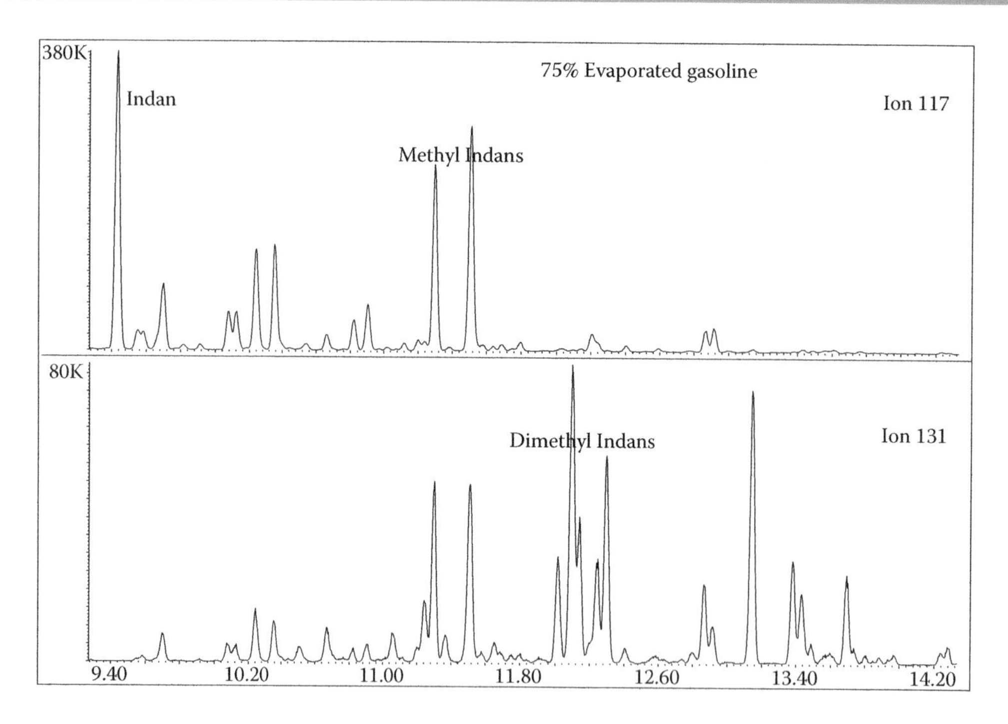

Figure 5.16 Ion chromatograms for indan, the methylindans, and dimethylindans found in 75% evaporated gasoline.

should be. The standard states, "Indan (dihydroindene) and methylindans are usually present." The first tall peak in the 117 chart is indan, appearing at 9.4 min. Figure 5.16 shows the peak groupings characteristic of gasoline for the indans, methylindans, and dimethylindans. Like the C_3 alkylbenzenes, the indans and alkyl-substituted indans can be found in roughly the same proportions in gasoline that is 98% evaporated. Other petroleum products, particularly the distillates, contain indans and methylindans, but their mass chromatographic patterns are more complex than the ones for gasoline.

The next group of compounds that should be present in a sample identified as containing gasoline is the *methyl-* and *dimethylnaphthalenes.* Naphthalene is also present but it is so common that its presence in a sample is all but meaningless. 2-Methylnaphthalene elutes before 1-methylnaphthalene and is almost always more abundant than 1-methylnaphthalene. In any case where this ratio is reversed, the extract should be considered suspect. The boiling point of the methylnaphthalenes is high enough that the ratio, unlike xylenes or C_3 alkylbenzenes, is unlikely to be affected by evaporation. All petroleum products are likely to contain both the naphthalenes and the dimethylnaphthalenes, and the ratios should be highly comparable to each other.

There are some *aromatic products* in the marketplace that meet almost all of the ASTM criteria for the identification of gasoline. Such products are used as solvents for stains, insecticides, and industrial and commercial products. The way to differentiate these aromatic products from gasoline is to look for the presence of alkanes. All gasolines contain high percentages of alkanes when fresh, but even highly evaporated gasolines contain some branched alkanes, such as the trimethylpentanes used as octane boosters. Straight-chain hydrocarbons, if they are present at all, are at very low concentrations. This is because they are undesirable components that cause knocking when burned in gasoline engines. The trimethylpentanes can be viewed by examining ions 57 and 71.

5.6.1.2 Identification of distillates

Most distillates that one is likely to encounter in fire debris are straight-run distillates from crude oil. They have not been subjected to cracking or reforming, so they lack the high aromatic content found in gasoline, but there are still aromatics present. The distillates are characterized by an abundance of normal alkanes, along with branched alkanes and cycloalkanes.

Distillates are usually easy to find, except for light petroleum distillates (LPD), which might be missed unless the analyst checks for them in each sample before calling it negative for ILR. Because of their high volatility, they tend to be present at low concentrations in fire debris samples and are typically found on the left side of the chart, where they might be mistaken for decomposition products. This is a particular hazard when the higher boiling components are decomposition products. LPDs do not generally exhibit the Gaussian distribution so frequently seen in MPDs and HPDs. When the peaks that elute prior to 8 min are examined, and the analyst sees a mixture of cycloalkanes, branched alkanes, and perhaps some normal alkanes in the C_7 through C_9 range, an LPD should be suspected. The analyst should have in the library as many LPDs as possible because, unlike the heavier distillates, the patterns tend to vary from one to the other. Figure 5.17a shows three different brands of cigarette lighter fluid, each exhibiting a different pattern of peaks. The recommended extracted ion profiles for LPDs are 57, 55, 83, and 91, which can be seen in Figure 5.17b. This particular LPD is unusual in that it contains xylenes, but no toluene is present. Because of their high volatility, LPDs are not persistent in the environment and are not usually expected to be found as background material. There are some cleaning agents and automotive products that contain LPDs, but if the sample was collected from somewhere other than the garage, the workshop, or under the kitchen sink, a finding of LPD generally indicates the presence of a foreign ignitable liquid.

This is not the case with medium or heavy petroleum distillates, which are far more common in our environment [25]. A typical medium petroleum distillate, Sparky Charcoal Lighter Fluid, is shown in Figure 5.18. The ions necessary to make a good identification of a medium petroleum distillate are 57 and 83. The analyst should also check for ion 105 or 91, to be certain that the aromatics have not been removed, in which case the proper classification of the residue would be as a de-aromatized distillate. One does not notice much difference between the TIC and the ion 57 EIC because distillates are dominated by the normal and branched alkanes. The *alkylcyclohexanes*, which elute about halfway between the normal alkanes, present an overall appearance similar to that of the alkane chromatogram. If the normal alkanes are not present, or if they are present at approximately the same concentration as the branched alkanes, an *isoparaffinic product* should be suspected. When the cycloalkanes are present at a concentration greater than about 20%, a *naphthenic–paraffinic* source is indicated. In most distillates, the cycloalkanes are present at about 5% to 10% of the concentration of the normal alkanes.

Medium petroleum distillates cover a wide range of products and can be used as fuels, such as lamp oil or charcoal starter, or as solvents, such as mineral spirits or insecticide carriers. This wide range of products and formulations requires that the library of reference materials include numerous MPDs.

HPDs include kerosene and diesel fuel. Except for HPDs that are formulated for specific applications, such as jet fuel, the carbon number range of heavy petroleum distillates can vary. Thus, unless one has a sample of the unevaporated

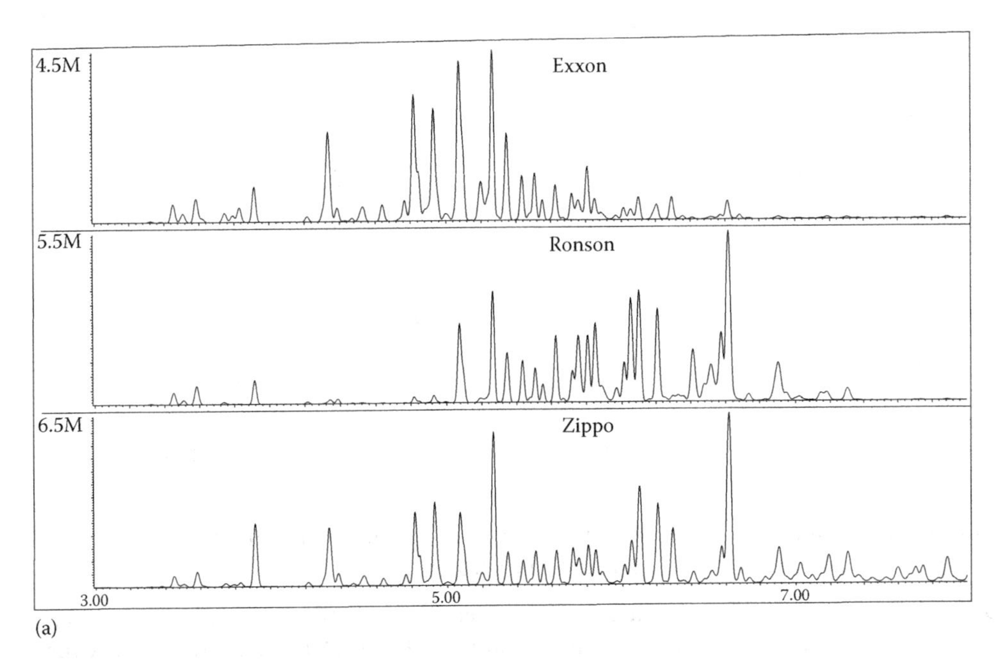

Figure 5.17 (a) Total ion chromatograms of three different brands of cigarette lighter fluid.

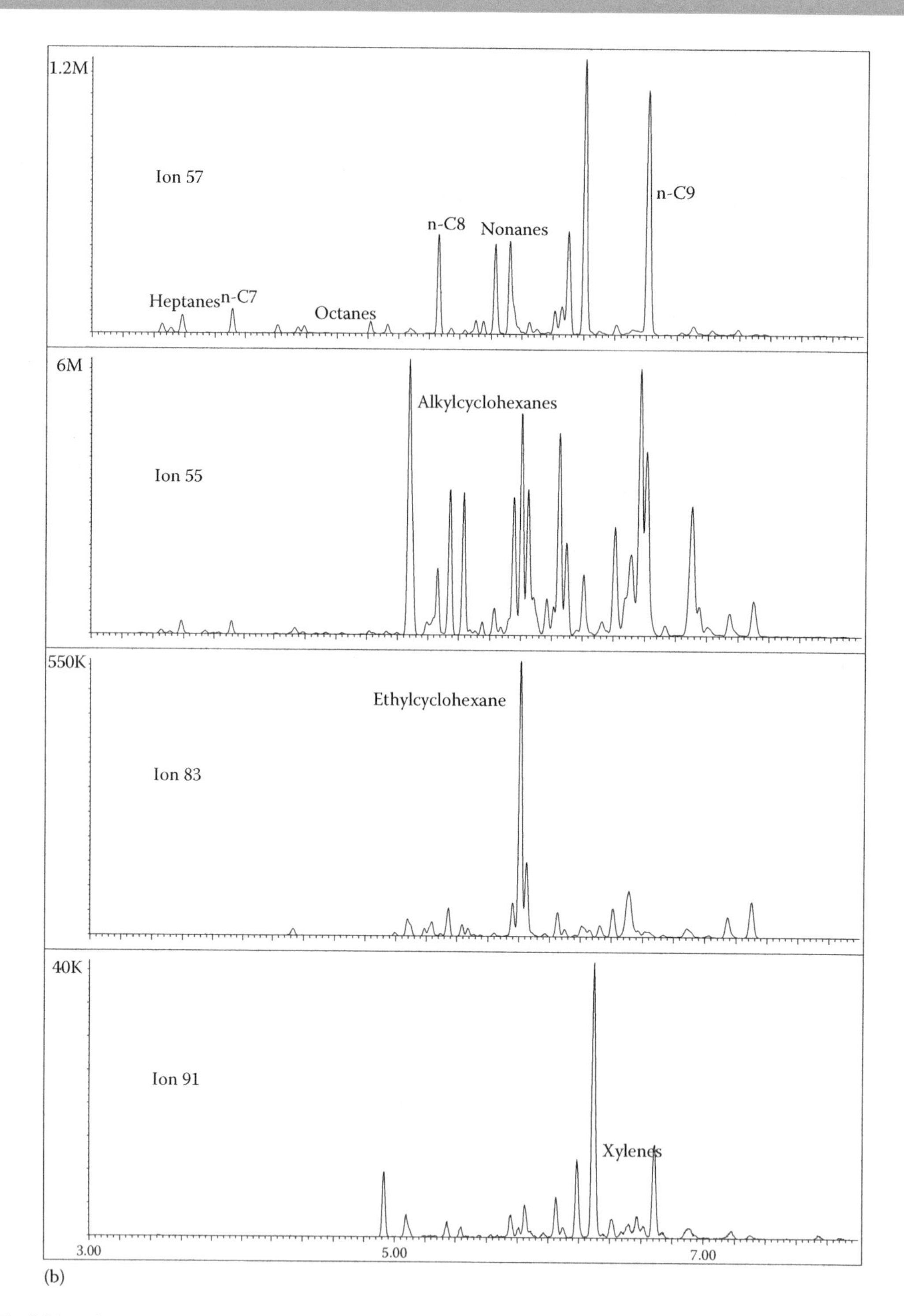

Figure 5.17 (b) Ion chromatograms of ions 57, 55, 83, and 91 from a standard of Ronson Lighter Fluid.

liquid, it is difficult to determine the degree of evaporation of the residue isolated from a sample. In northern parts of the country, diesel fuel sold in the winter may actually be kerosene.

Figure 5.19 shows a comparison of kerosene and diesel fuel. These distillates come with internal carbon number markers, in the form of pristane and phytane. Pristane is 2,6,10,14-tetramethylpentadecane ($C_{19}H_{40}$), and it elutes immediately after normal heptadecane. Phytane is 2,6,10,14-tetramethylhexadecane ($C_{20}H_{42}$), and it

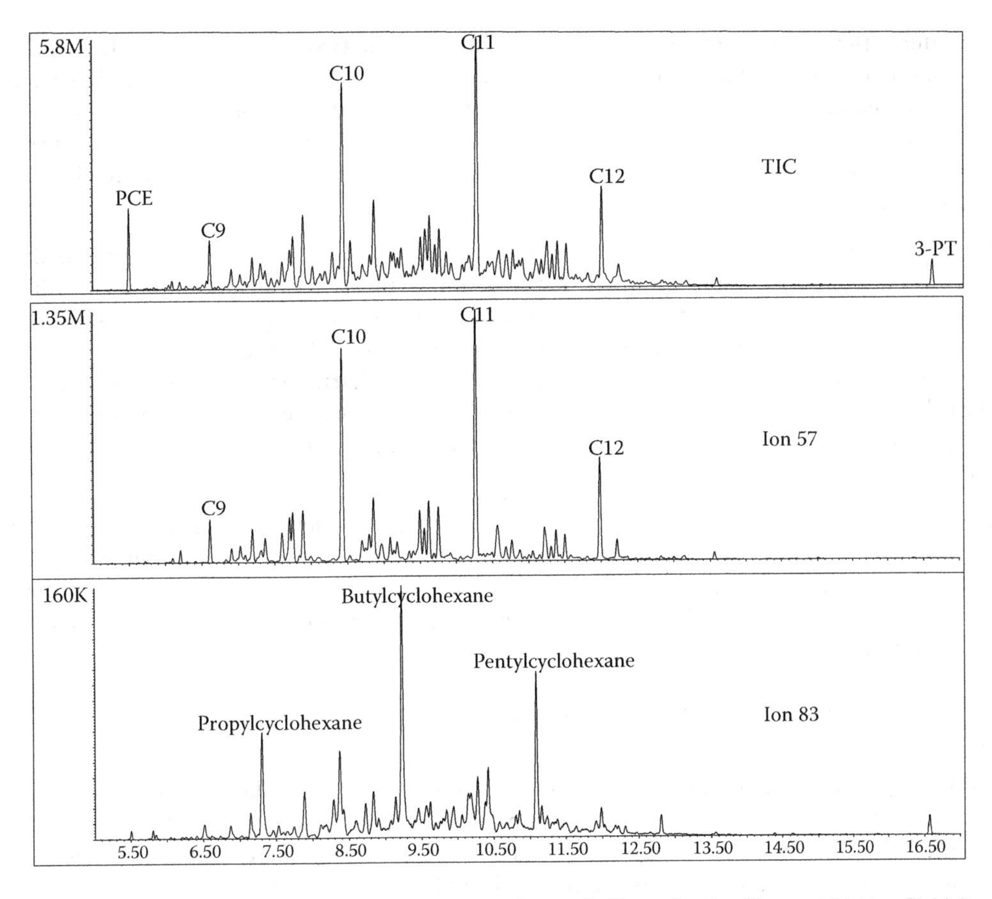

Figure 5.18 Total ion chromatogram of a typical medium petroleum distillate, Sparky Charcoal Lighter Fluid (top chart), and extracted ion profiles for ions 57 and 83.

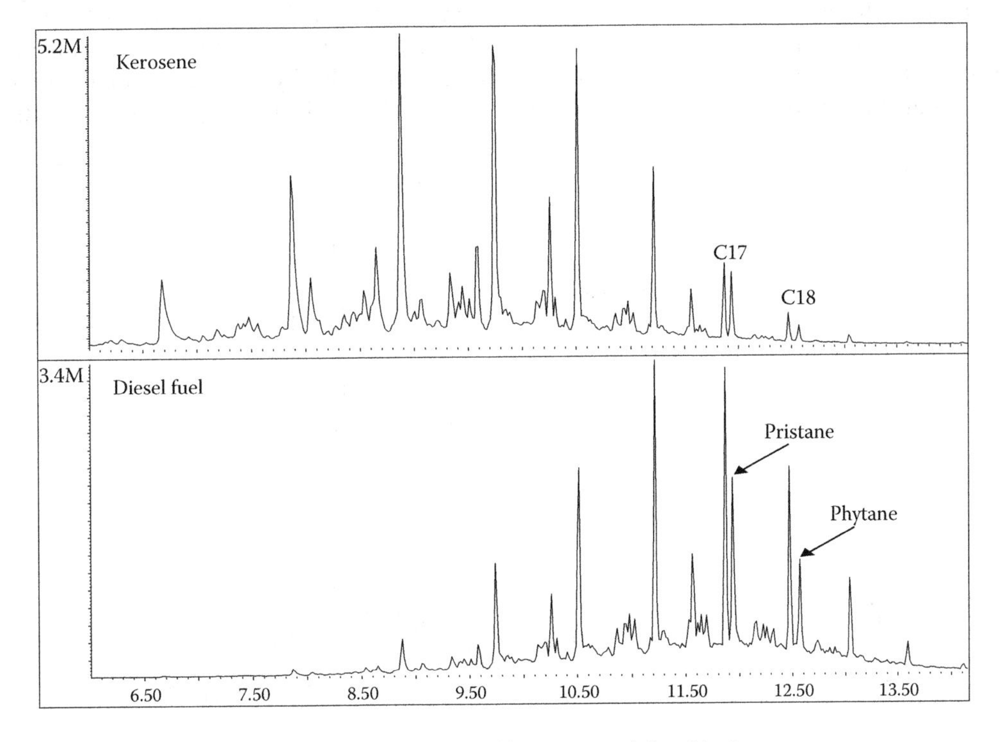

Figure 5.19 Total ion chromatograms showing a comparison of kerosene and diesel fuel.

elutes immediately after normal octadecane. Thus, we look for the two doublets on the high side of the bell-shaped curve and can count carbons up and down from there. In Figure 5.19, these doublets occur at 12 and 12.5 min. We can see that the kerosene, which is a known standard that has been evaporated to 50% of its original volume, covers the range from C_{11} to C_{19}, while the diesel fuel range is from C_{12} to C_{21}. If the kerosene is evaporated further, it will look more like the diesel fuel. Had the diesel fuel been evaporated less, it would look more like the kerosene.

As with the medium petroleum distillates, interpretation of a finding of HPDs should be approached with caution. Numerous household products contain HPDs, including many of the same kinds of products in which MPDs are found. The safer charcoal lighters are made from kerosene rather than mineral spirits. They are safer because of their higher flash point. Figure 5.20 shows a total ion chromatogram of lemon oil furniture polish. This polish has a carbon number range from C_{12} to C_{22}, with C_{17} being the tallest peak on the chart. The limonene peak at the left side of the chart could easily be attributed to a pine substrate. Pinenes and limonene are very common in samples containing structural (coniferous) wood. This particular furniture polish, if found in a fire debris sample, could easily be reported as diesel fuel.

It is not necessary for a liquid to be present in order for a distillate to be detected. Figure 5.21a shows the total ion chromatogram of a piece of pine wood that was stained with Minwax finish 10 months before it was subjected to headspace concentration. The naturally occurring terpenes α- and β-pinene and D-limonene are the dominant peaks on the chart, but the mineral spirits solvent is still clearly visible. When ions 57 and 83 are extracted, the terpenes disappear to yield the charts shown in Figure 5.21b.

> These results show the critical necessity of asking for comparison samples, particularly when samples of flooring are submitted for analysis.

The flooring does not need to be recently painted in order to exhibit the distillate solvent used to apply the floor coating. This author has reported finding distillates 24 months after application [26] and has detected distillates in samples of finished flooring and furniture up to 10 years old. There is no reason to believe that these solvents do not persist indefinitely, trapped in either the wood matrix or the polymer coating matrix.

In addition to true distillates, there are some distillate-like residues produced as the result of decomposition of other products. *Asphalt* is what is left at the bottom of the distillation pot after all of the volatiles have been distilled from crude oil. It contains hydrocarbons ranging from C_{30} to C_{60}. When these long-chain hydrocarbons undergo pyrolysis, they do so in much the same way as the long-chain hydrocarbons in polyethylene (i.e., via random scission). This results in the production of normal alkanes in the range of C_9 to C_{18} [27]. When these pyrolysis products are present

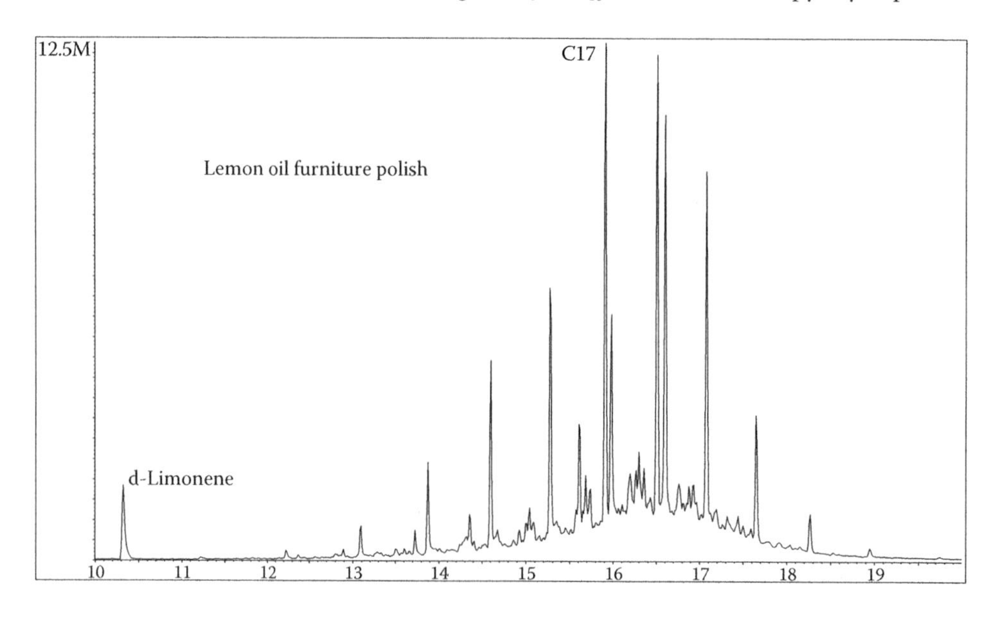

Figure 5.20 Total ion chromatogram of lemon oil furniture polish. This sample could easily be reported as diesel fuel.

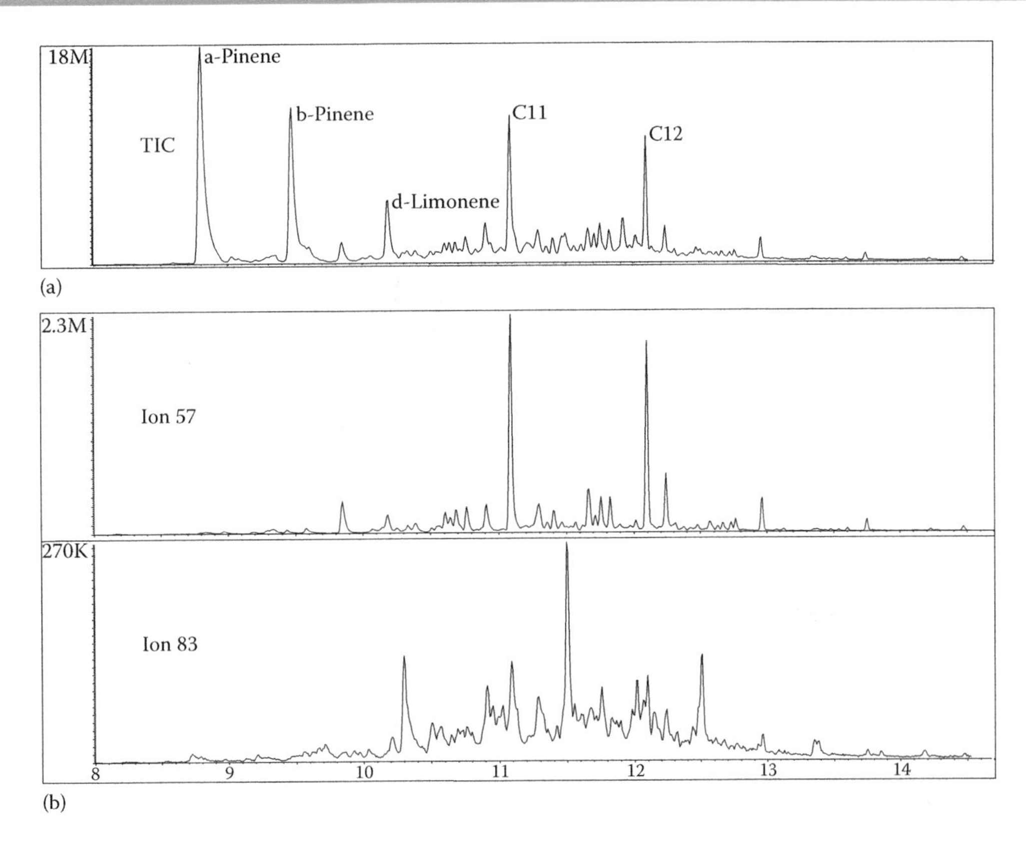

Figure 5.21 (a) Total ion chromatogram of a piece of pine wood stained with Minwax Finish 10 months prior to its analysis. (b) Extracted ion profiles for ions 57 and 83.

in fire debris and subjected to headspace concentration, they produce a chromatogram that can be and has been mistaken for the chromatogram of a heavy petroleum distillate. Such a chromatogram is shown in Figure 5.22a. In 1982, it was reported that roof shingles could produce "accelerant-like residues," and at that time, there existed no reliable method for distinguishing HPDs from asphalt shingle residues [28]. The increased use of capillary columns allowed for the occasional visualization of a *double-peak kerosene*, sometimes called "pseudo kerosene," but even with capillary columns, it was not uncommon for asphalt residue to be misidentified as a liquid petroleum distillate. This author learned how to make the differentiation in 1995 in connection with the investigation of an insurance claim that had been erroneously denied because of a finding of HPDs where none should have been present (This case, *Clark v. Auto Owners*, is discussed in detail in Chapter 9). Figure 5.22b shows how the distinction is made. One compares ion 57 with ion 55. If a second peak appears in front of the n-alkane peak, or if a small peak grows larger, we can conclude (particularly after we collect the mass spectrum) that the second peak is the 1-olefin. A sample such as this must be classified as asphalt smoke condensate or asphalt decomposition residue. In the presence of such an asphalt residue, the presence of gasoline in a sample should be interpreted with caution. If the building in question had an asphalt-covered roof, the possibility that "cutback asphalt" was used on the roof should be explored. Cutback asphalt is asphalt that has been thinned with a volatile solvent such as gasoline. The specifications for this "rapid curing" asphalt allow for up to 45% solvent [29].

The amount by which the olefin peak "grows" varies, depending on the sample. When one looks at kerosene and diesel fuel, the only difference in appearance between the ion 55 chart and the ion 57 chart is that the abundance of ion 55 is lower. The relative abundances of the individual peaks with respect to each other do not change.

Another way to distinguish asphalt decomposition products from HPDs is to look for the cycloalkanes. They are not present. There is a Gaussian pattern observed when ion 83 is extracted, but this is due entirely to the presence of the olefins. The ion 83 and ion 55 ion chromatograms show the same peaks. Asphalt smoke condensates also contain little, if any, pristane and phytane.

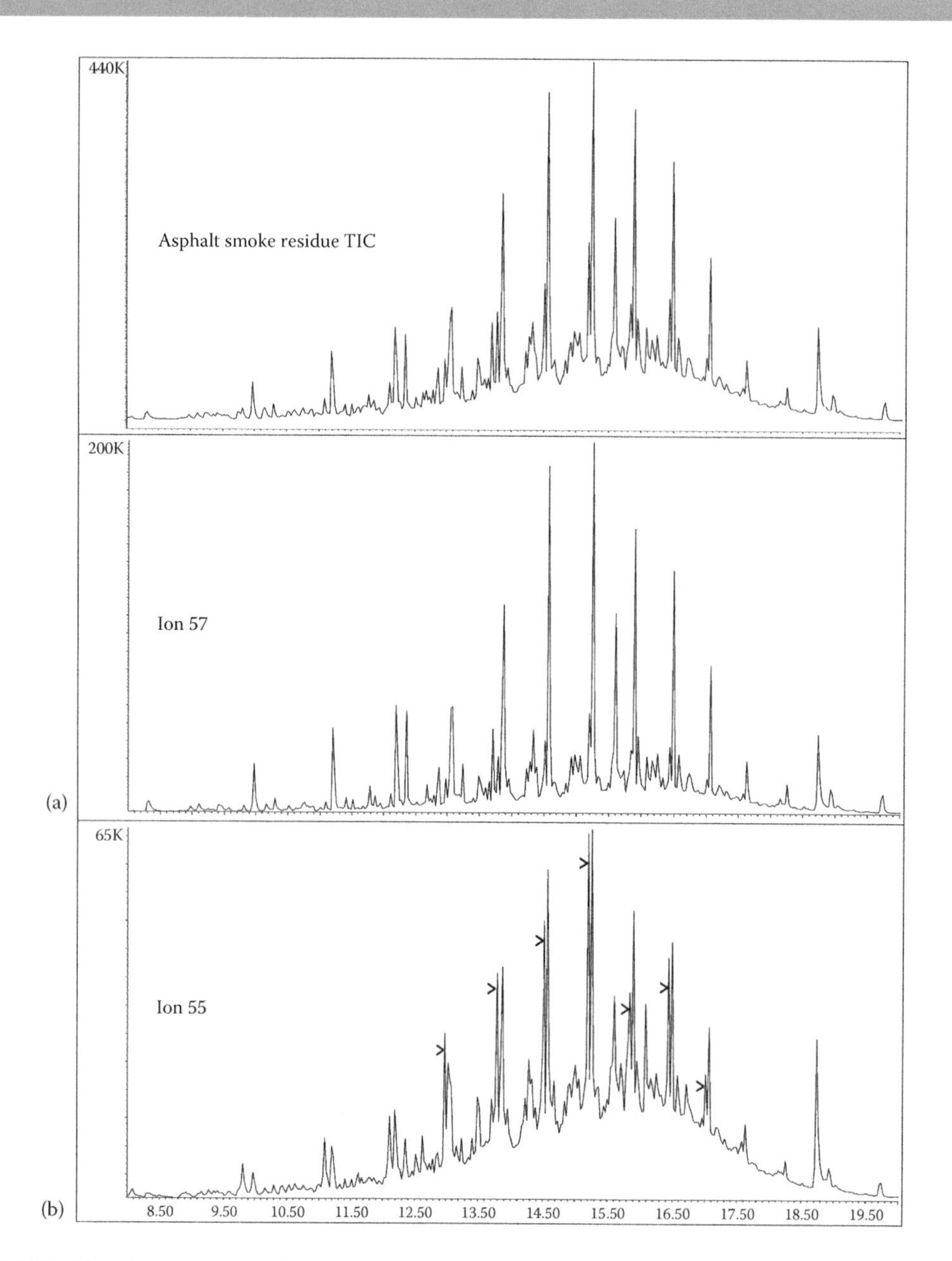

Figure 5.22 (a) Total ion chromatogram of asphalt smoke residue, of the type that can be mistaken for a petroleum distillate. (b) Extracted ion profiles for ions 57 and 55. The "growth" of the peak in front of the normal alkane when comparing 55 to 57 demonstrates the presence of *alkenes*, which allows the identification of the residue as a decomposition product rather than a foreign petroleum distillate.

Polyethylene is another substance that produces a distillate-like appearance in the chart, but if the capillary column has any resolution at all, it will be obvious that one is looking at polyethylene residue rather than at a distillate. Figure 5.23 shows the total ion chromatogram of polyethylene smoke condensate. All the peaks are doublets, and most are actually triplets, owing to the presence of the 1, (n-1)-diene in the mixture. When one looks at ion 57 and compares it with ion 55, the growth in the olefin peak is obvious relative to the alkane peak. The diene peak also grows because the dienes contain more ion 55 than ion 57. Lubricating oils are subject to the same decomposition processes as asphalt and polyethylene, and the smoke condensates of lubricating oils appear similar.

A careful examination of the chromatographic and mass spectral data prevents the analyst from misidentifying decomposition products as distillates, but automation of pattern recognition is available and presents some dangers if not

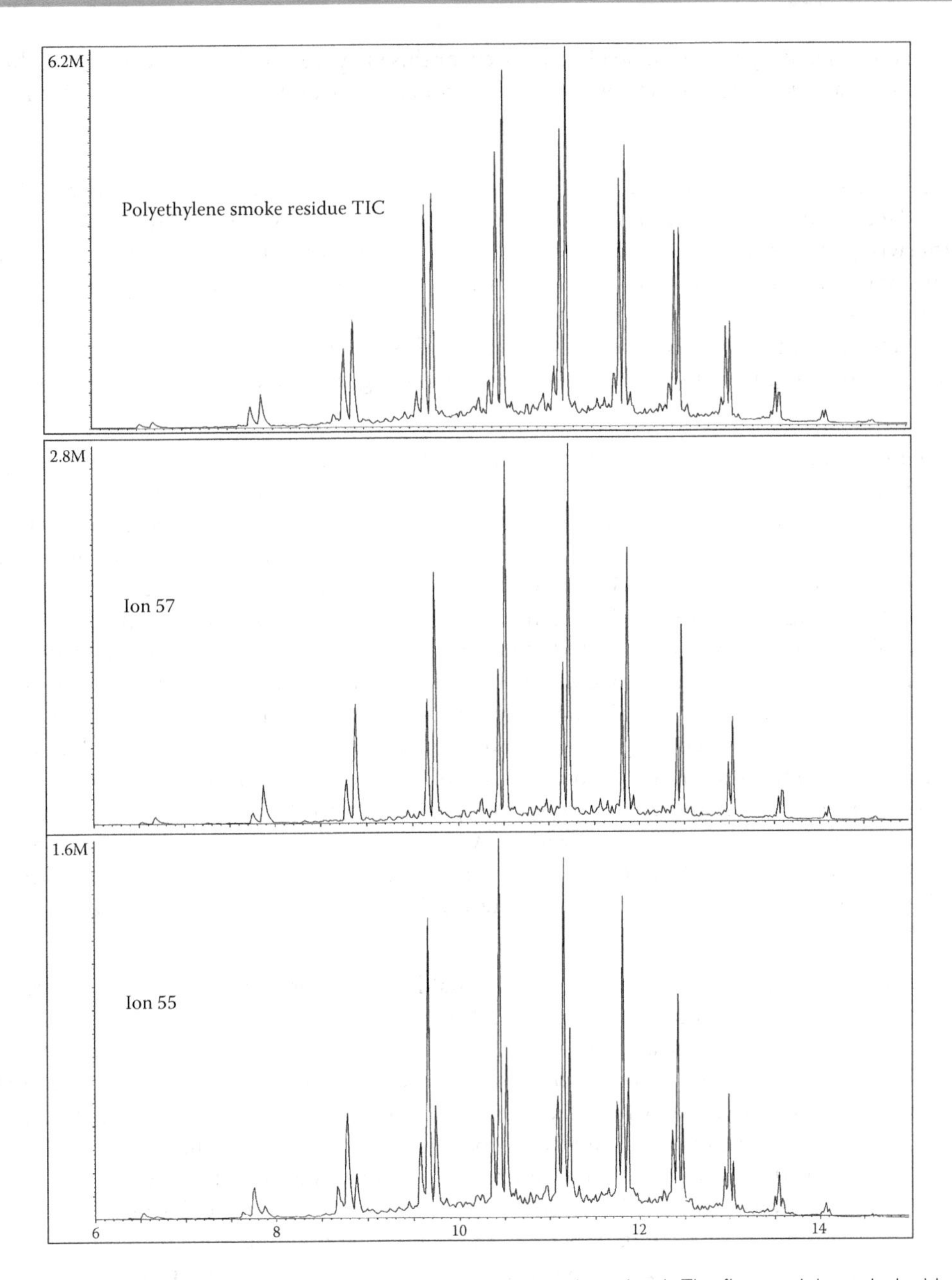

Figure 5.23 Total ion chromatogram of polyethylene smoke condensate (top chart). The first peak in each doublet is the alkene; the second is the alkane. Pristane, phytane, cycloalkanes, and aromatics are all absent. The middle chart is the ion 57 EIC, and the bottom chart is the ion 55 EIC. In the ion 55 chromatogram, a third peak, representing the n, (n-1)-diene, appears.

handled properly. As with all computer "answers," one should always do a "reality check." The same pattern recognition software that allows matching a mass spectrum from an unknown compound against the spectra of 250,000+ compounds[2] can be applied to chromatograms. The chromatogram is converted using Microsoft Excel into a bar graph that has the same general appearance as a mass spectrum. This graph can then be compared against a library of known ignitable liquids that have been similarly transformed. Like the mass spectral libraries, "extra" peaks do not necessarily keep the library from recognizing a "match." The polyethylene smoke condensate TIC shown in Figure 5.23, matched against a library that contains only ignitable liquids, yields a match of exceptionally high quality for diesel fuel. This is

[2] As of January 2018, the NIST mass spectral library contained spectra for 267,376 chemical compounds.

why it is necessary to populate ignitable liquid libraries used for this purpose with known background chromatograms. The chart would make an even better match for polyethylene smoke condensate, but only if it is in the library.

5.6.1.3 Identifying other classes of products

The remaining classes of ignitable liquid residues can frequently be identified by what is and is not present. A *de-aromatized distillate,* for example, has a signal from the aromatics that is less than 1% of the signal from the aliphatics; otherwise, it looks the same with respect to both the alkanes and the cycloalkanes. Because of environmental regulations governing the aromatic content of distillates, there has been, since 1973, an ASTM standard for this determination. ASTM D3257-06 (2012) is entitled *Standard Test Methods for Aromatics in Mineral Spirits by Gas Chromatography* [30] and uses a specified test blend for calibration. If the calculation of aromatic content is an issue, a forensic scientist might avoid a *Daubert* challenge by using this established method rather than devising a new one.

The *normal alkane* products are very easy to recognize as long as one makes sure that one is looking at a homologous series of normal alkanes, as opposed to a homologous series of aldehydes (commonly found in sweat) or some other group of homologs that are pyrolysis or decomposition products. Groups of compounds that differ from each other only in that they have one additional CH_2 group look pretty much the same as a series of normal alkanes and require some caution. Figure 5.24 shows a series of normal alkane products marketed by Exxon, as well as a sample of Lamplight Farms Ultra Pure Lamp Oil, which also consists of normal alkanes. One common source of normal alkanes is carbonless forms (NCR paper), and the bottom chromatogram in Figure 5.24 shows the chromatogram of concentrated headspace vapors from a 2 × 2 inch square of a carbonless form. These forms contain microspheres filled with normal alkane solvent that, when broken, causes color to develop in the ink. There are only a few microliters per square meter, but this concentration is easily detectable. Figure 5.25 is a scanning electron micrograph that shows the bottom side of a carbonless form. We should note that the alkanes found in carbonless forms are usually accompanied by a pair of substituted biphenyls. Normal alkane products are also found in some brands of vinyl floor covering. Anytime a sample is collected from a floor likely to have vinyl in its structure (e.g., kitchen, bathroom, or laundry room), a finding of normal alkanes is probably not meaningful. Some vinyl floor coverings may also contain HPDs. A comparison sample is an absolute necessity in such cases.

Isoparaffinic hydrocarbons are made by removing the normal hydrocarbons with a molecular sieve. These liquids are becoming more and more common as petrochemical manufacturers move to more environmentally friendly and odor-free replacements for the straight-run distillates such as mineral spirits. Similar to the isoparaffinic hydrocarbons are the *naphthenic–paraffinic products,* which are characterized by an abundance of cycloalkanes. Whereas one might expect to find cycloalkanes present at less than 5% in the isoparaffinic hydrocarbons, they may be present at up to 30% in naphthenic–paraffinic products. The distinction requires looking at the ion profiles for the cycloalkanes versus the alkanes, as well as looking at the abundance numbers on the left side of the chart. Figure 5.26 shows a comparison of the ion 57 and ion 83 chromatograms from Isopar H. Compare this to Figure 5.27, which compares those same ion chromatograms from a naphthenic–paraffinic solvent, Vista LPA 170. Instead of having an ion 83 chromatogram that is 2% the height of the ion 57 chromatogram, in the naphthenic–paraffinic solvent, the ion 83 chromatogram is more than 30% of the height of the ion 57 chromatogram. Also note that the tall peaks in the ion 57 chromatogram from the naphthenic–paraffinic product are not normal alkanes but are branched alkanes.

Another way to make the distinction between the isoparaffinic products and the naphthenic–paraffinic products is to take an average mass spectrum. Representative isoparaffinic products are shown in Figure 5.28. A mass spectrum of any of these looks very much like the mass spectrum of a single alkane. Ion 57 is the base peak and the other fragments are spread out in a Gaussian distribution with a spacing of 14 mass units between them. A typical isoparaffin average mass spectrum, that of Isopar L, taken over a 3-minute portion of the chromatogram, is shown in Figure 5.29.

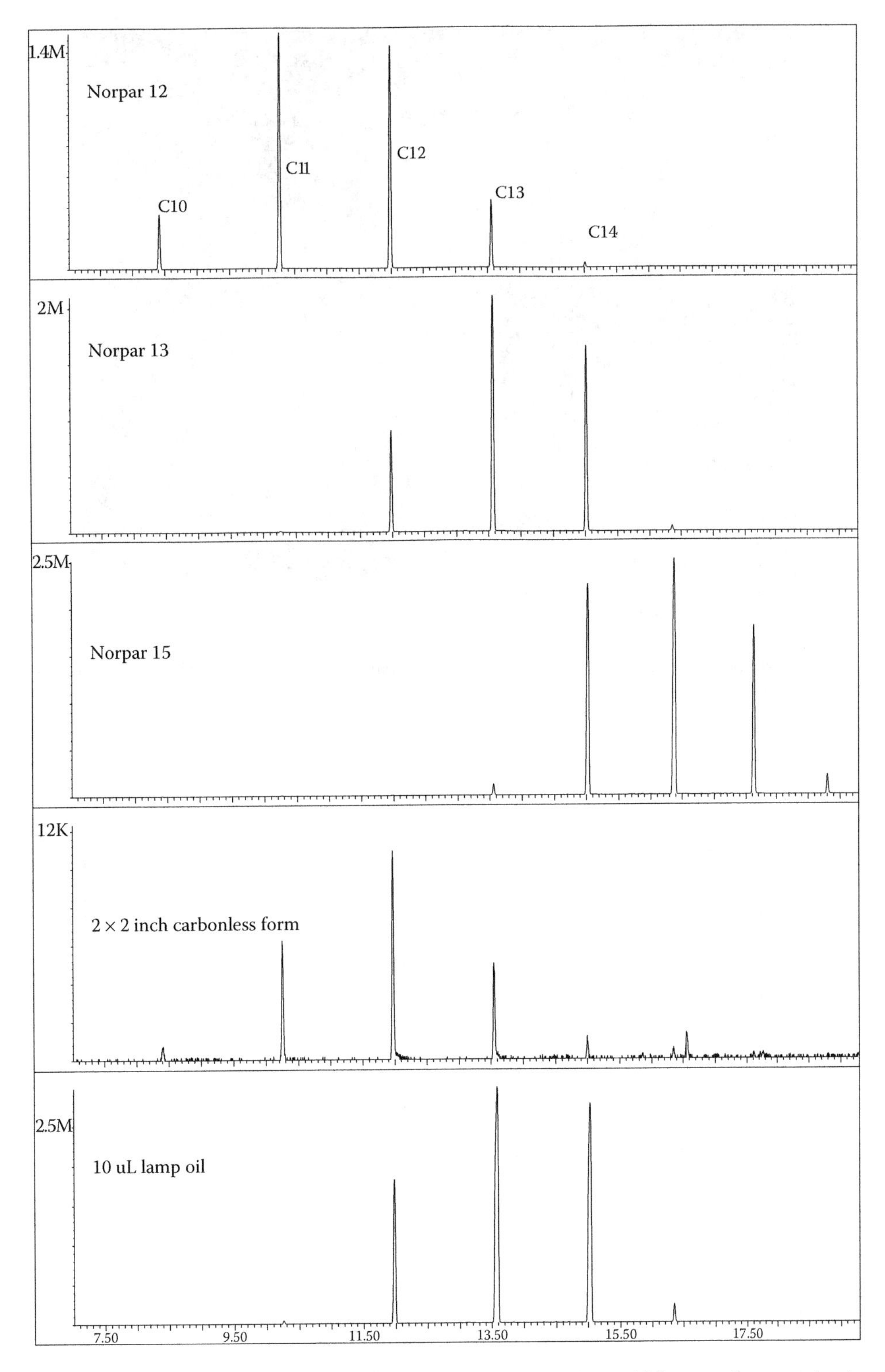

Figure 5.24 Ion 57 chromatogram of three Exxon Isopar products, compared with an ACS extract from a carbonless form, and the ACS extract of a 10-µL sample of Lamplight Farms Ultrapure Lamp and Candle Oil.

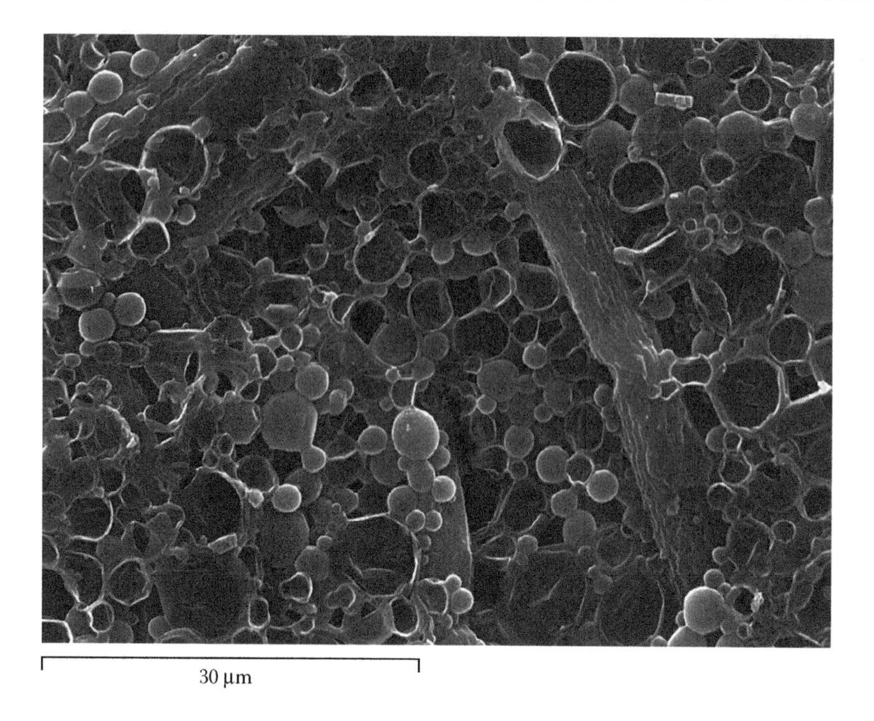

Figure 5.25 Scanning electron micrograph of the underside of a carbonless form. The micro-spheres range in size from 2 to 10 μm and are filled with normal alkanes.

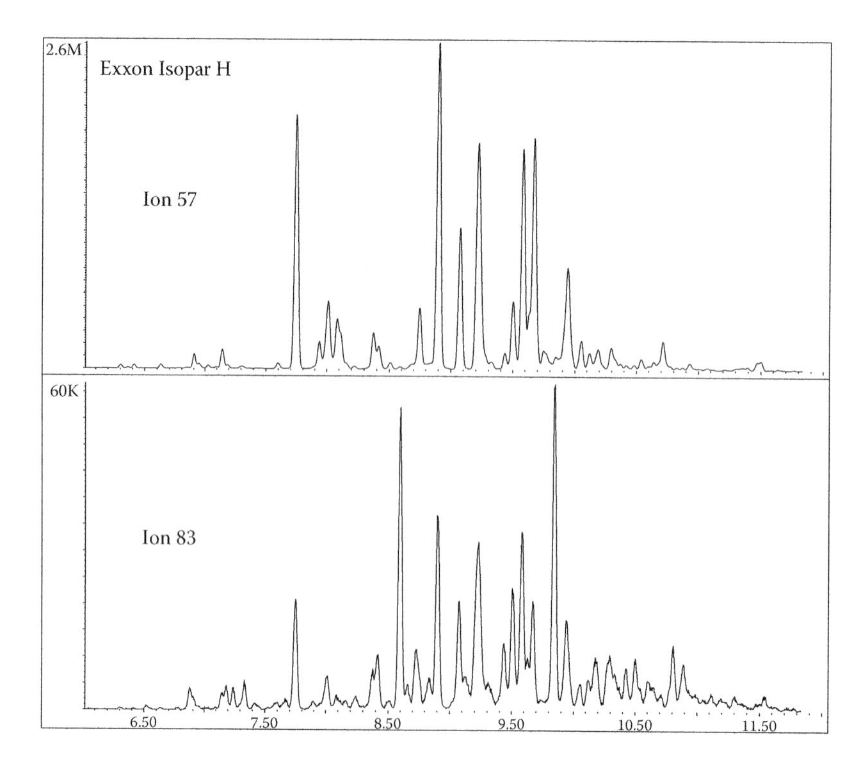

Figure 5.26 Comparison of the alkane (ion 57) and cycloalkane (ion 83) chromatograms from Exxon Isopar H. The cycloalkane peaks are only about 2% of the height of the branched alkane peaks.

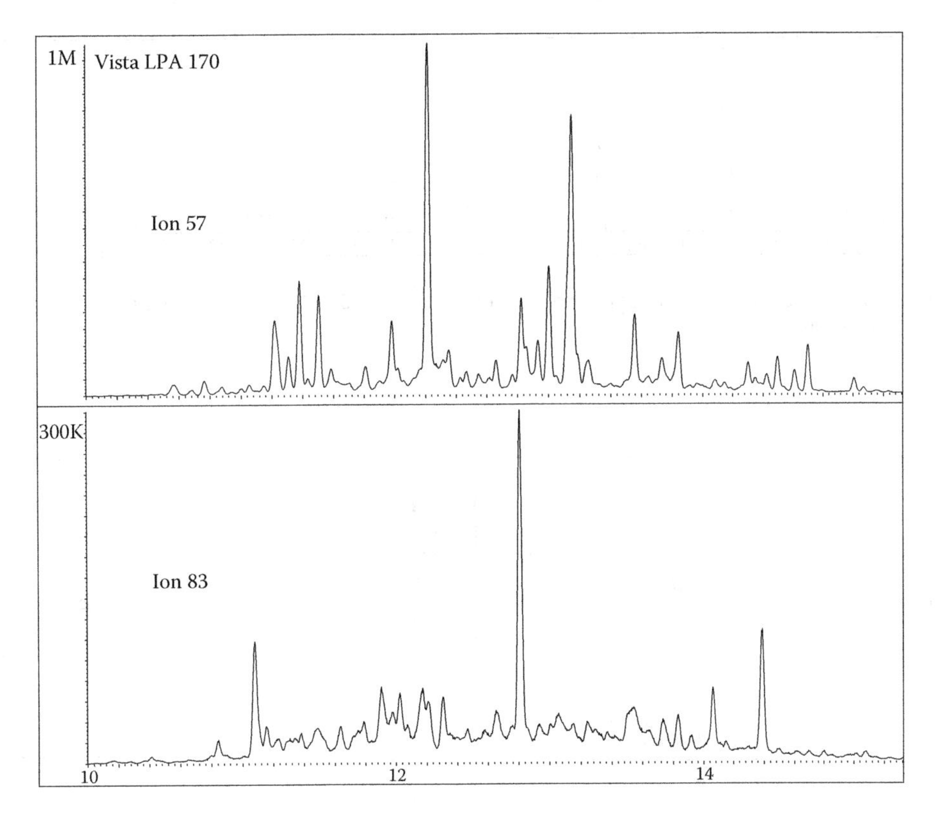

Figure 5.27 Comparison of the alkane (ion 57) and cycloalkane (ion 83) chromatograms from Vista LPA 170, a naphthenic–paraffinic product. The height of the cycloalkane peaks is about 30% of the height of the branched alkane peaks.

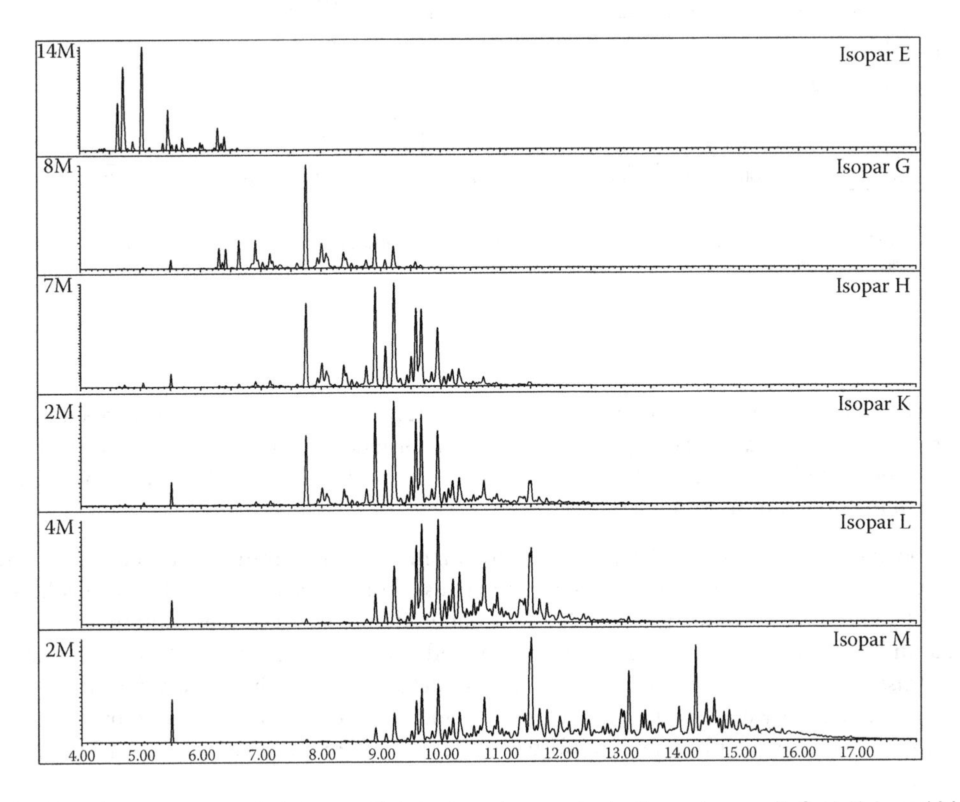

Figure 5.28 Total ion chromatograms of six isoparaffinic hydrocarbon products: Exxon Isopars E, G, H, K, L, and M.

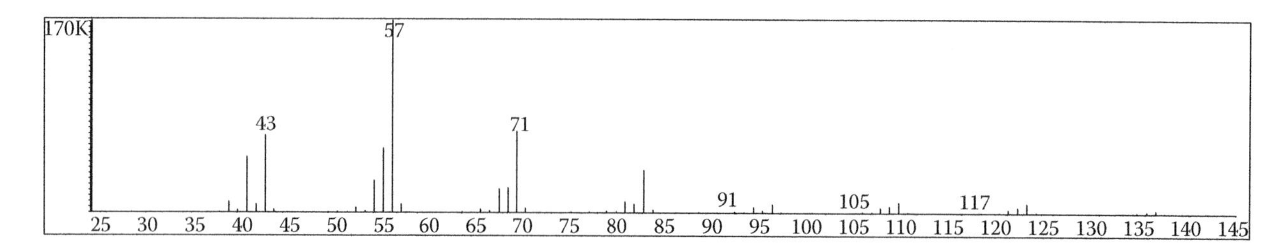

Figure 5.29 Average mass spectrum taken from Isopar L, over the range of 9 to 12 min. This spectrum has a very similar appearance to that of a branched alkane.

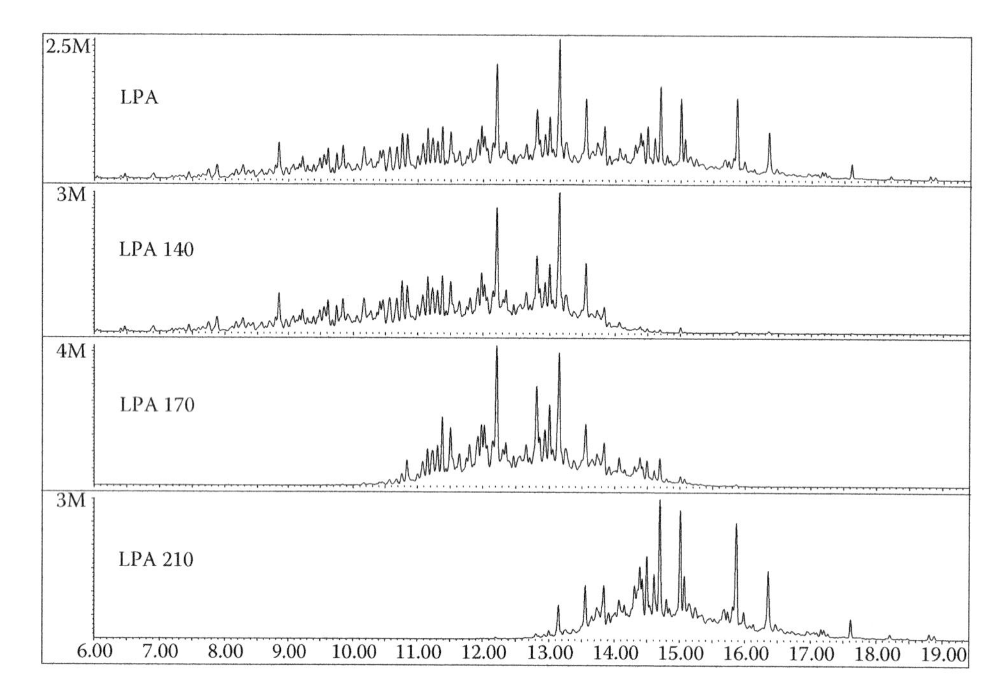

Figure 5.30 Total ion chromatograms of four naphthenic–paraffinic products: Vista LPA, LPA 140, LPA 170, and LPA 210. The numbers in the product names approximately correspond to flash points.

The same exercise can be performed on the naphthenic–paraffinic products, several of which are shown in Figure 5.30. The average mass spectrum of a typical naphthenic–paraffinic product (shown in Figure 5.31) looks quite different from that of the isoparaffinic product. While it still exhibits a base peak of 57, there is a greater abundance of ions 55, 69, and 83 than in the isoparaffinic product.

The identification of other ignitable liquid residues requires an individual examination of the peaks in the chromatogram. It is possible to find just about any single compound in just about any sample. Findings of alcohols, turpentines, aromatic solvents, and other "flammable" liquids should be undertaken with great caution. ASTM E1618 recommends not making an identification of these single compounds unless they are present in such concentrations that the signal is *at least two orders of magnitude greater than the background.* Just about every fire debris sample contains methanol, and just about every fire debris sample contains toluene. Unless these substances are present at concentrations sufficiently high for the analyst to feel comfortable saying they are not native to the background, they should not be reported.

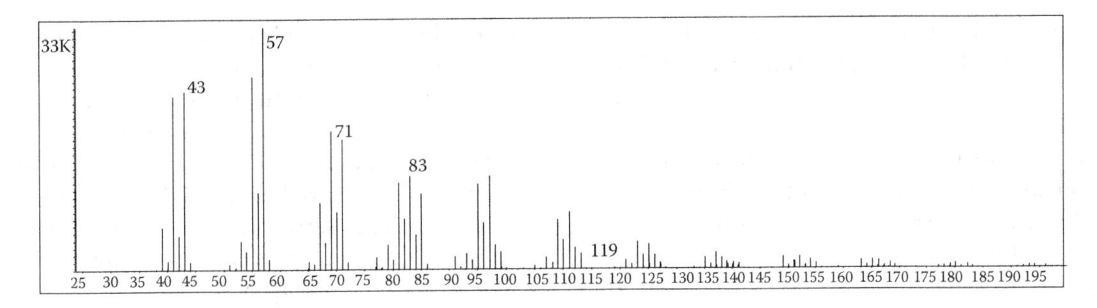

Figure 5.31 Average mass spectrum of Vista LPA, taken from 8 to 17 min.

5.6.2 Improving sensitivity

The techniques routinely used for isolating and analyzing ignitable liquid residues are quite sensitive. In samples with nontenacious substrates, it is possible to detect as little as one-tenth of a microliter of ignitable liquid residues such as gasoline in a kilogram or more of debris. The task gets a little more difficult when there are active surfaces in the fire debris, such as charcoal, but an activated carbon strip is more tenacious than most debris one is likely to encounter.

Before attempting to improve the sensitivity of the detection method, it is necessary to get a handle on how much sensitivity one already has. Although fire debris analysis is generally considered a "qualitative" exercise, it is really semi-quantitative, not so much as it pertains to the amount of ILR in a sample but as it pertains to the *relative* concentrations of the individual components. A laboratory's validation documents should contain detection limits of whole liquids. This information is best obtained by preparing serial dilutions of an ignitable liquid with an internal standard and comparing the signals to those generated by extracts from known amounts of ignitable liquid run through the laboratory's adsorption/elution process. In this author's laboratory, we typically use 10 μL standards. The liquid is applied to a piece of filter paper in a 1-quart can, and the filter paper is subjected to ACS headspace concentration, just like any debris sample. When exposed to 10 μL of 75% evaporated gasoline, a 10 × 10 mm carbon strip eluted with 500 μL diethyl ether spiked with 100 ppm perchloroethylene yields a concentration of approximately 1,000 ppm. This is easily detectable; 1 μL in the sample can yield a concentration of approximately 100 ppm in the eluate—again, very easily detectable. At 0.1 μL (10 ppm in the eluate), the signal is noisy and it becomes difficult to extract acceptable ion chromatograms except for ion 105, showing the C_3 alkylbenzenes. To obtain better ion chromatograms, a number of different strategies can be employed.

The lower limit of detection is formally defined as that concentration of analyte that gives a signal that is twice the background level; for this analyst, the detection limit is that concentration of analyte that produces a signal large enough to feel comfortable about a determination. We can lower the detection limit by either increasing the signal or by decreasing the background noise. Increasing the signal can be accomplished in a number of ways. One obvious way is to increase the concentration of the analyte. In the case of fire debris eluates from C-strips, this is accomplished by evaporating some of the solvent. This strategy is not without its costs. It is time-consuming and must be done very carefully if one is to avoid losing the analyte along with the solvent. No heat should be used in this process. At some point, the volume of the solution becomes too small to be handled by an autosampler, and this requires manual injection. While it can be done, easier ways exist to increase the signal. Another obvious strategy is to simply increase the size of the injection. Normally, we inject 1 μL; when we are looking at samples of low concentration, we inject 2 μL.

One way to dramatically increase the signal is to run the gas chromatograph in the *splitless mode.* (Running splitless is one reason that SPME techniques are so sensitive.) Most laboratories use a split ratio of 20:1 to 50:1. This dilution of the sample improves resolution. Turning off the split increases the amount of analyte presented to the detector, and the signal increases by up to a factor of 10. The only cost is likely to be a small decrease in resolution.

Another way to increase the signal from the ions of interest in order to produce better ion chromatograms is to increase the dwell time (i.e., the time that the detector spends looking at a particular ion). This is accomplished by using the selected ion monitoring (SIM) mode on a quadrupole instrument. In the full scan mode, we typically look from 33 to 300 amu, resulting in a dwell time of less than 1 msec per ion. If we ask the instrument to look only at the ions in which we are interested, we can increase the dwell time dramatically. The detector response increases in proportion to the square root of the dwell time; that is, if we increase the dwell time by a factor of 4, the signal strength increases by a factor of 2. Our instrument performs 5.24 scans per second in the full scan mode, resulting in a dwell time of 0.7 msec. In the SIM mode, there is only about one scan per second, resulting in somewhat rougher peaks, but with a dwell time of 50 msec (an increase of about 70×), the signal increases by about a factor of 8. It is not the increasing signal that is the most attractive feature of SIM, however; it is the reduction in noise. "Changing channels" 267 times per scan (1,400 times per second) generates an abundance of noise in the instrument. This is known as "housekeeping noise." If we only look at the 23 ions listed in Table 5.4, we reduce the noise by a factor of 60 or more. Figure 5.32 shows a comparison of the noise generated in the full scan mode versus the noise generated in the selected ion monitoring mode. It is easy to see that, even with a much smaller signal, the peaks of interest are plainly visible.

The price to pay for this increase in sensitivity is a decrease in specificity (there is no such thing as a free lunch). The mass spectra produced in the SIM mode do not contain enough ions for a library to match. The analyst can, however, make certain that the ion of interest is the base peak for a particular compound and also make sure that there is a "qualifier ion" present. The abbreviated spectrum from a SIM peak can be compared with the SIM spectrum from a known compound—but that is as far as it can go. For this reason, the analyst should apply more rigid criteria when trying to determine whether a SIM pattern exhibits "sufficient similarities" to a standard pattern.

Table 5.4 Ions used for SIM

Ion	Compounds
31	Methanol
43	Alkanes
45	Ethanol
55	Alkenes, cycloalkanes
57	Alkanes
69	Alkenes
71	Alkanes
83	Cycloalkanes, alkenes
85	Alkanes
91	C_1, C_2 alkylbenzenes
104	Styrene
105	C_3 alkylbenzenes
117	Indan, methylindans
118	Methylstyrene
119	C_4 alkylbenzenes
120	C_3 alkylbenzenes
128	Naphthalene
131	Methyl, dimethylindans
133	C_5 alkylbenzenes
134	C_4 alkylbenzenes
142	Methylnaphthalenes
156	Dimethylnaphthalenes
168	3-Phenyltoluene

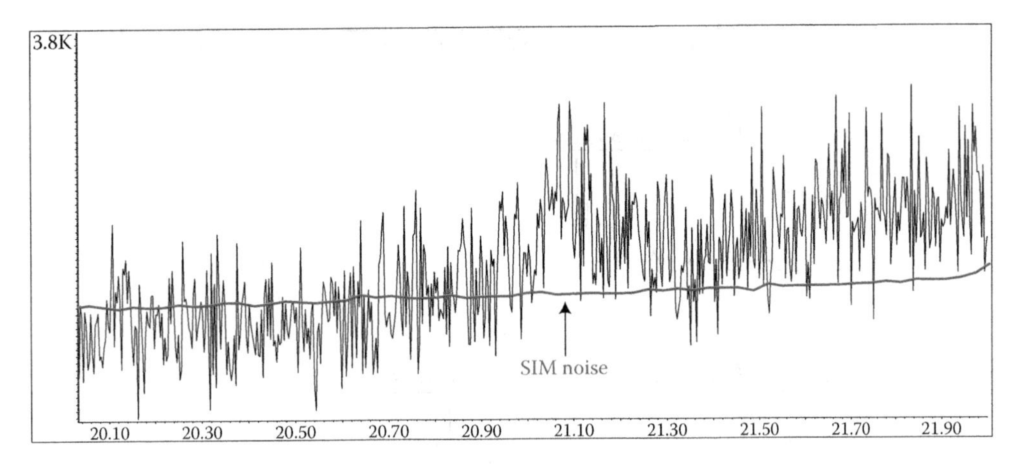

Figure 5.32 Comparison of the baseline noise generated by a full scan with that generated in the selected ion monitoring (SIM) mode.

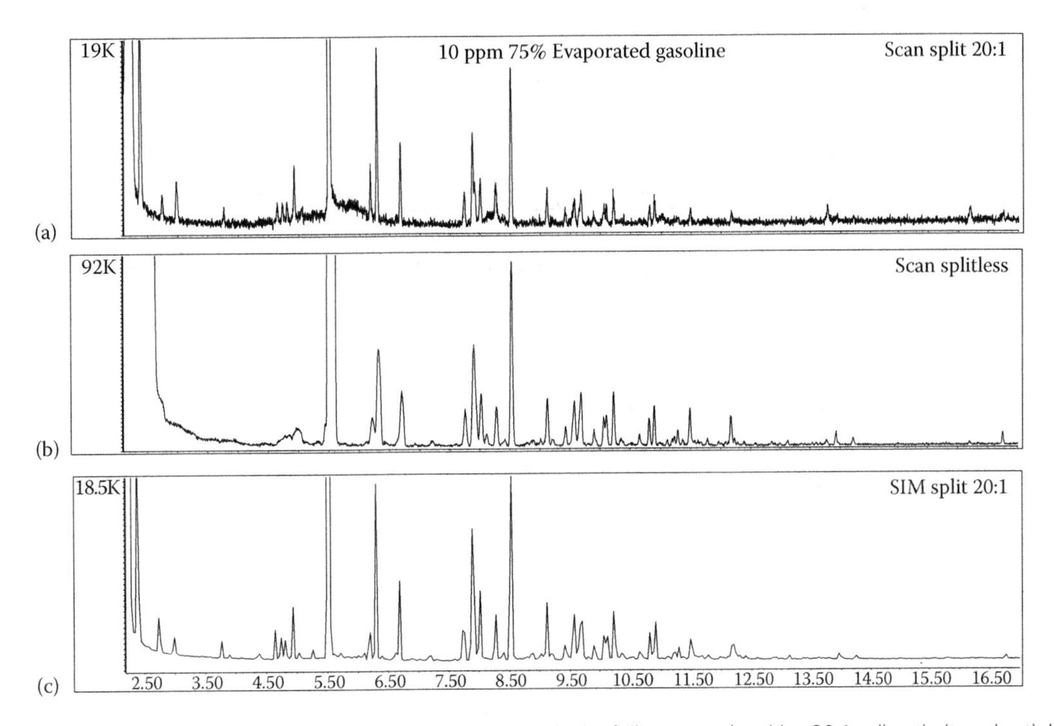

Figure 5.33 (a) A 10 ppm solution of 75% evaporated gasoline, run in the full scan mode with a 20:1 split ratio (top chart). In each of these three charts, the off-scale peak at 5.5 min is the perchloroethylene internal standard. (b) A 10 ppm solution of 75% evaporated gasoline, run in the full scan mode with a splitless injection (middle chart). The abundance is increased by about a factor of 5 over the split injection. (c) A 10 ppm solution of 75% evaporated gasoline, run in the SIM mode with a 20:1 split ratio. The abundance is not increased compared to the scan mode, but the noise is significantly reduced.

Figure 5.33 shows a comparison of a 10 ppm standard of 75% evaporated gasoline run in the full scan mode, in the full scan splitless mode, and in the SIM mode. Looking at just the total ion chromatograms, one might conclude that the SIM mode does not add value compared to the splitless mode, but when one generates the six extracted ion profiles for gasoline, the SIM chromatograms clearly outshine those produced in the full scan mode, even with no split. This is demonstrated in Figure 5.34.

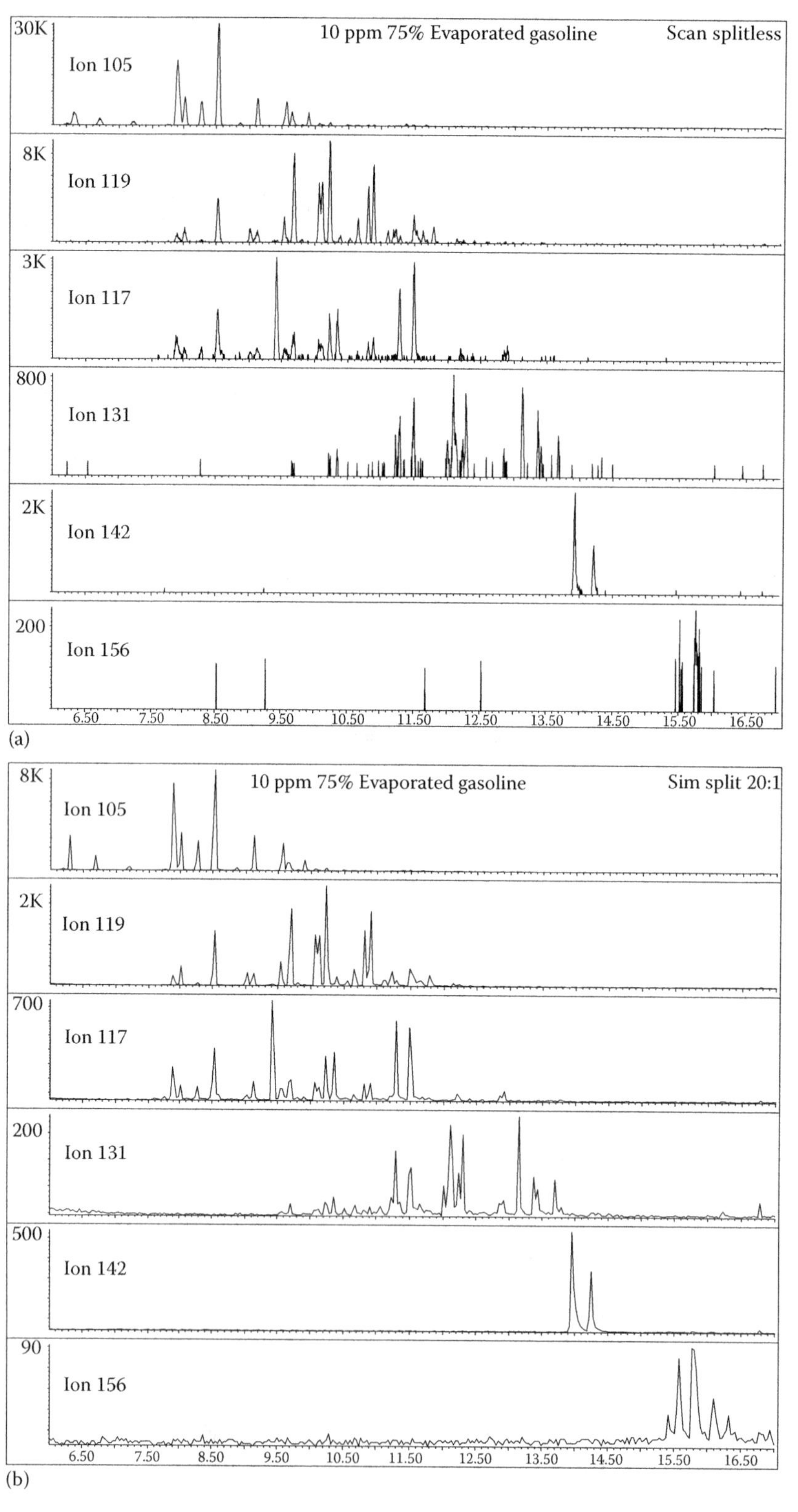

Figure 5.34 (a) Six ion chromatograms produced by a 10 ppm solution of 75% evaporated gasoline run in the full scan mode, with splitless injection. (b) Six ion chromatograms produced by a 10 ppm solution of 75% evaporated gasoline run in the SIM mode, with a 20:1 split ratio.

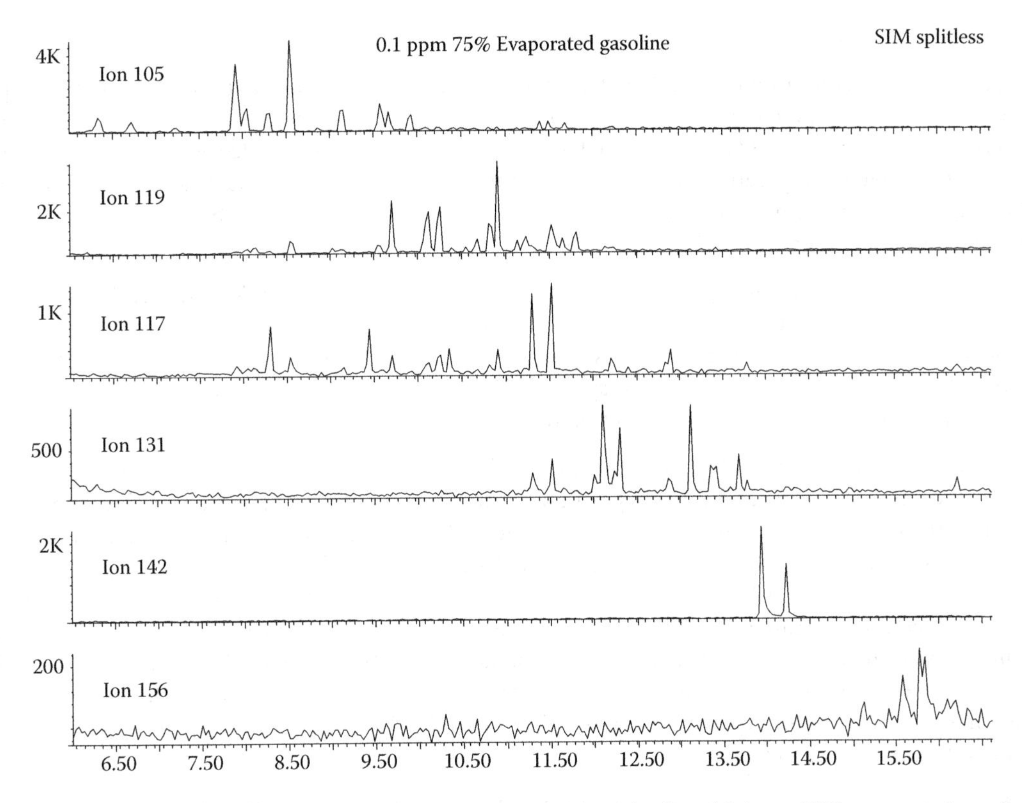

Figure 5.35 Six gasoline ion chromatograms produced by a 2-μL splitless injection of 0.1 ppm 75% evaporated gasoline.

This process of increasing sensitivity can be taken even one step further by using a splitless injection and selected ion monitoring, although in doing so, we are approaching territory where findings of lower and lower amounts may not be meaningful. Figure 5.35 shows the six gasoline ions from a solution of 0.1 ppm standard of 75% evaporated gasoline. This is roughly equivalent to a sample containing 0.001 μL or 1 nanoliter of gasoline. Even at a concentration in the eluate of 0.01 ppm (10 ppb), the baseline does not go completely flat in the SIM splitless mode.

By simply using the tools available, we can see that the sensitivity of the analysis can be increased by four orders of magnitude. It does not seem to this analyst that lowering the detection limit further than this would be worthwhile. Table 5.5 shows a rough order of magnitude calculation of the detection limits available using the techniques described earlier.

Table 5.5 Detection limits for four instrumental configurations

Method	LLD (solution)[a]	LLD (sample)[b]
Full scan, split 20:1, 1 μL injection	100 ppm	1.0 μL
Full scan, splitless, 1 μL injection	10 ppm	0.1 μL
SIM, split 2:1, 2 μL injection	1 ppm	0.01 μL
SIM, splitless, 2 μL injection	0.1 ppm	0.001 μL

Note: LLD = lower limit of detection.

[a] Concentration at which this analyst feels comfortable calling a sample positive.

[b] Assumes little or no competition from the sample, and a 5% recovery rate.

5.6.3 Estimating the degree of evaporation

Estimating the degree of evaporation is a task that an analyst is occasionally asked to perform. Usually, evaporation is not an issue, and frequently, even if it is an issue, estimating how much of the original volume of a sample might have evaporated is difficult. Work in 2017 by Birks et al. has indicated that the apparent degree of evaporation may change depending on the ambient temperature at which the evaporation took place. For example, 90% evaporated gasoline (10% of the original volume remaining) weathered at 500°C might be indistinguishable from 70% evaporated gasoline weathered at room temperature [31].

Unless we know what the ignitable liquid was to start with, it is next to impossible to determine how much it has evaporated. This is particularly true for distillates, which can range from C_7 through C_{12} in one batch and C_9 through C_{13} in the next batch of the same product. Likewise, kerosene and diesel fuel formulations change with latitudes and seasons. So, unless there is a container that is suspected of being the source of the residue, this is a task that simply cannot be accomplished (with the possible exception of gasoline).

Fresh from the pump, traditional reformate gasolines look pretty much alike, particularly with respect to major components, and they change in a reasonably predictable fashion as a result of exposure to fire. The issue of evaporation frequently arises in the context of trying to decide whether gasoline was present in a sample before the fire or it was introduced during or after the fire. It sometimes happens that firefighters bring gasoline-powered equipment to fire scenes, and it is sometimes necessary to eliminate such equipment as the source of residue found in a debris sample. A number of different techniques can be brought to bear on this issue. The simplest way to make the estimation is to compare the total ion chromatograms. In fresh gasoline, toluene is the tallest peak, and there are numerous light (C_5–C_8) alkane peaks to the left of the toluene peak. By the time gasoline is 25% evaporated, many of these light alkane peaks have disappeared. Also, in a 25% evaporated gasoline sample, the tallest peak is the m- and p-xylene peak.

The analyst's library of gasolines should include gasoline in many stages of evaporation, as well as fresh gasoline. In our laboratory, we keep gasoline standards on file at 0%, 25%, 50%, 75%, 90%, 98%, and 99% evaporated. Performing a rough estimation of the degree of evaporation happens every time we identify gasoline, simply because it is necessary to put a copy of at least one standard in the file [32], and it is desirable for the reference chromatogram to match the sample chromatogram as closely as possible.

By the time gasoline is 50% evaporated, the toluene peak is significantly shorter than the m- and p-xylene peak and, in fact, is shorter than the ethylbenzene peak. The pseudocumene peak, meanwhile, has grown so that it is almost as tall as the m- and p-xylene peak. At 75% evaporation, the pseudocumene peak is the tallest peak in the chart, and the toluene peak has almost disappeared. At 90% evaporation, the toluene peak is gone, and the xylenes are far shorter than the C_3 alkylbenzenes. The C_4 alkylbenzenes, meanwhile, have grown as tall as most of the C_3 alkylbenzenes, but pseudocumene is still the tallest peak in the chart. By the time the sample is 98% evaporated, the C_4 alkylbenzenes are taller than the C_3 alkylbenzenes.

A more quantitative technique for estimating evaporation involves the use of the mass spectrometer. Collecting the average mass spectrum from C_6 to C_{13} (about 3 to 13 min in our laboratory) yields an average that changes predictably from one stage of evaporation to the next. In fresh gasoline, ions 91 and 105 are the tallest, followed by ions 119, 43, 57, and 71. Ion 43 is produced by the light branched alkanes to the left of toluene. At 25% evaporation, the 91 and 105 ions still dominate the mass spectrum, but the 57 ion is now taller than the 43 ion. At 50% evaporation, ion 91 is no longer equal to ion 105, and ions 119 and 120 exceed the height of ion 57.

At 90% evaporation, the 119 ion is almost equal in size to the 105 ion, and ion 134 is beginning to catch up with ion 91. At 98% evaporation, ion 119 is the tallest, followed by ions 105, 134, and 91. Figure 5.36 shows these six average mass spectra. Keep in mind that the substrate can contribute to the spectra, thus making this approach more difficult in some cases.

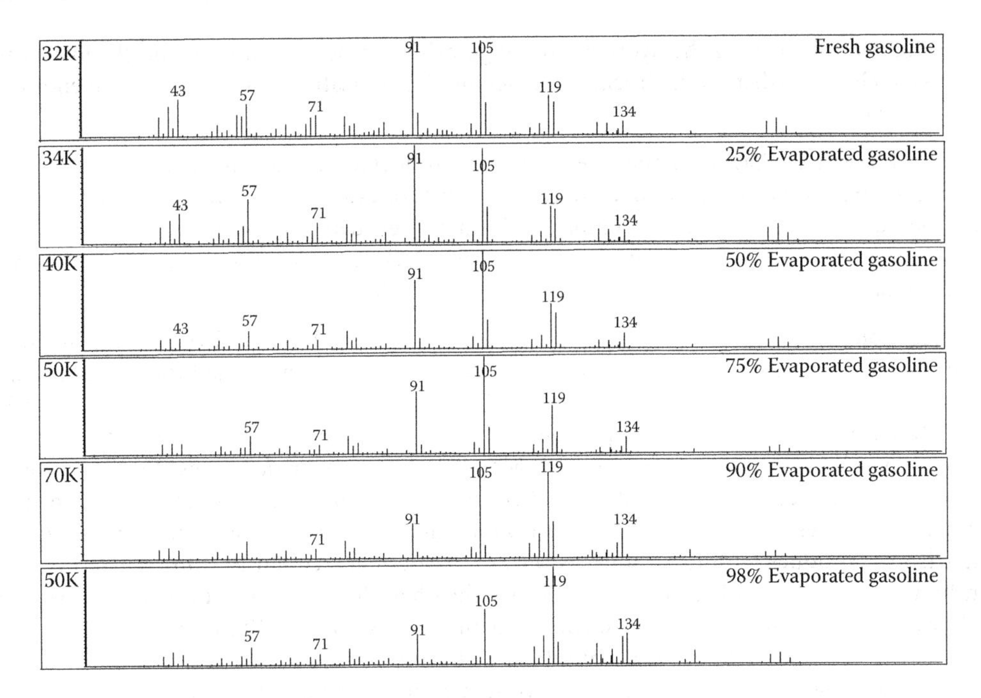

Figure 5.36 Average mass spectrum from 3 to 13 min of fresh gasoline, followed by the same spectra taken for gasoline evaporated to 25%, 50% 75%, 90%, and 98%.

Any attempt to estimate evaporation should be done by comparing "apples to apples." While it is hoped that the analyst has optimized the isolation process so that an eluate from an ACS looks like the liquid in solution that has not gone through the ACS process, standards should be eluates rather than solutions prepared by simply diluting the liquids. Further, samples and standards should have approximately equal concentrations. More concentrated samples tend to appear more evaporated due to the effects of displacement [33].

Certain substrates preferentially hold certain compounds, and substrate effects must be considered before reporting a sample as having a problematic evaporation level. Concrete, for example, tends to retain lighter hydrocarbons, making residues appear less evaporated than the fire scene conditions would indicate.

5.6.4 Identity of source

The ability to match a weathered ignitable liquid residue with a proposed source of unburned ignitable liquid has eluded fire debris analysts until just recently. The environmental forensics community has developed some tools to identify the sources of petroleum spills, but they have the advantage of having huge quantities of material available for characterization. Petroleum hydrocarbons contain trace quantities of *biomarkers*, substances that have changed little since they were first synthesized inside a living organism. The relative amounts of these biomarkers frequently allow for the identification of the source of a major spill. Polycyclic aliphatic and aromatic hydrocarbons are also useful in "fingerprinting" hydrocarbons. Examination of additives, such as oxygenates, or in the case of old gasolines, the alkyl lead compounds, can also provide clues as to the identity of the source of a spill.

There are three levels of control on the composition of petroleum. The primary control is the genesis of the petroleum (i.e., its "genetic" features), including its geographic location, the original source of the fuel (terrestrial or marine), and the conditions in the rock where the oil resided for millions of years.

The second level of control affecting the hydrocarbon fingerprint is processes imposed on the hydrocarbon by the refiner. Activities such as distillation, hydrocarbon cracking, isomerization, alkylation, and blending change the composition of the fuel.

The tertiary controls are those that occur after the petroleum product leaves the refiner. The most important tertiary control of interest to fire debris analysts is, of course, weathering, or evaporation. Mixing in the service station tank is also critical. Because the tanks are seldom empty, each time a new delivery is made, the fuel service station tank assumes a new temporary identity. For a more extended discussion of chemical fingerprinting in the environmental forensic science arena, see *Introduction to Environmental Forensics* [34].

Some of the most comprehensive work to identify the source of ignitable liquid residues from fire debris was performed by Dale Mann in 1987 [35,36]. Using a 60-meter column, Mann learned that it was possible to make comparisons on samples of fresh gasoline and correctly identify the source by examining the relative ratios of the light hydrocarbons, ranging from n-pentane to n-octane. Exactly these hydrocarbons are lost, unfortunately, before a gasoline is even 25% evaporated. Mann's second paper describes the limitations of making comparisons between ignitable liquid residues and fresh samples due to the contamination of residues with pyrolysis products, changes introduced by the isolation method, and the loss of volatiles resulting from weathering. Comparison of residues with fresh sources was, therefore, not frequently studied after Mann's extensive work. Later, however, Dolan and Ritacco reported that they had devised a way to measure the relative abundances of 20 peak pairs in gasolines and were able to identify the source of 30 samples that were evaporated to 25% and 50%. This process uses the relative abundances of branched alkanes that occur generally in the center of the gasoline chromatogram. Because the peaks are sequential, eluting only a few seconds apart, the relative ratios are not significantly affected by evaporation. The peaks examined are all minor components. The major components of gasoline are similar enough to each other that they are not useful in discriminating between sources [37].

Following up on this work, Wintz and Rankin applied principal components analysis, only to learn that while the ratios within the pairs show little if any change between unevaporated and 50% evaporated gasoline, only a few of those pairs were actually useful in distinguishing between one gasoline source and another, and some of those pairs occurred in the light petroleum distillate range and thus are unlikely to be found in gasolines evaporated to more than 50% [38].

Further work by Barnes and Dolan identified six ratios of sequentially eluting minor components in 50% evaporated gasoline, and four ratios in 75% evaporated gasoline. Using these ratios, they were able to successfully discriminate among 16 gasoline samples evaporated 50%, and 10 gasoline samples evaporated 75%. They used a blind study to confirm the ability of the technique to correctly identify the source of three gasolines, matched against a library of ten candidate sources. The technique uses a 60-meter column and is automated using a target compound program. The program takes the abundance of the base peak from the mass spectrum of each compound of interest, and divides the base peak abundance of the later eluting compound by the base peak abundance of the earlier eluting compound. As well as this technique appears to work, the authors state that it is most useful for *eliminating* a source and caution that the analyst's statement of conclusions reflect that a common origin is indicated but is not a certainty [39]. Despite these cautions, it appears that Barnes and Dolan may have solved the problem that Mann started investigating 20 years earlier.

A somewhat simpler approach, using an analysis of naphthalene, methyl, and dimethylnaphthalenes, has been reported by Sandercock and Du Pasquier in a three-part series published in 2003 and 2004. The column used was an everyday 30 m HP-5 column of the type used for routine fire debris analysis. A SIM method was used, and in order to have all of the peaks capable of being integrated at the same approximate level of precision, the signal for naphthalene and the methylnaphthalenes was selectively reduced by monitoring ions 127 and 129 rather than 128 for naphthalene, and 141 and 143 rather than 142 for methylnaphthalene. The base peak of 156 was used for the dimethylnaphthalenes. Sandercock and Du Pasquier started their research by pretreating the gasoline with an alumina column to separate out both polynuclear aromatic hydrocarbons and the polar alkylphenols, but they learned that the alkylphenols did not provide much help in discriminating among different samples. The polycyclic aromatic hydrocarbon (PAH) ratios do not change significantly upon evaporation and seem to be a characteristic imparted by the refinery [40].

One way to compare any type of ILR to a suspected source or to see if two ILRs within the same class came from the same source is to compare the average molecular weights. When exposed to a fire, ignitable liquid evaporates, and its average molecular weight necessarily increases. No predictable experience causes the average molecular weight of the fire-exposed liquid to decrease. If the suspected source of an ignitable liquid, which was not exposed to the fire, exhibits a higher average molecular weight than the residue extracted from samples collected at the scene, the suspected source can be conclusively eliminated.

In doing comparisons of gasolines or any ignitable liquid, the best approach takes the proposed source liquid, evaporates some of it so that it matches the residue in terms of carbon number range, and then subjects it to the same separation procedure that was applied to the questioned sample. Once this is accomplished, a detailed examination of the finer points in the chromatogram can be carried out.

A really good match between chromatograms not only matches the peaks but also matches the valleys. Examining multiple extracted ion chromatograms for both quality and quantity is a necessary step when doing comparisons. This practice has been codified with respect to the analysis of distillates heavier than kerosene in ASTM D5739-06, *Standard Practice for Oil Spill Source Identification by Gas Chromatography and Positive Ion Electron Impact Low Resolution Mass Spectrometry* [41]. The approach in this method compares 24 extracted ion chromatograms of polynuclear aromatics and certain biomarkers. While this method is not applicable to kerosene and lighter mixtures like gasoline, the general approach seems workable. Given the state of the art, it is probably a good thing that comparisons are not frequently requested.

5.7 REPORTING PROCEDURES

The analyst's report can be one of the most important documents generated during the investigation of a fire. The report, therefore, should be written carefully enough so that readers are not misled. The objective of fire debris analysis is to determine whether there is any *foreign* ignitable liquid residue present in a sample.

The report from a forensic science laboratory is supposed to be a *scientific* report. As such, it should include an introduction, a section detailing the test methods and results, and a discussion and conclusion section, if necessary. **The laboratory report should also state what was done with the evidence.** There is an unfortunate tendency in some agencies to provide checklists and merely state "positive for gasoline" or "negative for accelerants" rather than preparing a real laboratory report. Reading such a report, a reviewer cannot tell whether the analyst used a GC-MS or a Ouija board.

Given that it is now possible to store templates with all possible results in them, the excuse that it is "too time-consuming" to prepare an understandable narrative report is unacceptable. Although laboratory reports are intended to assist investigators, they should also be understandable to individuals requested to review that work. At a minimum, a report should include the following:

- An identification of the fire in question.
- A description of how the sample was delivered to the laboratory, when, and by whom.
- A description of the samples, including container size, substrate material, and a reported location from where the sample was collected.
- A description of the isolation procedure used to separate the ILR from the sample substrate (ASTM separation procedure).
- A description of the analytical technique applied to the sample extract (ASTM E1618).
- The results of the analysis of the data.

- A discussion of the meaning of the results, if there is any chance of misinterpretation or misuse. In the discussion section, the analyst can provide examples of potential sources for whatever ignitable liquid residues may have been identified. This is also an appropriate place to put in a *disclaimer* about the possibility that the ILR may not be foreign to the background or a disclaimer that states that negative results do not eliminate the possibility that ignitable liquids were present at the fire scene.
- A conclusion or bottom line, understandable to even an attorney, is helpful
- A sentence stating what has happened to the sample (returned to the submitter or placed in storage)

If a substance is found that is natural or incidental to the background, it is the analyst's duty to say so. ASTM E1618 allows for the inclusion of disclaimers, on both positive and negative reports. With respect to negative reports, a disclaimer to the effect that negative results do not preclude the possibility that ignitable liquids were present at the fire scene is suggested as an aid to help readers avoid misunderstanding the report. Likewise, in the case of a positive report, the standard states, "It may be appropriate to add a disclaimer to the effect that the identification of an ignitable liquid residue in a fire scene does not necessarily lead to the conclusion that a fire was incendiary in nature. Further investigation may reveal a legitimate reason for the presence of ignitable liquid residues." Certainly, a finding of gasoline in the living room is noteworthy. It is not so noteworthy to find gasoline in the basement near the chain saw.

On occasion, it seems that the fire debris analyst is too eager to "help" with an arson investigation. Reporting a sample as being "positive" when all the analyst has identified is background compounds is neither helpful nor morally defensible. Karen and Paul Stanley of Akron, Ohio, were falsely accused of setting the fire that killed their infant son. The fire debris samples were submitted to a laboratory that reported finding turpentine in a sample consisting of charred yellow pine. The laboratory analyst had absolutely no business reporting turpentine, or at the very least, should have reported the finding of turpentine with a huge disclaimer stating that the turpentine was indistinguishable from naturally occurring turpentine found in coniferous woods. The case was dismissed after the prosecutor understood the significance (or lack thereof) of the chemist's findings [42].

A similar case of an analyst being too "helpful" occurred in a homicide case in Georgia, where the laboratory analyst reported finding "toluene, a flammable liquid" on a suspect's "clothing." Actually, the "clothing" included a pair of tennis shoes, and the examination of a new pair of tennis shoes directly from the shoe store revealed the presence of a high concentration of toluene. A closer examination of the first analysis revealed that, in addition to toluene, diethylene glycol and butylated hydroxytoluene (BHT) were also present in the suspect's shoes and the exemplar shoes in identical relative proportions. Less than a year earlier, there had been a presentation on the analysis of suspects' shoes in arson cases at an American Academy of Forensic Sciences (AAFS) seminar [43]. Perhaps if the analyst had attended that presentation (or read the proceedings), the toluene would have been properly reported as native to the sample background or, more appropriately, not reported at all.

When shoes are submitted for analysis, it is important to keep in mind the high concentration of hydrocarbons used in their manufacture. Shoes should always be analyzed separately, so that one shoe can act as a comparison sample for the other. If there is ILR on a shoe due to an arsonist spilling ignitable liquid on him- or herself, it is unlikely that he or she spilled exactly the same amount on each shoe. Unless both shoes reek of gasoline, only when the concentration of ILR is orders of magnitude higher on one shoe should the results be considered meaningful.

A typical positive report and a typical negative report from our laboratory are shown in Figure 5.37.

SCIENTIFIC FIRE ANALYSIS, LLC

88005 Overseas Highway, #10-134 Islamorada, FL 33036

e-mail: scientific.fire@yahoo.com, website: www.firescientist.com

CHEMICAL ANALYSIS REPORT		
File No. Lab File #	**Date** January 18, 2018	**Page** 1 **of** 1

Fire Investigator Agency
Address
City, State ZIP

Subject

Incident # (Claim # 78910)
Your File # 23456.
Victim (Insured): Name.
Date Of Loss: January 8, 2018.
Analysis of Fire Evidence.

Background

On January 15, 2018, John Lentini of Scientific Fire Analysis (SFA) received from client via UPS the following:

Item 1. A one-quartcan containing burned carpet and pad identified as removed from the north end of the south bedroom.

SFA was requested to analyze the sample for ignitable liquid residues.

Test Methods and Results

The sample was separated according to ASTM Practices E 1412-16, and analyzed according to ASTM Standard Method E 1618–14.

Gas chromatographic/mass spectrometric (GC/MS) analysis of concentrated headspace vapors from Item 1 reveals the presence of components having retentiontimes and mass spectra characteristic of components of known weathered gasoline.

Conclusion

Gasoline was present in the sample.

Sample Disposition

The sample has been returned to client via UPS.

Prepared by ____ John J. Lentini, F-ABC

Reviewed by ____ P. E. Rogers, Senior Chemist

(a)

Figure 5.37 (a) Typical positive report from our laboratory.

SCIENTIFIC FIRE ANALYSIS, LLC

88005 Overseas Highway, #10-134 Islamorada, FL 33036
770-815-6392
e-mail: scientific.fire@yahoo.com, website: www.firescientist.com

CHEMICAL ANALYSIS REPORT		
File No. Lab File #	**Date** January 18, 2018	**Page** 1 **of** 1

Fire Investigator Agency
Address
City, State ZIP

Subject

Incident # (Claim # 78910)
Your File # 23456.
Victim (Insured): Name.
Date Of Loss: January 8, 2018.
Analysis of Fire Evidence.

Background

On January 15, 2018, John Lentini of SFA received from client via UPS the following:

Item 1. A one-quart can containing burned carpet and pad identified as removed from the north end of the south bedroom.

Item 2. A one-quart can containing debris identified as removed from the center of the living room.

SFA was requested to analyze the samples for ignitable liquid residues.

Test Methods and Results

The samples were separated according to ASTM Practices E 1412-16, and E 1386-15, and analyzed according to ASTM Standard Method E 1618-14.

Gas chromatographic/mass spectrometric (GC/MS) analysis of concentrated headspace vapors from Items 1 and 2 fails to reveal the presence of any ignitable liquid residues.

These results do not eliminate the possibility that ignitable liquids were present at the fire scene.

Sample Disposition

Samples have been returned to client via UPS.

Prepared by ______ John J. Lentini, F-ABC

(b) Reviewed by ______ P. E. Rogers, Senior Chemist

Figure 5.37 (b) Typical negative report from our laboratory.

5.8 RECORD KEEPING

Each case file that includes a positive identification of an ignitable liquid residue should contain not only the sample charts but also a standard to which that sample can be compared. This means that both the sample and standard should be printed with the same time scale so that the data can be easily reviewed. Even if there is no criminal prosecution, the analyst should be aware that many fire cases involve civil litigation, so the case file should be kept for a reasonable period of time.

Electronic data files should also be protected and stored. There are only a few versions of GC-MS software in widespread use, so it is possible for one analyst to review another's raw data. Such reviews should be facilitated by the preservation of the data. Instrumental data, such as tune reports or spectrum scans, should be kept, if only to keep track of the performance of the instrument. Certainly, if a case file has matching standards and samples, there is no need to review the tune report for that particular week. No matter what happened, the instrument tune parameters would not cause a false positive. Blanks, on the other hand, should be stored somewhere, particularly for those cases where the detection limit of the procedure is being reduced.

5.9 QUALITY ASSURANCE

The factors affecting the overall quality of a laboratory's work product are numerous. For this reason, accreditation programs have been set up so that an outside agency can verify that a laboratory is performing appropriately. Each laboratory should have a detailed written procedure for the examination of fire debris, and this procedure should be based on recognized standards. The procedure used in our laboratory appears at the end of this chapter in Sidebar 5.1. As part of the quality assurance program, fire debris analysts should participate in proficiency testing at least once a year. This can be internal or external, but certainly external proficiency testing carries more credibility. Proficiency tests are available from a number of commercial suppliers, but it is not necessary to actually purchase proficiency tests. Analysts can form round-robin groups and take turns preparing samples for each other. Because it is not necessary to demonstrate one's ability to place a strip in a can, samples can be prepared by exposing 10 strips to the same sample and then just mailing the strips. By doing regular proficiency tests or round-robins, analysts can monitor their own performance, and laboratory directors can be assured that their analysts are performing adequately.

Individual professional development and continuing education are an important component of any quality assurance program. Analysts should take the time to keep up with the literature and should attend professional meetings when possible. Although the science of fire debris analysis is reasonably settled, this is an interesting field with a large cadre of professionals who perform and publish research on a continuing basis. Keeping up with this research is a professional responsibility.

The ultimate sign that an analyst is keeping up with the profession is the decision to become certified. The American Board of Criminalistics offers certification in fire debris analysis. Certification can only be maintained by continuing education and annual participation in external proficiency testing. Becoming certified is a way for an analyst to demonstrate that he or she cares about professional development. Supporting individual certification is a way that a laboratory director can assure his or her employees that the agency also cares.

SIDEBAR 5.1 Procedure for the analysis of fire debris samples

Proc: 801
Rev: 1
Date: 4/14/17
Page:
Approval:

1. SCOPE
   This procedure covers the analysis of fire debris samples by gas chromatography–mass spectrometry.
2. REFERENCED DOCUMENTS
   Scientific Fire Analysis (SFA) Procedure 800, Handling of Fire Related Evidence
   ASTM E1388-17, Standard Practice for Sampling of Headspace Vapors from Fire Debris samples.
   ASTM E1412-15, Standard Practice for Separation of Ignitable Liquid Residues from Fire Debris Samples by Passive Headspace Concentration with Activated Charcoal
   ASTM E1618-14, Standard Test Method for Ignitable Liquid Residues in Extracts from Fire Debris Samples by Gas Chromatography/Mass Spectrometry
   ASTM E1386-15, Standard Practice for the Separation of Ignitable Liquid Residues from Fire Debris by Solvent Extraction
   ASTM E1459-13, Standard Guide for Physical Evidence Labeling and Related Documentation
   ASTM E1492-11 (2017), Standard Practice for Receiving, Documenting, Storing, and Retrieving Evidence in a Forensic Science Laboratory
   ASTM E2451-13, Standard Practice for Preserving Ignitable Liquids and Ignitable Liquid Residue Extracts from Fire Debris Samples

3. PURPOSE
   The purpose of this procedure is to determine whether a sample contains detectable quantities of ignitable liquid residue and to identify the liquid.
4. PROCEDURE (These steps to be followed in conjunction with SFA Procedure 800)
   4.1 Separation
      4.1.1 Passive headspace concentration.
         4.1.1.1 This procedure will be used for most of the samples submitted to the SFA laboratory for analysis of ignitable liquid residues.
         4.1.1.2 Visually examine the sample to make sure that the container is properly labeled per ASTM E1459 and the evidence transmittal form (or work order if the sample was collected by SFA staff) accurately describes the contents of the container.
            4.1.1.2.1 Resolve any discrepancies with the submitter prior to issuing a report.
         4.1.1.3 Determine whether the sample contains a large concentration of hydrocarbons detectable by odor.
            4.1.1.3.1 If the sample exhibits a strong odor, remove it to the hood and perform a solvent extraction, as described in Section 4.1.2.
         4.1.1.4 Determine whether the sample contains a large amount of water.
            4.1.1.4.1 In the event the sample is very wet, puncture a small hole in the sample can lid and seal with cellophane tape.
         4.1.1.5 Add 20 μL of internal standard to the sample container.
            4.1.1.5.1 The internal standard is a 2.5% solution of 3-phenyltoluene in diethyl ether.
            4.1.1.5.2 Prepare this internal standard solution by adding 200 μL technical grade 3-phenyltoluene to 8 mL diethyl ether. Keep the internal standard solution in the refrigerator.
         4.1.1.6 Place a "C-strip" absorbent package into the sample container and close the container.
         4.1.1.7 Place the container into the convection oven.
         4.1.1.8 Prepare a system blank by placing a piece of filter paper in the bottom of an empty can and adding 20 μL of internal standard solution. Analyze in the same manner as samples.
            4.1.1.8.1 Run a system blank each day that samples are run.
         4.1.1.9 Be certain that the temperature is set to 80°C and the timer is set for 16 h.
         4.1.1.10 At the conclusion of the 16-hour adsorption phase, remove the sample from the oven and place the C-strip into an appropriately labeled 2-mL septum seal vial.
            4.1.1.10.1 Label the vial with the following information: SFA Job Number, Sample Serial Number, Date, and Initials.
         4.1.1.11 Add 0.5 mL eluting solvent to the septum seal vial.
            4.1.1.11.1 The eluting solvent consists of diethyl ether spiked with 100 ppm tetrachloroethylene.
         4.1.1.12 Seal the vial and place it in the autosampler tray of the GC-MS.
      4.1.2 Solvent Extraction.
         4.1.2.1 This separation procedure is typically used on liquid samples, on small portions of very concentrated samples, on very small samples, or on empty containers.
            4.1.2.1.1 In the case of organic liquid samples, add 10 μL liquid to 1 mL diethyl ether eluting solution, label the vial, seal, and place in the GC-MS autosampler tray.
            4.1.2.1.2 If an aqueous solution is submitted, perform a liquid/liquid extraction using a minimum quantity of diethyl ether eluting solution. Place 1 mL of the eluting solution in a septum seal vial, seal, label, and place in the GC-MS autosampler tray.
            4.1.2.1.3 In the case of solids, these will generally be from samples containing a high concentration of ignitable liquid. Select a small sample to be extracted and place it in a small beaker. Cover the sample with diethyl ether eluting solution and allow 5 min for dissolution of soluble hydrocarbons. No evaporation step should be necessary.
            4.1.2.1.4 For empty containers or very small samples, use the minimum quantity of ether necessary to either cover the sample or rinse the interior of the container, usually about 10 mL.
               4.1.2.1.4.1 Evaporate the solution gently, using no heat, to a volume of approximately 1 mL.
         4.1.2.2 In some cases, it is necessary to filter the extract to remove large particulates.
         4.1.2.3 Prepare blanks for solvent extraction by evaporating 20 mL solvent to 1 mL prior to analyzing.
   4.2 Analysis
      4.2.1 Tune the mass spectrometer each Monday morning, using "standard spectra autotune."
         4.2.1.1 Place the tune report in the instrument logbook.
         4.2.1.2 Run a spectrum scan daily except Monday. Print the scan report and place it in the instrument logbook.
      4.2.2 Load the sequence entitled "Default" and then select the menu item "Edit Sample Log Table."
      4.2.3 Enter the appropriate information into the sequence log table.
         4.2.3.1 Necessary information includes the sample vial location; a data file location, which should be the same as the job number followed by the serial number; sample information to include job number, sample serial number, separation method, and solvents; and miscellaneous information to include a sample description and location.
      4.2.4 Enter the appropriate method for the analysis.
         4.2.4.1 Use the full scan method, FCLR, for the first analysis of an extract.
         4.2.4.2 If there are still peaks eluting from the column at the end of the run, extend the runtime. The sample may need to be rerun using the method FCLRLONG.

4.2.4.3 If the extract exhibits a low concentration of residue, run the sample again in the splitless mode, FCLRSPTL.
4.2.4.4 Use a selected ion monitoring (SIM) method if still more sensitivity is required.
4.2.4.4.1 Using a SIM method prevents the collection of complete mass spectra.
4.2.4.5 Create a directory for the day's analyses and run the sequence.
4.2.4.6 When a sample run has been completed, load the data file for the sample and examine the total ion chromatogram.
4.2.5 Obtain ion chromatograms for the classes of organic compounds frequently found in petroleum hydrocarbons.
4.2.5.1 For all samples, obtain ion chromatograms for 57, 71, 83, and 105.
4.2.5.2 For all samples, obtain ions 105, 119, 117, 131, 142, and 156.
4.2.6 Compare the patterns produced by the sample with patterns produced by known ignitable liquids. Print any patterns that are considered to match.
4.2.6.1 Conduct the comparison according to ASTM E1618.
4.2.6.2 If the patterns exhibit sufficient similarities to the standards, classify the ignitable liquid residue into the appropriate category.

5. ALTERNATE PROCEDURE FOR LIGHT ANALYTES
5.1 The solvent peak may obscure some low molecular weight analytes. In such cases, it is necessary either to use a different solvent or to sample in the vapor phase.
5.2 Use perchloroethylene as an alternate solvent and select an analytical method that turns the detector off while the PCE is eluting. Such methods are designated with the letters PCE in the method name. Carry out the analysis as described in Section 4.
5.2.1 If the necessity for an alternate solvent is anticipated, place two ACS strips in the sample can during the adsorption step.
5.3 It is sometimes necessary to sample the vapor phase because of the low affinity of the sample for the ACS strip. Withdraw 500 μL of vapor from the heated sample can and inject manually. It is necessary to select a method that designates manual injection. Use the analytical protocols for analysis described in Section 4.
5.3.1 Run a blank consisting of 500 μL of lab air prior to each headspace analysis.
5.3.2 Refer to ASTM E1388 for guidance on the use of headspace techniques.

6. DOCUMENTATION
6.1 Prepare a report describing the samples, the methods employed, and the results of the analysis, using a standard report template.
6.1.1 Place a copy of the report and a copy of all of the data printed during the analysis into the job file.
6.1.1.1 Place other documentation describing the handling of the evidence, as described in SFA Procedure 800, in the job file.

7. STORAGE OR RETURN OF SAMPLES AND EXTRACTS
7.1 Depending on the submitter's instructions, either store the sample in the evidence room, or return it to the submitter.
7.1.1 Follow ASTM E1492 in all procedures involving storage or return.
7.2 For all cases in which at least one sample tested positive for the presence of ILR, store all of the sample vials containing the carbon strips, including negatives, in a refrigerator for at least one year. Sample vials may be moved to room-temperature storage facilities after one year, if space is required.
7.2.1 Follow ASTM E2451 in all procedures involving archiving sample strips.
7.3 Archiving can also be applied to liquid samples for comparison or extracts created by solvent extraction.

5.10 CONCLUSION

The isolation and identification of ignitable liquid residues from fire debris samples are an important part of the fire investigation process. The fire debris analyst should be familiar with the common techniques of fire investigation and understand the language used by investigators. The analyst must also understand the stakes involved. If an ignitable liquid residue is identified, a hypothesis that a fire was incendiary may be supported, and a long litigation process may ensue, either in criminal or civil court. The laboratory results are often the deciding factor in a prosecutor's decision to indict or in an insurance company's decision to resist a claim.

The technology for fire debris analysis has improved dramatically over the past 40 years, to the point where our methods are now as sensitive as they need to be, and possibly more sensitive than they need to be. The analytical procedure requires not only focus and creativity but also adherence to the generally accepted criteria for making an identification. The analysis should be conducted so that it is capable of being reviewed by another analyst. There is no reason that a reviewing analyst should reach a different conclusion, and such reviews should not be only expected but also welcomed. Science is based on "multiple witnessing." Fire debris analysts must be acutely aware of the stakes involved in what they are doing and of the need to communicate effectively about the meaning of their findings.

Review questions

1. What is the advantage of using steam distillation over the carbon strip method for separating ignitable liquid residues from fire debris samples?
   a. It has more sensitivity.
   b. It produces a visible extract that can be shown to the jury.
   c. It is less time consuming to use.
   d. It is nondestructive.
2. Which of the following substances are NOT normally found in the background of the living space in a residence?
   a. Kerosene
   b. Toluene
   c. Gasoline
   d. Mineral spirits
3. What is the routine lower limit of detection for ignitable liquid residues in a competent forensic science laboratory?
   a. 1 drop, or 50 microliters in a kilogram of debris
   b. 1/10 of a drop, or 5 microliters in a kilogram of debris
   c. 1/500 of a drop, or 0.1 microliters in a kilogram of debris
   d. 1/10,000 of a drop, or 5 nanoliters in a kilogram of debris
4. When submitting a sample of hardwood flooring to a fire debris laboratory, why is a comparison sample necessary?
   a. Submitting a comparison sample provides the appearance of objective science.
   b. Hardwood floor coatings often contain residual solvent.
   c. There is often not an opportunity to go back and collect another sample.
   d. All of the above
5. Which of the following statements about fire debris analysis laboratories are true?
   I. A positive report does not necessarily mean the fire was set.
   II. Drug and environmental chemists have the same equipment and the same skill set as fire debris chemists.
   III. As long as the equipment is properly calibrated, the lab results will be reliable.
   IV. A certified analyst in an accredited laboratory following standard methods is more likely to be correct than a chemist without these advantages.
   V. A negative report is inconclusive, and does not necessarily mean there were no accelerants used.
      a. All of the above statements are true.
      b. None of the above statements are true.
      c. Only I, IV and V are true.
      d. Only III and IV are true.

Questions for discussion

1. Why is an unconfirmed canine alert not suitable for use as evidence in a trial?
2. Even though the analysts use the same instrument, a gas chromatograph–mass spectrometer, environmental analysis, drug analysis, and fire debris analysis require different skill sets. Discuss why this is so.

3. Why does a negative sample not necessarily indicate the absence of accelerants, and why does a positive sample not necessarily mean a fire was intentionally set?
4. How can the degree of evaporation be used to include or exclude a potential source of an ignitable liquid residue?
5. What are the benefits of using an accredited laboratory with analysts certified in fire debris analysis?

References

1. Kirk, P. (1969), *Fire Investigation*, John Wiley & Sons, New York, p. 153.
2. Chrostowski, J., and Holmes, R. (1979) Collection and determination of accelerant vapors, *Arson Analysis Newsletter*, 3(5):1–17.
3. Juhala, J. A. (1982) A method for adsorption of flammable vapors by direct insertion of activated charcoal into the debris samples, *Arson Analysis Newsletter*, 6(2):32.
4. Dietz, W. R. (1993) Improved charcoal packaging for accelerant recovery by passive diffusion, *Journal of Forensic Science* 38(1):165.
5. Waters, L., and Palmer, L. (1991) Multiple analysis of fire debris using passive headspace concentration, *Journal of Forensic Science* 36(1):111.
6. Fultz, M., and Wineman, P. (1982) AANotes, "Accelerant Classification System," *Arson Analysis Newsletter*, Systems Engineering Associates, Columbus, OH, 6(3):57–56.
7. Midkiff, C. (1982) Arson and explosive investigation, in Saferstein, R. (Ed.), *Forensic Science Handbook*, Prentice Hall, Upper Saddle River, NJ, p. 225.
8. ASTM E1387-01 (2001) *Standard Test Method for Ignitable Liquid Residues in Extracts from Fire Debris Samples by Gas Chromatography*, Annual Book of Standards, Volume 14.02, ASTM, West Conshohocken, PA.
9. Stone, I. C. (1976) Communication to *Arson Analysis Newsletter, 1*(1), Systems Engineering Associates, Columbus, OH, 5.
10. Nowicki, J. (1990) An accelerant classification scheme based on analysis by gas chromatography–mass spectrometry (GC-MS), *Journal of Forensic Science* 35(5):1064.
11. Smith, R. M. (1982) Arson analysis by mass chromatography, *Analytical Chemistry* 54(13):1399.
12. IAAI Forensic Science Committee (1988) Guidelines for laboratories performing chemical and instrumental analyses of fire debris samples, *Fire & Arson Investigator* 38(4):45.
13. ASTM E1618-01 (2001) *Standard Test Method for Ignitable Liquid Residues in Extracts from Fire Debris Samples by Gas Chromatography–Mass Spectrometry*, Annual Book of Standards, Volume 14.02, ASTM, West Conshohocken, PA.
14. U.S. Department of Justice, Office of Law Enforcement Standards, (1999) *Forensic Sciences: Review of Status and Needs*, Washington, DC, 38.
15. Fire Debris and Explosives Analysis Subcommittee (2017) Position Statement on E1618-14. Available at www.nist.gov/sites/default/files/documents/2017/05/19/osac_fde_subcommittee_-_e1618_position_statement.pdf (last visited January 16, 2018).
16. AAAS Working Group on Fire Investigation (2017), Forensic Science Assessments, A Quality and Gap Analysis: Fire Investigation, AAAS, Washington, DC, 33. Available at https://www.aaas.org/report/fire-investigation (last visited January 16, 2018).
17. Newman, R., and Dolan, J. (2001) Solvent options for the desorption of activated charcoal in fire debris analysis, in *Proceedings of the American Academy of Forensic Sciences*, (*AAFS*) February 2001, Seattle, WA, p. 63.
18. Armstrong, A., and Lentini, J. (1997) Comparison of the eluting efficiency of carbon disulfide with diethyl ether: The case for laboratory safety, *Journal of Forensic Science* 42(2):307.

19. CDC (2016), NIOSH Pocket Guide to Chemical Hazards. Available at https://www.cdc.gov/niosh/npg/default.html (last visited January 16, 2018).

20. Newman, R., Gilbert, M., and Lothridge, K. (1998) *GC-MS Guide to Ignitable Liquids*, CRC Press, Boca Raton, FL.

21. Dolan, J. (2004) Analytical methods for the detection and characterization of ignitable liquid residues from fire debris, Chapter 5 in Almirall, J., and Furton, K. (Eds.), *Analysis and Interpretation of Fire Scene Evidence*, CRC Press, Boca Raton, FL, p. 152.

22. Peschier L. J., Grutters M. P., and Hendrikse J. N. (2017) Using alkylate components for classifying gasoline in fire debris samples. *Journal of Forensic Sciences* 63(2):420–430. doi:10.1111/1556-4029.13563.

23. Almirall, J., and Furton, K. (2004) Characterization of background and pyrolysis products that may interfere with the forensic analysis of fire debris, *Journal of Analytical and Applied Pyrolysis* 71:51–67.

24. Keto, R. O. (1995) GC/MS data interpretation for petroleum distillate identification in contaminated arson debris, *Journal of Forensic Science* 40(3):412–423.

25. Lentini, J., Dolan, J., and Cherry, C. (2000) The petroleum-laced background, *Journal of Forensic Science* 45(5):968.

26. Lentini, J. (2001) Persistence of floor coating solvents, *Journal of Forensic Science* 46(6):1470.

27. Lentini, J. (1998) Differentiation of asphalt and smoke condensates from liquid petroleum distillates using GC-MS, *Journal of Forensic Science* 43(1):97.

28. Lentini, J., and Waters, L. (1982) Isolation of accelerant-like residues from roof shingles using head-space concentration, *Arson Analysis Newsletter*, 6(3):48.

29. Layven, P. (2003) *Asphalt Pavements: A Practical Guide to Design Production and Maintenance for Architects and Engineers*, CRC Press, Boca Raton, FL, 6, 7.

30. ASTM International (2012) ASTM D 3257-06 (2012), *Standard Test Methods for Aromatics in Mineral Spirits by Gas Chromatography*, Annual Book of Standards, Volume 6.04, ASTM, West Conshohocken, PA

31. Birks, H., et al. (2017) The surprising effect of temperature on the weathering of gasoline, *Forensic Chemistry* 4:32–40.

32. ASTM International (2014) ASTM E1618-14, *Standard Test Method for Ignitable Liquid Residues in Extracts from Fire Debris Samples by Gas Chromatography-Mass Spectrometry*, Annual Book of Standards, Volume 14. ASTM, West Conshohocken, PA, Paragraph 9.1.6.

33. Newman, R. T., Dietz, W. R., and Lothridge, K. (1996) The use of activated charcoal strips for fire debris extractions by passive diffusion. I. The effects of time, temperature, strip size, and sample concentration, *Journal of Forensic Science* 41(3):361.

34. Stout, S., et al. (2002) Chemical fingerprinting of hydrocarbons, in Murphy, B., and Morrison, R. (Eds.), *Introduction to Environmental Forensics*, Academic Press, San Diego, CA, 140.

35. Mann, D. C. (1987) Comparison of automotive gasolines using capillary gas chromatography. I. Comparison methodology, *Journal of Forensic Science* 32(3):606.

36. Mann, D. C. (1987) Comparison of automotive gasolines using capillary gas chromatography II: limitations of automotive gasoline comparisons in casework, *Journal of Forensic Science* 32(3):616.

37. Dolan, J., and Ritacco, C. (2002) Gasoline comparisons by gas chromatography–mass spectrometry utilizing an automated approach to data analysis, in *Proceedings of the American Academy of Forensic Sciences Annual Meeting*, Atlanta, GA, February 16, 2002, 62.

38. Wintz, J., and Rankin, J. (2004) Application of principal components analysis in the individualization of gasolines by GC-MS, in *Proceedings of the American Academy of Forensic Sciences*, Dallas, TX, February 2004, 48.

39. Barnes, A., Dolan, J., Kuk, R., and Siegel, J. (2004) Comparison of gasolines using gas chromatography–mass spectrometry and target ion response, *Journal of Forensic Science* 49(5):1018.

40. Sandercock, M., and Du Pasquier, E. (2004) Chemical fingerprinting of gasoline: 3. Comparison of unevaporated automotive gasoline samples from Australia and New Zealand, *Forensic Science International* 140:71.

41. ASTM International (2013) *D5739-06 Standard Practice for Oil Spill Source Identification by Gas Chromatography and Positive Ion Electron Impact Low Resolution Mass Spectrometry*, Annual Book of Standards, Volume 11.02, ASTM, West Conshohocken, PA.

42. Trexler, P. (2002) Prosecution expert rejects short as cause, *Akron Beacon Journal*, Akron, OH, February 8, 2002. Available at http://truthinjustice.org/stanleys.htm (last visited January 16, 2018).

43. Cherry, C. (1996) Arsonist's shoes: Clue or confusion, in *Proceedings of the American Academy of Forensic Sciences*, Nashville, TN, February 1996, AAFS, 20.

CHAPTER 6

Evaluation of ignition sources

The use and operation of an appliance should be well understood before it is identified as a fire cause.

—NFPA 921

LEARNING OBJECTIVES

After reviewing this chapter, the reader should be able to:

- Understand the kinds of physical evidence that are often examined after the scene examination, in a location away from the scene
- Describe the steps taken in the laboratory examination of potential ignition sources
- Understand typical failure modes for common potential ignition sources
- Write a protocol for the joint examination of evidence, and organize the examination
- Design and carry out a physical experiment to test a hypothesis

6.1 INTRODUCTION

With the legal system demanding ever-increasing levels of sophistication in the analysis of fire evidence, the fire investigation laboratory is now required to perform all kinds of examinations and experiments beyond the detection of ignitable liquid residues. The tests routinely conducted in the modern forensic laboratory with respect to fire evidence break down generally into two categories: (1) the examination of physical evidence, and (2) the testing of fire scenarios. As with other aspects of fire investigation, this phase of the investigation requires planning, cooperation among various parties, the involvement of several technical disciplines, and a written protocol. At this stage of the investigation, documentation is as important as ever, particularly when technical experts representing different parties get together to perform joint inspections, and the destructive disassembly of items of evidence is necessary. Inspections lasting 2 or 3 days, involving up to 10 experts (with 5 or 10 lawyers standing off to the side), are not uncommon in fires involving multimillion-dollar losses or the loss of life. This chapter deals with what happens to the evidence after it has been identified at the fire scene and taken to a laboratory for further inspection.

While most public sector laboratories have ignitable liquid residue detection capabilities, they typically do not have the equipment or the staff to conduct most of the examinations that will be described in this chapter. Even so, it is sometimes necessary in criminal cases to have certain potential sources of ignition examined and ruled out as the cause of the fire, so these examinations may apply in both civil and criminal venues.

Both the NFPA and the Consumer Product Safety Commission (CPSC) use the National Fire Incident Reporting System (NFIRS) to keep track of fire causes. The top four accidental fire causes, according to NFIRS, are cooking, heating, wiring (including lighting), and smoking. Much of the time, these "causes" are the result of poor interactions between people and appliances, but defective products do cause a large number of fires. Therefore, the examination of products suspected of causing fires requires the attention of fire investigators.

6.2 JOINT EXAMINATIONS OF PHYSICAL EVIDENCE

Examinations of evidence that are preliminary and nondestructive do not require the presence of all interested parties. Depending on the results of the inquiry, however, one can be amazed at what might be construed as "destructive" by a party that wishes to keep examination results from being admitted into evidence. Given that spoliation motions are becoming more common, providing notification is easier than explaining why notification was not made. It is therefore necessary to make certain that all interested parties have been identified and can be represented at the inspection. (One hopes that these parties were all invited to the fire scene as well, but sometimes information develops after the scene has been investigated.) Even if everyone is cooperating, it is not unusual for a joint laboratory examination to take a month or more to arrange. As people arrive at the host facility, they should all be required to identify themselves and the party who retained them. This is accomplished with a sign-in sheet, a copy of which should be provided to everyone at the end of the inspection.

It is first necessary to agree on a protocol, and this is best accomplished in advance of the examination. This protocol will necessarily be flexible because during the examination, evidence may be uncovered that causes the investigation to turn in an unexpected direction. Equipment should be available for most of the foreseeable tasks that need to be accomplished. Most of the participants will bring their own cameras, but beyond that, the host facility should be prepared to provide test equipment such as a stereo microscope, an X-ray capability, videography,[1] an ultrasonic cleaning bath, and whatever chemical and microscopic analytical equipment can be reasonably anticipated. This usually includes a Fourier transform infrared spectrophotometer and a scanning electron microscope with energy dispersive X-ray capabilities (SEM/EDX). Whenever possible, inspections should be scheduled in facilities that have this equipment. In contrast to the fire scene, it is usually possible to conduct evidence examinations in a comfortable environment. Steps should be taken to make this happen. Examining appliances in the parking lot of a rental storage facility is generally not conducive to having a productive day.

In many cases, there will be more than one item of evidence or potential ignition source to be examined. Participants should agree on the order of the examination and should stick to that agreement.

6.3 APPLIANCES AND ELECTRICAL COMPONENTS

In the examination of manufactured items, everyone should have the opportunity to photograph all surfaces of the piece of evidence. At this point, the item should be examined to conclusively determine who manufactured it. Before the case of an appliance or suspected appliance is opened, it is useful, and sometimes indispensable, to X-ray the device. This can be accomplished using film X-ray or real-time X-ray. Each has advantages and disadvantages. The major disadvantage of film X-ray is that it takes a considerable period of time to expose and develop the film, and then it may be necessary to shoot the device again from a different angle. Film X-ray, however, has a very high level of resolution. Real-time X-ray, on the other hand, allows a device to be examined from many angles in a short period of time. The output of the real-time X-ray machine can be converted to a .jpg file and stored on a computer. The capabilities of real-time X-ray units have increased, and the costs have decreased, so these devices are more common than they once were. Additionally, it is now possible (but still expensive) to use computed tomography (CT) to examine evidence.

Prior to opening the case of an appliance, the exterior should be examined for evidence of prior opening or repair. It is also exceedingly useful to have an exemplar for an examination of this type. There are times when obtaining exemplars is simply not possible or is cost prohibitive, but consideration should always be given to examining an unburned exemplar prior to opening a burned product.

The opening of an appliance should be undertaken with the minimum amount of force possible. There are times, particularly when plastic cases have melted down, when it is necessary to physically break the case off the inner parts, and this method of disassembly requires videotaping. Some plastic cases are so tough that opening them requires an overnight bath in hot methylene chloride. Fortunately, these are the exceptions.

[1] It is preferable to turn off the audio recording at a joint inspection to allow for a more open exchange of ideas. If audio is being recorded, everyone in attendance should be aware of that.

Once an appliance is opened, the individual components all require examination, testing, and more documentation. Many times, additional X-rays prove useful at this stage. It is also at this point that items might be discovered that require identification, chemical analysis, microscopic examination, or SEM/EDX examination. For example, the end of a wire might exhibit a rounded appearance, suggesting that the wire was arc-severed. Sometimes this can be clearly established by microscopic examination, as shown in Figure 6.1. At other times, a chemical analysis is required because the rounded appearance might be due to the melting of solder rather than the melting of copper. To keep the globule intact, this analysis must be done nondestructively, as with an SEM/EDX. SEM/EDX is also frequently used to examine steel objects such as nails or staples to determine whether they might have inadvertently become part of an electrical circuit.

The cause of a fracture or discontinuity in metal objects frequently becomes an issue. It is useful to have the services of a metallurgist available. Metallography can also reveal the heating history of a particular piece of metal. Metals have a granular microstructure, which changes in appearance as a result of exposure to heat. Metals that have melted and resolidified exhibit a different microstructure than the unmelted metal. "Cold working" during manufacture imparts a definite orientation to grains in many alloys. This orientation can become randomized due to annealing as a result of heating. It is sometimes possible to distinguish different heating histories in two identical parts, as shown in Figure 6.2. In this investigation, one of two clamps was suspected of overheating; when the bronze bolts on those clamps were examined, one showed evidence of cold working acquired in the manufacturing process, while the second showed evidence of annealing. Local heating also resulted in visible changes to the grain structure of one of the two brass clamp bodies, as well as to the steel rod to which one of the clamps attached. Because both assemblies were exposed to exactly the same fire, we were able to determine that one of the clamps had, in fact, overheated as a result of electrical activity and not as a result of the fire. Had the fire been capable of causing these metallurgical changes, they would have been manifested in both assemblies.

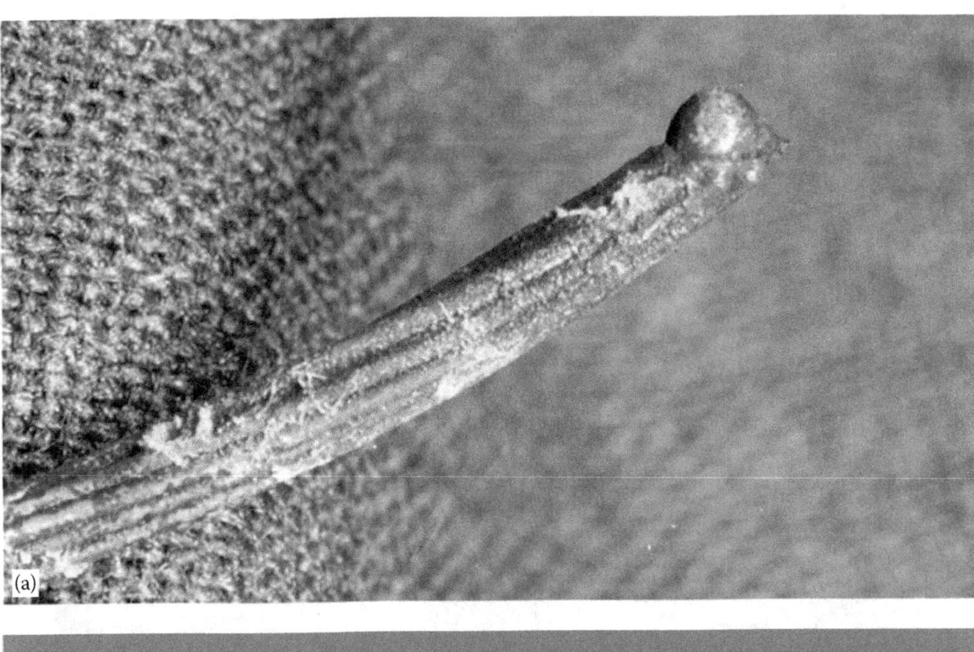

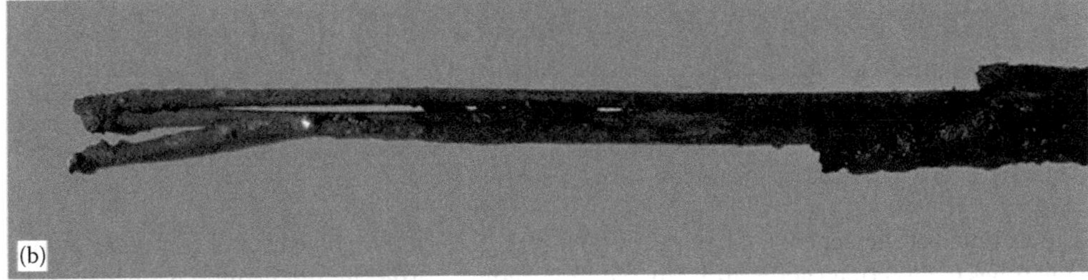

Figure 6.1 (a) Photomicrograph of an arced-severed wire. The dispositive test for determining whether a wire was damaged by fire melting or electrical activity is the finding of a sharp line of demarcation between the melted and unmelted parts of the wire. (b) Photograph of wires melted by a fire. This wire came from a mobile home burned in a test fire. There was no power on the house wiring.

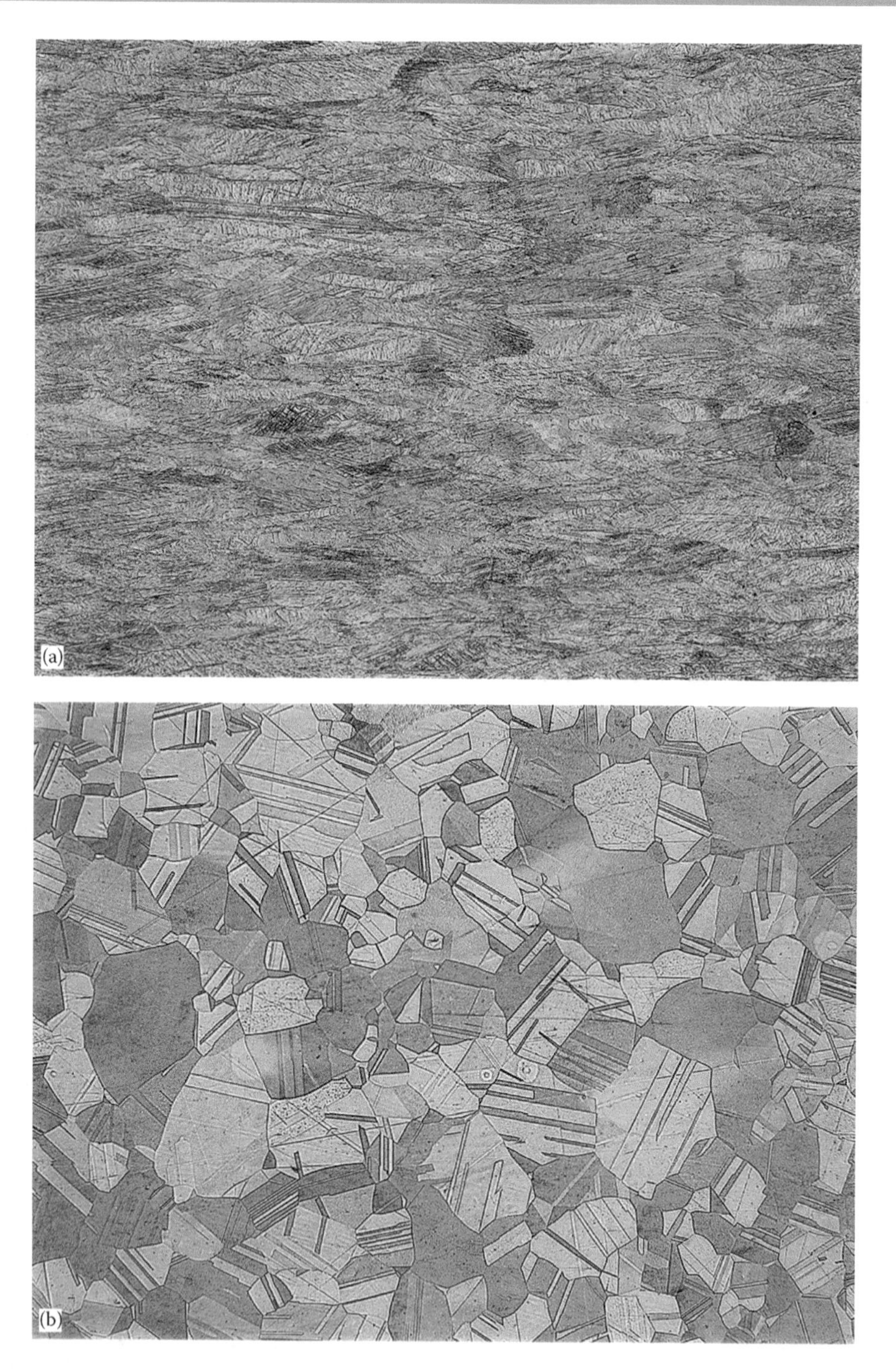

Figure 6.2 Metallographic evidence of overheating. (a) Photomicrograph showing cold working in the grain structure of the bronze bolt. This is how it left the factory. (b) Photomicrograph showing the homogeneous grain structure resulting from prolonged heating at temperatures in excess of 1000°F. Because both bolts were exposed to the same fire, it was concluded that electrical resistance heating was responsible for the condition of the annealed bolt.

Figure 6.3 is an example of the benefits of having an exemplar. This case involved small computer monitors containing cathode ray tubes (CRTs) that caught fire. The monitor manufacturer did not want to believe there was anything wrong with the product; with only the heavily damaged unit shown in Figure 6.3a, it would have been difficult to convince them that their product was anything but a victim of an advancing fire. Fortunately, there were two smaller fires found that allowed investigators to pinpoint the exact origin of the fire, which turned out to be at the location of a particular capacitor. Figure 6.3b shows the moderately damaged unit, and Figure 6.3c shows a unit disconnected when the operator noticed it was smoking. Further examination of the capacitors in two of the three monitors revealed an internal failure. The third capacitor was never recovered, but the fire patterns indicated that the fire originated in the same location on the circuit board.

(a)

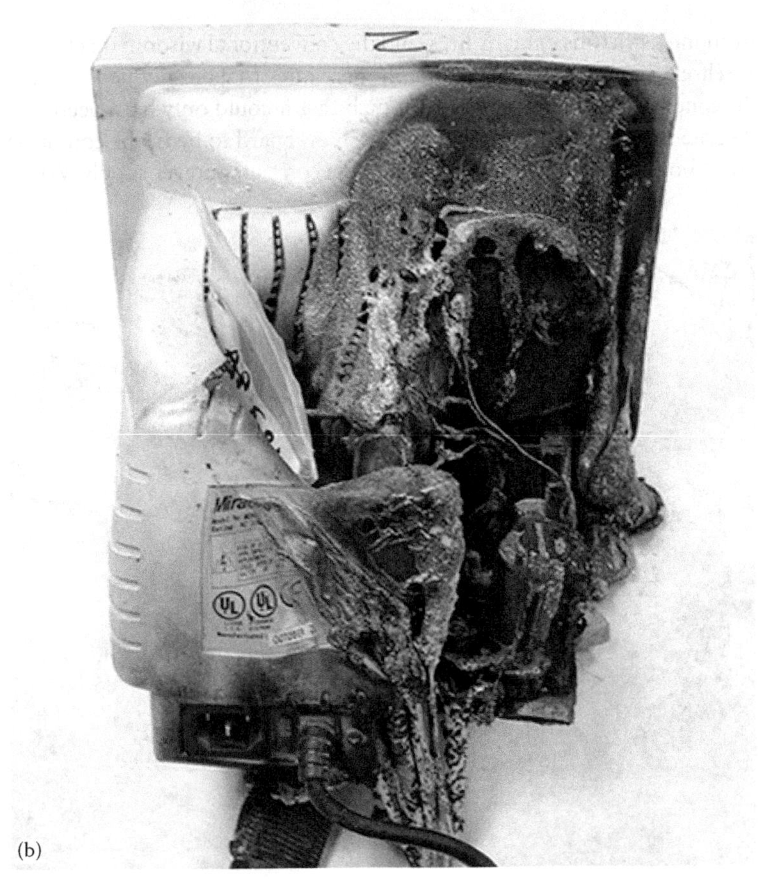

(b)

Figure 6.3 (a) This heavily damaged monitor shows more damage on one side of the unit, but a definitive origin determination is elusive without the other two monitors. (b) This unit sustained far less damage.

Figure 6.3 (c) This monitor never caught fire. It was seen smoking and was disconnected. The damage on the circuit board confirmed that a failed capacitor was the cause of all three events.

Printed circuit boards are found as victims of many fires, and the conventional wisdom on circuit board fires has been that because the fiberglass is "self-extinguishing," circuit board fires are rare. That has not been our experience in recent years. There are times when the condition of the circuit board is such that it could only have been the point of origin. In such cases, there exists no alternate scenario that would cause the circuit board to be in the condition in which it was found. Figure 6.4a shows the circuit board removed from the interior of a computer power supply. What exactly went wrong was

Figure 6.4 (a) Circuit board from a computer power supply. No scenario other than this being the origin of the fire can account for this isolated damage on the bottom of the board. The fire vented out the back of the power supply at the location of the cooling fan.

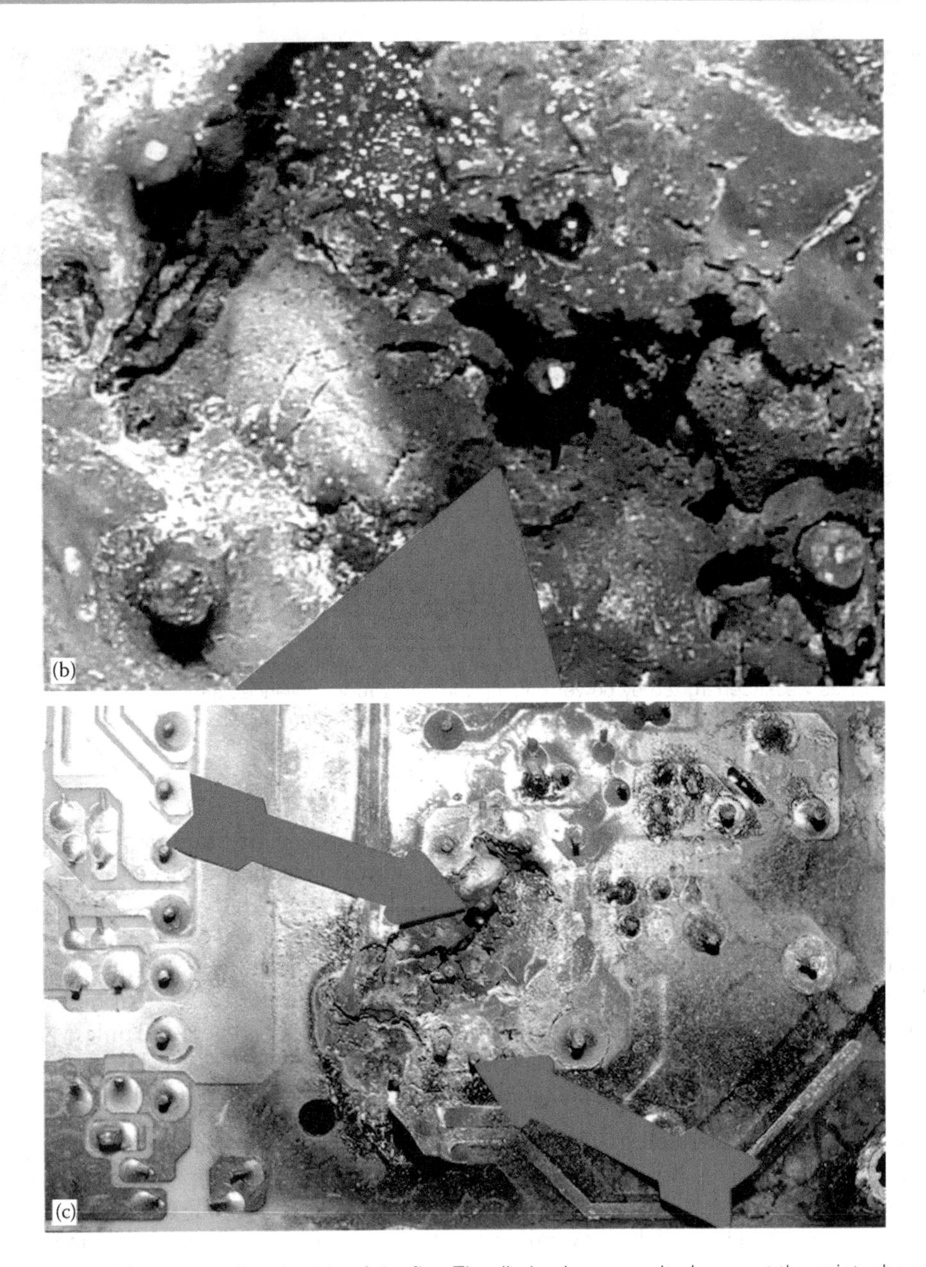

Figure 6.4 (b) Close-up of the exact point of origin of the fire. The diode pin was melted away at the point where it was soldered to the board. This was the result of a damaged solder joint that caused locally high resistance. (c) Comparison of the connections at the two ends of the diode. Only one pin was damaged, demonstrating that the fault was not an overheated diode but a locally high resistance at the point of connection.

not clear at first, but a careful examination of the CPU showed arcing on its power cord, opening of its fuse, and evidence of electrical activity inside the case. This could only happen with the fire starting on the inside of the case and moving out. A single sprinkler head controlled the fire, so the CPU and its monitor were the only items burned. A closer inspection of the area of heaviest damage on the circuit board revealed one end of a diode connection at the center of the action. This pin can be seen in Figure 6.4b. The pin at the other end of the diode, in contrast, showed no evidence of heating at its connection to the board, as shown in Figure 6.4c. This allowed us to eliminate the possibility that the diode had overheated and to conclude that a poor connection at one end of the diode generated a locally high resistance and set the circuit board on fire. After the catastrophic failure, the cause of the poor connection was not evident. There could have been a poor connection from the beginning, but it is more likely that the joint fractured some time later. Such fractures can result from the cycle of heating and cooling as the device was turned on and off. This can result in low cycle fatigue.

Figure 6.5 Exemplar circuit board made to ignite when the computer was connected to a disconnected leg of a service panel. The disconnected leg was weakly energized by power coming through 240-V appliances attached to the other leg. The low voltage caused overheating in the power supply at the resistor in the center of this photograph. The resistor was originally striped like the others in the row. The fiberglass circuit board base around the resistor and the three adjacent transistors began to char and eventually ignited.

Another example of a computer fire is shown in Figure 6.5. In this case, the fire damage was limited to a bedroom, and there were several devices near the origin of the fire. The examination of these devices was not conclusive because three of them (a television, a printer, and a computer) all burned heavily enough that they could not be eliminated as potential causes. Because the residence was not heavily damaged, an electrician was brought in to power up those parts of the house that could be safely used, and it was discovered that only one leg of the service entrance was hot. The house had several 240-volt (240 VAC) appliances, including air conditioners and water heaters that came on during the day. When there is a loss of leg, one of the two buses in the electric service panel is not energized unless a 240-volt appliance such as a water heater is activated. When that occurs, the de-energized bus becomes partially energized, delivering anywhere from 5 to 50 volts to anything that is plugged in, depending on how many 240-volt appliances are operating.[2]

We obtained an exemplar computer and supplied it with low voltage, expecting the fan motor or printer motor to overheat. Instead, we observed a fire breaking out on a circuit board inside the computer, at a resistor near the point where three transistors were located. It turned out that an underground electric cable (the service lateral between the transformer and the house) had been nicked several years prior to the fire when the underground natural gas supply was installed. The corrosion of the service entrance cable took several years and caused intermittent problems for the homeowners, who thought they were receiving poor power service because they lived at the end of an island. The gas company eventually accepted responsibility for the fire.

The experiment with the computer was less of an examination of the physical evidence and more of an examination of a proposed ignition scenario. At the fire scene, it is not uncommon to simply collect any devices in the area of origin that could have started the fire. It is only when these devices are examined in a laboratory setting that ignition scenarios can be proposed and tested. The laboratory offers the advantages of good lighting, protection from the elements, and the ability to observe the finer details that can support or disprove a hypothesis.

There have been significant developments in the understanding of circuit board failures in recent years. Richard Vicars, a colleague who specializes in printed circuit board failure analysis, has contributed the following information dealing with the reliability of electronic devices.

---

[2] It is actually possible to use an electric range as a "dimmer" when there is a loss of leg. Turning a burner up causes lights powered by the dead bus to glow brighter.

6.3.1 Electronic device reliability and failure modes

Electronic components, used in vast quantities in consumer electronics, can fail for a variety of reasons. The failure rate of electronics, expressed as failures/million hours of operation (λ), can typically be characterized by the shape of a bathtub, shown in Figure 6.6, where failures occur at a high yet rapidly decreasing rate very early in a product's life; decrease substantially throughout its useful life, where the failure rate is pretty much constant; and then increase in the wear-out cycle (end of life).

Electronic devices, being composed primarily of semiconductors and other, nonmechanical components like those shown in Figure 6.7, do not wear out in the traditional mechanical sense, like the brake pads on a car, which sacrifice pad material until the linings reach the minimum required thickness to safely stop the car.

There simply are no mechanical surfaces or moving parts inside the majority of electronic components to mechanically wear (with relays, switches, and motors and fans being the exceptions). Electronic devices and components fail for other nonmechanical reasons that this section will describe in detail. Before this discussion can occur, a brief tutorial on the reliability bathtub curve for electronic devices is required.

6.3.1.1 The burn-in phase

A new product that literally fails right out of the box, in the context of reliability engineering, demonstrates a phenomenon known as infant mortality. Infant mortality failures, or dead-on-arrival devices, are perhaps the most frustrating failures that a consumer can experience. Not only do such early failures send the message to the consumer that the design may be defective, but they also cause the consumer to question his or her brand loyalty.

In an attempt to prevent infant mortality failures from reaching the consumer, manufacturers deploy many extensive processes to ensure that these weaker products are, at a minimum, weeded out before they reach the field. A process known as environmental stress screening (ESS), or burn-in (originally developed by defense contractors to age newly manufactured electronics into the useful life phase of the product life cycle), involves stressing the product using thermal cycling and vibration prior to final acceptance testing and shipment. The extreme thermal excursions of the temperature chamber and the aggressive vibration and shock of the slip table impart just enough stress to fail those electronic assemblies that would have found their way into consumers' hands, without applying too much stress to take appreciable life out of the product. Although this is a good process to age the unit just enough to screen out infant mortality, newer processes that allegedly involve more stringent control and testing at the tier 1 supplier component level have theoretically replaced burn-in as the cost-effective acceptance test of choice.

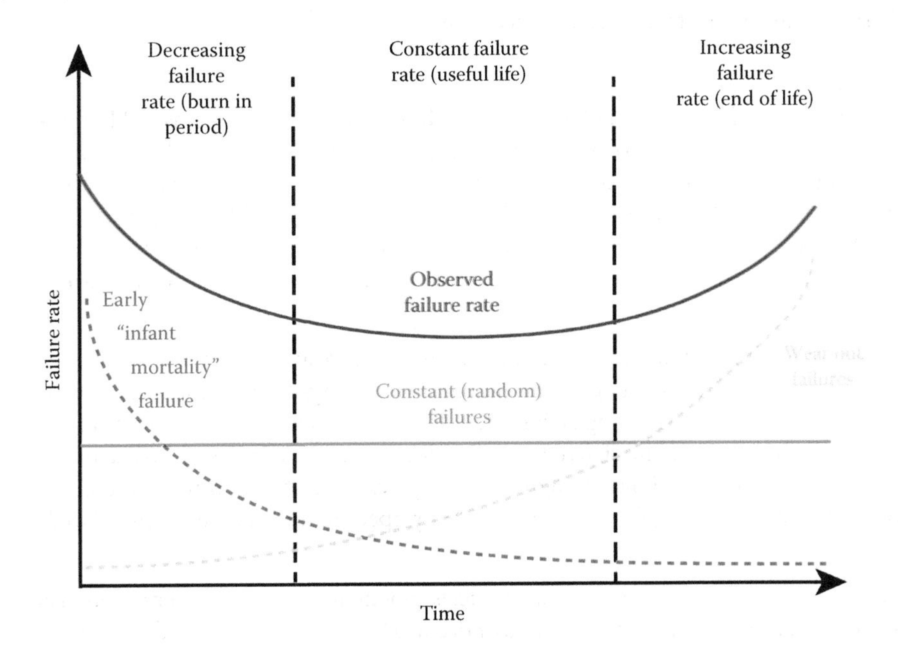

Figure 6.6 Reliability bathtub curve.

Figure 6.7 Surface mount technology (SMT) printed circuit board (PCB). (Courtesy of Richard Vicars, Jensen Hughes, Baltimore, MD.)

Once the product has been aged into the flat portion of the reliability bathtub curve (useful life), it should be experiencing only the normal, or as-designed, failure rate. This is a consumer-friendly reliability level. There is a common misconception in industry that products are designed to be free of defects. Although this is an admirable goal, it is simply not possible to achieve. Consequently, during the design and development process, engineers set predetermined reliability specifications with which to design and test the product. Designing with a failure rate in mind does not imply that the product is poorly designed. On the contrary, it means that the engineering team consciously designed the product to meet or exceed a desired reliability by using various engineering tools and processes. In other words, the engineering team knows in advance, during the design phase, how the product will perform once in the hands of customers.

6.3.1.2 The useful life phase

If an electronic assembly accumulates enough operational time to make it into the useful life phase of the reliability bathtub curve, then only normally occurring component failures, theoretically, will cause product unreliability or failure. Every electronic assembly has a failure rate, and this is also true for the components that comprise them. As electronic components like capacitors, resistors, diodes, microprocessors, and LEDs operate, they fail in accordance with their designed-in failure rates or those failure rates defined by the physics of the device materials. These failures occur throughout the useful life phase, but they are far less frequent than during the burn-in phase, where infant mortality occurs.

One of the most common failures of the electronic components is the result of electrical overstress—that is, operating the component up to or over its rated voltage, current, or power ratings—either intentionally or due to an extreme environmental event (power surge, lightning strike, etc.). Figure 6.8 shows the heating effects of a 7200 VAC surge on a 120 VAC dimmer switch. Notice the charred and discolored areas of the circuit board and evidence of reflowed solder as a result of the heat generated by this significant transient overstress. Moreover, the light blue component, which is a carbon composition resistor, failed in an electrically open manner as it succumbed to the excessive heating of this event.

Other common failure modes are due to various quality defects (not design defects) by component manufacturers who are running manufacturing processes that are not optimized or deemed "out of process control." Such failures usually mean that

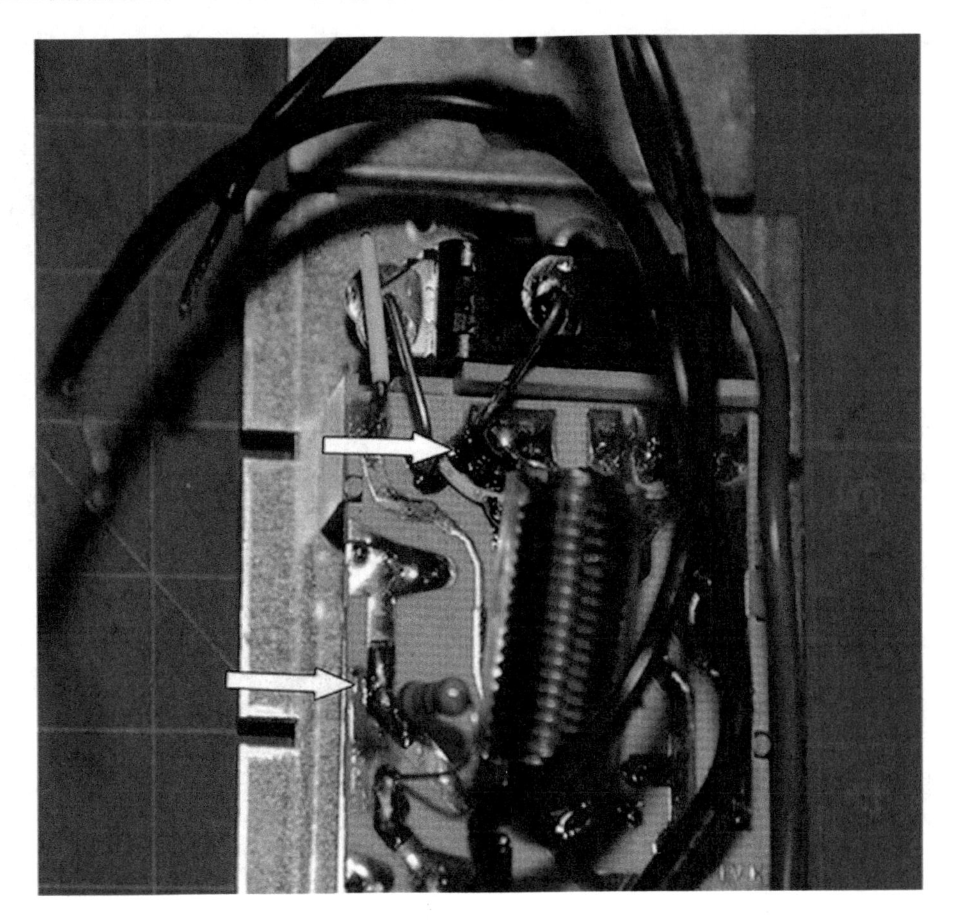

Figure 6.8 Voltage overstress on dimmer switch electronics (white arrows). (Courtesy of Richard Vicars, Jensen Hughes, Baltimore, MD.)

our cell phones, tablets, e-readers, or televisions simply stop working with no advance notice or symptoms. In some instances, the user may observe a wisp of smoke emanating from the device or, in a worst-case scenario, the entire device can catch fire.

6.3.1.3 The wear-out phase

As an electronic assembly ages, it moves from the useful life phase of the reliability bathtub curve to the wear-out phase. Since solid-state electronic devices generally do not wear out in the traditional sense, the wear-out phase includes failure of mechanical aspects of the components, solder joints, or solder connections; the continued aging effects of power on/off cycles; and so on.

It is the mechanical aspects of these electronic components that may wear out, depending on the robustness of the design (i.e., how the components were soldered to the printed circuit board). If little attention was paid to the circuit board layout and the physical geometries of the solder attachments from a mechanical perspective, then these connections and interconnections will ultimately fail from creep or fatigue failure mechanisms (or user abuse) that will ultimately compromise the electrical connections to the components, causing product failure and, in some cases, arcing, ignition, and fire at this interface, as shown in Figure 6.9.

6.3.1.4 Electronic device failure causes

There exist a variety of reasons why electronic devices fail within the useful life phase of the reliability bathtub curve, other than the inherent and additive failure rates of the individual components. For example, using electronic devices in environments for which they were not designed can create reliability problems. The moist environment can cause

Figure 6.9 Fire on PCB from an arcing at a failed connection to the board.

unintentional “sneak” circuits and corrosion and can even activate preexisting manufacturing contaminants that not only can cause malfunction of the device but also produce localized ignition and fire.

The primary causes for electronic device failure within the useful life phase are:

1. Humidity and contamination
2. Salt air
3. Extreme temperatures
4. Mechanical—extreme vibration/shock
5. Radio frequency interference (RFI) and conducted emissions (EMI)
6. Electrostatic discharge (ESD)

Contamination comes in all forms. For example, a microwave oven can ignite during operation as a result of food contamination and residue that accumulates in the waveguide/stirrer mechanism, creating a localized hot spot, as shown in Figure 6.10. As the contamination continues to heat from the RF energy emitted by the magnetron tube, the waveguide will eventually burn or melt, as shown in Figure 6.11. Depending on the unit’s materials and construction, a fire can result.

A circuit board, such as the electronic control in a microwave oven, however, responds to contamination in a different, perhaps more subtle and insidious way. A printed circuit board consists of hundreds or thousands of discrete electronic components soldered to (SMT) or through (through hole) a printed circuit board. The circuit traces on and within the layers of the printed circuit board act as conductors for the various voltages and signals that allow the individual components to interact properly.

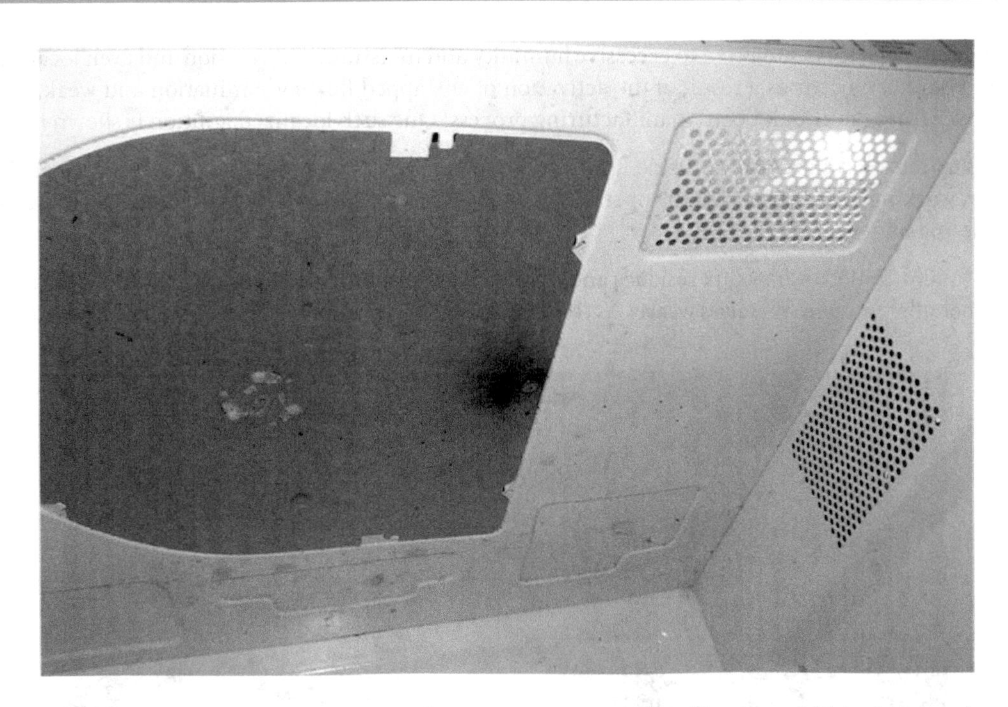

Figure 6.10 Charring of food debris and grease on microwave oven waveguide. (Courtesy of Richard Vicars, Jensen Hughes, Baltimore, MD.)

Figure 6.11 Burned plastic waveguide cover in microwave oven due to contamination accumulation. (Courtesy of Richard Vicars, Jensen Hughes, Baltimore, MD.)

When the electronic device is subjected to excessive humidity and moisture, malfunction and even localized ignition of the circuit board can occur as a result of the activation of entrapped flux contamination and weak organic acids (WOAs) left over from the circuit board manufacturing process. One such localized ignition is shown in Figure 6.12.

Figure 6.13 illustrates localized, bottom-side ignition between two solder joints that were hand retouched. When soldering or reworking boards by hand, the operator may apply too much flux and then fail to clean the flux residue once the soldering has been completed.

Figure 6.14 is an illustration of how flux residues and WOAs become entrapped within the circuit board substrate that consists of paper filler through fiberglass weave. As the holes in which the components are inserted through the PCB are

Figure 6.12 Localized board ignition due to humidity activated contamination on board surface (SMT). (Courtesy of Richard Vicars, Jensen Hughes, Baltimore, MD.)

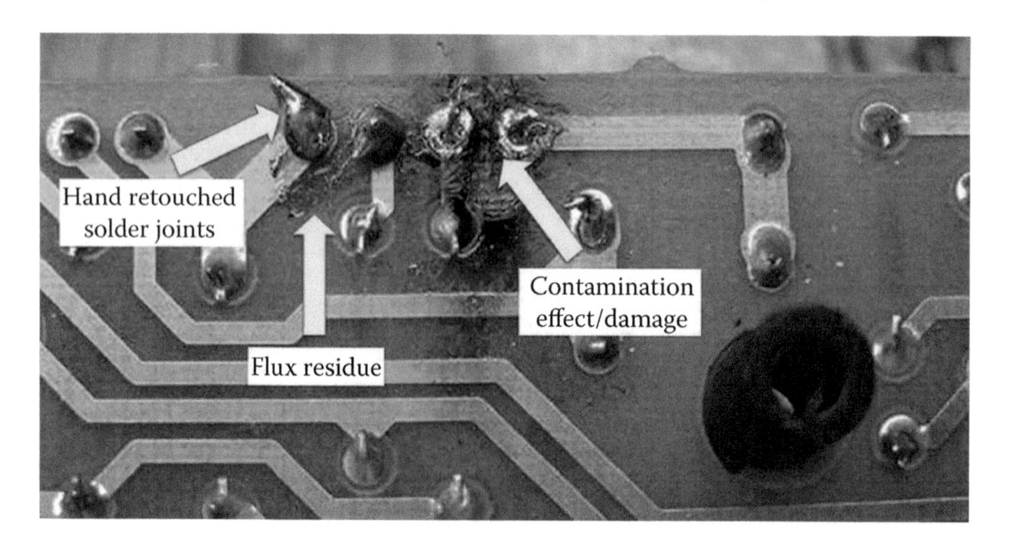

Figure 6.13 Electrical activity due to hand-applied flux contamination. (Courtesy of Richard Vicars, Jensen Hughes, Baltimore, MD.)

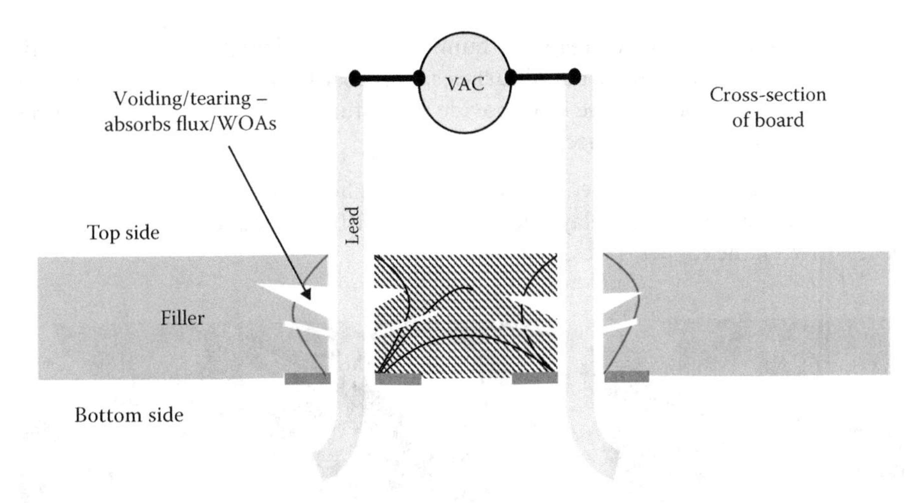

Figure 6.14 Inner board flux penetration. (Courtesy of Richard Vicars, Jensen Hughes, Baltimore, MD.)

created at the board manufacturer via a punching (or drilling) operation, a tearing action through the substrate occurs as the tools age. This tearing action creates voids adjacent to the holes, which create entrapment areas for flux residues.

The entrapped residues produce inner-board conductive paths that create localized inner board heating and ignition of the substrate between the oppositely charged component leads. If the plastic case that contains the circuit board is not sufficiently fire retardant, the consumption of the entire device can occur, even when powered only by common household batteries. Television remote control devices are not commonly thought of as ignition sources, but, as shown in Figure 6.15, even a device powered by only two AA batteries is capable of igniting when the PC board is contaminated.

Figure 6.15 Remote control ablaze. (Courtesy of Richard Vicars, Jensen Hughes, Baltimore, MD.)

In order to allow the electronic assembly to operate in humid or even condensing environments, the manufacturer may apply a layer of varnish or silicone to the board surfaces. This coating, known as a conformal coating, is shown in Figure 6.16. A conformal coat's sole purpose is to provide a moisture barrier for the copper traces and a laminate of the printed circuit board. In general, it is successful.

However, conformal coat is not intended to be, nor can it be, 100% hermetic. As it is somewhat hygroscopic (i.e., it absorbs water), in some instances, it merely delays the ingress of moisture to the sensitive electronic components and the very small spaces between the copper traces.

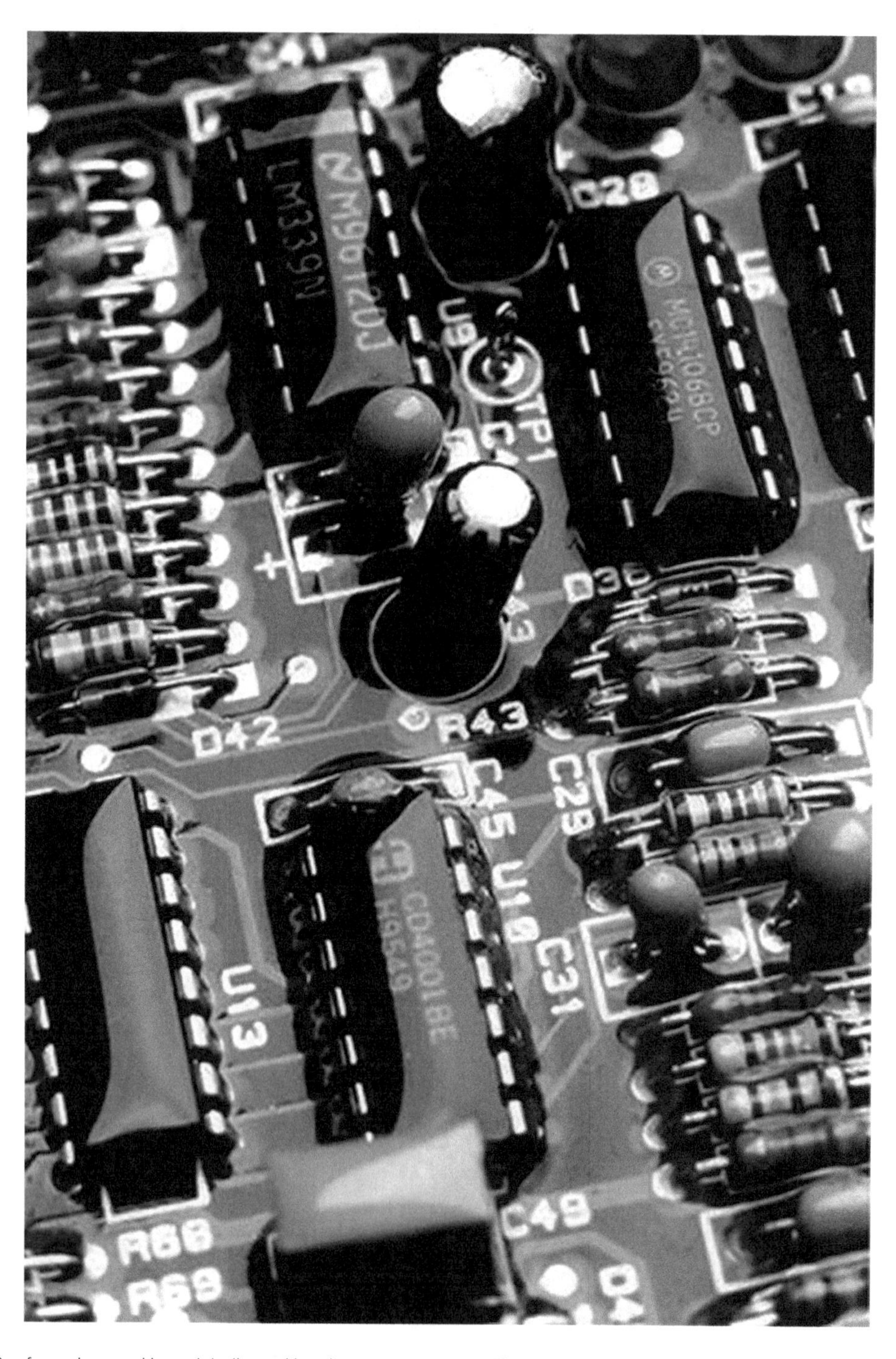

Figure 6.16 Conformal coated board, indicated by glossy appearance. (Courtesy of Richard Vicars, Jensen Hughes, Baltimore, MD.)

SMTs are capable of growing root-like structures known as dendrites. Dendrites can grow beneath the "protective" conformal coat layer. A dendrite is a conductive metal filament or crystalline structure that grows in the presence of contamination, moisture, and voltage bias. It creates an unintended current path (short circuit). The moisture from ambient humidity can migrate through the coating over time, activating the residual contamination across the bias (voltage nodes). This shorting, even across such a small component, can be a competent ignition source, especially as the dendrite continues to regrow and thicken, carrying increased amounts of fault current upon each regrowth cycle.

6.3.1.5 Case study—Fire remote from root cause

A more insidious and less obvious failure mode due to the ingression of moisture onto and into the circuit board's inner workings is the activation of manufacturing-introduced and environmentally introduced contamination residues causing a fire remote from the malfunctioning product. The fire's origin was not in the same location as the fire's cause.

In one example, the fire investigator selected a severely burned stereo speaker as the fire's origin, but speakers are typically passive unpowered devices. Further, the homeowner reported that the stereo was turned off. The investigator chose a reasonable origin based on the fire patterns and relevant damage, but an extensive series of empirical tests and analyses was required to establish the true cause. Normally a speaker fire would require the stereo to have been powered on and playing music, probably very loudly, at the time of ignition.

Most speaker assemblies utilize UL (Underwriters Laboratories) 94 V flame retardant rated materials; however, the performance of many materials is unpredictable at best, especially with the introduction of the reduction of hazardous substances (RoHS) requirements and the difficulty in controlling quality for products originating from the Pacific Rim. Assuming the speaker met the UL requirements, if the speaker were to be overdriven due to excessive volume, any excessive heating of the respective coil wire would not, in theory, be able to ignite the cone material (or if it did, the flame would promptly self-extinguish). There would need to be continuous amperage applied to the speaker in order to achieve sustained combustion.

It is known that unpowered speakers do not spontaneously combust. If the statement by the homeowner regarding the amplifier having been powered off were true, then it is difficult to understand how the speaker could have caused the fire. Fortunately, the *entire* contents of the room, including the amplifier, were harvested. When the amplifier was unwrapped for documentation, the engineers invited to the inspection snapped a few perfunctory photos and promptly moved on to the next piece of evidence. They were not interested in this product from a causation standpoint because it exhibited no fire damage.

With the agreement of all in attendance, the amplifier was subsequently powered up by pressing the "soft start" ON button while connected to exemplar speakers taken from the same home. The amplifier produced clear and audible music, even at full volume, at least for a period of time. The amplifier was then powered off by pressing the soft OFF button. Note that even when the power button on the amplifier is in the OFF position, it is not actually powered down. Many electronic products are never really powered off because the microprocessor in the device is always powered up and waiting for inputs from the user. One of those inputs is the recognition of a depressed "power on" button on the front panel of the amplifier. So power is usually flowing through the product.

The subject speaker was subsequently analyzed and exhibited an overcurrent or melting open (like a blown fuse filament) of one of two speaker coil wires. The wire was clearly electrically overstressed, creating a length of wire that is fused open due to localized internal heating, not fire melting.

At this point in the investigation, the team was still trying to understand how an unpowered speaker could spontaneously ignite. More attention was focused on the amplifier, even though it was remote from the area of origin. Knowing that microprocessor-controlled electronics can misbehave for a variety of manufacturing or environmental reasons, it was decided to raise the humidity around the amplifier. Within seconds, a strange phenomenon occurred. As the relative humidity was raised just 5 points, from 35% to 40%, the amplifier actually turned on by itself.

The problem was not necessarily that the amplifier actually powered *on* spontaneously. The bigger problem was *why* the amplifier powered on spontaneously. This is a clear indication of a software or environmentally induced malfunction, suggesting that the microprocessor was "confused" and received an errant command. This allowed the music to continue to play through all of the exemplar speakers, but the left front exemplar speaker (the same channel that was wired to the subject speaker) started to distort ever so slightly after a period of time. Note that the humidity was raised to achieve this result. This is a key point in the investigation, as embedded software (firmware) is not affected by humidity.

While the speaker was playing distorted music, a spectrum analyzer was connected to the speaker output the amplifier on the subject channel to study the signal more closely. Something unusual was observed on the instrument display. The measurement revealed that the speaker was seeing not only the appropriate audio signal but also an inappropriate, constant 12 VDC (volts direct current)—the equivalent of connecting a car battery directly across the speaker. At about 30 seconds into the testing, the music started to distort and the DC voltage proceeded to overheat the exemplar speaker coil, which subsequently erupted into flames. The fire eventually consumed all of the materials within the speaker basket before it was extinguished by the investigation team.

The main circuit board, also containing the amplifier section, was removed and inspected. A white film was observed atop certain areas of the circuit board, as shown in Figure 6.17. These were visible signs of contamination (most contamination is not visible to the naked eye). The board was sent to a chemical laboratory for ionic contamination testing. Ionic contamination levels from manufacturing by-products of greater than 10 $\mu g/in^2$ were detected via ion chromatography, creating unintentional circuit paths, or "sneak circuits," to the power-on circuit as well as the audio processor. This was the root cause of the fire.

Given that contamination this subtle can occur on a device like a stereo speaker, causing a propagating fault that creates a fire in a product across the room, consideration must be given to other electronics as well.

6.3.1.6 Hardware versus software

The most benign failures are mere nuisances that require rebooting, for example, computers "locking-up" or displaying the infamous *blue screen*. In such a case, there is no life-threatening situation. These are instances of a "fault" or preexisting error in millions of lines of embedded software—the instructions executed by the microprocessor (at a defect rate measured in parts per million)—that manifest in a particular operational sequence to cause a functional problem noticeable to the end user.

The consequences of a software error in a consumer electronics device such as a cell phone can also have severe, even life-threatening consequences. The millions of lines of code in a microprocessor instruct the many subsystems of the associated on-board hardware electronics. The microprocessor communicates with memory, displays, batteries, chargers, RF modules, and more. As it executes its commands, sending and receiving signals at nearly the speed of light, an error in the embedded software can actually cause the operation of an unintended feature. For example,

Figure 6.17 Board contamination > 10 $\mu g/in^2$. (Courtesy of Richard Vicars, Jensen Hughes, Baltimore, MD.)

various on-board electronic components could be called to turn on simultaneously, causing excessive heating of the board as the system attempts to source the required current from the battery.

Another example is any device that is used to turn on an electric motor through a relay. Consider a garage door opener. When the user presses the remote control to open the garage door, the opener receives an RF signal from the transmitter. The microprocessor reads this signal; translates it to binary code; executes its embedded software; and finally, once it has determined that it is supposed to open the door, sends a signal to a relay to close and supply a higher voltage at the motor. If the microprocessor experiences a software error, it could:

1. Fail to respond to the remote control—a benign failure
2. Open or close the garage door when not instructed to do so—a more serious failure
3. Ignore the safety limits and sensors when someone is passing under the door as it is opening—a life safety failure
4. Send a series of high-speed on/off commands to the relay that turns the motor on and off—a reliability failure or a life safety failure

Failure modes 1, 2, and 3 are concerning in their own right, but for the purpose of fire causation, failure mode number 4 can produce ignition and fire in the garage door opener as follows.

If the microprocessor erroneously instructs the relay to turn the motor on and off at a fast duty cycle, the relay contacts, which are rated for light duty, begin to heat. As the contacts heat within the case of the relay, the contacts themselves can weld or degrade to the extent that they cause the plastic relay case to pyrolyze and burn.

Still another failure mode is possible. As the relay contacts chatter, they do not have enough "off time" to sufficiently cool per their design specification. The heat generated at the contacts begins to rise dramatically and is conducted down the pin from the contact area and into the circuit board/solder joint. If the temperature rises above the melting point of the solder, the current-carrying capacity of the solder connection is reduced, and the temperature at the connection point of the relay begins to rise, causing the solder to flow. Tin–lead solder melts at 360°F–370°F. Lead-free solder (increasingly mandated due to its compliance with RoHS requirements) melts at a much higher temperature, that is, 430°F.

Arcing between the copper-clad hole on the circuit board and the pin on the relay begins as the cross-sectional area of the solder joint changes or voids completely. This arcing represents a competent ignition source.

The circuit board epoxy and laminate, in the presence of this arcing, become the first fuel in the presence of this competent ignition source. Even though the board material may be UL 94-V0 flame rated (i.e., it must self-extinguish within ten seconds following the removal of flame), it will continue to burn in the presence of an energy source and, in some instances, it will burn violently.

Further, not all 94-V0 rated board manufacturers are producing circuit board laminate that, when actually tested to 94-V0 test procedures, performs to the V0 specifications. This is occurring for a variety of reasons, including aggressive cost reduction and RoHS initiatives. Formerly, brominated compounds were used extensively as flame retardants in printed circuit boards. But as is the case with the lead in solder, brominated flame retardants are now considered unsafe (they are suspected carcinogens), and their use is being substantially reduced.

6.3.1.7 Beware of red phosphorous—A popular fire retardant

Red phosphorous is a low-cost, highly abundant fire retardant with an environmentally friendly profile (Figure 6.18). As such, red phosphorous has become the fire retardant of choice in many electronic devices, especially as RoHS compliance becomes more burdensome to the product manufacturers.

One of the most popular uses of red phosphorous is its incorporation in the plastics of IEC-type connectors and power cords. IEC (International Electrotechnical Commission) connectors and powers cords are commonly used in a wide variety of consumer electronic products (TVs, computers, laptops, countertop appliances, and more). As such, virtually any consumer electronics device that can be plugged into an outlet can have red phosphorous in its power cord and connector socket.

Red phosphorous is an effective fire-inhibiting additive when combined with flammable plastics and resins. This fire inhibitor functions by forming a flame-resistant intumescent layer of char during combustion of the plastic. This property is in part based on its ability to form phosphoric acid through a reaction with water and heat during combustion.

Figure 6.18 Red phosphorous. (Courtesy of Richard Vicars, Jensen Hughes, Baltimore, MD.)

Unfortunately, it is this same property that also permits the slow evolution of phosphoric acids during exposure to elevated temperatures in the presence of moderate to high humidity. In some plastics such as nylon, the characteristic surface morphology is porous, and moisture can percolate within the surface and react with the red phosphorus additive. On close examination, particles of red phosphorous can be seen on the surface of the plastic when observed with a hand lens. The red phosphorous is even more visible in cross-section (Figure 6.19), where the red particles can be viewed with the naked eye.

Exposure of red phosphorous impregnated plastics to a warm and humid environment results in the formation of conductive, fluid-deposited beads of phosphate salts on the surface. These conductive salts of sodium and magnesium

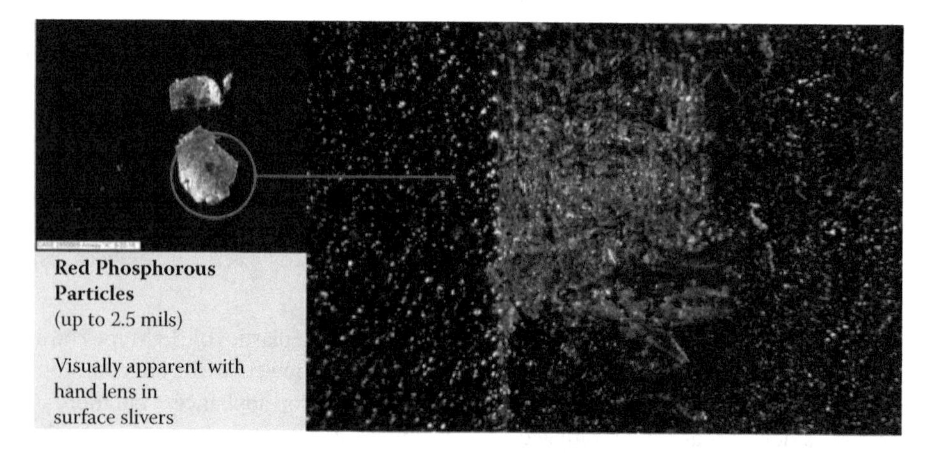

Figure 6.19 Red phosphorous particles visible in the cross-section of an IEC connector. (Courtesy of Richard Vicars, Jensen Hughes, Baltimore, MD.)

phosphates form in situ through a reaction between the induced phosphoric acid and the mineral fillers used in the plastic formulation. This condition results in the formation of dangerous conductive salt bridges and dendrites, which ultimately create electrical shorting paths (arc-tracking) between adjacent pins, conductors, or traces. Once the shorting path is complete, ignition occurs (sometimes with violent arcing activity) at this precise location (Figures 6.20 and 6.21).

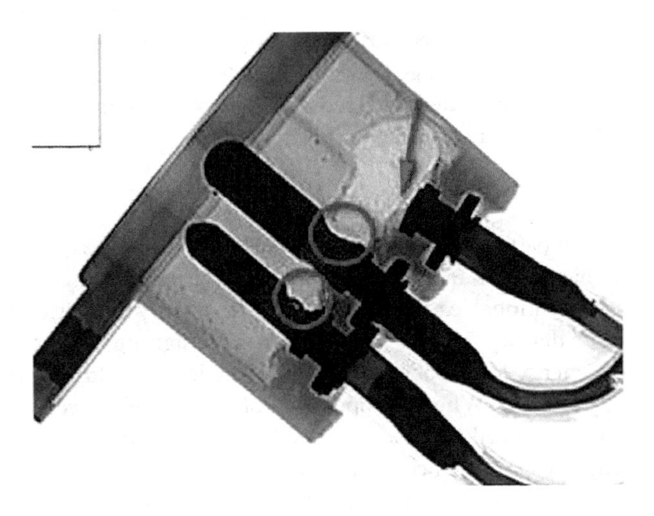

Figure 6.20 Erosion of conductor pins caused by arc tracking. The short-circuit current flowed over a conductive phosphate salt bridge. (Courtesy of Richard Vicars, Jensen Hughes, Baltimore, MD.)

Figure 6.21 Flaming line cord at IEC interface. Ironically, red phosphorous flame retardant was the root cause of the fire. (Courtesy of Richard Vicars, Jensen Hughes, Baltimore, MD.)

This is why it is of the utmost importance for the fire investigator to collect any and all power-carrying plastic connectors and line cords, as potential ignition sources, when processing a fire scene, including exemplar products. Subsequent root cause analysis, looking for the presence and reaction of red phosphorous, can be performed by a qualified chemist or material scientist on both subject and exemplar products. Using a carefully designed environmental test, exemplar products can also be driven to failure, and fire, providing verification of the investigator's fire cause hypothesis in accordance with the requirements of NFPA 921.

In summary, the failure of any electronic component or assembly, even those using low voltage, can start a fire. When components experience failure modes as described earlier, are mounted to printed circuit boards that do not perform to the flame ratings as specified, and are packaged within plastics that have significantly reduced (or completely eliminated) flame retardants, or are impregnated with red phosphorous as a fire retardant, a fully involved product fire, and resulting compartment or structure fire, can ensue.

6.3.2 Lithium ion batteries

The greatest strength of lithium ion batteries is also their greatest weakness. The chemistry of the Li-ion cell allows a battery to be constructed that has an extremely high *energy density*, being measured in the range of .93–2.34 MJ/Liter. By way of comparison, a conventional lead-acid battery has an energy density of .54 MJ/Liter, and a battery of Ni Cd construction has an energy density of .18 to .52 MJ/Liter. In designing battery systems, engineers try to use a topology that allows the most energy to be packed into either the lowest volume (volumetric energy density) or lowest mass (gravimetric energy density). When the operation of a given cell is well controlled, the battery system is very reliable. However, when a given cell is operated outside its usual parametric envelope (to include external insult or internal manufacturing defect), the uncontrolled discharge of energy can result in extreme temperatures, flame, or both. Li-ion batteries do not respond well to either overcharging or undercharging [1]. Although there are exceptions for large applications (airplanes and automobiles), the phone, computer, and appliance batteries that an investigator is likely to encounter are either a set of "18650" cells or pouch-type batteries. These batteries, examples of which are shown in Figure 6.22, contain charge controllers and safety devices in addition to the chemical cells that provide the current.

As the use and fielded population of lithium ion batteries increases, even if the battery technology's failure rate (estimated at 1 in 10 million cells) remains constant, the likelihood of an investigator encountering a lithium ion battery failure and fire statistically increases simply because the population of such battery-powered devices continues to grow over time. Manufacturers know what causes Li-ion batteries to fail (Figure 6.23) and are taking steps to design safer ones. One new design uses a nonflammable electrolyte, an obvious risk reducer. Solid electrolytes are also on the horizon. Another design incorporates fusible microspheres of polyethylene. When the temperature in the cell reaches the melting point of the polymer, the now-liquid plastic seals the pores in the separator, shutting down flow of electrons. Although there will be safer designs in the future, today's fire investigators must deal with the ones currently in service.

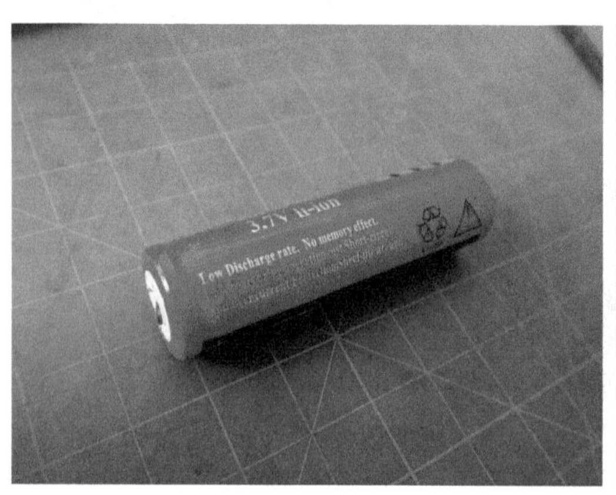

Figure 6.22 A single 18650 cell (left) and a lithium ion pouch battery of the type used in tablets, laptops, and cell phones (right). (Courtesy of Richard Vicars, Jensen Hughes, Baltimore, MD.)

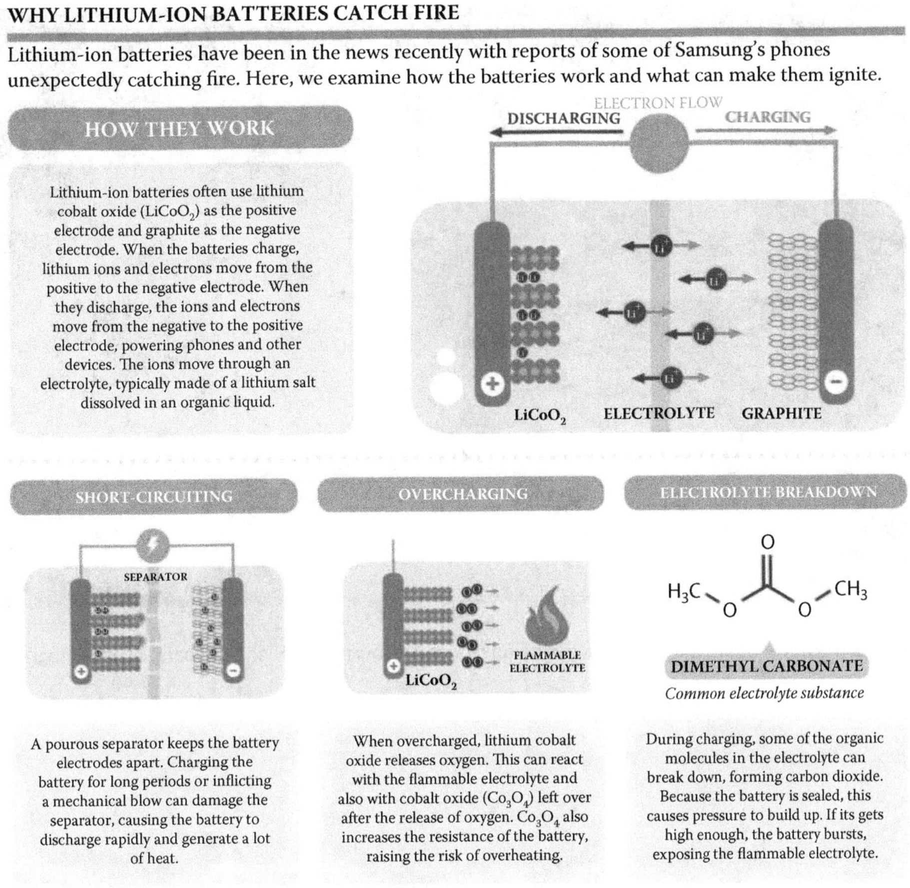

Figure 6.23 Diagram showing the various failure mechanisms for Li-ion batteries. (Reprinted with permission from *Chem. Eng. News*, November, 14, 2016, 94(45), 33. Copyright 2016 American Chemical Society.)

The failure modes of Li-ion batteries can be seen before a fire, but afterward, the cell where the failure started is likely to be so totally destroyed that not much can be said, other than, hopefully, "This is the cell that failed." The ensuing fire may damage adjacent cells to the point where all that can be said is "This battery pack failed." Thus, as with so many issues in fire investigation, the chicken-and-egg question arises. "Was this battery pack the cause of the fire, or did the fire damage the battery pack?" In some cases, both questions may be answered in the affirmative.

Figure 6.24 shows a hoverboard fire that occurred when the lithium ion cells, while being charged, went into thermal runaway.

Any time a lithium ion–powered device is the suspected cause of a fire, a fire investigator must work diligently to harvest all of the components. This includes finding all of the constituent pieces of the battery cells—cans, end caps, straps, copper jelly roll, and so on, or pouches, chargers, and the battery management circuit boards. To avoid spoliation, special attention must be given to the harvesting and storage of the steel cans of 18650 cells as the cans quickly rust, corrode and ultimately disintegrate in evidence containers not properly prepared. As such, the optimum method of harvesting and storing these cans for long-term storage prior to a laboratory evidence exam is in a sealed container with a desiccant package (Figure 6.25).

6.3.3 Metal oxide varistors

Metal oxide varistors (MOVs) are incorporated into relocatable power taps (RPTs; also known as power strips) and ground fault circuit interrupters (GFCIs). An MOV is a dime-sized or smaller disc of zinc oxide treated with small quantities of

Figure 6.24 Remains of a hover board that ignited while being charged. (Courtesy of Richard Vicars, Jensen Hughes, Baltimore, MD.)

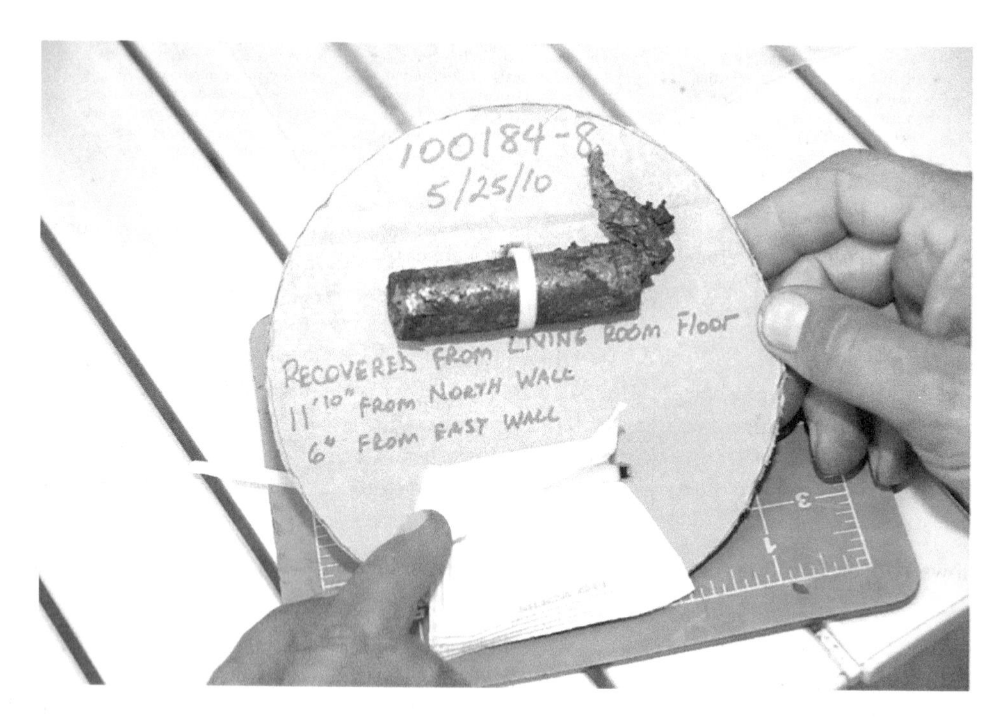

Figure 6.25 Proper method for harvest/storage of lithium ion cells. (Courtesy of Richard Vicars, Jensen Hughes, Baltimore, MD.)

other substances and is designed to protect electrically "downstream" devices by diverting excess voltage that results from a power surge. Power surges result from lightning, utility switching, and inductive loads inside a structure.

The disc has two leads and is encased in epoxy. During operation with nominal line voltage (120 VAC) no current passes through the zinc oxide disk. When voltage exceeds the MOV's "clamping voltage," energy passes from a conductor on one side of the disk through the zinc oxide to the opposing conductor. Typically this causes no problems, but when excessive voltage or repeated spikes are experienced, the MOV can fail. Failure occurs in one of three ways:

1. Conductors inside the MOV weld together forming a short circuit. This action should cause the integral circuit breaker to open, preventing the unit from working. When this happens, the suppressor must be replaced, typically by getting a whole new device.
2. Conductors inside the MOV completely separate, preventing electrical flow across the MOV's components. Though the outlet strip continues to function, no surge suppression capability is present. Unless the unit is equipped with an indicator light, the user cannot know that no protection is present.
3. Conductors form a circuit with high resistance. In this case a heater is created. Depending on characteristics generated, this is the result that initiates fires.

When finding an RPT at the origin of a fire, the investigator may be tempted to count the number of appliances plugged in, assume they were all drawing current at once, and declare that the fire resulted from an "overload." While not impossible, if the RPT were truly overloaded, the circuit breaker should have tripped. It is always necessary to carefully collect all of the pieces of the RPT. This can be difficult because MOVs are fragile and may be missed. If the MOV looks like the one shown in Figure 6.26, it can be concluded that the MOV was the likely ignition source. The melting point of zinc oxide is 1975°C (3587°F), so one can be sure it required an electrical source to melt.

The RPT shown in Figure 6.27a was found at the origin of a bedroom fire. Because of the destruction caused by the fire, neither the switch nor the MOV was collected. Melting on the brass neutral bus and on one of the other buses, shown in Figure 6.27b provided sufficient evidence to conclude that the RPT was the ignition source. As can be seen, the evidence was quite subtle and the melting escaped the notice of the first investigators, so they did not look at the RPT. As with RPTs, any time a GFCI receptacle is found at the origin of a fire, an MOV failure should be considered as the possible ignition source.

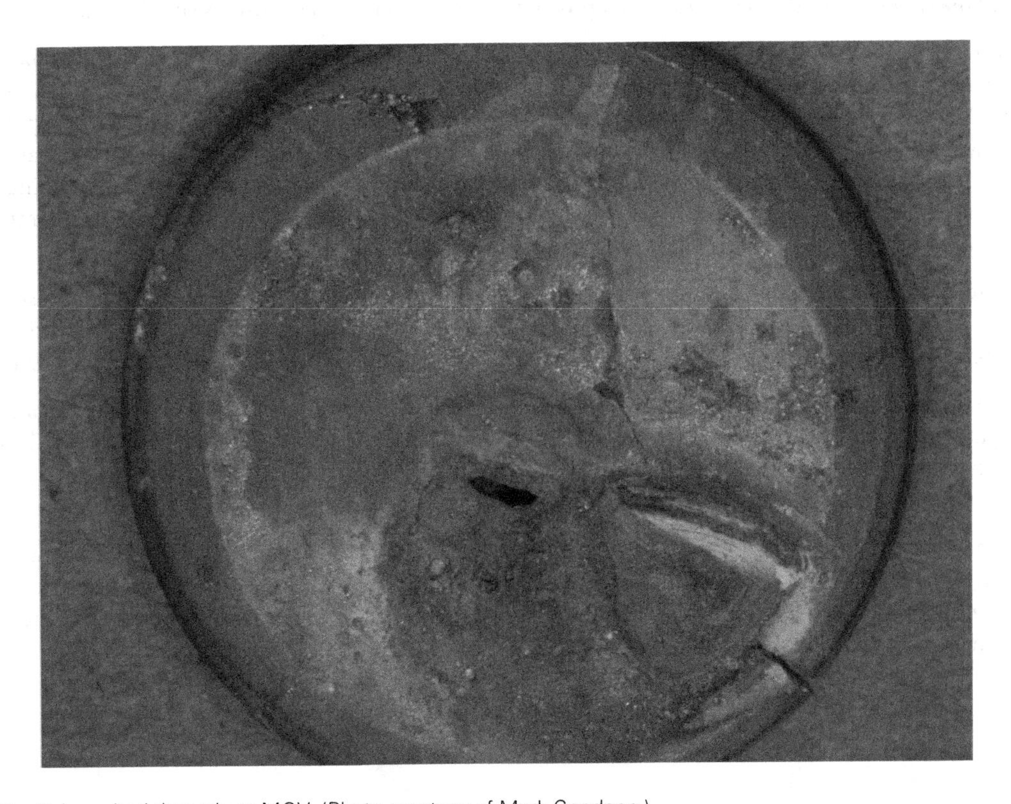

Figure 6.26 Hole melted through an MOV. (Photo courtesy of Mark Goodson.)

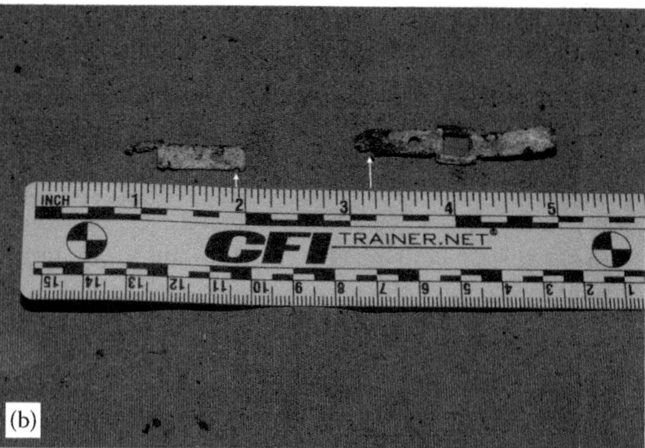

Figure 6.27 (a) Overview of an RPT found at the origin of a bedroom fire. Neither the switch nor the MOV was collected. (b) Close-up view of the neutral and one other bus. Arrows show localized melting, which was not seen by the original investigators.

6.3.4 Kitchen ranges

According to NFPA, cooking equipment was involved in 46% of reported home fires, 19% of home fire deaths, and 44% of reported home fire injuries. Ranges, with or without ovens, account for the majority (62%) of reported home structure fires involving cooking equipment and even larger shares of associated civilian deaths (88%) and civilian injuries (79%) [2]. Because the range is the ignition source for so many kitchen fires, it is often necessary to rule the range in or out as the source of the fire. This is usually accomplished by examining the burner control knob posts or the position of the cams inside the controls. If one of four posts is in a different orientation, it usually means that a burner was on, but this can be confirmed by laboratory examination. The wire terminals can be nondestructively tested for continuity using an ohmmeter. X-rays sometimes reveal the position of contacts inside the phenolic control case, but opening the case and making the visual observation often provide superior data. Even so, X-rays should be taken, if possible.

Even if it is determined that a burner was left on, that does not tell the whole story. It should be possible to leave an electric or gas burner on indefinitely without consequence, as long as there is no food or other nearby fuel to be ignited. Even if there is a pan on the stove, a scenario for the ignition of the kitchen must be developed. The pan shown in Figure 6.28 was found on a burner that was determined to have been inadvertently left on, but the pan was used exclusively for boiling water. After the water boiled away, how could the heat be transferred to the surrounding structure? The answer lies in the construction of the pan. It is a stainless steel vessel with a ⅛ inch cladding of aluminum on the bottom. The aluminum, in turn, is covered by another sheet of stainless steel.

Once the water boils away, the temperature of the aluminum increases until it melts in the center of the clad disc. The melted area expands radially until it reaches the edge, at which point the molten aluminum squirts out with sufficient energy to ignite any combustible surface on which it lands. Neither an all-aluminum pan nor an all-steel pan exhibits such behavior.

Finding that a range burner was inadvertently left on and unattended almost always ends the investigation. Even if the investigator is unable to develop a satisfying ignition scenario, such a finding effectively eliminates the possibility of proving that another appliance caused the fire. The possibility of a coincidence of another appliance malfunctioning at the exact same time the burner was left on simply lacks credibility.

In any fire in which the kitchen **or** the room next to it has become fully involved, it is necessary to eliminate the range. Work by Carman on elevated origins (the Z-factor) has demonstrated that it is possible for a range fire to spread to an adjacent room without first bringing the kitchen to flashover. The hot gas layer may descend and smother the fire on the range while simultaneously spreading the fire to the next room [3].

Figure 6.28 A common pan design, consisting of an aluminum plate *sandwiched* between the bottom of the vessel and a sheet of stainless steel. When inadvertently left on a hot burner, even when empty, this pan can spread a fire by squirting molten aluminum onto combustible surfaces.

6.3.5 Coffeemakers

During the 1980s, many kitchen fires were started by malfunctioning coffeemakers. When the main thermostat fails closed, there exists a safety device designed to sense the heat and open the circuit. This device may open as a result of bimetallic movement or, more commonly, it will melt open. These devices are known as thermal cutoffs (TCOs). Unfortunately, a large number of fuse-type TCOs were defective and failed to open when the need arose. One manufacturer, after being held responsible for more than 400 serious fire losses, finally agreed to a recall, and others followed suit. The industry responded by placing two TCOs in series, but even that was a risky strategy if the TCOs both came from the same bad batch. The more savvy manufacturers placed two TCOs with different temperature ratings in series, thus ensuring that they came from different batches. As a result of the industry response, the number of coffeemaker fires has drastically declined, but coffeemakers are still carefully scrutinized. In any case where a coffeemaker is suspected, one hypothesis that should be carefully examined is the possibility of intentional sabotage. The redundant safety features incorporated into coffeemakers in the last 20 years make it very hard for them to start a fire unless the safety features have been overridden.

Determining *whether* a coffeemaker was the cause of a fire (as opposed to determining *why* it failed) is a straightforward exercise. If the coffeemaker has overheated, the aluminum case around the heater melts, but only in those places where there are actual heater coils. The ends of the horseshoe-shaped heater usually survive. Any time a substantial puddle of molten aluminum is generated, it is a very competent ignition source. Consequently, the countertop under the coffeemaker is likely to be severely burned. Conversely, the survival of a plastic case bottom is persuasive evidence that the coffeemaker did not cause the fire. Figure 6.29 shows a burn in a countertop caused

Figure 6.29 Origin of a fatal fire where a failed coffeemaker was found. The kitchen counter burned through directly underneath the coffeemaker. The board at the back of the counter had been located directly over the coffeemaker and exhibits an oval-shaped area of damage.

by a coffeemaker that started a fatal fire. Figure 6.30 shows a comparison of the coffeemaker heater that started this fire and another heater that did not. Examination of the TCOs in a coffeemaker (or any other appliance) usually begins with electrical continuity testing. One-time fuses are likely to test "open" because of their exposure to the fire. Resettable TCOs may be found closed if they were not severely damaged. X-rays of the heater and the TCOs are often quite useful. Close examination of the contact surfaces in a bimetallic TCO often reveals whether the contacts have been cycling.

6.3.6 Deep fat fryers

Electric and gas-fired deep fat fryers are the ignition source for many restaurant fires. These devices are, for the most part, well designed and only cause fires when they are abused. The exception is the electric fryer that contains a single control circuit and a single power circuit. The control circuit operates a single relay, which turns power off to the element when the thermostat is satisfied. Contacts in the control circuit open, removing power from an electromagnet, and the spring-loaded relay opens the main power contacts. In the event of a relay failing closed (contacts stuck together), the spring in the relay is unable to separate the contacts. When the temperature increases, a second, high-limit thermostat again tells the relay to open, but because power in the control circuit has already been removed, nothing happens. This is simply a poor design that fails to take into account a foreseeable failure mechanism. Properly designed electric fryers incorporate a second relay or switch *in series* with the first so that if the first one fails, the power can still be disconnected by the high-limit thermostat.

Most of the time, a laboratory examination of a fryer that started a fire reveals that a human has defeated a safety device or done something to allow grease to accumulate in the burner area. Most fryers operate with two capillary-type thermostats. These consist of a sealed reservoir of fluid (usually xylene) connected to a diaphragm

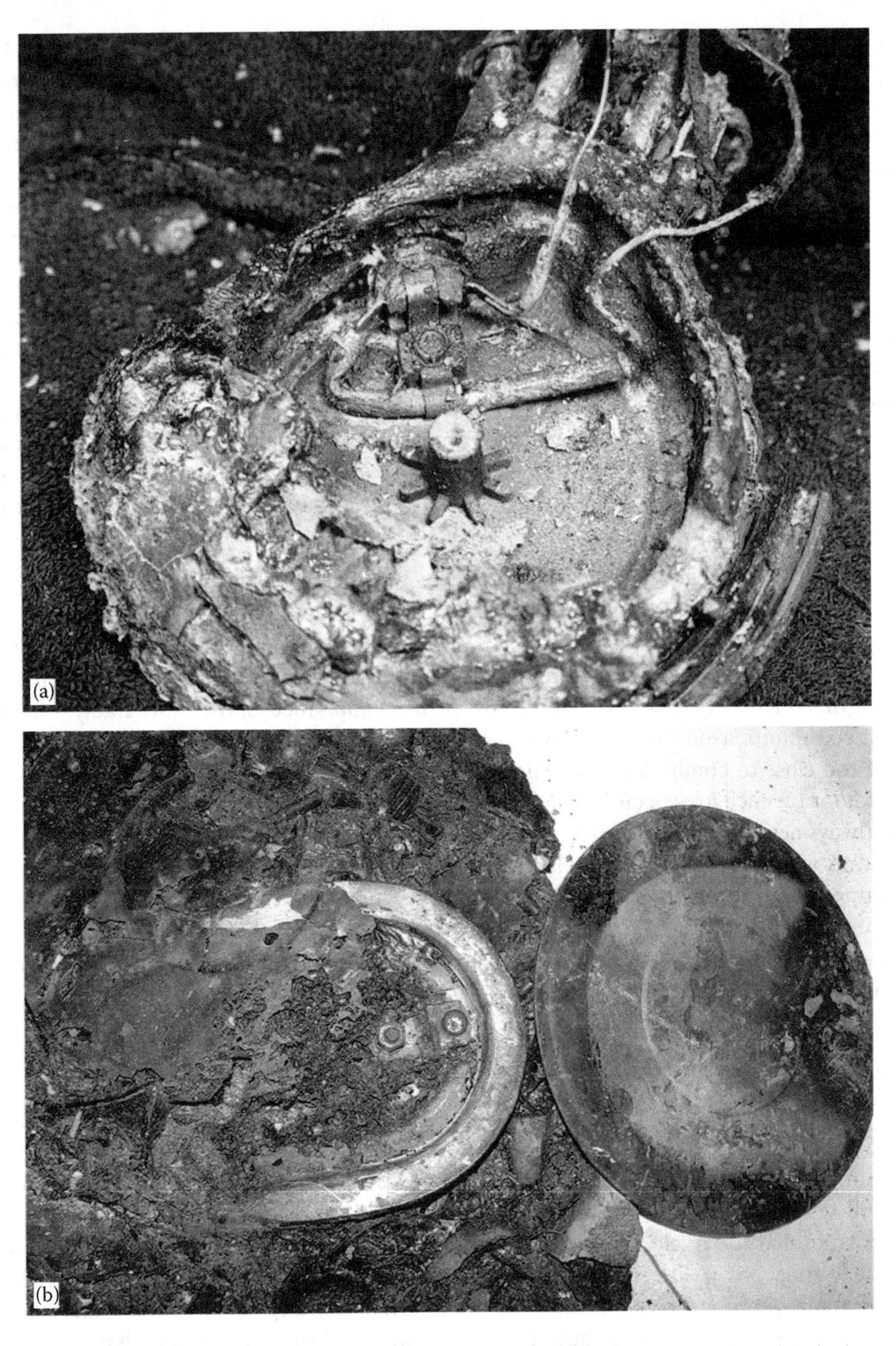

Figure 6.30 (a) Coffeemaker remains from the unit that started the fire shown in Figure 6.29. Note that the heater is melted in the area where the coils are located but not in the area containing only connecting wires. That area, at the upper right, is still intact. This design (which was recalled 2 weeks after the fire) had a single TCO in the heat-resistant sleeve at the center of the heating plate. (b) "Innocent" coffeemaker. The aluminum heater still maintains its original shape. The pan shown in Figure 6.28 was at the origin of the fire that damaged this coffeemaker.

and switch by a narrow (and consequently fragile) copper tube. As the temperature increases, the liquid expands, causing movement of the diaphragm, which in turn pushes on a small switch and opens a set of contacts. In electrical units, these contacts are in the control circuit and cause a relay to open. In gas units, the contacts are in series with the thermocouple. Opening the contacts has the same effect as no pilot—an electromagnet is de-energized and a spring closes the main valve. The capillary tubes are subject to breaking, and when that happens, the fluid leaks out and the thermostat no longer responds to heat. The fryer keeps working, however, because the contacts remain closed. The temperature is then under the control of the high-limit thermostat. The oil may smoke, but many restaurant operators continue cooking if they can. If it is designed properly, the high-limit thermostat includes a "negative biased" reservoir that is slightly below atmospheric pressure so that if the capillary tube breaks, the loss of negative pressure opens the circuit. The fryer cannot operate in this condition, but it takes only a minute to move the wire from the terminal at one side of the microswitch to the other side, effectively taking the safety device out of the circuit. If it is the high-limit capillary that breaks first and it is defeated, the operating thermostat will control the oil temperature just fine until its capillary breaks, and then the result will be a big fire.

Once a vat full of oil is raised above its autoignition temperature, control is very difficult. Dry chemical extinguishers blanket the burning oil with a layer of sodium bicarbonate, which forms a soap film and smothers the fire. Wet chemical extinguishers also contain soap-forming compounds and provide a cooling effect. Because there are usually people around when these fryers catch fire, they can intervene and frequently prevent the automatic extinguishing systems from working. Placing a cookie sheet over a burning fryer to smother the flames might seem like a good idea, but it keeps the agent from interacting with the oil; as soon as enough air gets to the surface, the overheated oil is on fire again. An examination of the extinguishing system usually accompanies the examination of any commercial cooking appliances.

6.3.7 Space heating appliances

Space heating appliances are responsible for a large number of fires, and their condition should be observed and documented at all fire scenes. Space heaters frequently require laboratory examination as well if they are found in or near the origin. As a general rule, the smaller the heating appliance is, the more likely it is to be an ignition source. The most common reason for a fire caused by a space heater is misuse on the part of the consumer, who places the heater too close to combustibles or allows combustibles to fall too close to the heater. (Wall-mounted space heater and floor furnace fires occur most frequently at the start of the heating season. In this author's practice, there were always new assignments arriving after the first cold snap of the year.) Small electric heaters next to beds often start fires when a blanket or sheet is kicked onto them. There is a class of wall-mounted electric space heaters that is impossible to turn off; they can only be turned down. These obsolete heaters are frequently abandoned in place, and combustible items are placed near them. In the event that the main heating system allows the temperature inside to drop below about 50°F, these heaters come on and ignite whatever is nearby. A typical heater with this design is shown in Figure 6.31.

The examination of heating devices in the laboratory can reveal the orientation of a device at the time of the fire. This can be important in cases where the floor collapses and the orientation is an issue. Liquids that fall or drip move straight down and can be used to help orient objects such as space heaters. These liquids might be melted plastics, melted metals, or simply condensation or extinguishment water.

Because of the number of safety devices in their design, central furnaces cause a smaller number of fires than portable or small wall-mounted heaters, and many of these fires are the result of misuse or old age. Electric heaters have their elements located in largely inaccessible cavities. Fires in these units are likely caused by poor connections to the house wiring. Gas-fired central furnaces can cause fires when their heat exchangers rust out or when they are deprived of air. "Combustibles too close" to a gas furnace (or any gas appliance) cause a fire only if those combustibles starve the furnace for air or cause the fire to roll out of the combustion chamber. Figure 6.32 shows the protection pattern on the floor in front of a furnace, caused by the presence of a stack of boxes during the fire. Laboratory examination of this furnace revealed that there was nothing wrong with the unit, once it was supplied with adequate air.

Any gas appliance whose air supply is blocked is a dangerous one. Air-starved gas burners produce soot, which can coat the inside of the combustion chamber and catch fire. Air-starved burners also produce high levels of carbon monoxide, which can kill in the absence of a hostile fire. The laboratory examination of any gas appliance should include the inspection of the burner ports, the inside of the combustion chamber, and the flue to determine if there is evidence of soot build-up.

Figure 6.31 This wall-mounted electric heater that ignited the chair placed in front of it has no "off" position. When the central heat failed, this heater came on and ignited the chair. The homeowners thought the heater was turned off.

Figure 6.32 Protection patterns on the floor indicate that boxes were placed in front of this furnace. Another utility room had boxes stacked 4 ft high in front of a similar furnace.

Furnaces installed in confined spaces are required by the *National Fuel Gas Code* (NFPA 54) to be provided with sufficient air for combustion and ventilation.[3]

Blocked air inlets can result in both a fire and an asphyxiation hazard. Blocked air inlets can result in the circulating fan creating a vacuum inside the appliance enclosure, causing the flue to run backward. In any carbon monoxide poisoning investigation, it is imperative that the appliances be tested *in place*.

When a heat exchanger in a gas furnace rusts out, the room air that is blown over the exchanger can penetrate through the rusted holes. This can result in flames rolling out of the draft hood or the burner chamber. If there are combustibles nearby, then there is a risk of ignition. When examining electric furnaces, it is frequently possible to test the control and safety devices, as they are designed for high temperature and often survive fires. In gas furnaces, the temperature sensing and switching devices might survive, but the aluminum control valves frequently melt during the fire.

6.3.8 Water heaters

Gas-fired water heaters cause more fires than space heaters. Electric water heaters do not cause fires. This is a bold statement, but after tearing apart more than 50 electric water heaters found near the origins of fires, this author has not seen a single fire that can be attributed to one. The result of a failure in an electric water heater is cold water. Goodson reports similar experiences, having investigated 40 fires alleged to have been caused by electric water heaters and confirming none [4]. Because the typical water heater draws around 5 kilowatts, it is surprising that they do not cause fires, but they are only likely to be involved in ignition if there is a loose connection to the house wiring and there are combustibles close to that failing connection.

Gas water heaters, on the other hand, have a number of failure modes. One common fire cause is the introduction of flammable liquid vapors near the flame. Any time a gas-fired water heater is suspected as the ignition source, this scenario must be ruled out. Incidents involving flammable liquid spills usually involve large quantities of liquid, so making a determination as to the viability of this scenario is usually not difficult.

The replacement of an electric water heater with a gas-fired one has resulted in several fires examined by this author. Many electric water heaters are installed in confined spaces, and putting a gas water heater in that same space requires that the space be ventilated. Failure to ventilate causes the flame to produce soot that coats the inside of the flue and the combustion chamber. When someone opens the door to check on the quality of the flame, however, there will be plenty of air and an efficient blue flame will be observed. If allowed to burn in an unventilated confined space for even a few weeks, a fire can occur. Figure 6.33 shows the results of starving a water heater for air. The installer knew he was required to add ventilation to the closet where the heater was installed and promised the homeowner that he would come back and do it, but he did not get back to the residence in time to do the job. This heater had been installed for only 6 weeks. It should have been ventilated before it was turned on.

The two failure scenarios described earlier involved human failures, but there have been numerous water heater fires resulting from product failures. Most of these are the result of the gas control valve failing to close properly. A failure to seat properly can allow a small amount of gas to pass to the main burner. This results in *candling*, a phenomenon wherein the flame burns directly above the main burner orifice rather than at the burner ports. In normal operation, the gas is too rich and moving too fast to burn in this location. From the main orifice, gas is propelled to the underside of the burner and is deflected radially to the burner ports, where the fuel/air mixture supports combustion. When the burner valve fails to completely close, candling occurs and a small yellow flame exists beneath the burner. Because it is not properly mixed with air, this flame produces a significant amount of soot, resulting in the burner ports becoming clogged. When the main valve again opens fully, the gas cannot be directed to the burner ports, and the flame may roll out of the combustion chamber. The underside of the burner assembly in such a case will be coated with soot. Several recalls of gas valves prone to this failure mechanism have

[3] A confined space is defined as a space whose volume is less than 50 ft^3/1000 Btu/h of the aggregate input rating of all gas appliances installed in that space.

Figure 6.33 (a) Closet fire origin where an electric water heater was replaced with a gas water heater 6 weeks before the fire. There was no ventilation in the closet.

Figure 6.33 (b) Heavy carbon deposit on the bottom of the pressure vessel of a water heater starved for air. The carbon ignited and fell down, igniting the floor of the closet.

occurred over the years [5]. Figure 6.34 shows such a water heater. Although the failure of the valve to close caused the heater to ignite its surroundings, the valve itself was only superficially damaged, and it was possible to reconnect a gas supply and re-create the failure scenario. Failure of the main valve to close completely can occur because of a misalignment or because of the introduction of particles into the gas stream.

Gas furnaces and water heaters are required to have a "sediment trap" to catch particles in the gas stream [6]. These traps are nothing more than a "T" fitting and a short length of pipe installed just upstream of the appliance. Sediment traps have been shown to be especially effective at trapping particles of copper sulfide, a compound that forms inside copper pipes carrying "sour gas,"[4] which has a high concentration of corrosive hydrogen sulfide. Copper and brass piping is prohibited for gas that has more than a "trace" of hydrogen sulfide, defined as 0.7 mg per 100 L [7]. Note that clothes dryers, ranges, and outdoor grills are specifically exempted from the requirement for a sediment trap.

As a result of the number of fires caused by water heaters igniting flammable liquids, gas-fired appliances installed in residential garages or in spaces that open to residential garages are required to be elevated so that the flame is at least 18 inches above the floor. Some new designs exist that prevent gas water heaters from igniting flammable liquid spills through the use of a flame arrester. The 18-inch elevation requirement is waived for such appliances under the provision that states, "Gas appliances shall not be installed in areas where the open use, handling, or dispensing of flammable liquids occurs, unless the design, operation, or installation reduces the potential of ignition of the flammable vapors [8]."

[4] Hydrogen sulfide (H_2S) is typically scrubbed from gas that is transported across state lines. "Sour gas," therefore, is only likely to be encountered in states where the gas is produced.

6.3.9 Clothes dryers[5]

There are more than 15,000 dryer fires each year in the United States. The failure mechanisms that cause these fires are not all well understood, but one common feature is poor ventilation or no ventilation. Dryer manufacturers require that exhaust vents be made of smooth metal and be as short and straight as possible, yet many dryers are vented with corrugated plastic or metal vent pipes that are prone to clog up with lint (and stores continue to sell these unsuitable materials to vent dryers).

Figure 6.34 (a) Burn pattern on the side of a water heater that failed due to blockage of the burner ports by soot from "candling" at the main burner orifice.

[5] Disclosure: Much, but certainly not all, of the author's experience with clothes dryer fires was obtained while serving as a retained expert for a major dryer manufacturer.

Figure 6.34 (b) Soot build-up on the burner of the water heater shown in Figure 6.34a. (c) Laboratory test showing the behavior of the main burner as a result of blockage by soot.

Dryer fires either start in the drum or in the other spaces. A drum fire requires an investigation of its contents. Because most dryers have an operating thermostat and at least two additional high-limit thermostats, multiple device failures are required to set the clothing on fire by simple overheating. The clothing must be checked for the presence of substances that would account for the fire. While the clothing should be examined for the presence of common ignitable liquids, such liquids are not likely to cause a dryer fire. First, clothing contaminated with ignitable liquids is likely to be much less contaminated after it is washed. Second, because of the airflow through the dryer, it is very difficult to build up sufficient vapors to reach the lower explosive limit, at least while the dryer is running and a source of ignition is present.

Many investigators make the mistake of asking their laboratory to look only for ignitable liquid residue (ILR). What the laboratory analyst needs to know is that the debris is from a dryer fire, and in addition to ILR, there should be a check for oils subject to spontaneous heating.

At ordinary dryer temperatures, about 150°F, many vegetable oils will undergo spontaneous heating. Because the dryer is tumbling and there is an abundance of ventilation, spontaneous ignition of the clothing only takes place *after* the dryer stops. Delays of 5 or 6 hours are not uncommon. If the clothing is removed from the dryer and immediately placed into another container, or if it is folded and stacked, the clothing can ignite outside the drum [9].

Spontaneous heating fires in dryers are common in commercial occupancies that use large amounts of oils, such as restaurants and health spas. Such fires are less common in residences, but if there is furniture or floor refinishing, or any activity involving drying oils or vegetable oils going on, the possibility of this ignition mechanism should be examined. One artifact that is sometimes found in spontaneous heating fires is a "clinker." This is a hardened mass of fabric that has become hard as a result of the polymerization of the vegetable oil. Most of the burned fabric retains its original softness, but the finding of a lump of hardened, stiff material is a sign that spontaneous heating has occurred, and the clinker is actually the first fuel ignited. A careful sifting of the drum contents is necessary, as the clinker may be inadvertently crushed. Figure 6.35 shows a large clinker removed from a dryer that ignited in the dryer several hours after the business, a skin care salon, had closed. Later chemical analysis revealed the presence of vegetable oils on the dryer contents.

For fires beginning outside the drum, several failure mechanisms are possible. One common fire cause involves the connection of the electric power supply cord. A strain relief device is required at the point where the cord passes through the rear bulkhead of the dryer, shown in Figure 6.36. This device is often omitted. Even a well-built power cord erodes after a few years of vibrating against an unprotected metal bulkhead. If there is easily ignited lint behind the dryer, and there often is, a fire results. The owner of the dryer shown in Figure 6.37 routed the power supply cord under an access panel rather than through the round hole in the bulkhead. Laboratory examination of the dryer revealed that the metal access panel had eroded the insulation until the conductor arced to the cover. Faulty connection of the power supply is the most common cause of residential dryer fires seen by this author.

The possibility that the fire began in the control cavity should be examined. Almost all the wires in this cavity have the capacity to become an ignition source and are subject to being pinched during assembly and abraded afterward. A dryer control cavity or drum cavity is one of the few places where a short-lived electrical event is likely to find the necessary fuel to start a fire.

Figure 6.35 Clinker removed from a dryer in a health spa where a fire occurred several hours after the business closed. (Courtesy of Ron Ready, Seattle Fire Department, Seattle, WA.)

Figure 6.36 Power cord passing through a bulkhead without the required strain relief. This could have started a fire, but this dryer burned because of spontaneous heating.

Figure 6.37 Dryer power cord that started a fire because of improper installation. Rather than routing the cord through the bulkhead and using a strain relief device, the homeowner passed the cord under an access panel. Abrasion of the insulation caused the cord to arc to the panel cover.

Figure 6.38 Half-inch-thick layer of lint inside the cavity of a dryer believed to be responsible for a large fire loss.

It is necessary to remove the drum to examine the motor and the heater. The area around the drum should be checked for excessive lint build-up. There should be very little lint in this space if the seals around the drum and the lint screen have not been compromised. A build-up of lint like that shown in Figure 6.38 may result in a fire. This dryer was found at the origin of a large fire, and the lint build-up was believed to be meaningful, but to this author's knowledge, an ignition scenario sufficiently credible to explain the fire (and pass a *Daubert* challenge) has not been demonstrated. One possible explanation is that a piece of the lint was drawn into the heater and ignited, and then blown into the dryer drum where it ignited the clothing. There are several problems with this scenario. First, it requires that the lint is somehow drawn into the heating device, ignited, and then delivered to the clothing in the drum, which must necessarily be dry. The lint has very little mass and would be expected to burn itself out quickly. The burning ember must penetrate a screen that limits its size to about 1 cm. If the clothing is not completely dry, the burning lint will not be able to ignite it. That is not to say that this scenario would not explain many dryer fires. There are on the order of 10 billion loads of laundry dried each year. If this was a 1 in 1 million occurrence, that would still account for 10,000 fires, but the chances of re-creating the event in a laboratory are quite small. It is important to keep in mind that, although the *rate* of failures of a particular design might be exceedingly small, the *number* of failures can be quite large. Common household appliances are produced by the millions. A catastrophic failure rate of only 1 in 100,000 will result in 10 fires in 1 million products.

Once a dryer is opened, it is frequently possible to remove and test the safety devices. Figure 6.39a shows the operating thermostat from a burned dryer in which a fire occurred in the drum. The operating thermostat was too badly damaged to test. Figure 6.39b shows a one-time fuse that tested open. A second such fuse, shown in Figure 6.39c, was mounted on the heater and also tested open. It was possible to test the high-limit thermostat shown in Figure 6.39d using a hair dryer, a thermocouple, and a continuity tester. This thermostat still functioned and opened at 175°F, which means that it was not possible for the drum fire to have been caused by a failure of the temperature control system.

Figure 6.39e shows the actual cause of the fire in the drum. About 1 mL of vegetable oil was extracted from a 6 × 12 inch sample of cloth towel material that had supposedly been laundered. Apparently, the washing machine failed to remove sufficient oil to prevent the fire. In this case, and in other cases investigated by the author, the unburned cloth in the dryer drum had the distinct odor of cooked vegetable oil.[6]

[6] The odor is the same as that on the clothing of a typical fast food establishment employee.

Figure 6.39 (a) Operating thermostat from a dryer that experienced a drum fire. This device was too badly damaged to test. (b) Fuse mounted next to the operating thermostat. The fuse tested open, most likely as a result of exposure to the fire.

Figure 6.39 (c) Fuse mounted on the heater enclosure. This fuse also tested open, most likely as a result of exposure to the fire. (d) High-limit thermostat mounted on the heater case. This thermostat remained connected to the bimetallic sensor and was capable of being tested using a hair dryer and an ohmmeter.

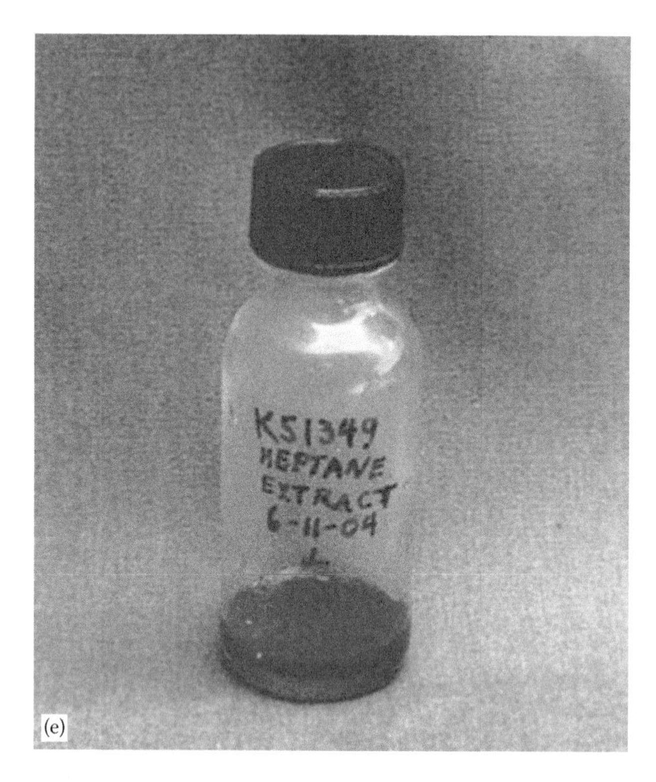

Figure 6.39 (e) Vegetable oil extracted from a 6 × 12 in. piece of unburned fabric from the dryer drum.

In many cases, the clothing in the drum will be too badly burned to test for vegetable oils. In those cases, the water in the drain hose from the washing machine might be the only place to look for oils. If there has not been another load of laundry washed, the water will be from the final rinse of the clothing in the dryer. This rinse water should be secured at the fire scene. The protocol for the detection of vegetable oils described in Sidebar 6.1 is sensitive enough to easily detect 10 mg of oil.

There is one design of electric dryer, easily recognizable by the large triangle on the back of the drum, which is exquisitely prone to failure. The drum is mounted to the dryer case by a single bearing, which rests in a metal slot. If this bearing fails, the drum can contact the heating element, resulting in arcing between the element and the drum. The heating element would be arced open, as shown in Figure 6.40a, and there would be a corresponding bead on the rear of the steel drum, as shown in Figure 6.40b.

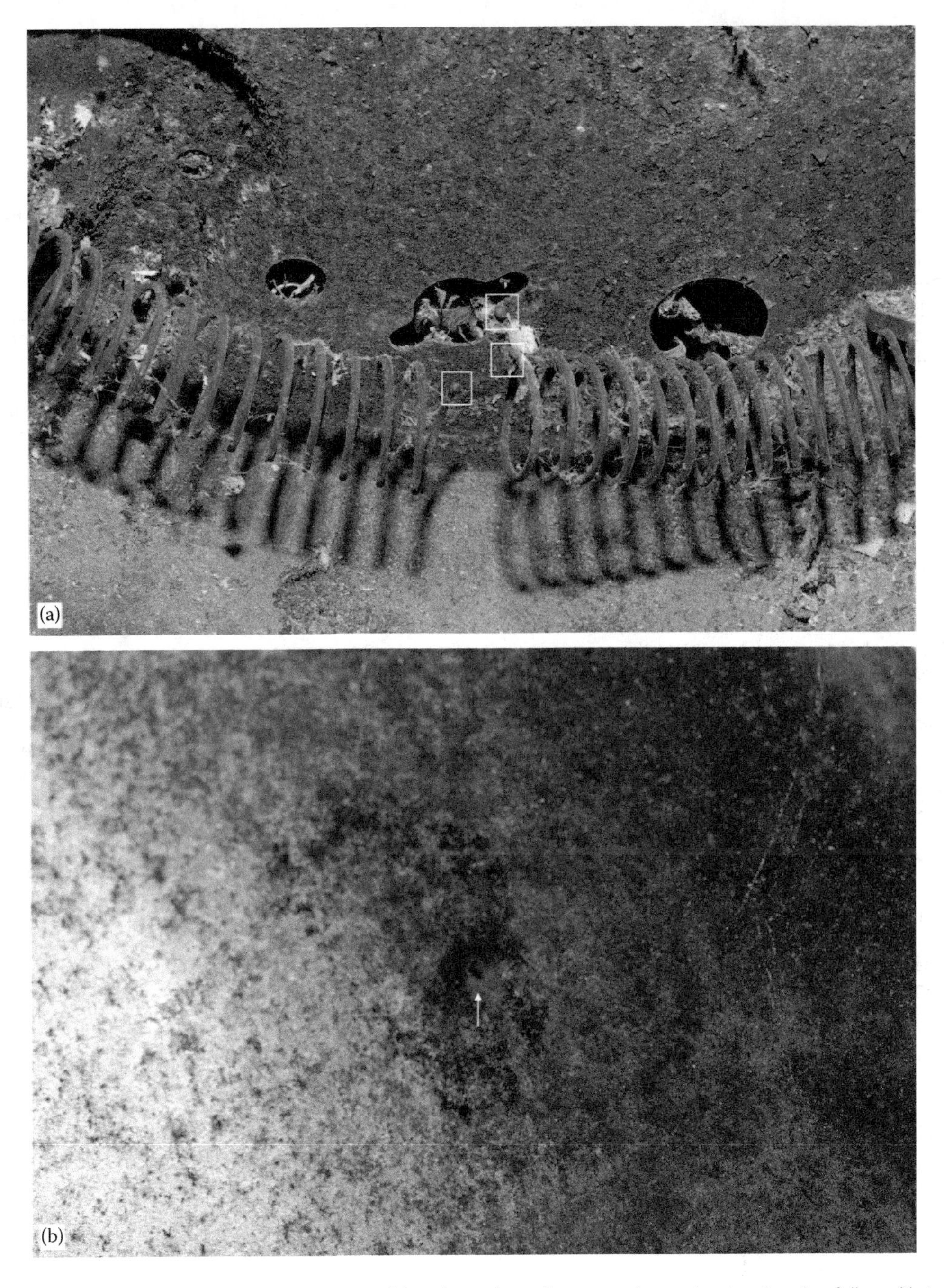

Figure 6.40 (a) Open heater coil caused by contact with a dryer drum that came loose due to a bearing failure. Note the bead of splattered metal on the coil support structure. (b) Small (1–2 mm) bead on the rear of the drum where it contacted the heater coil.

SIDEBAR 6.1 Detection and analysis of vegetable and drying oils in fire debris

Proc: 805
Rev: 1
Date: 8/15/04
Approval:

1. SCOPE
   This procedure covers the analysis of nonvolatile oils, such as cooking oils and drying oils, by extraction with heptane, derivatization with potassium hydroxide, and identification of fatty acid methyl esters by GC-MS.
2. REFERENCE DOCUMENTS
   ATS Procedure 803, GC/MS Analysis
   Badings, H. T., and DeJong, C., "Glass capillary gas chromatography of fatty acid methyl esters: a study of conditions for the quantitative analysis of short- and long-chain fatty acids in lipids," *Journal of Chromatography*, 279, 1983, 493–506.
3. PURPOSE
   The purpose of this procedure is to qualitatively determine whether a sample contains oils subject to spontaneous heating, and to characterize the most likely source of the oils. Such analyses are important in investigations of fires where it is necessary to test a hypothesis that spontaneous heating may have occurred.
4. PROCEDURE
   - 4.1 Obtain a sample of the suspected oil weighing approximately 100 mg.
     - 4.1.1 If the sample is an aqueous liquid, extract with n-heptane in a separatory funnel and dry.
     - 4.1.2 If the sample is solid, extract with n-heptane and dry.
     - 4.1.3 Evaporate or add heptane so that there is approximately 100 mg sample in 10 mL heptane, in a 20 × 150 mm test tube.
   - 4.2 Add 100 μL of 2N KOH (potassium hydroxide).
   - 4.3 Use the "vortex" mixer to agitate the sample in the test tube for 60 seconds.
   - 4.4 Centrifuge the sample for three minutes at 2,500 rpm.
   - 4.5 Open the GC-MS "top" program and load the sequence entitled "default," and select "Edit Sample Log Table" from the sequence menu.
   - 4.6 Enter the sample location; the name of the data file, which should be the ATS Job Number followed by the sample number; the method of analysis, "FAME1"; and the sample description, to include the ATS Job Number and sample number.
     - 4.6.1 If there is only a trace amount of oily residue, or if the residue is darkly colored (indicating heavy oxidation), use the method "FAME2," which is a more sensitive, splitless method.
     - 4.6.2 If the chromatogram is too noisy due to small sample concentration or contamination, use the selected ion monitoring program "FAMESIM."
   - 4.7 In the "Miscellaneous Information" field, enter the sample description.
   - 4.8 Run the sequence and collect the data.
   - 4.9 Load the data file and compare the total ion chromatogram (TIC) with the TIC of reference samples of animal and vegetable oils stored in the data file entitled "FAME."
     - 4.9.1 If a reference oil chromatogram matches the sample in question and can be distinguished from other potential reference oils, it is not necessary to rerun the reference oil.
     - 4.9.2 Print a copy of the TIC for the entire run, and for the period from 5 to 15 minutes, for both the sample and the reference oil that it matches most closely.
     - 4.9.3 Run a library search to identify the fatty acid methyl esters in the sample.
5. DOCUMENTATION
   Place a copy of the report and all data, including a copy of the reference oil chromatograms, into the job file.

6.3.10 Fluorescent lights

Between 2 and 4 billion fluorescent light fixtures are currently in use in the United States. Every day, about 50,000 of the ballasts in these fixtures reach the end of their useful life. The overwhelming majority of these do so quietly, as they flicker and die or simply fail to start. It is possible, however, for a ballast to overheat or for a lamp holder spring to become loose, creating the possibility of an arc. Since 1968, there has been a requirement that fluorescent lamp ballasts be equipped with a TCO, and most manufacturers provide an automatically resettable bimetallic switch to keep ballast temperatures below 94°C (194°F). A typical TCO is seen in Figure 6.41.

Because of their widespread use, fluorescent light fixtures are likely to be found within 10 feet of the origin of most commercial fires. Far more fluorescent fixtures are blamed for fires than actually cause fires, but because of the knowledge that these fixtures are potential ignition sources, they must be ruled out in cases where another ignition source is suspected.

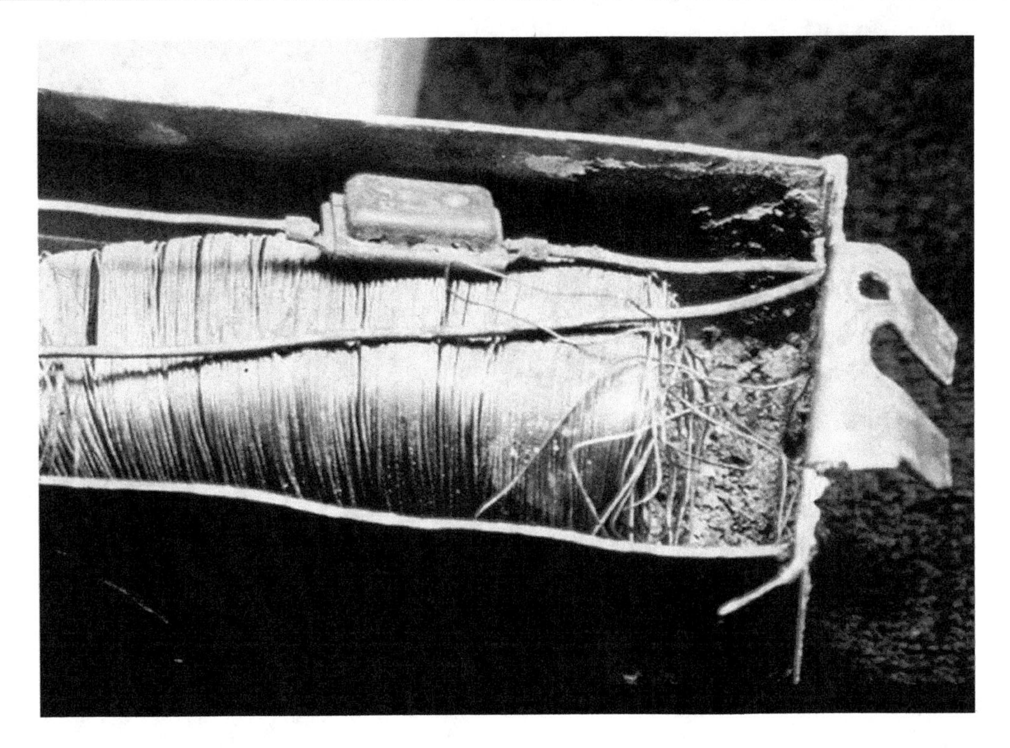

Figure 6.41 A TCO in a fluorescent light ballast that has had all of its pitch burned out.

The failure modes of fluorescent ballasts that cause fires are limited. The conventional magnetic ballast consists of an iron core transformer and a capacitor potted in pitch, an asphalt-like paste, which quiets the annoying buzz the transformer would make without it. While the presence of pitch makes the ballast combustible, it also excludes oxygen from the case, making ignition unlikely. If the ballast case is only partially filled with pitch, ignition can result, but only if conditions are perfect. Even then, it is difficult for the fire to move outside the sealed metal case. Overheating of ballasts usually occurs when there is a short circuit on one of the coils, reducing the resistance and increasing the current. This is the condition that the TCO is designed to detect. If the TCO fails, or if the incoming 120-V wire short-circuits, conditions may deteriorate to the point where arcing occurs through the case of the ballast. When there is a hole arced in the ballast case, such as the one shown in Figure 6.42, it can be said to be a credible ignition source. (Even then, it is necessary to make certain that the light can be shown to have been energized prior to the fire because the failure could have occurred long before the fire.) As with any proposed electrical ignition source, verification of a power supply is essential.

Another mode of failure of fluorescent fixtures involves a loose connection at the lamp holder. This may cause flickering with conventional ballasts. Electronic ballasts, however, function at 20,000 cycles per second or higher. At this frequency, the operation of the light appears normal, and the user may not be aware that there is a "series arc,"[7] such as the one shown in Figure 6.43, occurring between the lamp holder and the lamp pins until the arc ignites the holder. For this reason holders, known as "tombstones" because of their shape, should be made of ceramic, but not all of them are. Laboratory examination may or may not reveal evidence of this condition after a fire.

The ballast is usually confined inside a metal enclosure that runs the length of the fixture. The 120-V supply and 600-V fixture wiring in this channel might arc to the enclosure, providing some sequential data about when the fixture was involved in combustion. As with many electrical ignition scenarios, arcing within the enclosure of an appliance suggests that the appliance was involved early in the fire and may well be the ignition source.

---

[7] Series arcing is arcing within the intended path of the current, as opposed to line-to-line-or line-to-ground arcing. Series arcing most often occurs because of a gap or a loose connection.

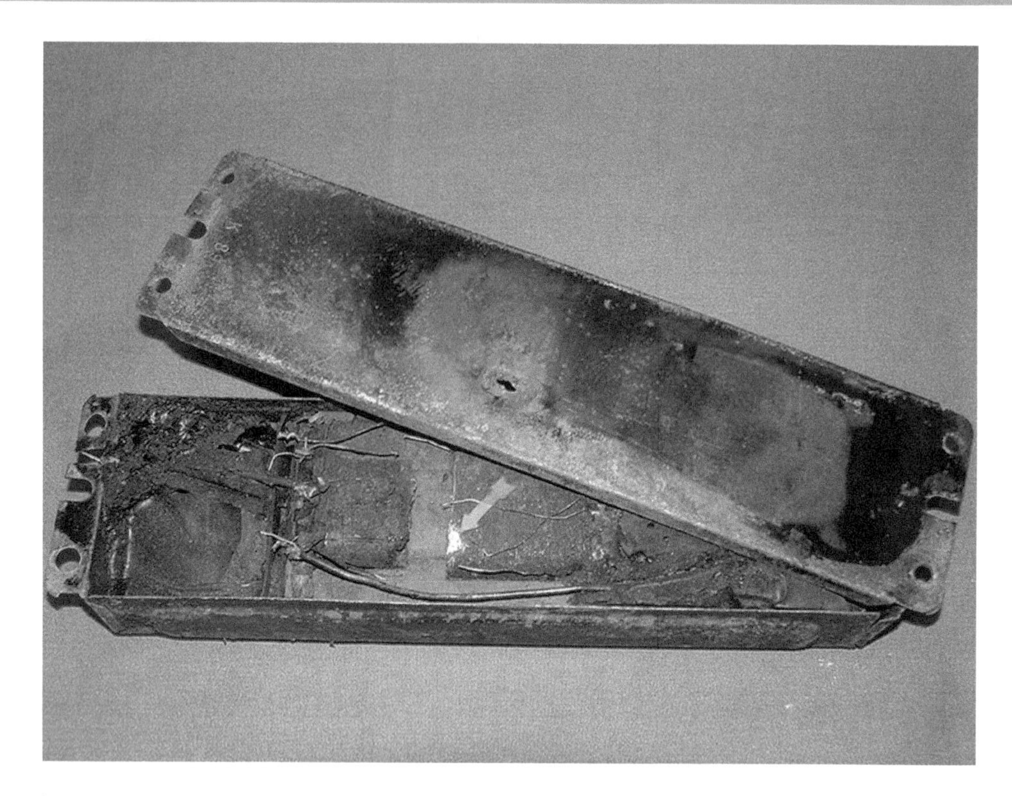

Figure 6.42 A conventional magnetic fluorescent light ballast that was positively identified as the cause of the fire. There is a hole arced through the steel case. The arrow shows the point on the windings where the arc occurred.

Figure 6.43 Series arc in a fluorescent fixture with an electronic ballast. Because of the high frequency, there is no discernible difference in the output of the bulb. (Photo courtesy of Mark Goodson, Goodson Engineering, Denton, TX.)

A more powerful variety of fluorescent lighting is the metal halide discharge lamp. These lamps use bulbs that are known to fail catastrophically, showering the area beneath them with hot glass fragments. These lamps, unless properly shielded, are only suitable for use over places where there is no combustible stock below, as in a concrete block factory or above a tennis court or other sports arena. Bulbs are available with an extra envelope, but these cost more than those without this protection, and the fixtures accept either type. Fixtures are available with shielding, but these cost more and the fixture manufacturers are loathe to warn against using unshielded fixtures. The worst of all designs is the metal halide discharge lamp with a combustible shield. Instead of dropping hot fragments when the bulb fails, it drips flaming molten plastic droplets.

6.3.11 Recessed lights

Recessed lights have the capability to cause a fire when they are improperly surrounded with insulation, are overlamped, or both. Some recessed lights are rated for direct contact with insulation and are equipped with overheat protective devices. These lights incorporate a bimetallic, automatically resetting TCO in the "can" portion of the fixture that surrounds the lamp. If the user installs a higher wattage bulb than the fixture is designed to use and the fixture is unable to radiate the heat away, the TCO causes the lamp to cycle off and on to prevent overheating and alert the user to the problem. A second kind of "thermally protected" fixture is not really thermally protected. These lamps use a 1- to 2-W heater and a TCO mounted on a metal junction box outside the can. These devices are more accurately described as "insulation detectors." Because it is the bulb that produces the heat that might cause a fire, one might think that it is bulb heat, not the output of an additional heat source, that should be sensed by a thermal protector. Despite the drawbacks of proxy sensing, however, the second design is allowed under current codes and standards (NFPA 70 and Underwriters Laboratories [UL] 1598.) Figure 6.44 shows the TCO in a fixture rated for direct contact with insulation, and Figure 6.45 shows the "insulation detector" TCO. The UL test for recessed light fixtures requires that shredded cellulose (blown-in) insulation be uniformly distributed over the fixture, and the TCO must operate within 3 hours and before the temperature of the fixture exceeds 320°F (160°C). Both fixture designs pass this test, but if paper-backed fiberglass batt insulation is substituted for blown-in insulation, it is

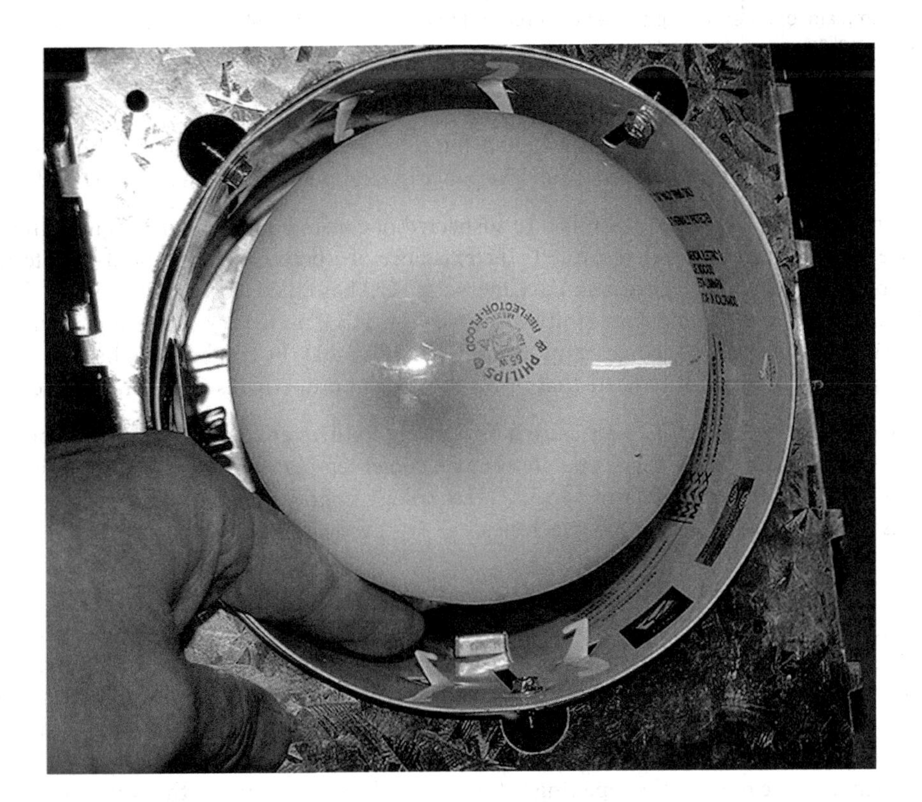

Figure 6.44 A recessed light fixture with a resettable TCO inside the can, where it senses actual temperatures generated by the bulb.

Figure 6.45 A recessed light fixture with the TCO mounted on the junction box. The only temperature this sensor measures is that generated by an internal heater. This sensor will operate if the junction box is insulated, but it will not operate if only the light can is insulated. Exemplars for recessed lights are best obtained from the fire scene, where fixtures of the same manufacture and vintage were likely installed.

possible to have the lamp can temperature exceed 320°F but the sensor on the junction box not respond. The author has seen several fires caused by this configuration. The paper insulation backing was the first fuel ignited.

A laboratory examination of a recessed light fixture suspected of causing a fire should begin with an examination of the bulb to see if its wattage can be determined. The existence of a power supply should be determined, and the type of overheat protection should be documented. Unless the TCO has failed closed, it is unlikely that a fixture with the TCO in the can represents a competent ignition source. The TCO, no matter where it is mounted, should first be tested for continuity, X-rayed, and then carefully disassembled so that the contacts can be examined. Evidence of repeated operation of the TCO might indicate that the fixture was improperly insulated and/or overlamped.

With the decline of incandescent lamps, fires caused by lighting fixtures are likely to decline. Compact fluorescent (CFL) and light emitting diode (LED) lamps operate at much cooler temperatures, and their failure modes tend to be less energetic than incandescent lamp failures. As LED technology improves, it is likely that even CFLs will become a thing of the past, like compact discs and fax machines.

6.3.12 Exhaust fans

Exhaust fans usually use shaded pole induction motors, in which a rotating magnetic field is created and the rotor tries to keep up with the field rotation. The rotation of the magnetic field happens in the coils surrounding the rotor. The resistance of the wire in these coils causes the coils to heat up, but in normal operation, particularly in the operation of a fan motor, the rate of heat dissipation greatly exceeds the rate of heat generation. When the motor stalls, the current does not increase, but the loss of air movement reduces the dissipation of the heat. Even when one of these motors fails, the usual failure mode is the "opening" of the circuit by arc-severing of the coil wire in a brief, relatively low-energy event.

Like many appliances, however, millions of these fans exist, and some are in operation almost constantly. Fans tend to gather dust, and if the motor stalls and is covered with dust, heat may fail to dissipate quickly enough to prevent the temperature of the dust from rising until ignition occurs. Alternatively, the brief spark that occurs when the coils open may be sufficient to ignite the accumulated dust. Because of their location, fans sometimes catch fire, but only the immediate surroundings ignite or the fire spreads up and away from the fan, leaving it sufficiently intact to conduct a meaningful examination. It is situations like the one shown in Figure 6.46 that allow for a detailed study of the failure mechanism that can lead to a fire.

Until recently, shaded pole induction motors required no auxiliary overheat protection. The fan industry convinced standards development organizations such as NFPA and UL that such safety precautions were unnecessary. (The argument was that these devices were "impedance protected"—an oxymoron.) In 1994, however, after a number of catastrophic fires were attributed to exhaust fans, a requirement for the inclusion of a TCO was adopted. Most manufacturers use a one-time thermal fuse in the circuit. Although the presence of a TCO greatly reduces the possibility of an exhaust fan catching fire, it does not entirely eliminate it. Careful examination of the motor requires that the coil be unwound to look for evidence of short-circuiting. Sometimes, it is possible to see melted windings metal in the interior of the coil, which proves the motor as the cause. The motor shown in Figure 6.47 was equipped with a TCO, but evidence of arcing was found near the point where the power supply was attached to the coil. In the motor shown in Figure 6.48, melted aluminum can be seen in the center of the coil, which is in much worse condition than the surface windings. In such a case, the motor cannot be said to be a victim of an advancing fire because the wires on the surface of the coil would necessarily be attacked first.

There is often more than one exhaust fan installed at a particular site. Because new exemplars may be difficult to obtain, and because burned fans are difficult to identify as to manufacturer, the fire site is the best place to look for a fan of the same manufacture and vintage.

Up until the turn of the century, fan manufacturers routinely accepted responsibility for losses caused by their products. If you determine that an exhaust fan was at the origin of a fire today, however, be prepared for stiff resistance and a challenge to the admissibility of your opinion.

Figure 6.46 The fire started by this exhaust fan/bathroom light combination spread mainly up the underside of the roof deck, making for a straightforward determination of the origin and cause.

Figure 6.47 Despite the fact that the fan motor design included a TCO, a loose connection between the pigtail power supply and the motor windings caused sufficient heating to start a fire.

Figure 6.48 An exhaust fan motor showing melting beneath the surface of the coil, indicating that the fire began inside the unit.

6.3.13 Service panels

Electric service distribution panels frequently exhibit evidence of intense electrical activity but are only rarely the cause of fires. When approached by an advancing fire, the service entrance cables inside the panel typically arc to the case, and because there is only limited protection on these cables, the arcing can be quite extensive.

The key to determining whether a service panel started a fire is the existence of arcing downstream of the panel. If there is arcing downstream, the panel can be eliminated. Once a panel starts a fire, and the fire has had time to get outside the panel, all downstream conductors are expected to be de-energized.

If no downstream arcing is present, attention should focus on two connection points inside the panel. The first connection point to examine is the stab where the circuit breaker attaches to the bus. In residential applications, two buses with interwoven stabs are present. A 120-V circuit is protected by a circuit breaker that is attached to one stab on one bus. A 240-V circuit is protected by double breakers that attach to two adjacent stabs. Loose connections between the breaker and the stab can cause overheating, which might lead to a catastrophic failure. Heat from this loose connection can be conducted to the surface on which the panel is mounted, but the usual failure of this type leads only to a loss of power on the affected circuits. Figure 6.49 shows a damaged service panel in which heating occurred at the point where the breakers mounted on the bus. The heating at the stab melted surrounding thermoplastic insulation. This panel was replaced before it started a fire.

The second point of interest is where the service entrance cable attaches to the main breaker. Loose connections in this area can lead to loss of power throughout the house (or, more likely, on half of the circuits in the house) and

Figure 6.49 Circuit breaker panel that was replaced *before* it caused a big fire. The panel reportedly gave off a buzzing sound and an odor of burning plastic. Insulating thermoplastic had melted, but the insurance company would not cover the panel replacement. (They would have covered it if the building had burned down, however.)

usually cause flickering lights and damage to the incoming cable. Some panel designs, however, are less forgiving[8] than others when loose connections occur at the service conductor lugs.

The panel shown in Figure 6.50a was found at the bottom of a perfect V-pattern. The fire spread upward, leaving the offending panel in surprisingly good condition but causing extensive damage to the rest of the dwelling. This panel design used an independent lug, separate from the main breaker, to attach the service entrance conductors. The lug was supported in a thermoplastic (polyphenylene oxide) channel, and when a loose connection caused the lug to

Figure 6.50 (a) Electric service distribution panel at the origin of a large fire. The fire spread upward, leaving the "smoking gun" of the damaged panel lug at the point of origin.

[8] "Less forgiving" is a polite way of saying that the design responds inappropriately to a 100% foreseeable event.

Figure 6.50 (b) Electric service distribution panel at the origin of a somewhat more devastating fire. The panel was damaged more severely, but there was no downstream arcing. The failure scenario was again proved by the location of the hole burned in the steel directly behind the lug where the service entrance cable connected. (c) Electric service distribution panel at the origin of a fire that burned the house to a pile of rubble in the basement. A combination of eyewitness statements, the lack of downstream arcing, and this evidence proved that the panel was responsible for the fire.

overheat, the lug migrated through the plastic, touching the back of the panel and setting the wall on fire. Since identifying the failure mode in this fire, the author has seen it in six other fires.

For a fire to start in this scenario, several circumstances must occur in sequence. First, excessive heat must be generated. This requires the service entrance cable to become loose at the lug, a process that may take several years after installation to occur. Additionally, the service entrance conductor must be mechanically loaded so that it has an impetus to move backward when the plastic supporting the lug melts or softens. (If the cable is loaded to move toward the cover, or not loaded at all, no arcing is likely.) Finally, there must be a fuel supply close to the point where the arcing takes place on the back of the panel. The event is sufficiently energetic that it will ignite a plywood wall or dimensional lumber. The diagnostic feature for identifying this failure mode is a hole arced through the back of the panel directly behind the location of the service entrance cable lugs. Figure 6.50b shows a panel from another fire caused by this unforgiving panel design. Figure 6.50c shows still another panel where the damage was much more extensive, but the hole in the case was directly behind the main lug. This evidence, in combination with an absence of downstream arcing and credible eyewitness statements, finally convinced the manufacturer to accept responsibility.[9] The manufacture of panels with this design has been discontinued, but there are more than 15 million of these panels in current use.

6.3.14 Oxygen enrichment devices

As the population ages, more individuals will rely on supplemental oxygen to prolong and improve the quality of their lives. Supplemental oxygen is provided in three ways: (1) as compressed gas, (2) as oxygen-enriched air, and (3) as liquid oxygen. With the use of a continuous oxygen supply, people with impaired lungs can lead long and normal lives. A patient with 50% of his original lung capacity needs a supply of air containing 40% oxygen to make up for the lung deficiency. Fire investigators are increasingly likely to encounter oxygen enrichment devices. Because of the enriched oxygen atmosphere around these devices, they are likely found in a severely damaged state, which makes them suspects as the cause of the fire. Unfortunately, the tobacco habit that caused many patients to require oxygen enrichment in the first place is now a much more dangerous habit. When smokers fail to take precautions (e.g., turning off the oxygen while they have their cigarette), the results can be disastrous. Some people involved in such fires fail to take responsibility for the event.

Oxygen enrichment devices react differently than other electric appliances when exposed to a fire, but they behave in predictable ways. Just as an overloaded power cord can have successive arcs back toward its power source, an oxygen tube, once ignited, burns back toward the source of oxygen, whether it is a tank, a concentrator machine, or a liquid oxygen dispenser. Because the tubing runs to the inside of the machine, the fire is likely to be carried inside, resulting in arcing within the enclosure and casting suspicion on the machine. While arcing inside an enclosure usually provides grounds for suspicion, such is not the case with these devices. Generally, after a fire, the examination of an oxygen concentrator reveals a path of the fire from the outside to the inside, following the path of the oxygen tubing. Figure 6.51 shows the interior of one heavily damaged concentrator, which exhibited arcing on its power cord inside the cabinet. At the scene, evidence was found that the fire had followed the path of the oxygen tubing through two rooms and an intervening hallway.

Figure 6.52 shows an oxygen supply tube on fire, having been ignited by the author. When the power supply to the oxygen concentrator was turned off, the tubing ceased burning within 20 seconds. Finding evidence of a fire following the path of the tubing outside the machine therefore allows one to conclude that the machine was operating properly and the ignition source was not the machine but something else, even if electrical arcing is found inside the enclosure.

Oxygen concentrators work by passing air through an alternating pair of molecular sieves, which remove nitrogen from the air and deliver 90% to 95% oxygen at up to 5 liters per minute. One sieve releases oxygen while the other exhausts nitrogen. They then switch roles. The air is filtered before it is sent to a compressor upstream of the sieve beds. If a patient has been smoking while the machine has been running, the filter (if it survives) frequently tests positive for nicotine. One such filter tested by the author was negative for nicotine but positive for cannabinols.[10]

[9] For a time, the defendant service panel manufacturer's retained expert proposed that the fire was actually a result of a "power surge" from a defective utility transformer, but that assertion was abandoned when it was proved that the same transformer was still providing uneventful service 3 years after the fire.

[10] More disclosure: all of the author's experience investigating oxygen concentrator fires was gained as a retained expert for an oxygen concentrator manufacturer. To date, that manufacturer has never accepted responsibility for causing a fire.

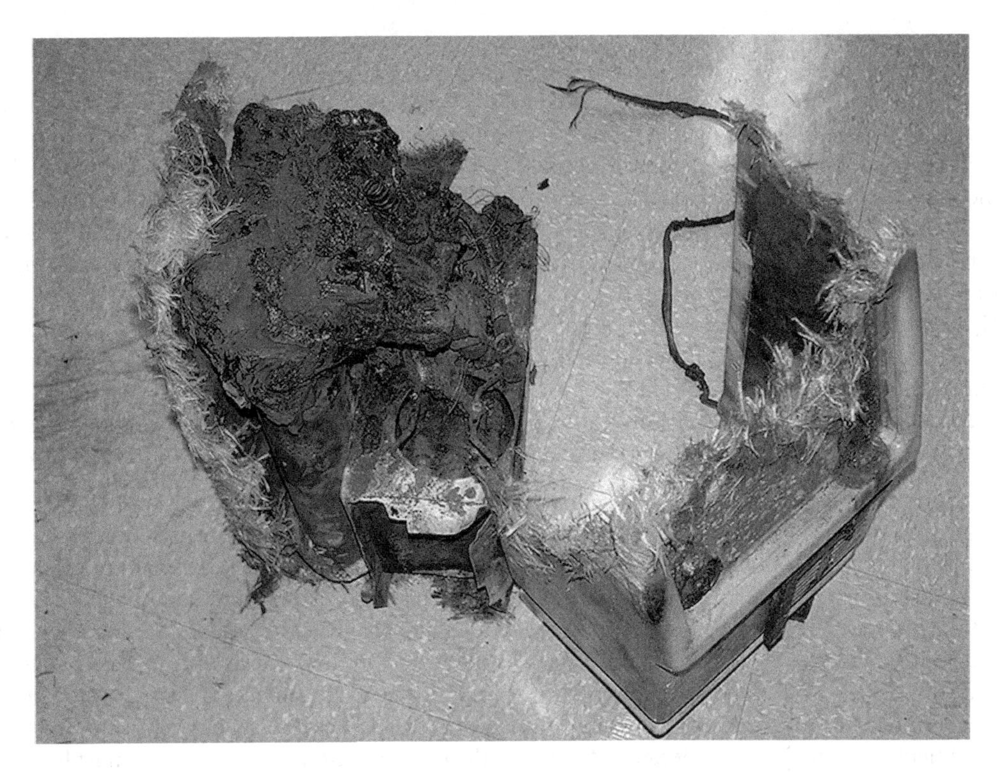

Figure 6.51 Oxygen concentrator, with an arc inside its enclosure, found in a seriously fire-damaged residence. A trail of burning coincident with the location of the oxygen tube proved that the concentrator was functioning normally until attacked by the fire.

Figure 6.52 Oxygen supply tube on fire. The fire self-extinguishes within 20 s of the concentrator shutting down.

6.4 TESTING OF IGNITION SCENARIOS

The laboratory analyses described thus far have related mostly to the examination and interpretation of physical artifacts, in an attempt to find sufficient data to develop a credible ignition scenario. Many times, the correctness of the scenario is self-evident; that is, no other scenario accounts for the condition of the appliance or other evidence. Other times, it is necessary to prove that the proposed scenario is actually possible. Generally, at least one party working on an investigation hopes to disprove the possibility of a particular failure scenario. The experimental design, therefore, should be critically considered because even the most careful designs will be subjected to intense scrutiny. CPSC staff members were roundly criticized for testing the concept of lint being ignited by an electric dryer heater and then subsequently igniting clothing in a dryer drum. Because they proposed a low-frequency occurrence, and did not have the budget to run a million tests, the staff set up conditions to favor ignition. The work was intended to improve dryer safety. Despite the disclaimer at the front of the report that reads "The experiments described in this research report were undertaken to support future advances in clothes dryer safety. This report should not be used to suggest that current clothes dryers are unsafe or defective," the report has been characterized as "Bad Science from Big Brother [10,11]."

NFPA 921, in describing hypothesis testing, was changed between the 1998 and 2001 editions. During the deliberations preceding the issuance of the 2001 edition, some misinformed sentiment was expressed that a "scientific" analysis must include a re-creation of the event. This is no truer for a fire than it is for a plane crash or other catastrophic event. The proponents of a particular hypothesis for a plane crash are not required to obtain a new plane and cause it to fail, and the proponents of a hypothesis of fire causation are not required to acquire another identical structure and burn it down. To clear up the apparent confusion, the paragraph on hypothesis testing was changed to include the sentence, "This testing of the hypothesis may be either cognitive or experimental." The scientific method allows a "thought experiment" as a means of attempting to disprove a hypothesis. Like physical experiments, however, thought experiments must be carefully designed to answer the right question, and the results must be persuasive.

Physical experiments are generally considered more persuasive because "seeing is believing." When an ignition scenario, particularly a novel ignition scenario, is put forward, unless the physical evidence is unequivocal, nothing matches the credibility of an actual experiment. Experiments can be designed by either the proponent of an ignition hypothesis or by one who believes that it is incorrect. The interest of the person designing the experiment must be kept in mind when reviewing the results.

To enhance their credibility, experiments should be designed to favor an outcome beneficial to the *adverse* party. Getting experimental results admitted into evidence is no easy task. Whichever party is prejudiced by the results will object that the experiment "does not substantially reflect the circumstances" of the event and therefore is not relevant. For this reason, experiments should be kept as simple as possible while still providing a relevant test of the hypothesis.

It is unwise to perform experiments on actual evidence, particularly if there is a risk of damaging the evidence by heating or energizing it with electrical current. As a rule, exemplars should be obtained for this work. For example, a small-scale experiment might test whether a particular appliance is capable of producing sufficient heat to ignite a particular target.

Both a thought experiment and a physical experiment were conducted to test a hypothesis that a torchiere lamp could ignite a wooden wall. In the thought experiment, the following conditions were set.

- The bowl of the fixture was touching the wall, placing the bulb 5.5 inches away from the target fuel. (The bowl may have been a few inches away from the wall.)
- The bulb had an output of 500 W. (It may have been only 300 W.)
- All of the bulb's energy was directed upward and spread out over a hemisphere. (Some of it was certainly directed downward.)
- There were no energy losses to convection. (There certainly were some losses to convection.)

Doing a little math, the target surface area of a hemisphere ($4\pi r^2/2$) with a radius of 5.5 inches is 190 in.^{2} or 0.12 m^2; 500 watts (0.5 kW) spread over that surface area yields an incident radiant heat flux on the target of 4 kW/m^2—not nearly enough for ignition to occur, even under a worst-case scenario. The physical experiment, shown in Figure 6.53, confirmed what the thought experiment predicted.

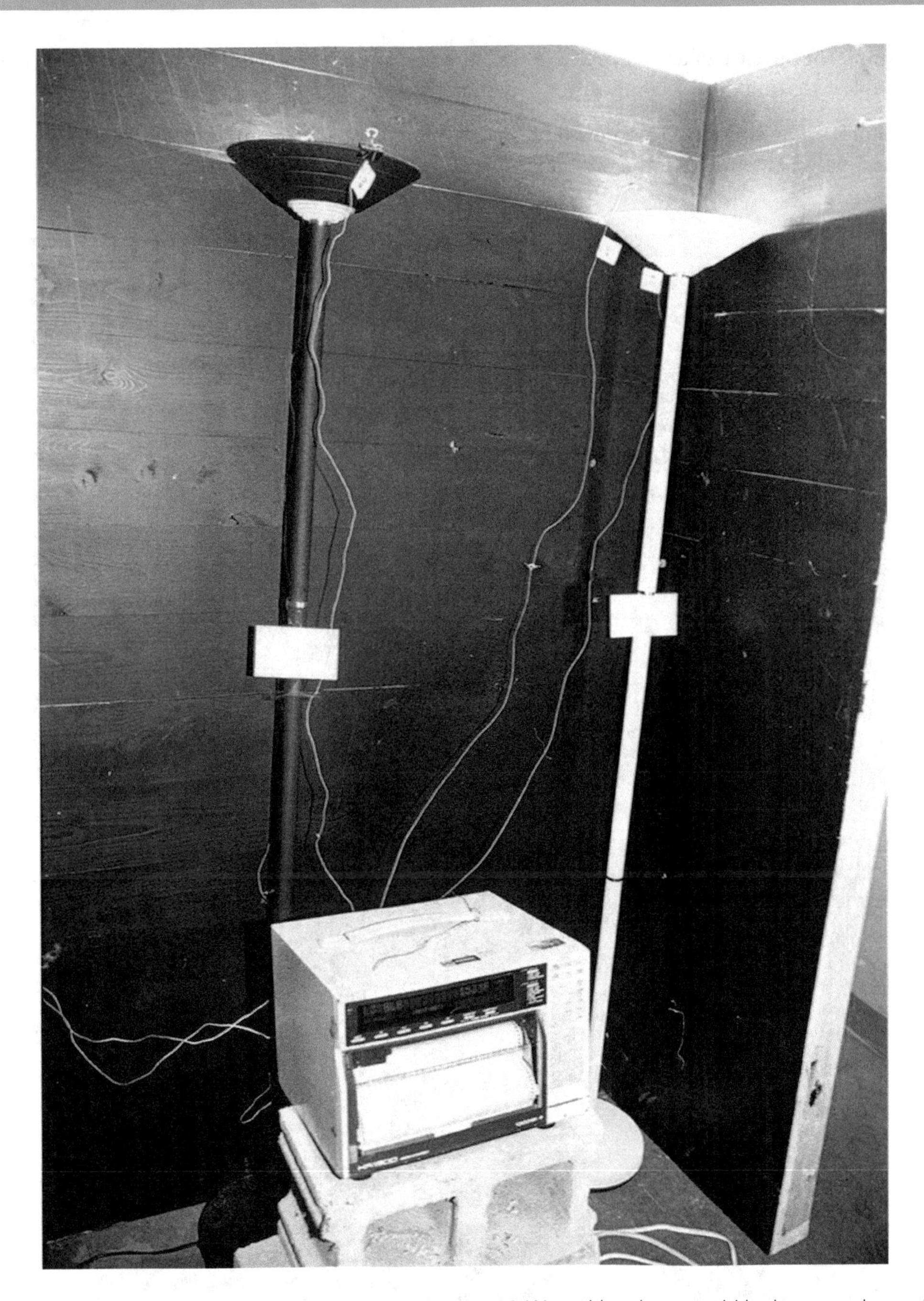

Figure 6.53 A physical experiment designed to test whether a 500-W torchiere lamp could ignite a wooden wall by radiation alone.

The conditions of the physical experiment were also designed to favor ignition. The pine wall was painted flat black, which raised the temperature at the surface by about 30°F. Lamps (500 W), with their overheat protective devices removed, were used. One lamp was placed in a corner to reduce radiative losses and air entrainment. The fire in question reportedly occurred after a few hours of exposure to the light, so this test was run for 14 days continuously.[11]

[11] There was no argument about the fact that torchiere lamps without guards are unsafe. These lamps have caused many fires. The question being addressed here was whether the radiant ignition scenario was correct.

As another example, in the case of an oxidizer fire, an experiment might be required to prove that a particular contaminant is capable of setting off a decomposition reaction.

Larger experiments can be conducted to show how a particular fuel package might burn. In one such case, the sofa shown in Figure 6.54 was identical to the one that burned in a two-room hotel suite. Understanding how this sofa

Figure 6.54 (a) A test of an exemplar sofa for resistance to cigarette ignition. (b) The same sofa was set on fire to determine its burning characteristics.

burned aided in understanding why two individuals were unable to escape the room. Two experiments were conducted. First the sofa was subjected to a cigarette ignition test and then it was set on fire.

Figure 6.55 shows an experiment conducted to examine artifacts left on a gypsum drywall ceiling by a smoke detector. It was alleged that the smoke detector had failed to function, but this experiment disproved the hypothesis that the smoke detector was in place. It had been taken down prior to the fire, as indicated by the absence of any protection

Figure 6.55 (a) This "hallway" was built to test how smoke detectors exposed to a small fire would behave. Several detectors were mounted on the ceiling.

Figure 6.55 (b) Melted smoke detectors, with protection patterns left on the ceiling. The results disproved the hypothesis that the smoke detector in question was in place prior to the fire.

patterns at the fire scene. It was easy enough to *say* that there should have been a pattern on the ceiling, but the experiment proved that some patterns should have been found if the device were present.

The difference between cause and effect on a kerosene heater was tested in the experiment shown in Figure 6.56. The unlit kerosene heater was surrounded with a wood crib and exposed to a fire of about a megawatt. A second fire was lit with the heater functioning, and the results were compared. Due to the presence of heat shielding inside the heater, the patterns produced upon exposure to an external fire differed, depending on whether the heater was on or off. Comparison with the heater from the fire scene revealed that it was on. The photographs of the test heaters were admitted into evidence.

The largest experiments to test ignition and/or spread hypotheses are full-scale fire tests. Such tests are prohibitively expensive and difficult to get into evidence, even if extreme precautions are taken.[12] All relevant parameters must be demonstrably the same as during the fire incident in question, an often impossible requirement. A simple error, such as leaving a door open when it should be closed, or vice versa, can skew the results dramatically. Tests of this size are not unlike rocket launches. Once the test fire is ignited, it is too late to change parameters. The Lime Street Fire tests conducted by this author and John DeHaan in 1991, and shown in Figure 6.57, are an example of a full-scale experiment designed to test an ignition and spread scenario.

[12] Even if the scene is re-created exactly, Murphy's law predicts that there will be a strong wind from the wrong direction, rain, snow, or some other confounding influence.

Figure 6.56 One of two kerosene heaters exposed to a 1-megawatt fire. Heaters were exposed while off and on, and the fire patterns on the exterior were compared.

Figure 6.57 Full-scale house fire test. This residence was located a few doors from the one that burned in the fire at issue. It had an identical floor plan, and the room of origin was furnished with exactly the same furniture and interior finishes as the original scene.

6.4.1 Spontaneous ignition tests

The testing of spontaneous ignition hypotheses is not very difficult if the scenario is sufficiently specific. Usually, the product in question already comes covered in warnings, so the question of configuration or ambient temperature needs testing. ASTM D6801-07 (2015) is the *Standard Test Method for Measuring Maximum Spontaneous Heating Temperature of Art Materials* [12]. A more widely applicable test is the United Nations (UN) Test for Materials Liable to Spontaneous Combustion. This test has also been adopted by the US Environmental Protection Agency (EPA) [13]. Each test has advantages and drawbacks.

The UN/EPA test uses 100- and 25-mm stainless steel mesh cubes and an exposure temperature of 140°C (284°F) for 24 hours. If the sample temperature exceeds the oven temperature by 60°C in the 100 mm cube, the test is repeated in the 25 mm cube. Materials that react positively in the small cube are assigned a more dangerous classification than those that react only in the larger cube. The test is repeated at 120°C and if the sample still tests positive, it is repeated at 100°C. This test is really designed for granular materials or powders; however, any liquid that has even a slight tendency to self-heat when on a cloth surface will likely ignite or at least smolder when exposed to this test. It is not a very realistic test to determine if a particular configuration of rags could be an ignition source, but if the test is negative, a particular scenario can be excluded. After a positive test, if the cloth is not consumed, it will be hardened, and a clinker of the type frequently seen in spontaneous combustion dryer fires will be produced. Figure 6.58 shows the results of a 100-mm cube UN/EPA test.

In the ASTM art materials test, the test liquid or paste is placed on a nonwoven paper substrate, and a manganese-based drier is added to improve the chances for spontaneous heating. This test is carried out at 70°C (158°F) and is rigidly structured but has the advantage of having undergone a series of interlaboratory tests to measure reproducibility and repeatability.

Babrauskas [14] provides an extensive discussion of various tests for self-heating and their history and development. The bottom line is that, aside from pass/fail tests, there is no uniform "scale" of the relative tendencies to self-heat among various substances. Any time a test is conducted to aid a fire investigation, it will be useful to test some commonly recognized substances, as well as the substance in question. Thus, when trying to characterize the self-heating tendency of a coating, running the same test on pure linseed oil and olive oil will provide some insight into the relative risks of ignition under the conditions of the test.

In many cases, the fire scenario requires a more "customized" approach to make a test substantially similar to the facts of the case. It is useful to run the standardized tests and then compare the results with a test setup that more closely reflects the actual conditions.

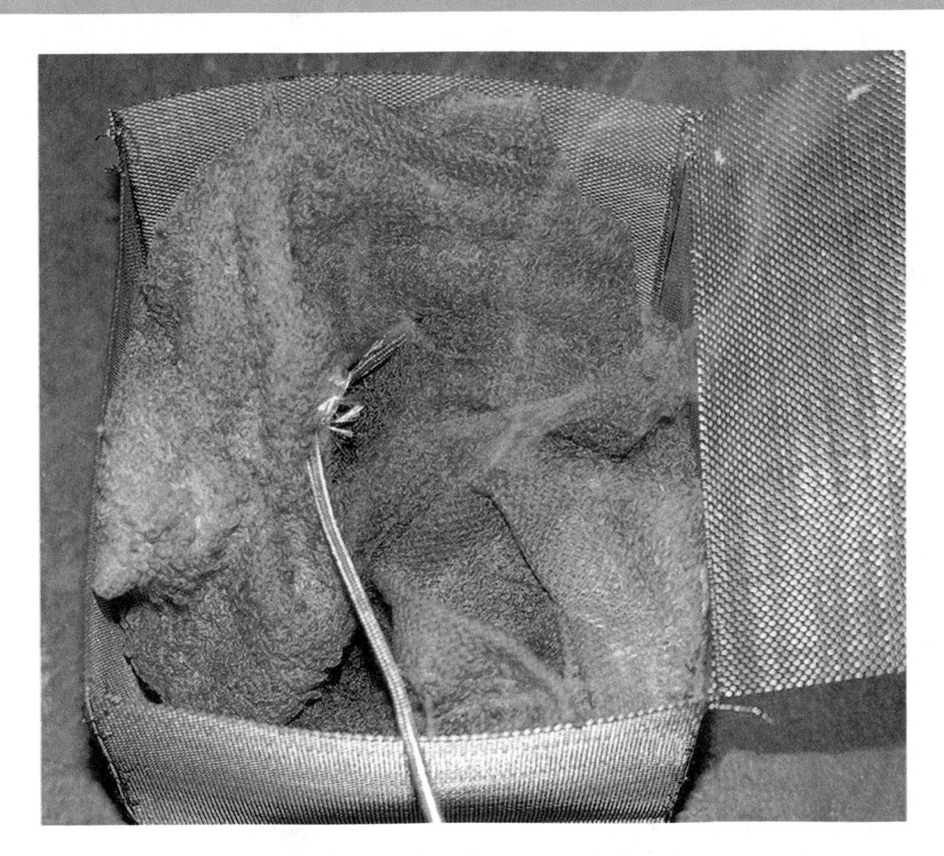

Figure 6.58 The 100-mm cube test for spontaneous ignition. The material in the cube is terrycloth cotton fabric saturated with peanut oil. Although the test is designed for solids, not liquids, just about any liquid capable of self-heating will ignite after a few hours of exposure to the high temperature called for in this test.

Sometimes it is necessary to generate the material for testing. Sanding dust is a particular hazard but may require days to generate sufficient sample material. Dust collected in batches should be kept in a freezer between collection and testing because it might slowly react if left at room temperature. Paint overspray particles are another material subject to spontaneous ignition that require generation from a sprayer tip that produces the same size droplets as the suspect sprayer. It is also important to match temperature and humidity conditions. Many commercial fire losses have resulted from the improper disposal of overspray or paint filters with overspray particles on them, but re-creating the particles for testing is a challenge. Dusts and small particles can undergo spontaneous heating at a very rapid rate compared to liquids on cloth. Many liquids include solvents that must evaporate before any significant heating can take place. The solvents use up energy in the evaporation process, and this loss of energy can make it impossible for the sample to reach its ignition temperature.

Once suitable materials have been obtained, the testing can take place. It is useful to record the results using a multichannel temperature recording device and, if possible, time-lapse photography. A fuel package undergoing spontaneous heating can take hours or days to develop into a smoldering fire, and frequently, the temperature rises a few degrees above ambient and then falls back down before rising again. While the results of tests of spontaneous heating hypotheses can be quite compelling, generating such results is often about as exciting as (literally) watching paint dry.

In one investigation conducted by the author, an allegation was made that a fire in a residence under construction was caused by improper storage of stain-soaked rags. The stain label included warnings about spontaneous heating, but a test conducted by the defendant contractor revealed no ignition in a 24-h period. The author's tests confirmed this result, and there seemed to be little support for the spontaneous ignition scenario. When discarded testing materials were seen on fire some 50 hours after the rags were initially exposed to the stain, however, a different test was conducted. First, more information was collected from the painter about the location of the box of rags and its size.

Next, a long-term experiment, complete with a sprinkler head in case the rags ignited overnight, was set up. The test setup is shown in Figure 6.59a. Eighty-eight hours after the test started, the result shown in Figure 6.59b was

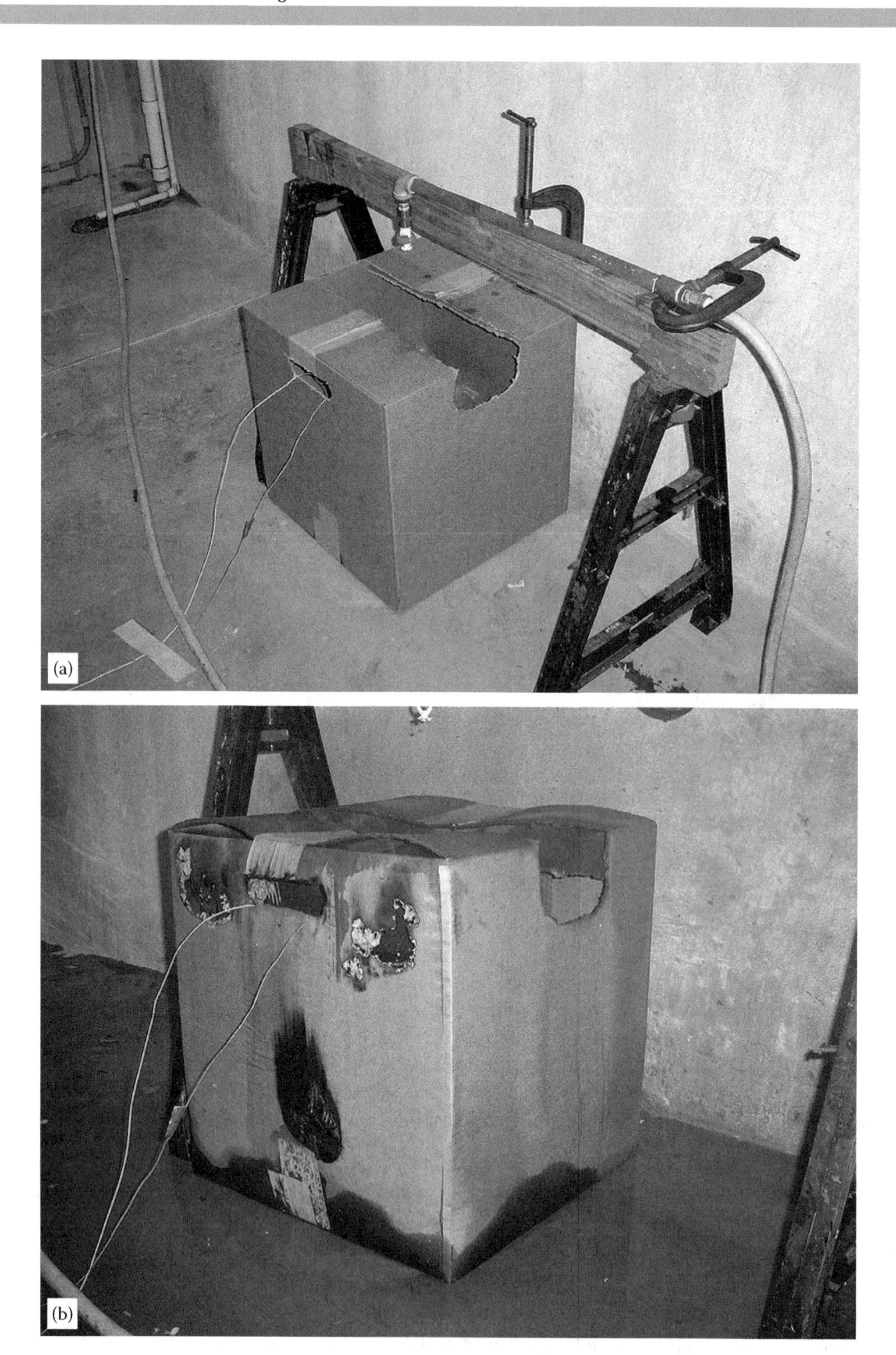

Figure 6.59 (a) Spontaneous ignition test setup. This box was configured as described by the worker who had been applying stain to a log house. Two thermocouples were placed in the pile of rags, which were sun-dried before placement in the box, as described by the worker. The sprinkler head allowed the test to be left unattended. A video camera recorded images every thirty seconds. (b) Spontaneous ignition test results. The rags began flaming combustion 88 hours after being placed in the box.

obtained. The case quickly settled, and the widely held notion that spontaneous heating fires can only occur if they happen within a few hours was put to rest.

6.5 FOLLOWING UP

Once laboratory testing has allowed for the development and testing of an ignition hypothesis, it is useful (often necessary) to research whether this is a one-time event or has happened before. In the case of a product believed to be defective, it is defective either because of a manufacturing defect (also known as a unit defect) or because of a design deficiency.[13] Certainly, if there is something wrong with the design that causes fires, other fires will have occurred. The CPSC has conducted thousands of recalls and collected "epidemiological" data on numerous events that have not resulted in recalls. Much information about recalled items is available at the CPSC website. The Freedom of Information Act can be used to obtain additional data from the CPSC, although this can be a painfully slow process. Manufacturers can object to the release of information, and such objections can take 6 months or more to resolve. It is sometimes quite instructive to type "[Product name]+fire" into a Google search box.

Other sources of information about particular failure scenarios are publications such as the *Fire and Arson Investigator.* Internet bulletin boards can also provide leads, but one must always find the documentation on which those leads are based. "I read it on the Internet" is unlikely to impress the judge hearing a *Daubert* motion.

If a poorly performed service or installation is believed responsible for the ignition, the *National Fire Codes* are the first place to look for the standards of care that might have been violated. Chances are that the person who caused the fire was not the first to engage in a particular series of actions or omissions. Hot work is covered by NFPA 51b, *Standard for Fire Prevention during Welding, Cutting, and Other Hot Work.* The application of spray paint and proper disposal of waste materials and filters are covered by NFPA 33, *Standard for Spray Application Using Flammable or Combustible Materials.* Electrical installation is covered in the *National Electrical Code,* NFPA 70. Gas appliances and installation are covered in the *National Fuel Gas Code,* NFPA 54. Construction and demolition are covered in NFPA 241. Codes exist for just about everything that can possibly cause a fire. It is unusual for an investigator to come across an act of negligence that is not proscribed somewhere in the *National Fire Codes.*

6.6 CONCLUSION

Finding an appliance or an electrical or gas system component at the hypothetical origin of a fire is frequently insufficient, in and of itself, to prove the cause of the fire. The device must be shown to be a competent ignition source or capable of becoming one, and it must be shown to have been the ignition source for the fire in question. A careful examination of the proposed ignition source might lead to the development of a testable hypothesis as to the failure mechanism, or the failure mechanism might become self-evident on close inspection. It is desirable, but in many cases difficult or impossible, to conclusively prove the root cause of the failure due to destruction of the evidence by the fire. Alternatively, a close examination of a proposed ignition source or scenario may allow it to be excluded as the cause of the fire.

The devices discussed in this chapter are those commonly examined in the context of fire investigations and have typical failure modes that cause fires. These typical failures must either be ruled in or ruled out. Once a failure scenario has been developed, it is important to learn whether similar fires have been reported, or whether any standards of care have been violated.

[13] The design of a product also includes its packaging, instructions, and warnings, but those aspects of the design are beyond the scope of this volume. For an excellent discussion of these and other product liability issues, see T. F. Kiely, *Science and Litigation: Products Liability in Theory and Practice,* CRC Press, Boca Raton, FL, 2002.

Review questions

1. Which of the following electrically powered devices are the exception to the rule that if there is arcing inside the appliance, it is likely that the fire started there?
   a. Fluorescent light ballast
   b. Oxygen concentrator
   c. Microwave oven
   d. Computer power supply
2. Which of the following statements regarding spontaneous combustion are true?
   I. A spontaneous combustion fire in a clothes dryer is unlikely to happen until the drum stops turning.
   II. Spontaneous combustion is usually a very fast process.
   III. Spontaneous combustion is not going to occur if it has not happened within 24 hours.
   IV. Petroleum oils, such as motor oil or kerosene are not likely to undergo spontaneous combustion.
   V. If no clinker is found, spontaneous combustion can be ruled out.
      a. I and IV only
      b. II, IV, and V
      c. I, II, and IV
      d. III only
3. Which of the following statements about ignition sources are true?
   I. Printed circuit boards are made from fire resistant fiberglass and will not sustain a flame.
   II. Battery operated devices do not have sufficient energy to ignite plastics.
   III. Impurities on a printed circuit board can cause short circuits.
   IV. Failures can occur in devices that appear to function properly for years.
   V. Once a device has been "burned in," it is less likely to fail.
      a. All of the above are true.
      b. I, III, and V
      c. III, IV, and V
      d. I and IV only
4. When designing an experiment to test a hypothesis, what is the MOST important element of the design?
   a. The experimental setup should exactly mimic the actual scene.
   b. The design should be absolutely objective.
   c. The design should be skewed toward producing results that are adverse to your hypothesis.
   d. The uncertainty of measurements should be known to $\pm5\%$.
5. According to NFPA, what were the two leading fire causes from 2010 to 2014?
   a. Smoking and intentional
   b. Heating and cooking
   c. Cooking and electrical
   d. Intentional and electrical

Questions for discussion

1. You are the host of a joint inspection of a dryer suspected of causing a fire. Write a protocol for the examination.
2. Explain the components of a reliability bathtub curve.
3. Why is an oxygen concentrator an exception to the general rule that arcing inside an appliance enclosure means the fire likely started there?
4. Why should a physical experiment be designed to favor the hypothesis of the adverse party?
5. What risks are associated with a homeowner switching from an electric water heater to a gas water heater?

References

1. Jordan, J. (2012) Batteries under fire, *Fire and Arson Investigator* 63(2):12.
2. Ahrens, M. (2016) *Home Fires Involving Cooking Equipment*, National Fire Protection Association, Quincy, MA. Available at NFPA.org.
3. Carman, S. (2011) Investigation of an Elevated Fire—Perspectives on the "Z-Factor," *Presentation to the 2011 Conference on Fire and Materials*, San Francisco, CA. Available at www.carmanfireinvestigations.com (last visited January 17, 2018).
4. Goodson, M. (2000) Electric water heater fires, *Fire and Arson Investigator* 51(1):17.
5. CPSC (1996) White-Rodgers Announce Gas Water Heater Temperature Control Recall, Consumer Product Safety Commission, Release # 96-070, February 15, 1996.
6. NFPA 54, (2015) *National Fuel Gas Code,* National Fire Protection Association, Quincy, MA, 2015, 9.6.8.
7. NFPA 54, (2015), 5.6.2.3.
8. NFPA 54, (2015), 9.1.9.
9. Monroe, G., and Wuepper, J. (1992) Spontaneous combustion of vegetable oils on fabrics, *Appliance Engineer* 54(8):14.
10. Lee, A. (2003) *Final Report on Electric Clothes Dryers and Lint Ignition Characteristics*, USCPSC, May 2003. Available at www.cpsc.gov (last visited January 17, 2018).
11. Gamse, B., McDowell, J., Nolen, D., Camara, N., et al. (2004) Bad science from Big Brother, in *Proceedings of the AAFS Annual Meeting*, Dallas, TX, February 2004, 138.
12. ASTM D6801-07 (2015), *Standard Test Method for Measuring Maximum Spontaneous Heating Temperature of Art Materials*, Annual Book of Standards, Volume 6.02, 2015.
13. US EPA, (2007) Method 1050 Test Methods to Determine Substances Likely to Spontaneously Combust. Available at https://www.epa.gov/sites/production/files/2015-12/documents/1050.pdf (last visited January 17, 2018). Until 1991, this test was published in 49 CFR Part 173, Appendix E, Division 4.2, 1991.
14. Babrauskas, V. (2003) *Ignition Handbook*, Fire Science Publishers, Issaquah, WA, SFPE, 405.

CHAPTER 7

Some practical examples

In theory, there is no difference between theory and practice, but in practice, there is.

—Jack Handy

LEARNING OBJECTIVES

After reviewing this chapter, the reader should be able to:

- Identify some common characteristics of different kinds of fires
- Understand the causes of variability in the experiences of fire investigators
- Appreciate the role of the fire investigator in different kinds of litigation

7.1 INTRODUCTION

The abbreviated fire scene inspection reports presented in this chapter are a selection of actual fires investigated or reviewed during the past few years. These are cases over which this author had either primary responsibility for determining the origin and cause or a supervisory role, or in some cases was asked by the original investigating agency to provide a technical review. These particular cases have been selected either because they are especially instructive, demonstrate typical fire behavior, or are more interesting than most. In addition to the technical description of the fire investigation, a short epilogue follows each case study. The epilogue describes the practical or legal consequences of the investigation.

The character of the cases present in an investigator's portfolio is controlled by several factors. The first is the investigator's clientele. Public-sector investigators are often called to investigate a fire when suspicious circumstances exist. In cases where the fire is known to be accidental, law enforcement authorities often do not have the resources, the interest, or the jurisdiction to conduct an in-depth investigation. Most private-sector investigators receive the majority of their case assignments from insurance carriers that want to know whether the fire was accidental or incendiary, and if incendiary, whether their insured was involved in setting the fire. In the case of an accidental fire, the insurance carrier wants to know if any avenues exist for subrogation, that is, finding a third-party provider of a product or service that may have caused the fire. Insurance carriers and law enforcement agencies represent the bulk of the clientele that employs fire investigators.

Because the system of justice in the United States is an adversarial one, the next largest groups of clients are those opposed to the first two. Criminal defense attorneys require the services of fire investigators to provide people accused of setting fires with effective assistance. In fact, the assistance of a competent fire investigator has been held to be an essential component of a defendant's right to effective assistance of counsel [1]. Civil litigators may represent individuals who are at odds with their insurance company over a denial of a claim, uninsured or self-insured individuals or corporations attempting to recover fire losses from third parties, or persons injured in a fire. Civil litigators may also be defending subrogation or personal injury claims. Fire investigators who are called in by this third group of clients are almost never the first on a scene because they are responding to another investigator's allegation as to the cause of a fire.

> Defense counsel failed to conduct an adequate background search for a competent expert, permitting false scientific testimony to go unchallenged during trial, and defense counsel failed to call an expert witness to rebut the testimony of the state's experts.
>
> Under the facts of this case, the failure of Richey's defense counsel to call an expert witness to rebut the false scientific testimony of the state's experts deprived Richey of the effective assistance of counsel.
>
> 6th Circuit, *Richey v. Bradshaw* August 10, 2007

In addition to clientele, a fire investigator's portfolio is influenced by geography. This is mainly due to the differences in climate, construction practices, the age of the building stock, and the type of heating system, all of which vary from region to region.

This author's practice was based in the southeastern United States, and the clientele for scene investigations was composed largely of insurance carriers, manufacturers defending themselves from insurance carriers, trial lawyers representing injured persons, civil attorneys defending clients from injured persons, criminal defense attorneys, and civil attorneys litigating first-party claims against insurance companies. Occasionally, the client is a prosecutor, but in those cases, the author is never the first responder but rather has been called in to review the initial determination. Cases in this chapter are presented alphabetically by cause. The author's current practice consists almost entirely of reviewing the work of other investigators and "preaching the gospel" of scientific fire investigations.

7.2 ARSON

The term *arson* is used to indicate that a crime has occurred, while the term *incendiary* is used to indicate that the fire was intentionally set. While it is possible for incendiary fires to be the result of stupid but not criminal acts, there is usually a large overlap between incendiary and arson fires. The terms are used interchangeably with respect to the cases presented here.

Generally, arson fires are not difficult to detect. Arsonists typically use an ignitable liquid, which is detectable after the fire is out. In those cases where this author has been asked to evaluate claims that a particular fire was intentionally set, the majority of those determinations were based on the "reading of burn patterns" or the presence of certain artifacts, but not on the finding of ignitable liquid residue in the fire debris because the samples all tested negative. There are, unfortunately, a large number of fire investigators who believe that most fires are arsons and that the finding of an ignitable liquid residue is the exception rather than the rule. That has not been this author's experience. For the first 15 years of this investigator's career, the clients were bulk insurance carriers with suspicions about their insureds having caused the fire. Perhaps hundreds of arsons were missed, but hundreds were found, and they were generally not subtle. All that was required was an excavation of the scene and proper reconstruction. Usually the cause of the fire became clear. Burning one's residence or business to defraud an insurance carrier is generally the act of a desperate person, and desperate people are seldom subtle (or clever). It has been said that proving a fire to be incendiary is not that difficult, but proving who did it presents problems. That is also frequently not true because, at least when motivated by fraud, the fire setter will behave in such a way as to make clear who the perpetrator is.

7.2.1 Arson fire 1: The fictitious burglar

In the spring of 2001, Applied Technical Services (the author's former employer, hereafter referred to as ATS) was requested to investigate the origin and cause of a fire that seriously damaged an insured residence, located in southeast Georgia. A Georgia deputy state fire marshal had already investigated the fire. The fire alarm was turned in at 2:34 a.m., and the fire department arrived on the scene at 2:37 a.m. to find smoke coming from the residence. The front of the residence is shown in Figure 7.1a, and a floor plan is shown in Figure 7.1b. The only damage visible from the outside was just above the southwest bedroom window (Figure 7.1c), from which the fire auto-ventilated when that room went to flashover.

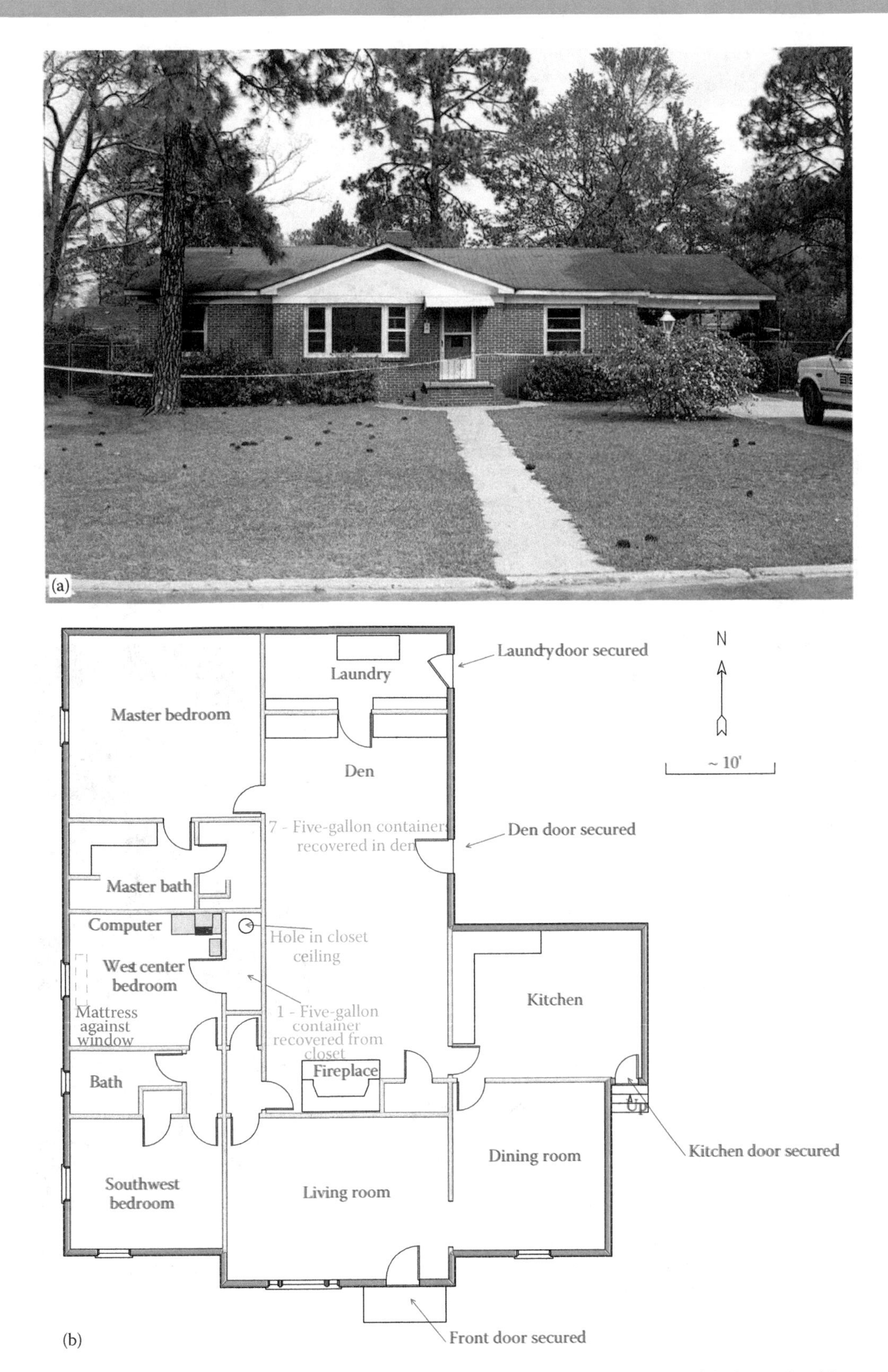

Figure 7.1 (a) Southeast Georgia residence set on fire on a cool spring morning. (b) Floor plan of the insured residence.

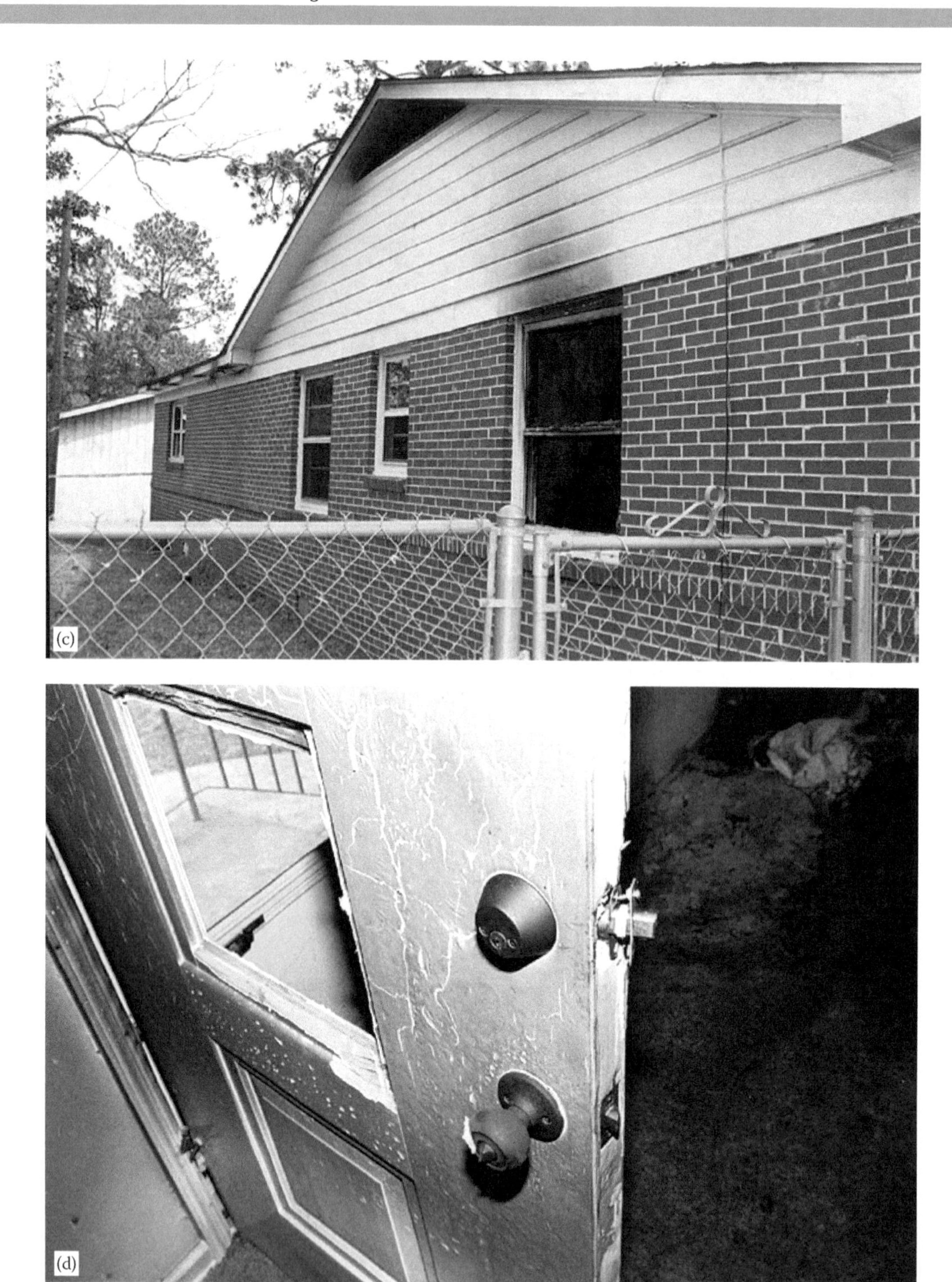

Figure 7.1 (c) Fire damage extending out the southwest bedroom window. This was the only window that broke during the fire. (d) Evidence of forcible entry by the fire department on the front door. Three other doors exhibited similar damage, characterized by breakage of the door and frame, and a lack of smoke deposits on the fracture surfaces.

Figure 7.1 (e) Exterior of the kitchen door where a key was broken off in the deadbolt. (f) Keys recovered on the doormat just inside the kitchen door.

Figure 7.1 (g) Flammable liquid fire patterns on the living room floor. (h) Narrow trail pattern on the living room floor.

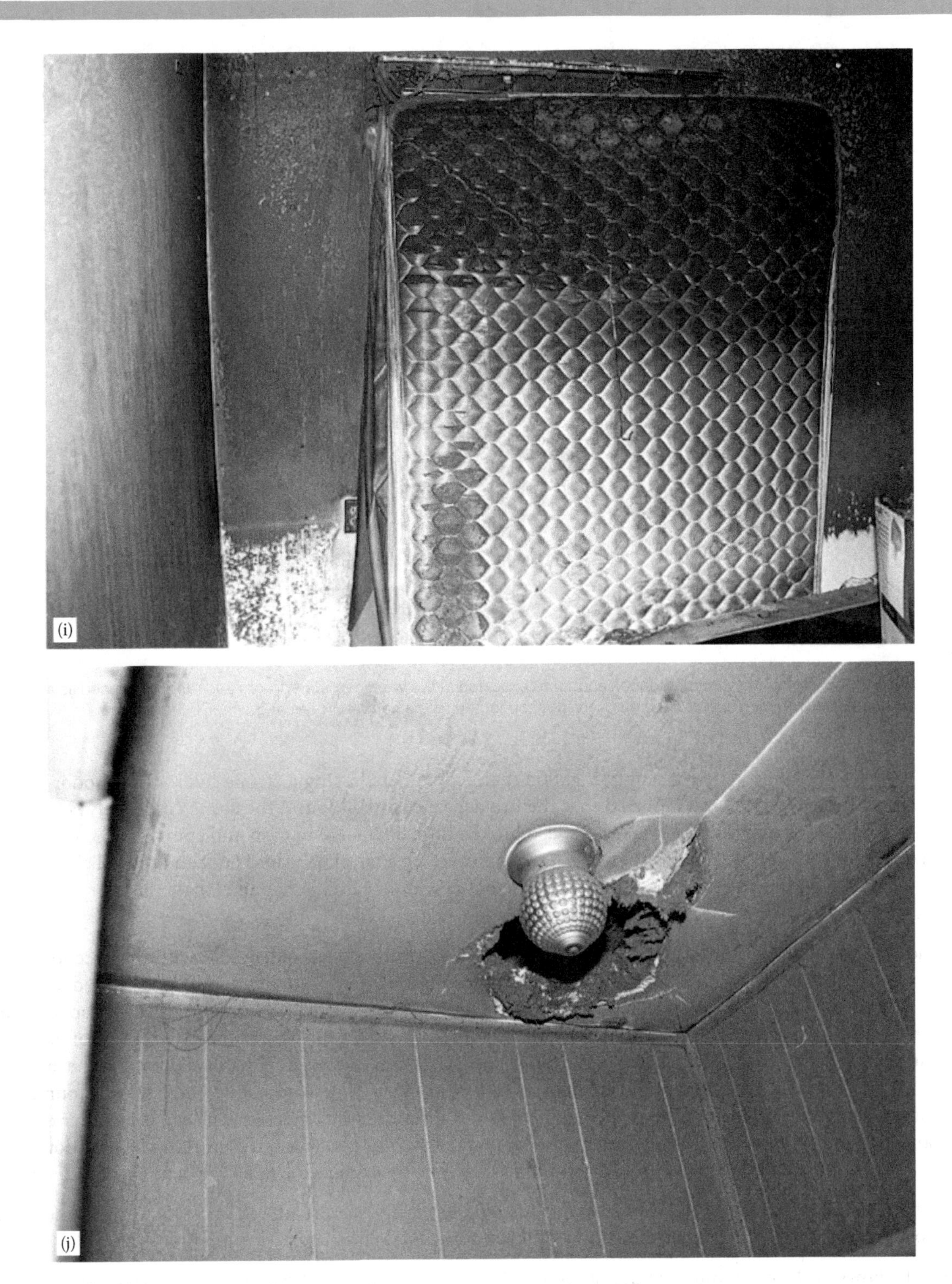

Figure 7.1 (i) Mattress propped up against a window in the west center bedroom. (j) Hole punched in the west center bedroom closet ceiling.

Figure 7.1 (k) Computer with keyboard, monitor, and printer, located in the west center bedroom and reportedly used as an office by the homeowner. The power cord and peripherals were not connected to the central processing unit (CPU).

The fire marshal had removed eight empty 5-gallon (gal.) containers having a strong hydrocarbon odor from the residence prior to our arrival and reported that the fire department had found the scene completely secured. The homeowners were claiming that the house had been broken into, $8,000.00 in cash and coins and $3,000.00 worth of silverware had been stolen, and the fire had been set to cover up a burglary. This is not an unusual claim in cases where the fire fails to destroy the evidence of its cause but it seldom actually happens that a burglar wants to call attention to the place he has just robbed by setting it on fire.

The condition of each of the entrances to the residence was carefully documented, and with the exception of the kitchen door, all exhibited damage similar to that shown in Figure 7.1d. Firefighters were interviewed, and it was possible to find the individual responsible for opening each of the doors. The kitchen door was never opened and is shown in its as-found condition in Figure 7.1e. The remains of a key were broken off in the exterior side of the kitchen deadbolt. (Smoke is present on the outside of this door because of the location of the storm door, which trapped the smoke as it came around the edges of the kitchen door.) A set of keys, one of which was broken, was found on the doormat just inside the kitchen door, in such a position as to indicate that they were placed there after entry had been gained through another door. The kitchen door would not swing open without disturbing the position of these keys. The edges of the broken key (Figure 7.1f) exhibited a physical match to the key in the exterior deadbolt.

Inside the residence, furniture was overturned, clothing and papers were scattered about, and the place literally reeked of kerosene. Figure 7.1g and h show some of the representative damage.

Figure 7.1i shows evidence of unusual but not atypical behavior on the part of the fire setter. The mattress from the west center bedroom was placed in an upright position over the window. This is the kind of evidence referred to in NFPA 921 in Section 24.3, "Potential indicators not directly related to combustion [2]." Given that this is a known incendiary fire, the most likely explanation for the presence of this mattress in front of the window is that whoever set the fire hoped to delay its detection. The author had previously investigated a fire in a brick veneer house where this behavior was the only clue as to the cause of the fire. Everything inside the house had burned to powder; had it not been for the mattress springs hanging over the window ledges, it might have been necessary to write off that fire as "undetermined."

Additional evidence of incendiary activity (also not directly related to combustion) was found in the west center bedroom closet, where a hole had been punched in the ceiling, as shown in Figure 7.1j. The attic pull-down steps were also found open and completely extended. (The springs that hold attic steps in place frequently lose their temper and open during fires, but in those cases, the steps remain folded.)

The homeowner was present during our inspection and pointed out the location of two shoeboxes that allegedly each contained $2,500.00 in cash, as well as the location of a jar that contained $3,000.00 in silver quarters. No evidence of any money was found in the house. (Banks were still paying interest on deposits in 2001.) The location of four pistols and a long gun was also pointed out, and again, it was clear that if these weapons were ever present in the house, they were removed prior to the fire. A VCR was also reportedly stolen from the living room, but the televisions were left behind. Still, there was no evidence to tie the homeowner to the disappearance of these items.

Then she made a tactical error. The homeowner pointed out a computer that was set up in the corner of a bedroom that she said she used as an office area. The computer, shown in Figure 7.1k, was particularly damning evidence because it was clearly not a functional machine. When the back of the CPU was examined, there were no peripherals connected nor was there a power cord. This was clearly a setup, and a lazy one at that. Had she bothered to make the connections, there would have been nothing suspicious about the computer. Had the house burned to powder, there would have been nothing suspicious about the computer. The computer represents the kind of evidence that tends to show that the homeowner had *prior knowledge* that this fire was going to happen. NFPA 921 describes replacement of contents as a strong indicator of prior knowledge. This was almost certainly a replacement computer.

Despite the overwhelming evidence of the presence of ignitable liquids, four samples were collected and returned to our laboratory for analysis, and all four were positive for the presence of kerosene. Because of its relatively high flash point, kerosene makes a poor accelerant, particularly in the early morning hours of a cool spring day.

The inventory of contents revealed a lack of clothing and other ordinary household items, but the homeowner explained the absence of these contents. She had another residence and admitted that she was in the process of making that her full-time residence, which explained the absence of clothing in the dresser drawers, food in the refrigerator, and dishes in the kitchen cabinets. While this explained the lack of personal property, it also provided a motive for ridding herself of this house. The insurance claim for this fire was successfully denied.

Epilogue: This fire was interesting not because the cause was difficult to determine but because of the existence of several indicators not directly related to the fire, which provided clues as to the motivations and identity of the perpetrator. There were two competing hypotheses: either the fire was set by the homeowner (or someone acting at her behest) or by someone else. The perpetrator gained entry without breaking a door and left behind a set of keys. He (or she) carried 40 gal. of kerosene inside and spread it around. A mattress was placed against a window. Efforts were made to spread the fire to the attic by opening the stairway and poking a hole in the ceiling. All of this activity, in conjunction with the evidence that the computer was where it was simply to add value to the claim, led investigators to believe that a fraudulent arson by the homeowner was the more likely hypothesis.

7.2.2 Arson fire 2: Three separate origins

The author reviewed the report of this fire as part of his participation in the Texas Fire Marshal's science advisory workgroup (SAW). This fire occurred around midday in September 2014 in east Texas and was reported by a worker who was mowing the grass alongside a highway. He reported smoke coming from under the eaves. The house had been previously damaged by a small bedroom fire in June, and the homeowner, who was uninsured, was performing repairs. There was reportedly nobody at home at the time of the fire.

There was very little damage visible from the exterior (Figure 7.2a). A small amount of damage from the June fire was still visible in a room under repair, as shown in Figure 7.2b. A floor plan of the house is shown in Figure 7.2c.

Examination of the house revealed at least three separate, uncommunicated areas of burning. Origin 1 was in the living room and is shown in Figure 7.2d. The sofa was almost completely consumed. There was an area of burning on the nearby loveseat, shown in Figure 7.2e, and this burning may represent a separate ignition, but because there were already three clearly separate ignitions, it was not necessary to go out on a limb and call this a fourth point of origin.

Origin 2 was across the living room from the sofa, where a box of clothing was lit on fire. This is shown in Figure 7.2f. This fire was short-lived as evidenced by its triangular shape. The plume never matured into a V-pattern. Origin 3,

Figure 7.2 (a) Exterior of the residence, which showed very little actual damage. (b) Location of the June fire, which was being repaired by the homeowner when the September fire occurred.

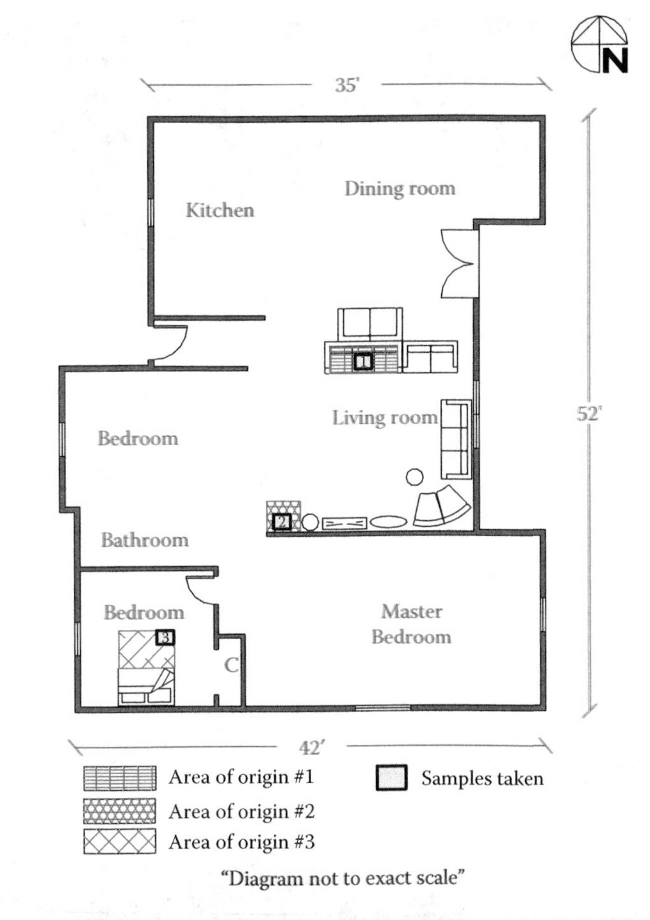

Figure 7.2 (c) Floor plan of the residence, showing the three points of origin. (d) Origin 1 on the sofa in the living room.

Figure 7.2 (e) Area of burning on the arm of the loveseat located behind the sofa. This may represent yet another origin, but because of its small size and its proximity to the sofa, the investigator refrained from declaring it to be so. (f) Origin 2, where a box of clothing was ignited.

Figure 7.2 (g) Origin 3 showing damage to the bed in the southwest bedroom.

shown in Figure 7.2g, was located in the southwest bedroom. The bed was about half consumed. None of the burning fuel packages drove either room to flashover, so this was a relatively straightforward fire to investigate.

Epilogue: Although this was clearly a set fire, no suspect was ever identified. The uninsured homeowner had no apparent motive. The June fire had resulted in the removal of the window from the bedroom, so the house could not be secured. A vandalism fire could not be eliminated. The purpose of including this fire here is to demonstrate what multiple origins actually look like. There is an unfortunate tendency on the part of some investigators to find multiple origins, when there is, in fact, only one. Two-dimensional thinking, discussed in Chapter 9, causes investigators to focus only on the floor when looking for connections between apparent multiple origins. NFPA 921 in Section 24.2.1.2 cautions that there are at least 10 different ways that apparent multiple origins can be created from a single fire [3].

7.2.3 Arson fire 3: Unpleasant neighbors

ATS was requested to investigate the origin and cause of a fire that occurred in late September 2000 and that destroyed three occupancies in a strip mall located in southern Kentucky and damaged three more. The two most heavily damaged occupancies were a print shop and a branch dental office, both insured by the same carrier. Because of the numerous interests involved, separate investigators, in addition to the author, were retained for the insurance carriers for the building owner and the health club owner. A deputy state fire marshal had ultimate control over the scene. As a fair amount of heavy lifting was involved, the fire marshal was happy to have the assistance of the other investigators.

The west end of the damaged shopping center is shown in Figure 7.3a, and an overhead view, taken from a hill behind the shopping center, is shown in Figure 7.3b. The fire damage seemed more complete toward the rear of the building, and based on a preliminary inspection as well as an estimation of the degree of difficulty, all the investigators agreed that it would be best to begin clearing debris from the dental office, starting in the rear. Figure 7.3c is a view from rear to front, or south to north, in the dental office; Figure 7.3d is a similar view looking into the print shop. As can be seen in these photographs, the stud wall between the two occupancies recorded burn patterns that indicated movement in both directions. This fire was reported just before 4:00 in the morning, and the fire department had a difficult time getting it under control. Because of the extent of destruction, the security of the doors could not be verified.

Figure 7.3 (a) West end of the burned shopping center, view from the front. The print shop is at the center of this photograph, and the dental office is at the far right. (b) Overhead view of the dental office and the print shop.

Figure 7.3 (c) Fire damage in the dental office, view from south to north (rear to front). (d) Fire damage in the print shop, view from south to north (rear to front). Note that the burn pattern on the stud in the foreground indicates movement of the fire from right to left, while movement from left to right is recorded on the very next stud.

Figure 7.3 (e) Newly installed air handler at the southwest corner of the building. This unit was eliminated as the cause of the fire.

Figure 7.3 (f) Burned studs between the print shop (left) and the dental office (right), indicating movement from the print shop to the dental office, probably a result of late extinguishment.

Figure 7.3 (g) Trail of burning uncovered in the rear room of the dental office.

The investigators did not consider the burn patterns on the stud walls significant (which is a good thing because they were contradictory). The fire moved through the wall more than once, but it was not possible to determine which side burned first until the floor was cleared.

It was reported that the dentist had recently installed a new air handler in the closet at the southwest corner of the building, in the area shown in Figure 7.3e. This was one of the first areas excavated, and the new furnace was carefully examined but was found to be unremarkable.

As some of the deeper debris was removed, there were additional fire patterns found on the stud wall closer to the floor, in the area shown in Figure 7.3f. The print shop is on the left side of the wall shown in this photograph, and the beveling clearly shows the fire burning from the print shop toward the dental office. We knew, however, that the dental office fire was extinguished first, and that this directional indicator on the stud wall might be no more meaningful than the patterns observed previously. A beveled pattern on a single stud this close to the floor is unlikely to be meaningful.

A pattern that everyone agreed was meaningful was revealed only when the floor was cleared off. The pattern extended over 40 feet (ft) and is shown in Figure 7.3g and h. This pattern extended through two doorways and made what appeared to be a loop at the south end, although the destruction was sufficiently complete at the south end that we were only able to see two sides of the roughly circular pattern. Figure 7.3i shows a sketch of the pattern that was finally uncovered.

Samples of the carpet exhibited a strong odor of gasoline and, in fact, tested positive for gasoline. Everyone involved in the investigation was allowed to collect samples.

Examination of the fire damage in the print shop revealed that all of it was caused by exposure to the fire from the dentist's office. Naturally, attention focused on the dentist. This was one of three branch offices that the dentist or an associate visited. The dentist visited this particular office on only one day per week. The fire occurred on a Thursday evening, and the dentist said that he had appointments for the next day at this office. During the insurance company's investigation, the financial condition of the dentist was investigated thoroughly, as some suspicion

Figure 7.3 (h) Continuation of the trail of burning running between two doorways.

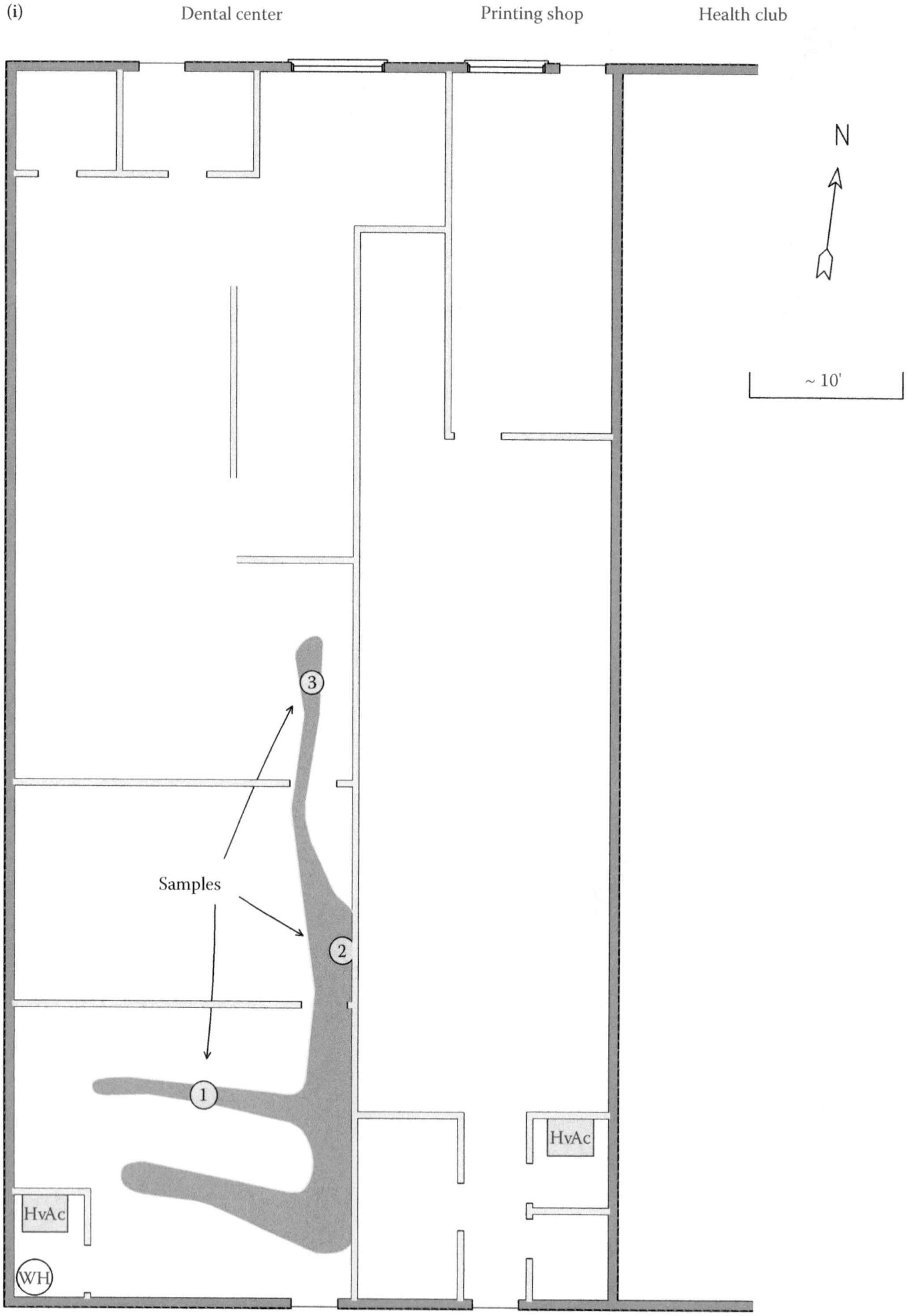

(i) Dental center Printing shop Health club

N

~ 10'

Samples

1 2 3

HvAc HvAc WH

Figure 7.3 (i) Sketch showing the location of the trail of burning, and location of samples. All three samples tested positive for gasoline.

existed that he might have overextended himself in opening so many branch offices. None of this analysis actually revealed a motive. In fact, the dentist provided services to many patients *pro bono*.

Epilogue: Once the floor was cleared, the cause of the fire was clear. As is often the case, determining *what* caused the fire was easier than determining *who* caused the fire.

During the routine processing of the claim from the print shop, it was discovered that the inventory was grossly inflated, with claims being made for machines that were not found during the inspection (large machines that would have been found had they been present). Further investigation revealed severe financial distress.

The investigation continued until it ended with a confession by the owner of the print shop that he had set his neighbor's office on fire, knowing that his shop would be destroyed and that suspicion would be directed at his neighbor.

7.3 DRYER FIRES

A significant number of fires are started by clothing dryers or their power supplies every year, but there are a surprisingly small number of fires that start because the heat from the dryer sets the clothes on fire. The typical dryer failure occurs because of poor installation practices, misuse, or spontaneous ignition of oil-containing laundry. The reason that so few fires begin in the clothing itself is that dryer manufacturers incorporate three or four safety devices to prevent that from happening. Even when the dryer fails because of a manufacturing or design defect, the redundancy in temperature controls usually prevents the clothing from being ignited. The following are typical examples of fires caused by dryers.

7.3.1 Dryer fire 1: Misrouted power cord

ATS was requested to investigate the origin and cause of a fire that destroyed the insured residence, an A-frame dwelling located in northwest Georgia. The fire occurred in late October 2001 while the homeowners were enjoying a crisp fall afternoon playing with their children in the front yard. One of the children noticed the smoke but her mother thought that one of the neighbors was burning trash. About 2 minutes (min) later, the volume of smoke increased and could be seen coming out from under the eaves. The homeowner approached the house and opened the kitchen door, but by that time, the kitchen was full of black smoke. The homeowner stated that she was not cooking but that the washer and dryer were on. The front of the residence is shown in Figure 7.4a, and a floor plan is shown in Figure 7.4b. Extensive damage was present throughout the residence, including the upper floor, but the heaviest damage appeared to be on the ground floor around the doorway to the large bathroom, which also served as a laundry room; it is shown in Figure 7.4c. The interior finish was stained wood paneling on both walls and ceilings everywhere in the house except for the kitchen and bathroom. This difference explains why there was heavier damage in rooms distant from the origin. Figure 7.4d shows the washer and dryer at the origin of the fire. Both of these appliances were returned to the ATS laboratory for further examination. Washing machines almost never cause fires—they cause floods, but failure to preserve the washing machine provides the dryer manufacturer or installer with an argument that they have somehow been prejudiced by the inability to examine the washing machine, and thus spoliation of evidence has occurred.

It was noted that there was no strain relief on the power cord, and, in fact, the homeowner had routed the power cord directly into the area behind the access panel where the power cord lugs connected to a terminal block. The dryer was designed to route the power cord through a round hole in the bulkhead below the terminal block, and the access panel was designed to allow access to the terminal block, not the passage of the power cord. The homeowner compounded the error by putting the access panel back in place, where it would squeeze the flat power cord. The cord, as it appeared after we removed the access panel, is shown in Figure 7.4e. The bathroom in this house was equipped with an electric range receptacle, designed to carry 60 amperes (amps), rather than a dryer receptacle, which is designed to carry 30 amps. The homeowner obtained a range power cord and attached it to the dryer. That would have made no difference if he had properly installed the cord, but more than 2 years of vibration while being squeezed by the access panel cover led to the arcing shown in Figure 7.4f, which ignited lint behind the dryer.

Figure 7.4 (a) Northwest Georgia residence burned in a dryer fire. (b) Floor plan of the residence.

Figure 7.4 (c) Fire damage in the living room. The doorway on the left leads to the bathroom where the washer and dryer were located. Note the heavier damage immediately around the doorway on the ceiling and the upper wall. (d) Washer and dryer at the origin of the fire.

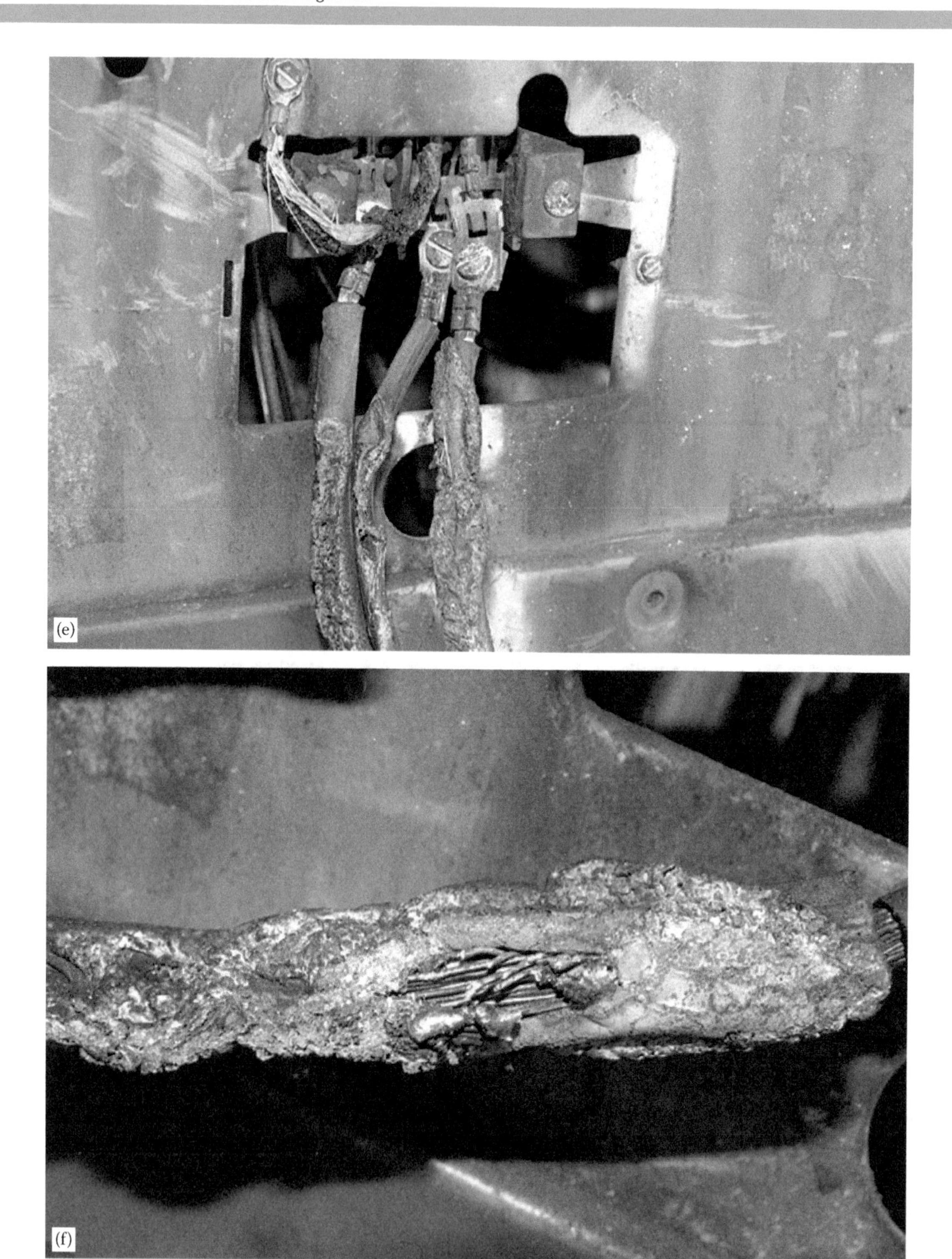

Figure 7.4 (e) Power cord misrouted through the rear bulkhead at the location of an access panel. (f) Location where the dryer power cord arced to the access panel cover.

Epilogue: Improper installation of the power cord is a common cause of dryer fires. Failure to use a strain relief device is a common error—one that dryer manufacturers know about and warn against. The combination of errors in this installation was unique, but the ignition source, a power cord arc, was the same as is commonly seen when the sole cause of the fire is the lack of a strain relief device.

7.3.2 Dryer fire 2: Cross-threaded electrical connection

ATS was requested to investigate the origin and cause of a fire that damaged an apartment building located in north metropolitan Atlanta. The fire occurred in mid-February of 2002 and was reported just before 1:00 p.m. The tenant was home at the time of the fire and reported to fire department investigators that she had seen it start in her utility closet. A brief initial inspection of the apartment building shown in Figure 7.5a revealed unequivocal evidence that this was a fire that began at the rear of the dryer. The origin of the fire is shown in Figure 7.5b.

The dryer had been purchased new from a local hardware store within the previous year, and the retailer had provided the installation. Therefore, both the manufacturer of the dryer and the hardware store were put on notice of a potential claim before the dryer was moved. The manufacturer responded that it would not send an investigator to the scene because the size of the loss was under its $50,000 limit for engaging its own independent investigator. It did, however, send a two-page letter requesting that all evidence be preserved and that it be provided with all reports and photographs. The manufacturer further requested that the inspections be carried out according to ASTM E860, ASTM E1188, and NFPA 921. The letter advised:

> These procedures may be accomplished with minimal or no added burden to you and with no adverse effects on your investigation. In any event, it is essential that these procedures be followed to ensure fair trial of any litigation that may arise and failure to follow them would likely lead to judicial finding of improper spoliation of evidence. See, for example, *Allstate Ins. Co. v. Sunbeam Corp.*, F.3d 1995 W.L. 242567 (7th cir. APR 27, 1995).

As it turned out, the manufacturer did not have any liability in this case, but its response was instructive. The owner of the hardware store did contact his insurance carrier, who sent out an independent adjuster with whom our office had worked on many occasions, and he agreed to allow the site investigation to proceed. He also agreed to the examination of the appliances at our laboratory.

Debris was cleared from the utility room, and the appliances were put back in place as shown in Figure 7.5c. After the appliances were removed, the fire pattern clearly showed an origin behind the dryer, as can be seen in Figure 7.5d.

The laboratory examination revealed several interesting findings, the first of which was a build-up on the lint filter, the likes of which this investigator had never seen. Figure 7.5e shows the multilayered lint blanket that resulted from the user having *never* cleaned the lint screen. She had no knowledge of it, having never read the instruction manual. This misuse, while interesting, ultimately proved irrelevant to the cause of the fire, except to the extent that there was plenty of easily ignited fuel in the area. The clothing inside the dryer drum was slightly smoked but still intact. It was also noted that, astonishingly, no vent was connected to the dryer exhaust; thus, before the lint screen became completely blocked, an excessive amount of lint had likely accumulated behind the dryer. This lint may have been the first fuel ignited.

An examination of the installation method revealed that the installer had used a strain relief device, as shown in Figure 7.5f, but it was noted that the power cord had arced to the device. A close-up of that arcing is shown in Figure 7.5g. This is the kind of arcing that one expects to see when an advancing fire approaches a dryer. In this case, however, we clearly had an origin located behind this dryer. Further examination revealed that the arcing shown in Figure 7.5g was an effect of the fire, *not* the cause. The terminal block connection point for the dryer power cord is shown in Figure 7.5h. The screw holding in the right-hand lug was slightly cocked, and a close examination of the screw, shown in Figure 7.5i revealed that it was cross-threaded. As such, the screw was incapable of holding the power cord lug tightly against the terminal block, resulting in localized heating, as shown in Figure 7.5j.

Figure 7.5 (a) Apartment building damaged by a dryer fire. (b) Fire damage at the utility closet where the fire was first seen by the tenant. This is a classic case of a "clearly defined origin."

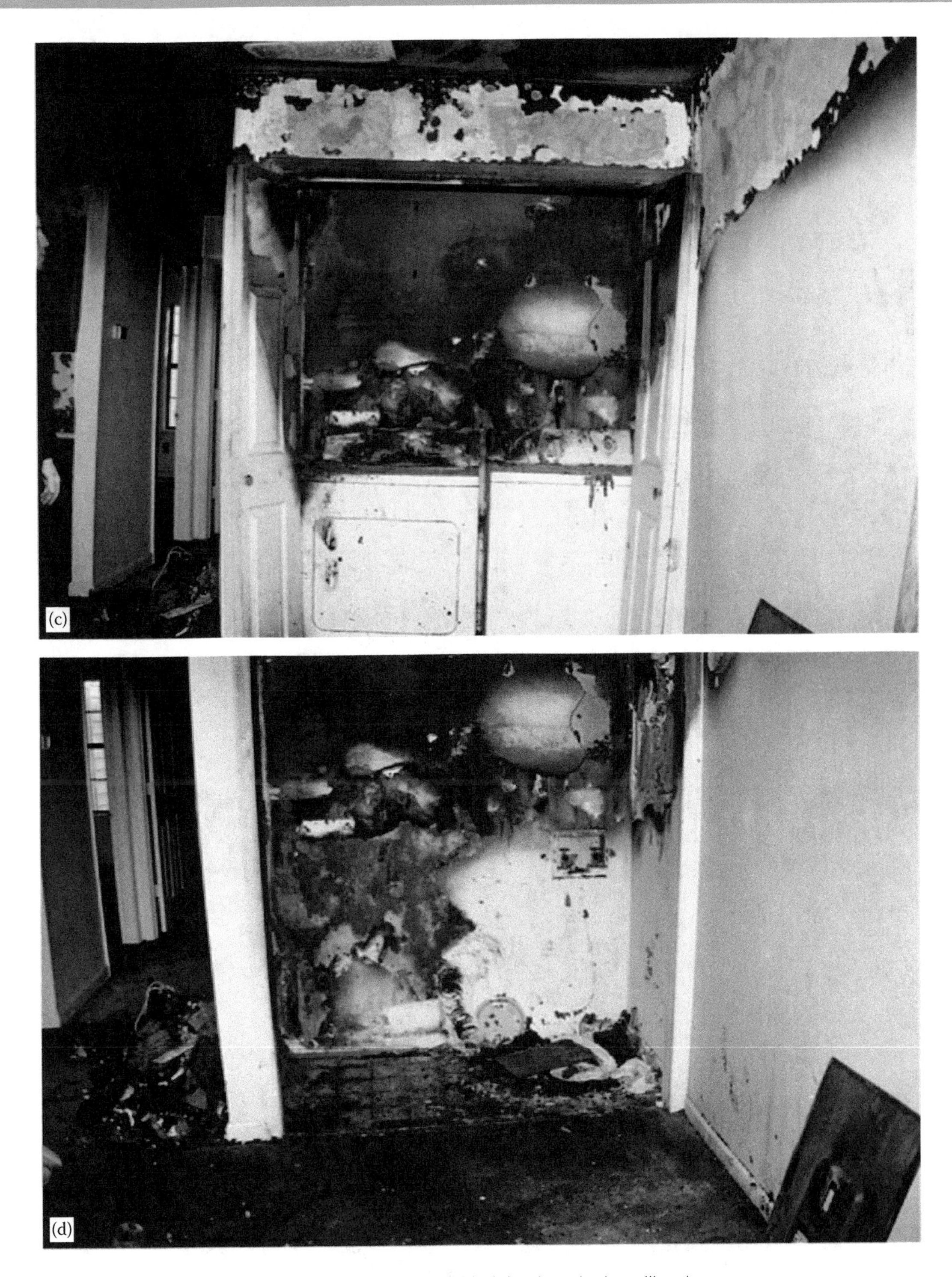

Figure 7.5 (c) Reconstructed utility closet. (d) Fire pattern behind the dryer in the utility closet.

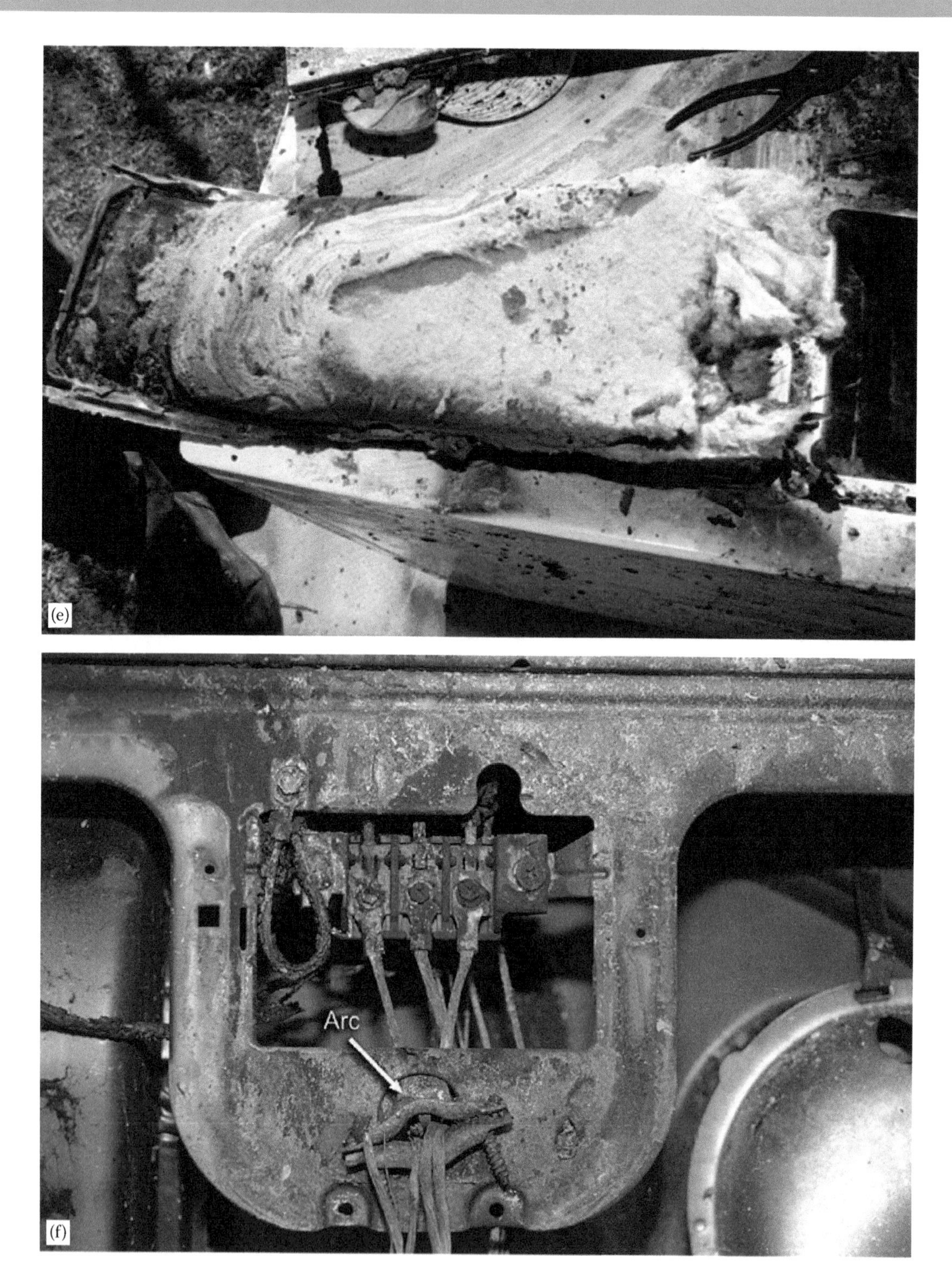

Figure 7.5 (e) Multilayered and multicolored lint build-up on the lint filter. The lint screen had never been cleaned. (f) Strain relief device used to install the power cord, as required by the manufacturer's instructions.

Figure 7.5 (g) Close-up of the electrical activity on the conductor where it arced to the strain relief clamp. (h) Terminal block connection point for the dryer power cord. The screw at the right side of this photograph was installed crookedly.

Figure 7.5 (i) Close-up of the cross-threading of the connection at the right side of the terminal block.

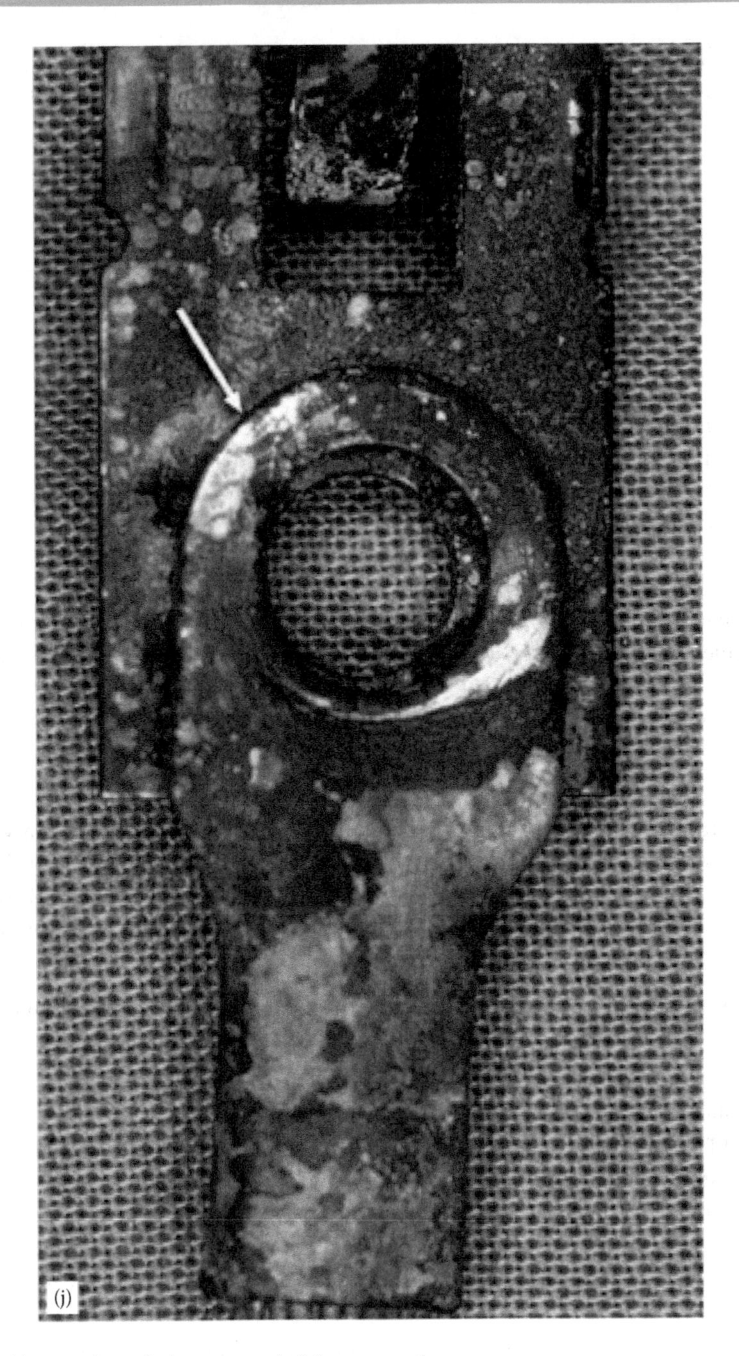

Figure 7.5 (j) Overheated lug as viewed after removal of the connection screw.

As with most electrical fires, the ignition source was not a short circuit or a ground fault, although a ground fault did occur. The cause of this fire was the flow of current through its intended path across the high-resistance point where the power cord attached to the terminal block.

Epilogue: This dryer presented two artifacts that could have confounded the determination. The lint build-up, while interesting and an extreme example of a common finding in dryer fires, did not play a role in this fire. The finding of excessive lint is a common *feature* of dryer fires but not a common *cause*. It may be that this frequently seen coincidence is a result of the lint supplying a finely divided fuel that can be ignited by a short-lived low-energy ignition source. The installer's failure to attach the dryer to a vent is also a unique finding.

The arcing between the power cord and the strain relief clamp might have been mistaken for the primary ignition source, but arcing at this location is almost always the effect of an advancing fire.

The hardware store's insurance carrier sent two investigators, and the hardware store sent two managers to observe the detailed inspection of the dryer connections. After seeing the evidence, these individuals all agreed that a poor job of installing the power cord was, in fact, responsible for causing the fire. The insurance carrier for the hardware store accepted liability for this claim.

7.3.3 Dryer fire 3: Spliced power cord

ATS was requested to investigate the origin and cause of a fire that occurred in late March 2000, in a rental mobile home located in north metropolitan Atlanta. The fire resulted in eight fatalities. Figure 7.6a shows the exterior of the mobile home, and Figure 7.6b shows a U-shaped burn pattern next to a dryer that was purportedly in use at about 4:30 a.m. when the fire broke out. The fire blocked the exit out the front door. A bed that had been placed in front of it blocked the back door. The victims were all found in that back bedroom near the door. All of them died as a result of smoke inhalation. A floor plan of the mobile home is shown in Figure 7.6c.

Figure 7.6d shows the dryer in place, as viewed from the living room. Figure 7.6e is a view down the hallway toward the bedroom area from the location of the dryer, and Figure 7.6f shows a natural gas-fired water heater located in that hallway. Extensive damage was noted to the roof structure directly above the water heater, as shown in Figure 7.6g. A fire pattern around the combustion chamber suggested a possible origin at the water heater, as shown in Figure 7.6h.

The water heater baffle was removed and found to exhibit a thin carbon deposit, but it could not be determined initially whether the deposit shown in Figure 7.6i was laid down during the fire or was the result of a sooty flame during regular operation.

This investigation involved several parties who were present to witness all of the inspection at the mobile home and the testing at the ATS laboratory. The manufacturer of the dryer sent an investigative team, despite the fact that the dryer was at least 20 years old, and under Georgia law, manufacturers cannot be sued for product defects after the unit is 10 years old. The owner of the mobile home was also represented, as was the manufacturer of the water heater valve, which, it turned out, was under a recall because some of the valves failed to close properly (see the section on water heaters in Chapter 6 for a detailed description of the failure mechanism that caused this recall).

It was possible to test the water heater by providing it with a gas supply. The test is shown in Figure 7.6j. A less than optimal flame was observed, and while it eventually shut down when the thermostat was satisfied, it did take a long time to do so. Examination of the underside of the burner, shown in Figure 7.6k, revealed some evidence of candle-like burning, as well as a light deposit of carbon directly above the orifice. It was possible, therefore, to show that this water heater was defective, but there did not appear to be a causal link between the marginal defect and the fire.

Attention next focused on the dryer and its power cord. Figure 7.6l shows two holes in the floor next to the dryer. One hole had served as the exhaust vent, and the second hole was located under a cardboard box. The initial investigators

Figure 7.6 (a) Rented mobile home that was the site of a dryer fire that caused eight fatalities. (b) U-shaped burn pattern on the skin of the mobile home, directly behind an electric dryer.

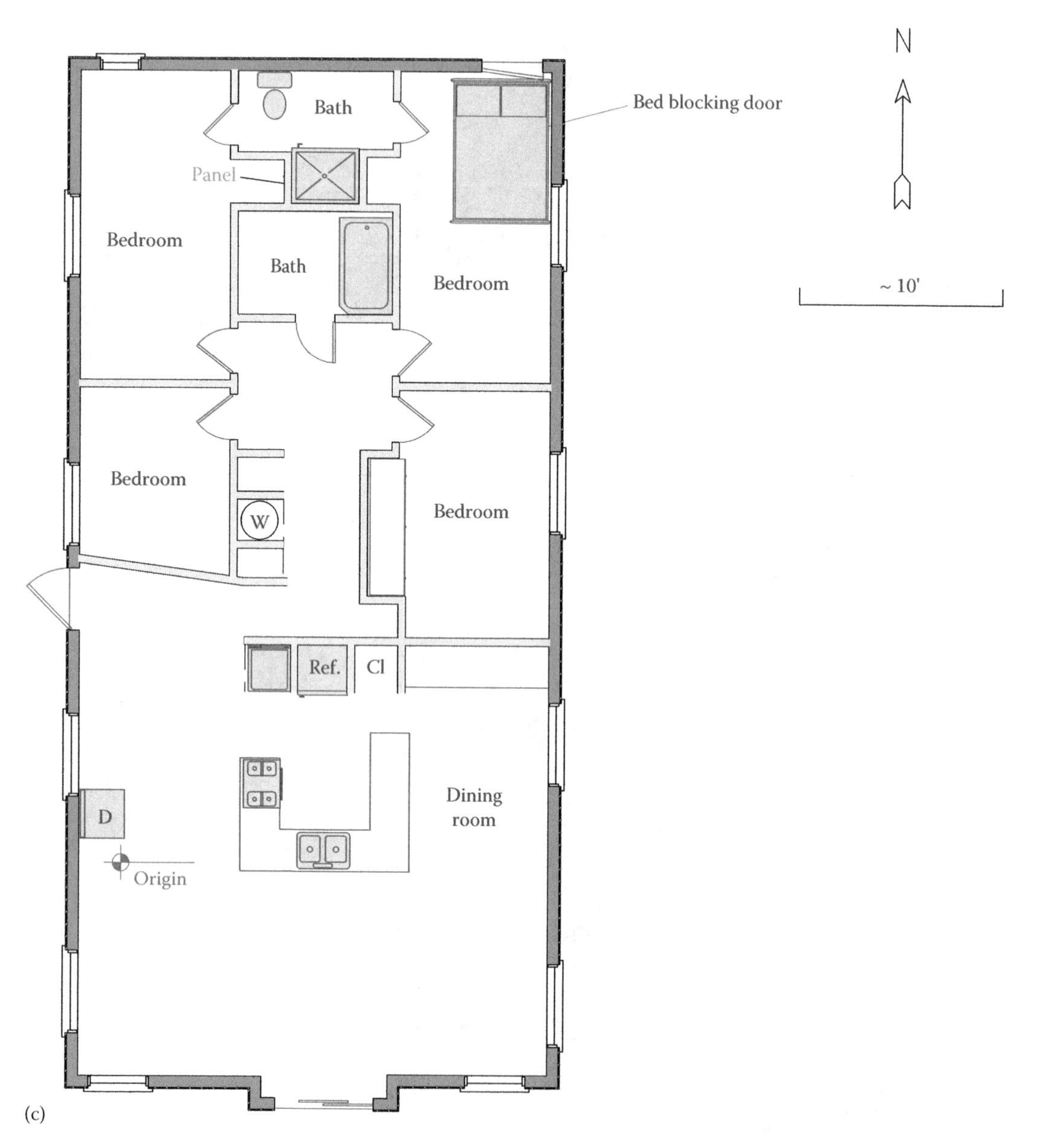

Figure 7.6 (c) Floor plan of the mobile home.

Figure 7.6 (d) View of the dryer from the living room. (e) View from the location of the dryer, looking down the hallway toward the bedrooms.

Figure 7.6 (f) Water heater that was examined as a potential cause of the fire.

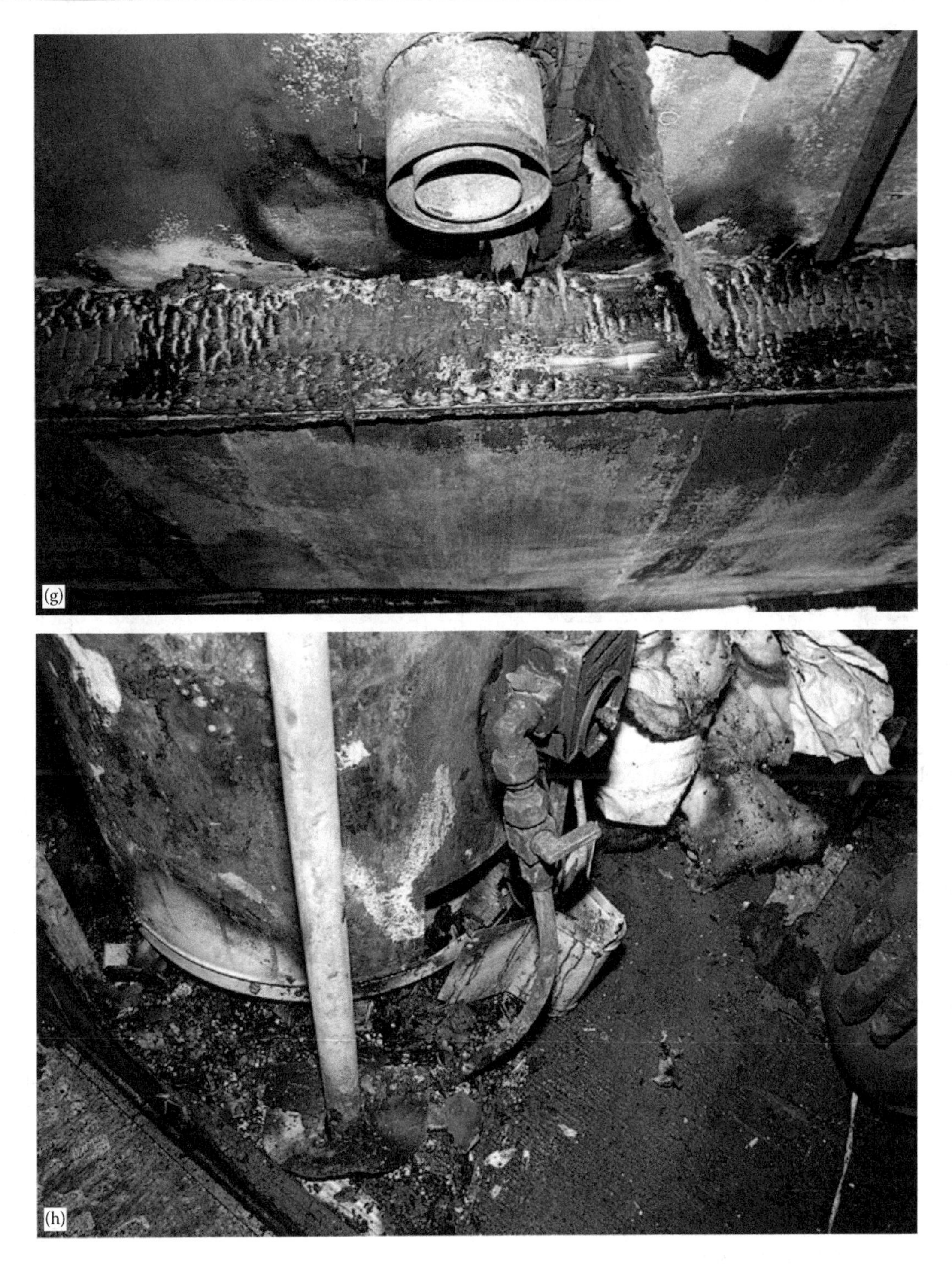

Figure 7.6 (g) Damage to the roof structure directly above the water heater. (h) Fire pattern around the combustion chamber. The undamaged paint was protected by clothing that was jammed into the water heater closet between it and the wall.

Figure 7.6 (i) Carbon deposit on the baffle from the water heater flue.

from the county fire department had moved the power cord but not before properly documenting its location. There were splices in all three conductors directly over the hole under the box. Figures 7.6m and n shows one of these splices, which exhibited localized melting of the copper conductor. This spliced power cord, which presented a point of high resistance in a high-current application, represented a competent ignition source in the presence of a fuel supply that was located in an area that could reasonably be determined to be the point of origin. This splice was ultimately determined to have been the cause of the fire.

The initial media reports stated that the people who died in this fire did so because there was no smoke detector in their mobile home, but the owner of the mobile home insisted that he had provided two smoke detectors. One of them, shown in Figure 7.6o, was found during the reconstruction of this fire scene. There is no evidence that the smoke detector failed, and it is probably what alerted the residents to the existence of the fire, but they had no viable means of escape.

Epilogue: This fire presented two potential ignition sources: (1) the dryer cord and (2) the water heater. The defect in the water heater required careful study, but the typical results of the defect were not evident in this fire scene.

As often happens, the victims of this fire were at least partially responsible for the outcome. Using a spliced cord in a high-current application was unwise, as was the blocking of the secondary means of escape. Such poor choices are frequently made because they are expedient, and the people involved lack the resources to make safer choices. In the end, no litigation was joined.

Figure 7.6 (j) Water heater, as it was set up for a functional test in the ATS laboratory.

Figure 7.6 (k) Underside of the water heater burner, showing some evidence of malfunction. Compare this burner to the one shown in Figure 6.34b. (l) Two holes burned in the floor at the location of the dryer. The hole against the wall was the vent outlet, and the hole at the left side of this photograph was at the location of the splice in the dryer power cord.

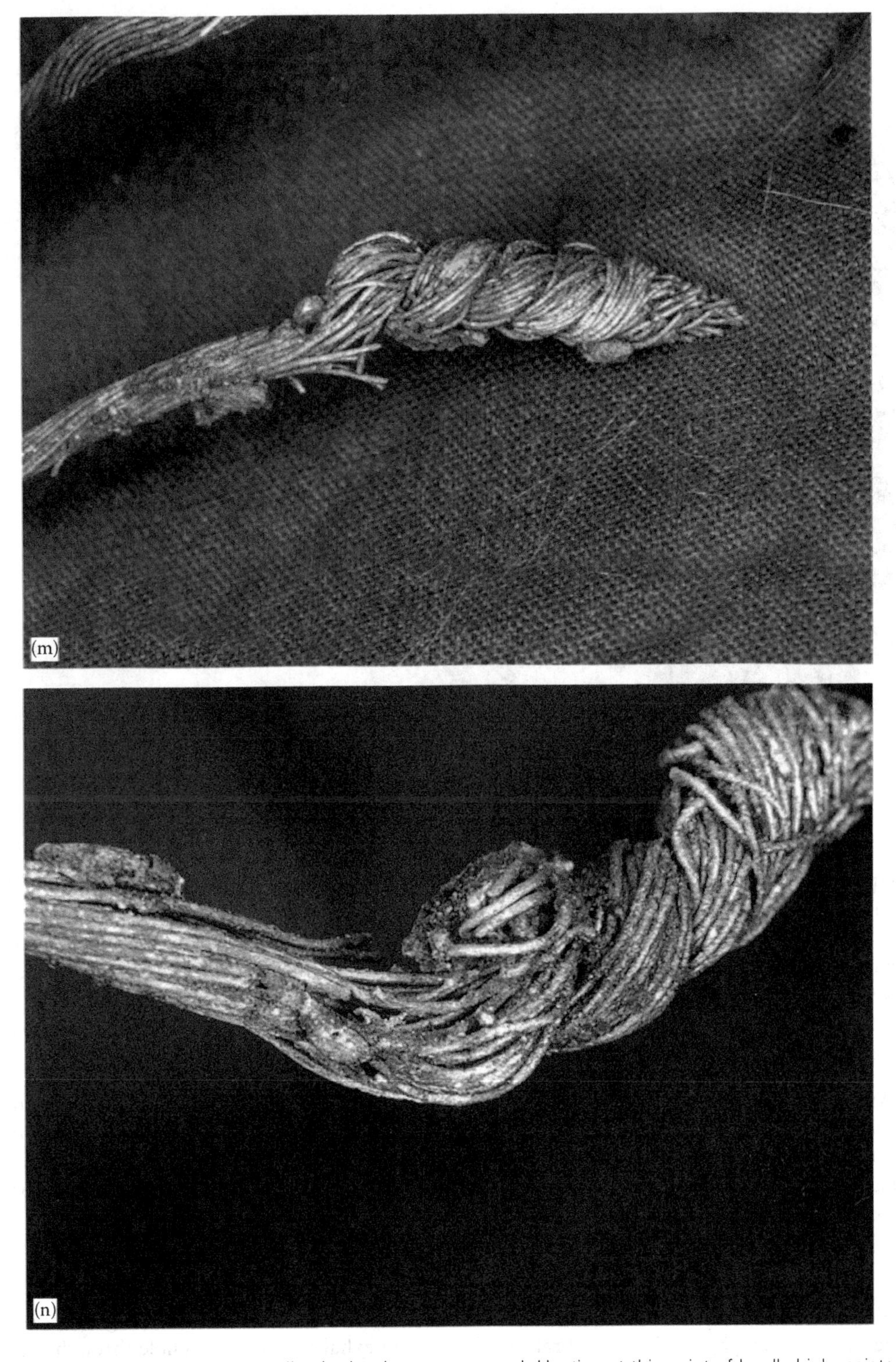

Figure 7.6 (m) Copper bead located at a splice in the dryer power cord. Heating at this point of locally high resistance was the ignition source for the fire. (n) Close-up view of the melted splice at the origin of the fire.

Figure 7.6 (o) Smoke detector found during the reconstruction of the fire scene.

7.3.4 Dryer fire 4: Internal power wire comes loose

ATS was requested to investigate the origin and cause of a fire that seriously damaged a residence in upper east Tennessee. The fire occurred in early March 2002 in mid-afternoon while the homeowner was sitting in the den with her son. Figure 7.7a shows the most heavily damaged part of the residence: the master bedroom located at the southeast corner. A floor plan of the residence is shown in Figure 7.7b. Using movement and intensity patterns, the fire was traced back to its origin in a laundry room at the southwest corner of the master suite. The washing machine and dryer are shown in Figure 7.7c. When these appliances were pulled away from the wall, the V-shaped fire pattern shown in Figure 7.7d was revealed.

The homeowner stated that she had been drying a load of clothing when the fire broke out, so both the washing machine and dryer were returned to the ATS laboratory for further inspection. Because the dryer was known to be more than 10 years old, and because it had been "worked on" several years earlier, the manufacturer was not put on notice of a potential claim. Figure 7.7e shows the dryer at the beginning of the inspection, and Figure 7.7f shows the interior of the dryer with the back panel removed. It was noted that the power cord was properly routed through the bulkhead using a strain relief device, but arcing was noted on the power supply cord downstream of the terminal block, as shown in Figure 7.7g.

The cause of the fire is shown in Figure 7.7h. Apparently, sometime during the previous few years, the plastic support that held the internal power cord in place either fell out of its hole or cracked, allowing the power cord to lie on top of the exhaust vent. Eventually, the vibration of the warm exhaust vent wore a hole through the power cord, resulting in the damage shown in Figure 7.7h.

Epilogue: The wires inside dryers are frequently supported with plastic clamps, which are subject to embrittlement and cracking. This author has previously seen two nearly identical dryer fires that resulted when internal power conductors dropped from their clamps and were abraded by the rotation of the dryer drum, resulting in the deposition of copper droplets around the circumference of the drum. The abrasion or pinching of internal power conductors is the manufacturing defect that this author has seen more frequently than any other.

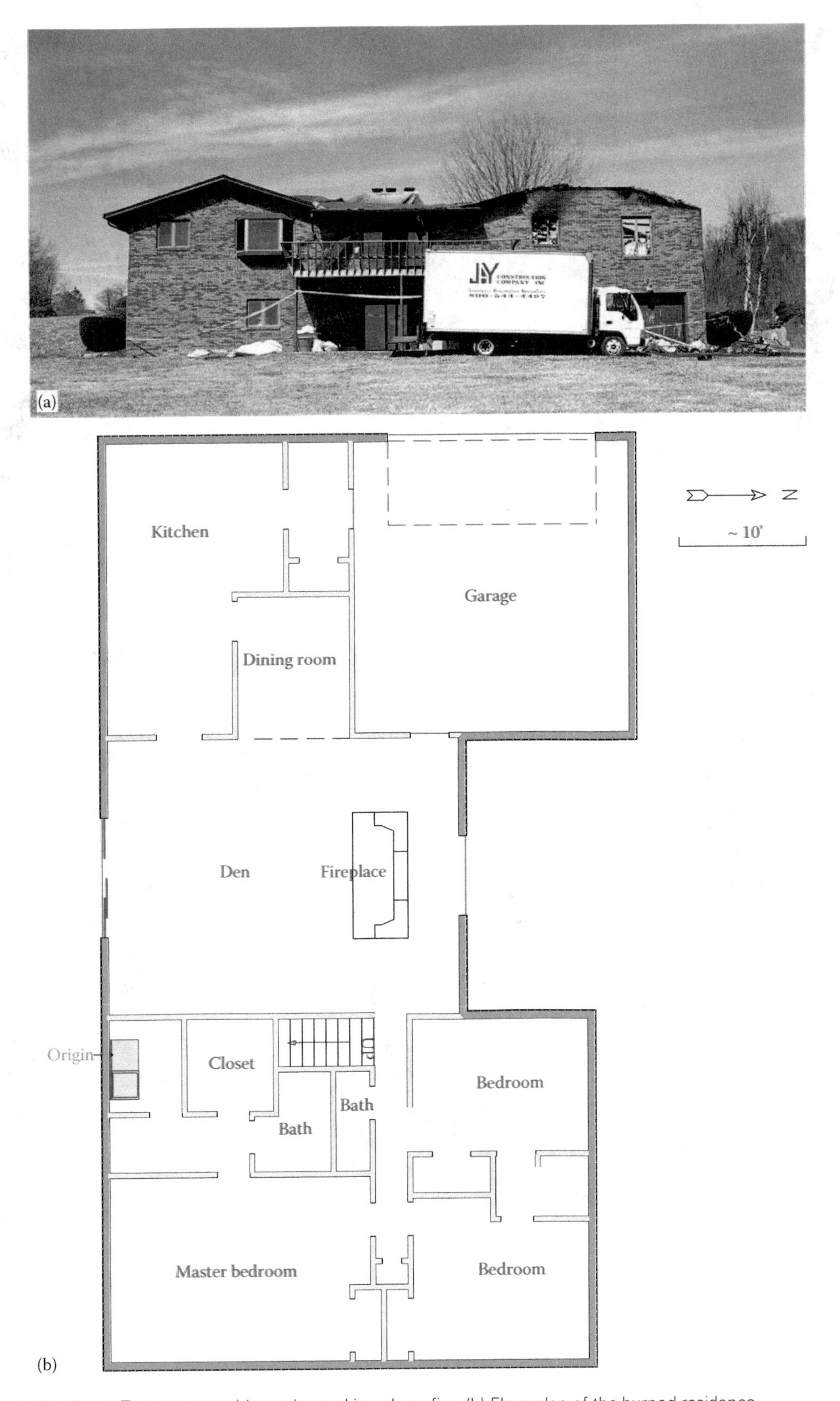

Figure 7.7 (a) Northeast Tennessee residence burned in a dryer fire. (b) Floor plan of the burned residence.

Figure 7.7 (c) Washing machine and dryer at the origin of the fire. (d) V-shaped burn pattern behind the location of the dryer.

Figure 7.7 (e) Dryer, after removal to the laboratory facility.

Figure 7.7 (f) Removal of the back panel allowed inspection of the interior of the dryer.

Figure 7.7 (g) Pigtail strain relief and terminal block. The power cord to the right was damaged at the point where it passed over the exhaust duct. (h) Hole arced in the exhaust duct because of contact with an energized conductor.

7.4 ELECTRICAL FIRES

The loss of electrical power that follows a natural disaster such as a hurricane makes us realize just how much we depend on electricity as an energy source. At any given moment, millions of megawatts of power are flowing through billions of conductors and connections. Each one of these connections represents an opportunity for a malfunction, so it is not surprising that many fires of undetermined cause have been written off as "electrical." Given the number of opportunities for malfunctions, however, it is surprising that the number of electrical fires is not much greater. In general, our electrical systems are sufficiently well designed that electrical mishaps result not in fires but in the opening of a circuit, which temporarily removes the danger. Even when the electrical event is itself a competent ignition source, frequently no fuel is available, so the only thing damaged is the part of the electrical system where the malfunction occurs. The fires reported in this section are cases in which the protective devices were not sufficient to prevent a fire or where the event was not the kind for which protective devices are designed.

7.4.1 Electrical fire 1: Energized neutral

ATS was requested to investigate the origin and cause of a fire that seriously damaged a residence located in middle Tennessee. The fire occurred in March 2002 at about 4:00 in the morning and resulted in damage to two residences, both served by the same pole-mounted transformer. One couple was awakened by what they believed to be an explosion, and their neighbor happened to be awake, watching television, when she heard a loud "bang" in her neighbor's backyard. She looked out her bedroom window and saw a glow around the transformer on the utility pole. When she looked back at the TV, it was off. There was no lightning that evening, although high winds were prevalent.

The seriously damaged residence is shown in Figure 7.8a. Very heavy damage was present on the main level of the house, as can be seen in Figure 7.8b, but an analysis of the fire patterns led to an origin in the basement near the washing machine, which is shown in Figure 7.8c. After the debris was removed and the washing machine was pulled away from the wall, an area of arcing was seen on the back panel of the washing machine control cavity, shown in Figure 7.8d. Damage was seen on the washing machine power cord and the receptacle where it was plugged in, which included melting of the ground pin at the point where it plugged into the receptacle (Figure 7.8e). The fire progressed upward after laundry, and other materials stored in the area were ignited, and it burned through the floor directly above the washing machine, which allowed the fire to spread to the main level. The hole in the floor directly above the dryer is shown in Figure 7.8f.

The damage in the neighbors' house was considerably less severe, although all of their electronics were rendered inoperative by the event. A nightlight exploded, and the receptacle where the nightlight was plugged in was scorched. Figure 7.8g shows the remains of the nightlight, and Figure 7.8h shows the receptacle.

The only explanation for this series of events is that the neutral circuit became energized as a result of contact between an uninsulated high-voltage conductor and an uninsulated neutral conductor upstream of the transformer. A branch falling from the tree shown in Figure 7.8i and causing the two conductors to momentarily contact each other most likely caused this to occur. This would account for the explosions heard by both neighbors.

Epilogue: This was a very unusual case and one of the few that this author has seen involving a failure of the power supply *upstream* of the meter. The damage was similar to that sometimes seen in lightning fires but considerably less violent. The fact that two adjacent homes supplied by the same transformer were damaged by the same event proved that the failure was in the external power supply and not in one of the residences.

7.4.2 Electrical fire 2: Worn-out outlet

ATS was requested to investigate the origin and cause of a fire that seriously damaged a four-unit apartment house located in upper east Tennessee. This building was originally a large, single-family dwelling that was converted into four apartment units, two upstairs and two downstairs. The fire occurred in late April 2001 and was reported at 11:30 a.m. by a passing motorist. One of the downstairs apartments was vacant, and all three of the current tenants were at work when the fire broke out. The front of the building is shown in Figure 7.9a. Pinpointing the room of origin was not difficult. The front room at the right side of the house was the only room that became fully involved, and all

Figure 7.8 (a) Tennessee residence damaged by a fire that started when the neutral circuits in the house were energized. (b) Area of heaviest fire damage on the main level, directly above the laundry room.

Figure 7.8 (c) Washing machine at the origin of the fire. (d) Hole melted in the rear panel of the washing machine control cavity.

Figure 7.8 (e) Electrical melting on the ground pin and receptacle behind the washing machine.

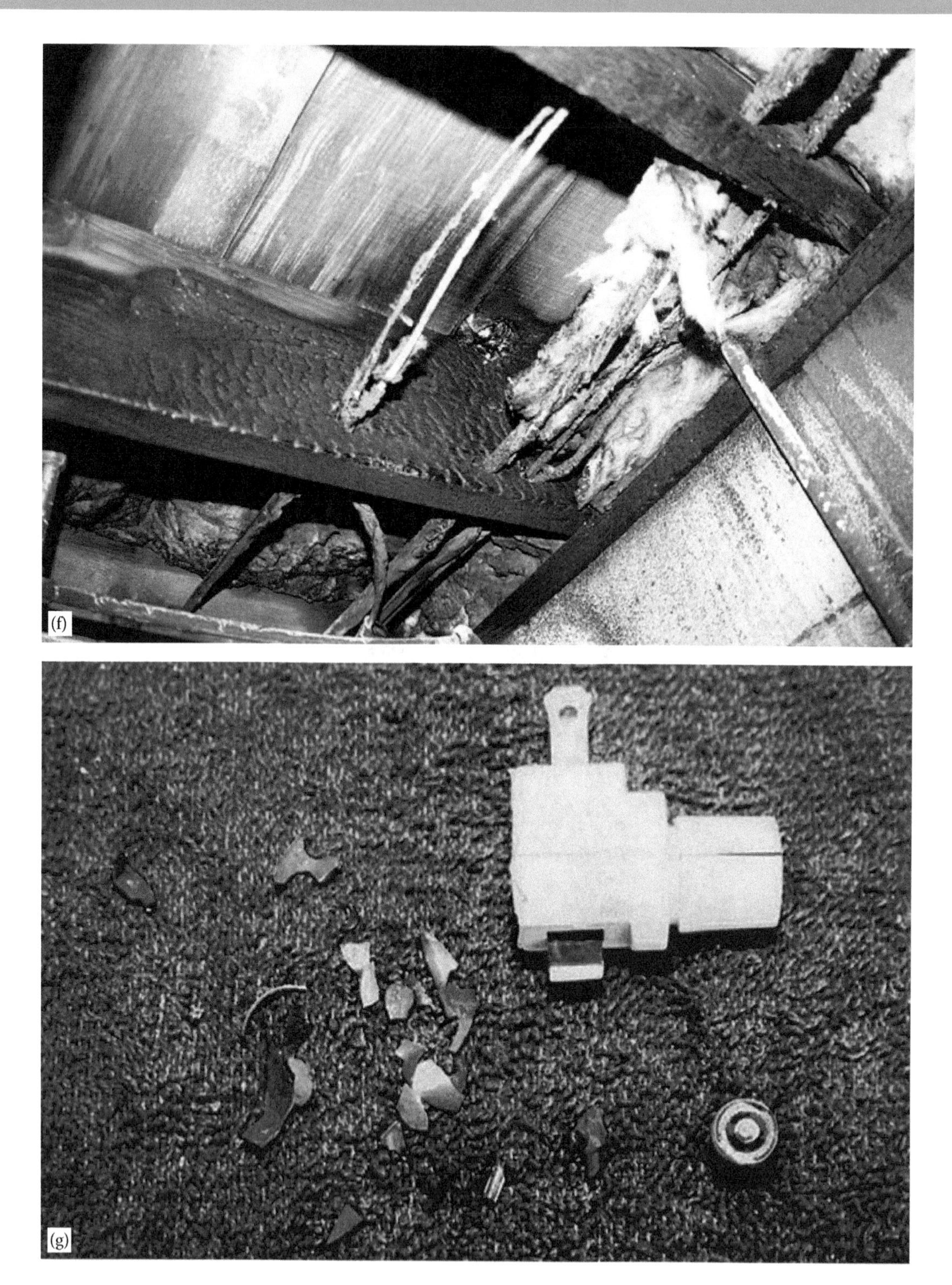

Figure 7.8 (f) Area of fire damage directly above the washing machine where the fire penetrated to the first floor. (g) Remains of an exploded nightlight from the residence next door.

Figure 7.8 (h) Damage on the receptacle where the nightlight had been located. (i) Utility pole where the transformer was mounted.

Figure 7.9 (a) East Tennessee apartment house damaged by a fire that started at an electrical outlet. (b) View into the living room where the fire started.

Figure 7.9 (c) The origin of the fire was under the boarded-up window on the right. (d) Torchiere lamp and portable space heater found in the front yard.

Figure 7.9 (e) Remains of the outlet where the fire started. (f) Heater control turned to the "high" position, showing welded contacts.

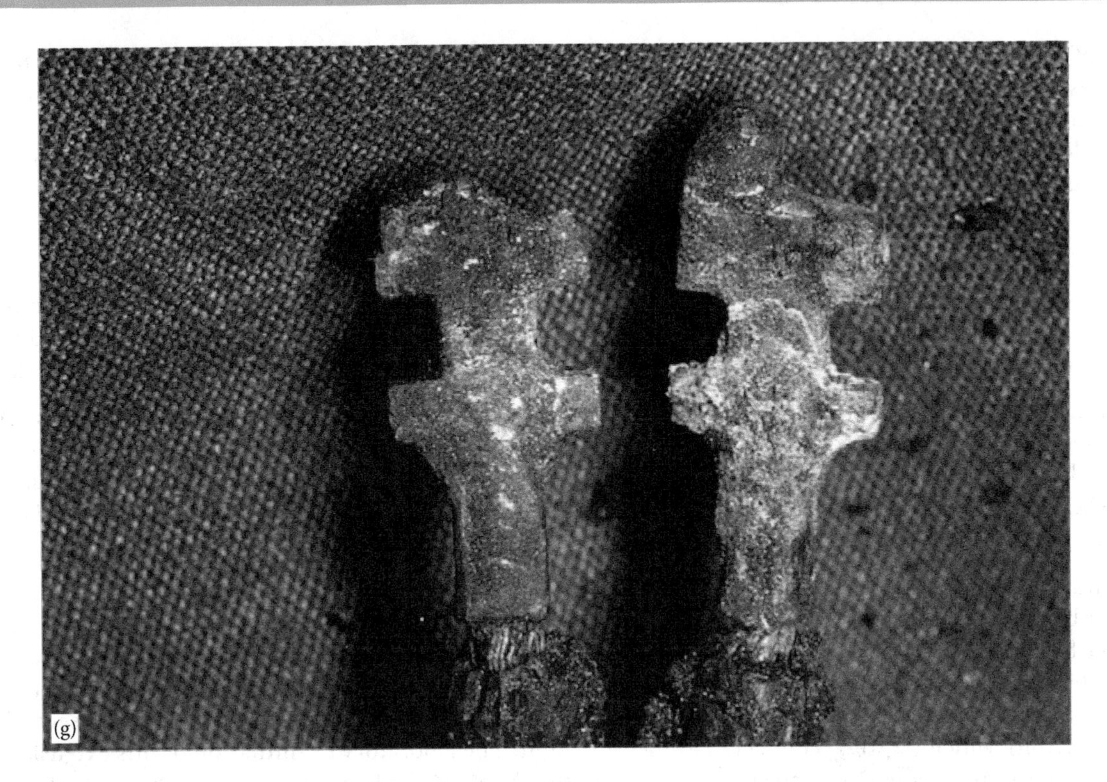

Figure 7.9 (g) Melted ends of the male prongs on the heater plug.

of the damage could be accounted for in terms of an origin in that room, which is shown in Figures 7.9b and 7.9c. Because of the extensive damage, however, pinpointing the cause was more problematic. *Every* potential ignition source in the room required examination. Two potential ignition sources were of particular interest: a space heater and a torchiere lamp thrown out onto the front lawn by firefighters, shown in Figure 7.9d. The occupant of the apartment had provided a written statement to city fire investigators to the effect that she had turned everything off prior to leaving for work. Certainly, it would have been obvious to her if the torchiere lamp were on, but it would have been less obvious if the space heater were inadvertently left on. A single circuit supplied the living room. Only the power cords for the space heater and the torchiere lamp exhibited any evidence of electrical activity. Figure 7.9e shows the remains of the duplex receptacle to which these devices were connected. The left receptacle was completely destroyed. The right receptacle showed evidence of arc-severing of the torchiere lamp cord, indicating that when the fire reached it, this receptacle was still energized.

The heater and lamp were returned to the laboratory for a closer examination. Figure 7.9f shows the temperature control for the heater, which indicates that it was turned on high. Figure 7.9g shows the ignition source for this fire. These prongs are from the male end of the heater power cord, which was still attached to the heater. A loose connection existed between the prongs and the very old receptacle, resulting in overheating at the connection. The sofa in front of the receptacle provided a very large and convenient fuel package.

Epilogue: Most electrical fires are caused by loose connections, and old receptacles are a common location where loose connections can occur.

None of the tenants had renter's insurance, and because this fire was the result of a defect in the building, the landlord was concerned that the tenants might look to him for compensation. In this state, however, for a tenant to sue the landlord, the tenant must prove that the landlord was on notice of a defective condition and failed to repair it. No evidence was brought forward to indicate that the landlord had any idea that the old receptacles in the house were unsafe.

7.4.3 Electrical fire 3: Makeshift extension cord

ATS was requested to investigate the origin and cause of a fire that seriously damaged a residence located in suburban Atlanta in mid-summer 2001. The fire was reported at about 1:00 p.m., and the homeowner stated that he had come home from work for lunch and had left at about 12:15 p.m. A neighbor, who called 911, reported the fire.

The front of the residence is shown in Figure 7.10a, and a floor plan is shown in Figure 7.10b. The fire clearly started in the basement, where the homeowner had been installing some new wiring that was not yet hooked up to the service panel. The fire spread to the upper level, mainly via a plumbing and HVAC (heating, ventilation and air conditioning) chase behind the west wall of the kitchen, shown in Figure 7.10c and d. The fire burned a hole in the roof above this chase.

This was not the only avenue by which the fire spread upward. Figure 7.10e shows a minor penetration of the master bedroom closet from below. This unusual burning focused our attention on the area directly below the master bedroom closet, shown in Figure 7.10f. A shop light was located in this area, and because the wiring in the basement was incomplete, the light was powered through an orange extension cord that ran through a doorway to the front basement room (Figure 7.10g). The orange extension cord was connected to a second cord (Figure 7.10h), which was then connected to the shop light. The laboratory examination of this extension cord proved instructive. Figure 7.10i shows the area where the shop light had been located and the heavy damage to the floor structure directly above it.

The orange extension cord was found plugged in, and one end of it was found to have arc-severed, as shown in Figure 7.10j. The homemade extension cord, shown in Figure 7.10k, was made with bell wire, a thin solid conductor that is wholly unsuitable for use in 120-V applications, particularly as an extension cord. The termination of one of these wires at the receptacle is shown in Figure 7.10l, and Figure 7.10m shows one of several loosely twisted connections in the power cord. Examination of the shop light showed no evidence of overheating or other malfunction. The fire is believed to have started at one of the loosely twisted connections, and the first fuel ignited was the moisture barrier on the insulation in the ceiling.

Epilogue: Household electrical systems are not very complicated. Attaching makeshift devices, however, evidences a lack of appreciation of the potential hazards of 120-V systems.

7.4.4 Electrical fire 4: A failed doorbell transformer

ATS was requested to investigate the cause of a fire that started in the attic over the kitchen in a middle Tennessee residence. The homeowners first noticed that the lights went out in their kitchen. They found a tripped circuit breaker, which they reset immediately before they noticed that the attic was on fire. Despite a rapid response from the fire department, tens of thousands of dollars in damage occurred to the roof structure and the living space. Figure 7.11a shows the origin of the fire over the kitchen, and Figure 7.11b is a closer view of the origin. The attic was filled with blown-in cellulose insulation. Buried in that cellulose insulation were a transformer and a junction box. The transformer took its power supply from the junction box but was not mounted on the junction box, as required by the *National Electrical Code*. Both of the wires from the junction box were arc-severed at the location of the transformer. Figure 7.11c shows the junction box and associated wiring after removal from the ceiling. Figure 7.11d shows the arc-severed end of one of the transformer leads.

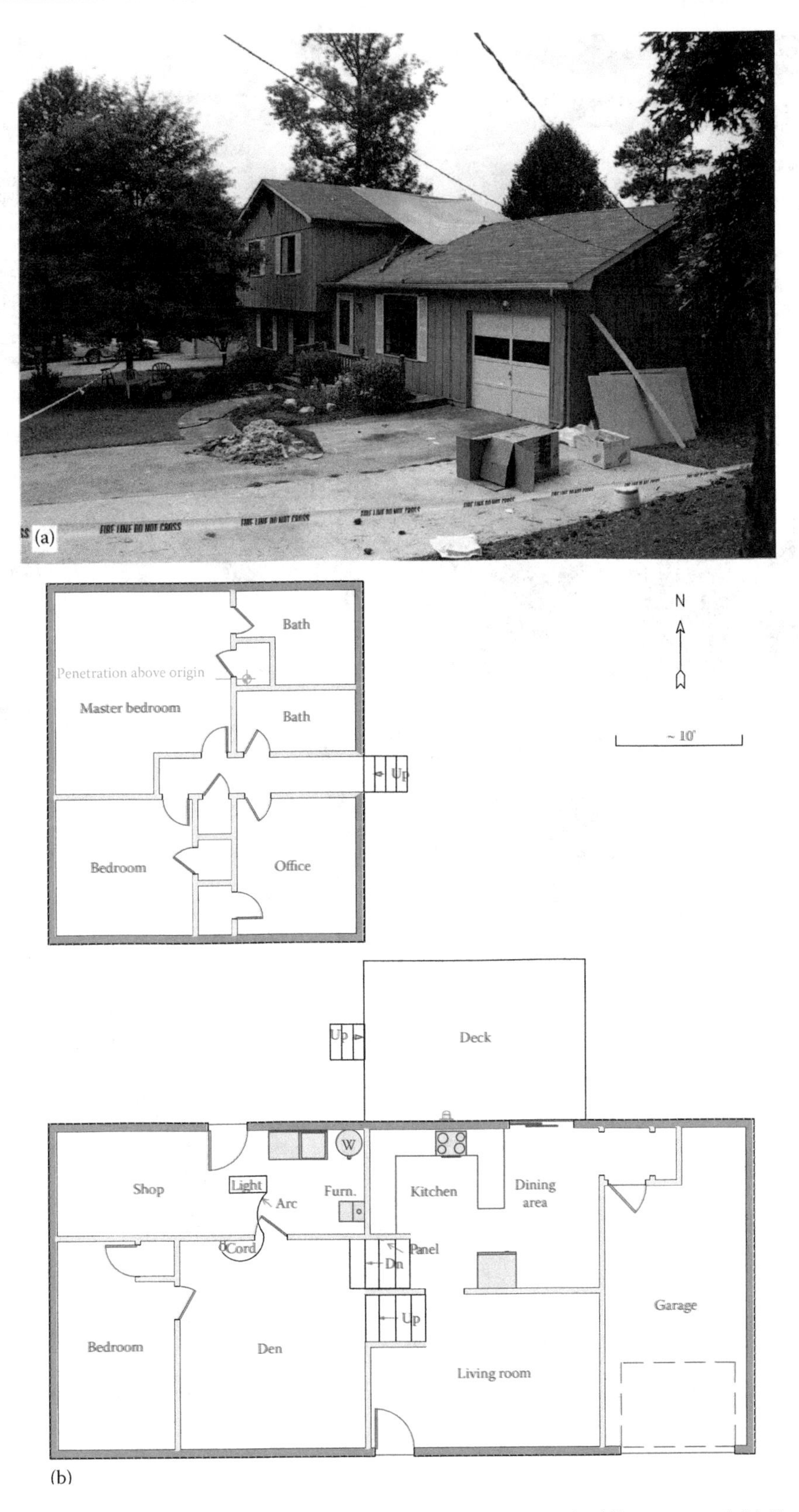

Figure 7.10 (a) Suburban Atlanta residence damaged by a fire that started on a makeshift power cord. (b) Floor plan of the damaged residence.

Figure 7.10 (c) Path of the fire behind the kitchen wall. (d) Close-up of the damage in the chase behind the kitchen wall.

Figure 7.10 (e) Fire damage in the master bedroom closet, directly above the origin. (f) Origin of the fire in the utility room.

Figure 7.10 (g) Path of the orange extension cord run through a doorway. (h) Shop light plugged into the makeshift extension cord.

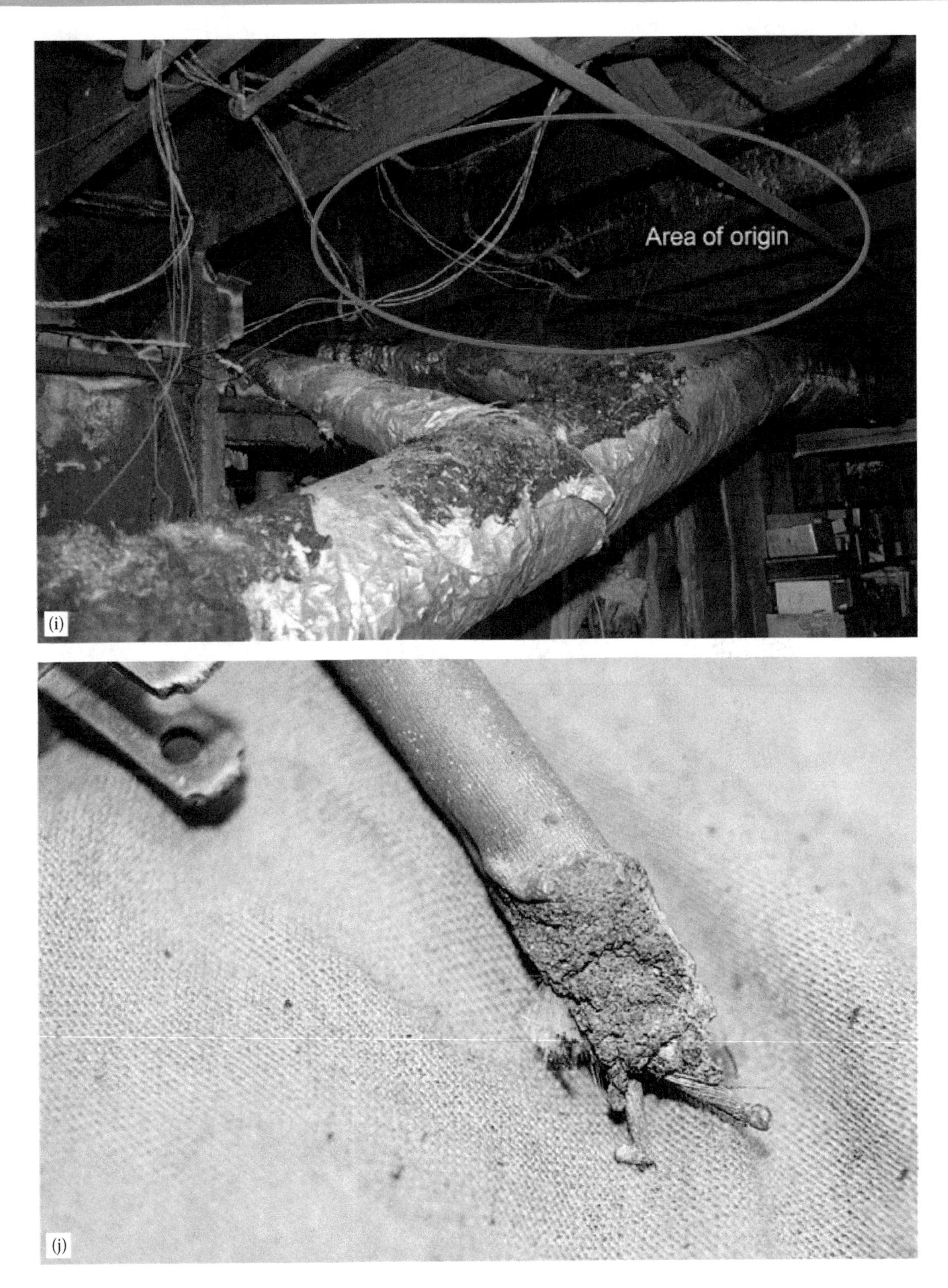

Figure 7.10 (i) Damage above the location of the shop light, which was removed prior to this photograph being taken. (j) Arc-severed end of the orange extension cord.

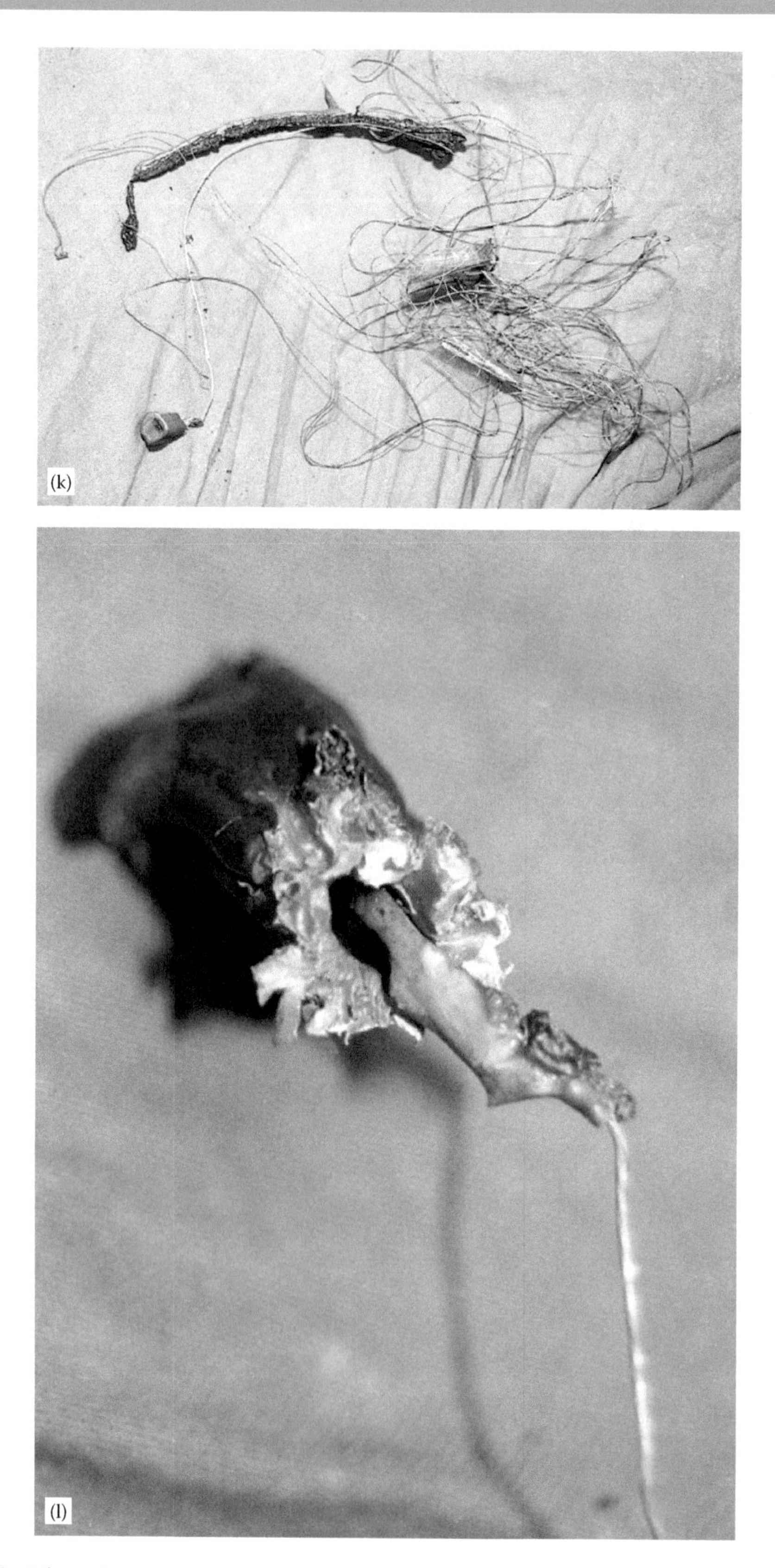

Figure 7.10 (k) Makeshift conductors used as a cable to supply the shop light. (l) One of several loosely twisted connections in the makeshift power supply cord.

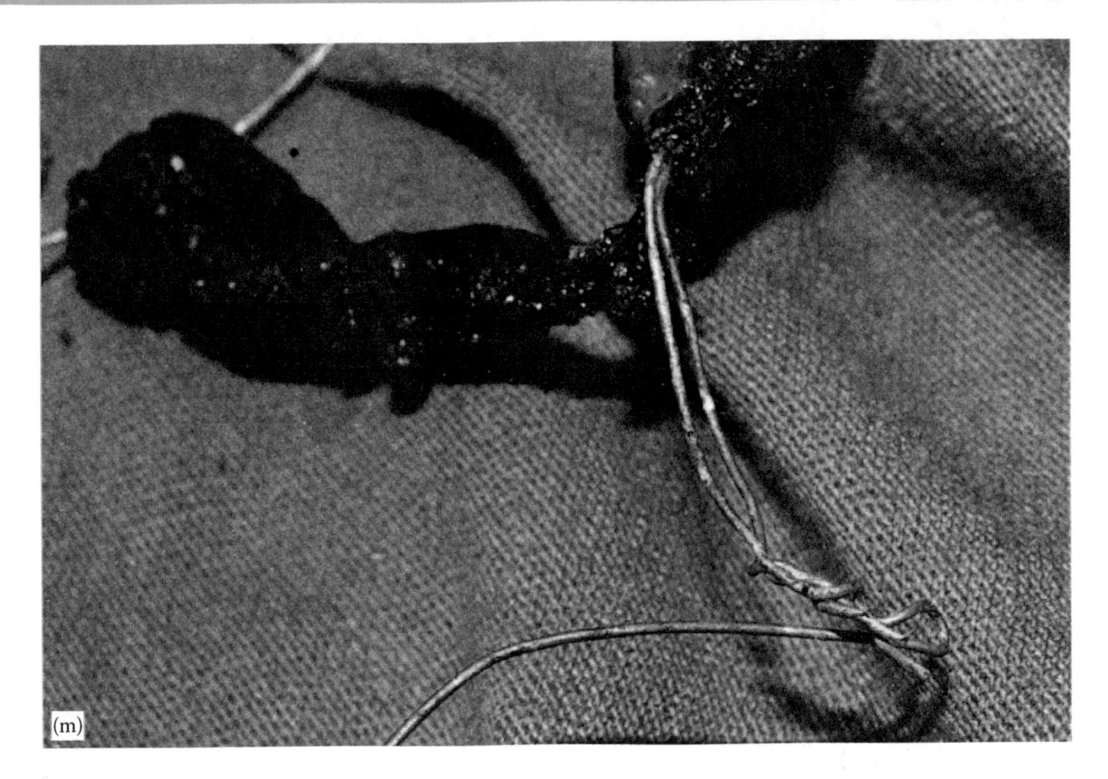

Figure 7.10 (m) Termination of one of the bell wires at the cord receptacle.

The transformer was taken back to the laboratory and disassembled, and the cause of the fire could be seen. Figure 7.11e shows an area of shorting of the transformer windings. The most likely cause of the failure was deterioration of the varnish insulation on the windings. The deterioration may have been hastened by the presence of the cellulose insulation. Any heat produced by the idle transformer would have been retained in the windings, instead of being radiated away. The arc-severing of the transformer pigtail was probably what caused the breaker to trip, but by that time, the insulation was most likely already burning. Had the transformer been mounted on a junction box, as it should have been, it would have been above the level of the insulation, and the only consequence of the transformer failure would have been the loss of the doorbell. Another contributing factor in this fire may have been the location of the transformer near the dormer/roof joint. This was the site of a roof leak several years prior to the fire. Blown-in cellulose insulation is required to be treated with a flame retardant, but this flame retardant is usually boric acid, which is soluble in water. Cellulose insulation that has been wetted is less likely to contain the required amount of flame retardant.

The doorbell chimes were connected to another transformer and found to work perfectly. The possibility that a short circuit on the doorbell wire caused the transformer failure was eliminated by testing the doorbell chimes with the exemplary transformer. When energized continuously for 30 min, the chime solenoid warped permanently. Thus, because the chime was in working order when found, the possibility that a continuously energized solenoid caused the transformer failure can be excluded.

Epilogue: Further investigation revealed that the doorbell transformer was original equipment supplied when the house was new, some 20 years earlier. There was thus no possibility of subrogation.

Figure 7.11 (a) Origin of a fire caused by an overheated transformer. (b) Transformer and junction box, covered with blown-in cellulose insulation.

Figure 7.11 (c) Area of origin with some of the insulation removed. (d) Junction box and transformer wire stubs, after removal from the ceiling.

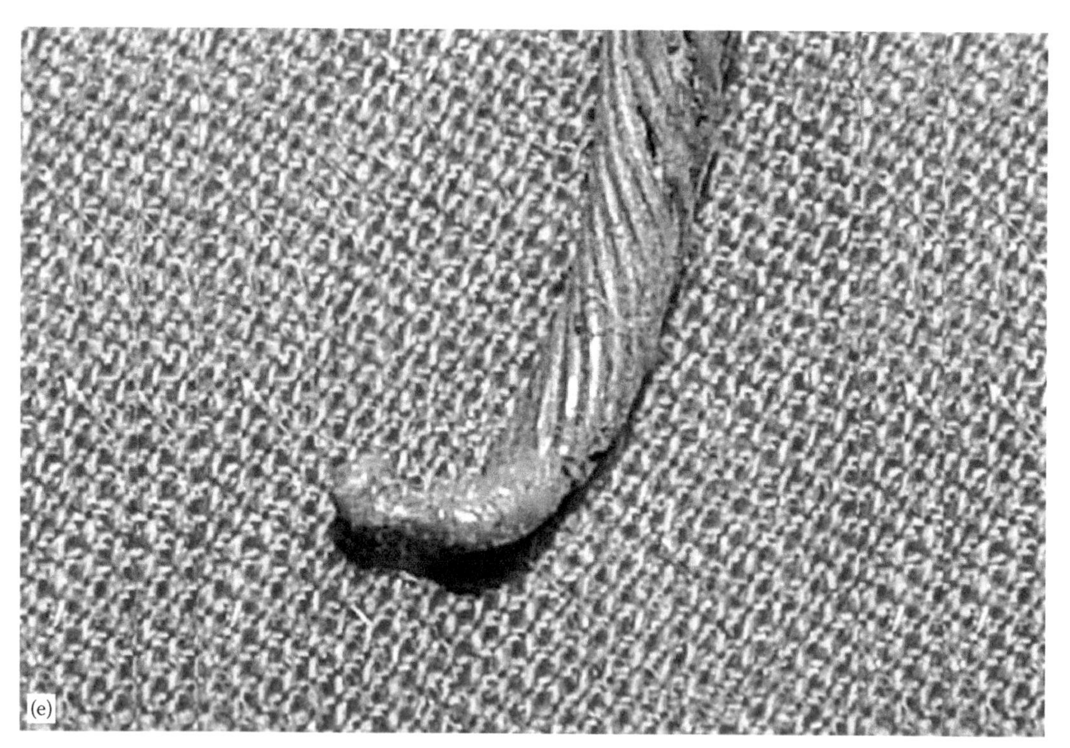

Figure 7.11 (e) Arc-severed end of one transformer lead. (f) Transformer windings, after removal from the case. The oval indicates the location of a short in the windings.

7.4.5 Electrical fire 5: The elusive overdriven staple

ATS was requested to investigate the origin and cause of a fire that occurred in a house under construction, located in northeast Georgia. The fire occurred in late October 2001 when two electricians were on the scene troubleshooting a tripped circuit breaker. The electricians identified the breaker as one protecting the refrigerator circuit, on which there were no known loads at the time. They reset the breaker and it tripped again, so the electricians began looking for a fault. Observing none when they examined the receptacle, they assumed that the circuit breaker was faulty, so they replaced it. When the new breaker was reset, it tripped again. At this point, the electricians hypothesized that a screw used to mount the kitchen cabinets had penetrated the circuit. They removed the screw and reset the breaker. This time, the breaker held for about a minute, and then it tripped again. At that point, three other breakers in the service panel tripped and the electricians noticed smoke, whereupon they flipped the refrigerator circuit breaker and the main breaker to the off position and called the fire department.

A joint inspection of the scene was conducted with investigators retained by the general contractor, the general contractor's insurance company, and the electrical subcontractor in attendance. Also present were the owner of the electrical subcontracting company, as well as the general contractor.

The front of the residence is shown in Figure 7.12a, and the area of damage visible from the outside is shown in Figure 7.12b. The top of the roof had burned off, and heavy damage was sustained by the upstairs "bonus room" shown in Figure 7.12c. Based on the eyewitness accounts and a careful search for electrical activity, the origin of the fire was determined as the front wall of the bonus room, located directly above the foyer.

Examination of the electric service distribution panel revealed that three breakers, one 15-amp breaker and two 20-amp breakers, were in the tripped position, and the 20-amp breaker for the refrigerator circuit was in the off position. The wires from all four circuits were traced, and to the surprise of everyone present, the arcing events that caused each of them to trip were identified. The refrigerator circuit exhibited two areas of arcing, while the other three circuits exhibited one apiece. These areas were marked by tying a piece of ribbon around the damaged area of the wire, as shown in Figure 7.12d.

Figure 7.12 (a) Front of the residence under construction, damaged by a fire caused by an overdriven staple.

Figure 7.12 (b) Rear of the residence, showing damage to the upper level. (c) Fire damage at the front of the bonus room on the upper level.

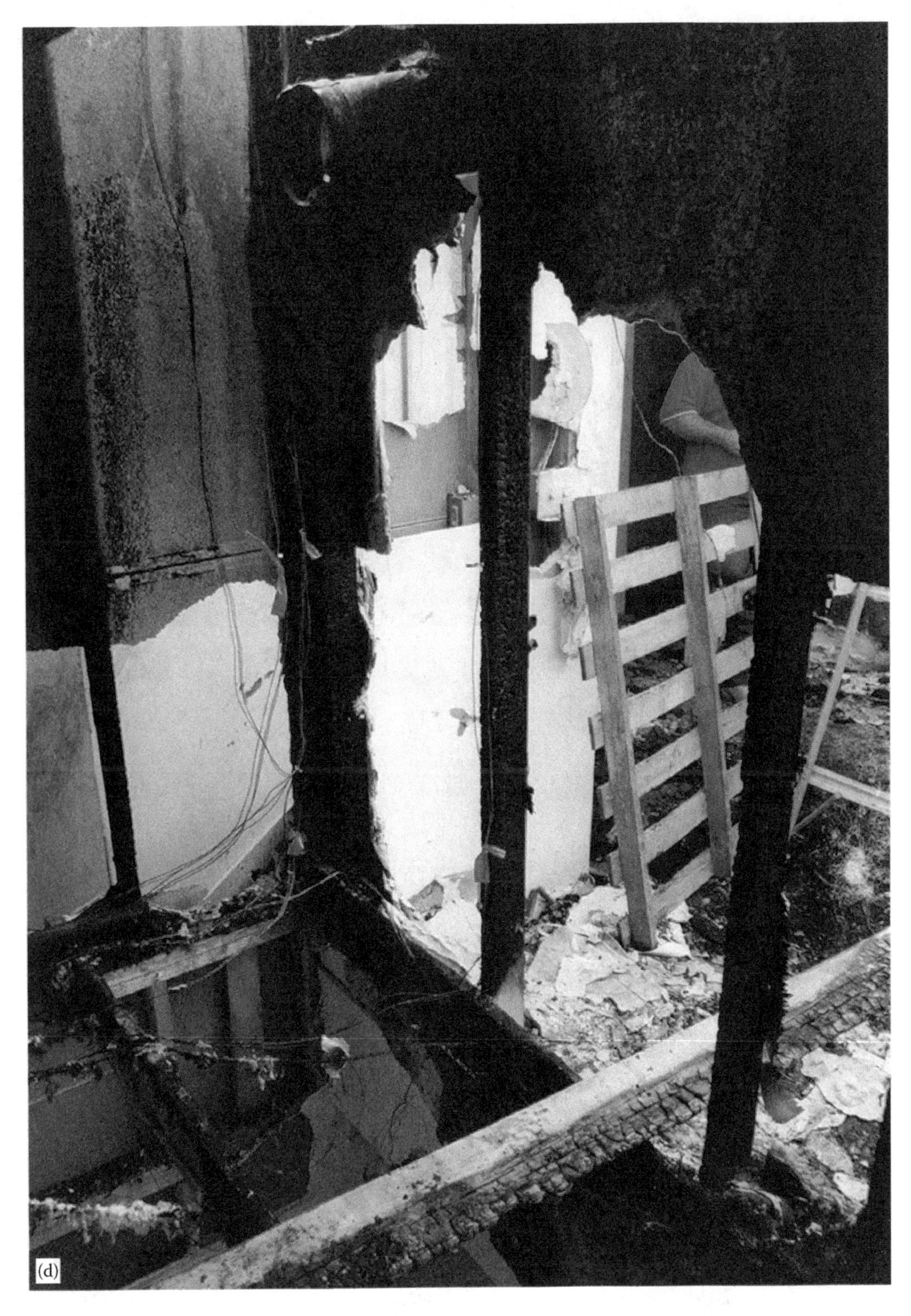

Figure 7.12 (d) Area of origin with five points of arcing damage marked by ribbons tied around the conductors.

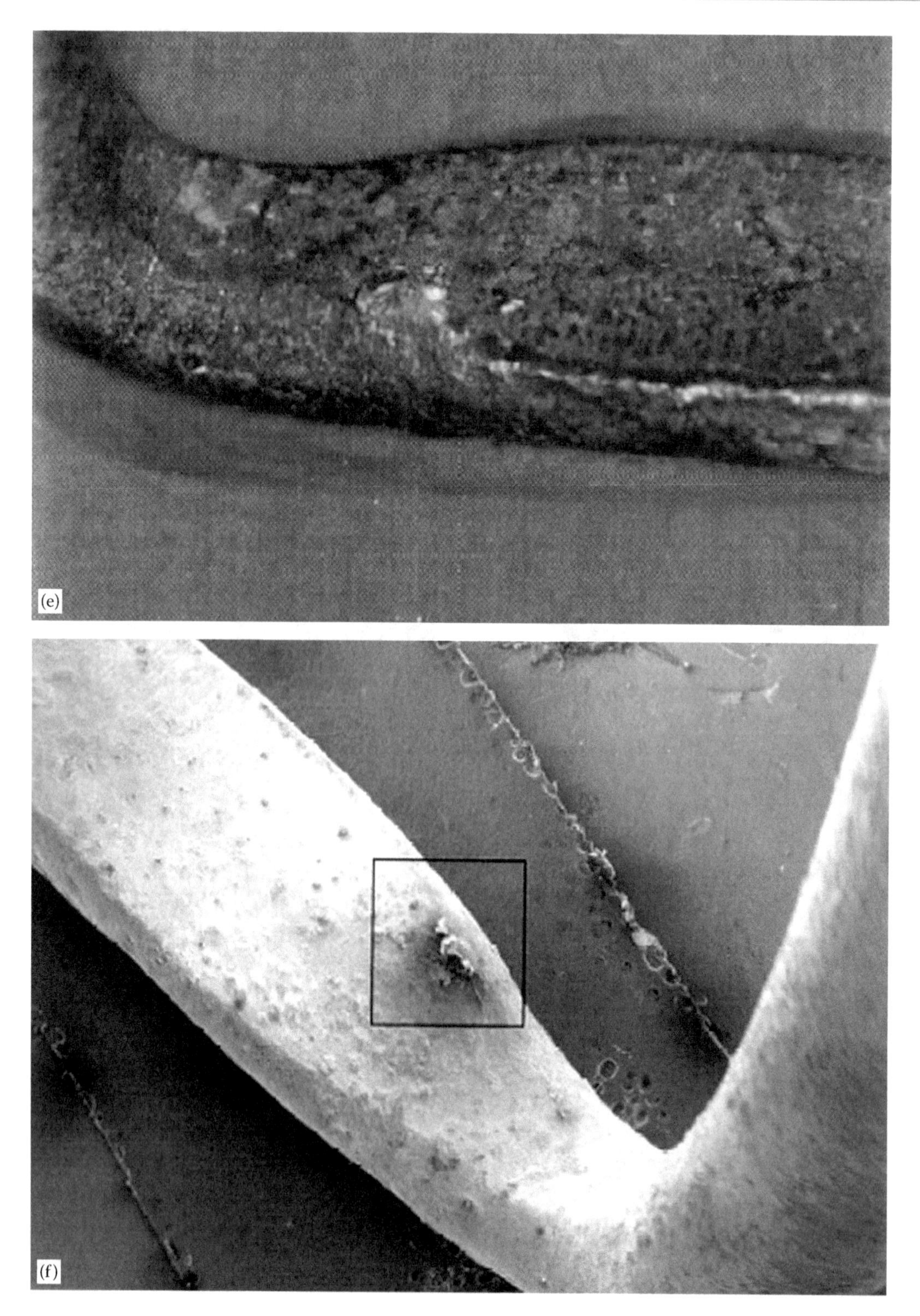

Figure 7.12 (e) Photomicrograph of the deposit on one of the staples found in the foyer. (f) Scanning electron micrograph of the staple, with a box around the area of the deposit.

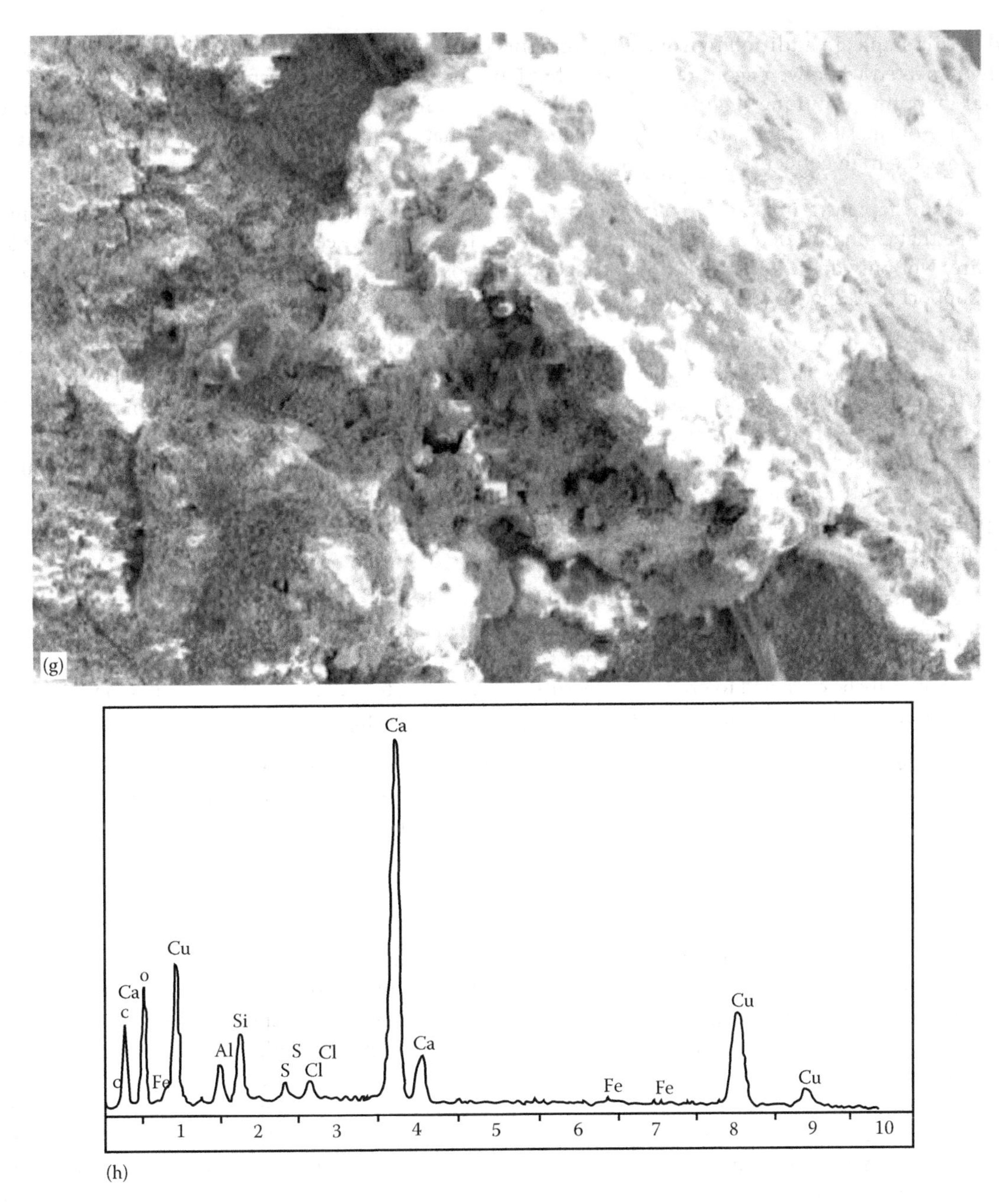

Figure 7.12 (g) Elemental map showing copper-rich area of the staple in green. (h) EDS spectrum of a copper-rich area of the deposit found on the staple.

An examination of the wiring technique in those parts of the house that were not damaged by the fire revealed several instances where staples were severely overdriven, causing damage to the insulation. The investigators hypothesized that a similar event had caused the failure of the refrigerator circuit, but several of the staples from the origin area had fallen out, landing on the floor of the foyer below. Using a magnet, we were able to recover three staples from the floor, and when viewed with a stereomicroscope, one of those staples exhibited a copper-colored deposit in the area shown in Figure 7.12e. Because small copper deposits and small deposits of iron oxide (rust) look very much alike, further analysis was required. Figure 7.12f is a scanning electron micrograph of the deposit seen under the stereomicroscope. Energy dispersive X-ray spectroscopy (EDS) was applied to this staple; in addition to iron, large amounts of calcium, oxygen, and silicon were observed. This is normally expected when we encounter samples that have been exposed to burned gypsum drywall. Using an elemental mapping technique (Figure 7.12g), we were able to locate an area where copper was present on the inner surface of the steel staple, and the EDS spectrum of that area is shown in Figure 7.12h. This part of the deposit showed a high concentration of copper and a very low concentration of iron.

The overdriven staple is a fire cause that is far more frequently hypothesized than demonstrated (and probably causes some that cannot be demonstrated); but in this case, the evidence seems to indicate no other possible cause for this fire. Not only does the staple exhibit the copper deposit but also the circumstances surrounding the fire strongly support the hypothesis.

In 1994, Ettling [4] conducted numerous experiments wherein he tried to drive staples through energized Romex wires and was unsuccessful in starting a fire. Babrauskas lists several possible mechanisms for an overdriven or otherwise improperly installed staple to cause a fire but makes clear that these mechanisms are "not all fully elucidated [5]." Two of the mechanisms described—(1) driving the staple so that one leg contacts the hot conductor and the other touches the ground and (2) driving the staple through a conductor so that it becomes part of the circuit—result in a fire almost immediately. A third mechanism, involving creep of the staple through insulation over a period of years, has been hypothesized but not yet demonstrated. The longest period reported between overdriving a staple and witnessing a fire is 15 hours.

The overdriven staple as a fire cause is only likely to be proved when the building is new and only when the actual staple, showing evidence of the ignition event, can be examined. The possibility that an advancing fire could produce the artifact we found in this case (i.e., copper deposited on the flat part of the staple) is not a likely one. The staple would initially be expected to provide a measure of protection to the wire underneath it. One would actually expect to see arcing on the wires in areas that were not protected by a staple.

Epilogue: Despite the abundance of evidence in this case, the attorneys retained on behalf of the electrical subcontractor mounted a vigorous defense at first. After some months of litigation, but before depositions were necessary, however, the liability carrier for the electrical subcontractor settled the claim.

7.5 FLUORESCENT LIGHT FIRES

Ordinary fluorescent lights generally fail uneventfully, but occasionally they start fires. They are accused of setting far more fires than they actually set, simply because they are ubiquitous. A fluorescent light is likely to be found within a few feet of the origin of any commercial fire, and most residences contain at least a few, particularly in workshops and utility areas. The most common failures involve either overheating of the ballast or overheating of the lamp holders. Since 1969, ballasts have been required by the *National Electrical Code* to be equipped with an overheat protective device. Most of the ballasts manufactured prior to 1969 have since gone out of service, so it will be unusual for a fire investigator to encounter a nonprotected ballast. Ballasts can fail just like any other transformer through an insulation failure and subsequent shorting of the windings, and it is possible for localized overheating to escape detection by the thermal cutoff (TCO). Unlike most transformers, however, potting material (pitch) surrounds ballasts to reduce the irritating buzzing sound they make. This potting material can provide a fuel source.

Lamp holders can overheat due to locally high resistance between the spring-loaded socket conductors and the pins of the fluorescent tube. This resistance can result from the spring losing its tension, from physical damage that causes the spring to come loose, or from a slight misalignment of the tube. Because fluorescent lights operate at 600 Volts (V), a user might not notice a high-resistance point in the circuit because the bulb continues to function, though a 60-cycle flicker may be noted in lamps supplied by magnetic ballasts. In electronic ballasts, which function at frequencies over 20,000 cycles per second, no such flickering is likely (Figure 6.43).

Another means by which lamp holders fail is through arc-tracking. This requires the presence of a conductive medium on the surface of the lamp holder that provides a path to a grounded surface. One type of each failure is described in the following sections.

7.5.1 Fluorescent light fire 1: A ballast failure

ATS was requested to investigate the origin and cause of a fire that occurred at a strip shopping center in eastern Tennessee in the summer of 2001. An employee standing at a cash register of the thrift store located in the middle of the shopping center saw a fluorescent light directly over her begin to shoot sparks. She ran to the back of the store and turned off the power and then called 911. The shopping center is shown in Figure 7.13a, and the interior of the thrift store is shown in Figure 7.13b. The fluorescent lights were mounted under a dropped ceiling, above which there was a layer of paper-backed insulation. The insulation quickly caught fire, and the fire spread throughout the concealed space above the ceiling. Firefighters had a difficult time getting water on the flames. They cut a hole in the roof above the rear of the thrift shop at the point shown in Figure 7.13c, and the increased ventilation resulted in more intense burning in that location. At one area, roof tar melted and flowed through a seam, landing on top of a steel duct, which resulted in a locally intense fire in the area shown in Figure 7.13d. Upon examining this area, it became clear that there was an electrical code violation (connections made outside a junction box), but this was just a code violation—not the cause of the fire.

Figure 7.13 (a) Eastern Tennessee shopping center damaged by a fire caused by a ballast failure. (b) Interior of the thrift shop.

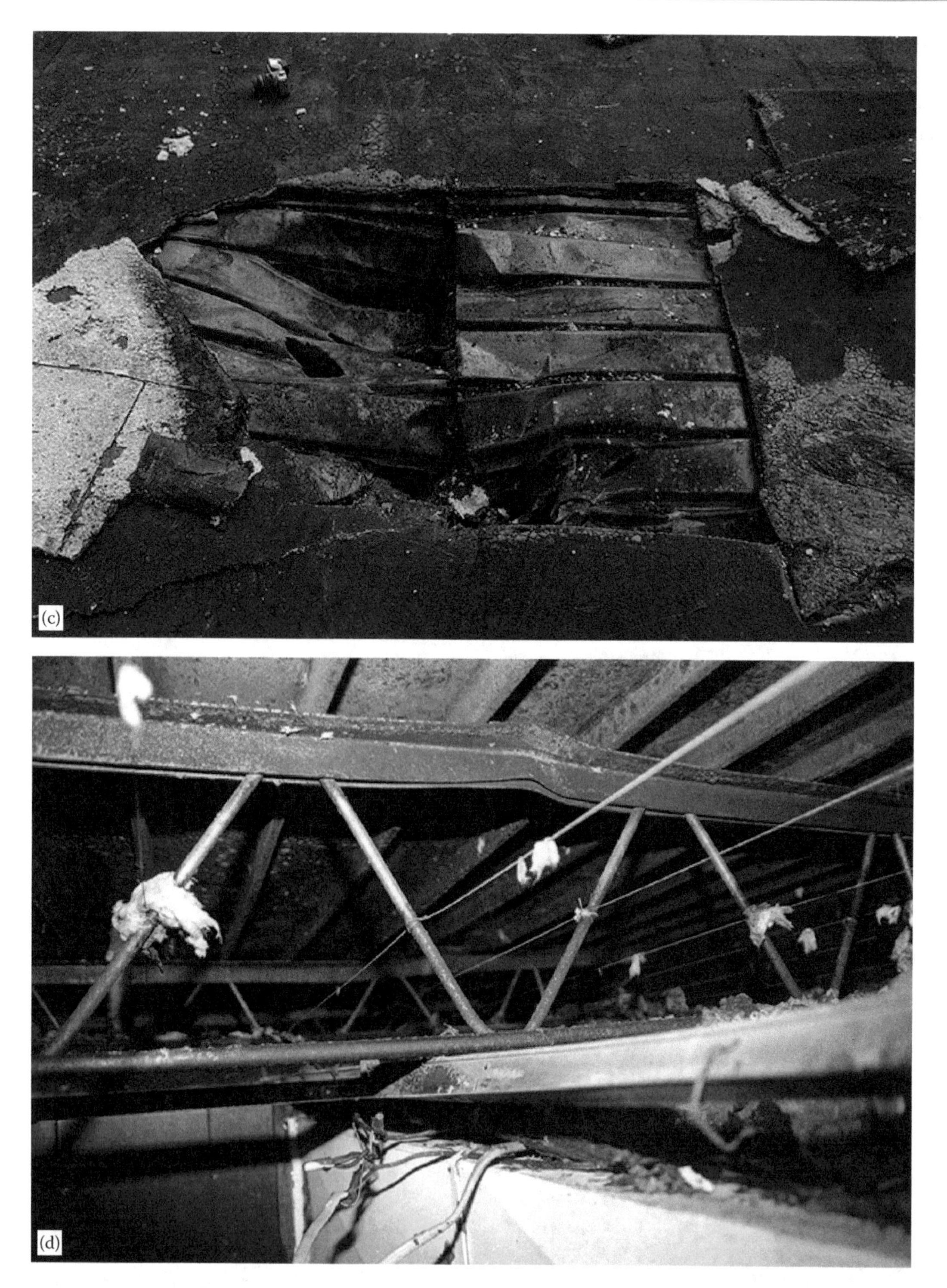

Figure 7.13 (c) Hole cut in the roof by firefighters above the rear of the thrift shop. (d) Area of localized heavy fire damage away from the origin. Note the electrical connection made without benefit of a junction box.

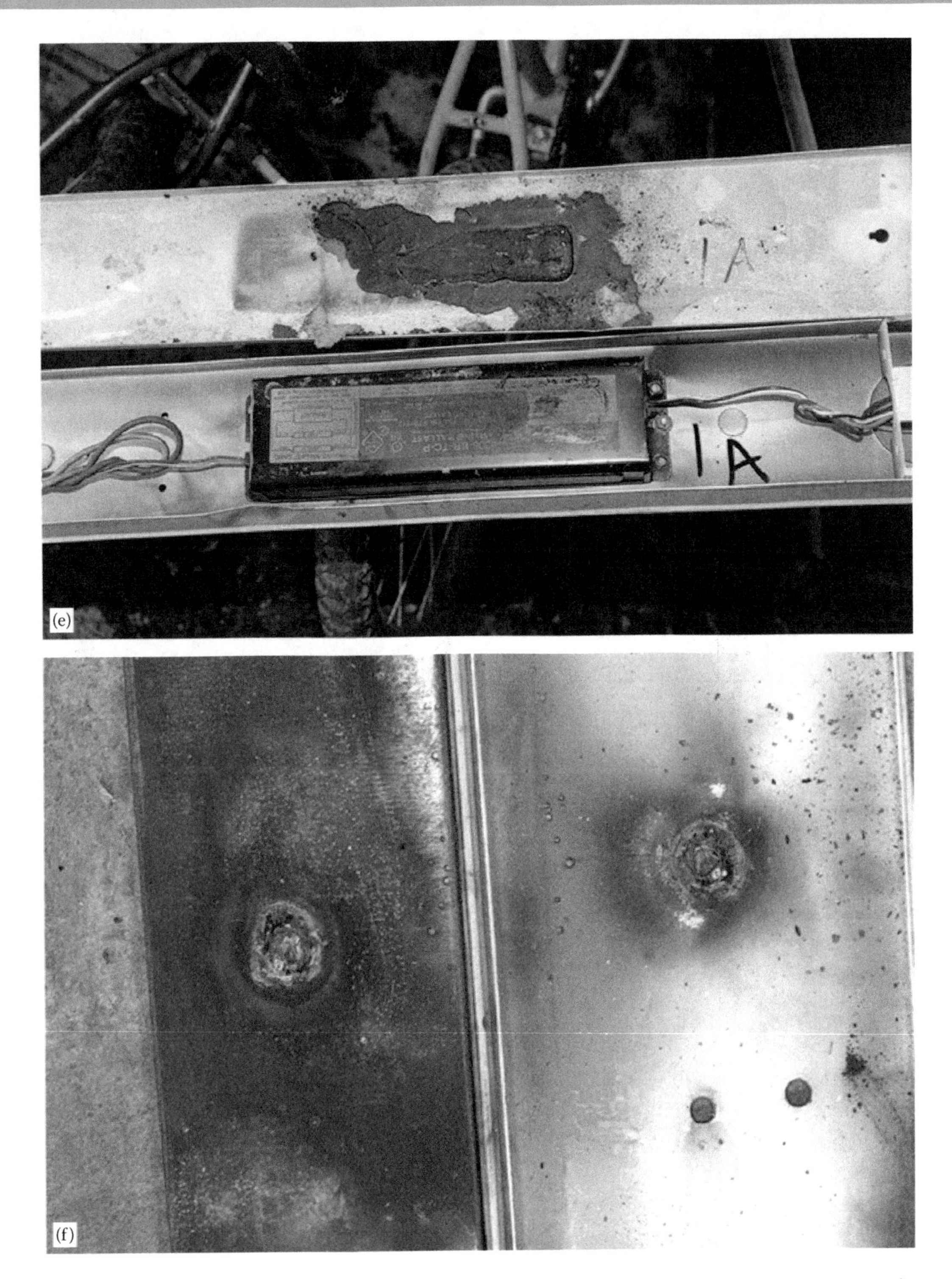

Figure 7.13 (e) Fluorescent light ballast that started the fire. (f) Holes arced through the ballast case and through the fixture case. This arcing took place in direct contact with the dropped ceiling tile.

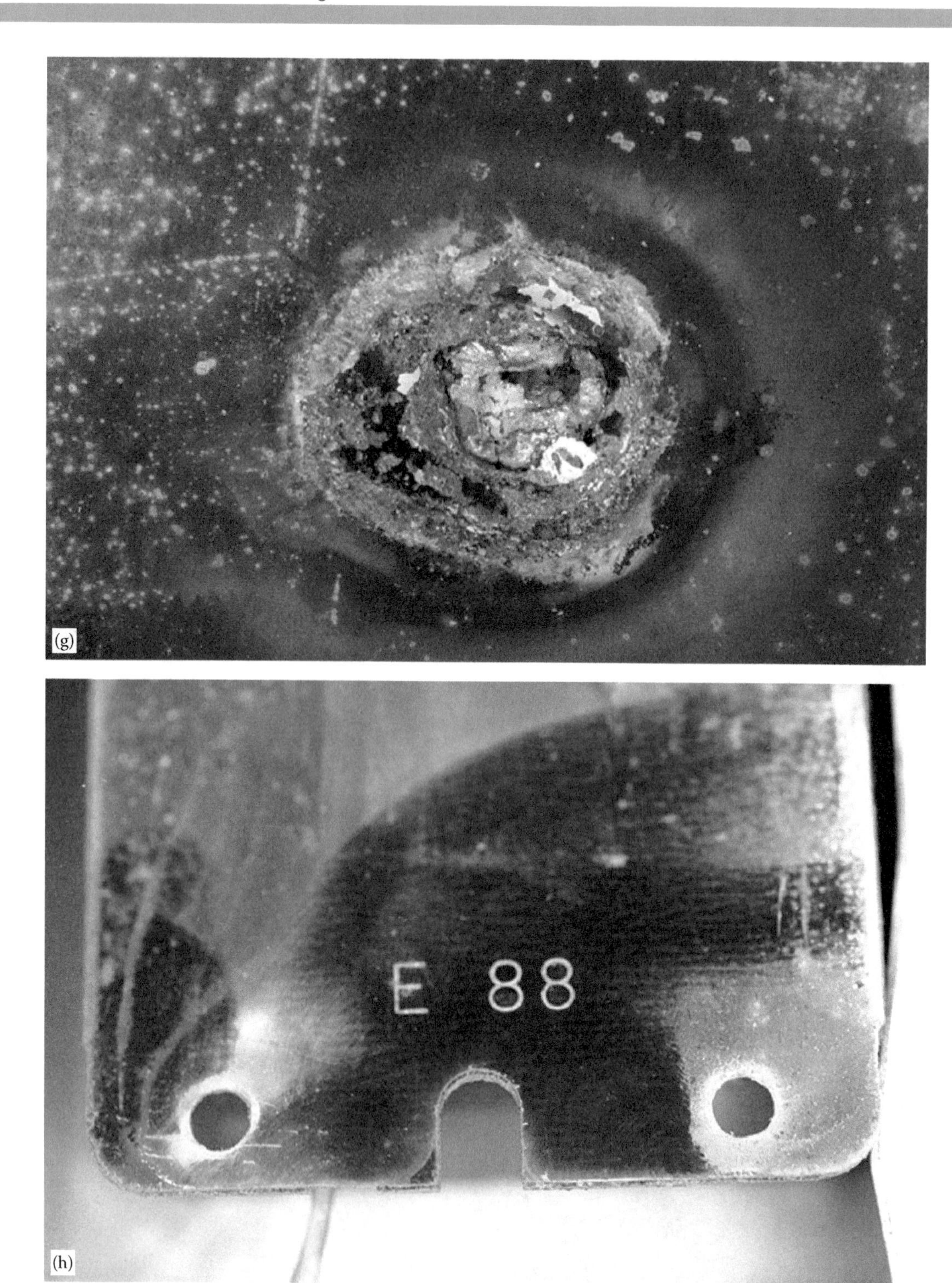

Figure 7.13 (g) Close-up view of the hole arced in ballast case. (h) Manufacturer's date code, indicating that the ballast was 13 years old when it failed. The date was made more visible by rubbing a piece of chalk over the stamped code.

Heavier burning, therefore, occurred in areas away from the origin than at the origin. As with most multi-tenant occupancies, the investigation of this fire involved investigators retained by several different insurance carriers. Scheduling the inspection took some time, but a cordial and cooperative atmosphere prevailed when the inspection convened.

Figure 7.13e shows the ballast from the fluorescent light at the origin, which still had an unburned paper label attached to it. Very little damage was visible until the fixture was turned over. Figure 7.13f shows clearly how the fire got out of the fixture. A hole had melted through both the ballast case and the fixture case, a close-up of which is shown in Figure 7.13g.

Manufacturers of fluorescent ballasts typically stamp a date code onto the bottom of the ballast; Figure 7.13h shows the manufacturer's date code, indicating that the ballast was manufactured in 1988. (Rubbing a piece of chalk—gypsum drywall also works—over the back of the ballast enhanced the visibility of this date code.)

Epilogue: Without the eyewitness account of the fixture shooting sparks, the cause may not have been determined. This fire was clearly caused by a defective product; because this ballast was 13 years old when it failed, however, the manufacturer could not be held liable for the fire under Tennessee law.

7.5.2 Fluorescent light fire 2: An overheated lamp holder

ATS was requested to investigate the cause of a small fire accompanied by a large amount of smoke damage that closed a public school in middle Tennessee. Figure 7.14a shows the school, and Figure 7.14b shows a sample of the smoke damage in the lobby. The only fire damage was located across the lobby and was confined to the soda-vending machine, shown in Figure 7.14c. A limited inspection was conducted on the first of three days required to complete this investigation. The plastic front panel was held back in place for a photograph, shown in Figure 7.14d. Although the owners of the vending machine were represented at the initial inspection, they insisted on coming back another day with an independent origin-and-cause investigator. That inspection took place a week later, at which point all parties agreed that the machine had caused the fire. An attempt was made to open the machine, but because the lock mechanism had melted, it was determined that the machine would need to be cut open. Counsel for the machine owner then decided that cutting the lock open would be "destructive" and we should come back yet another day when "all parties could be represented."

Four weeks later, the same parties that were present at the previous inspection met and cut the lock so the machine could be opened. Figure 7.14e shows the fire patterns on the machine, visible after the outer door was opened. Further examination revealed numerous areas of arcing on the conductors inside the vending machine, and an examination of an exemplar revealed that the arcing shown in Figure 7.14f was probably the first 120-V arcing that occurred. Figure 7.14g shows the lower left-hand lamp socket bracket, close to the point of arcing. Based on the history of these machines and the condition of this one, we (ATS) hypothesized that the arcing resulted from an overheating of the lamp socket. The likely cause of the overheating at the lamp socket was arc-tracking. Because of the orientation of the lamp socket, spilled soda can get into this area, with arc-tracking a likely result. Contamination such as that can cause arc-tracking to be initiated even at 120 V. In fluorescent lamps, however, the voltage is five times as high, so the likelihood of an unwanted conductive path being set up is greater [6]. Actually proving that arc-tracking has occurred, by showing the path of the current over a surface, requires a much less heavily damaged appliance.

Epilogue: Our client, the school's insurance carrier, was not happy that it required three days to investigate the fire loss, but the current climate regarding spoliation gave it few other options if it wanted to pursue its claim. Eventually, the insurance carrier recovered its loss from the vending machine owner, but it was not reimbursed for investigative expenses.

Figure 7.14 (a) Public school damaged by a fluorescent light fire in a soda-vending machine. (b) Smoke damage in the school lobby.

Figure 7.14 (c) Fire-damaged vending machine at the origin of the fire. This was the only item burned.

Figure 7.14 (d) Limited reconstruction allowed on the first day that an inspection took place.

Figure 7.14 (e) Fire pattern inside the vending machine visible after the outer door was opened.

Figure 7.14 (f) Arcing damage to conductors routed through a hole in the outer door. (g) Location of a fluorescent lamp socket directly below the farthest downstream arcing.

7.6 GAS FIRES

Gas systems and appliances are frequently involved in fires and are sometimes responsible for starting them. As with electrical fires, a *chicken-and-egg* problem exists in determining whether the open gas line found in a burned structure was opened before the fire or opened as a result of the fire. Sometimes, such as when the fire starts with a major explosion, this determination is not difficult. Other times, there is room for argument.

In the many gas fires, a recent change in the configuration of the system has been made. It may be a new installation or there may have been a recent delivery after an *out-of-gas* situation. Minor leaks usually cause fires, not explosions. When an explosion occurs, it is almost always because of a major leak involving tens to hundreds of cubic feet per hour.

Both natural and liquefied petroleum (LP) gas systems and appliances are capable of starting fires, but LP gas fires seem more numerous. This is certainly a result of a combination of factors. First, LP gas is heavier than air, while natural gas is lighter than air. Fugitive LP gas is more likely, therefore, to collect inside a structure, whereas natural gas is more likely to disperse (although it can collect in high places, such as attics). As a result of its tendency to disperse more readily, the odorant in natural gas is more likely to be detected than the odorant in LP gas. Because most gas-burning appliances are located below nose level, we are more likely to smell natural gas on the way up than we are to smell LP gas on the way down.

A cubic foot of LP gas contains 2.5 times as much potential energy as a cubic foot of natural gas, so a leak of the same magnitude is going to be more dangerous if LP gas is leaking. Finally, the quality of installation and maintenance of LP gas systems is often not as robust as the quality found in natural gas systems, where the gas lines are typically part of the original house construction. The portability of LP gas in containers provides additional opportunities for failure and additional failure modes.

7.6.1 Gas fire 1: Leak in a corrugated stainless-steel tubing line (and failure to inspect)

ATS was requested to investigate the cause of an explosion that destroyed a vacation rental cabin in eastern Tennessee. The cabin blew up on the morning of September 14, 1999, on the second day of use. The cabin was owned by a couple who contracted with a resort developer to build it for them, and their plan was to use it for 2 weeks per year and have the developer rent it out for the balance of the year. The first tenants in the cabin, a couple on vacation, noticed that they were unable to light the gas logs in the glass-enclosed fireplace soon after they arrived. They called the resort office, which sent a technician to check out the problem. The technician learned that the LP gas was turned off at the tank. A few minutes after he turned the gas on, the technician was able to light the fireplace, which had an automatic timer and shut itself off after about an hour. The following morning, while the tenants were cooking breakfast, the cabin exploded. Amazingly, both survived, although the wife lost a foot and the husband lost his spleen as a result of injuries sustained in the explosion. Fortunately, the fire was not large or sustained. It took rescuers only 20 min to remove the wife from the scene, but her husband was trapped in the wreckage for 5 hours.

Site inspections were conducted on two different days and were attended by investigators retained by the gas company, the general contractor who built the cabin, the developers, the homeowners, the tenants, the fireplace manufacturer, the manufacturer of the corrugated stainless-steel tubing (CSST) used in the gas line, and the furnace installer. Figure 7.15a shows the cabin as it appeared immediately after the explosion. The cabin was situated on the side of a steep hill, with an enclosed cinder block basement, having walls 8 ft tall on the uphill side and 20 ft tall on the downhill side. Examination of the structure revealed that the explosion had pushed the main floor upward as a result of an accumulation of gas in the basement. This accumulation would have begun at the bottom of the hill and gradually worked its way upward until it encountered an ignition source.

The ignition source for this fire is believed to be the electric furnace. The furnace should not have been a competent ignition source, but one of the power supply conductors was loose at the lug where it attached to the terminal block. Consequently, there would have been a small spark produced every time the furnace came on or turned off. The furnace, located at the upper end of the basement, is shown in Figure 7.15b. The gas line, where it connected to the fireplace, is shown in Figure 7.15c.

Figure 7.15 (a) Cabin that exploded 12 hours after the LP gas was turned on. (b) Furnace on the uphill side of the basement. The red power lead was found disconnected, although there were no apparent mechanical stresses on this conductor. It is believed that the conductor was loose, and when the furnace came on, this loose connection provided the ignition source for the fugitive LP gas.

Figure 7.15 (c) Point of connection between the CSST gas line and the glass-enclosed fireplace.

Figure 7.15 (d) Misaligned fitting just upstream of the gas fireplace cutoff valve. Note that the installer had inappropriately used joint compound on the flare fitting. (e) Gnarled end of the CSST line. There is no way this end could ever make a tight seal with a flare fitting.

The gas line from the tank to the house was copper, which connected to a second-stage regulator. A section of black (steel) pipe was connected to that regulator, and CSST was routed from the end of the black pipe to the fireplace. Used properly, CSST makes a versatile gas line that is easily bent without kinking and can stretch for very long distances. It is quite difficult to form proper joints in this material, however, and in fact, the manufacturer requires that the installer must be certified at a school run by the manufacturer before it will sell the product to an installer. The installer of this CSST line did, in fact, take and pass the certification course; however, the joint shown in Figure 7.15d falls far short of the quality workmanship taught in the certification course and will not hold even low pressure. In addition to the difficulty encountered in forming joints, CSST is also exquisitely sensitive to being punctured by direct or indirect lightning strikes. This aspect will be discussed later.

Some 5 months after the explosion, when everyone could agree on an inspection date, the evidence was examined in the ATS laboratory. When the joint was disassembled, the gnarled end of the CSST pipe was visible and showed clearly that the installer failed to form a serviceable flare.

This poor workmanship should not, however, have resulted in the explosion. The *National Fuel Gas Code* (NFPA 54) required that all gas systems be tested for leaks prior to first being put in service. In the operative (1998) edition the following instruction appears at Section 7.2.2:

> Before gas is introduced into a system of new gas piping, the entire system shall be inspected to determine that there are no open fittings or ends and that all valves at unused outlets are closed and plugged or capped.

The Code goes on to require in Section 7.2.3:

> Immediately after the gas is turned on into a new system or into a system that has been initially restored after an interruption of service, the piping system shall be tested for leakage. Where leakage is indicated, the gas supply shall be shut off until the necessary repairs have been made [7].[1]

Note that the Code fails to specify *who* is responsible for conducting the required inspection and testing. Although both the installer and the gas company insisted that they performed the required check, it is clear that these assertions are untrue. They may have *thought* they did the check or they may have checked the gas line upstream from the leak, but the required 10-minute pressure check—which involves attaching a pressure gauge at the tank, then opening the valve to pressurize the system, and then closing it—would have easily detected a leak of this magnitude. Several hundred cubic feet of gas leaked out overnight.

Epilogue: After all the evidence had been developed, all of the parties involved except the retailer of the LP gas accepted a share of responsibility and settled the claim brought by the injured tenants. The retailer, curiously, despite the fact that it claims to participate in the Gas Appliance System (GAS) Check Program, denied that it had any responsibility for actually checking to see whether the gas systems it supplies are safe. The GAS Check Program is sponsored by the Propane Education and Research Council (PERC) and requires retailers that adopt the program to follow it. A tape-recorded script that is supposed to be presented to homeowners when they first begin service describes the inspection process as follows: "During the gas check, we have inspected your entire propane system from the container to the appliance burner." Unlike the *National Fuel Gas Code,* this voluntary program places responsibility for inspection squarely on the retailer. The retailer's retained expert opined in his report, however, that the retailer "does not have a duty nor is it reasonable for the gas company to provide building code inspections on customer systems that are not installed or maintained by the gas company." The safety program in this case appears to be simply a marketing scheme. (About a year after the first edition of this book was released, when it became clear that a trial could not be otherwise avoided, the gas retailer finally agreed to settle the case.)

7.6.2 Gas fire 2: A leak at a new flare fitting

ATS was requested to investigate the origin and cause of a small but very expensive fire in a historical house located in eastern Georgia. The homeowners had contracted with a local heating and air conditioning company to replace their old electric furnace with a new LP gas furnace. The furnace was installed in the crawl space under their home. This involved setting the furnace in place, connecting new ducts, setting a new gas tank, and running a gas line and a flue. On the day the furnace was first fired up in early February 2000, the homeowner reported an odor of LP gas but was told that it was "normal" for someone unaccustomed to having an LP gas furnace to smell the odor initially. The odor continued on the second day. On the third day, the owner called back to the installer and again requested assistance. The sales manager at the HVAC company said that he would call back but failed to do so. The next evening, while eating dinner, the homeowner again smelled the LP gas odor and made plans to call the HVAC contractor again, but at just about that time, his smoke alarm activated because there was a fire in his crawl space.

[1] The current (2015) edition of the Code treats testing and inspection as follows, and still does not name the party responsible for either:
8.2.2 Turning Gas On. During the process of turning gas on into a system of new gas piping, the entire system shall be inspected to determine that there are no open fittings or ends and that all valves at unused outlets are closed and plugged or capped.
8.2.3* Leak Check. Immediately after the gas is turned on into a new system or into a system that has been initially restored after an interruption of service, the piping system shall be checked for leakage. Where leakage is indicated, the gas supply shall be shut off until the necessary repairs have been made.

This author (working on behalf of the homeowner's insurance company) and a mechanical engineer (working on behalf of the HVAC contractor) conducted a joint inspection of the site and later conducted laboratory analyses of gas and electrical equipment removed from the area of origin. The home is shown in Figure 7.16a. The origin of the fire was directly below the heart pine floor in the area shown in Figure 7.16b. It was hoped that the floor could be repaired in place, but there was no way to complete this investigation without cutting through the floor. Replacement of the floor was an expensive proposition.

Figure 7.16 (a) House burned as a result of a leaking flare joint in a new LP gas installation. The newly installed tank is visible at the right. (b) Origin of the fire, on the underside of the heart pine floor, directly over the gas line to the newly installed furnace.

Figure 7.16 (c) View of the origin area through the opened floor. (d) On-site soap bubble test of the LP gas line.

Figure 7.16 (e) Electrical junction box that represents a code violation. This code violation had existed without incident for more than 20 years. (f) Fire test, demonstrating the magnitude of the LP gas leak.

The opened floor is shown in Figure 7.16c. Once the floor was opened, we turned the valve on at the gas tank and soon smelled the odor of LP gas. A soap bubble solution was applied to the joint shown in Figure 7.16d, and the presence of a large leak was indicated. It was noted that the leak in the gas line was located below the most heavily damaged area, but the fittings themselves were not visibly damaged.

This seemed like a straightforward determination, but there was a "complication." Figure 7.16e shows an extended electrical junction box that has clearly been installed in violation of the *National Electrical Code*. There were no "connectors" (strain relief devices) protecting the wires as they entered the box, and there was no cover on the box. A second box had been inverted and used instead of a bottom cover, and wires were routed in the space between the two boxes. Because of this code violation, the engineer for the HVAC contractor decided that this must have been an electrical fire. Both the electrical box and the leaking joint were carefully marked and then carried back to the ATS laboratory for further examination. One of the wires in the box appeared to have a more rounded end than the others, and the engineer speculated that this may have been the result of electrical arcing, although what fuel might have been ignited by such an arc (and what the wire may have arced to) was not clear. Upon closer inspection of the end, however, he had to agree that it was, in fact, cut and not melted.

The pipe joint was connected to an LP gas supply, the soap bubble test was repeated and then the fugitive gas was ignited to determine whether the leak was of sufficient magnitude to damage the floor above it. Figure 7.16f shows this functional test. Although LP gas is heavier than air, burning LP gas is not, and the fire shown in this test is exactly the right size to cause the damage seen at the residence. Finally, a rotation test was conducted to determine just how loose the joint was. One-third of a turn was required to make the joint tight. The HVAC contractor's expert said that he believed that the flare fitting must have come loose as a result of the fire, as such joints are known to do [8].

The installer from the HVAC contractor insisted that he had run the required 10-minute pressure test and that no leaks were detected. However, a shutoff valve was located just upstream of the leaking joint; thus, if the test was conducted, it was probably conducted with that valve closed.

Epilogue: The HVAC contractor's engineer continued to insist that this was an electrical fire, despite the lack of evidence for it. He took the position that the leaking gas line was the result of the fire and could not be proved to be the cause of it. There was, in fact, a code violation in the electrical box, but that violation had existed without incident for 20 years. Further, a hypothetical electrical fire could not account for the homeowner's report of smelling gas for 3 days prior to the fire. Fortunately, the installer's claims adjuster was quite knowledgeable and reasonable. The claim was eventually settled.

7.6.3 Gas fire 3: Overfilled cylinders

ATS was requested to investigate the origin and cause of a fire that destroyed a suite of temporary offices located on the ground floor of a parking garage under construction. The fire occurred in early March 2002. The construction company that built and occupied the offices had, since January, been heating the office suite with three LP gas-burning radiant heaters mounted directly on the collars of propane cylinders. The fire occurred on a Monday, shortly after three new cylinders were brought into the office. A gas company employee had filled these cylinders on Friday afternoon. Fortunately, the employees in the office were all out to lunch when the fire occurred. One cylinder brought in on Monday morning began to vent soon after it was brought inside, and it was taken outside and replaced with another cylinder. One cylinder was brought in but left in place because the cylinder that had been in use since January was not yet empty. At the time of the fire, a total of six cylinders were present in the office area, and five of them vented. Needless to say, extensive damage occurred. Figure 7.17a shows the location of the offices, and Figure 7.17b shows the fire damage near the entrance to the offices. Some shoring up had taken place between the time of the fire and the time of our inspection. Much of the office furniture and building materials had been dragged out of the office area, so considerable reconstruction was required. Figure 7.17c is a view of one of the offices prior to any reconstruction efforts.

Investigators retained by the gas company's insurance carrier were present during the inspection. All the cylinders and the damage around them were examined, and we determined that the fire originated in the office shown in Figure 7.17d. This was based on the extent of the damage, including spalling of the concrete ceiling. Note that the cylinder in the foreground in Figure 7.17d has a cooling line approximately one-third of the way up. This indicates

Figure 7.17 (a) Parking deck where the temporary office structure was located. (b) Fire damage at the entrance to the offices.

Figure 7.17 (c) Fire damage in the office where the fire started. (d) Burned LP gas cylinders in their pre-fire locations. Note the cooling line on the cylinder in the foreground. Both cylinders were empty after the fire. The cylinder in the background vented first. Because it was not full of liquid, it took considerably longer for the pressure in the cylinder in the foreground to exceed the set pressure of the relief valve.

Figure 7.17 (e) Reconstruction, showing the pre-fire locations of the water cooler, coffeemaker, and microwave oven, all of which were ruled out as potential causes of the fire. (f) One of the overfilled cylinders venting through its overpressure relief valve in response to being warmed by sunshine. Cold LP gas causes frost to form on the valve. Five cylinders vented the day this photo was taken. They vented in order of the most overfilled to least overfilled, as determined by weight.

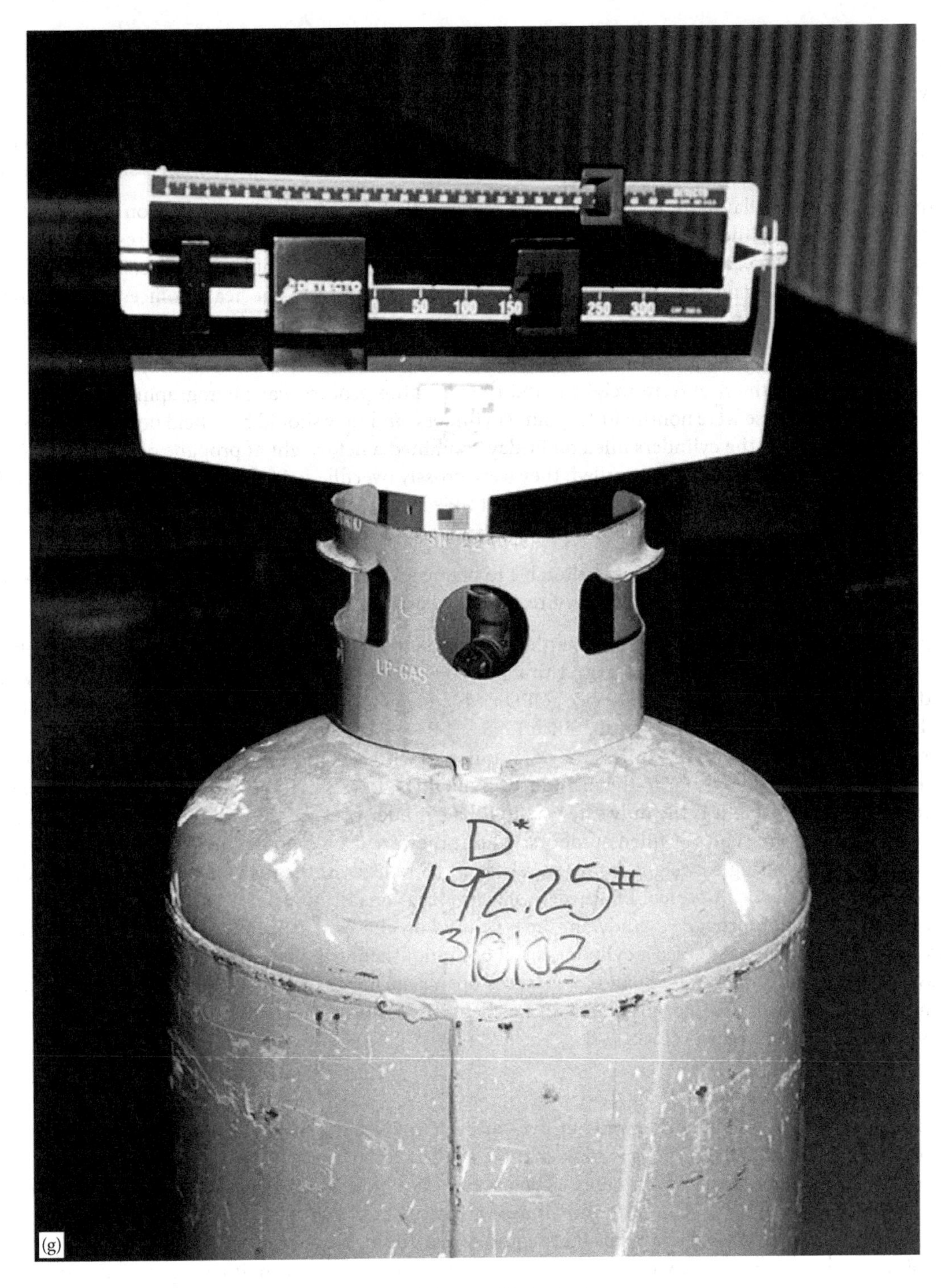

Figure 7.17 (g) Documentation of the weighing of one of the overfilled cylinders.

the liquid level in the cylinder at the beginning of the fire. It was this cylinder in the foreground that had a heater attached at the time of the fire. It was our hypothesis that the heater on the cylinder in the foreground ignited fugitive gas venting from the cylinder in the background.

Additional reconstruction was carried out to eliminate all other potential sources of accidental ignition in the office area, including a water cooler, coffeemaker, and microwave oven. Part of the reconstruction is shown in Figure 7.17e.

The cylinders filled on Friday were moved from the loading dock to a holding area, and on the first day of our inspection, one of these cylinders began venting due to exposure to the afternoon sun. This resulted in frosting on the overpressure relief valve, as shown in Figure 7.17f. This was the cylinder that had been removed from the office area when it started leaking on Monday morning. Bringing the cylinder into the heated offices had the same effect as exposing it to sunlight.

A calibrated scale was brought to the scene; all the cylinders that had been filled on the Friday before the fire (except those that discharged in the fire) were weighed, and the weighing process was photographically documented, as shown in Figure 7.17g. These were nominal 100-pound cylinders, and they should have held no more than 80 pounds of liquid propane. None of the cylinders filled on Friday contained a net weight of propane of less than 110 pounds, which means that they were not only overfilled; they were grossly overfilled. After they were weighed, the cylinders were taken outside and placed in the sun, at which point they began to warm up. The heaviest cylinder vented first and was followed in order by the second heaviest, third heaviest, fourth heaviest, and so on.

Epilogue: After it was satisfied that it would be unable to blame anything other than its own employee's negligence, the gas company accepted responsibility for the fire, and its insurance carrier paid off.

Overfilling of LP gas cylinders is a problem that has caused many fires, but the problem has been largely solved, though the solution would not have prevented this fire. Upright cylinders of 4- to 40-pound capacity are now equipped with an overfilling prevention device (OPD), which is essentially a float that prevents more liquid from being delivered once the cylinder is 80% full. Since 1998, NFPA 58, the *Liquefied Petroleum Gas Code,* has prohibited the refilling of portable cylinders unless they are equipped with an OPD [9]. Cylinders equipped with an OPD have a triangular valve handle. Prior to the introduction of OPDs, LP gas cylinders could be filled either by weight or by volume. Filling by weight is the only safe way to fill a cylinder because the volume of the fluid changes with temperature. Cylinders are typically filled outdoors; thus, if they are used indoors, the liquid is likely to expand, as it did in this case. Even filling by weight, however, relies on the reflexes of the individual filling the cylinder rather than on a mechanical device. A typical LP gas delivery nozzle is capable of delivering a gallon of the liquid fuel in just 5 seconds. A 20-pound propane cylinder, of the type used on charcoal grills, is completely liquid-filled when it contains 24 pounds of propane. Once a cylinder is full of liquid, there is no gas to compress, and liquid cannot be compressed. As the liquid warms, it expands and forces open a spring-loaded pressure relief valve, even one set at 325 pounds per square inch. Cox has presented an excellent discussion of the issues involving overfilling of LP gas cylinders and the solution to the problem [10].

7.6.4 Gas fire 4: New installation, open line

On December 31, 1991, a young mother and her five children arrived at their rural south Georgia home, unaware that their newly installed 250-gal. LP gas tank had been filled for the first time approximately 6 hours earlier. They were only in the house for a few minutes when an explosion that could be heard 20 miles away flattened the house. Portions of the roof blew 50 ft away and landed in the cemetery across the street. Miraculously, everybody lived, although they sustained horrific burns that required years of medical treatment. It is likely that the immense pressure of the explosion forced the victims to exhale, thus preventing them from inhaling the flames. The remains of the house are shown in Figure 7.18a and b.

Figure 7.18 (a) South Georgia residence flattened by a propane explosion, view from the front. Six people were inside this house when it blew up. All survived. (b) Exploded residence, view from the rear. Massive pieces of the roof were blown 50 ft away. Smaller pieces were found 200 ft away.

Figure 7.18 (c) Open flare fitting. Gas flowed from this open line for about 6 hours.

The fire was only burning in one place when the fire department arrived on the scene. They reported a "torch" burning in the area of the open line, shown in Figure 7.18c. The fire went out when the valve on the gas tank was turned off.

Gas company employees had moved a space heater from the kitchen to the living room and plugged the gas line in the kitchen, but they had not performed any pressure check because there was no tank installed when they finished the modification. The gas company employee who set the tank in place did not perform any pressure test because the tank was empty. The gas company employee who delivered the gas did not perform the pressure check for unknown reasons. The company's safety manual did set forth a policy that the delivery person should have conducted the pressure check (this company also subscribed to the GAS Check Program) or at least locked the tank valve in the off position so that the test could be performed when the residents returned home.

Although it was an open gas line that provided the fuel, it was the delivery person's failure to conduct a necessary test that caused the explosion. This situation actually required a complete inspection in addition to the leak test.

Epilogue: The local sheriff considered filing criminal charges but did not do so because of the vague provisions of the *National Fuel Gas Code* (adopted into law in Georgia and most states), which fails to state *who* is responsible for testing the system. On the civil side, this case moved at record speed. Depositions were not necessary, and the gas retailer's insurance carrier settled the claim of the injured family for $40 million.

7.7 HEATER FIRES

Space heaters, particularly floor furnaces and portable electric heaters, are responsible for a disproportionate number of fires. These fires can be the result of manufacturing defects but are much more frequently the result of misuse or abuse. Many heater fires occur on the first cold day of the year or when something in the environment changes, such as tenants moving in or out of a residence. Many heating devices are sufficiently robust that they survive a fire, and some can even be tested afterward. This robustness often makes it difficult to demonstrate anything more than that the heater was located at the origin of the fire. This is particularly true when the heater acts as an ignition source for combustibles placed too close to it. The heater is not likely to change as a result, nor is it likely to exhibit any defects when examined after the fire. The following three fires are examples of typical heater fire scenarios.

7.7.1 Heater fire 1: Combustibles on a floor furnace

ATS was requested to investigate the origin and cause of a fire that destroyed a rental house and resulted in three fatalities. The fire occurred in the spring of 1999 in an east Atlanta residence. The tenants were in the process of moving out and had placed many of their belongings in corrugated cardboard boxes. The fire occurred in the middle of the night while five people were asleep in the house. Two survived, and shortly after the fire, began seeking compensation for their injuries and for the loss of their loved ones.

The local gas utility had been replacing gas mains on this street about 3 months prior to the fire, which required that all gas service be turned off for a period of time. A contractor for the gas company restored service and relit pilots. Individuals representing the estates of the deceased persons claimed that the gas company should not have relit the pilots because of alleged code violations in the gas system.

The inspection of the site took place on two separate days, and two additional days were required for laboratory examination of the gas-fired floor furnace. Investigators representing the estates of the deceased persons, the landlord, the gas company, and the gas company's contractor were all involved in the inspection and testing.

The residence is shown in Figure 7.19a, and a floor plan is shown in Figure 7.19b. Based on an examination of the fire patterns and eyewitness statements, the origin of the fire was determined to be in the living room at the center of the house where a floor furnace, shown in Figure 7.19c, was located. There was unanimous agreement among the experts that the floor furnace was the ignition source. The floor furnace, which fell into the basement, is shown in Figure 7.19d.

The cause of the fire appeared to be straightforward. Figure 7.19e is a closer view of corrugated cardboard debris found immediately next to the furnace grille. Numerous items of personal property were found beneath the furnace and inside it. It appears that personal property was stacked, at least partially, over the furnace grille.

The furnace was returned to the ATS laboratory for testing with the agreement of all parties. Because the fire did not significantly damage the parts of the furnace below the floor, it was possible to connect a gas supply and test the

Figure 7.19 (a) Atlanta residence where three tenants died in a fire started by a floor furnace.

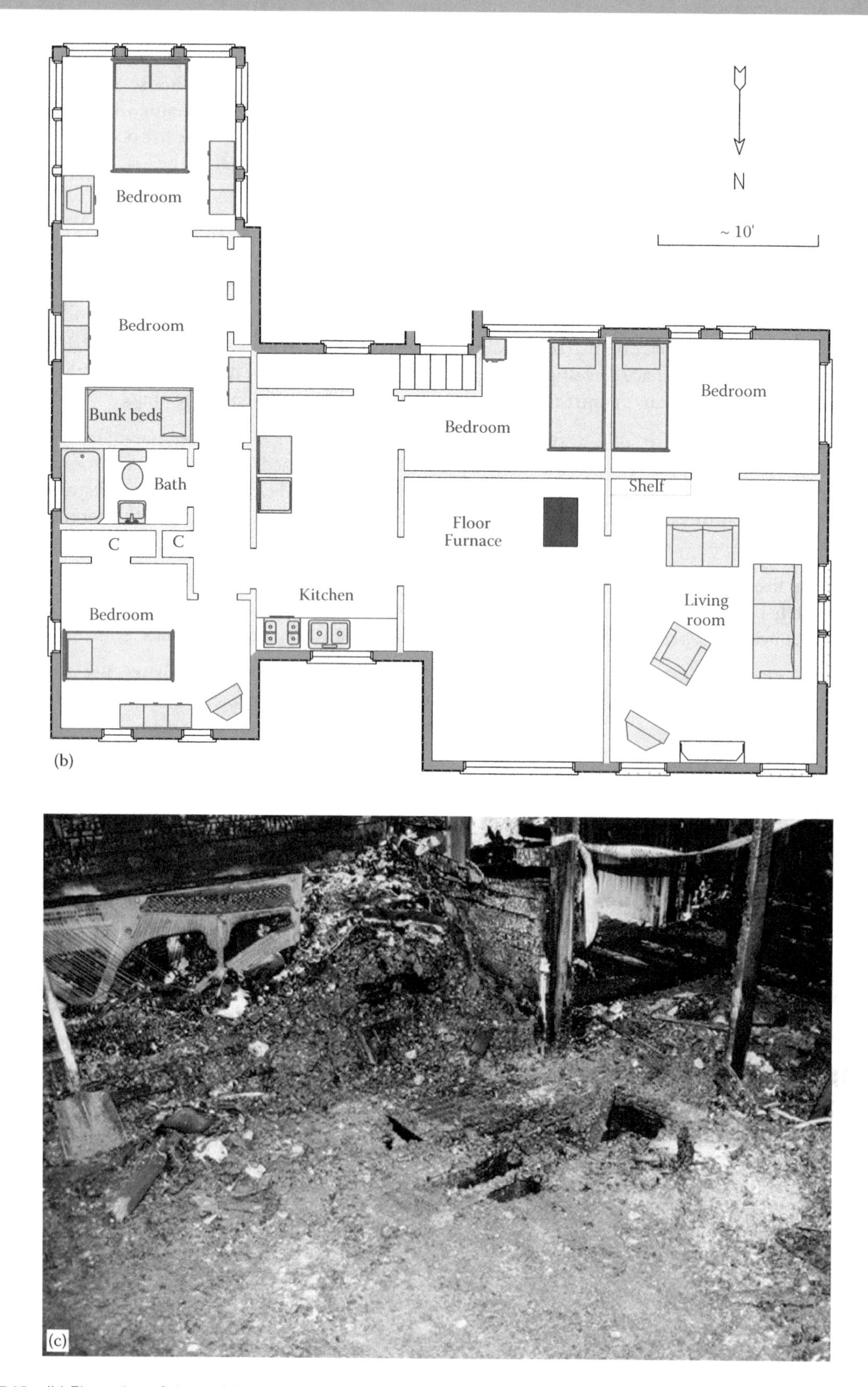

Figure 7.19 (b) Floor plan of the residence. (c) Floor furnace at the origin of the fire.

Figure 7.19 (d) Closer view of the furnace. Corrugated cardboard, some of which can be seen at the far side of the furnace grille, was likely the first fuel ignited. (e) Floor furnace in the basement below the origin of the fire.

Figure 7.19 (f) Laboratory testing of the floor furnace.

operating characteristics of the furnace. Both gas consumption and temperatures at various locations were monitored. The test setup is shown in Figure 7.19f. No abnormal operation was observed, but temperatures in excess of 900°F (approximately 480°C) were measured on the top of the heat exchanger, which was approximately 2 inches below the steel grille. Temperatures in excess of 700°F (370°C) were measured at the center of the grille, but at the edge of the grille, the temperature did not exceed 200°F (approximately 93°C).

Epilogue: Based on the results of the testing, one expert for the family of the deceased alleged that the furnace ran "too hot." No expert was able to develop a theory of liability stating that the person who lit the pilot 3 months before the fire should have been aware of this characteristic of the furnace. No litigation ensued.

7.7.2 Heater fire 2: Portable heater ignites cardboard

ATS was requested to investigate the origin and cause of a fire that all but destroyed a central Georgia residence. The fire occurred at about 4:30 a.m. on a cold January night. The homeowner was awakened by a noise and saw a glow like a flashlight in her kitchen. She stated that when she opened the back door, she saw that the garage was fully engulfed in flames. She ran to wake her husband and they barely escaped with their lives. Figure 7.20a shows the front of the house, and Figure 7.20b shows the garage.

The fire moved rapidly from the garage into the kitchen, which became fully involved, then into the dining room and the living room at the center of the house, which also became fully involved. Only the bedrooms at the far end of the house from the garage were not completely gutted. A view from the kitchen into the garage is shown in Figure 7.20c.

Excavation of the garage and the two vehicles required two men working for about 4 hours. Examination of the electrical circuits in the garage revealed no evidence of arcing, other than on the service panel located on the east wall of the garage, which exhibited arcing at its lower right corner as well as on the bottom. The service entrance cables were routed through the bottom of the panel, as shown in Figure 7.20d. The cables were routed up each side to the top of the panel, where they curved around to connect to the main breaker lugs. One would normally expect that if one of the vehicles had caught fire, the hot gas layer would build down from the ceiling and attack the panel at the top first. These findings led to the hypothesis that the fire may have originated below the panel. Figure 7.20e shows the excavation of the east wall of the garage in progress. A steel plant stand was found directly below the electric panel and directly in front of the wall stud that burned to the lowest level. A close-up of the area behind the plant stand is shown in Figure 7.20f. A hole was burned in the kitchen floor joist, and the molding at the bottom of the wall was burned away in a shallow V-shape. (Note that a V-pattern this low and this shallow is almost certainly not a result of the fire plume intersecting the wall, but it may reflect a longer burning time.)

The service panel was put back in place for the reconstruction shown in Figure 7.20g. The final step in the reconstruction is shown in Figure 7.20h. The homeowner had placed a small space heater in the garage on that night for the first time to keep the plants warm. Pieces of corrugated cardboard were found adhering to the front of the heater. The placement of the heater was the only change that took place in the garage on the night of the fire.

Epilogue: Because of the extent of the damage, we are left with a hypothesis that is supported by the fire patterns, but the level of certainty is such that the most we can say is that this heater cannot be excluded as the ignition source.

Figure 7.20 (a) Central Georgia house damaged by a fire that started at a heater in the garage.

Figure 7.20 (b) View of the fire-damaged garage. (c) View of the fire-damaged garage from the kitchen.

Figure 7.20 (d) Electric service distribution panel, as found.

Figure 7.20 (e) Plant stand at the point of lowest burning on the east wall of the garage. (f) Shallow V-shaped burn, hole in the floor joist, and burned-off wall stud on the east wall of the garage.

Figure 7.20 (g) Reconstruction, showing the location of the service panel above the plant stand.

Figure 7.20 (h) Portable electric space heater with cardboard adhering to its face.

7.7.3 Heater fire 3: Contents stacked in front of the heater

ATS was requested to investigate the origin and cause of a fire that damaged a four-unit apartment building in central Tennessee. The fire occurred in mid-December 1999 at approximately midnight, while the apartment where the fire started was unoccupied. The city fire department was able to keep the fire damage confined to the apartment of origin, but the apartment directly below it sustained significant water damage. The front of the apartment building is shown in Figure 7.21a, and a floor plan of the affected apartment is shown in Figure 7.21b. Most of the fire damage was confined to the west side of the

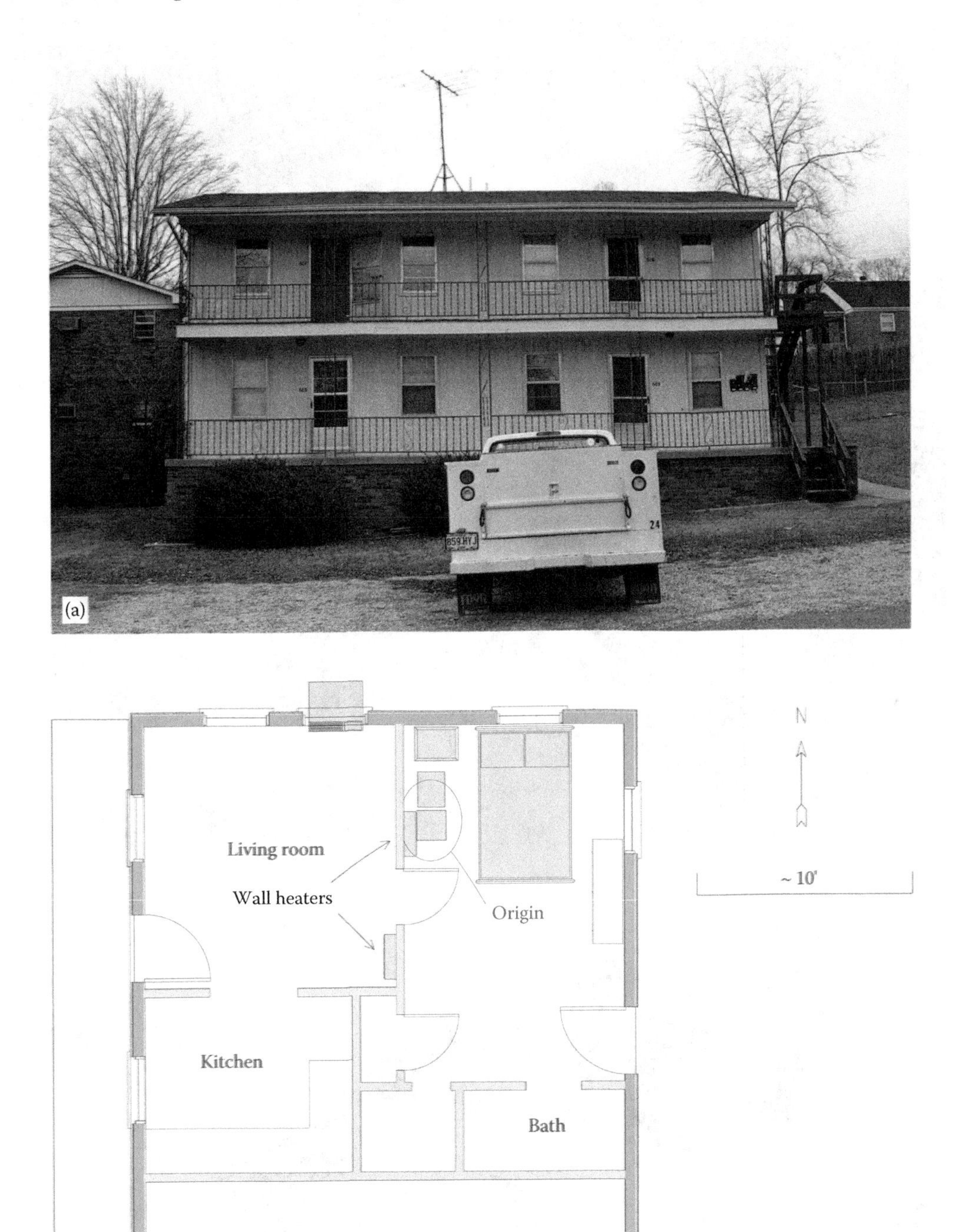

Figure 7.21 (a) Apartment house damaged by a heater fire. (b) Floor plan of the affected apartment.

Figure 7.21 (c) Fire damage in the room of origin.

Figure 7.21 (d) Origin of the fire in front of the wall-mounted space heater.

Figure 7.21 (e) Area between the wall and the bed, after being cleared of debris. There is a rectangular footprint of a box located directly in front of the heater.

Figure 7.21 (f) Burned cloth adhering to the upper right-hand side of the wall-mounted space heater. (g) Position of the heater control knob. It should not have been possible to put the heater knob in this position.

one bedroom in the apartment, as shown in Figure 7.21c. A wall-mounted space heater was located in the middle of the most heavily damaged area, and a large amount of personal property was found on the floor in front of the space heater, between the wall and the bed. This accumulation of clothing and other personal property is shown in Figure 7.21d.

Excavation of this area revealed the rectangular footprint of a corrugated cardboard box directly in front of the wall heater as shown in Figure 7.21e, and a close inspection of the front of the heater revealed a piece of cloth adhering to the upper right-hand side of the grille, as shown in Figure 7.21f.

The tenant stated that he had turned the heater off prior to leaving the apartment (and prior to placing anything in front of the heater), but examination of the heater control knob post revealed that the heater was, in fact, turned beyond “high.” Figure 7.21g shows the heater at the origin with a knob removed from an exemplary heater in the living room. The “high” position on the heater is at exactly 3 o'clock, and the living room heater knob could not be made to turn past 3 o'clock. Unlike the wall-mounted heater described in Chapter 6, this heater had a positive “off,” but apparently, the detent became worn so that when the tenant turned this heater off, he actually turned it beyond the “off” position into the position in which it was found, resulting in the heater being set to the highest temperature.

Epilogue: This fire was the result of a combination of factors, including negligence on the part of the tenant, who piled combustible materials in front of a heating device, and a failure of the detent in the heating device itself. No litigation resulted from this fire loss.

7.8 INDUSTRIAL FIRES

A fire in an industrial plant is likely to present the fire investigator with challenges not found in ordinary business or residential fires. The plant may contain unique machines and systems whose intricacies are known only to the employees of the plant, whose cooperation is therefore essential. If the fire was limited in scope, however, cooperating with a fire investigator is probably the last thing that most employees and their managers want to do. Their goal will be to get the plant back up and running as quickly as possible, and evidence might be destroyed in the process. Companies purchase insurance to get them over situations such as a fire, and they are frequently not interested in dwelling on the root cause, although the smarter ones will want to know how to prevent such a costly event in the future.

In addition to getting back up and running as quickly as possible, plant employees may have other reasons to avoid cooperating with an investigation. Fires in industrial facilities are often the result of bad decisions made by someone, and that person may not be overly eager to take responsibility for those decisions. People in this position may be evasive or downright hostile if they fear that the investigation is likely to lead to a conclusion that a decision made by the plant management months or years ago was a bad one. Such decisions may include the decision to forego a fire suppression system because of the expense involved. The damage and downtime caused by a fire often dwarfs the estimate for a sprinkler system. In other cases, the fire may be the result of someone's decision to override a safety device or failure to take the time to do a procedure correctly in order to increase production. By far, the most common cause of industrial fires is poor housekeeping, and the plant manager, who is usually the individual responsible for that function, may not want to hear that a clean factory is a safe factory.

In the industrial setting, the investigator should become familiar with the processes taking place in the part of the plant where the fire occurred. The investigator should learn the nature of the raw materials, and how and where they are stored prior to processing. The raw materials are then transported to the processing area where processes are applied to them, and the investigator should become familiar with these processes. Finally, the finished product comes out of the process, and is packaged and stored until shipment. Prior to beginning an investigation, the investigator should request an interview with the plant manager or with someone familiar with all these issues so that he or she has an idea of the kind of things that might go wrong. The following are examples of typical industrial fires.

7.8.1 Industrial fire 1: Machine shop spray booth

ATS was requested to investigate the origin and cause of a fire that seriously damaged a tool and die shop located north of Nashville, Tennessee. The fire occurred in mid-July 2000 while the business was closed. The alarm was turned on at approximately 10:30 p.m.

The machine shop produced large steel brackets to hold fans and motors, marine parts, railroad parts, and conveyers with industrial applications. They bought steel shapes, then cut, welded, painted, and sold them. The business was under new management and, several months before the fire, had engaged a professional loss control consultant to review their operation and make recommendations for a safety program. A written program had, in fact, been adopted, but the fire prevention and control program took up only two pages of the document and failed to recognize several major hazards.

The front of the plant is shown in Figure 7.22a. Most of the employees were involved in cleanup operations on the date of this inspection, but the plant manager did take the time to walk this investigator through the processes that took

Figure 7.22 (a) Tennessee machine shop, damaged by a fire in the spray booth room. (b) Smoke and heat damage above the doorway to the spray booth area.

Figure 7.22 (c) Fire damage at the west side of the spray booth area.

Figure 7.22 (d) Fire damage around the vent on the south wall of the spray booth.

Figure 7.22 (e) Improperly stored containers of flammable paints in the spray booth area.

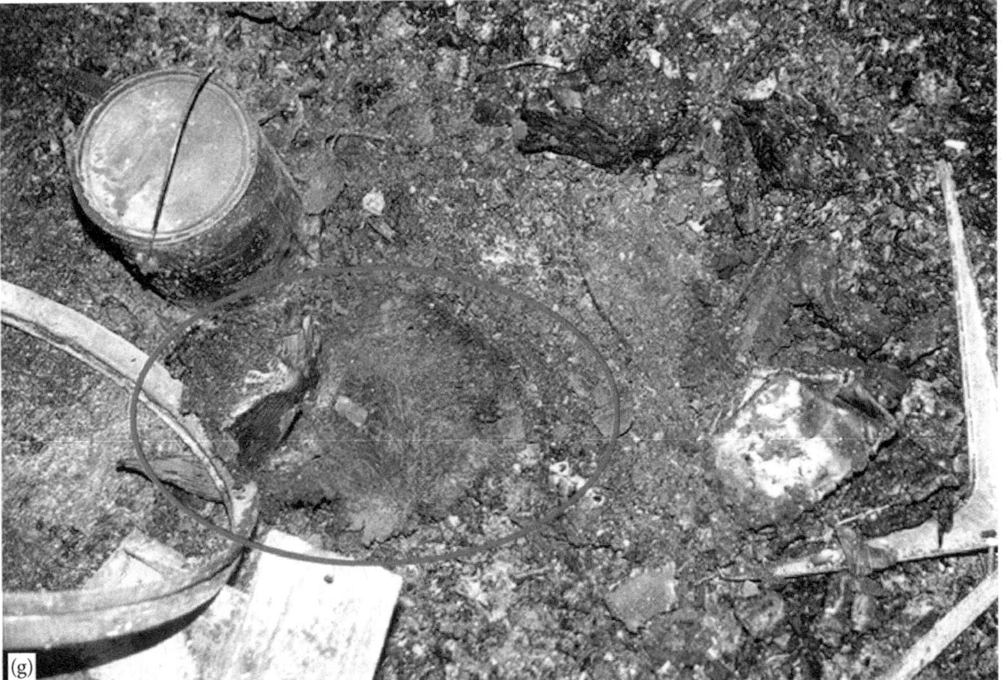

Figure 7.22 (f) Fire pattern on the side of the pallet stack facing the trash barrel. (g) Filter material found in the remains of the trash barrel. This material is known to be subject to spontaneous ignition. OSHA rules and NFPA 33 require that it shall be removed immediately to a well-detached location or placed in a water-filled metal container and disposed of at the close of the day's operation unless maintained completely submerged in water.

place. The exterior of the building was largely intact, although the interior was heavily smoke damaged. At the center of the building, some sagging of the roof was noted directly over the paint spray area, which was the area of origin. Figure 7.22b shows a pattern of heat and smoke coming out of the top of the doorway to the spray booth area, and fire damage in the spray booth is shown in Figures 7.22c through e. This was the most heavily damaged area of the building and the only place where any serious fire damage occurred. Note the improperly stored containers of flammable and combustible liquids in Figure 7.22e.

A stack of oak pallets was located in the doorway of the paint spray booth area, and charring was seen on one side of the pallets in the direction of a fiber trash barrel. The trash barrel was completely burned, but it was possible to see the remains of the fiberglass filters used in the paint spray area. The pallets are shown in Figure 7.22f, and the remains of the trash barrel are shown in Figure 7.22g.

The employee responsible for spray painting was interviewed, and he stated that he had done some painting on the day of the fire and had placed some used filters in the trash barrel before leaving.

Overspray particles collected on paint filters are a known hazard for spontaneous ignition. This hazard is so well known that the Code of Federal Regulations at 29CFR Section 1910.107 sets forth specific rules for the disposal of filters from spray booths. Most of the provisions in this Occupational Safety and Hazard Administration (OSHA) document were lifted directly from NFPA 33 *Standard for Spray Application Using Flammable or Combustible Materials*. NFPA 33 specifically requires that spray areas, which may include by definition any associated exhaust plenums and exhaust ductwork, any particulate filters, any solvent concentrator units, any recirculation air supply units, and mixing rooms shall be protected with an approved automatic fire protection system [11]. There was no fire protection system anywhere in the building.

Section 10.4.2 of NFPA 33 requires that

> [a]ll discarded overspray collector filters, residue scrapings, and debris contaminated with residue be immediately removed to a designated storage location or placed in a water-filled metal container and disposed of at the close of the day's operations unless maintained completely submerged in water.

This fire clearly originated in the fiber drum and was caused by the spontaneous ignition of the filters coated with overspray particles.[2]

Epilogue: This plant manager had endeavored to make his operation safer, but his loss prevention consultant made a serious omission in failing to consider the well-known hazards of spray painting. The plant manager and the consultant were close personal friends, however, and the company was therefore not enthusiastic about pursuing subrogation when the insurance company suggested that avenue. A sprinkler system was installed when the spray booth was rebuilt, and the employees now pay much more careful attention to housekeeping.

7.8.2 Industrial fire 2: Waste accumulations on the roof

ATS was requested to investigate the origin and cause of fires that were extinguished on two different days, three days apart. The fires occurred in the early summer of 2001 while the plant was in operation. This plant received raw material in the form of large rolls of heavy Kraft paper, which it made into sheets of corrugated cardboard. The cardboard was then cut into shapes to produce corrugated cardboard boxes. Several cardboard cutting operations were performed during this process, which generated many small pieces of cardboard and much fine dust. This waste material was collected by a vacuum system and sent to two cyclones on the roof.

[2] At the time of the fire, the operative edition of NFPA 33 was the 1995 edition. The 2016 edition is substantially the same with respect to the requirements cited.

These cyclones were continuously emptied into a hopper and the collected materials sent to another plant for recycling.

The first fire was detected at approximately 2:30 p.m. on Tuesday. The fire started on the roof outside one of the cyclones, but the unit had been shut down and opened because it had clogged. The employees working to unclog the cyclone had taken a break and were not on the roof when the fire started. Because the cyclone was open, the fire extended to the fuel in the interior of the unit. The fire department responded and extinguished the fire. Sprinkler heads inside the cyclone also contributed to the extinguishment. A second fire occurred on the following Friday at 4:30 a.m., in the vicinity of the second cyclone. The Friday fire did not penetrate into the cyclone.

One of the motors that powered the blower on the first cyclone had been recently replaced, and the plant manager suspected that it might have been involved in causing the first fire. Figure 7.23a shows the two cyclones on the roof of the plant. The Tuesday fire began near the cyclone shown on the left. The fire damage from the Tuesday fire, some of which had been cleaned up by the time of this inspection, is shown in Figure 7.23b, while the fire damage from the Friday fire is shown in Figure 7.23c. This author was at the plant on Friday afternoon, and only minor cleanup had taken place. The Tuesday fire caused damage inside the plant when the contents of the cyclone ignited and fell down a chute, as shown in Figure 7.23d.

As the investigation into the fire progressed, a fire pattern was noted that indicated a fire inside one of the cyclones with significant air movement. Because we knew that the cyclone had been shut down at the time of the Tuesday fire, this was a curious finding. It was also noted that some of the bolts on the cyclone that caught fire on Tuesday had been painted with silver paint, as shown in Figure 7.23e. When asked about these conditions, the plant manager remembered a fire in this same cyclone approximately 18 months earlier. Not only were we being asked to separate the artifacts of the Tuesday fire and the Friday fire but also leftover artifacts from an 18-month-old fire.

One of those artifacts may have included the arcing event shown in Figure 7.23f, found on the load side of a disconnect box, which may (or may not) have been used to supply the recently replaced motor. We also learned that there had been a severe thunderstorm on Tuesday, so yet another variable was added to the problem.

A considerable accumulation of cellulose waste was found around the cyclones, and it appeared sufficiently deep to allow a burrowing fire to exist for three days. Finely divided cellulose, when stacked deep enough, is capable of smoldering for extended periods of time beneath the surface and only manifesting itself as a flaming fire when it breaks through the surface and gets sufficient oxygen. It was likely that the Friday fire was simply a rekindle of the Tuesday fire, but because of all of the confounding influences, the best this investigator could do was to make recommendations to prevent future fires.

Two steps can be taken to prevent fires such as these: Remove sources of ignition and remove sources of fuel. Figure 7.23g shows one of the ducts leading to one of the cyclones, indicating numerous dents caused by impacts of metal objects against the inside of the duct. A fan housing, shown in Figure 7.23h, exhibited similar dents from the inside out, some of which had actually resulted in tears that were welded closed by plant maintenance personnel. The author made a recommendation to install magnetic filters to remove the ferrous materials getting sucked into the exhaust system. This would reduce the chances of an ignition inside the ducts or the cyclones.

A more important recommendation was to establish a cleaning routine to remove the accumulations of finely divided cellulose from around the cyclones. A daily removal was suggested, with the possibility that this could be reduced to a weekly removal, depending on the rate of waste accumulation. Because no previous program existed for cleaning the area around these cyclones, the rate of waste accumulation was simply not known. What was known was that there was too much waste there at the time of the fire.

Epilogue: The evidence did not allow for the categorical elimination of the possibility that the two recent fires had been intentionally set, but the employees who had access to the area were long-term, loyal employees with no motive that anyone could discern. There have been no fires since the housekeeping changes were instituted.

Figure 7.23 (a) Two cyclones damaged in two separate but related fires. (b) Damage on the cyclone that caught fire Tuesday, indicating that the fire spread to the inside.

Figure 7.23 (c) Fire damage on the exterior of the cyclone that was damaged on Friday. There was no damage on the inside of this cyclone.

Figure 7.23 (d) Fire damage around a chute below the cyclone that caught fire on Tuesday.

Figure 7.23 (e) Silver paint on some of the bolts of the cyclone that caught fire on Tuesday. This paint was applied after a fire that occurred 18 months previously. (f) Arcing damage found on the load side of a disconnect box that may have been abandoned in place 18 months previously or may have been damaged in the Tuesday fire.

Figure 7.23 (g) Dents caused by mechanical impact of metal objects on the inside of a duct. (h) Dents caused by mechanical impacts in a fan housing. This fan housing was repaired in several areas by welding where it had been penetrated by flying debris.

7.8.3 Industrial fire 3: A design flaw in a printing machine

ATS was requested to investigate the cause of a fire and explosion in a small printing machine designed to produce labels for building materials. The machine sustained an explosion and fire while in use, but employees of the print shop were able to bring the fire under control with a hand-held extinguisher. This machine was a new purchase, and the owner wanted to know the cause of the fire and how a future loss could be prevented.

The printer is shown in Figure 7.24a. The interior is shown in Figure 7.24b, and it is possible to see the dry powder extinguisher in this photograph. This machine uses two reservoirs of highly flammable ink having a flash point of 24°F. The bottom of the unit is shown in Figure 7.24c. Evidence of burning underneath was seen. Apparently, one of the ink reservoirs leaked, and some of the ink managed to flow into the electronics cavity. Figure 7.24d shows the electronics cavity where the ink is routed in steel tubes to prevent leakage. The electronics cavity is also protected from ink vapors by fans that blow from the electronics cavity across the ink reservoir cavity, but the bulkhead between the two cavities is not sealed against liquid penetration.

Figure 7.24e shows the corner of the bulkhead through which the ink flowed. The small cutout seen in the photo was probably designed for ease of assembly. It would have been a simple matter to seal this cutout with a drop of silicone.

Once the fugitive liquid vapors in the electronics cavity were ignited, they flashed back to the ink cavity, producing the pattern shown in Figure 7.24f.

Epilogue: The printing machine in question failed because of a design defect. The designer knew the hazards of flammable vapors and took steps to avoid those hazards. No provision was made, however, for the foreseeable occurrence of a spill of the highly flammable ink.

Figure 7.24 (a) Printing machine in which an explosion occurred.

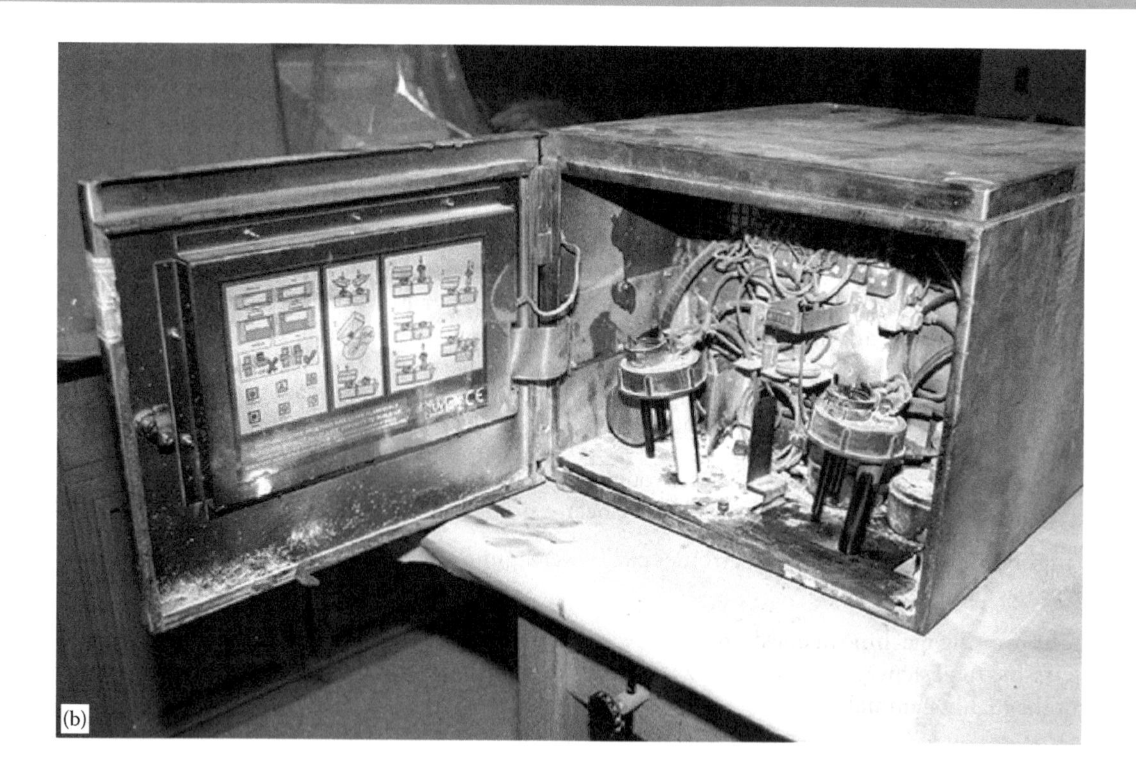

Figure 7.24 (b) Cavity where two reservoirs of highly flammable ink were located. The light-colored powder is dry chemical fire-extinguishing agent. (c) Fire pattern on the bottom of the printer.

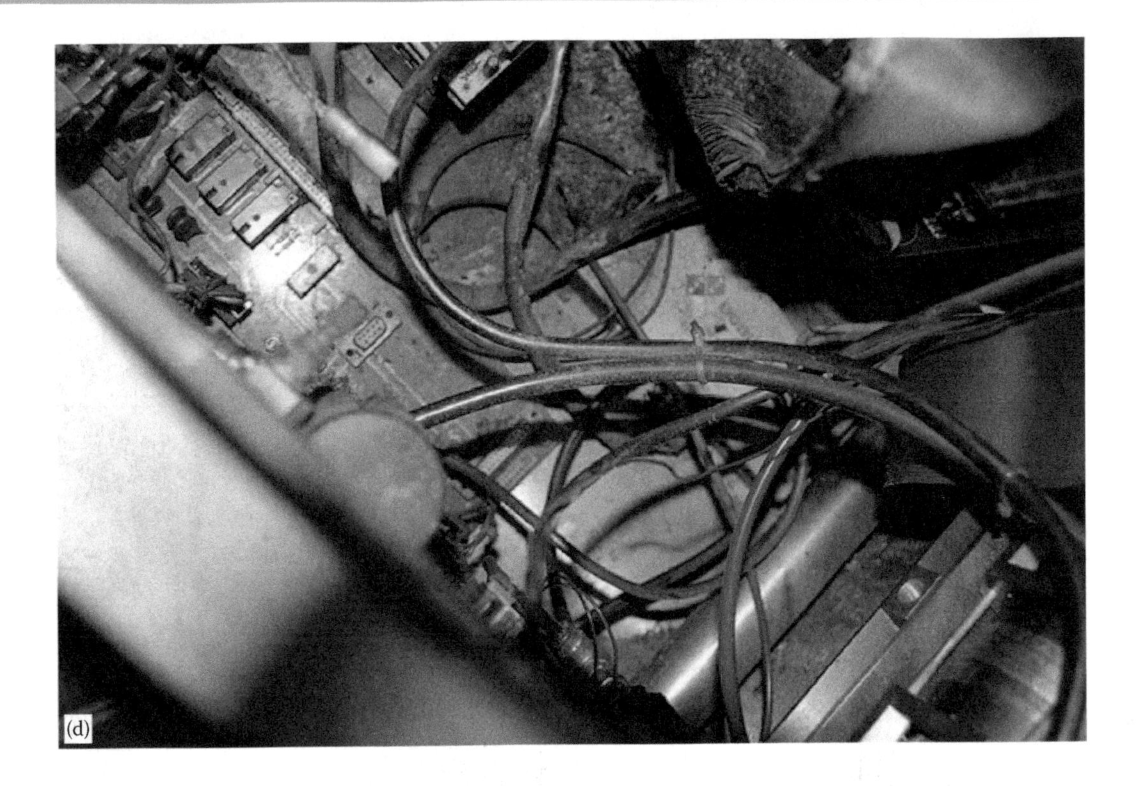

Figure 7.24 (d) Electronics cavity. The steel tubes carried the ink hoses through this cavity. (e) Location where ink spilled in the next cavity flowed through a cutout in the bulkhead.

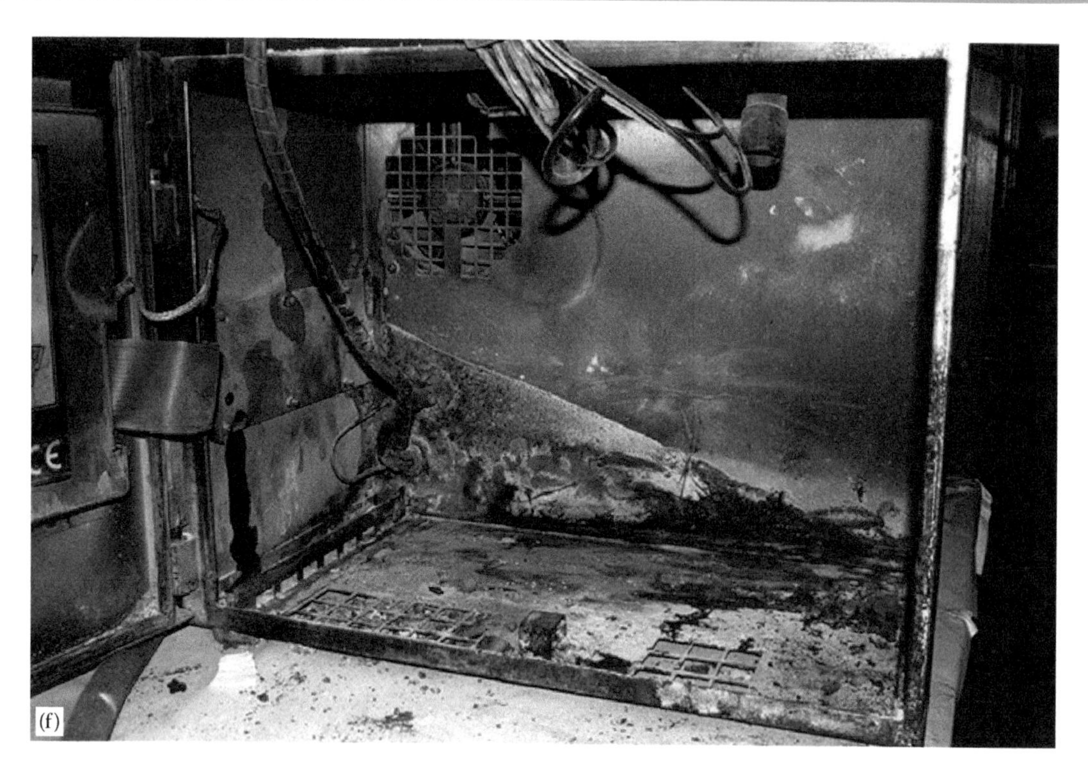

Figure 7.24 (f) Fire pattern in the ink cavity, moving from the lower right to the upper left at the location of the cutout.

7.8.4 Industrial fire 4: Hydraulic fluid fire

ATS was requested to investigate the origin and cause of a fire that seriously damaged a chicken-processing plant located in northern Georgia. The fire occurred as employees were setting up a large deep-fat fryer for the day's operations after the maintenance shift had concluded. The fire damaged the refrigeration system at the plant, and several hours passed before the ammonia from that system could be reduced to levels that allowed safe entry into the burned area. The plant is shown in Figure 7.25a. Although some damage was visible on the exterior skin, the roof above the origin of the fire exhibited much more damage, as shown in Figure 7.25b. This damage distribution is typical in large metal structures. Origin determination was straightforward as there were numerous eyewitnesses. The fire began at a frying machine, used to "par fry" the seasoned coating onto chicken parts that would later be shipped to many different fast-food restaurants. The machine was supposedly protected by a large carbon dioxide system in a hood that lowered to a position close to the surface of the oil; however, the oil was not the first fuel ignited. The output end of the fryer is shown in Figure 7.25c. The fire originated at the input end, where a bank of three hydraulic valves controlled the speed of the conveyor and the position of the hood. The middle valve shown in Figure 7.25d was identified as the source of the fuel (pressurized hydraulic fluid) when the hydraulic system valve just upstream of the machine was opened. The fluid came out of a hole designed to accommodate one of two guide pins, which slipped into notches at either end of the movable valve seen in Figure 7.25e. We found half of one guide pin, and a metallurgist determined that it had broken due to an overload failure. A scanning electron micrograph of the guide pin is shown in Figure 7.25f. The cause of the failure of the guide pin turned out to be an accumulation of metallic debris in the hydraulic system. The debris, shown in Figure 7.25g, introduced friction, which the operator overcame by forcing the valve lever. When presented with this debris, the plant manager remembered a hydraulic pump failure two weeks earlier, which had shut down the plant and required replacement of the pump. After such a failure, which involved the compressor vanes flying apart, it is necessary to drain and clean the hydraulic system to remove the debris produced by the pump failure. The plant owners instead decided to try to remove the debris by filtration, which involved less downtime but ultimately turned out to be a very costly decision.

In following up on the history of this fryer, we learned that there had been several major fires caused by hydraulic failures, and we even assembled a group of investigators to compare notes.

Figure 7.25 (a) Chicken-processing plant that sustained a fire when a hydraulic valve failed and sprayed atomized hydraulic fluid into a large natural gas flame under a fryer. (b) Damage on the roof over the origin of the fire.

Figure 7.25 (c) Output end of the fryer. "Par-fried" chicken parts were conveyed on a steel belt through the oil. They were then moved into the freezer, the entrance of which is at the left side of this photograph. The fire originated just below the light in the background. (d) Bank of three hydraulic valves at the origin of the fire. A guide pin in the middle valve broke, and a stream of pressurized hydraulic fluid was sprayed into the natural gas flame under the fryer.

Figure 7.25 (e) Valve spool from the middle hydraulic valve. Metal debris from a hydraulic pump failure jammed between the guide pins that attached to the end of this spool and the valve body, resulting in excessive load on the pins when the valve lever was moved. (f) Scanning electron micrograph of the broken end of one of the guide pins from the failed valve. The small box was identified as the origin of a crack caused by overloading the pin in tension.

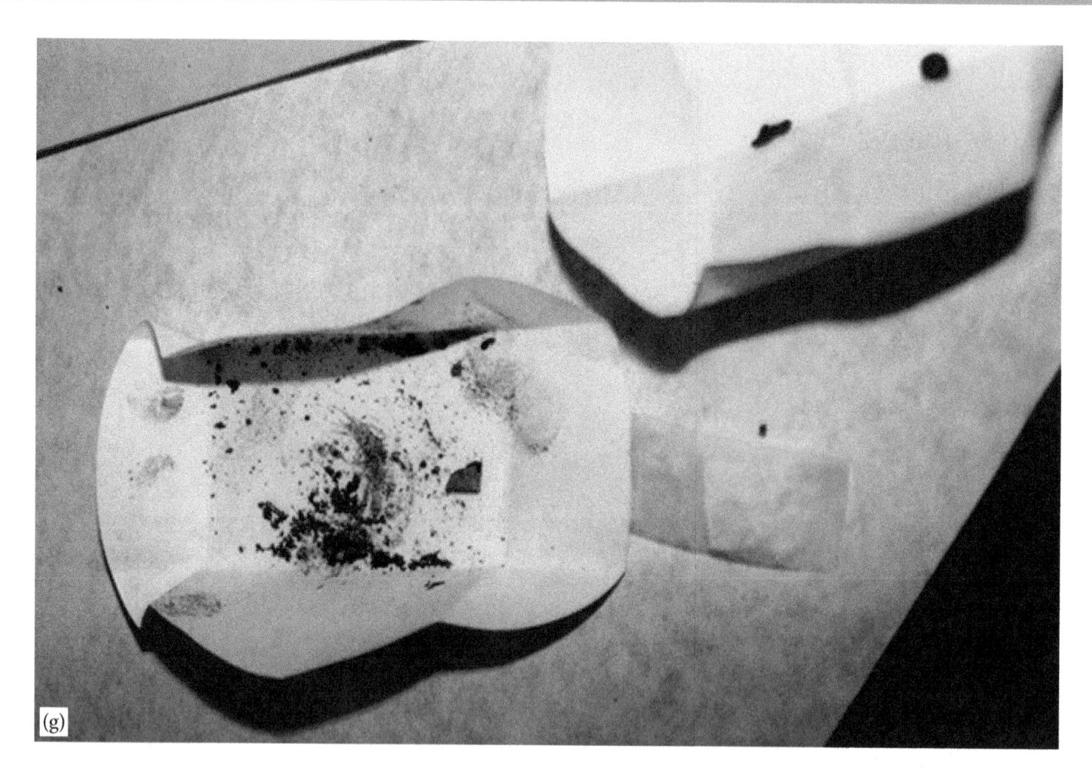

Figure 7.25 (g) Metal debris, identified as the remains of failed compressor vanes. This debris should have been cleaned out of the hydraulic system, but plant management made the decision to restart the plant without adequate cleanup.

Fires of this type were well known to the manufacturer, who had devised a solution involving the provision of a small hydraulic fluid reservoir for each individual fryer rather than connecting the fryer's hydraulic system to the plant hydraulic system. Thus, in the event of a hydraulic fluid leak, there would be 10 gal. of fuel instead of hundreds.

Epilogue: Subrogation against the manufacturer of the fryer by the insurance carrier for the factory was unsuccessful due largely to the fact that the factory owners purchased a new unit of identical design from the same manufacturer. They did not even spend the additional money for the safer hydraulic system. This made it difficult for their insurance carrier's attorney to argue that the plant owners, the nominal plaintiffs, actually believed that the product was unsafe.

7.8.5 Industrial fire 5: Another chicken story

ATS was requested to investigate the origin and cause of a fire that destroyed a chicken plant located in central Alabama. The damage was so extensive that it was necessary to tear the plant down and rebuild from the bare slab. Figure 7.26a shows the plant immediately after the fire. Not much damage was visible from the outside, but the inside was devastated. The origin of the fire was a deep-fat fryer, but in this case, the fire was caused by overheating of the oil. The fryer is shown in Figure 7.26b. This was an electric fryer with tubes containing the heating elements that were immersed in the oil. The main temperature sensor is seen in Figure 7.26c. If the oil level drops below the level of this sensor, the sensor measures air temperature, not the oil temperature. Because the air is cooler than the oil, the sensor does not stop the current. Further, the tubes overheat because air is a poor conductor compared to the oil, so there can be local overheating of the oil should the heating tubes become exposed. A second set of sensors was attached to a float in a sump to the side of the main oil reservoir, designed to cut off the heaters if the oil level either gets too high or too low. These sensors are shown in Figure 7.26d. The manufacturer of the fryer was put on notice and allowed to inspect the fire scene. A joint inspection was conducted on site, and then the controls were removed for laboratory testing. Figure 7.26e shows the tags placed on the fryer to indicate where the devices were removed. A series of experiments was conducted in the laboratory to design a testing protocol for an examination that would take place on the subject fryer, which had been removed to a warehouse. Figure 7.26f shows an experimental setup that allowed us to heat a small volume of oil to test the thermostatic controls. Once a consensus protocol was accepted, a field test was conducted using the fryer controls. In the experiment shown in Figure 7.26g, we were able to reproduce the overheating of the oil. When we traced out the circuitry in the control panel, we learned that two safety devices had been defeated, as demonstrated by the wire nuts shown in Figure 7.26h.[3]

Once the cause of the fire was determined, we examined the failure of the fire suppression system to extinguish it. The oil vat was "protected" by a carbon dioxide (CO_2) gas extinguishing system. Two T-cylinders of carbon dioxide

Figure 7.26 (a) Chicken-processing plant that was completely destroyed by a cooking oil fire. (b) Deep-fat fryer at the origin of the fire.

[3] At least they used wire nuts rather than just twisting the wires together and taping them.

Figure 7.26 (c) Temperature sensor located just above the heating tubes. If the oil level drops, this sensor senses air temperature, not the temperature of the oil. (d) Sump for measuring the oil level. The float switch installed here shuts off current to the heaters if the oil level is either too high or too low.

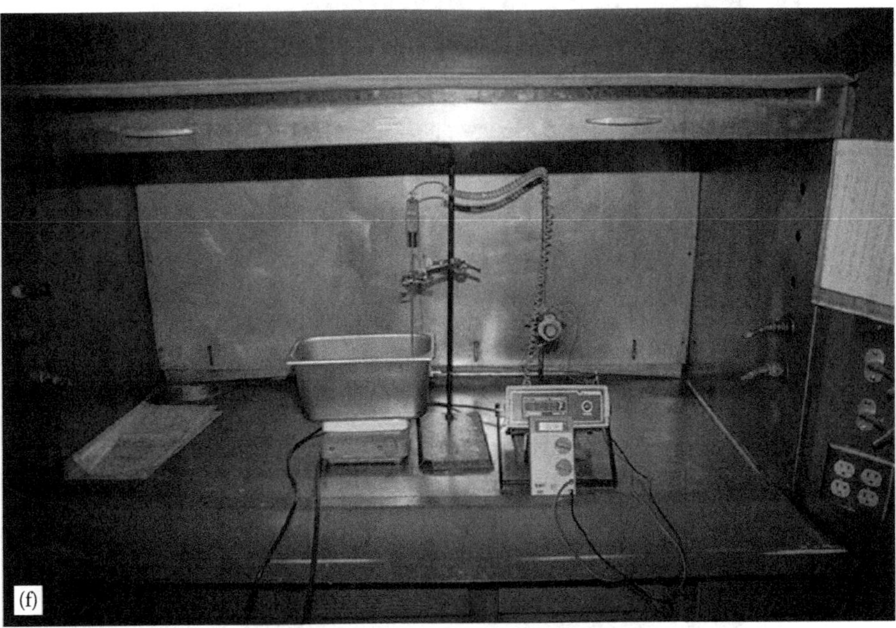

Figure 7.26 (e) Tags showing the date of removal of sensing devices. These were taken back to the laboratory for testing and for the development of a testing protocol for the control panel. (f) Laboratory test of the thermal sensor. This test also validated the design of an experiment to test the control panel.

Figure 7.26 (g) Field test that allowed us to heat a small quantity of oil to test the relays and other components in the control panel. (h) Wire nuts used to remove two separate safety devices from the circuit in the control panel. An expedient measure such as this is a common cause of industrial fires.

Figure 7.26 (i) 10,000-gal. CO_2 tank that might have made the extinguishing system more effective.

were attached to the system, which proved wholly inadequate. When the system first discharged, the flames died down, but a gaseous CO_2 system does not do anything to cool the oil nor does it do anything to shut off the hood fan to prevent fresh air from getting to the oil. The CO_2 dispersed and the fire flared back up. Plant employees were able to connect two more cylinders to the system, with similar results—the fire died down and then flared back up. There was a 10,000-gal. tank of liquid CO_2 on the premises and within easy reach of the fryer. The CO_2 was used to flash-freeze the chicken after it came out of the fryer. Perhaps if this tank, shown in Figure 7.26i, had been connected to the extinguishing system, it might have been possible to control a fire in the oil, but it would probably have killed all of the employees in the plant by displacing all the oxygen. Gaseous CO_2 is a useful extinguishing agent for small fires because it leaves no residue. It is not useful for large oil fires when the oil has been heated above its ignition point, particularly with a fume hood running that helps disperse the extinguishing agent.

Epilogue: The fryer manufacturer was clearly not at fault in this case. Plant personnel had taken the expedient step of wiring around two safety devices rather than fixing the problem that ultimately led to the fire that burned the plant down. The design of the extinguishing system was so severely inadequate, however, that the company that put it together agreed to accept some liability for the fire. Unfortunately, their insurance policy was capable of covering only about 5% of the loss.

7.9 LIGHTNING FIRES

Lightning produces some of the most interesting artifacts a fire investigator is ever likely to encounter. It is often said that the only predictable thing about lightning is that it is unpredictable. While that is generally a true statement, lightning does produce certain artifacts when it interacts with a structure, and an investigator will come to associate certain fire patterns with lightning. Because lightning is the ultimate accidental fire cause, it is not investigated nearly as often as it could or should be.

Certain steps can be taken to reduce the likelihood that a lightning strike will produce serious damage. Chief among these is bonding the electrical system to metal piping and framing systems in a building. The fire protection value of bonding is sufficiently well recognized that it is required by the *National Electrical Code,* NFPA 70, at Article 250 and Article 820 [12]. Even when lightning is the cause of a fire, the failure of the electrical system or gas system or cable TV system installer to bond these systems together may be the actual cause of the damage. This author has been involved in several subrogation cases where the insurance carrier for the residence was successful in seeking compensation from the individual or company that should have bonded the systems but failed to do so.

Lightning artifacts are often destroyed by the ensuing fire. The investigator is left with the option of checking nearby trees and tall structures for evidence of lightning in the area or relying on eyewitnesses. One artifact that lightning leaves behind permanently is a record of its electronic signature. This signature is recorded by (and can be retrieved from) a lightning detection network. The fire investigator provides the address (or the coordinates) of the fire scene, and within 24 hours, a report can be provided stating whether lightning occurred within the time period of interest within a 5- or 10-mile radius of the location. For an additional fee, "error bars" can be included; if there is a lightning strike indicated within 500 yards of the fire scene, the investigator will have a very difficult time excluding lightning as the cause. Lightning can begin up to 5 miles high in the atmosphere, and a single strike can hit in several places [13].

The voltage of a lightning strike can exceed 1 million volts, and it has the capacity to cause damage at a distance through the "transformer effect." The lightning strike can induce a very large current in conductors that are parallel to the direction of the lightning strike, generally vertical. This can cause all kinds of damage to electronic equipment connected to the circuit and may even cause fires. Further, lightning is capable of causing perforations in gas lines, resulting in subsequent fires. The three fires reported here are typical lightning fires.

7.9.1 Lightning fire 1: Be careful what you wish for!

ATS was requested to investigate the origin and cause of a fire at a residence in southern Georgia. The fire occurred late one July afternoon while the homeowner's daughter was alone at home. The insurance carrier learned that the homeowner had divorced his wife about 18 months prior to this fire and had been trying to sell the house ever since. Because the fire was so "convenient," the insurance company elected to have the scene investigated. Figure 7.27a shows the right rear corner of the house, where the daughter stated that she heard the lightning strike. Figure 7.27b shows the reason for the investigation, a "For Sale" sign that had been located in front of the property.

Figure 7.27 (a) South Georgia house reportedly struck by lightning.

Figure 7.27 (b) The "For Sale" sign that prompted the investigation.

Figure 7.27 (c) Fire damage in an upstairs bedroom, all caused by dropdown from above. (d) Fire damage above the center of the house, showing that the fire burned in the attic space.

Figure 7.27 (e) Exploded section of closet wall in an area unaffected by the fire.

Like many lightning fires, this was a high fire that burned in the attic. Figure 7.27c shows an upstairs bedroom that was heavily damaged, but the damage was due to a falling roof structure. No evidence existed that the fire was confined in this room at any point. Figure 7.27d shows a more typical ceiling in the house. No damage whatsoever was seen below the ceiling line. Figure 7.27e shows a typical lightning artifact: a closet on the first floor where a section of wall had simply blown out. Several areas like this were found in the residence.

Epilogue: All of the evidence pointed to this being a lightning fire, and no evidence existed to contradict this inference. An inventory of the contents revealed no evidence that any items of personal property had been removed from the house prior to the fire. Further, it would take a cold individual, indeed, to set his house on fire with his daughter in it. The homeowner, after an interview with this author, had a parting comment, "John, I don't know if you're a praying man but if you ever ask the Lord for something, be specific."

7.9.2 Lightning fire 2: Lightning opens a gas appliance connector

This fire occurred in the middle of the night in mid-summer of 2003, during a thunderstorm. The homeowners were away for the weekend, and the fire was not reported until a neighbor saw it venting out the living room window. There was a classic V-pattern on the side of the house, as shown in Figure 7.28a. This fire appears to have started in the living room, auto-ventilated through the window, and spread up the outside of the dwelling. A closer examination revealed, however, that the fire actually started below the living room, where the flexible stainless steel gas line shown in Figure 7.28b was located. This flexible line supplied a set of artificial logs in the fireplace. It opened up and burned through the floor directly underneath the sofa. The reason for the gas line opening up was apparent when the chimney cap, which had been blown into the driveway, was examined. Figure 7.28c shows a hole almost exactly the size of a quarter (24 mm) burned into one edge of the chimney cap. Figure 7.28d is a close-up view of that hole, which is almost perfectly round. A lightning strike energized the metal chimney and the attached firebox, and the current jumped to a section of black pipe that supplied the fireplace logs, which in turn was connected to the flexible stainless steel tubing. A close-up of the melted end of the stainless steel tubing is shown in Figure 7.28e. This tubing appeared yellow because it was exposed to heat, and when it was brought back to our

Figure 7.28 (a) V-pattern around the living room window of a house ignited by lightning.

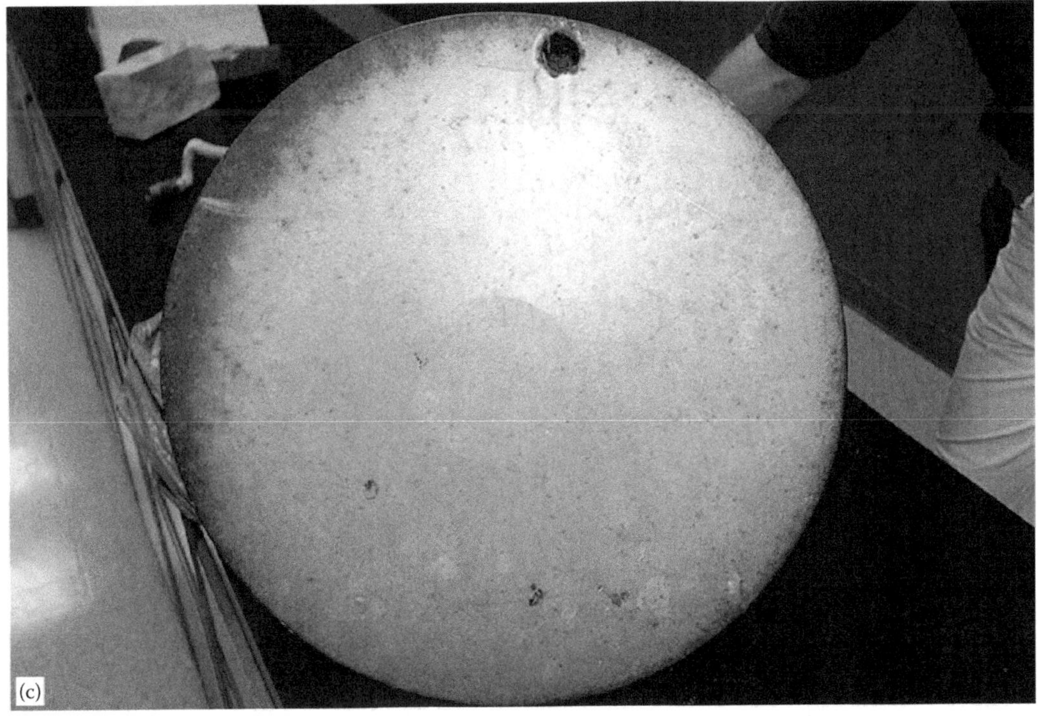

Figure 7.28 (b) Flexible stainless steel gas line that burned open as a result of being energized by lightning. (c) Chimney cap, showing a quarter-sized hole burned in one edge.

Figure 7.28 (d) Close-up of the hole burned in the chimney cap. (e) Close-up of the stainless steel tubing, where it burned open.

Figure 7.28 (f) Brass nut to which the stainless steel tubing was connected.

laboratory, the author was required to conduct a chemical test to prove to the investigator who brought it in that the tubing was, in fact, stainless steel and not brass. The brass nut on the tubing is shown in Figure 7.28f. It also melted when the current reached a point of locally high resistance. When the tubing opened up, a very competent ignition source in the form of molten metal was already available to ignite the natural gas. The gas probably burned under the floor for 15 min or more before it penetrated into the living room, igniting the sofa.

Epilogue: The artifacts left by the lightning strike were classic. Even if this house had been located far out in the country and burned to powder, the hole in the chimney cap and the melted gas line would have been enough to determine the cause of the fire.

7.9.3 Lightning fire 3: Nearby lightning strike causes perforation in a CSST line

This fire was the result of lightning causing the penetration of CSST. This mass-produced, thin-walled gas piping was introduced into the stream of commerce in the late 1980s with minimal testing and no apparent thought to what might happen if the residence where it was installed was subjected to either a direct or indirect lightning strike [14]. Its resistance to damage is far less than either copper tubing, which has thicker walls, or black pipe [15]. It is much easier to install, however, than either of these older alternatives. (The exception being that joints are difficult to make properly, as described earlier in this chapter.)

The home shown in Figure 7.29a exhibited no visible damage on the exterior, but it sustained major damage to the ceiling in the basement. An indirect lightning strike was responsible. There was an invisible fence installed in the yard to keep a dog in, and where that fence crossed over a polyvinyl chloride (PVC) gas line equipped with a metal tracer wire, the energy was directed into the house and on to the CSST piping system. The point of crossover is shown in Figure 7.29b. The energized CSST opened up where it crossed a section of metal HVAC duct in the area shown in Figure 7.29c. The green ribbon in this photo is used to mark the hole. A close-up view of the hole is shown in Figure 7.29d.

Figure 7.29 (a) Residence damaged when an indirect lightning strike caused a CSST line to be perforated. (Courtesy of Mark Goodson, Goodson Engineering, Denton TX.) (b) Underground crossover point where lightning energy on an invisible fence jumped to a tracer wire on the gas line. (Courtesy of Mark Goodson.)

Figure 7.29 (c) Origin of the fire above the CSST line under the basement ceiling. Green ribbon shows the location of the perforation in the gas line. (Courtesy of Mark Goodson.) (d) Close-up view of the perforation in the CSST gas line. (Courtesy of Mark Goodson.)

7.10 WATER HEATERS

The failure modes of water heaters were discussed at length in Chapter 6. The fire presented below is intended to illustrate the point that, while it is never a good idea to violate the National Fire Codes, code violations, in and of themselves, are frequently nothing more than code violations. It would be difficult to conduct a thorough inspection of any residence and not detect at least one violation of a fire code, whether it be the *National Electrical Code*, the *National Fuel Gas Code*, or the *Flammable and Combustible Liquids Code*.

7.10.1 Water heater fire 1: A code violation that did not cause the fire

ATS was requested to investigate the origin and cause of a fire that damaged a residence located in northern Georgia. The fire occurred shortly after the teenage son arrived home from a fishing trip and piled his gear in the storage room at the rear of the garage. The door to this room was normally left open. The teenager's mother was unhappy about the mess that he had made in the room, so she closed the door. Figure 7.30a shows the side of the house with fire damage above the carport. Some smoke damage was noted inside the house as a result of the fire penetrating through the kitchen doorway, shown in Figure 7.30b. The fire clearly originated in the storage room, as indicated by the V-pattern at the doorway of the storage room, shown in Figure 7.30c. An LP-gas-fired water heater was located in the rear corner of this room, as shown in Figure 7.30d.

Because this room had a finished ceiling, it was immediately obvious to this investigator that a serious code violation occurred during the installation of the water heater. This room clearly met the definition of a "confined space" as set forth in NFPA 54, *The National Fuel Gas Code.* The total volume of the room was 374 ft^3. *The National Fuel Gas Code* defines a confined space as any room having a volume of less than 50 ft^3 per 1000 Btu/hr. This was a 40,000-Btu water heater and therefore required a 2000 ft^3 room, unless the room was ventilated. The installer of the water heater had made no effort to ventilate this room as required by the gas code. The installer was a contractor located in the same town as the fire, and he was immediately put on notice of a potential claim for liability. He stated that he had installed hundreds of water heaters just like this one and was not going to respond to any claims.

A fire pattern was located around the opening of the combustion chamber, shown in Figure 7.30e, which suggested that the fire might have come from there, but examination of the water heater did not reveal any excess soot on the burner or in the flue, which is a sign of air starvation. As the area was being excavated, a melted plastic container with a strong odor of gasoline, shown in Figure 7.30f, was found. Apparently, this container was knocked over when the gear was thrown into the storage room, and the fire developed about 10 min later when the gasoline vapors were ignited by the heater flame.

Epilogue: What had initially appeared to be a case where the insurance carrier might recover its loss through subrogation was finally determined to be the result of carelessness on the part of the homeowners.

Figure 7.30 (a) North Georgia residence damaged by a fire started by a water heater.

Figure 7.30 (b) Point where the fire moved from the carport into the residence.

Figure 7.30 (c) Fire pattern, showing the fire moving out of the storage room.

Figure 7.30 (d) Water heater installed in the unventilated storage room with an inadequate supply of air for combustion and ventilation.

Figure 7.30 (e) Fire pattern around the combustion chamber.

Figure 7.30 (f) Melted plastic gasoline container.

7.11 CONCLUSION

In many of the cases cited here, the determination of the circumstances that brought fuel, air, and heat together was straightforward. Fire investigators are frequently asked, however, to go beyond that narrow determination of cause. For example, the spread of the fire, rather than its cause, may be more important in understanding the damage. The cause of fatalities frequently has more to do with the arrangement of exits than the source of ignition. What our clients are often seeking is a determination of *responsibility* for the fire in addition to the origin and cause. The physical evidence and eyewitness accounts often allow an investigator to make such determinations.

Additional data, not directly related to the fire, often must be considered to evaluate whether the investigator's hypotheses make sense in the real world. The individual who acts as the *principal investigator* is charged with putting all the evidence together into a coherent story. Applying the scientific method to questions of responsibility is as important (and often more difficult) than applying it to origin and cause determinations.

Review questions

1. Although a build-up of lint is a common *feature* of dryer fires, what is the most common actual *cause* of dryer fires?
   a. Overheating of clothes in the drum
   b. Improper connection of the power source
   c. Spontaneous heating
   d. Contamination with ignitable liquids
2. Which of the following documents can be said to govern a fire investigator's duties with respect to preservation of evidence?
   a. ASTM E 860 *Standard Practice for Examining and Testing Items That May Become Involved in Litigation*
   b. ASTM E 1188 *Standard Practice for Collection and Preservation of Information and Physical Items by a Technical Investigator*

c. NFPA 921, *Guide for Fire and Explosion Investigations*
d. NFPA 1033 *Standard for Professional Qualifications for a Fire Investigator*
e. All of the above

3. What is the most common root cause of electrical fires?
a. Overcurrent
b. Overdriven staples
c. Short circuits
d. Loose connections

4. Why are problems with LP gas systems and appliances more dangerous than problems with natural gas systems and appliances?
a. LP gas contains more energy per unit volume than natural gas.
b. LP gas systems operate at a higher pressure than natural gas systems.
c. LP gas is heavier than air and more likely to accumulate than natural gas.
d. All of the above

5. What is the most common cause of industrial fires?
a. Expedient decision making to avoid costs or to keep the plant running
b. Labor unrest
c. Defective machine installation
d. Careless smoking

Questions for discussion

1. What accounts for the differences between a public-sector investigator's portfolio of cases and that of a private investigator?
2. Explain why the presence of residual vegetable oil is more likely to cause a dryer fire than the presence of residual ignitable liquid.
3. What are the relevant factors to understand when investigating an industrial fire loss?
4. Why is it important for a fire investigator to be familiar with the fire codes?
5. What are examples of recent changes to a structure that might cause a fire?

References

1. 6th Circuit, (2007) *Richey v. Bradshaw*, 498 F.3d 344.
2. NFPA (2017) NFPA 921, *Guide for Fire and Explosion Investigations*, National Fire Protection Association, Quincy, MA, p. 258.
3. NFPA (2017) NFPA 921, *Guide for Fire and Explosion Investigations*, National Fire Protection Association, Quincy, MA, p. 256.
4. Ettling, B. V. (1994) The overdriven staple as a fire cause, *Fire and Arson Investigator* 44(3):51.
5. Babrauskas, V. (2003) *Ignition Handbook*, Fire Science Publishers, Issaquah, WA, p. 790.
6. Sanderson, J. (Ed.) (2000) Carbon tracking: Poor insulation combined with contaminants is potential fire cause, *Fire Findings* 8(3):1.
7. NFPA (1998) NFPA 54, *National Fuel Gas Code*, National Fire Protection Association, Quincy, MA, 7.2.
8. Sanderson, J. (Ed.) (1997) Loose gas pipe fittings: Physics basics may aid your investigation, *Fire Findings* 5(2):1.
9. NFPA (2017) NFPA 58, *Liquefied Petroleum Gas Code*, National Fire Protection Association, Quincy, MA, 5.9.3.

10. Cox, D. E. (1997) LP gas cylinders, *Fire Findings*, 5(3):7.

11. NFPA (2016) NFPA 33, *Standard for Spray Application Using Flammable or Combustible Materials*, National Fire Protection Association, Quincy, MA, 9.1.

12. NFPA (2017) NFPA 70, *National Electrical Code*, National Fire Protection Association, Quincy, MA, Articles 250 and 820.

13. NASA (2003) Lightning really does strike more than twice. Available at https://www.nasa.gov/centers/goddard/news/topstory/2003/0107lightning.html.

14. Goodson, M., and Icove, D. (2016) Electrical characterization of corrugated stainless steel tubing components and systems, *Proceedings of the 7th International Symposium on Fire Investigations Science and Technology* (*ISFI*), NAFI, Sarasota, FL.

15. Goodson, M., and Hergenrether, M. (2005) Lightning Induced CSST Fires, *Fire and Materials Delegate Handbook*, Interscience Communications, London, UK.

CHAPTER 8

The mythology of arson investigation

For the great enemy of truth is very often not the lie—deliberate, contrived and dishonest—but the myth—persistent, persuasive, and unrealistic. Too often we hold fast to the clichés of our forebears. We subject all facts to a prefabricated set of interpretations. We enjoy the comfort of opinion without the discomfort of thought.

—John F. Kennedy
Yale University Commencement, June 11, 1962

LEARNING OBJECTIVES

After reviewing this chapter, the reader should be able to:

- Avoid the use of arson myths when investigating fires
- Understand how arson myths developed
- Understand how arson mythology in the fire investigation literature has influenced the development of the fire investigation profession
- Recognize arson myths when reviewing the reports of other fire investigators or reading outdated fire investigation texts

8.1 DEVELOPMENT AND PROMULGATION OF MYTHS

The first edition of this text was written in 2004. At that time, the myths described in this chapter were believed by a significant number of fire investigators. Over time, retirements and the effect of texts like National Fire Protection Association (NFPA) 921 and others, as well as court rulings overturning cases based on myths, have reduced the prevalence of the myths in current practice. For fire investigators who began their careers after 1992, NFPA 921 has always been a fact of life, although the *Guide* cannot be said to have gained general acceptance before 2001. There are still incarcerated citizens who are appealing convictions that occurred before 2001, but today it is unusual to see a report containing conclusions based on myths. Many of the texts containing the myths are still in print, but fire investigation texts published after 2001 are likely to be myth-free or nearly so. That is not the case with some general investigation or criminal justice texts. In 2015, the author found it necessary to contact the writers of two widely used texts, and request that they fix their chapters on fire investigation in their next editions.[1]

[1] The books were *Criminal Investigation*, 10th edition by Orthmann and Hess (Delmar, Clifton Park, NY, 2013), and *Encyclopedia of Security Management*, 2nd edition, by Fay Butterworth-Heinemann, Waltham. MA, 2007). Both authors have promised corrections in the next edition.

Despite the myths' waning popularity, an exposition of their history is still worthwhile. The introduction and persistence of mythology in fire investigation are an unfortunate part of the history of the discipline and an area that many fire investigators do not like to think about. Some would like to keep these dirty little secrets locked away in a closet in the hope that people will gradually forget about them and they will not be a problem anymore. It is this failure to address a serious problem in the training and education of fire investigators that caused the myths to persist. The unfortunate consequence is that innocent lives are destroyed by well-meaning but ignorant investigators. The purpose of this chapter is to attempt to understand why the myths came into being and why some of them still persist. The hope is that new investigators, or those considering entering the discipline, can be spared the necessity of having to "unlearn" things that are simply not true.

Just as in the study of Greek or Roman mythology, no single reason explains why a myth develops. Certainly, no reason exists to believe that any investigator deliberately sets out to promulgate something that is not true. It is likely that most myths came about as a result of unwarranted generalizations. For example, an investigator might observe that in a garage fire, a pattern of spalling surrounds the remains of a gasoline container and makes an association of gasoline with spalling. The next time he sees spalled concrete, he infers that gasoline must have been involved.

Some myths arise because of intuitively obvious "deductions." The notion that gasoline burns hotter than wood is appealing; as anyone who has ever started a wood fire knows, it is much easier to start it with liquid fuel, and certainly after a short time, a fire started with, for example, gasoline is throwing off much more heat than the fire burning wood only. Therefore, the flame temperature must be higher, right? Wrong! But even Paul Kirk, arguably one of the finest forensic scientists of his time, bought into this notion. In the first edition of *Kirk's Fire Investigation* (1969), he described the utility of examining melted metals:

> Whenever any residues of molten metal are present at the fire scene, they will reliably establish a minimum temperature for the point of their fusion in the fire. The investigator may use this fact to advantage in many instances, because of the differences in effective temperature between simple wood fires and those in which extraneous fuel, such as accelerant, is present [1].

To this day, investigators sometimes infer the presence of accelerants when they observe a melted aluminum threshold.

The notion that crazed glass indicates that the glass was rapidly heated was appealing enough that Brannigan, Bright, and Jason, three respected researchers at the National Bureau of Standards (now the National Institute of Standards and Technology [NIST]), included it in the *Fire Investigation Handbook* (1980). Some authors have declared that crazed glass is sufficiently useful that the size of the crazing cracks can indicate proximity to the area of origin [2].

It is the publication and continued promulgation of myths that ensure their longevity. If an "arson school" decides to use a text in its training courses, hundreds of investigators can be exposed to this false "gospel." Those who take few refresher courses, fail to keep up with the literature, and attend few meetings may never be exposed to updated ideas and new research.

The question naturally arises as to why fire investigation espouses (or has espoused) such a wide variety of myths, whereas DNA analysis, a forensic discipline derived from molecular biology, has many fewer myths to expunge. To some extent, the answer lies in the nature of the practitioners. In forensic DNA, the leaders in the field are trained scientists. If someone told them that crazed glass resulted specifically from rapid heating, they might remember an experiment in an undergraduate chemistry lab that they tried to save from overheating by adding some water, only to watch the glass beaker craze when the water touched it. Thus, they might consider an alternate explanation for the observation of crazing. During their training, scientists are supposed to acquire what Carl Sagan delicately referred to as a "baloney detector," otherwise known as natural scientific skepticism. However, one need not possess a science degree to be appropriately skeptical. Sagan wrote:

> The tenets of skepticism do not require an advanced degree to master as most successful used car buyers demonstrate. The whole idea of a democratic application of skepticism is that everyone should have the essential tools to effectively and constructively evaluate claims of knowledge. All science asks is to employ the same levels of skepticism we use in buying a used car [3].

Presented with the notion that large shiny alligator blisters occur only on wood surfaces that have been rapidly heated, a scientist will say, "Show me the data!" while an apprentice fire investigator will absorb the "knowledge" from his experienced mentor. When someone with an advanced degree publishes the myth and maybe even an apparent explanation for why it is so (albeit with no real data), the apprentice internalizes the fallacy as fact, making retraining difficult. Once the fire investigator uses the myth to send someone to prison, he or she is extremely reluctant to question the myth's authority lest he or she be forced to admit to an unspeakable error.

That is not to say that fire investigation myths have never been challenged. In 1979, Harvey French's *Anatomy of Arson* took on many of the "old wives' tales." In 1984, Bruce Ettling, writing in the *Fire and Arson Investigator,* asked, "Are we kidding ourselves?" Ettling listed many of the myths that will be explored in this chapter. He wrote:

> Some of the "old firemen's tales" that need to be discontinued are: that spalling of concrete indicates that an accelerant was used; that big rolling blisters on burned wood indicates a fast hot fire and probably accelerants; that black sooty smoke indicates petroleum products; that an accelerant will burn a big hole in a floor; that loose insulation on a wire indicates internal heating; that the standard time–temperature curve indicates how hot a fire ought to be at a given time; that PVC insulation after being warm for some time becomes brittle and allows shorts; that very hot fires can be obtained only with accelerants; and that a momentary arc at several thousand degrees can ignite solid fuels [4].

In 1986, in the third edition of *Scientific Evidence in Criminal Cases,* the authors noted the dearth of research to back up the conventional interpretation of "arson indicators" with the following indictment:

> Many of the arson indicators which are commonplace assertions in arson prosecutions are deficient for want of any established scientific validity. In many instances the dearth of published material in the scientific literature substantiating the validity of certain arson indicators should be sufficient grounds to mount a challenge to the general scientific acceptability of such indicators. It is clear, from the cases, however, that arson indicators are given a talismanic quality that they have not earned in the crucible of scientific validation [5].

Much of the mythology about fire investigation was collected by the Aerospace Corporation, under a contract awarded to the Law Enforcement Assistance Administration (LEAA), in a 1977 booklet entitled "Arson and Arson Investigation: Survey and Assessment." (Apparently, Aerospace had insufficient work from its primary client, the Department of Defense, so it "branched out" into criminal justice.) To their credit, the authors of this survey pointed out, "Although burn indicators are widely used to establish the causes of fires, they have received little or no scientific testing." They recommended "a program of carefully planned scientific experiments be conducted to establish the reliability of currently used burn indicators. Of particular importance is the discovery of any circumstances which cause them to give false indications (of, say, a fire accelerant)." In a remarkably prescient statement, they added, "A primary objective of this testing would be to avert the formidable repercussions of a court ruling on the inadmissibility of burn indicators on the grounds that their scientific validity had not been established." Moenssens et al. repeated this thought 9 years later, but serious challenges to the myths did not become common until NFPA 921 was published and the *Daubert* decision required that expert testimony be reliable.

The LEAA study is one of the oldest and most comprehensive listings of "indicators" and provides as good a starting point as any for the study of the myths of fire investigation. Here is the list from the survey:

- *Alligatoring effect:* Checking of charred wood, giving it the appearance of alligator skin. Large rolling blisters indicate rapid intense heat, while small flat alligatoring indicates long, low heat.
- *Crazing of glass:* Formation of irregular cracks in glass due to rapid intense heat—possible fire accelerant.
- *Depth of char:* Depth of burning of wood—used to determine length of burn and thereby locate the point of origin of the fire.
- *Line of demarcation:* Boundary between charred and uncharred material. On floors or rugs, a puddle-shaped line of demarcation is believed to indicate a liquid fire accelerant. In the cross section of wood, a sharp distinct line of demarcation indicates a rapid, intense fire.
- *Sagged furniture springs:* Because of the heat required for furniture springs to collapse from their own weight (1,150°F) and because of the insulating effect of the upholstery, sagged springs are believed to be possible only in

either a fire originating inside the cushions (as from a cigarette rolling between the cushions) or an external fire intensified by a fire accelerant.

- *Spalling:* Breaking off of pieces of the surface of concrete, cement, or brick due to intense heat. Brown stains around the spall indicate the use of a fire accelerant [6].

In addition to the misconceptions listed in the LEAA report, the following myths have also been widely promulgated:

- *Fire load:* Knowing the energy content (as opposed to the energy release rate) of the fuels in a structure was believed to allow an investigator to calculate the damage that a "normal" fire should produce in a given time frame.
- *Low burning and holes in the floor:* Because heat rises, it was widely believed that burning on the floor, particularly under furniture, indicated an origin on the floor.
- *V-pattern angle:* The angle of a V-pattern was supposed to indicate the speed of the fire.
- *Time and temperature:* By estimating the speed of a fire, or establishing the temperature achieved by a fire, it was believed that an investigator could determine whether it was accelerated.

Many of the myths about fire investigation were addressed in the first two editions (1992, 1995) of NFPA 921. In the chapter on fire patterns, there were several paragraphs entitled "Misconceptions about ______." While the technical committee felt it important to shine a spotlight on these myths, many in the fire investigation community railed against the notion that any of them had ever harbored any misconceptions about anything. They insisted, and the committee acquiesced, to a change in the 1998 edition that changed the section titles to "Interpretation of ______" as if removing the word *misconception* would remove the misconception. The optimistic thought was that the earlier editions (to which many investigators still objected) had relieved the discipline of its mythology. In discussing the origins and spread of myths, it will be necessary to name names (i.e., cite the sources where the myths are repeated) before citing the source that debunks the myth.

8.2 ALLIGATORING

After its citation in the LEAA survey, we find mention of alligatoring in the *Fire Investigation Handbook,* a mostly useful book published by the NBS (now NIST). The *Handbook* states:

> In determining whether the fire was a slowly developing one or a rapidly developing one, the following indicators may be used: (a) Alligatoring of wood—slow fires produce relatively flat alligatoring. Fast fires produce hump-backed shiny alligatoring [7].

The 1982 International Fire Service Training Association (IFSTA) manual unequivocally states:

> If alligatoring is large, deep, and shiny, the fire spread extremely rapidly. Large alligatoring should be considered an indication of the nearby presence of a flammable or combustible liquid [8].

Nowhere is it stated what the difference is between a "fast" fire and a "normal" fire. The lack of a definition of these subjective words not only renders the "indicators" of a fire's progress meaningless but also makes it nearly impossible to design an experiment that tests the indicator's usefulness. The Army's Field Manual 19–20, *Law Enforcement Investigations*, provides a slightly different interpretation of alligatoring when it states:

> When wood burns, it chars a pattern of cracks which looks like the scales on an alligator's back. The scales will be the smallest and the cracks the deepest where the fire has been burning the longest or the hottest. Most wood in structures char at the rate of 1 inch (in.) in depth per 40–45 minutes of burning at 1400 to 1600°Fahrenheit (°F)—the temperature of most house fires [9]. [*Thus combining three misconceptions in a single paragraph!*[2]]

[2] This document was originally published in 1985 and reprinted in 1995, 2005, and 2013. Sadly, the 2013 version still contains misinformation about alligatoring, the behavior of glass, and the angle of the V. It now references NFPA 921 and has copied some of the language from the 2000 U.S. Department of Justice (DOJ) publication *Fire and Arson Scene Evidence: A Guide for Public Safety Personnel.*

O'Connor's *Practical Fire and Arson Investigation* (1986) stated:

> Deep alligatoring (large rolling blisters) on an exposed wooden surface ordinarily indicates an intense, rapidly moving body of flame. This condition may be associated with the use of an accelerant [10].

The second edition of the book is far more cautious, the authors having been brought up to speed on this subject. The newer text states:

> It has been suggested that the presence of large shiny blisters (alligator char) and the surface appearance of char, such as dullness, shininess, or colors, have some relation to the presence of liquid accelerant as the cause, but no scientific evidence substantiates this. The investigator is advised to be very cautious in using wood char appearance as an indicator of incendiarism [11].

They have not completely given up on the myth, however. The 1997 text shows a photo of "a heavy rolling char … caused by the rapid intense movement (extension) of heat and flame [12]."

Noon, in his 1995 *Engineering Analysis of Fires and Explosions*, wrote:

> In the same way that a hunting guide interprets signs and markers to follow a trail of game, a fire investigator looks for signs and markers which may lead to a point of origin. For example, a fast, very hot burn will produce shiny type wood charring with large alligatoring. A cooler, slower fire will produce alligatoring with smaller spacing and a duller appearing char.

Noon then went on to explain "scientifically" why this should be so:

> As heat impinges on the piece of wood, the water in the surface material will evaporate and escape from the wood. The rapid loss of the water at the surface is also accompanied by a rapid loss of volume, the volume which the water formerly occupied. The wood surface then is in tension as the loss of water causes the wood to shrink. This is the reason why wood checks or cracks when exposed to high heat or simply dries out over time. Of course, if the heat is very intense, more of the water "cooks" out, and the cracking or alligatoring is more severe [13].

The scientific-sounding explanation (although it is rubbish) lulls the reader into believing that the author actually knows what he is talking about. This kind of exposition in many books that repeat the myths has enhanced the credibility of the myths and thus their longevity. (Both Noon's and O'Connor's books were published by a highly respected distributor of scientific books—CRC Press, which publishes this text.)

There were doubts about the shiny alligator theory even back in 1979. Harvey French, in *The Anatomy of Arson*, wrote:

> Due to limited scientific experimental study on these particular phenomena in relation to wood and fire, there is no present reliable data tending to identify the size of the alligatoring (its coarseness or fineness), its gloss, or other visual appearance with either rapid temperature rise or with the presence or use of flammable accelerants, such as gasoline, acetone, thinner or other volatile liquids or incendiary materials [14].

The final word on this and most other myths is NFPA 921. Here is what it says about alligatoring:

> **6.2.4.3 Appearance of Char.** In the past, the appearance of the char and cracks had been given meaning by the fire investigation community beyond what has been substantiated by controlled testing. The presence of large shiny blisters (alligator char) is not evidence that a liquid accelerant was present during the fire, or that a fire spread rapidly or burned with greater intensity. These types of blisters can be found in many different types of fires. There is no justification for the inference that the appearance of large, curved blisters is an indicator of an accelerated fire. Figure 6.2.4.3, showing boards exposed to the same fire, illustrates the variability of char blister.
> **6.2.4.3.1** It is sometimes claimed that the surface appearance of the char, such as dullness, shininess, colors, or appearance under ultraviolet light sources, has some relation to the use of a hydrocarbon accelerant or the rate of fire growth. There is no scientific evidence that such a correlation exists, and the investigator is advised not to claim indications of accelerant or a rapid fire growth rate on the basis of the appearance of the char.[3]

[3] The following applies to all quotations from NFPA 921 (except for reference to historical editions) cited in this chapter: Reprinted with permission from NFPA 921-2017, *Guide for Fire and Explosion Investigations*, Copyright © 2017, National Fire Protection Association, Quincy, MA. This reprinted material is not the complete and official position of NFPA on the referenced subject, which is represented only by the standard in its entirety.

The referenced figure is a photograph taken by Monty McGill, reproduced here as Figure 8.1a, which was first shown in the second edition of *Kirk's Fire Investigation*. It is the definitive evidence that debunks the myth of the shiny alligator. Figure 8.1b is a similar photograph taken by the author. Although McGill's photo is famous, the phenomenon that it shows is not unique. For our friends in the United Kingdom, we note the fact that Dougal Drysdale prefers the term *"crocodiling."*

Figure 8.1 (a) Photograph of different-sized char blisters produced on the same wall by the same fire. (Courtesy of LaMont "Monty" McGill.)

Figure 8.1 (b) Photograph of different-sized char blisters produced on the same wall in the same fire in one of the author's fire scenes.

8.3 CRAZED GLASS

It is unclear why anyone ever thought that crazing of glass indicated rapid heating. Perhaps a piece of crazed glass was observed near the known origin of a fire, and one influential investigator reached the wrong conclusion and repeated it to a large group of seminar attendees. However the notion began, it achieved widespread acceptance. Unlike most myths, this one has proved especially amenable to testing, but until 1992, nobody bothered to make the effort.

The NBS *Handbook* stated, "Window glass fragments in large pieces with heavy smoke deposits usually indicates slowly developing fires. Crazed or irregular pieces with light smoke deposits indicate a rapid buildup of heat [15]." Both sentences are false, but crazing is our focus for now.

The Army's Field Manual, *Law Enforcement Investigations*, states, "As a general rule, glass that contains many cracks indicates a rapid heat buildup. Glass that is heavily stained indicates a slow, smoky fire [9]."

IFSTA's *Fire Cause Determination* stated:

> A window with small crazing (minute cracking), and perhaps with light smoke accumulation, is probably near the point of origin, its condition suggesting intense and rapid heat buildup. Large crazing and a heavy smoke accumulation suggest slow heat buildup and remoteness from the point of origin [16].

The IFSTA manual may have been the source used in *Practical Fire and Arson Investigation* (O'Connor, 1986; O'Connor and Redsicker, 1997), which repeats the notion that crazing implies a "rapid and intense" heat buildup, and that if the crazing is "small," it is close to the area of origin. A larger crazing pattern, on the other hand, "implies that it may have been located in an area some distance away from the point of origin." The misconception about crazing follows an extensive discussion of the types of glass that an investigator may encounter, complete with softening points, chemical compositions, and applications. The reader thus is led to believe that the writers know all about glass [2]. DeHaan, who today lists crazed glass under "Myths and Misconceptions [18]," still believed the myth in 1991. He stated then, "Crazed glass, where the fractures or cracks resemble a complex road map in the glass, is certainly indicative of a very rapid buildup of heat sometime during the fire." Getting closer to an understanding of the true cause of the phenomenon, he went on to state, "Small cratering or spalling of the glass is more likely due to a spray of water hitting a hot pane of glass during suppression [19]." By the time the fourth edition of *Kirk's* was published in 1997, DeHaan acknowledged the work by this author that proved that crazing is *only* the result of rapid cooling [20,21].

In the study organized by the author following the urban wildland fire in Oakland, California, in 1991, crazed glass was one of three "indicators" examined. We observed that all the crazing occurred at those parts of the fire where there had been active suppression efforts, suggesting that water was associated with crazing.

Moving from field observations to laboratory experiments, this author tested that hypothesis. The experimental results should have put an end to the myth once and for all, by comparing the effects of "slow" heating and "rapid" heating on glass. (While definitions are lacking for "rapid" and "slow" heating, when the rates of heating are taken to extremes, little room for argument exists.) Window glass samples of various thicknesses were either put into a cold oven, and heated slowly to 1,500°F (approximately 800°C), or plunged into the same oven preheated to the same temperature. To simulate an even "faster" fire, the glass samples were heated to over a 40,000 British thermal unit (Btu) (approximately 12 MW) propane burner. While some cracking was observed with the propane flame, in no case did the rapid heating result in crazing. In every case in which the glass reached a temperature of at least 500°F (260°C), whether heated rapidly or slowly, the application of water invariably resulted in crazing. The crazing only occurred where the water was applied. By using a wet cotton swab, it was possible to write on the glass as shown in Figure 8.2.

The misinterpretation of the significance of crazed glass played a major role in the trials of Han Tak Lee in Pennsylvania (1989)[4] and of Ray Girdler in Arizona (1983). Crazed glass also played a role in the analysis of the Paul Camiolo case in Pennsylvania (1997). In that case, an expert brought in to bolster the Commonwealth's case was the first to point out that crazed glass was visible in one of the photographs. His report stated, "A photograph of melted 'crazed' glass also indicates a very rapid buildup of heat in the family room. This indicated a very rapid spread of the fire, unlike

[4] The presence of crazed glass on a bedroom window allowed a witness to "eliminate" a smoldering fire because the crazed glass indicated a rapid heat buildup. The condition of mattress springs also factored into that "elimination."

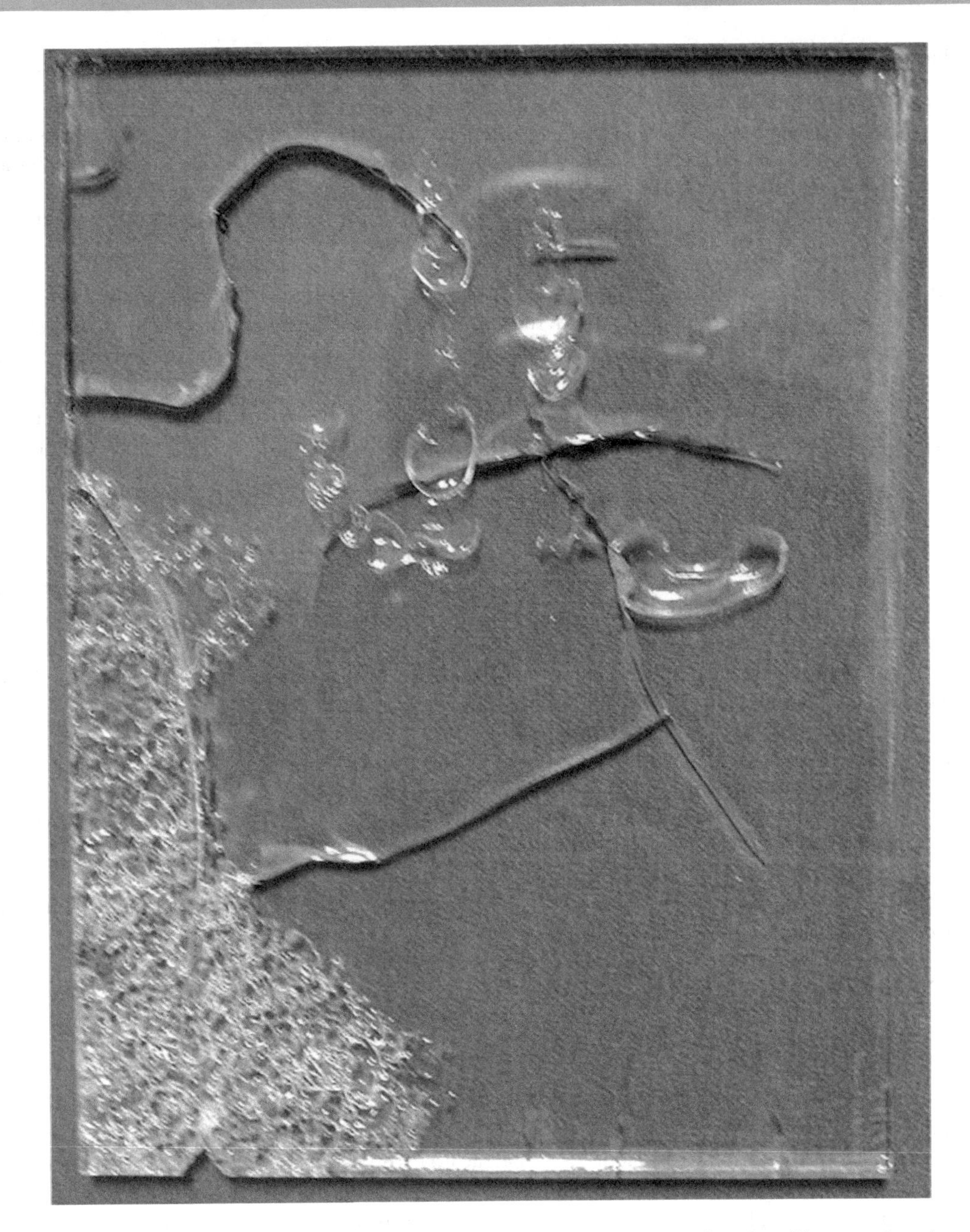

Figure 8.2 Author's initials written in crazed glass by applying water on a cotton swab to hot glass. The area of crazing at the lower left was caused by the application of water from a spray bottle.

a cigarette fire which burns much more slowly." This expert's use of a discredited indicator in testimony before the grand jury resulted in his being named as a defendant in a civil suit after the criminal charges were dismissed. In his deposition, he allowed that there were two ways glass could be crazed, by rapid heating or rapid cooling, but relied on logic that can most kindly be described as circular to support the former contention. He stated, "Well, as we spoke earlier in this deposition, we talked about the two ways you could get crazing. If I am to believe that the burning patterns on the floor are caused by a flammable liquid that was present, then I have to believe that the crazing was also caused by that condition itself. Had we not had the burning patterns on the floor, I probably would have looked more toward the crazing being caused by water which was put onto the glass [22]." It is possible that this particular individual has learned his lesson, but to this day, there are fire investigators who believe (and teach!) that crazed glass indicates (or may indicate) rapid heating.

NFPA 921 has the following comment on crazing:

> **6.2.13.1.4** Crazing is a term used to describe a complicated pattern of short cracks in glass. These cracks may be straight or crescent-shaped and may or may not extend through the thickness of the glass. Crazing has been claimed to be the result of very rapid heating of one side of the glass while the other side remains cool. Despite widespread publication of this claim, there is no scientific basis for it. In fact, published research has shown that crazing cannot be caused by rapid heating, but can only be caused by rapid cooling. Regardless of how rapidly it was heated, hot glass will reproducibly craze when sprayed with water.

It is interesting to note that crazing of glass as an indicator of rapid heating is a myth that never caught on in the United Kingdom. This is almost certainly because in the United Kingdom, the most widely read fire investigation text, *Principles of Fire Investigation*, correctly identified "the appearance of many small conchoidal fractures on one surface of the glass" as being the result of rapid cooling from extinguishment water [23]. The authors of that text did not use the term *crazing*. The absence of the crazing myth in the United Kingdom lends credence to the proposition that it is *publication* in apparently respectable texts that is responsible for the perpetuation of the mythology of arson investigation.

8.4 DEPTH AND LOCATION OF CHAR

The insurance industry has the greatest financial stake in accurately determining the cause of fires and in controlling the problem of arson fires. Thus, it is no surprise that insurance companies have historically assumed a leading role in fighting arson; even today, insurance companies provide the funding for almost all private-sector investigations. They have also contributed to the propagation of myths. In 1979, Aetna Life & Casualty published a brochure-style handbook, authored by John Barracato, which espoused as many myths as any publication ever printed on the subject. On the subject of depth of char, the booklet entitled "Fire … Is It Arson?" states:

> The speed at which a fire burns is an important indicator of its cause. A fire not involving accelerant (such as gasoline or other flammable liquid) burns at the rate of ¾ in. per hour (hr) into pine wood. The investigator should ask the fire department how long and intensely the fire burned, then carefully inspect any charred wood to see if there is a reasonable correspondence between the length of time the fire burned and the degree of damage it caused [24].

Exactly this type of analysis was put forward in the case *of Commonwealth v. Han Tak Lee* (1989). The investigator in that case made the following observations:

- The fire burned for a total of 28 min.
- Fire burns 1 in. in 45 min. (Note: this is a more commonly cited charring rate than Barracato's ¾ in./hr.)
- 2 × 10 s were completely consumed.

Therefore, the fire *must* have been accelerated because it would take 4.5 hr to burn through a 2 × 10. (This, of course, assumed that the fire would only burn in one dimension, as opposed to attacking the wood from both sides. This unique concept of fire spread is shown in Figure 8.3.) Barracato estimated the time required to burn a 2 × 4 at 1 hr 43 min at 1,780°F. The investigator was misled by two myths: (1) that the depth of char could be used to reliably determine the time of burning and (2) a uniquely held belief that only one dimension of the wood would be attacked by the fire. If we assume, for the sake of argument, that fire burns 1 inch in 45 min, then it should take only 34 min to burn through a 1.5-inch piece of wood, assuming it is attacked from both sides. This investigator had both his premise and his implementation of that premise wrong. Here is what NFPA 921 says about the rate of charring:

> **6.2.4.4 Rate of Wood Charring.** The correlation of 2.54 cm (1 in.) in 45 min for the rate of charring of wood is based on ventilation-limited burning. Fires may burn with more or less intensity during the course of an uncontrolled fire than in a controlled laboratory fire. Laboratory char rates from exposure to heat from one side vary from 1 cm (0.4 in.) per hour to 25.4 cm (10 in.) per hour. Care needs to be exercised in solely using depth of char measurements to determine the duration of burning.
>
> **6.2.4.4.1** The rate of charring of wood varies widely depending upon variables, including the following:
>
> 1. Rate and duration of heating
> 2. Ventilation effects
> 3. Direction, orientation, and size of wood grain
> 4. Species of wood (pine, oak, fir, etc.)

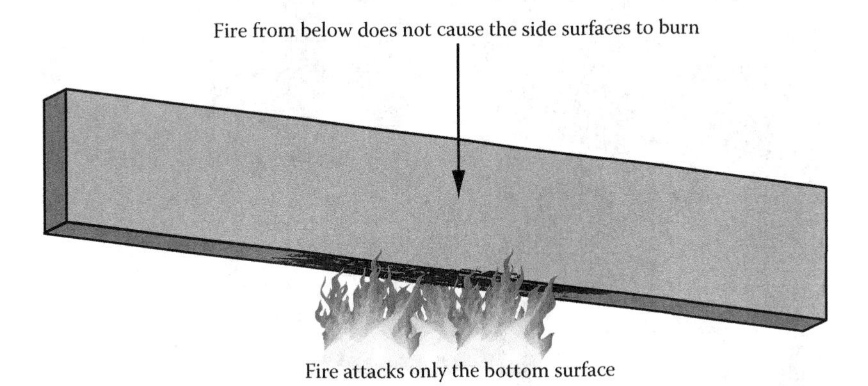

Figure 8.3 Conceptual drawing of how it would take more than 4 hours to burn a 2 × 10. One-dimensional thinking such as this caused an erroneous determination of arson.

5. Wood density
6. Moisture content
7. Nature of surface coating
8. Oxygen concentration of the hot gases
9. Velocity of the impinging gases
10. Gaps/cracks/crevices and edge effect of materials

8.5 LINES OF DEMARCATION

This is one of the more complex myths in fire investigation because, in some instances, lines of demarcation can be used to tell exactly what happened, whereas in other instances, lines of demarcation are just lines. The threshold question is whether the compartment where the lines occur experienced full room involvement. Let us be clear. There are times when a fire pattern is so obviously caused by an ignitable liquid that further analysis truly is "the icing on the cake." Unfortunately, once a fire progresses to full room involvement, it is no longer valid to make a determination using visual cues alone, and there are some who maintain this should never be done. Patterns like the one shown in Figure 8.4, however, when viewed in a similar context, are simply not worth arguing about. The arguments begin when

Figure 8.4 A rare case of an "obvious pour pattern." Only the carpet burned. There was no other fuel present, and the room showed no evidence of radiant heat. The carpet shows a "doughnut" pattern, and a sample of the carpet tested positive for the presence of a medium petroleum distillate.

Figure 8.5 Alternating areas of burning separated by sharp lines of demarcation between burned and unburned areas. No ignitable liquids were used in this test fire. These patterns are the result of areas of exposure to and areas of protection from the intense radiation that occurred when the room became completely involved. The white wires are thermocouple leads.

the room becomes fully involved (ventilation controlled) and the investigator still wants to be able to make a call based on the shape of the pattern alone. Such investigators are asking for trouble. The lines of demarcation seen in a 1992 test fire in Tucson, Arizona, shown in Figure 8.5, gave pause to the author and his colleagues conducting the test. No efforts were made to produce the pattern (later used in NFPA 921), but it looked to all present like something resembling a flammable liquid pour pattern had been created.

Sharp, continuous, irregular lines of demarcation between burned and unburned areas are frequently cited as evidence of the use of ignitable liquids. It is true that ignitable liquids can produce such patterns on carpeting, and many arson seminars include staged fires that are extinguished early so that investigators can learn to recognize "pour patterns." What is not evident from these incipient test fires is what happens after the room becomes fully involved. The carpet in the living room shown in Figure 8.5 failed in strips that resulted in alternating areas of exposure and protection. When the flashover and subsequent full room involvement occurred, the radiant heat burned the exposed floor and did not burn the protected floor. What resulted was an area of burning with sharp, continuous, irregular lines of demarcation between burned and unburned areas. On smooth surfaces such as tile or finished hardwood, ignitable liquids may burn off so quickly that they leave no patterns at all. (See the story of the Herndon case in Chapter 9.)

Another influence on the production of floor patterns of burning is carpet underlayment. Figure 8.6 shows a nearly straight-line burn from the top down that goes across the room to a sofa. This pattern was the source of much puzzlement and was declared by the original investigators to have been caused by the application of a flammable liquid. (As with many questionable determinations, the laboratory testing revealed no ignitable liquid residue [ILR].)

Lines of demarcation can also occur for no apparent reason. The intensity of radiation falls off as the square of the distance from the source to the target, so at some point, perhaps a sharply defined point, insufficient energy exists to maintain combustion. This property, as well as the random nature of some burning, can result in sharp lines of demarcation like the ones shown in Figure 8.7. This piece of carpet was exposed to a controlled source of radiant heat in a radiant panel test, but the edges of the burned area look remarkably like they were influenced by the presence of an ignitable liquid.

Figure 8.6 Straight-line pattern caused by a seam in the underlayment below the carpet. Holes lined up with each other, but there was very little burning under the floor. This linear pattern was originally and erroneously thought to be the result of an ignitable liquid.

Figure 8.7 This carpet specimen was exposed to a controlled radiant heat source in a flooring radiant panel test per ASTM E648 (also known as NFPA 253).

Figure 8.8 "Trailer" pattern created by clearing a path through clothing on the floor of a mobile home burned in a test fire. Close examination of the edges of the pattern reveals the true nature of the sharp lines, but the pattern has the appearance of a trailer. The key to distinguishing real trailers lies in understanding that the unburned sides of the trail should have burned in this fully involved compartment. Rather than trying to explain the burning, the investigator should explain the lack of burning. In this fire, the only valid explanation is that something protected the unburned areas.

Protection patterns can be produced by irregularly shaped pieces of gypsum drywall that fall from the ceiling and provide protection to whatever portion of floor they land on. Clothing on the floor has also been known to produce alternating areas of exposure and protection. Figure 8.8 shows a pattern intentionally generated in a test fire using clothing on either side of the "trailer." A real-world fire, the Lime Street fire that occurred in Jacksonville, Florida, in 1991, contained a similar pattern in the front hallway at the living room doorway. The pattern shown in Figure 8.9 was originally thought to result from gasoline burning on the floor in the doorway.[5]

Figure 8.9 The burn pattern cut out of the doorway to the room of origin in the Lime Street fire. The pattern was thought to result from the presence of gasoline but was later determined to be the result of radiation. The unburned areas on either side of this "trail" were protected by plastic bags containing clothing.

[5] This was an understandable misinterpretation because a chemist (who has since found other work) had erroneously identified gasoline in the sample collected from this pattern.

Figure 8.10 "Puddle-shaped" pattern produced by burning a cardboard box on an oak parquet surface. While it has the shape of a puddle, this pattern actually does not resemble a real flammable liquid pour pattern, but its shape is what many investigators *expect* to see as a result of burning liquids. Those expectations need to be calibrated.

In a groundbreaking study of fire patterns caused by burning pools of gasoline and kerosene, Putorti [26] demonstrated that even on wood and vinyl floors, the edges of the patterns produced are not necessarily all that sharp. The only definitive pattern he found that could reliably be associated with the use of ignitable liquids was the "doughnut" pattern on carpeting, caused by protection at the center of the pattern by the presence of liquid fuel that had not yet evaporated. The "obvious pour pattern" shown in Figure 8.4 exhibits a doughnut pattern.

Radiation is capable of making sharp, "puddle-shaped" patterns, even in cases lacking full room involvement. Figure 8.10 shows a pattern produced by the author (which also appears in NFPA 921) that was generated by the radiant heat from a burning empty corrugated cardboard box on an oak parquet floor. Compare the sharpness of the lines of demarcation with Putorti's pattern, which was produced by a liter of gasoline, shown in Figure 8.11.

Lines of demarcation in the cross section of charred wood have been cited since 1980 as an indicator of the speed of a fire. The *Fire Investigation Handbook* stated, "A distinct line between charred and uncharred portions indicates a rapidly developing fire. Lack of a distinct line usually indicates a slow, cooking process, thus, a slowly developing fire [7]."

O'Connor (1986) and O'Connor and Redsicker (1997) both provide a diagram of a cross section of a piece of lumber showing a sharp line of demarcation indicating a rapid spread, and a gradual line of demarcation indicating a slow-burning fire [27,28].

DeHaan (2011) states, "One indicator that is more useful [than the surface appearance of char] is the appearance of the charred wood in cross section… When a charred beam is cut lengthwise, the gradation between the charred layer and underlying undamaged wood is more gradual with a prolonged, slowly developing fire [29]." He provides a perfectly reasoned analysis of why this should be so, but, like O'Connor, provides neither data (although he does provide a drawing) nor a definition of what is meant by "sharp," "gradual," "fast," or "slow." It seems to be a case of "I know it when I see it." To his credit, DeHaan cautions that a fast-developing fire may or may not be accelerated. Nonetheless, this is the type of "data" that an investigator can use to "eliminate" incorrectly a smoking fire because smoking fires are not "fast developing." The usefulness of a cross section was finally excised from *Kirk's* in the eighth edition (2018, now authored by Icove and Haynes). This is one indicator that was always illustrated by drawings instead of photographs because no photographs exist.

Figure 8.11 A flammable liquid pour pattern on oak parquet produced in a test fire. A liter of gasoline was poured on the floor and ignited to make this pattern. (From Putorti, A., *Flammable and Combustible Liquid Spill/Burn Patterns,* NIJ Report 604-00. With permission.)

Some of the more consistently debated sections of NFPA 921 deal with determinations made by observing lines of demarcation. While it is appropriately silent on the observation of cross sections, the document contains an entire section devoted to caution in the interpretation of irregular fire patterns on the floor:

> **6.3.7.8 Irregular Patterns.** Irregular, curved, or "pool-shaped" patterns on floors and floor coverings should not be identified as resulting from ignitable liquids on the basis of visual appearance alone. In cases of full room involvement, patterns similar in appearance to ignitable liquid burn patterns can be produced when no ignitable liquid is present.
> **6.3.7.8.1** The lines of demarcation between the damaged and undamaged areas of irregular patterns range from sharp edges to smooth gradations, depending on the properties of the material and the intensity of heat exposure. Denser materials like oak flooring will generally show sharper lines of demarcation than polymer (e.g., nylon) carpet. The absence of a carpet pad often leads to sharper lines.
> **6.3.7.8.2** Irregular patterns are common in situations of post-flashover conditions, long extinguishing times, or building collapse. These patterns may result from the effects of hot gases, flaming and smoldering debris, melted plastics, or ignitable liquids. If the presence of ignitable liquids is suspected, supporting evidence in the form of a laboratory analysis should be sought. It should be noted that many plastic materials release hydrocarbon fumes when they pyrolyze or burn. These fumes may have an odor similar to that of petroleum products and can be detected by combustible gas indicators when no ignitable liquid accelerant has been used. A "positive" reading should prompt further investigation and the collection of samples for more detailed chemical analysis. It should be noted that pyrolysis products, including hydrocarbons, can be detected in laboratory analysis of fire debris in the absence of the use of accelerants. It can be helpful for the laboratory, when analyzing carpet debris, to burn a portion of the comparison sample and run a gas chromatographic–mass spectrometric analysis on both samples. By comparing the results of the burned and unburned comparison samples with those from the fire debris sample, it may be possible to determine whether or not hydrocarbon residues in the debris sample were products of pyrolysis or residue of an accelerant. In any situation where the presence of ignitable liquids is suggested, the effects of flashover, airflow, hot gases, melted plastic, and building collapse should be considered.
> **6.3.7.8.3** When overall fire damage is limited and small, or isolated irregular patterns are found, further examination should be conducted for supporting evidence of ignitable liquids. Even in these cases, radiant heating may cause the production of patterns on some surfaces that can be misinterpreted as liquid burn patterns.
> **6.3.7.8.4** Pooled ignitable liquids that soak into flooring or floor covering materials as well as melted plastic can produce irregular patterns. These patterns can also be produced by localized heating or fallen fire debris.

> **6.3.7.8.5** The term *pour pattern* implies that a liquid has been poured or otherwise distributed, and therefore is demonstrative of an intentional act. Because fire patterns resulting from burning ignitable liquids are not visually unique, the use of the term *pour pattern* and reference to the nature of the pattern should be avoided. The correct term for this fire pattern is an *irregularly shaped fire pattern*. The presence of an ignitable liquid should be confirmed by laboratory analysis. The determination of the nature of an irregular pattern should not be made by visual interpretation of the pattern alone.

NFPA 921 contains more cautions on this subject than on any other. The reason for the abundance of cautions on the subject of interpreting lines of demarcation is simple—the errors caused by this particular misinterpretation have been legion.

8.6 SAGGED FURNITURE SPRINGS

The Aetna booklet "Fire … Is It Arson?" (1979) advised fire investigators to photograph furniture springs "because their appearance can help the investigator document the area of origin. Severely sagging springs can indicate that a flammable liquid was involved and created heat intense enough to cause the springs to sag [30]."

Carter, on the other hand, writing in *Arson Investigation* (1978), stated that collapsing all or part of a coil spring was an indication of a cigarette starting the fire [31].

In the Han Tak Lee case, smoking in bed was ruled out because the bedsprings had lost their temper. Clearly, this is an area of much confusion. In 1989, Tobin and Monson, two FBI laboratory scientists, subjected furniture springs, both loaded and unloaded (with and without weights on them), to different fire conditions and basically concluded that the condition of the springs is of little probative value in fire investigation [32]. DeHaan correctly states that varying degrees of spring damage can provide some insight about the progress of a fire but cautions that the collapse of springs cannot be reliably used to determine whether a fire was incendiary [33]. The investigator needs to consider the complexity of the multiple factors affecting spring conditions. Cushions, for example, may at first provide protection but later provide additional fuel. The portion of a mattress closer to a doorway may experience higher temperatures due to ventilation. NFPA 921 provides the following guidance on the subject:

> **6.2.14 Collapsed Furniture Springs.** The collapse of furniture springs may provide the investigator with clues concerning the direction, duration, or intensity of the fire. However, the collapse of the springs cannot be used to indicate exposure to a specific type of heat or ignition source, such as smoldering ignition or the presence of an ignitable liquid. The results of laboratory testing indicate that the annealing of springs, and the associated loss of tension (tensile strength), is a function of the application of heat. These tests reveal that short-term heating at high temperatures and long-term heating at moderate temperatures over 400°C (750°F) can result in the loss of tensile strength and in the collapse of the springs. Tests also reveal that the presence of a load or weight on the springs while they are being heated increases the loss of tension.
> **6.2.14.1** The value of analyzing the furniture springs is in comparing the differences in the springs to other areas of the mattress, cushion, or frame. Comparative analysis of the springs can assist the investigator in developing hypotheses concerning the relative exposure to a particular heat source. For example, if at one end of the cushion or mattress the springs have lost their strength, and at the other end they have not, then hypotheses may be developed concerning the location of the heat source. The hypotheses should take into consideration other circumstances, effects (such as ventilation), and evidence at the scene concerning duration or intensity of the fire, area of origin, direction of heat travel, or relative proximity of the heat source. The investigator should also consider that bedding, pillows, and cushions may shield the springs, or provide an additional fuel load. The portion with the loss of spring strength may indicate more exposure to heat than those areas without the loss of strength. The investigator should also consider the condition of the springs prior to the fire.

8.7 SPALLING

There exists no more misunderstood and misused indicator than concrete spalling. It has been the pivotal "indicator" in many major fire cases and has been the subject of numerous contentious articles in the *Fire and Arson Investigator*. There are still a few investigators who believe that spalling indicates the presence of an ignitable liquid, but their ranks are thinning because of retirement.

Kennedy's "Blue Book" (1977) had the following comments on spalling:

> Spalling caused by flammable liquids burning is usually found at low levels because the flammable liquid vapors are heavier than air and tend to go down.
>
> We often find spalling on concrete blocks and foundations under wooden flooring when flammable liquids are applied to the wooden floor or floor coverings and seep down under the flooring.

> Spalling or fusion requires an explanation. When the fire investigator finds fused materials or spalling at a fire scene, he should be alerted and must demand an explanation or determine the reason.
>
> Fusing or spalling mean that high temperatures were reached during the fire. These high temperatures must be explained. How and why did these high temperatures occur? The answer to this question may explain the cause and origin of the fire... .
>
> The fire investigator is not particularly concerned with the different types of concrete, stone or bricks involved in the fire. Regardless of the composition of the concrete or brick, the indicator is the spalled area or areas indicating the burning of accelerants... .
>
> The spalling temperatures are usually much higher than the temperatures found in the normal dwelling or commercial building fire. Therefore, we know that accelerants were used [34].

A different Kennedy (no relation to the author of "Blue Book") wrote a somewhat ambivalent article on spalling in 1982, stating, "Don't be led to believe that spalling only occurs in accelerated fires. Testing by the American Society for Testing and Materials (ASTM), the Portland Cement Association Research Department, Frederick Smith, Jack Mitchell and many others may prove you wrong." He went on to state, however, that spalling under washers, dryers, or furnaces was a "strong indicator" of the use of flammable liquids and that a pink to orange-brown discoloration around the spalled area was a sure sign of flammable liquid use [35].

IFSTA's *Fire Cause Determination* provided the following statement on spalling in 1982:

> Concrete floors and assemblies that have spalling should be examined closely. The spalling may be an indicator of the use of accelerants. If the accelerants had adequate time to soak in before ignition, the spalling will follow the flow pattern of the liquid. Spot spalling is not a clear indicator of the use of accelerants. Further, it is not unusual for spot spalling to result from severe fire exposure [8].

This semicautious language is typical of what has been written about spalling. Skeptics have always questioned the relationship between ignitable liquids and spalling. In 1979, French wrote:

> There is no reliable scientific basis for the contention that spalling in concrete ... denotes either the presence of hydrocarbon flammables or any other volatile accelerant, or that spalling indicates criminal or incendiary origin.... Spalling is common where concrete surfaces are exposed to heat generated in a fire, regardless of the cause of the fire and when water is applied to hot surfaces during extinguishment causing rapid temperature change and cooling [36].

Fred Smith and Jack Mitchell conducted some of the first experiments that tended to discredit this popular myth in 1981. These researchers lit some experimental fires on concrete using gasoline and wood. They observed spalling in four of six test fires, and of these, the deepest spalling was observed in the two fires involving no ignitable liquids [37]. Shortly after Smith and Mitchell's work was published, this author examined a fire scene where spalling was found up and down the aisles of a commercial establishment. Samples of the concrete tested positive for the presence of kerosene. Using what, in hindsight, was arguably circular logic, an article entitled "A Documented Case of Accelerant Induced Concrete Spalling" was published to "balance" Smith and Mitchell's article [38].

One of the largest bad faith insurance awards in Alabama history was the result of a fire investigator who relied on a "trail of spalling" in addition to other "indicators" to conclude that the cause of a fire was arson. It did not help that the fire chief stood on the "trail" before the fire reached the basement. Neither was the court impressed with the shape of the "trail" when it learned that its shape resulted from the investigator shoveling a trail.[6] When the slab was completely cleared, it was found that the entire slab had spalled, and no "trail" of any kind had ever existed [39].

Fire investigators have argued endlessly about the characteristics of an accelerant-induced spall as opposed to a naturally occurring one. A brownish or pinkish halo around the hole was thought to indicate the presence of burning hydrocarbons.[7] Numerous slides and photos were exchanged, but in the end, the consensus was that French was right.

---

[6] The court's characterization of the testimony of the investigator is instructive. "The court concludes that not only is [the investigator's] testimony as a whole completely void of credibility, but the presentation of his testimony borders on the perpetration of a fraud upon this court. For [the investigator] and USAA to present to this court a case so heavily dependent upon "spalling" as this case, when it is indisputable that [the investigator] selectively cleared only those areas of the floor which supported this incredulous theory is reprehensible."

[7] Cook and Ide (1985) reported that the color change was probably a result of the dehydration of yellow-colored hydrated iron oxides, which turned pink or reddish brown at about 300°C.

Charles Midkiff reviewed the literature in 1990 and called for an end to the use of spalling as an indicator of the presence of ignitable liquids [40]. Smith performed a similar review in 1991 [41], and Beland repeated the exercise in 1993 [42]. Most of the books on fire and arson investigation address the issue and advise caution in varying doses, but some fire investigators still fail to take these cautions seriously. DeHaan in 1991 wrote, "As a fire indicator, spalling can indicate the presence of such suspicious sources of localized heating as a chemical incendiary or a volatile petroleum liquid [43]." In the next edition, the language was moderated to the following: "As a fire indicator, spalling can indicate the presence of a significant fuel load of ordinary combustibles, as well as the presence of suspicious sources of localized heating as a chemical incendiary or a volatile petroleum liquid [44]." DeHaan had conducted some experiments in the meanwhile and stated, "Experiments by this author have shown that concrete spalling is very unlikely (although it can take place) when a volatile liquid alone is ignited on its surface."[8] He then goes on to state the obvious fact that as long as there is a liquid on a surface, that surface is not able to achieve a temperature much in excess of the boiling point of the liquid. By the fifth edition of *Kirk's*, spalling had made DeHaan's list of "Myths and Misconceptions [45]."

This fact that a burning pool of ignitable liquid is incapable of reaching a temperature greater than the boiling point of the liquid was apparently lost on Randall Noon, who opens his discussion of spalling with the following statement: "If an accelerant has been poured on a concrete floor and ignited, it sometimes causes spalling and other temperature-related damage to the concrete." He then shows a drawing of how a flammable liquid might appear on a cross section of concrete and presents a series of equations to determine the increase in temperature (ΔT) required to cause a given area of concrete to spall. He also suggests that the size of the spall is related to this ΔT: "large pieces can spall when the temperature difference is great, and small pieces can spall when the temperature difference is small." Noon does discuss heat transfer but only in the context of a sealant or coat of paint that might make for poor heat transfer. The possibility that the liquid itself might inhibit heat transfer was not discussed. There is a mention that "spalling can also be caused by non-incendiary effects," but this seems to be presented as the exception rather than the rule [46].

Noon notwithstanding, most of what is written about spalling wishfully refers to misconceptions formerly held. NFPA 921 has, since its inception, warned about misinterpreting spalling. Guidance on the interpretation of spalling as it appears in the 2017 edition reads as follows:

> **6.2.5 Spalling.** Spalling is characterized by the loss of surface material resulting in cracking, breaking, and chipping or in the formation of craters on concrete, masonry, rock, or brick.
> **6.2.5.1 Fire-Related Spalling.** Fire-related spalling is the breakdown in surface tensile strength of material caused by changes in temperature resulting in mechanical forces within the material. These forces are believed to result from one or more of the following:
>
> 1. Moisture present in uncured or "green" concrete
> 2. Differential expansion between reinforcing rods or steel mesh and the surrounding concrete
> 3. Differential expansion between the concrete mix and the aggregate (most common with silicon aggregates)
> 4. Differential expansion between the fire-exposed surface and the interior of the slab
>
> **6.2.5.1.1** A mechanism of spalling is the expansion or contraction of the surface while the rest of the mass expands or contracts at a different rate; one example is the rapid cooling of a heated material by water.
> **6.2.5.1.2** Spalled areas may appear lighter in color than adjacent areas. This lightening can be caused by exposure of clean subsurface material. Adjacent areas may also tend to be darkened by smoke deposition.
> **6.2.5.1.3** Another factor in the spalling of concrete is the loading and stress in the material at the time of the fire. Because these high-stress or high-load areas may not be related to the fire location, spalling of concrete on the underside of ceilings or beams may not be directly over the origin of the fire.
> **6.2.5.2** The presence or absence of spalling at a fire scene should not, in and of itself, be construed as an indicator of the presence or absence of liquid fuel accelerant. The presence of ignitable liquids will not normally cause spalling beneath the surface of the liquid. Rapid and intense heat development from an ignitable liquid fire may cause spalling on adjacent surfaces, or a resultant fire may cause spalling on the surface after the ignitable liquid burns away.
> **6.2.5.3 Non-Fire-Related Spalling.** Spalling of concrete or masonry surfaces may be caused by many factors, including heat, freezing, chemicals, abrasions, mechanical movement, shock, force, or fatigue. Spalling may be more readily induced in poorly formulated or finished surfaces. Because spalling can occur from sources other than fires, the investigator should determine whether spalling was present prior to the fire.

[8] No better way to learn about the behavior of fire exists than to participate in controlled live burn experiments.

Figure 8.12 Deep spalling on the ceiling of a temporary office setup in a parking garage.

This language, although "toned down" and made more "user friendly" since the 1992 and 1995 editions of NFPA 921 in which the cautions were entitled "Misconceptions about Spalling," is essentially unchanged. In this author's view, the move away from a title including the word *misconception* reflected an overly optimistic view on the part of the Technical Committee that the earlier editions of the document had cleansed the fire investigation community of the misconceptions. Warnings about misinterpretation of spalling dating back to French in 1979 and through nine editions of NFPA 921 continue to be ignored to this day. One can only wonder what investigators who believe that spalling always indicates the presence of accelerants think about patterns such as the one shown in Figure 8.12. Certainly, no pool of liquid existed on the concrete ceiling.

In a 1998 fire report, a fire investigator with over 20 years of experience combined an observation of spalling with four (unconfirmed) canine alerts to conclude that a fire had been set. He wrote: "The extent and location of spalling patterns are indicative of accelerant use." When he was deposed on the subject, in 2000, the following exchange took place:

Q: Do you have any indication that these would have been hydrocarbons or some type of accelerant on the floor?

A: It's my belief that the accelerant was placed on the floor in the areas outlined in blue.

Q: Based solely on the spalling areas?

A: Based on my experience investigating fires for 20 years, yes.

Q: What in those 20 years would lead you to believe that the spalling area was from accelerants?

A: I suppose it would have to be my conversations with people.

Q: Here on this accident, now, I mean this fire?

A: You're talking about the last 20 years, that's your question. What in my experience?

Q: Yes.

A: In my experience over 20 years of talking with people, investigating fires and learning what flammable and combustible liquids were in the area of fire origin, or in the fire scene itself... The only time I have seen evidence of spalling is when a combustible or flammable liquid is in contact with the concrete surface during the fire [47].

This investigator, as is typical of investigators who believe in myths, relied not only on the observation of spalling but also on his belief that the fire had "spread laterally too fast." The four unconfirmed canine alerts should have been irrelevant to his scientific conclusions. Thus, one meaningless indicator was used to bolster other meaningless indicators. This investigator's client was the same insurance carrier involved in the Alabama "trail of spalling" case cited earlier. Getting tagged with a $2 million bad faith judgment had apparently failed to get their attention, or perhaps the effect had worn off after 10 years. Not only were they relying on the same kind of discredited evidence to bolster an arson defense, but they even engaged the same defense attorney whose presentation of the trail of spalling evidence had been found to "border on the perpetration of a fraud upon the court." This time, however, the insurance company finally abandoned the unsupportable defense and settled the case.

The following is an excerpt from a report of a 2001 fire that resulted in three deaths and an indictment for capital murder: "[The Deputy Fire Marshal] showed me locations of the spalding [*sic*][9] of the concrete. At this time [the Detective] came to the scene with a video recorder to film the areas of spalding. Due to spalding being found on the slab, [the Major] advised me that [the DA investigator] advised that a search warrant would be required before completion of the search of the slab." (The laboratory detected no ignitable liquid in the concrete sample, but the lab report did note, "This sample was taken due to the spalling of concrete." At least the chemist had the terminology correct.)

In a 2003 fatal fire, an indictment was obtained on the strength of the testimony of a sheriff's detective who claimed at one time to have seen "probably 25" arson/homicides in his 25-year career in this rural Georgia county (where an average of less than one fire death per year is reported). This detective testified as follows at the preliminary hearing (and presumably in his sealed grand jury testimony as well):

Q: What did you see on the floor around this body?

A: A fresh crack, very severe crack associated with this type of scene. It's typically associated directly underneath that of a body.

Q: And this crack is located directly underneath the body?

A: Yes, sir. It's referred to as a "spaulding crack."

Q: "Spaulding crack." What is the cause of a "spaulding crack" such as this?

A: It's the generation of heat that's held underneath a body after it's been subjected to a flammable liquid and ignited. The liquid pulls toward the back and it holds the heat down like making a seal and it starts to evaporate because of the heat, the water density in the concrete, and that's what causes the crack.

Q: Could you tell whether that crack existed prior to this fire? Is there anything in the crack that would indicate it had been there a while?

A: No, there was not. There was no—it was fresh powder. Even when the body was moved there was nothing that suggested that it was there before. It has the characteristics of a "spaulding pattern" associated with an arson/homicide... And you could see an outline of the victim.

The detective then proceeded to demonstrate his prowess in reading patterns other than "spaulding" patterns. In a breathtaking display, he opined on pour patterns, the self-extinguishing properties of bodies, and the significance of the pugilistic position.

Q: And what made up this outline of the victim?

A: A pattern that's typically associated with that of a flammable liquid of petroleum products, what we refer to as pour patterns.

[9] Some people misspell it "spalding" and some misspell it "spaulding." The origin of the misnomer is probably the past tense of the word *spall*. One sees spalled concrete. People who do not know any better add the "d" and call the process spalding or spaulding. Such people have apparently never read a text or even an article on the subject.

Q: What would be your conclusion then, that you in your investigation reached and would be the opinion you had of what happened here from having seen the body and these pour patterns around the body?

A: My opinion is that he was down on his back and someone poured him with a petroleum product at least twice ... we found patterns that were commonly found in arson/homicide that could only have been left by a petroleum flammable product.

Q: And you said it had been poured at least twice. What is your basis of saying that?

A: A body just doesn't burn that quick. It's hard to burn a body. A body that is not repeatedly doused will put itself out. I mean, we're made up of liquid. You know, the majority of us are made up of water and we self-extinguish.

Q: So is what you're saying is that the body had to have been poured, lit, and then re-poured in order to have had the extent of damage that you observed?

A: That's correct.

Q: Do you know what "pugilistic" means?

A: I know what the term means in relation to arson.

Q: What does it mean?

A: It means that you draw back up into a fetal position. It's the act or the need to draw up into the fetal position to try to get as small as you can to get away from the fire. You're not going to make yourself bigger to get to the fire sooner, you're going to make yourself smaller [48].

The body in question was in a textbook pugilistic attitude, as reflected in the autopsy report. The testimony beyond the "spaulding" discussion is presented here to demonstrate that when an investigator carries around one myth, his "toolbox" is likely to include a whole collection of them.[10] Also, despite the fact that some of the individuals who continue to use these myths to prosecute innocent victims of fire are otherwise dedicated public servants, as fire investigators, those who fail to even learn the correct terminology demonstrate a disturbing lack of professionalism. Such individuals do not meet the minimum requirements of NFPA 1033.

8.8 FIRE LOAD

In one of the early attempts to bring a quantitative approach to the practice of fire investigation, French (1979), who has so far been held out in this discussion as a model of appropriate skepticism, described the process by which an investigator could calculate whether a fire was "normal." Closely related to the calculation of fire load was the comparison of the expected behavior of that fire load in a real fire to the "standard time/temperature curve." French described the process as follows:

> The heat energy production of fuels is extremely important to any competent fire investigator in determining fire load in the premises or equipment under investigation, again in respect to its potential in affecting temperature rise and spread and the time spectrum...
>
> Fire load of a given space may be established by knowing the type of combustibles in storage, their calorific heat-producing capacity in Btus per lb., the total weight of the combustibles in storage, and the square foot capacity of the space.
>
> The formula is as follows: multiply the calorific contents in Btus per lb. by the total weight of the contents or materials in pounds. Then, divide the result by the area in square feet. The answer is fire load per square foot.
>
> National Bureau of Standards and American Standards as well as the National Fire Protection Association and British time/temperature curves are in general agreement as to what temperature rise may be expected in various occupancies, with known fire loads, particularly during the first two hours of combustion.
>
> For example, with sufficient oxygen to support continuing combustion, fires in buildings may be expected to attain 1000°F to 1200°F during the first 5–10 min, accelerating on the curve to approximately 1500°F in the first half hour and with temperatures reaching on the order of 1700°F at one hour [49].

[10] The district attorney actually took this case to trial in the spring of 2004. After a 2-week trial, the jury returned a verdict of not guilty in less than 4 hours. It did not help his credibility with the jury when the deputy testified that the state fire marshal and an insurance investigator agreed with him. Both individuals had called the fire "undetermined" in their written reports and in their trial testimony.

Carroll, writing in *Physical and Technical Aspects of Fire and Arson Investigation,* adopted a similar approach but, instead of fire load, urged investigators to use the flame-spread index described earlier in his text. The vaguely defined process was described in two paragraphs as follows:

> Using this Flame-Spread Index (available from Underwriters Laboratories, Inc.), a fire investigator can determine the comparative rate of how fast a fire should or should not have spread under normal circumstances by comparing the burning rates of known fires and the Standard ASTM (American Society of Testing and Materials) Time/Temperature Fire Exposure chart shown in Figure 10.
>
> Figure 10 shows the temperature acquired as a function of time, which has been found to be the average temperature eight feet off the floor.

Carroll also wrote, "By knowing the fire load of the building, i.e., the material available for the creation of heat, a reasonable approximation of the highest temperatures attained can be made and compared with temperatures to be expected had an accelerant been used [50]."

The ASTM Time/Temperature Exposure Curve referenced by Carroll is actually a specification for how to run a fire resistance test, described in ASTM E119. This curve does not come close to approximating real-world fires, and is of no use in determining whether a fire behaved "normally." Figure 8.13 compares the ASTM E119 curve with five test fires, all of which exceeded the temperatures specified in the standard.

Comparison of fire behavior to the Standard Time Temperature Curve was used with tragic results in the case of *Commonwealth of Pennsylvania v. Han Tak Lee.*[11] Because the fire lasted only about 30 min, the jury heard testimony that it could not possibly have reached temperatures sufficiently high to melt copper (which the

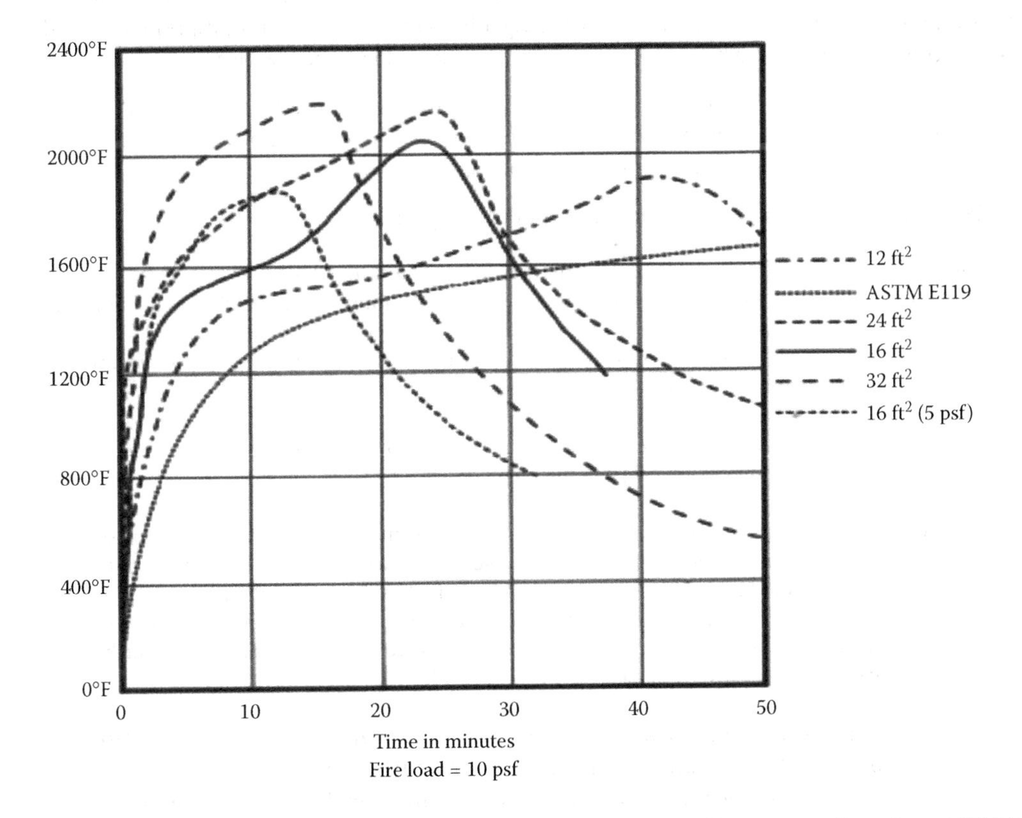

Figure 8.13 Comparison of Data from five test fires compared to the Standard Time Temperature Curve from ASTM E119, which is the bottom curve. All five test fires exceeded the standard curve by hundreds of degrees at 10 minutes.

[11] Mr. Lee's conviction was vacated and he was released from custody on August 22, 2014, after having served 25 years in prison.

Standard Time/Temperature curve says will not melt for over 3 hours in a "normal" fire), but melted copper was found. An expert for the Commonwealth calculated fuel loads and Btus expected, and seeing more Btus than he expected, calculated the excess fuel that must have been present in the form of flammable liquid. This author has reviewed the calculation and listed the incredible number of myths that went into this so-called analysis in "A Calculated Arson [51]."

The calculation of fuel load to determine whether a fire was "normal" fortunately never enjoyed widespread acceptance. Its use was common enough, however, that it has been addressed in NFPA 921 in the chapter on basic fire science. It states quite simply at 5.6.3.1, "Total fuel load in the room has no bearing on the rate of growth of a given fire in its pre-flashover phase."

8.9 LOW BURNING AND HOLES IN THE FLOOR

A common misconception is that because hot air rises, fire burns up and out and will not burn downward unless it has "help." This simplistic explanation of fire behavior has formed the basis of many an arson investigator's determination of an incendiary cause and plays very well with a jury that knows that hot air rises but has no knowledge of flashover.

In Carroll's 1979 text, he discusses multiple low points and states that "the discovery of a low point should not be considered the end of the search, since more than one low point may be discovered. This is particularly true in arson fires." He states further, "every effort should be made to determine whether multiple low spots are accidental or deliberate. If they have been set in an incendiary effort, these would be considered evidence of arson [52]."

Carroll states that significant differences in char depths at two different low points would indicate an accidental low point. If the low points or holes in the floor are all about equally charred, the use of multiple holes to indicate multiple points of origin could arguably be justified by referring to this alleged learned treatise.

The IFSTA manual, *Fire Cause Determination* [53], espoused a similar misinterpretation when it stated, "Low levels of charring are good indicators of flammable liquid having been used. For example, accidental fires are unlikely to burn the bottom edge of furniture or the bottom edge of a door."

The Army's 1995 Field Manual succinctly restated the popular myth:

> Liquid accelerants leave evidence of low burn. That is, they show burning on the floor of the structure. A normal fire chars only the upper portion of a room. Floor damage in natural fires is usually limited to about 20 percent of the ceiling damage. Low burn, shown by complete charring of large areas of the floor or the baseboards, is not natural.
>
> Nor is fire burning downward natural. Fire burning downward is a prime indicator of use of a flammable accelerant. Patterns burned in wood floors or holes in a floor may show that an accelerant was used [54].

Kirk was one of several workers who disagreed with the notion that holes observed in a burned floor necessarily indicated the presence of an ignitable fluid. In 1969, he wrote:

> In many instances, the lowest burn is a floor surface or a region directly under a floor. These points are sometimes difficult to evaluate and often lead to error in interpretation. For example, there is a hole burned in the floor in a region away from all walls or other objects that could carry fire upward by providing fuel in the path of the flames. It is not uncommon for the investigator to assign the cause to the use of a flammable liquid. Such an interpretation is more often incorrect than otherwise. On a tight floor, it is always incorrect, unless holes or deep cracks are present. Lacking such conditions, *flammable liquids never carry fire downward.* (Emphasis in the original.) [55]

NFPA 921 has dealt with this myth in a straightforward admonition at 6.3.3.2.5: "Holes in floors may be caused by glowing combustion, radiation, or an ignitable liquid. The surface below a liquid remains cool (or at least below the boiling point of the liquid) until the liquid is consumed. Holes in the floor from burning ignitable liquids may result when the ignitable liquid has soaked into the floor or accumulated below the floor level. Evidence other than the hole or its shape is necessary to confirm the cause of a given pattern." Figure 8.14a shows a floor burned in a test fire involving no ignitable liquids. Using a PowerPoint presentation, a deluded investigator could produce the kind of exhibit shown in Figure 8.14b, showing all of the "suspicious" aspects of this innocent burn pattern.

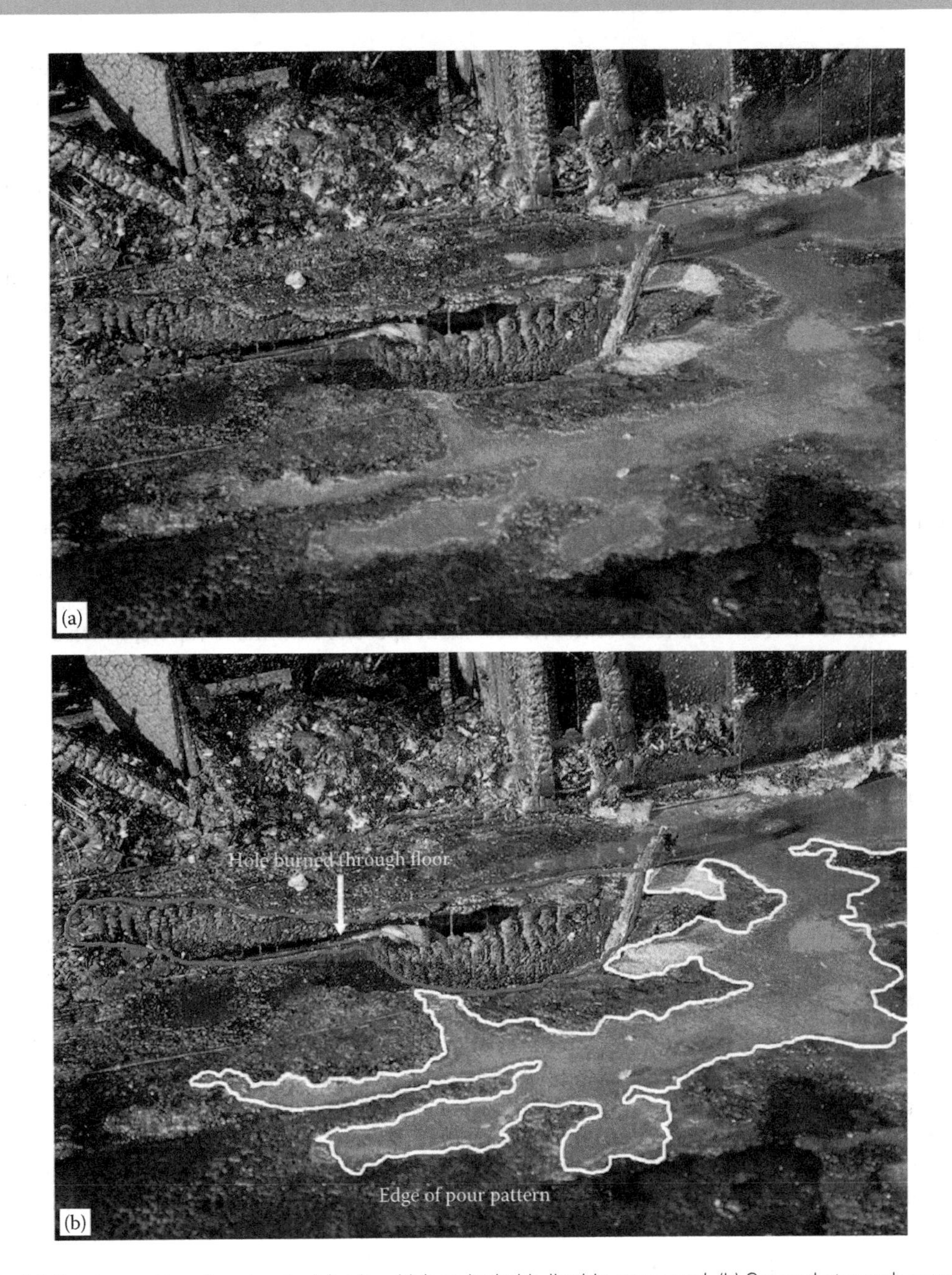

Figure 8.14 (a) Hole burned in a floor in a test fire in which no ignitable liquids were used. (b) Same photograph annotated for use in a PowerPoint program. A practiced expert witness showing this annotated photo to a jury would likely convince them that they were looking at a "flammable liquid pour pattern."

8.10 THE ANGLE OF V

The V-angle myth has been published by many authors. It goes like this: The sides of a "normal" conical fire plume are angled 15° from vertical. The faster a fire burns, the slower it will spread laterally and the closer the angles will be to vertical. Conversely, the slower a fire burns, the further the angle will tilt from vertical. This author has reviewed reports of investigators who saw a fire burning straight up (never mind that it was constrained by 2 × 4 studs) and concluded that not only was this the point of origin, but the fire had to have been set because of its speed. No scientific support exists for this myth. Like most of the myths presented in this chapter, it is a deceptively appealing notion.

The *Fire Investigation Handbook* stated (in the same section that described the meaning of alligatoring) that the V-pattern should be examined to determine whether the fire was a slowly developing one or a rapidly developing one. Rather than defining rapid and slow, the intent was apparently to let the indicator do the defining. The authors stated:

> Fire patterns—a wide angle or diffuse “V” pattern generally indicates a slowly developing fire. A narrow sharply defined “V” pattern generally indicates a fast-developing, hot fire [56].

The Army’s Field Manual (even the 2013 edition!) states, “Fire burns up and out. It leaves a V-shaped char pattern on walls and vertical structures. A fire which is hot and fast at the point of origin will leave a sharp V-pattern. A slow fire will produce a shallow V [57].”

Carroll was more “quantitative,” at least as far as the angles are concerned:

> A normal fire, consuming wood, plastic or electrical insulation, would burn with a “V” pattern of approximately 30° measured vertically. If an accelerant was used, or if highly combustible material was involved, the “V” would be narrower as the temperature of the fire increased, due to the additional heat content of the accelerant or flammable liquid. This would cause a faster rise of heat and flame, resulting in a “V” pattern of approximately 10°, depending upon the heat flux generated by the accelerant [58].

Noon devotes an entire section of his book to burning velocities and V-patterns. (There is nothing wrong with that, unless one tries to prove a relationship where none exists.) In Noon’s view, shallow V-patterns actually indicate downward burning. He relates narrow V-patterns to the vertical flame speed of flammable vapors, which are much greater than the vertical flame speed of the underlying combustible wall. Typically, Noon’s discussion is accompanied by equations to help the investigator determine the ratio of the upward burning rate to the lateral burning rate by finding the tangent of the angle of the V [59].

O’Connor (1986) and O’Connor and Redsicker (1997) repeat the story. “The breadth or width of the V (also called the *funnel pattern)* is affected by (and hence, indicative of) the buildup, progression, speed and intensity of the fire. An intense rapidly moving fire produces a narrow V-pattern, whereas a slow, less intense fire produces a wide V-pattern. The angles of the V boundaries average between 10° and 15° [60,61].”

DeHaan urged caution in the interpretation of V-pattern angles in the third edition of *Kirk’s*, where he stated, “Although it is sometimes claimed that the more vertical the sides of the V, the faster the initial fire (and therefore the more suspicious), one can appreciate that the nature of any wall covering and conditions of ventilation have important effects on the shape of the pattern, and must be taken into account [62].” By the fourth edition, he came right out and said, “The angle and width of the V are not dependent on the rapidity of ignition of the fuel [63].”

NFPA 921, since its first edition in 1992, has contained a section on the interpretation of V-patterns and described the equating of angles with speed as a “misconception.” The current edition, which eschews the word *misconception* for fear of offending its readership, simply makes the following flat statement at 6.3.7.1.2: The angle of the borders of the V pattern does not indicate the speed of fire growth or rate of heat release of the fuel alone; that is, a wide V does not indicate a slowly growing (“slow”) fire and a narrow V does not indicate a rapidly growing (“fast”) fire.

8.11 TIME AND TEMPERATURE

A fire that burns “hotter than normal” or “faster than normal” is thought to indicate a set fire. Actually, fire temperature and the perceived speed of the fire are not valid indicators of a fire’s cause.

A major misconception underlying many calls of arson is that the temperature achieved by a particular fire can help an investigator evaluate whether a fire was “normal” or “abnormal,” with an abnormal fire being attributed to incendiary activity. We have already covered Paul Kirk’s published misconception.

Kennedy, discussing the fusion of copper, wrote:

> Copper fuses at 1980°F, which is very high. Therefore, if we find fused copper or beaded copper wires, we are immediately alerted because we have an unusually high temperature that must be explained. The normal burning of a structure should not cause temperatures in the 2000°F range, which is necessary to fuse or melt the copper.
>
> What could cause copper to fuse or melt? The burning of an accelerant such as flammable liquids, natural or LP gases or electrical shorts or arcing are some of the causes for “high heat”—meaning excessive temperatures [64].

Barracato summed up the time and temperature equation as follows:

> Fires which burn through entire floor sections, destroy large support beams in a relatively short period of time, or melt articles located in the area of origin such as metal, copper, aluminum or glass, are unusual. It takes tremendous heat to cause such damage and unless there is a rational explanation for the heat build-up—for example, if the room was used to store highly flammable material—it's very probable that the fire was intentional and an accelerant was used [26].

Carroll, as discussed previously under "8.8 Fire Load," presented the standard time/temperature curve from ASTM Standard E119, *Standard Test Methods of Fire Tests of Building Construction and Materials,* and stated that it can be used as a basis in comparing the burning rates of structures [65]. Carroll also stated, "If accelerants or other chemicals are present, temperatures can reach higher than on the *Standard Fire Curve* [66]."

Based on research that began in the late 1970s and continues until today, it is now well understood that there is no such thing as "normal" fire spread and also that the ASTM time/temperature curve has little relation to the behavior of a "normal" fire. Babrauskas, however, has recently compiled considerable data to support the contention that the standard time/temperature curve can be used to judge the minimum length of a fire's duration, but *only after flashover,* and *only if there are no gaps* in the surface under consideration. Using test data from ASTM E119 furnace tests, he reported that the charring rate of wood ranges from 0.5 to 0.8 mm/min (0.7–1.2 in./hr) if there are no gaps in the wood. With gaps, the rates of char after flashover more than double, to 1.6 to 2.1 mm/min (3.8–5 in./hr) for oak and 3 to 4 mm/min (7–9 in./hr) for softwoods. Babrauskas argues that if an investigator finds ceiling joists (a "no-gap" situation) with 20 mm of char, he can reasonably conclude that the fire burned for 25 min after flashover [67].

In the second edition of *Kirk's* (1983), DeHaan somewhat moderated Kirk's enthusiasm for interpreting melted metals but still leaves the readers with the suggestion that an abnormal fuel load, such as provided by an accelerant, increases temperatures.

> While melted metals cannot and should not be used as proof that the fire was incendiary, the fire investigator should note their presence, extent, and distribution. Such information can be of help in establishing differences between normally fueled and ventilated accidental fires and those produced by enhanced draft conditions or unusual fuel loads from accelerants in incendiary fires [68].

Although the third, fourth, and fifth editions of *Kirk's* include a discussion of the fact that gasoline burns at essentially the same temperature as wood, it still states that temperature can be used to determine the presence of "enhanced draft conditions *or* unusual fuel loads," when the data only support the former. Nonetheless, the modern text of *Kirk's* at least recognizes what blacksmiths and metallurgists have known for centuries: that increased ventilation causes increased temperatures.

NFPA 921, beginning with the 1992 edition and continuing until today, contains an admonition about placing too much stock in the perceived temperature of a fire. In the 2017 edition of the document, both the temperature and speed of the fire are addressed, and investigators are warned to be cautious when interpreting temperature and speed of the fire. The following language appears in the current edition:

> **6.2.2.2** Wood and gasoline burn at essentially the same flame temperature. The turbulent diffusion flame temperatures of all hydrocarbon fuels (plastics and ignitable liquids) and cellulosic fuels are approximately the same, although the fuels release heat at different rates. Burning metals and highly exothermic chemical reactions can produce temperatures significantly higher than those created by hydrocarbon- or cellulosic-fueled fires.

Despite this language having appeared in all nine editions of NFPA 921, there are still investigators today who will testify that a melted aluminum threshold indicates the presence of a burning ignitable liquid.

The speed at which a fire progresses is frequently used to imply that the fire is incendiary. While it is true that an accelerated fire burns faster than an unaccelerated fire, at least in its initial stages, serious caution is required when confronted with information about how fast a fire burned. Most observations about the "speed" of a fire are provided by eyewitnesses, but there have been reported instances of an investigator looking at a destroyed structure and knowing the time from alarm to extinguishment, opining that the amount of destruction could not have occurred in that time frame unless it had "help." These conclusions are usually based on a misconception that wood has a fixed burning rate, such as the oft-quoted "1 inch in 45 minutes."

The 1995 Army Field Manual advises investigators: "Note precisely the time you arrive at the scene. Then quickly note whether the fire is burning slow or fast. If the fire is burning fast, is this consistent with the type of fire load present in the building? If not, suspect the use of an accelerant [9]." No advice is given, however, to help the investigator decide what speed constitutes a "slow" fire or a "fast" one. NFPA 921 warns:

> **5.10.1.4** The rate of fire growth as determined by witness statements is highly subjective. Many times witnesses are reporting the fire growth from time of discovery, which cannot be directly correlated to ignition time. The rate of fire growth is dependent on many factors besides fuel load, including fuel configuration, compartment size, compartment properties, ventilation, ignition source, and first fuel ignited. Eyewitnesses reporting a rapid rate of fire growth should not be construed as data supporting an incendiary fire cause.

A study by the editors *of Fire Findings*, published in 1995, revealed that witnesses might make very different observations about a fire, even if their sightings are only a few minutes apart. This study, which involved the actual burning of a two-story house, resulted in the finding that fires may only appear to start rapidly and tended to dispel the widely held (but incorrect) belief that if a fire appears to start quickly, accelerants must have been involved. If an eyewitness only notes the existence of a fire at the point where it breaks out a window, the progress thereafter will be rapid indeed, regardless of the cause [69].

Additionally, NFPA 921, in its discussion of the progression of compartment fires, relying on studies conducted at the National Bureau of Standards states "Flashover times of 3 to 5 minutes are not unusual in residential room fire tests and even shorter times to flashover have been observed in nonaccelerated room fires [70]."

In *Kirk's Fire Investigation*, DeHaan published a time/temperature curve for a "typical furnished room" with no accelerants that shows flashover occurring just 210 seconds (3.5 min) after ignition [71]. This curve looks nothing at all like the "Standard Time/Temperature Curve" from the ASTM test.

Thus, while it is true that accelerants tend to make fires burn more rapidly, a rapidly burning fire does not necessarily indicate the presence of accelerants.

8.12 CONCLUSION

When the scientific community was first getting organized, it was understandable that misconceptions about fire, such as the phlogiston and caloric theories, should exist. What is surprising is that myths proliferated for much of the twentieth century. The sheer number of misconceptions and their widespread publication in learned and not-so-learned treatises contributed mightily to this proliferation of nonsense [72]. The publication of NFPA 921 and texts like the previous editions of this one have gradually reduced the acceptance of false knowledge and may one day result in the final elimination of arson myths.

Review questions

1. Which of the following statements about crazed glass is true?
   a. Crazed glass indicates locally rapid heating.
   b. Crazed glass can only be caused by the application of water to hot glass.
   c. Crazed glass with light smoke deposits was probably near the point of origin.
   d. Smaller cracks indicate that the glass was closer to the point of origin, while larger cracks indicate a greater distance from the origin.
2. Which of the following statements about spalling is true?
   I. Spalling is often caused by the burning of an ignitable liquid.
   II. A dark ring around the spalled area indicates a burning hydrocarbon.
   III. Spalling can be caused by the components of the concrete expanding at different rates.
   IV. Spalling is one of the most misunderstood of all the fire effects.

V. It is far easier to cause spalling by the sustained the burning of wood than by the short-lived burning of a pool of gasoline.
    a. I and II only
    b. I, II and III
    c. IV and V only
    d. III, IV and V

3. Why have so many arson investigation myths survived for so long?
    a. The myths have been published in many fire investigation textbooks.
    b. The myths are appealing notions.
    c. The myths have been embraced by highly respected organizations or authors.
    d. All of the above.
4. What do sagged furniture springs indicate?
    a. A slowly developing fire, such as one caused by a cigarette in the furniture cushions.
    b. A rapidly developing fire, such as one fueled by a liquid accelerant.
    c. A fire pattern that may show differential fire exposure.
    d. Sagged furniture springs are of no value to the fire investigator.
5. What does the standard time-temperature curve describe?
    a. An ASTM specification for running a standard fire-resistance test.
    b. The progress of a normally developing fire.
    c. The progress of a fire accelerated using natural gas.
    d. The progress of a fire that does not achieve flashover.

Questions for discussion

1. What fire myths have you seen in your own or a colleague's report?
2. How have fire myths adversely affected the professionalization of fire investigation?
3. Why was the fire investigation profession so reluctant to admit that certain indicators of arson were not valid?
4. Why is it not appropriate to use the standard time-temperature curve to determine if a fire's spread was faster than expected?
5. In the past, black smoke and orange flames were thought to be a sign of burning petroleum liquids. How has that changed?

References

1. Kirk, P. (1969) *Fire Investigation*, John Wiley & Sons, New York, p. 145.
2. O'Connor, J., and Redsicker, D. (1997) *Practical Fire and Arson Investigation*, CRC Press, Boca Raton, FL, p. 107.
3. Sagan, C. (1995) *The Demon Haunted World*, Random House, New York, p. 76.
4. Ettling, B. (1984) "Are we kidding ourselves?" *The Fire and Arson Investigator* 34(4):19.
5. Moenssens, A., Moses, R., and Inbau, F. (1986) *Scientific Evidence in Criminal Cases*, Foundation Press, Mineola, NY.
6. Bourdreau, J., et al. (1997) *Arson and Arson Investigation Survey and Assessment*, Aerospace Corporation, National Institute of Law Enforcement and Criminal Justice, LEAA, U.S. Department of Justice, 87. Available at https://www.ncjrs.gov/pdffiles1/Digitization/147389NCJRS.pdf (last visited on February 18, 2018).
7. Brannigan, F., Bright, R., and Jason, N. (1980) *Fire Investigation Handbook*, NBS Handbook, Vol. 134, US Department of Commerce, National Bureau of Standards, 6.

8. IFSTA (1982) *Fire Cause Determination*, 1st ed., Fire Protection Publications, Stillwater, OK, p. 48.
9. U.S. Army (1985) *Law Enforcement Investigations*, Field Manual 19–20, 220.
10. O'Connor, J. (1986) *Practical Fire and Arson Investigation*, CRC Press, Boca Raton, FL, p. 88.
11. O'Connor, J., and Redsicker, D. (1997) *Practical Fire and Arson Investigation*, CRC Press, Boca Raton, FL, p. 99.
12. O'Connor, J., and Redsicker, D. (1997) *Practical Fire and Arson Investigation*, CRC Press, Boca Raton, FL, p. 74.
13. Noon, R. (1995) *Engineering Analysis of Fires and Explosions*, CRC Press, Boca Raton, FL, p. 131.
14. French, H. (1979) *The Anatomy of Arson*, Arco Publishing, New York, 1979, p. 61.
15. Brannigan, F., Bright, R., and Jason, N. (1980) *Fire Investigation Handbook*, NBS Handbook, Vol. 134, US Department of Commerce, National Bureau of Standards, 5.
16. IFSTA (1982) *Fire Cause Determination*, Fire Protection Publications, Stillwater, OK, p. 46.
17. O'Connor, J. (1986) *Practical Fire and Arson Investigation*, CRC Press, Boca Raton, FL, p. 94.
18. DeHaan, J., and Icove, D. (2011) *Kirk's Fire Investigation*, 7th ed., Pearson, Upper Saddle River, NJ, p. 297.
19. DeHaan, J. (1991) *Kirk's Fire Investigation*, 3rd ed., Prentice Hall, Upper Saddle River, NJ, p. 129.
20. DeHaan, J. (1997) *Kirk's Fire Investigation*, 4th ed., Prentice Hall, Upper Saddle River, NJ, p. 171.
21. Lentini, J. (1992) "Behavior of glass at elevated temperatures," *Journal of Forensic Science* 37(5):1358.
22. *Paul Camiolo v. State Farm Fire & Casualty Co., Trooper Investigative Services et al.*, U.S. District Court for the Eastern District of Pennsylvania, Civil Action No. 00-CV-3696, Deposition Testimony of George Wert, OH, November 26, 2001, p. 66.
23. Cooke, R., and Ide, R., (1985) *Principles of Fire Investigation*, The Institution of Fire Engineers, Leicester, UK, p. 134.
24. Barracato, J. (1979) *Fire ... Is It Arson?* Aetna Life & Casualty, 15.
25. *Commonwealth of Pennsylvania v. Han Tak Lee*, Court of Common Pleas of Monroe County, 43rd Judicial District, No. 577 Criminal, 1989. Report of Daniel Aston, 1990.
26. Putorti, A. (2000) *Flammable and Combustible Liquid Spill/Burn Patterns*, NIJ Report 604-00, U.S. Department of Justice, Office of Justice Programs, National Institute of Justice. Available at https://www.ncjrs.gov/pdffiles1/nij/186634.pdf (last visited on February 18, 2018).
27. O'Connor, J. (1986) *Practical Fire and Arson Investigation*, CRC Press, Boca Raton, FL, p. 74.
28. O'Connor, J., and Redsicker, D. (1997) *Practical Fire and Arson Investigation*, CRC Press, Boca Raton, FL, p. 75.
29. DeHaan, J., and Icove, D. (2011) *Kirk's Fire Investigation*, 7th ed., 282.
30. Barracato, J. (1979) *Fire ... Is It Arson?* Aetna Life & Casualty, p. 23.
31. Carter, R. (1978) *Arson Investigation*, Glencoe Press, Encino, CA, p. 97.
32. Tobin, W. A., and Monson, K. L. (1989) "Collapsed spring observations in arson investigations: A critical metallurgical evaluation," *Fire Technology* 25(4), 317.
33. DeHaan, J. (2002) *Kirk's Fire Investigation*, 5th edition, Prentice Hall, Upper Saddle River, NJ, p. 212.
34. Kennedy, J. (1977) *Fire-Arson Explosion Investigation*, Investigations Institute, Chicago, IL, p. 392.
35. Kennedy, R. (1982) Concrete spalling and fire investigation, *National Fire and Arson Report* 1(3):1.
36. French, H. (1979) *The Anatomy of Arson*, Arco Publishing, New York, p. 64.
37. Smith, F., and Mitchell, J. (1981) Concrete spalling under controlled conditions, *The Fire and Arson Investigator* 32(2):38.
38. Lentini, J. (1982) "A documented case of accelerant-induced concrete spalling," *The Fire and Arson Investigator* 33(2):30.
39. *USAA v. Wade*, 5344 So. 2d 906 Ala. 1989.
40. Midkiff, C. (1990) Spalling of concrete as an indicator of arson, *The Fire and Arson Investigator* 41(2):42.

41. Smith, F. P. (1991) Concrete spalling: controlled fire tests and review, *Journal of the Forensic Science Society* 31(1):67.
42. Beland, B. (1993) Spalling of concrete, *The Fire and Arson Investigator* 44(1):26.
43. DeHaan, J. (1991) *Kirk's Fire Investigation*, 3rd ed., Prentice Hall, Upper Saddle River, p. 122.
44. DeHaan, J. (1997) *Kirk's Fire Investigation*, 4th ed., Prentice Hall, Upper Saddle River, p. 163.
45. DeHaan, J. (2002) *Kirk's Fire Investigation*, 5th ed., Prentice Hall, Upper Saddle River, p. 234.
46. Noon, R. (1995) *Engineering Analysis of Fires and Explosions*, CRC Press, Boca Raton, FL, p. 208.
47. *Rock Savage v. USAA, U.S. District Court*, District of South Carolina, Columbia Division Civil Action No. 3:99-893-19, Deposition Testimony of William P. Brooks, January 27, 2000, 66.
48. *State of Georgia v. Jean Long*, Magistrate Court of Butts County Case # 03-JT-019, Testimony of Detective Michael Overbey, February 20, 2003.
49. French, H. (1979) *The Anatomy of Arson*, Arco Publishing, New York, p. 36.
50. Carroll, J. (1979) *Physical and Technical Aspects of Fire and Arson Investigation*, Charles C. Thomas, Springfield, IL, p. 72.
51. Lentini, J. (1999) "A calculated arson," *Fire and Arson Investigator*, 49(3):20.
52. Carroll, J. (1979) *Physical and Technical Aspects of Fire and Arson Investigation*, Charles C. Thomas, Springfield, IL, p. 105.
53. IFSTA (1982) *Fire Cause Determination*, Fire Protection Publications, Stillwater, OK, p. 81.
54. U.S. Army (1985) *Law Enforcement Investigations*, Field Manual 19–20, 225.
55. Kirk, P. (1969) *Fire Investigation*, John Wiley & Sons, New York, p. 74.
56. Brannigan, F., Bright, R., and Jason, N. (1980) *Fire Investigation Handbook*, NBS Handbook 134, US Department of Commerce, National Bureau of Standards, 7.
57. U.S. Army (1985) *Law Enforcement Investigations*, Field Manual 19–20, 219.
58. Carroll, J. (1979) *Physical and Technical Aspects of Fire and Arson Investigation*, Charles C. Thomas, Springfield, IL, 103.
59. Noon, R. (1995) *Engineering Analysis of Fires and Explosions*, CRC Press, Boca Raton, FL, p. 104.
60. O'Connor, J. (1986) *Practical Fire and Arson Investigation*, CRC Press, Boca Raton, FL, p. 76.
61. O'Connor, J., and Redsicker, D. (1997) *Practical Fire and Arson Investigation*, CRC Press, Boca Raton, FL, p. 77.
62. DeHaan, J. (1991) *Kirk's Fire Investigation*, 3rd ed., Prentice Hall, Upper Saddle River, NJ, p. 109.
63. DeHaan, J. (1997) *Kirk's Fire Investigation*, 4th ed., Prentice Hall, Upper Saddle River, NJ, p. 148.
64. Kennedy, J. (1977) *Fire-Arson Explosion Investigation*, Investigations Institute, Chicago, IL, p. 396.
65. Carroll, J. (1979) *Physical and Technical Aspects of Fire and Arson Investigation*, Charles C. Thomas, Springfield, IL, p. 72.
66. Carroll, J. (1979) *Physical and Technical Aspects of Fire and Arson Investigation*, Charles C. Thomas, Springfield, IL, p. 54.
67. Babrauskas, V. (2004) Charring rate of wood as a tool for fire investigators, in *Proceedings of the 10th International Fire Science and Engineering (Interflam) Conference*, Interscience Communications, London, UK, 1155.
68. DeHaan, J. (1983) *Kirk's Fire Investigation*, 2nd ed., Prentice Hall, Upper Saddle River, NJ, p. 173.
69. Sanderson, J. (1995) "Fire timing test results: fires may only appear to start rapidly," *Fire Findings* 3(3):1.
70. NFPA 921, 2017 edition, Section 5.10.4.6, 49.
71. DeHaan, J. (2002) *Kirk's Fire Investigation*, 5th ed., Prentice Hall, Upper Saddle River, NJ, p. 42.
72. Lentini, J. (2017) Pernicious, pervasive and persistent literature in fire investigation, in Bartick, E. G., and Floyd, M. A. (Eds.), *Forensic Science Research and Evaluation Workshop*, USDOJ, OJP, NIJ. Available at http://www.ncjrs.gov/pdffiles1/nij/250088.pdf. (last visited on February 11, 2018).

CHAPTER 9

Sources of error in fire investigation

Learn from the mistakes of others. You can never live long enough to make them all yourself.

—Groucho Marx

LEARNING OBJECTIVES

After reviewing this chapter, the reader should be able to:

- Understand that error is part of being human, and errors occur in all forensic science disciplines
- List the common kinds of errors that lead to incorrect fire cause determinations
- Recognize an error when reviewing the work product of a fire investigator
- Understand how errors in fire investigations affect the lives of individuals after a fire

9.1 INTRODUCTION

Forensic science is a human activity, and as such, it is subject to error. Denying that error occurs is to deny reality. Errors should be recognized and embraced,[1] for they can help us learn what we are doing wrong and avoid making those same mistakes in the future. For the most part, the errors described in this chapter occurred not because investigators were evil or stupid, but because they made mistakes that were caused by cognitive biases, or the errors occurred due to a lack of knowledge, information, or experience. Some errors result from deviations from best practices. These errors are insidious. Small deviations seldom have any consequences, and that lack of consequences encourages continued and even larger deviations until a catastrophic event, such as a wrongful conviction, occurs. Four of the cases at the end of this chapter involved wrongful convictions.

Wrongful convictions have been described as "organizational accidents." According to the National Institute of Justice (NIJ):

> In an organizational accident, the correct answer to the question, "Who is responsible?" is almost invariably, "Everyone involved, to one degree or another," if not for making a mistake, then by failing to catch someone else's. In the instance of a wrongful conviction, "everyone" may include not only witnesses, police, forensic scientists and lawyers at the sharp end of the system, but also legislators, policymakers, funders and appellate judges who were far from the scene of the event but who helped design the system and dictated the conditions under which the sharp-end operators work [1].

While there are numerous players involved in a wrongful arson conviction, our focus here will be on the individuals whose errors got the process started: the fire investigators and engineers who incorrectly made the incendiary determination.

[1] Donald Berwick, one of the pioneers of the modern patient safety movement stated, "Every defect is a treasure."

It is disheartening to realize that so many erroneous determinations exist as to the origin and cause of fires that it is actually possible to categorize the types of errors that lead to those erroneous determinations. But looking at the big picture, it is not so surprising. More than 375,000 structure fires are reported in the United States every year. Of these, approximately 8% are classified as "intentional."[2] That means that in about 30,000 fires each year, fire investigators have the opportunity to make the most important call (or miscall). With accidental fires, no reason exists to believe that the error rate is any lower than it is with intentional fires. However, the consequences of an error in determining the cause of an accidental fire pale in comparison to the consequences of calling an accidental fire arson. If the error rate is 5% for those 30,000 intentional fires, the result is about 1,500 miscalled fire investigations every year.

Nobody knows the true error rate in fire investigation, but only the most optimistic investigator actually believes that the error rate is only 5%. This author has had the opportunity to discuss the issue in front of many audiences of investigators. It has been instructive to perform a survey by asking the audience first to stand up, then asking those who think that when they or a colleague call a fire arson, that person is right 100% of the time to sit down. Nobody sits. When the question is repeated with 95% as the number, only a very few sit. A few more sit down when asked if they think 90% of arson fires are called correctly. By the time those who say they believe that the call is correct 80% of the time have sat down, most of the audience is still on their feet. A 20% error rate translates into 6,000 erroneous calls every year. The collective opinion of the author's seminar audiences may or may not accurately reflect reality, but the opinion does not inspire confidence.

Aside from these admittedly speculative data, the National Registry of Exonerations lists 64 exonerations from arson-related convictions as of late 2017. Forty-two of these exonerations are classified as "no crime" cases,[3] meaning that the fire investigator(s) wrongly classified an accidental fire as incendiary [2]. Because of the stringent requirements for entry on this list, this is a low number and does not include any cases where the criminal prosecution was declined or dismissed prior to conviction, where the defendant was acquitted in his or her first trial, or where the defendant admitted to some other offense involving negligent conduct related to the fire but never admitted to setting the fire. One hopes that the safeguards built into our justice system result in the vast majority of incorrect arson determinations being declined by prosecutors or rejected by grand juries.

In many fields requiring specialized knowledge or skill, determining who is prone to making errors and who is not is a straightforward exercise. Pipes installed by incompetent plumbers leak; patients operated on by incompetent surgeons die; riverboats steered by incompetent harbor pilots run aground. No such real-world indicators exist to help us evaluate the work of fire investigators. In many fields of forensic science, it is possible to get a handle on the error rate through proficiency testing, wherein scientists are presented with samples of known origin, and either they get the analysis right or they get it wrong. If the proficiency tests are taken by a reasonable cross section of practitioners in the field, one can get a feel for how much confidence to place in a certain type of analysis, but not, of course, on a true "error rate." Neither will proficiency testing shed any light on the probability that a *particular* determination is sound. With fire investigation, the proportion of incorrect calls is really anyone's guess, but it certainly exceeds 5%. If one credits the data reported by Carman and others reported in Chapter 3, the error rate exceeds 5% by an order of magnitude. Thus, every year, a large number of accidental fires are erroneously called "incendiary." Most of these investigations probably go no further because no suspect exists, but many times, particularly in the emotionally charged context of a fatal fire, the obvious suspect is the person who survived the fire. The survivor tells a story that is at odds with the conclusion of the investigator, and because the investigator "knows" that the fire was intentionally set, the person "lying" about the fire's origin and cause must be the one who set it because no one else would have a "motive" to lie about the origin and cause.

It is hoped that the presentation of a listing of the sources of error in fire investigations will help readers avoid making these errors, or at least recognize the errors when they are reviewing an investigator's work product.

This author's study of fire investigation errors began in the early 1990s when, at the request of a superior court judge, the case of *Arizona v. Ray Girdler* was presented for review. Girdler had been convicted of arson and two counts of murder for setting the fire that killed his wife and daughter on November 20, 1981. Judge James Sult, hearing his first

[2] In 2000, the National Fire Protection Association (NFPA) reported that intentional fires accounted for 15% of 505,000 structure fires. The decline is probably due, at least in part, to the dropping of "suspicious" as a fire cause determination.

[3] In contrast, only two of the fires listed in the dedication of this text were actually incendiary. The rest were miscalled accidents.

capital case, sentenced Girdler to two life terms, sparing him the death sentence only because he believed that life in prison would be a more painful punishment than death. It was only after Girdler had served 8 years that he was granted a new trial because the fire marshal, who declared the fire to be arson, had testified that *only* the presence of accelerants could account for the condition of the burned-out mobile home. All the fire reports and all the expert testimony from the trial were reviewed with an eye toward understanding why Girdler was convicted, and what would happen if he were tried again. Since that time, reviewing cases of alleged arson has become a significant part of this author's practice. Sufficient data have been accumulated to define the following seven general categories of fire investigation errors:

1. Overlooking critical data
2. Misinterpreting critical data
3. Misinterpreting irrelevant data
4. Ignoring inconsistent data
5. Two-dimensional thinking
6. Poor communication
7. Faulty chemistry or engineering

In this last category, the fire investigator relies on a specialist who commits one or more of the errors listed earlier.

All the errors described in this chapter would likely have been caught had the investigators involved properly applied the scientific method. A careful test (in some cases, even a casual test) of their hypotheses would probably have let them know something was amiss. More important, had the investigators actively sought out alternate hypotheses that explained the data equally well, they might have been able to detect the errors before they committed to a final opinion.

9.2 OVERLOOKING CRITICAL DATA

Fire scenes, at least those where it is possible to overlook data, tend to be large, complicated, messy places. The important artifacts are often buried and are generally the same color as the rest of the environment, either black or gray. It is very easy to miss something. This is especially true if the fire investigator is in a hurry, is required to investigate the scene in the dark, or fails to keep an open mind. The finding of an ignition source, say a torchière lamp close to what appears to be a V-pattern, might be all that an overworked investigator needs to close his case and move on to the next one. The portable space heater 2 feet (ft) away might be overlooked, especially if the investigator thinks he has already figured out the cause of the fire. One of the best ways to avoid overlooking data is to use the classical approach of starting the inspection of the scene in the area with the least amount of damage and working toward the area with the greatest amount of damage. A fire investigator who works in the opposite direction is far more susceptible to developing a case of "tunnel vision," in which all of the data that he sees support his initial, prematurely formed hypothesis. The fire investigator who takes his time, familiarizes himself with the building, looks at everything before moving anything, and carefully documents the initial conditions is less likely to overlook a critical piece of data.

Sometimes a piece of data only becomes "critical" when a fire investigator makes it so. In one of this author's first experiences working for an accused individual, a homeowner was accused of setting her house on fire on the basis of some holes in the floor and the disappearance of a large set of silverware. This was back in the days when the price of silver was skyrocketing, so it would have made sense for a financially strapped homeowner to take it out prior to a fire. The initial fire investigator used the "disappearance" of the set of silverware as part of the basis for his determination that the fire was intentionally set.

Communication with the homeowner revealed that the box containing the silverware was located in a particular room in the basement into which several tons of rubble had fallen. It took the better part of a day and a crew of four with wheelbarrows and shovels to clear out the room, but the silverware was found. The original investigator's credibility was seriously compromised by the fact that he had written in his report that the homeowner had removed the silverware prior to the fire, then committed insurance fraud when she claimed its loss.

A similar set of circumstances occurred many years later when a former firefighter was accused of setting the fire that killed his wife. Hundreds of hours were expended trying to find his beloved electric guitar, which the police theorized he had removed from the premises before setting the fire. The defendant denied removing the guitar from the house, so finding it at a buddy's house or in a storage locker would have proved the case, or so it was thought. Unfortunately, the police were looking in all the wrong places. A careful sifting of the debris from the very place where the accused said he had left his guitar revealed the remains of strings and tuning keys [3].

Sometimes it is simply the size of the evidence that makes it difficult to spot. The fire that Weldon Wayne Carr was accused of setting in order to kill his wife was actually caused by a malfunction in a light switch. A careful reconstruction of the scene revealed that the fire began in a wall between the breakfast area and the bathroom, and the fire eventually flashed over in the bathroom and came back into the breakfast area. The fire investigator was so captivated by the radiant heat pattern on the floor that he failed to look in the bathroom, where the damage was much more severe, but even when an electrical engineer was asked to look at the light switch to "eliminate it," he failed to note that one of the contacts in this three-way switch was considerably more damaged than an identical contact located on the same moving member. When the fire scene was reconstructed (post conviction), a detailed analysis of the fire patterns led to the switch as the cause, and when the switch was made available, the failure mechanism was evident.

Investigators should consider very carefully the implications of declaring something to be absent. It is much easier to prove that an item is present in a fire scene than it is to prove that it was removed prior to the fire. By far, the most common cause of overlooking critical data is a rushed and cursory investigation of the fire scene. The only way to prevent this kind of error is to take sufficient time to thoroughly process the fire scene. If an investigator or agency lacks sufficient time or resources to do a proper job, then it would be appropriate to request that the investigation be carried out by a different agency.

9.3 MISINTERPRETING CRITICAL DATA

Misinterpreting critical data commonly results from a failure to understand the behavior of fire, but it can also result from poor communication, another category of error that is discussed later. An example of misinterpreting critical data is the determination that a fire was set by trailing ignitable liquid through a doorway. This particular misinterpretation is fueled by a basic misunderstanding of fire behavior. A V-pattern is evident at a doorway. Everyone who looks at the doorway agrees that a clear V-pattern can be observed, with its apex (or nadir) at the floor in the doorway. One explanation would be that the pattern was produced by a trail of ignitable fluid. In fact, if someone trailed ignitable liquid through that doorway, this is exactly the kind of pattern we would expect to see. On the other hand, if the room on one side of that doorway has become fully involved and the fire was ventilation controlled, it would be surprising *not* to see a V-pattern at the doorway.

Misinterpretation of critical data occurs when a fire somehow does not live up to an investigator's "expectations." If those expectations are not properly "calibrated," numerous misinterpretations will occur in the investigator's career because a constant mismatch exists between expectation and reality. As stated previously, innocuous or extraneous data can become "critical" as a result of nothing more than the investigator's thought processes. When reviewing a photograph taken during a fire, this author saw in the photograph (Figure 9.1) a light that appeared to be on. Had the light actually been on, rather than merely reflecting the photographer's flash, it could have indicated that the fire started in the attic rather than in the room where the initial investigators said it started, where the switch for this light was located. Because of a written report that called the condition of this light a critical piece of evidence, this investigator was ultimately made to look foolish. During a truly humbling cross-examination, opposing counsel ignored all the brilliant points made on direct and kept returning to the light in a moderately successful attempt at impeachment.[4]

Many fire investigations go down the wrong track because of the misinterpretation of a single piece of critical data. When the firefighters arrived at Ray Girdler's mobile home in 1981, he was next door, sitting at a neighbor's kitchen table, fully dressed. Because it was after 2:00 in the morning, fire officials were naturally suspicious of this, but rather than questioning Girdler, they simply counted this as a suspicious fact. Had they asked, they would have learned that Girdler escaped his mobile home on that cold winter night barefoot and dressed only in his undershorts and that the clothing had been provided by the neighbor.

[4] The case ended in a mistrial (hung jury). Two subsequent trials also ended in mistrials, after which the judge dismissed the case.

Figure 9.1 An example of evidence that became "critical" only because this investigator declared it to be so—a misinterpreted photograph. The author, viewing this photograph in the light most favorable to his client, opined in a report that the light at the peak of the building was on. Later review revealed that the white color was a reflection of a photographer's flash. The report provided ample grist for cross-examination.

Maynard Clark's furniture store was equipped with fusible links that caused fire doors to drop in the event of a fire, thus preventing the spread of a fire from one compartment to the next. The investigator who found the screwdriver jammed into the roll-up door mechanism concluded that this was part of the business owner's scheme to make the fire spread more quickly. Had he made the effort to inquire, he would have learned that the fusible link on the fire door had melted several years prior to the fire and that the screwdriver had been jammed in there at that time. Certainly, damage to fire protective equipment is one of the "indicators not directly related to combustion" listed in NFPA 921 as evidence tending to show that someone had prior knowledge of the fire. Further, the investigator probably did not want to give Clark a chance to "make up an excuse" for the screwdriver's presence or even know that the investigator had found it. Having a first hypothesis of accidental cause, which the investigator clearly did not carry into this scene, would have required him to inquire about this apparently incriminating data. Instead, he chose to play a game of "gotcha," leading to an embarrassing misinterpretation.

Another kind of misinterpretation occurs because of *negative corpus* methodology. Failing to find a competent ignition source and fuel source at a hypothesized origin should not be interpreted as evidence of an incendiary fire cause, unless there is affirmative evidence of such a cause. The failure to find the ignition source or fuel source should send an investigator in search of an alternative origin hypothesis.

The only way to avoid misinterpreting critical data is to ask whether there are any other interpretations of the data that explain it equally well. Careful consideration of alternate hypotheses is the most difficult step in the scientific method, partly because there exists a natural tendency to become overly attached to the first hypothesis that explains the data. Failure to consider alternate explanations, however, is a frequent cause of incorrect determinations. Investigators who accept work from "both sides" may have a broader view of possible alternate hypotheses.

9.4 MISINTERPRETING IRRELEVANT DATA

The misinterpretation of irrelevant data remains one of the most common errors in fire investigation. "Irrelevant data" is the kind of myth described in the previous chapter. Irrelevant data can also be described as "epiphenomenal" data; they are real but they do not help us understand anything. Like any historical science, fire investigation presents too much data—too many "clues"—many of which are totally irrelevant to the cause of the fire. Fire investigators who interpret spalling on concrete floors as incontrovertible evidence of the presence of an ignitable liquid, or those who insist that crazing of glass indicates rapid heating nearby, or those who believe that melted aluminum thresholds mean something important will misinterpret irrelevant data on a consistent basis. The fact that these investigators have failed to keep up with the progress of the fire investigation profession unfortunately means that they are likely to engage in the same sort of lazy approach to fire investigation that results in overlooking critical data as well. These same individuals are also likely to commit all the other errors that cause fire investigators to make incorrect determinations.

The only way to avoid the misinterpretation of irrelevant data is to become equipped with current knowledge of what is and what is not meaningful. This requires a dedication to keeping current and a willingness to "unlearn" that which is false. Unfortunately, the corrective action for this class of error usually involves teaching the investigator the error of his ways the hard way—by discrediting the false hypotheses that the investigator puts forward.

9.5 IGNORING INCONSISTENT DATA

With respect to both origin *and* cause hypotheses, NFPA 921 states, "It is unusual for all data items to be totally consistent with the selected hypothesis. Each piece of data should be analyzed for its reliability and value.... Contradictory data should be recognized and resolved. Incomplete data may make this difficult or impossible. If resolution is not possible, then the cause hypothesis should be reevaluated [4].

Ignoring or discarding inconsistent data is one of the most insidious of errors and a classic symptom of a closed mind. Inconsistent data should be apparent to anyone who follows the scientific method and performs hypothesis testing using deductive reasoning. Investigators who ignore data that do not fit their hypothesis really should find another line of work because they are going to make serious errors.

Inconsistent data might also be described as "inconvenient data." Criminal defense attorneys refer to such inconsistent data as "Brady material." In the civil arena, attorneys representing insureds who have wrongly been denied their insurance proceeds refer to the discarding of inconsistent data as "bad faith."

Maybe sufficient inconsistent data exists so that a fire investigator is forced to call a fire "undetermined." Although "undetermined" is not an actual fire cause, some investigators fail to understand that "undetermined" is the appropriate determination when the data do not support any hypothesis that the investigator can formulate. There exists no way to avoid this category of error, other than keeping an open mind. In many cases, ignoring inconsistent data goes beyond simple error; it is self-delusion or outright deception.

9.6 TWO-DIMENSIONAL THINKING

Fire is an inherently three-dimensional process evolving in time (yet another dimension), but some investigators seem to think in only one or two dimensions. Even those who think in three dimensions sometimes fail to take time into account. Fortunately, the one-dimensional thinkers are very rare, such as the individual who testified that it would take 4 hours to burn a 2 × 10. However, a significant amount of two-dimensional thinking still exists. Because structure fires are surrounded by two-dimensional surfaces, walls, floors, and ceilings, and because most of the burn patterns are recorded on those surfaces, it is easy to fall into the trap of thinking in only two dimensions. Because ignitable liquids tend to land on the floor, fire investigators can become fixated on the floor and see "multiple origins" in a fully involved room: They are unable to see a connection *on the floor* between two holes. In fact, an incorrect finding of multiple origins is the most common outcome of two-dimensional thinking. This can even occur in separate rooms if the fire is traveling, for example, through an attic.

Mr. and Mrs. Johnson owned a residence next to their business, and a fire started in the living room. It penetrated the ceiling and traveled through the attic, eventually causing failure of the bedroom closet ceiling some 20 ft away. The fire extended downward into the closet, which was erroneously identified as a second point of origin by the

Figure 9.2 A fire that was miscalled as a result of two-dimensional thinking. This is a photograph of the burned-out cottage floor where Han Tak Lee's daughter perished. An investigator was able to discern *nine* separate points of origin in this room. (He also "calculated" that more than 60 gallons of flammable liquid were applied to the floor, although the samples were all negative, and there were no 60-gallon containers available.)

fire investigator. It took years of litigation to overcome the false accusation of arson that resulted from this two-dimensional thinking [5].

The case of *Pennsylvania v. Han Tak Lee* rested, to a large extent, on a finding of "nine points of origin" in a single room, as shown in Figure 9.2. All of the interior finish was consumed, as well as the ceiling and the roof. There is no credible way to find evidence of multiple origins in such a room, even if they existed, but the investigator was certain and testified persuasively. He even testified to the order in which the fires were ignited. Obviously, the fire in front of the exterior door had to have been set last [6].

The only way to avoid making the kind of error caused by thinking in two dimensions is to constantly remind oneself to think in three dimensions and to remember that fire patterns do not record the time (and usually not even the sequence) that they were produced.

9.7 POOR COMMUNICATION

Erroneous determinations based on poor communication are common. Sometimes this results from the division of investigative responsibilities between different agencies, for example, when the fire department is responsible for determining the cause of the fire and the police department is in charge of interviewing the witnesses. If these two agencies do not communicate well, an erroneous determination can result.

In a homicide case reviewed by this author, the defendant was accused of pouring gasoline in the stairway of a two-story townhouse, thus preventing the exit of the children who were trapped upstairs. The fire investigator used "spalding" *[sic]* and "heavy alligator charring" as the basis for his determination, but until the day this case came to trial, he was not apprised of the fact that the mother of the children told police investigators that she had been *upstairs* when the fire started. She would have had to run through the stairway, which was allegedly doused with gasoline, to escape the fire, and she certainly would have been burned. Good communication between the police and the fire departments might have prevented the fire investigator from stating that gasoline had been poured on the stairs when it clearly had not been. (He probably still would have called the fire incendiary, however, based on the "spalding.") [7]

Poor communication can also take the form of simply failing to talk to the first firefighter on the scene or relying on someone else to relay critical information. In reviewing a product liability case, this author learned that the fire department that responded to the Smith residence carried a video camera with them. Shortly after the kitchen fire was extinguished, the firefighter who made entry came out and spoke to the homeowners on camera and told them that he had found the stove on and turned it off. (The reaction of the teenager who had been cooking earlier was interesting to watch.) The fire investigator who came along later did not get this information and blamed the fire on a coffeemaker. This author, in the capacity of assisting the coffeemaker manufacturer, had the opportunity to review the videotape. Because of poor communication, the manufacturer was forced to spend a considerable sum of money investigating a fire in which its product was clearly not at fault.

One way to avoid errors caused by poor communication is to designate one technically qualified individual as the *principal investigator*. This person is charged with collecting all the relevant technical and other relevant data from everyone involved in the investigation. This person should *not* be the lawyer who is directing the investigation—it should be a qualified investigator intent on discovering the truth about the case.

9.8 FAULTY CHEMISTRY OR ENGINEERING

It is not unusual for fire investigators to form hypotheses that are disproved when those hypotheses are tested. One means of testing the hypothesis that a fire was intentionally set using an ignitable liquid is to collect a sample of the flooring where the liquid was apparently poured and submit it to a laboratory. If the laboratory follows appropriate procedures, it will not find ignitable liquid residues where none are present and will call for comparison samples when necessary. If, on the other hand, the laboratory is not following appropriate ASTM procedures, it may fail to disprove an incorrect hypothesis, potentially resulting in someone being falsely accused of setting a fire that was accidental. This happened in the Lime Street fire. The chemist who first examined the debris from the suspicious V-shaped burn pattern at the living room doorway reported finding gasoline in the sample. A review of the chromatography, and later a review of the evidence, revealed no such thing. It was simply a bad call made by a chemist with a history of bad calls. Inaccurate chemical analyses are serious errors that often change the whole complexion of a case. An attorney who believes he can overcome a fire investigator's interpretation of burn patterns is likely to have second thoughts when that interpretation is backed up by a finding of gasoline on the living room carpet. Most attorneys will be intimidated by such findings and feel powerless to refute them. When reviewing any fire case, it is a good practice to review the laboratory data because chemists are not infallible.

Electrical engineers are also subject to making errors. The electrical arcing on the wire shown in Figure 9.3 was found on the power supply for a motor that moved clothing in a dry-cleaning establishment. An electrical engineering firm had examined the motor and associated wiring and concluded that it did not cause the fire, but it made no mention whatsoever of any arcing in its report. In fact, the neutral wire and the ground wire that ran in the same cable as the arc-damaged wire exhibited no corresponding arc damage, indicating that this hot wire had arced to a grounded surface in the normally dusty area in the middle of the clothing conveyer. There was actually a "clearly defined" origin in this case, and the engineer's failure to find this ignition source led the investigator to conclude incorrectly that the ignition source must have been an open flame.

Errors made by engineers and chemists are difficult for the average fire investigator to detect, particularly when the engineer or chemist has produced a report that supports the investigator's hypothesis. Getting a second opinion from an equally qualified individual, or at least making sure that the report was technically reviewed, will help to reduce this class of error. If an investigation is conducted correctly, qualified independent experts should be reviewing the evidence on behalf of other parties to the case.

The "objective" nature of chemical or engineering analysis, especially when compared to the more subjective discipline of fire scene analysis, tends to make people more confident in the engineer's or chemist's findings. In fact, *more than a third of all of the fire cases listed in the dedication* of this text involved faulty chemistry or engineering.

Erroneous determinations in fire cases usually result from multiple errors of the kind listed earlier. It is almost always a combination of several different errors that leads to the incorrect determination.

Figure 9.3 An example of critical evidence overlooked. This conductor supplied a dry cleaner's conveyor motor. The report of the electrical engineer who examined the motor and associated wiring made no mention of any arcing and "eliminated" electricity as a possible source of ignition. As a result, the fire investigator incorrectly determined that the fire was intentionally set. The author tied a red ribbon around this artifact and sent it back. The case settled shortly thereafter.

9.9 EVALUATING ALLEGATIONS OF ARSON

Fire investigators, particularly those working in the insurance industry, learn about "red flags" that might indicate the possibility that arson fraud has been committed to collect an insurance settlement. The following is a different list of red flags that can be used to evaluate arson investigations in general. While certainly not infallible, this list is at least as reliable as the red flags used to determine arson fraud. Note that this approach is not designed to provide a road map for fire investigators, only for those who are asked to evaluate a fire investigator's work product. If any of the questions listed below are not answered satisfactorily, the fire cause determination deserves careful reexamination. The first group of questions deals with the scientific aspects of a case.

9.9.1 Is this arson determination based entirely on the appearance of the burned floor in a fully involved compartment?

If the fire investigator opines that, based on the finding of "pour patterns," "accelerants" were used, but the laboratory results are all negative, this determination lacks support. It is the consensus of the relevant scientific community that in a fully involved compartment, fire patterns on the floor should not be attributed to ignitable liquids on the basis of visual appearance alone [8].

9.9.2 Is this arson determination based on "low burning," crazed glass, spalling, "shiny alligatoring," a "narrow V-pattern," or "melted/annealed metal"?

Fires do burn downward without "help." Crazed glass means only that hot glass was rapidly cooled. Shiny "alligatoring" means nothing. A narrow V-pattern does not indicate a "rapid" fire (whatever that means). Melted metals are used to determine temperatures achieved by the fire, but a well-ventilated gasoline fire burns no hotter than a well-ventilated wood fire. Further, ignitable liquids cannot burn under an aluminum threshold due to the absence of oxygen there. Annealed bedsprings or furniture springs do not indicate the presence of accelerants, nor can they be used to determine whether a fire started with a cigarette. Reliance on any of these indicators demonstrates that the investigator's "toolbox" is full of broken tools. Some investigators counter that they are not using an indicator as the "sole basis" for their determination, but reliance on a multiplicity of erroneous beliefs is no more credible than reliance on a single erroneous belief.

9.9.3 Is this arson determination based on an unconfirmed canine alert?

Canines are a wonderful tool for selecting samples to submit to the laboratory. The use of canines to detect ignitable liquid residues is one of the most significant advances ever in the fight against arson, but dogs can only tell us where to collect samples, not what the samples contain. Although samples collected with the aid of a canine are significantly more likely to test positive, if the laboratory does not confirm the canine alert, *the sample is negative*. The NFPA Technical Committee on Fire Investigations, addressing the problem of unconfirmed canine alerts being presented to juries, wrote, "In essence, a fraud is being perpetrated upon the judicial system [9]." Unconfirmed drug or explosive detection canine alerts are routinely dismissed as canine errors. Some canine handlers insist, however, that their "canine partner" is never mistaken, and some courts have been persuaded to allow the jury to hear these mistaken opinions. This misguided insistence has been denounced by the Canine Accelerant Detection Association (CADA). In a 2012 policy statement, the CADA Board of Officers and Directors declared, "[N]o Prosecutor, Attorney or ADC Handler should ever testify or encourage testimony that an ignitable liquid is present without confirmation through laboratory analysis." This position was echoed in the *AAAS Forensic Science Assessments Quality and Gap Analysis on Fire Investigation*, published in 2017 [10].

9.9.4 Is this arson determination based on a fire that "burned hotter than normal" or "faster than normal"?

It is the amount of ventilation that determines the temperature of the fire, not the nature of the fuel. The "speed" of a fire is a highly subjective evaluation, and relying on a witness's perception of speed assumes that the witness saw the fire from its inception.

9.9.5 Do neutral eyewitnesses place the origin of the fire somewhere other than where the fire investigator says it was?

Investigators should consider all the data, including (especially) eyewitness statements. If an investigator has the origin wrong, he or she almost certainly has the cause wrong as well.

9.9.6 Is this arson determination based entirely or largely on a mathematical equation or a computer model?

If there is no demonstrable physical evidence of the fire's origin and cause, such determinations are nothing more than digitized speculation. Mathematical models, even the most sophisticated computer models, were not designed for the purpose of resolving issues of fire investigation. As shown in Chapter 3, fire models, and especially "hand" models, have more uncertainty than is acceptable for this application.

9.10 INVESTIGATIONS GONE WRONG

The following pages describe in some detail a number of investigations that resulted in innocent persons being accused of setting fires. In all of these cases, the fires were accidental but were incorrectly called intentional. In most of these cases, the defendant (often an accused homeowner) was present at the time of, or shortly before, the fire and would have been an obvious suspect had the fire actually been intentionally set. All of these cases have been settled or adjudicated.

There is, necessarily, a "war story" aspect to the retelling of these cases. The retelling will necessarily be somewhat one sided, told from this author's perspective as a retained expert working for the innocent accused. A considerable amount of background information is provided so that readers can understand the context of these cases as well as the consequences of these errors. Frequently, the conduct of the nontechnical players has a strong bearing on the outcome of a case. Sufficient information is provided so that readers can seek out the participants on the other side of the case, if they so desire, and hear the other side. The purpose of retelling these stories is neither to cause embarrassment nor to gloat, but to provide real examples of egregious errors so that readers can learn from the mistakes of others and avoid repeating them. These stories also make clear the stakes involved—this is not a sporting event. In most of these cases, the author was not retained until long after the fire had occurred. The shortest time interval was about 6 months and the longest was more than 25 years. Some of the cases were very well documented, others less so.

Some of the defendants were fine, upstanding citizens who were unlucky enough to draw an investigator with a closed mind or who worked with outdated knowledge. Other defendants were less savory or had personality traits that just did not sit right with the investigator or the jury. Most of these cases are criminal cases, and the person(s) making the errors are mostly public servants. No attempt is being made here to disparage the overall work of the public sector. Significant errors occur in the private sector as well. Some of the criminal cases evolved into civil cases after the charges were dismissed or the defendant was acquitted.

The lesson common to all these cases is that while other individuals (often attorneys) made errors in the conduct of these cases, they would not have had the opportunity for error—the cases would not have been presented for litigation—had it not been for errors made by fire investigators on the fire scene, or engineers or chemists in the laboratory. Space does not allow for all of the relevant cases to be described here. Some cases published in the second edition have been replaced with more recent cases. Those that are repeated here have either historical or technical significance. In cases involving wrongful convictions, it is important to note the date of exoneration rather than the date of the fire, as such cases take many years, even decades, to resolve. The cases that are repeated from the second edition are the ones that still offer relevant lessons as of this writing.

9.10.1 State of Wisconsin v. Joseph Awe [11]

Joseph Awe was the owner of a bar, JJ's Pub, located in Harrisville, Wisconsin, which was destroyed by fire on September 11, 2006. The building was over 100 years old and had electrical problems. The state's investigators found no ignition sources at the place where they thought (erroneously) that the fire started, used "negative corpus" methodology, and declared the fire to be incendiary. They also made an issue of a framed poster of Joe and colleagues on an Army Jeep in Kuwait at the end of Desert Storm. The image was once on the cover of *Life* magazine but in a later incarnation, it was used as part of a Miller Lite promotion. It burned in the fire, but the state's investigators said Awe had removed it prior to the fire because of its sentimental value. This was based on an outdated notion of the irreplaceable nature of photographs. Miller actually sent Awe a new poster within three weeks of the fire. There was $900 cash found in the till, presumably left there to mislead the investigators. Awe was 40 minutes away at the time of the fire, so the state proposed that he hired someone (never even tentatively identified) to set the fire. The insurance company "supported" the state's investigator with an electrical engineer and another fire investigator. It was the electrical engineer's elimination of electrical cause that caused the lead state investigator to write, "The conclusions section of the fire scene examination report should be amended to indicate that the official cause for this fire incident is now being classified as incendiary in nature. Electrical engineer Chris Korinek has eliminated electricity as a potential cause for this fire incident."

Awe's trial counsel also hired an electrical engineer, but he had never testified and he spent all of ten minutes examining the service panel. He was discredited, and Awe was convicted on December 20, 2007. He remained free on bond pending appeal until 2011, when, having exhausted all of his appeals, he began serving his three-year term of imprisonment.

The author was not contacted until the spring of 2012, when Awe's sentence was more than half over. When I queried the appellate attorney, Stephen Meyer, why he would expend such effort, he stated that his client wanted his reputation back.

Appellate counsel also retained electrical engineer Mark Svare to reexamine the electrical evidence.

Figure 9.4a shows the front of the bar, and Figure 9.4b shows the alleged point of origin. Figure 9.4c is a "before" shot of the area shown in the foreground of Figure 9.4b, showing the door to the storeroom and the framed poster that the state alleged had been removed in anticipation of the fire.

During the investigation, the state's investigators tested their origin hypothesis by conducting a reconstruction. There had been a shelf mounted on the wall directly above the area where the investigators initially hypothesized the fire began. Examining the shelf, however, revealed that it was not nearly as damaged as the interior of the wall. This indicated that it fell down early in the fire and was not exposed to a fire burning on the floor. Rather than looking for another origin, though, the investigators modified their hypothesis to one that put the origin *inside the wall*. What clearly happened here is that there was fall-down inside the wall, shown in figure 9.4d, a common feature of fires in buildings with balloon construction. The reconstruction is shown in Figure 9.4e.

Figure 9.4 (a) Front of JJ's Pub. (b) View of the storage room where the fire originated, view from the bar area.

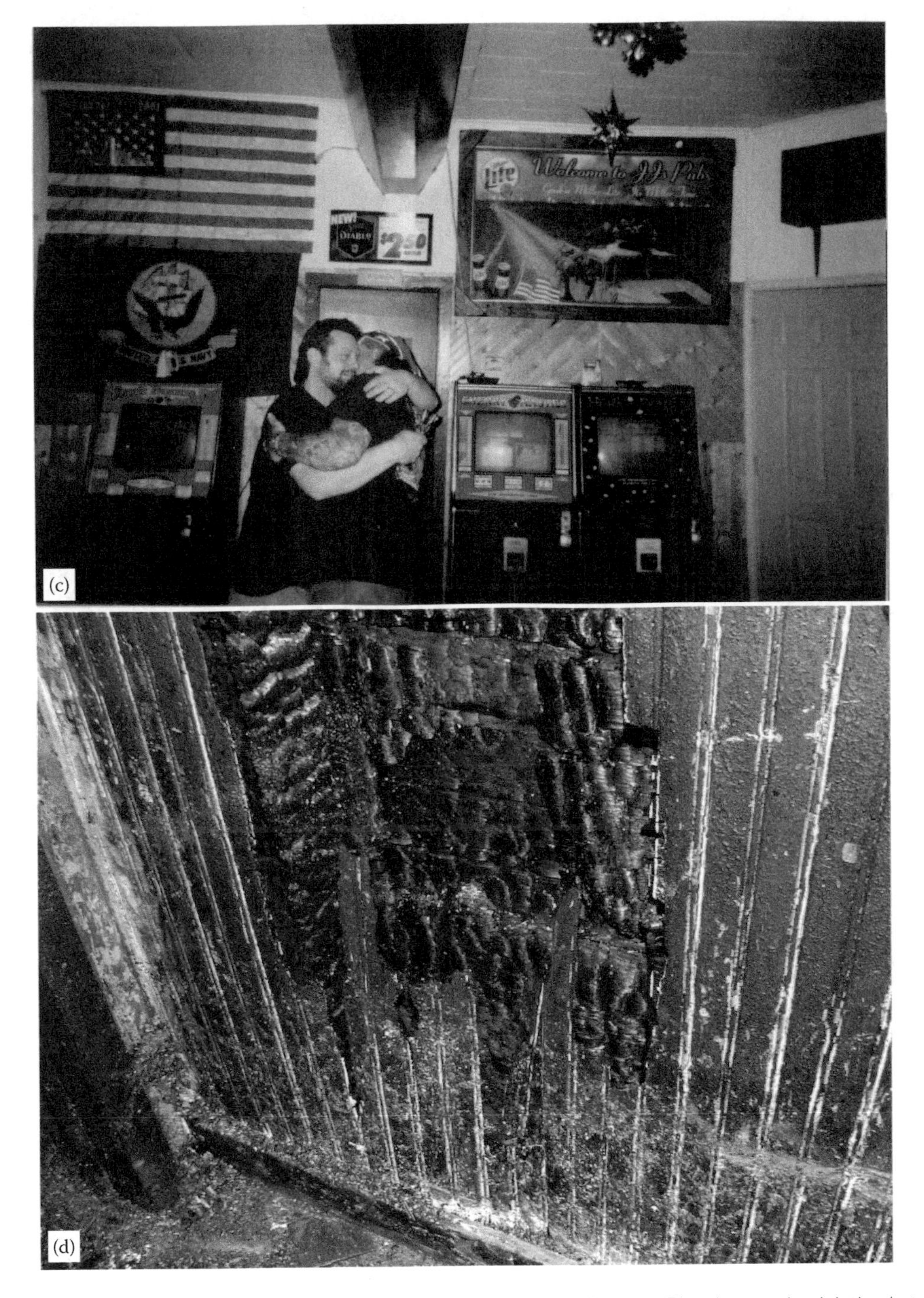

Figure 9.4 (c) Pre-fire view of the bar showing the "missing" (actually burned) poster. Blue door on the right leads to the storage room where the fire originated. (d) Base of the alleged area of origin prior to some of the wainscoting at the bottom being removed.

Figure 9.4 (e) Reconstruction of a shelf above the alleged area of origin. A fire set below this shelf should have caused far more damage to the exposed bottom side of the wood. (f) Base of the wall under the actual point of origin in the electric service panel mounted above.

Figure 9.4 (g) Electric service distribution panel in place after removal of some drywall. The cover had been removed from this panel before the fire to deal with electrical problems. (h) Electric service distribution panel, showing two circled areas where arcing through the steel case had occurred.

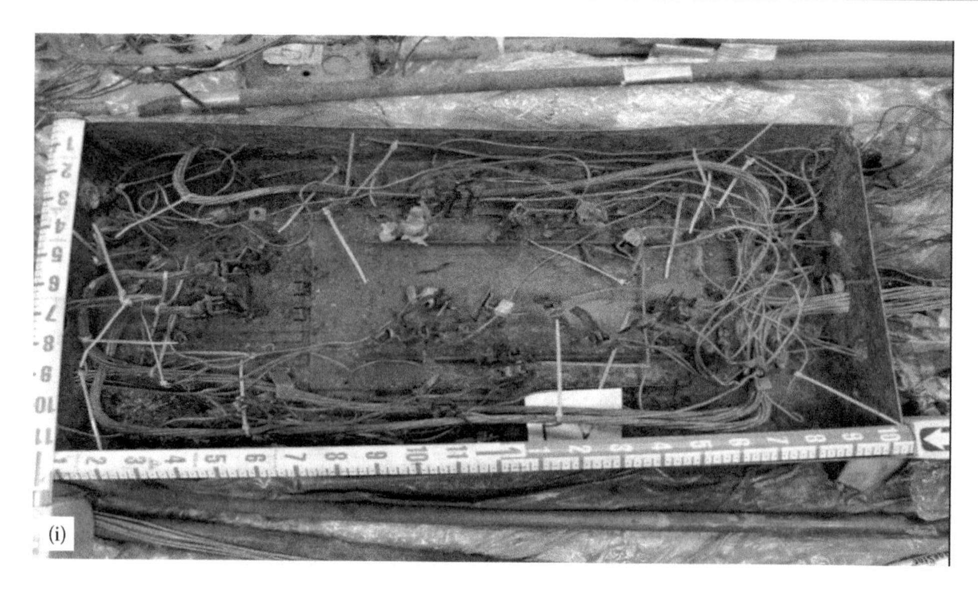

Figure 9.4 (i) Electric service distribution panel with orange zip ties showing the locations of the arcing events that occurred inside the panel.

This was the state investigator's testimony on direct:

> You have two by four studs within the wall there that, *once the fire breaks through that wainscoting or exterior surface and gets into the interior part of the wall* it's gonna act just like a chimney. Those of us who have fireplaces know what happens when the fire gets up into the chimney: It's gonna go to the least area of resistance. And all that available oxygen is up high so the fire is just gonna rise rapidly in between the studs and the wall. And in balloon construction, this type of construction, older construction, it's very common. *Once the fire gets into the walls* it goes right up to the second floor and into the attic immediately. And that's why you see such a distinct pattern here. And then the protected areas around it are like that because there was stuff stacked up around it protecting it. This was a storage room. There were all kinds of stuff stored on shelving and on the floor around it [12].

Now compare his testimony on direct with his testimony on *redirect* reflecting the change in origin to the inside of the wall:

> *A:* What it tells me, based on the elevation of the damage, the nature of the damage, and *how it got into the wall so quickly*, to me the likeliest scenario there is someone put a boot through there. Someone came up, gave it a good swift kick or knocked it out with a hammer or an axe or did something to damage it intentionally to get an opening into the wall. And it—I've seen it many times where that's what happens. People get upset and they just want to give something a good swift kick. And the height of it is consistent with where that would be. And it certainly would explain how it got into the wall as quickly as it did. But generally speaking, that's what I would have expected happened in this particular case.
>
> *Q:* And then whoever did that would have put some other type of material in there to—
> *A:* Absolutely. Absolutely. Would have used whatever was readily available. They could have brought certain things with them and took them when they left. The sky's the limit there. But—
> *Q:* And that's your—how you can attribute the damage that you see and—
> *A:* Absolutely.

The investigator apparently believed that the person kicking a hole through the wooden wall got very lucky and was able to avoid encountering a stud behind the wall [13]. When what really happened is that this section of wall became involved late in the fire after fall-down occurred inside the wall. The investigator also tightly embraced *negative corpus* methodology in this testimony:

> "One by one we try to address every single potential accidental ignition source *in our area of origin*. And once we are able to eliminate them all, there's absolutely nothing left that could have caused the fire except human involvement and that's the only way that we're able to do that [14]."

It was *exactly* this kind of flawed and unscientific thinking that caused the NFPA Technical Committee on Fire Investigations to disparage the use of negative corpus methodology in the 2011 edition of NFPA 921.

Failing to find a competent ignition source at the proposed origin should have led the investigators to look for a different origin, but these investigators apparently wanted to find an incendiary cause regardless of the facts. Ten feet away from the proposed origin, there was a very similar fire pattern behind the gypsum drywall, shown in Figure 9.4f, which coincidentally was directly below the electric service distribution panel, seen in Figure 9.4g. Here was a fire pattern that was nearly identical to the fire pattern selected as the origin, but with an actual potential ignition source above it.

Unfortunately, the insurance company's retained electrical engineer failed to recognize the significance of the evidence in this panel, from which the cover had been removed prior to the fire. The panel exhibited numerous arcing events inside, including two holes arced through the side of the steel case, as shown in Figure 9.4h.

Mr. Svare conducted his reexamination of the panel and other electrical evidence, he noted melted lugs, illegal wiring techniques, and numerous arcs, as shown in figure 9.4i. This panel was clearly the origin of the fire, and the first electrical engineer's failure to see that would prove costly later.

The fact that the state's and the insurance company's investigators determined both the wrong origin and the wrong cause had already been litigated, so it was *new evidence* that won the day for Awe. There was an evidentiary hearing held on September 17, 2012, at which both the author and Svare testified. The judge, Richard Wright, found that the changes in NFPA 921 between the time of the original trial in 2007 and 2011 constituted "new evidence." He even noted the similarities between arson cases and "shaken baby" cases, where the verdict is utterly dependent on the investigator's opinion as to whether a crime had even occurred. He wrote:

> Here the State's other evidence as to a set fire origin, to say the least, was weak; so this is not harmless. There was no direct evidence of the defendant's guilt. Circumstantial evidence of motive (financial hardship), and removal of keepsakes prior to the fire (just as likely consumed by the fire) added little. There was much argument over evidence regarding whether the electrical breaker box was involved and the location of the origin, and those arguments are not resolved by the new evidence. But the case was utterly dependent on the expert's opinion that this was a "set fire." Had the jury learned that the State's experts had used a methodology now disapproved by a mainstream arson investigation association, there is a reasonable probability it would have had reasonable doubt as to the defendant's guilt. The result would have probably been different.
>
> No one of the State's experts could determine a cause of the fire; they had theories as to potentialities, but no cause. From that they concluded, by elimination of hypothetical accidental causes, that the fire was not accidental. Without that opinion, there could hardly have been proof beyond a reasonable doubt. If it could have been accidental, it would have taken some strong evidence to point the finger at the defendant, and there was no such evidence.
>
> This is not the fault of the State's arson investigators, who were trained in the flawed methodology. It is the result of the maturation of the arson investigation field, a gradual process of taking a second look at the negative corpus thinking [15].

Judge Wright ordered a new trial and instructed that Joseph Awe be released from custody. The state declined to prosecute because they would again have to rely on negative corpus methodology. Awe was not content, however. He filed a lawsuit against his insurance carrier for promoting his prosecution as well as for the policy limits they owed. He also sued the electrical engineer for failing to properly interpret the evidence of electrical activity in the service panel. Both of these cases were settled before trial.

9.10.1.1 Error analysis

The electrical engineer misinterpreted critical data by attributing electrical arcing inside the panel to an external attack, and erroneously excluded the panel as the ignition source. He also relied on the fire investigators for their origin determination, and he was able to use circular logic to eliminate the panel because it was outside their area of origin.

> **This is not the fault of the State's arson investigators,** who were trained in the flawed methodology. It is the result of the maturation of the arson investigation field, a gradual process of taking a second look at the negative corpus thinking.
>
> **—Hon. Richard O. Wright, 2013**

The investigators also misinterpreted critical data. The obvious fall-down at their false origin was interpreted as the fire moving from the outside to the inside or, alternatively, starting inside the wall as a result of a deliberate act. They interpreted the absence of the burned poster as evidence of prior removal, when any reasonable hypothesis about how the fire progressed would not have resulted in any evidence of the poster being left behind. Further, they did not consider that photographs are no longer the irreplaceable sentimental items that they were before digital photography became the norm.

The use of the negative corpus methodology is a form of misinterpreting critical data. The absence of any fuel or ignition source at the proposed origin was misinterpreted as evidence of an incendiary fire rather than as evidence that another origin was possible. The investigators failed to give adequate consideration to the fire pattern directly beneath the electric service distribution panel, and went to great lengths to explain (incorrectly) that if the panel were at the origin, it would have exhibited much more damage.

The investigators also ignored inconsistent data. Awe was demonstrably 40 min away when the fire was discovered. He had an airtight alibi. The state and insurance company investigators "explained this away" by alleging, without evidence, that Awe had hired someone to set the fire. The reconstruction of the shelf provided inconsistent data, but instead of considering that data, the investigators simply moved the alleged origin to the interior of the wall and added the improbable and evidence-free explanation that someone kicked or cut a hole in the wall. They also ignored the finding of $900 in cash in the till. If Awe had a financial motive, why would he burn up so much money? Maybe $100 or $200 could have been left in the till to mislead the investigators, but $900 certainly seems excessive. Attributing inconsistent data to a diabolical plot by the arsonist to mislead investigators allows every fire to be called arson.

This combination of errors led to an innocent man being robbed of three years of his life and living under a cloud of suspicion for six. This case should never have been brought to trial.

9.10.1.2 Significance

This is the first (so far the only) case known to the author in which a change in the science of fire investigation was found to constitute "new evidence" that triggered the overturning of an arson conviction. This case also points out that photographs are no longer the irreplaceable items that they once were. Finally, the lawsuit against the electrical engineer demonstrates that errors can have real consequences for negligent fire investigators as well as for the falsely accused.

9.10.2 State of Georgia v. Weldon Wayne Carr [16]

Weldon Wayne Carr and his wife, Patricia, were asleep in the same bed on the evening of April 7, 1993, when they were awakened by smoke. Upon opening the bedroom door, Mr. Carr encountered thick black smoke blocking his access to the stairway, which was located down the hall and to the right. He attempted to find the chain escape ladder that had been kept under his bed but could not locate it. He made the decision that he and his wife were going to have to jump out the front bedroom window and was about to do so when his wife ran toward the hallway; Mr. Carr pursued her but was unable to find her in the smoke.

Mr. Carr jumped out the bedroom window, landing on and breaking a bush in the front yard and cracking a vertebra. In extreme pain, he hobbled across the street and pounded on the neighbor's door, breaking the door in the process. He screamed for the neighbor to call 911, and the fire department responded. Firefighters were able to retrieve Mrs. Carr and get her to a hospital, where she lived for 3 days. She never regained consciousness and died from complications of smoke inhalation. Mr. Carr, who was also hospitalized, was arrested in his hospital room and charged with murder the day after his wife died.

An anonymous phone call had been received by the Fulton County Fire Department, stating that the cause of the fire should be carefully investigated, and the captain in charge of the arson squad responded. The fire captain in charge of the investigation determined that the fire had been set intentionally using a trail of newspaper and "an accelerant" in the form of a liquid leather-finishing product known as Neat-Lac.[5] The trail of newspaper reportedly was spread from the kitchen, along a wall into the dining room, and around to the base of the stairs.

[5] According to the MSDS, Neat-Lac contains over 60% solvent, including toluene, isopropyl alcohol, isobutyl acetate, and light petroleum distillate.

The fire investigator sent out by Mr. Carr's homeowner's insurance carrier made the same observations and also found evidence of a "trail of newspaper" leading up the stairs. The *Atlanta Journal and Constitution* quickly jumped on this story and, within a week, Carr had been convicted in the paper. It came out during the investigation that there were marital difficulties and that Mrs. Carr had been having an affair with a neighbor. Mr. Carr had tapped the telephone and recorded conversations between his wife and her paramour. Mr. Carr had also recently updated his insurance coverage and he had sent certain sentimental items that had been in the house for several months to his mother. In addition, his car was parked at the end of the driveway rather than in the garage. All of this background information made Mr. Carr the obvious suspect if the fire actually had been set intentionally.

The investigation was aided by the use of an accelerant detection canine by the name of Blaze, who alerted 12 times in the Carr residence. Laboratory analysis failed to detect the presence of accelerants in any of the 12 samples collected. The prosecutor, Nancy Grace, went so far as to retain the services of a private laboratory to double-check the state's results, but the private laboratory's results were also negative.

Weldon Wayne Carr was a multimillionaire who owned a large nursery business, Hastings Nursery, in Atlanta, which was a nationwide mail-order business.[6] He retained a respected criminal defense attorney, Jack Martin. Martin, in turn, hired I. J. Kranats, who was later to become president of the International Association of Arson Investigators (IAAI), as an independent origin-and-cause investigator. During his investigation, Kranats persuaded Martin to retain another colleague, Ralph Newell, to look at the scene. Kranats and Newell, together, prevailed upon Kenneth Davis, a retired deputy state fire marshal, and Steven Sprouse, then a fire chief in nearby Douglas County and later the chief investigator for the Georgia State Fire Marshal's Office, to also review the evidence. All four of these individuals agreed that the fire captain had made a serious error in his determination.

As the case proceeded toward trial, Attorney Martin became concerned about his own inexperience with fire cases, and brought in Michael A. McKenzie, an experienced civil litigator in fire cases, to assist. Shortly after McKenzie was on board, he retained the author to assist him with testing some of the state's hypotheses. I was reluctant to enter the case at this late stage, only 3 weeks before the trial, but I agreed to participate in a limited capacity.

An alleged "pour pattern" was observed that covered approximately 4 ft^2 at the doorway between the kitchen and the hallway. A floor plan of the Carr residence is shown in Figure 9.5a. The alleged accelerant, Neat-Lac, was contained in an 8-ounce can. There was testimony that Mrs. Carr had used this can for several years, so it was probably half empty. There were still 2 ounces of Neat-Lac remaining in the container when the fire captain found it. A spill test showed that it would require 24 ounces of this liquid to cover the alleged pour pattern. However, as with many other aspects of the fire captain's opinion, the size of the pour pattern and the proportion covered by liquid accelerant on it changed from day to day. Whatever its size, there was not enough Neat-Lac in the house to create the pattern. No evidence suggesting the use of any other accelerants was ever developed.

The other aspect of this author's work in the original trial was to comment on the state of the art with respect to accelerant detection canines. Although Blaze had alerted 12 times, laboratory analysis revealed no evidence of ignitable liquid in any of the samples. Nevertheless, it was the state's intent to introduce the alerts as evidence that accelerants were present throughout the first floor. An investigation of Blaze's confirmation record, as documented by the staff of the Georgia State Crime Laboratory, revealed that when fire debris samples failed to exhibit an odor detectable to humans, Blaze's confirmation rate was *less than 50%*.

Two additional pieces of physical evidence, critical to the state's case, were not even tested until after the trial. A trail of newspaper was allegedly spread around the first floor to communicate the fire from the kitchen into the dining room and the hallway. The basis for this opinion was the finding of small pieces of newspaper in these rooms but, curiously, no damage was observed to the linoleum or hardwood floors where the "trail of newspaper" had allegedly been laid. A third aspect of the state's case was that accelerants somehow migrated into the wall (or were poured into the wall) between the kitchen and the bathroom, and then the vapors from those accelerants migrated *upward* and exploded, causing damage on the second floor.

[6] Like the plumber with leaky pipes or the cobbler's children with holes in their shoes, Carr's residence sported several very healthy poison ivy vines growing among the azaleas in the front yard. (Hastings Nursery closed in 2016 after 123 years in business.)

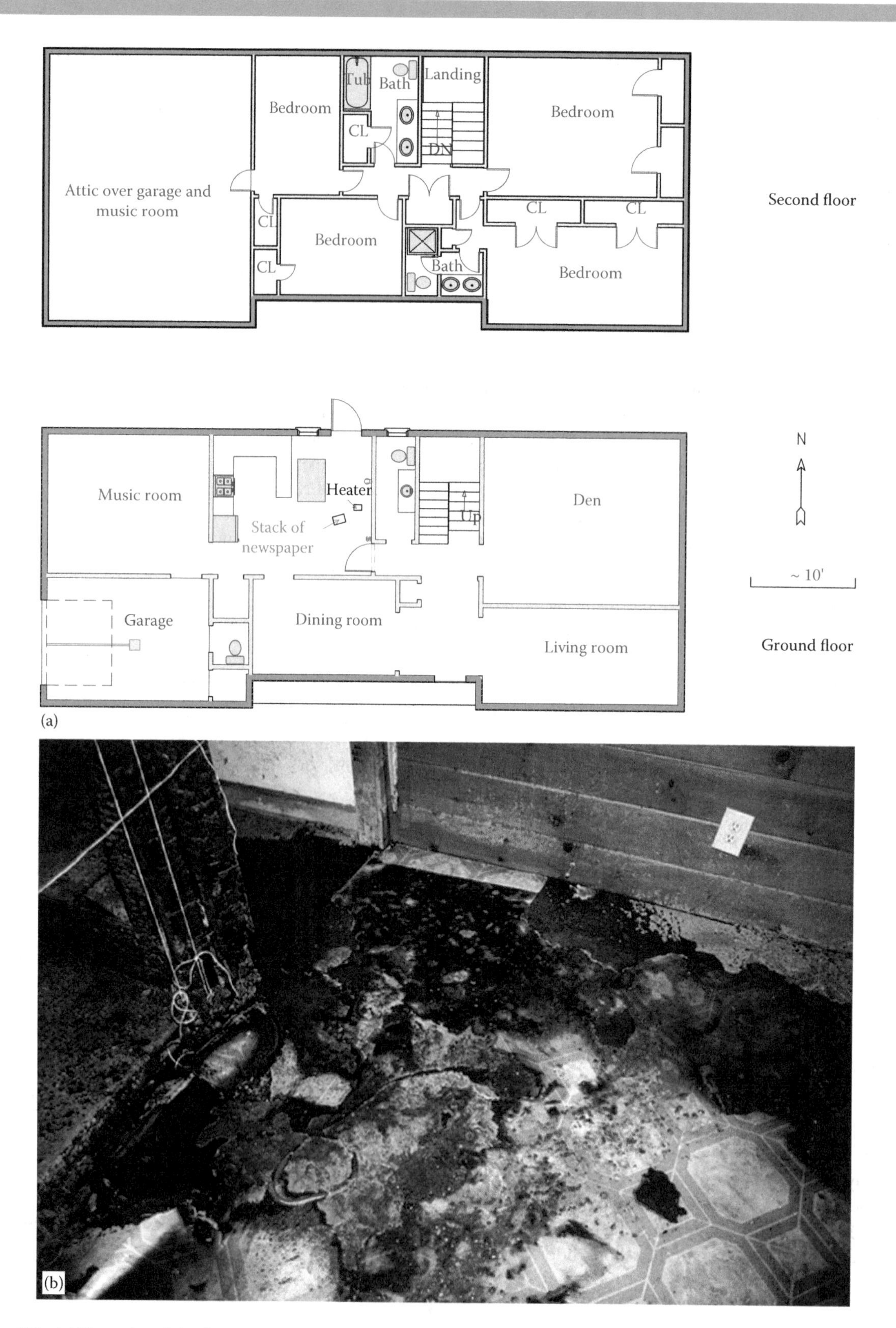

Figure 9.5 (a) Floor plan of the Carr residence. (b) Burn pattern at the doorway incorrectly characterized as having been caused by an ignitable liquid. Note the straight line on the floor at the base of the wall showing the position of the door when the radiant heat scorched the areas of the floor that were not protected.

Figure 9.5 (c) A test of the state's hypothesis that an ignitable liquid could burn on the floor without scorching the bottom of the door. (d) Fire pattern resulting from the fire test, showing that the door should have been scorched if there had been ignitable liquid burning on the floor.

Figure 9.5 (e) A test of the state's hypothesis that a trail of newspaper could burn the paper without damaging the floor.

Figure 9.5 (f) Paper found on the stairway in the Carr residence, falsely characterized as being part of a trail used to spread the fire. Note the floral pattern on the paper at the center of this photo.

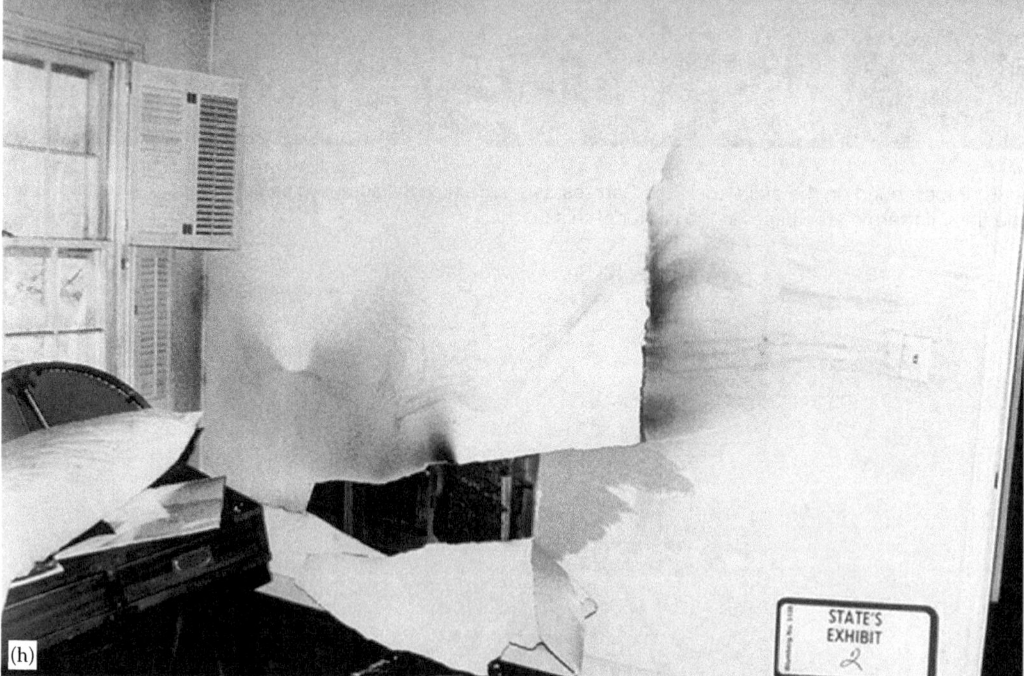

Figure 9.5 (g) Actual source of the paper on the stairs. (h) Blown-out section of wall on the upper level, between a bedroom and an upstairs bath, directly over the wall between the breakfast room and the bathroom. The state hypothesized that the explosion was caused by the ignition of vapors that somehow migrated upward.

Figure 9.5 (i) Apparatus for the test of the state's "vapor in the wall" hypothesis. The pressure rise was so slight that it could only be measured using inclined manometers.

Figure 9.5 (j) Upstairs floor structure directly over the wall cavity where the switch that caused the fire was located. A semicircular pattern is centered on the cavity. Other views of the charring in this area also showed that the wall cavity was the origin of the fire.

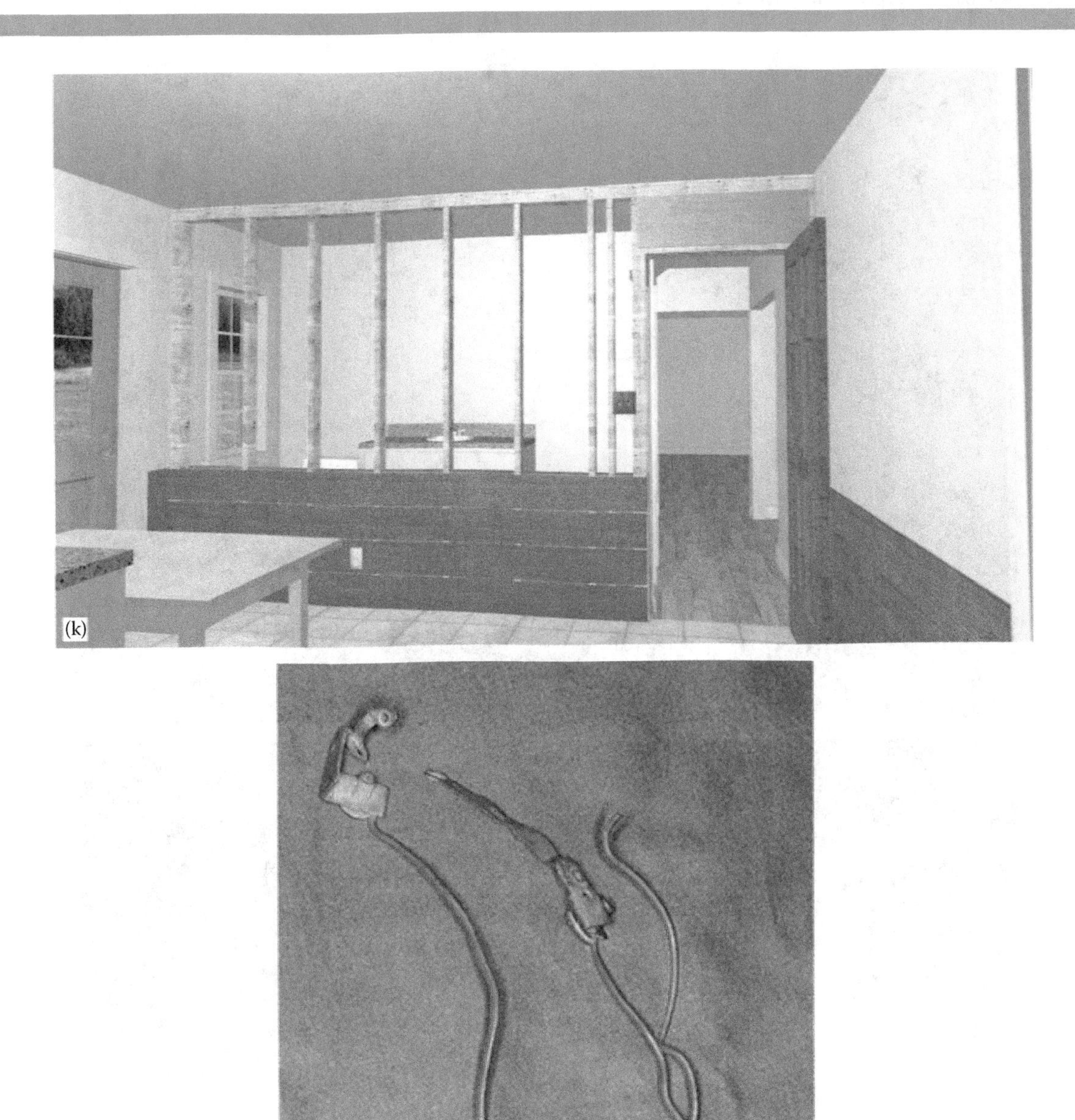

Figure 9.5 (k) Schematic view of the area of origin in the Carr fire. (l) Three-way switch and associated wiring from the wall at the origin of the fire.

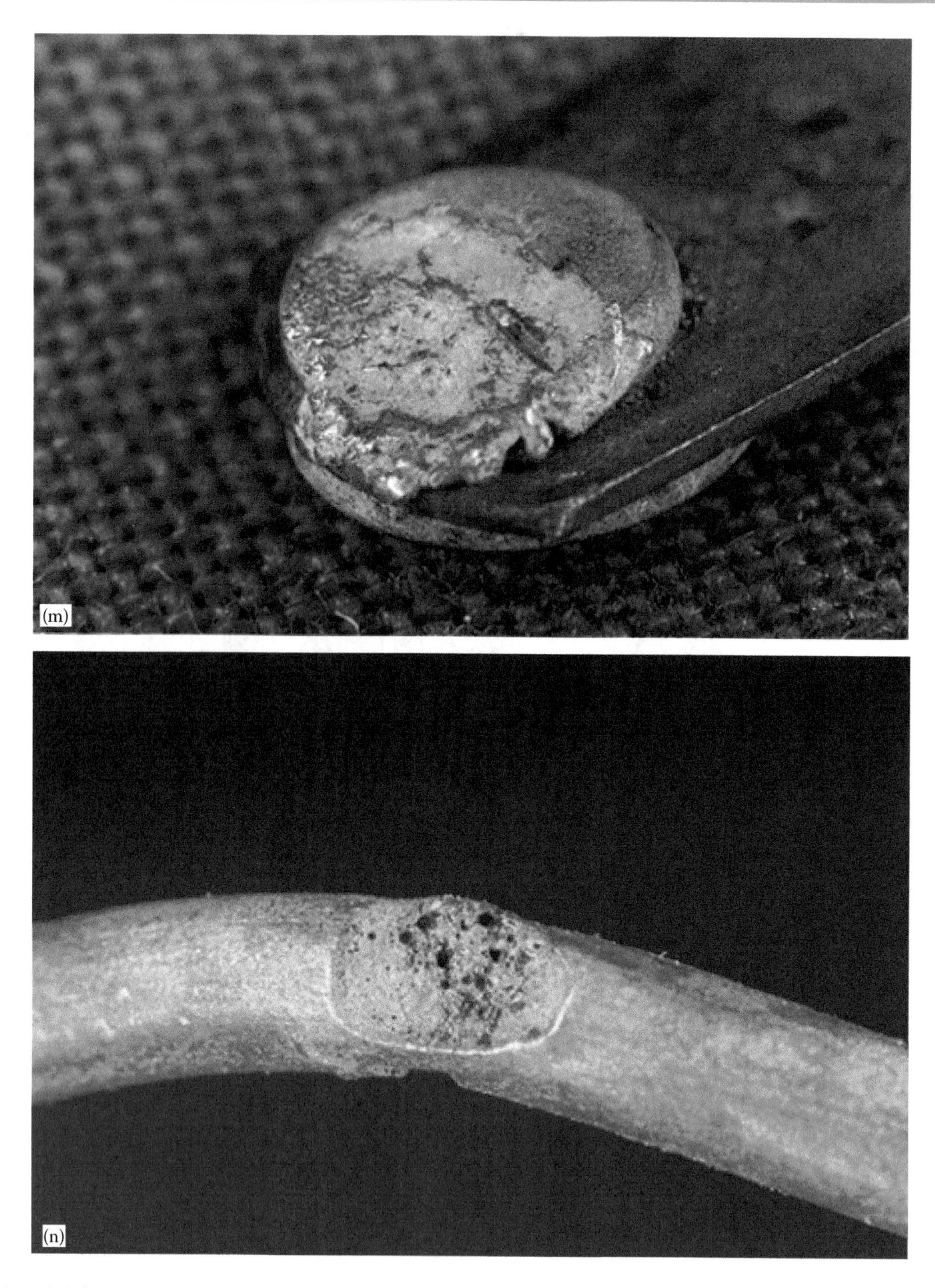

Figure 9.5 (m) Overheated contact pad on one side of the moving member of the three-way switch. This was not fire melting, nor could it be said to be a normal result of the switch's operation, because the other contact pad was in pristine condition. This condition had escaped the notice of the first two electrical engineers who had the opportunity to examine it. (n) One of two areas of arcing damage on the wire directly above the switch box where the fire started.

At the trial of the case, I was able to testify about the quantity of liquid that would have been required to cover the alleged pour pattern. However, the judge ruled that I was not qualified to testify about accelerant detection canines because I was not an accelerant detection canine handler. The canine handler put the unreliable evidence of the 12 alerts before the jury.

Mr. Carr had an explanation for each of the background circumstances that made it look like he had prior knowledge or a motive to set the fire, but counsel chose not to put him on the stand. He was the kind of witness who could not be controlled, and counsel was simply unprepared to subject him to Nancy Grace's cross-examination.

Mrs. Carr sustained a minor head injury as she was being transported down the stairs by firefighters. This led the state to charge Mr. Carr with assault, contending that he had struck his wife on the head prior to jumping out the window. The state attempted to introduce evidence that Carr was a wife beater, but there was absolutely no evidence from anyone on that score, although that did not stop the prosecutor from using a mannequin and beating up on it in her closing argument. Carr was acquitted of beating his wife, but he was convicted of setting the fire and of wiretapping. He was sentenced to life in prison.

It was only after the conviction that this author was allowed to seriously test the state's hypotheses. The defendant was under a court order to produce all results of any tests, favorable or not, to the prosecution prior to the trial.[7] As it was impossible to predict with certainty the results of any tests other than the one involving the determination of the coverage provided by the liquid Neat-Lac, no other testing was done pretrial. Once the conviction had occurred, Carr had nothing to lose by testing all of the state's hypotheses.

Figure 9.5b shows the doorway in front of which the alleged pour pattern existed. What actually happened is that the fire came through the breakfast area doorway after flashing over in the bathroom, and radiation burned the floor. This was obvious from an inspection of the pattern on the floor next to the door, which ended in a perfectly straight line. The straight line resulted from the door protecting the floor. This is a classic case of heat shadowing. There was no damage to the bottom of the door and, in fact, the lower half of the door was undamaged.

The first test we conducted was to build a mock-up of the doorway to demonstrate what would happen if there had, in fact, been a puddle of flammable liquid in the doorway. The test fire is shown in Figure 9.5c, and the resulting damage on the door is shown in Figure 9.5d. Of course, the door from the test fire looked nothing like the door from the Carr residence.

The next set of tests was conducted to determine whether it would be possible for a trail of newspaper to burn on a linoleum or finished hardwood surface, leaving no detectable damage. Actual flooring, both linoleum and hardwood, from the Carr residence was cut out and brought to the ATS laboratory. Trails of newspaper, both with and without Neat-Lac poured on them, were ignited. In every case, the burning newspaper trail caused significant damage to the floor. One such test is shown in Figure 9.5e.

The explanation for the newspaper found around the house was straightforward. There was a pile of newspapers on the floor in the breakfast room. When the firefighters arrived at the scene, their first attack was straight through the back door into this stack of newspaper, which was already on fire.[8] The extinguishment water carried the newspaper particles throughout the house. With respect to the newspaper found on the stairway, a careful examination of the paper showed that it was not newspaper at all but rather charred wallpaper. Figure 9.5f shows that the alleged newspaper used as evidence against Mr. Carr was nothing more than charred wallpaper. So anxious was the investigator to prove that the trail extended up the stairs that he failed to look up and see the obvious source of the paper, shown in Figure 9.5g.

---

[7] That order was later found by the Georgia Supreme Court to be a reversible error, and the court has since declared that such orders were a violation of a defendant's Sixth Amendment right to confront the evidence against him.

[8] Carr was not a tidy person. He would read the paper while sitting at the table in the breakfast room and then drop it on the floor. This behavior was one of the causes of tension in the Carr marriage. Carr stated that he was working on cleaning up his act when the fire occurred.

Still another test was conducted to examine the (truly strange, physically impossible) hypothesis that the accelerant vapors had somehow become lighter than air, causing an explosion of sufficient power in the wall cavity on the second floor that gypsum wallboard was dislodged. The broken wall is shown in Figure 9.5h. A mock-up of the first- and second-floor walls was built with a minimum of leakage so as to contain any pressure produced by an explosion, and quantities of Neat-Lac were introduced into the lower part of the mock-up. The test setup is shown in Figure 9.5i. The state's witnesses were never clear as to how someone would have introduced Neat-Lac into the wall. At one point, they suggested that it migrated there after being poured on the floor, and at another point, they said someone had cut or broken a hole in the wall to make it appear that the fire had started there. In the mock-up, we cut 4″ × 4″ holes in the wall cavity, added 2 ounces of Neat-Lac, and allowed it to vaporize for a minute prior to dropping in a lighted match. A small, low-order explosion occurred, but it was necessary to use an inclined manometer (a U-tube tilted almost horizontally to increase sensitivity to pressure changes) to measure the pressure. The pressure generated by the explosion in the upstairs cavity was 0.006 psi, or less than 3/10 of 1% of the 2 psi that is generally accepted as being required to break wallboard. In the second test, the Neat-Lac explosion produced 50% more pressure, all the way up to 0.009 psi. Clearly, the rupture of the wall was caused by the explosion of smoke rather than the explosion of flammable liquid vapors.

All of the test results were summarized in an affidavit that ran 147 paragraphs over 65 pages and was submitted as an exhibit attached to Carr's appeal brief.

A second avenue of appeal contested the admission of unconfirmed canine alerts being used as evidence. After the judge allowed the dog handler to testify, this author approached the Forensic Science Committee of the International Association of Arson Investigators and requested that they address the issue in a position paper on the appropriate use of canines. The paper [17] was published in late 1994 and caused quite a stir. Prosecutors and fire marshals wanted to know who these meddling chemists thought they were to question the authority of their highly trained dogs. The Forensic Science Committee position paper, along with information about the *Carr* case and a similar case in Iowa, where an appellate court had ruled that the admission of unconfirmed canine alerts was proper [18], was brought to the attention of the NFPA Technical Committee on Fire Investigations. The 1995 edition of NFPA 921 had already been finalized, but the Committee members believed that if they were to wait until 1998 to publish guidance on the proper use of accelerant detection canines, the case law that accumulated in the interim would render the Committee's guidance ineffective. The chair of the Technical Committee, Richard Custer, suggested, and the Committee agreed, that a Tentative Interim Amendment (TIA) would be appropriate in order to get the word out sooner rather than later. This met with some resistance in the accelerant detection canine community, some of the members of which hold the position, "Dog said it. I believe it. That settles it." A decision was made to host two separate meetings, one on the West Coast and one on the East Coast, to gather input from all relevant interested parties, but particularly from the community of accelerant detection canine handlers. Generally, responsible canine handlers did not believe that unconfirmed alerts constituted reliable evidence, and they said so. The Committee passed the TIA, but the NFPA Standards Council did not understand the emergency nature of the situation and requested further clarification. Generally, the Standards Council is concerned that failure to act quickly will result in additional injuries or deaths or property loss. When it was explained to the Standards Council that what concerned the Committee was wrongful convictions, and it was further explained that the emergency was of a judicial nature, they acquiesced and allowed the TIA to go into effect. Just days after the TIA went into effect, Carr's lawyers filed their appeal with the Georgia Supreme Court and pointed out that the relevant scientific community had a serious problem with unconfirmed canine alerts being considered conclusive evidence of the presence of accelerants. The author submitted an additional affidavit in support of this brief, but more importantly, five forensic scientists from the Georgia Bureau of Investigation Crime Laboratory submitted an affidavit stating that unconfirmed canine alerts were unreliable as evidence of ignitable liquids.

By this time, Carr had retained Millard Farmer as his lead appellate counsel, although Michael McKenzie stayed heavily involved in the technical aspects of the appeal. Farmer did not think that the appeal of the admission of the unconfirmed canine alerts would sway the court and was surprised when the justices requested additional briefing on that subject.

The court considered the case for several months. During that time, the prosecutor, Nancy Grace, spoke at a Georgia Fire Investigators seminar and introduced Blaze as her "star witness" in the *Carr* case. She stated, "A jury will believe a dog before they'll believe a man any day." She also voiced concern about the status of the appeal, as well she should have. A few months later, the Georgia Supreme Court overturned Carr's conviction on the basis that the judge had allowed the introduction of the 12 unconfirmed alerts. In enumerating six other major errors, the court held that it would have granted a new trial based on those points had they not already granted a new trial on the basis of the unconfirmed canine alerts being admitted. The court found it to be reversible error for the judge to have excluded the testimony of this author on the subject of canines. The court also found that Nancy Grace had violated Carr's Fourth Amendment rights by going back to his house many months after the fire with an outside expert and a news crew from CNN to conduct a "smoke test." Grace did not think it was actually necessary for her to obtain a real search warrant, and the court took strong exception to that. The court also took strong exception to Grace's conduct in several other areas and indicated that they would have overturned Carr's conviction on the grounds of *prosecutorial misconduct* had they not already overturned it for the wrongful admission of the unconfirmed canine alerts. Grace would later comment, "[The ruling was] one of the most painful incidents in my legal career. I was stunned by it—by the flogging they gave me and the judge [19]."

Although the conviction was overturned, Carr was still under indictment, and the Fulton County prosecutor indicated that he would be tried again. Consequently, he stayed in the penitentiary awaiting a second trial. Negotiations between Carr's defense team and the prosecutors resulted in an agreement that the state would hire yet another independent origin-and-cause investigator to evaluate the case. However, by the end of 1997, this had not happened and when Carr's attorneys moved to have him released, the motion was granted. Carr had served just over 3 years in the penitentiary.

As one assistant district attorney after another reviewed the physical evidence in the *Carr* case, they became persuaded that no proof existed that the fire was incendiary. The county finally hired an independent expert, and he concluded in a September 2001 report that the defense had made "a very convincing cause determination based on the fire scene." Consequently, nobody moved the case forward until Carr's attorneys, in late 2002, filed a motion that the case be dismissed, both because it lacked merit and because Carr's right to a speedy trial had been denied.

Early in the case, the state had retained an electrical engineer to eliminate all the potential accidental fire causes in the house, and he had obliged. Newell and Kranats had obtained the services of another electrical engineer who was unable to find any evidence of an electrical cause. In 2002, the evidence was made available to an ATS electrical engineer, Richard Underwood, and for the first time, a rational and complete ignition scenario was proposed. Underwood was the first electrical engineer of three who had looked at the switch from the wall between the breakfast area and the bathroom and noted that one of the contact pads had melted. Simultaneously with Underwood's evaluation, Jeff Morrill, also from ATS, was assigned to start over with a fresh reconstruction. Carr's residence was still available and the origin area was meticulously reconstructed. Small details, such as bent nails, were used to determine the exact location of the switch box, revealing the significance of the semicircular pattern above it, shown in Figure 9.5j. Numerous other photographs of the area at the top of the wall showed a clear pattern, indicating that the fire started in the wall cavity containing the switch. Figure 9.5k is a schematic view of the area of origin. There had never been any question in this author's mind that the fire burned quite vigorously for a considerable period of time in the bathroom before it ever entered the breakfast area, but this reconstruction brought the whole sequence of events into sharp focus. The three-way light switch and associated wiring are shown in Figure 9.5l. The overheated contact pad is shown in Figure 9.5m. The heating could not have been a result of the fire, as indicated by the pristine condition of an identical contact on the same moving member. If this overheated contact did cause the fire, then we would expect to see evidence of electrical arcing in the wall cavity containing the switch, and indeed there was evidence of arcing. One side of the arc damage is shown in Figure 9.5n. Eight years after the fire, Jeff Morrill and Dick Underwood had finally put it all together. The cause of the fire was the overheated light switch.

In December 2002, a hearing was held on the defense motions to dismiss. Several witnesses had moved away, some had died, and others' memories had faded. Judge Roland Barnes withheld adjudication on the merits of the case[9] but granted the motion to dismiss on the speedy trial grounds. This author attended the hearing and witnessed an exchange in which an assistant district attorney, having seen the evidence developed after the conviction, stated that it would be unethical to proceed with a prosecution. Unfortunately, the district attorney did not share that view and appealed the judge's ruling on the speedy trial motion.

[9] Thus allowing skeptical observers to believe that Carr "got off on a technicality." Defense counsel believed that the judge would have granted the motion on the merits of the case if the supreme court had not upheld the dismissal on speedy trial grounds.

> The dog handler's testimony concerning the presence of accelerants, based on the dog's alert, is conceptually indistinguishable from the other types of tests and analyses… . Therefore, the analysis and data gathering leading to the testimony should be subject to the same requirements of scientific verifiability applicable to the other procedures. To hold otherwise would make the dog alert procedure analogous to the earliest recorded lie detector, which involved touching a donkey's tail.
>
> **—Supreme Court of Georgia, 1997**

In the summer of 2004, the Georgia Supreme Court again reviewed the case of *Georgia v. Weldon Wayne Carr* and upheld the dismissal. Clearly, cause for suspicion existed in the case of the Carr fire, but this is a classic case of allowing suspicions to interfere with a careful examination of the physical evidence. Carr finally addressed each one of the suspicious circumstances in an interview with the *Fulton County Daily Report* in 2004. Mike McKenzie had previously addressed these circumstances with one assistant district attorney after another, which was one of the reasons they chose not to retry the case. Had the investigators kept an open mind, they would have taken the time to carefully examine the physical evidence and would have discovered the true cause of this fire long before 2002.

9.10.2.1 Error analysis

The fire investigators in this matter misinterpreted irrelevant evidence when they chose to make an inference from an irregular burn pattern on the floor. They also failed to adequately reconstruct the scene, and they failed to account for the fully involved conditions in the bathroom, ignoring inconsistent data. Further, they formed hypotheses that they did not test, particularly with respect to the ability of two ounces of liquid to cover a 4 ft^2 pattern, the ability of newspaper to burn on a combustible surface without damaging that surface, and the ability of heavier-than-air vapors to defy the laws of gravity. Nor did they test the hypotheses put forward by Carr's experts. The fire investigators simply let the background evidence overwhelm any reasonable interpretation of the physical evidence. The engineers who first looked at the electrical evidence overlooked critical data and, in so doing, gave unwarranted support to the incorrect hypotheses put forward by the fire investigators.

Some would argue that the background evidence—the failing marriage, the car being moved, the updated insurance coverage, and the shipment of photographs—constitute "data" that should be considered in determining the cause of the fire. The argument is that the investigator should take a "holistic" approach and consider the "totality of the evidence [20]," and failing to consider such "data" is "unscientific." This argument points out the difference between scientific "truth" and legal "truth." The argument is unlikely to be resolved anytime soon but, in this author's view, the background "data" will always lead to an incendiary determination, which will often be incorrect and not justifiable using science. For this reason, NFPA 921 states that background evidence such as motive and opportunity should only be considered *after* the physical evidence has been examined and the fire classified as incendiary [21].

9.10.2.2 Significance

The *Carr* case provided the impetus for both the IAAI Forensic Science Committee's Position Paper on Canines and NFPA 921's TIA on the subject, which later became part of the guide. Both of these papers helped to persuade the Georgia Supreme Court that unconfirmed canine alerts constitute unreliable expert evidence. The case has been cited on numerous occasions when prosecutors or insurance defense attorneys have attempted to misuse this valuable tool by introducing evidence of unconfirmed alerts.

This case was also Nancy Grace's last conviction. The Supreme Court (unanimously) found "that the conduct of the prosecuting attorney in this case demonstrated her disregard of the notions of due process and fairness, and was inexcusable," thus undermining the credibility of this television personality.

9.10.3 Maynard Clark v. Auto Owners Insurance Company [22]

On an April night in 1992, Maynard Clark's furniture store in Ruskin, Florida, burned to the ground during a severe thunderstorm. In examining a roll-up door that served as a fire barrier, the investigator found a screwdriver jammed into the gears, and from that point forward, everything he saw supported a hypothesis that Maynard

Clark had torched his own store. When the floor was excavated, spalling of concrete was seen. Twisted steel beams (described as "molten") were cited as evidence that the fire had exceeded 2,000°F, far higher than the 900°F that the fire investigator believed a "normal" fire would attain. Unlike in the *Carr* case, the fire investigator did not consider motive, because, as will be discussed, Maynard Clark had no motive to torch his store. The fire investigator also "eliminated" lightning, based on a lightning detection network report. The investigator's report stated, "The nearest lightning strike was 2.08 miles from the store." A lightning data report purchased by Clark's attorney from the same detection service for the same time period showed different results. Further, the lightning data report stated that the accuracy of the location of the flash was typically within 2 to 4 kilometers (km) (1.2 to 2.4 miles) of the actual location and claimed a detection efficiency of only 60% to 80%.[10] The graphic from the lightning detection report is shown in Figure 9.6a. After discounting the lightning report or, rather, turning it on its head, the investigator went on to discount more evidence. A young Hispanic boy, inexplicably outside at 2:30 in the morning did, in fact, see lightning strike the building. A parallel criminal investigation resulted in a sheriff's deputy pulling the boy out of school to interrogate him, in the process accusing him of lying about seeing the lightning. The boy was intimidated into retracting his story, although his statement, which was only revealed well into the litigation, gave no indication as to why he had made up a story about lightning or why he had retracted it. A second eyewitness, an elderly man who also said he saw lightning hit the building, was dismissed as not credible because he was thought to be senile.

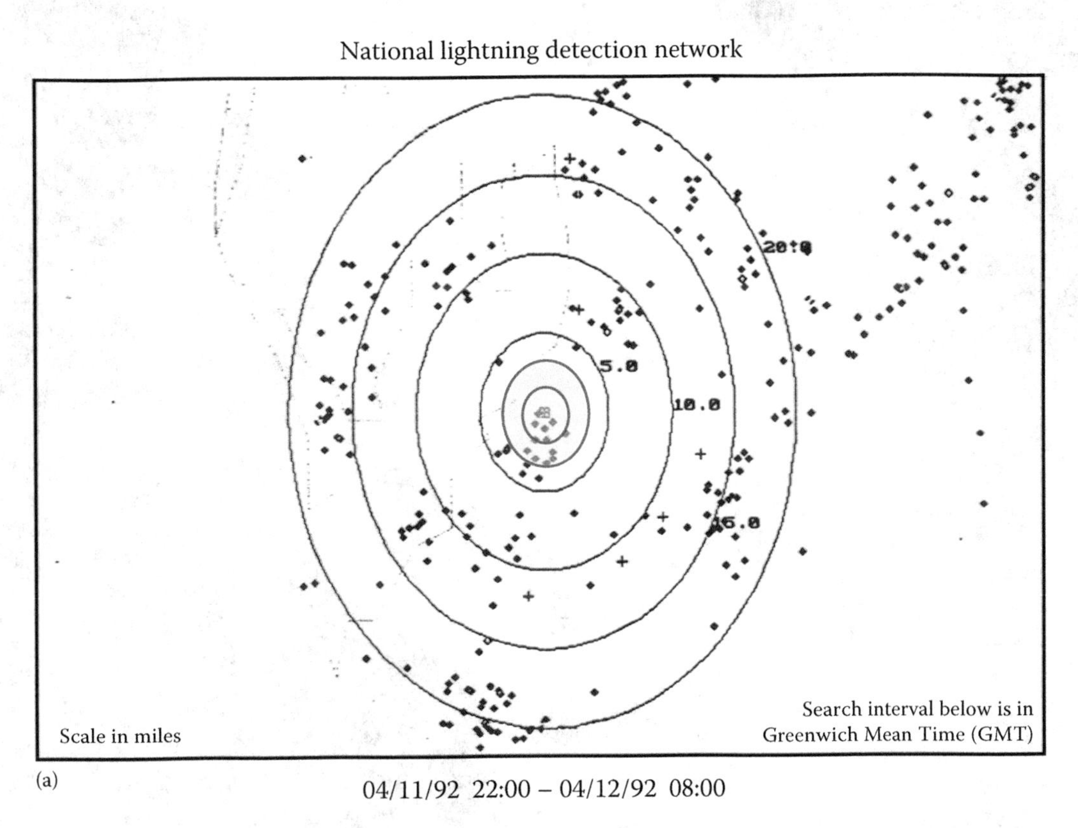

Figure 9.6 (a) Lightning data report with Maynard Clark's store at the center. The report had a "detection efficiency" of 60% to 80% (it missed 3 of every 10 strikes), and a location accuracy of ±2 to 4 km. Sixteen strikes were recorded in the hour before the fire. This report was cited as evidence that lightning had ***not*** struck the building. The "error bar" around the store is colored yellow. The inner circle around the furniture store at the center of the plot shows the minimum error of 2 km, or 1.2 miles, and the outer circle shows the maximum reported error of 4 km. Each strike indicated would also have a similar circle around it.

[10] Lightning data services now report a storm detection efficiency of greater than 99%, a flash detection efficiency of greater than 95%, and a median location accuracy of 250 m or better. Each lightning strike has its own unique "99% confidence ellipse," which is determined by several factors. The confidence ellipse is the equivalent of the "margin of error" for each individual lightning strike.

Figure 9.6 (b) Preparation of an asphalt sample during the development of a procedure to distinguish asphalt residue from heavy petroleum distillates.

Figure 9.6 (c) Condensation of asphalt smoke on a watch glass. Such condensates can be used to characterize the expected contribution of just about any combustible material to a fire debris sample.

This author was brought into the case to help resolve issues regarding the chemical analysis. The fire investigator had seen a "pour pattern" at the rear of the store where the tar-covered roof had collapsed. Samples were collected by both the Florida State Fire Marshal's Office and the independent investigator, who had sent his samples to his company's in-house laboratory in Ohio. The state fire marshal's laboratory reported finding diesel fuel in the sample, while the Ohio laboratory reported finding kerosene. When first asked to review the chemical data, this author made it clear to Clark's attorney, Barry Cohen, that making a distinction between kerosene and diesel fuel was not an easy task. The more salient question, however, was whether either lab had found anything other than roof tar. The private chemist had used a packed column and was therefore unable to resolve the peaks that might have alerted him in to the contribution of the asphalt to his results. The state chemist had used a capillary column but was not aware of the significance of the "extra" peaks in her chromatogram.

The private chemist's file (at least part of it) was produced at his first deposition, and his notes indicated that the sample contained "tar." In 1982, this author and Laurel Waters (now Laurel Mason) published an article discussing the chemical analysis of asphalt shingles and warning that the technology available to chemists at the time would not allow for a distinction between liquid heavy petroleum distillates (like kerosene and diesel fuel) and the residue from melted asphalt [23]. It was reported to counsel that, at the time, no valid technique existed to distinguish between asphalt and kerosene, or diesel fuel, in fire debris samples. Clark, having already rebuilt and reopened his furniture store with his own funds, was able to easily finance the time and materials required to conduct a research project to develop just such a technique. Clark was not in financial distress either before or after the fire.

The first phase of the research project involved burning asphalt and then sampling headspace for the resultant residue. Under these conditions, no kerosene-like substance was produced. About 3 weeks into the project, the author was preparing an asphalt sample by burning it in a small crucible, as shown in Figure 9.6b, and extinguished the fire by placing a watch glass over the crucible. The watch glass is shown in Figure 9.6c. As the fire went out, a condensate appeared on the inside surface of the watch glass, and it occurred to me that this was the substance that I wanted to test. By running either a solvent extraction or a headspace concentration on this asphalt smoke condensate, it was possible to see the kerosene-like substance that had been confounding chemists for decades. Using extracted ion profiling, it was also possible to determine that the large concentration of *alkenes* in the condensate, and the absence of cycloalkanes and two substances known as pristane and phytane, easily differentiated this smoke condensate from liquid kerosene or diesel fuel. The results were later presented to the American Academy of Forensic Sciences, published in the *Journal of Forensic Sciences*, and incorporated into ASTM Test Method E1618 for ignitable liquid identification. Thus, the fire debris analysis community benefited from Maynard Clark's adversity [24].

Shortly after the means of distinguishing asphalt from kerosene was discovered, the fire scene samples were made available to this author, and the residues detected by both laboratories were analyzed and conclusively identified as asphalt. In the meantime, the chemist from Ohio had been deposed and asked about communications he had with the Florida laboratory on the subject of the disparity of the results between the two labs. He denied any such communication.[11] Unfortunately for the Ohio chemist, the chemist from the Florida laboratory produced a ream of documents about 5 inches thick containing correspondence from the chemist in Ohio attempting to reconcile the two laboratories' findings with each other. When this correspondence was discovered, it was necessary for Clark's attorney to re-depose the Ohio chemist, who experienced a sudden refreshing of his recollection when confronted with the documents. He stated that he had "forgotten" all about the extensive correspondence when asked about it previously. Why the chemist's copies of that correspondence were not in his file remains a mystery.

Meanwhile, the misconduct of the law enforcement officer who intimidated the eyewitness into retracting his story came out in discovery, but only after the court ordered the sheriff's deputy to stop withholding evidence. Every time he was asked a question that made him uncomfortable, he responded, "Because there's an ongoing investigation, I choose not to elaborate on that." He was, however, unable to convince the judge that such an investigation actually still existed. After the law enforcement file was revealed, counsel that was overseeing this investigation on behalf of the insurance carrier became a defendant in the civil case. Shortly after new counsel was retained, the case settled for much more money than it would have cost the insurance company to pay for this simple lightning claim.

> 11.2.2 Extracts that meet the criteria for heavy petroleum distillates should be reviewed carefully for "extraneous components" that elute near *n*-alkanes and are the result of polyolefin or high molecular weight hydrocarbon (asphalt) decomposition. Peaks representing the corresponding 1-alkene or 1, (*n*-1) diene, and having an abundance near the concentration (within one-half an order of magnitude when viewed in the alkene profile) of the alkane, should be considered as indicating the presence of polyolefin or asphalt decomposition products rather than fuel oil products. Polyolefin decomposition products typically do not exhibit the same pattern of branched alkanes as fuel oils.
>
> **—ASTM E1618-14**

[11] The following exchange took place in the chemist's first deposition:

Q: Have you had occasion to correspond with [the chemist] in any fashion?
A: Well, I said that I communicated with her once, yes.
Q: Well, you mentioned a telephone conversation?
A: Uh-huh (affirmative).
Q: Was there written communication with her as well?
A: No, I don't believe so.
Q: Then the answer to my question, "Have you had any written correspondence with her," would be no?
A: No I have not had any written correspondence.

9.10.3.1 Error analysis

Some of the "mistakes" made by experts retained by the insurance company and other people involved in the case go beyond mere technical mistakes, although there were also plenty of those. Irrelevant data were misinterpreted when the investigators used "temperatures of 2,000°F" to infer the presence of accelerants and interpreted spalled concrete the same way. Some of the investigators' errors, however, sink to the level of misconduct. Using the categories of error we have outlined, we can place some of these (intentional or unintentional) errors into the category of ignoring critical eyewitness observations that were also "inconvenient." The investigators retained by the insurance company also disregarded the clear statements on the lightning detection report regarding the detection system's accuracy and precision. The communication in this case was not just poor; it was utterly lacking. Once the screwdriver was found in the fire door, the store manager or other employees should have been interviewed about it. The insurance company's investigator was, however, playing a game called "gotcha." When Cohen located former employees of the furniture store who had no particular loyalty to Clark and no "dog in the fight," they stated that the screwdriver had been jammed in the door for 5 to 7 years. Apparently, the fusible link that held up the door had melted in an earlier, much smaller fire, and the door had been pulled up and jammed open at that time. The screwdriver had been there ever since.

Additional errors were made by the chemists, whose reports made no mention of asphalt as a possible source of the "kerosene" or "diesel fuel," although their lab notes indicated that "tar" was present. The private chemist's "failure to recall" the extensive correspondence with the Florida chemist, even if it was an innocent mistake, would have been used in his cross-examination to destroy his credibility. All of these errors caused Clark to suffer through 4 years of litigation and cost his insurance company dearly.

9.10.3.2 Significance

The *Clark* case was significant in that it provided the data necessary to develop a methodology for distinguishing asphalt smoke condensate from petroleum distillates. The previous inability to make this distinction caused frustration in the community of fire debris analysts, and probably resulted in erroneous (false positive) findings both before and after this case. The methodology is now generally accepted and since 2001, it has been included in ASTM E1618, *Standard Test Method for Ignitable Liquid Residues in Extracts from Fire Debris Samples by Gas Chromatography–Mass Spectrometry.*

9.10.4 State of Georgia v. Linda and Scott Dahlman

Frances Dahlman suffered a severe stroke that left her paralyzed on the right side. She had initially been placed into a nursing home, but her son, Scott, and his wife, Valerie, were not pleased with the quality of care that Frances was being provided, so they had their basement refinished and built a mother-in-law suite for her. Additionally, they hired a nurse to stay with Frances for 12 hours a day and took care of her themselves for the balance of the time. Frances was in hospice care because, in addition to her stroke, she had a heart condition as well, and was not expected to live more than a few months. She had a standing "Do Not Resuscitate" order and her doctor had advised that her medications (she took both blood pressure and anti-seizure medicines) were optional. The stroke left her vulnerable to violent seizures, which caused her to flail her left arm.

There was a large supply of morphine sulfate in the house (a bottle was found on the table next to Frances's chair), so if Valerie or Scott had decided that it was time to end Frances's life, they could have done so easily, and there would have been no questions about the cause of death. In fact, there would have been no autopsy.

On the morning of August 31, 2008, at 5:27 a.m., there was a fire on and immediately next to Frances's chair, and she was badly burned. She was transported to Grady Hospital in Atlanta where she died of her injuries a few hours later.

Valerie Dahlman reported discovering her mother-in-law on fire and used a garden hose to extinguish her. But when the fire investigators first arrived at the scene, some of them were suspicious. Frances had a Perfect Chair, which was electrically powered to allow her to change position. The chair immediately after the fire is shown in Figure 9.7a. The Peachtree City Fire Marshal took the chair into custody, and the police sent other evidence to the Georgia Bureau of Investigation (GBI) Division of Forensic Services (DOFS).

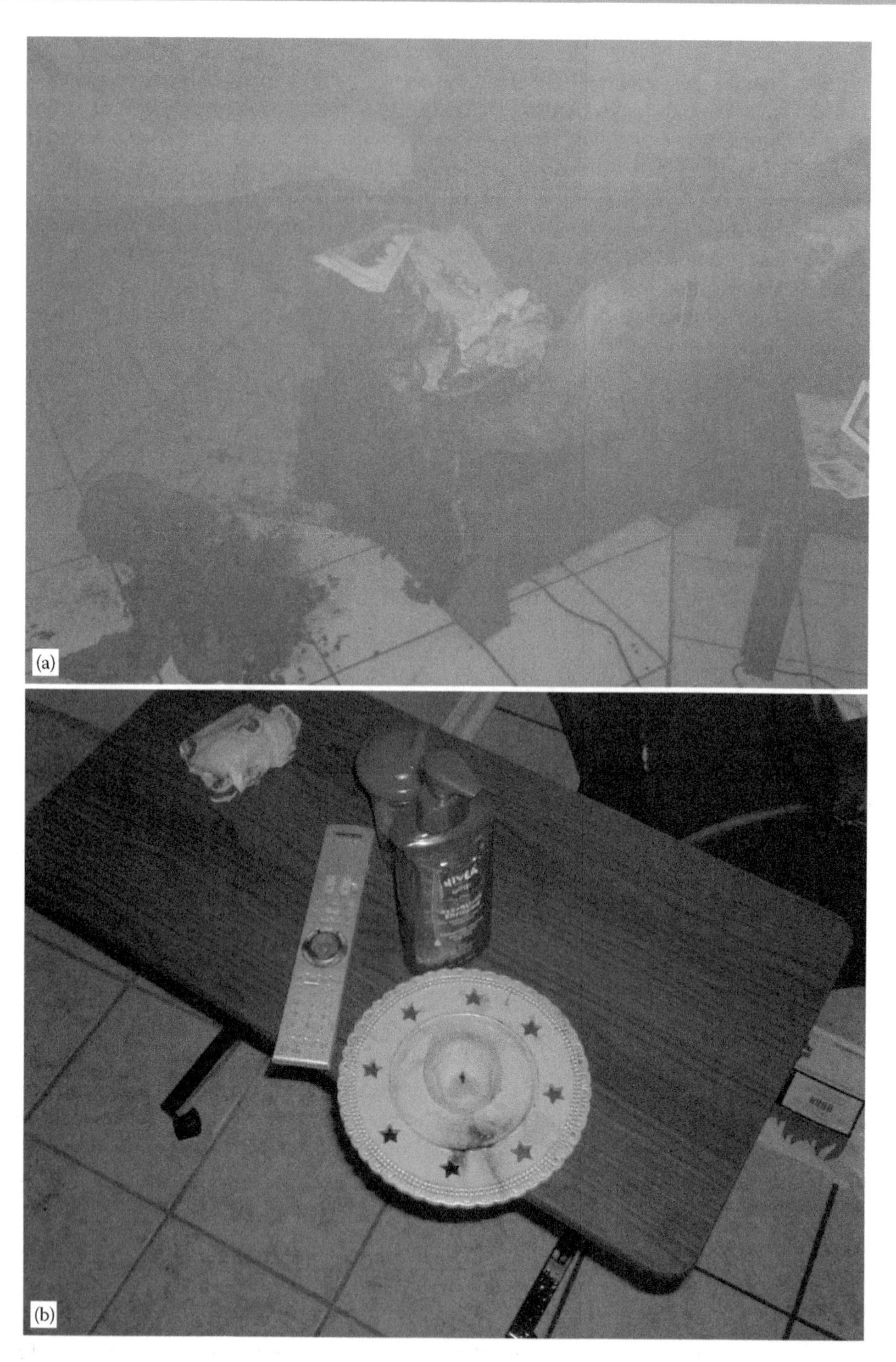

Figure 9.7 (a) Photo of the electrically operated Perfect Chair taken at 5:58 a.m., before the smoke had completely cleared. Note that the hospital table had been pushed aside by first responders. (b) Photo of the candle and other items on the hospital table, taken at 6:19 a.m. The objects clearly are arranged differently than they are in the next photo.

Figure 9.7 (c) Photo of a reconstruction of the origin area, taken at 7:22 a.m. The location and orientation of the objects on the table is unsupported by smoke patterns or any other evidence.

Examination of the chair by an electrical engineer retained by the city fire marshal revealed that there was burning on one of the 110 V cables in the chair, but this burning was almost certainly a result of exposure to an advancing fire. Because there was a death involved, on September 8 a deputy state fire marshal was called in. The fire marshal inspected the chair and photos of the scene, but apparently not all of them.

The state fire marshal prepared his first undated report, which was just over a page long. It is unknown when he decided that this fire was incendiary, but the medical examiner's autopsy report, dated April 7, 2009 (more than seven months after the fire), stated, "A thorough investigation failed to reveal the source of the fire; thus the manner of death is undetermined [25]." Using negative corpus methodology, this fire marshal concluded,

> The area of origin of the fire was determined to be within the area of the side of the chair and the blanket on the floor. Obvious sources of ignition in the area of origin were investigated. All accidental causes were eliminated as the possible cause of the fire. [**Emphasis** added.]
>
> Upon completion of examining all the photographs and evidence available of the fire scene and examination of the chair it is my opinion the chair Ms. Frances Dahlman was sitting in at the time of the fire was intentionally set on fire with criminal intent [26].

On November 30, 2009, some 15 months after the fire, Valerie was arrested, and ordered held without bond. On December 16, Scott was arrested as well. Despite the fact that he had offered to turn himself in, the police insisted on coming to his home and arresting him in front of his children.

The Dahlmans had recognized early on that they were the targets of a criminal investigation and retained highly respected defense attorneys Ed Garland and Richard Grossman. The author was retained by Grossman on September 12, 2008, made a trip to Atlanta to examine the chair on September 17, and concluded that the chair could not be blamed for the fire. After the arrests, I visited the Georgia State Crime Laboratory on March 10, 2010, where the evidence other than the chair had been taken. Among the items of evidence was a candle on a saucer that had been burning on a hospital table immediately to Frances's left. The fire pattern on the almost entirely burned candle and saucer suggested that perhaps an item of bed clothing such as a sheet or blanket had been inadvertently placed on the saucer and caught fire. The crime laboratory had identified fibers embedded in the candle wax, but concluded that these fibers did not match the blanket that was submitted. The candle was clearly visible in fire scene photos as shown in Figure 9.7b.

Throughout the pendency of these murder charges, I was curious as to why I had not seen a report from the Peachtree City Fire Marshal. He was one of the first people on the scene and had collected evidence and videotaped the scene. Unfortunately, his communication with the deputy state fire marshal was less than ideal, possibly because of a difference of opinion as to the fire cause. It was more than a year after the fire that the state fire marshal learned about the existence of the candle, an "obvious source of ignition in the area of origin."

In a preliminary proceeding, the judge was dismayed by the bare-bones quality of the state fire marshal's initial report and requested another one. In the second report, which revealed his thought processes more extensively, he wrote about the candle:

> There is no way to determine or confirm when the melting and blackened residue occurred because a subsequent interview with the caretaker stated that she, Ms. Dahlman's son, and her daughter-in-law would light a candle and leave it on the table to sooth Ms. Frances Dahlman. I did have the opportunity to examine the candle and saucer itself because they were taken as evidence. Closer examination of the blackened residue appeared to possibly be soot that would have landed on the saucer as the result of unburned by products of the chair floating down and landing on the saucer… .
>
> Through deductive reasoning I was able to eliminate the candle as being the ignition source of the fire. First observation was that the lotion bottle exhibited fire damage to the paper attached to the bottle on one side of the bottle. However, the fire damage of the paper on the lotion bottle was on the opposite side, away from the area of origin and the area of radiant heat and fire that would have been caused by the fire. I also observed the candle exhibited melting out of the far side of the candle, which again was further away from the fire. In conclusion the fire damage exhibited to the lotion bottle and candle is not indicative of the damage you would see if they had been close to the side of the electric chair. Therefore it was my determination through deductive reasoning that the candle was eliminated, was not close to the chair, and was not the ignition factor of the electric chair burning… .
>
> There was no fire damage found anywhere on the hospital table not even any soot, which would have been indicative of the table being next to the chair at the time of the fire. Therefore: it is my determination the table and the items on the table can be removed from the equation thus if the table was not close enough to the chair to sustain fire damage or receive any soot accumulation, the candle can be eliminated as the ignition factor of the fire… .
>
> As to the theory and elimination of another possible accidental cause which has been mention by the caretakers that Ms. Dahlman would with her mental status and condition have seizures and spells of wildly swinging her arms around and possibly threw the blanket over onto the candle which was actually located behind her chair. It is common sense this would not have been possible without knocking or dragging items off of the hospital table… .
>
> Upon completion of examining all the photographs and evidence available of the fire scene and examination of the chair and with the information received and the on-scene investigation completed, it is my opinion the cause of this fire is intentional and incendiary in nature [27].

Let us now consider the fire marshal's "deductive reasoning." First, he did not say *when* he examined the candle and saucer, or if he had actually seen them or just photographs. He stated that the soot on the saucer had landed there by "floating down." If that were the case, there should have been evenly distributed soot elsewhere, and there should have been a protected area under the saucer. No such evidence existed. So the smoke particles selectively landed on the saucer? Next, the observations about the items on the table were misplaced. The photograph in Figure 9.7c, from which these "observations" were made, was a *reconstruction*. The city fire marshal had requested the reconstruction from the first responders, but there is no way to tell how accurate their reconstruction was. They were focused on the patient, not the artifacts around her. The image of the reconstruction was made at 7:22 a.m. on the day of the fire. Figure 9.7a was shot at 5:58, while there was still smoke in the room, and Figure 9.7b was shot at 6:19 a.m. These photos were taken by Cpl. Hyatt of the Peachtree City Police Department between 5:57 and 6:25 a.m. On a second set of images, including the reconstruction, there is another photo showing the chair prior to the reconstruction, taken at 7:13 a.m. Even someone with minimal computer skills should be able to look at metadata to determine the provenance of an image. Due to the lack of smoke deposits on the table, there is no way to discern the location, much less the orientation of the objects in the photo. The fire marshal thus relied on demonstrably unreliable "data" to "eliminate" the candle as the ignition source. Not explained in the second report is how the fire on the chair could damage the "paper" (actually plastic) on the opposite side of the lotion bottle. Violating the laws of physics was no problem for this individual, because, despite being an IAAI-CFI, he did not understand the laws of physics.

When the city fire marshal was finally compelled to render an opinion, he stated that the cause of the fire was "undetermined." He knew about the candle, but because the state fire marshal, the police, and the prosecutor told him they had "something," he refrained from making a determination. After the arrests in December 2009, he had told the assistant district attorney that he did not believe this was an arson case. Further, the videotape he had made was made available to the state fire marshal and it showed a burned white sheet that had not been collected as evidence. The sheet, which may have been the source of the fibers that the GBI Laboratory found in the candle wax, certainly should have been tested for wax. The failure to collect the sheet was blamed for the case not being prosecuted.

The district attorney stated:

> The white sheet was not photographed, seized, entered on any chain of custody, secured, or properly maintained by the Peachtree City fire marshal and was thus not considered by the state fire marshal in his fire scene analysis. The white sheet is not now available. The state believes the above, when considered with other evidence available to the state at this time will create a reasonable doubt as to the guilt of the accused and that while there was probable cause for the arrest, there is insufficient evidence to prove guilt beyond a reasonable doubt [28].

The white sheet was certainly photographed in the video. The fire patterns did not show when it was ignited, but it could have been the first fuel ignited. There was, however, already plenty of reasonable doubt in this case. The DA dismissed the case because he was going to lose, with or without the loss of the white sheet.

Because he would not go along with the incendiary determination made by the state fire marshal, the city fire marshal was demoted and eventually left the fire service. To this day, John Dailey stands as an example of someone who would not sacrifice his integrity, even at the cost of his livelihood.

The state fire marshal, on seeing the videotape still insisted that the fire was incendiary, apparently because once he has made an incendiary determination, even with compelling new evidence, there is never any going back. The State Fire Marshal's office stated "… [O]ur opinion does not change that this fire was intentionally set," and "Hopefully, the DA will revisit this case [29]."

9.10.4.1 Error analysis

The city fire marshal surely should have photographed and retained the white sheet, and poor communication resulted in the state fire marshal not learning about the candle for over a year and the white sheet until almost two years after the fire. The state fire marshal, however, should have inquired about the evidence that was collected before he arrived on the scene. He also should have reviewed the first responder's photographs. Instead he relied on a reconstruction that could not be verified. There is no excuse for overlooking either the first photographs or the critical evidence that was in the possession of the GBI. He relied on negative corpus methodology to decide that the fire was incendiary. That fire marshal then decided to discard the data that was inconsistent with his determination, coming up with a physically impossible explanation for the smoke deposits on the saucer.

9.10.4.2 Significance

This case caused quite a stir in the local fire investigation community and resulted in a fire marshal losing his job for the supposed reason that evidence had been lost, but for the real reason that he would not go along with the prosecution. John Dailey set an example for people who are asked to do something that they know is wrong. Scott and Valerie Dahlman have resumed their lives after the trauma of losing Scott's mother and the even worse trauma of being accused of a horrific crime. This is one case where the background information should have persuaded even the most suspicious investigators that arson homicide was not a reasonable hypothesis. Under what circumstances would a couple having the means and opportunity to kill someone quietly and painlessly decide instead to burn the victim alive? The state fire marshal has retired.

> A new scientific *truth* does not triumph by convincing its opponents and making them see the light, but rather because its opponents eventually die, and a new generation grows up that is familiar with it.
>
> **—Max Planck**

9.10.5 State of Michigan v. David Lee Gavitt [30]

David Gavitt and his wife Angela were in bed at 10:30 p.m. on the night of March 9, 1985, when they were awakened by their dog and became aware of a fire in the living room. Angela ran down the hall, and David ran outside to attempt entry into the children's bedrooms through the windows. He cut himself badly in the process and was hospitalized for burns and severe lacerations.

By the next morning, investigators from the Michigan State Police (MSP) had concluded that the "distinctive flammable liquid burn patterns" and "multiple origins" in the house indicated that the fire had been set with ignitable liquids, and David was arrested.

The investigators collected 6 samples of debris on March 10, all of which tested negative. They returned to collect more samples on March 20 from the front yard where carpet had been thrown out during the initial scene excavation. The samples were submitted to the MSP laboratory, where they were analyzed by an individual who had experience as a food chemist but no serious experience as a fire debris analyst. He found that two of the samples, both from the front yard, were positive for "residues of highly evaporated gasoline."

The chemist ran an additional "flame spread test" involving an attempt to ignite the carpet with a Bunsen burner. Three times he applied the flame to the carpet and three times the fire went out shortly after removal of the flame. In a fourth test, he poured a small amount of gasoline on the carpet and lit it with a match. Unsurprisingly, the carpet stayed lit when the blowtorch was removed. This was "proof" that the carpet would not have burned had it not been for the application of gasoline.

The photographs of the scene clearly show that origin was in the living room (Figure 9.8a), but nothing in the photos suggested multiple origins.

In September 2010, the author was contacted by Michael McKenzie and asked to review the case. I told him that if gasoline really was present in the living room, there was not much that I could do. I was provided with the GC-FID chromatograms and was appalled by what I saw. The two allegedly positive samples matched neither each other nor the gasoline standard, and the quality of the work was unbelievably poor.[12] Figure 9.8b shows the comparison of the allegedly positive samples and the standard, and Figure 9.8c shows a publication quality chromatogram from 1977. The difference in resolution is quite remarkable.

Having determined that the chemistry was not credible, I continued my analysis of the fire scene investigation and found it to be equally unreliable. A timeline analysis revealed that the living room burned in a fully involved condition for an extended period of time, in excess of 13 min, so reading fire patterns was not reliable, although that was not known to the investigators of 1985.

I handed off the chemical analysis to a colleague, Craig Balliet of Barker and Herbert Analytical Laboratories, who agreed that the chemistry was abominable and prepared an affidavit to that effect. I prepared an affidavit about the fire scene analysis, and both were submitted to the court in support of a motion for relief from judgment.

The prosecutor attempted to have the chemistry reevaluated by the MSP laboratory. They reported, "Because the original sample processing methods are unclear and the instrumentation that was used is unavailable for evaluation of the reference samples, no definitive conclusion could be made regarding the presence or the absence of gasoline in the questioned carpet samples which were previously reported to contain highly evaporated gasoline." The ATF laboratory declined to reevaluate the data, stating that they had no protocols for doing so, having abandoned GC-FID 15 years ago. They also stated that because the data had already been reviewed by multiple experts, it was unlikely they would have anything to add. The prosecutor then turned to Dirk Hedglin of Great Lakes Analytical Laboratories. Hedglin agreed with the analysis conducted by myself and Balliet.

The prosecutor also reached out to ATF Special Agent Michael Marquardt to have him review the fire scene analysis. Marquardt reached the conclusion that there were "insufficient data (facts, circumstances, information, evidence, etc.) under current fire investigation accepted practices, procedures, and standards to clearly support an incendiary fire cause classification to a reasonable degree of scientific certainty."

---

[12] A colleague presented these chromatograms to an audience of forensic scientists at the Bureau of Criminal Apprehension Laboratory in Minnesota in 2012. She reported that there was an audible gasp when the scientists saw these charts.

Figure 9.8 (a) Two photos of the fire damage in the Gavitt living room.

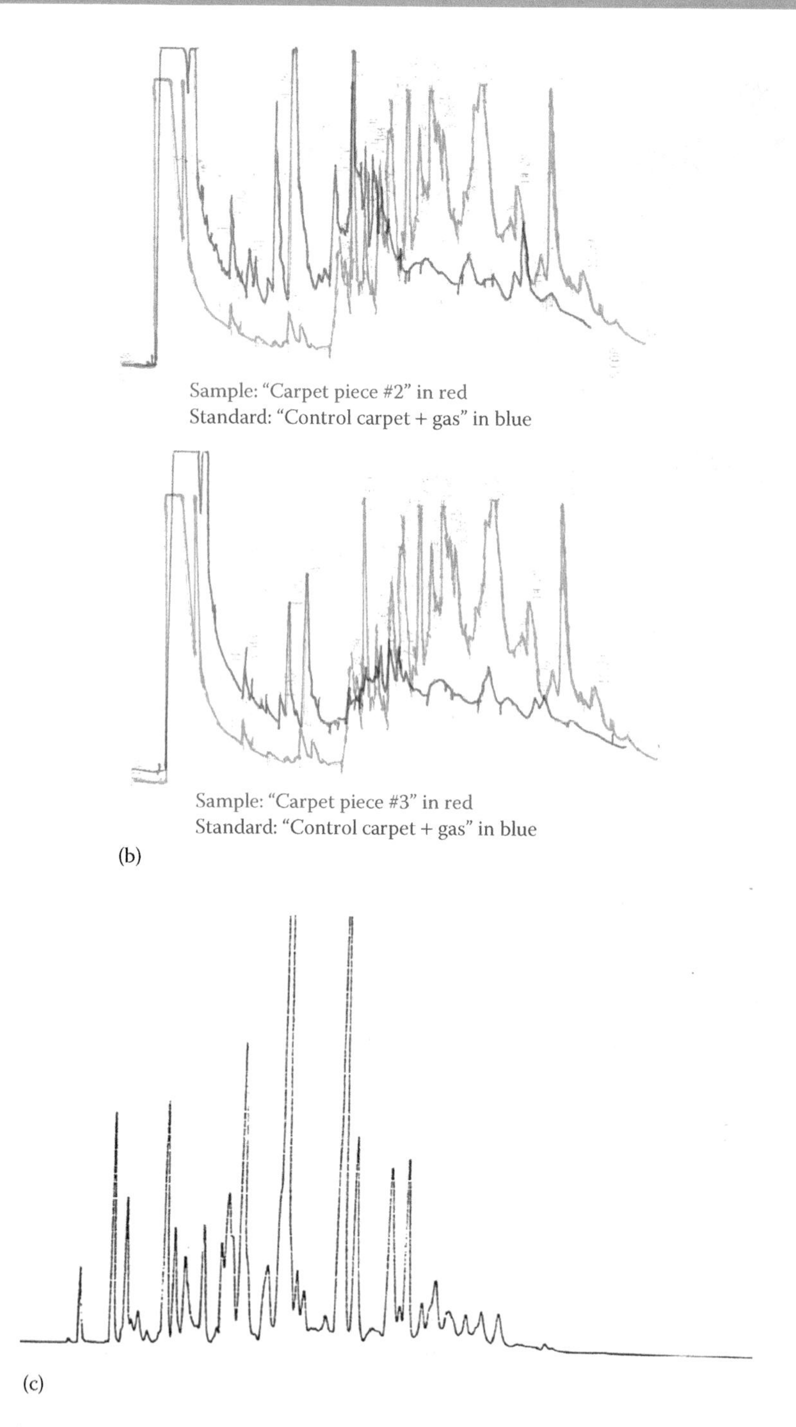

Figure 9.8 (b) GC-FID chromatograms from two samples incorrectly called positive for gasoline. In each chart, the sample chromatogram is gold and the gasoline "standard" is black. The samples match neither gasoline nor each other. (c) Publication quality GC-FID chromatogram of gasoline from 1977.

Figure 9.8 (d) David Gavitt leaves the penitentiary after serving 27 years for a crime that nobody committed. He is accompanied by his attorneys Imran Syed and Michael McKenzie.

The prosecutor delayed for an additional eight months after being told that the evidence against Gavitt was not supported by science, arguing *inter alia* that, although fire investigation practices had changed dramatically, Gavitt could have contested the faulty GC-FID analysis in his original trial. Despite the prosecutor's misgivings, he agreed not to object to a new trial. The prosecutor continued to maintain that he believed Gavitt was guilty, even though there was no evidence that indicated arson, and throughout the 27-year pendency of this case, nobody had even suggested what Gavitt's motive might have been. Because the evidence would not support a guilty verdict, on June 26, 2012, Gavitt was released from custody. He is shown, accompanied by his lawyers in Figure 9.8d. Gavitt's name has since been added to the National Registry of Exonerations. As is frequently the case with exonerations, no determination was ever made as to the actual cause of the fire. Seeing what they thought was evidence of arson, the investigators did not collect any data that would have indicated an accidental cause.

The *Gavitt* case formed the basis for a paper on the subject of what should happen when the science underlying a conviction changes. Imran Syed and Caitlin Plummer of the University of Michigan Law School's Innocence Clinic authored a paper called "Shifted Science and Post-Conviction Relief," which was published in 2012 [31].

> What can be done to cure the injustice of a conviction that was based on scientific testimony that may have been accepted in the relevant scientific community at the time of trial, but has since been completely repudiated?
>
> **—Plummer and Syed, "Shifted Science and Post-Conviction Relief," 2011**

9.10.5.1 Error analysis

The original fire investigators misinterpreted critical data by stating that the low burns were indicative of an incendiary cause (not at all unusual for 1985). Further, they engaged in two-dimensional thinking by stating that the lack of communication *at the floor level* indicated multiple origins.

The chemist (who has since died) was not properly trained or qualified in fire debris analysis, and the MSP was negligent in letting him conduct the analyses. He misinterpreted critical data by concluding that the chromatograms indicated the presence of gasoline when they clearly did not. The chemist's flammability test was beyond the pale and constituted irrelevant data. This chemist had no experience in conducting such tests and grossly overstated the meaning of his test with and without gasoline.

These errors cost David Gavitt 27 years of his life.

9.10.5.2 Significance

David Gavitt's wrongful conviction was a typical outcome of fire investigation in the 1980s. His case has encouraged discussion of how the law should respond when the science underlying a conviction changes.

9.10.6 State of Arizona v. Ray Girdler [32]

Ray Girdler lived with his wife and daughter in a mobile home near Prescott, Arizona. At about 2:45 a.m. on a cold November night in 1981, Girdler was sleeping in the bedroom when his cat's meowing awakened him. When he stood up, the room was full of smoke. He woke his wife, then ran down the hall and saw fire in the living room. He ran to the car to get a fire extinguisher, and his wife ran to the middle bedroom where their 2-year-old daughter was sleeping. A floor plan of the Girdler residence is shown in Figure 9.9a. When Girdler returned to the mobile home, the fire had grown to such an extent that he could not reenter. This was a 1964 model mobile home, and the interior finish was thin plywood paneling (flash paneling). Girdler was unable to save his family but managed to escape, barefoot and in his underwear, and ran to a neighbor's house to call the fire department. By the time the fire department arrived, the roof over the center of the home had collapsed.

Girdler's neighbor gave him some clothing to wear and a cup of coffee. When one of the firefighters went to the neighbor's house to interview Girdler and other witnesses, he saw Girdler fully dressed in the middle of the night and immediately became suspicious. His report stated, "… [T]he first thing I noticed was that he was dressed. He had on a T-shirt, pants, and in front of the couch next to him was a pair of socks and tennis shoes. He was drinking a cup of coffee and smoking a cigarette." It apparently did not occur to the firefighter that Ray had not yet put the socks and shoes on—he simply assumed that he had taken them off.

The local fire marshal came out and was able to discern "multiple points of origin" in the trailer, shown in Figures 9.9b and 9.9c. This "determination" was made at 4:00 a.m., *while it was still dark*. The Arizona state fire marshal was called, and the office sent out a relatively inexperienced investigator, who also "saw" evidence of an intentionally set fire: low burning, irregular patterns, holes in the floor, and crazed glass. Later, a senior Arizona deputy state fire marshal claimed to have reviewed the evidence, but in fact only talked to his subordinate by telephone. Three days after the fire, *before seeing the scene*, or even photos of the scene, before getting the (negative) results of the laboratory analysis of fire debris, and before the autopsy report was written, this supervisor told the district attorney that he was "absolutely certain" that this was an arson/homicide fire. Girdler was arrested the next day.

Girdler's wife and daughter had relatively high concentrations of carboxyhemoglobin in their blood (74% and 87%, respectively), and the county medical examiner opined that this was further proof that the fire was accelerated.[13] Given that Ray was in the house, he was the obvious suspect. If this was a set fire, Ray set it. This is the case with many of the false accusations the author has seen, and the main reason that a first hypothesis of accidental cause is warranted.

Girdler was brought to trial in front of Judge James Sult in June 1982 and was convicted of arson and two counts of homicide. During the trial, the fire marshal, who testified that he determined arson in 80% of the fires he investigated,

---

[13] All this really proves is that the fire was ventilation controlled, and therefore the products of combustion contained high levels (1% to 10%) of CO. See NFPA 921 at 25.2.1. There is an extensive discussion of carbon monoxide production and transport in the *SFPE Handbook*.

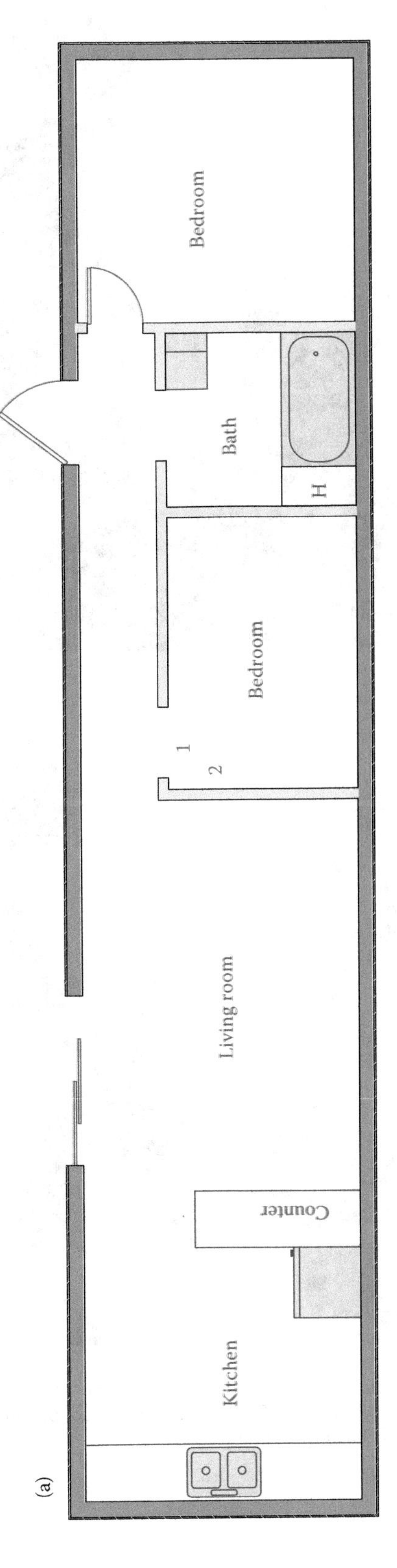

Figure 9.9 (a) Floor plan of the Girdler mobile home.

Figure 9.9 (b) Exterior view of the Girdler mobile home. A fire marshal had discerned multiple areas of origin before dawn broke on the morning after this fire. (Courtesy of David M. Smith, CFI, Associated Fire Consultants, Bisbee, AZ.) (c) Another view of the exterior of the Girdler mobile home. (Courtesy of David M. Smith.)

Figure 9.9 (d) Crazed and non-crazed glass outside the Girdler mobile home. The fire marshal testified that the explanation for the condition of the glass was that the defendant broke out the window "in a fit of rage" and then set the fire. Heat from the accelerant caused the glass to craze, but the previously broken out glass was protected. (Courtesy of David M. Smith.)

Figure 9.9 (e) Alleged "pour pattern" in the hallway of the Girdler mobile home. This floor would have presented a pattern of uniform charring were it not for the presence of boxes on the floor. (Courtesy of David M. Smith.)

Figure 9.9 (f) Continuation of the alleged pour pattern into the bedroom. The edges of the pattern are more irregular here than in the hallway because the objects providing the protection to the unburned areas were not boxes but items of clothing and bedding. (Courtesy of David M. Smith.)

repeatedly stated that nothing other than the presence of flammable liquids could account for the condition of the mobile home. About a hole burned in front of the sofa, he stated, "And this is an area without any possible natural or accidental fire cause." He further testified that it appeared that Ray had broken the leg off a coffee table and beaten his wife and daughter over the head with it. The injuries to the victims' skulls were actually thermal injuries, but that possibility was not considered by the fire marshal nor was it presented to the jury. The earliest fire reports describing the condition of the victims indicated "possible skull fractures."

In a remarkable example of the wide latitude afforded to experts, the fire marshal was allowed to opine that the crazed glass shown in Figure 9.9d had tremendous significance. Some of it was crazed, and some was not. The fire marshal's report stated:

> Much of this glass was heavily heat-crazed and free of smoke deposit, indicating a rapid buildup of intense heat and close proximity to initial fire, which is consistent with the expected results of a liquid accelerant fire. A significant volume of the arcadia door glass was found which was free of both heat and smoke damage in a configuration indicating breakage by physical force other than explosive force, prior to or in the very early stages of the fire [33].

The fire marshal elaborated on what he meant by "physical force" at the trial. With no objection from the defense, he testified, "I would say it was one of two things. Either there was a struggle or there was someone in there with a fit of rage."

During his trial testimony, the fire marshal stated that the wife and daughter "never had a chance" and were immediately overcome by this rapidly moving fire. After the conviction, there was a penalty phase, during which that same fire marshal testified that Mrs. Girdler and her daughter had died a "slow, painful death," suffered "excruciating pain," and had their lungs "severely seared." Actually, the autopsy found no evidence of damage to the trachea or lungs. These victims inhaled a few breaths of smoke that contained a high concentration of CO.[14]

[14] It would be revealed 8 years later that the fire marshal had testified at the first trial without the benefit of having actually read the autopsy reports.

In August 1990, in an extraordinary hearing that went on for several days in front of the same trial judge that oversaw the first trial, Ray Girdler, with the assistance of defense attorney Larry Hammond, was granted a new trial. The conviction was overturned and the sentence was vacated on the basis that "new evidence," in the form of new knowledge about the phenomenon of flashover, made the testimony that the jury heard about flammable liquids false.

Further, a respected medical examiner testified that the proposition that the high carboxyhemoglobin (COHb) levels found in the fire victims could only result from an accelerated fire had no scientific basis. The original medical examiner remained adamant but, under cross-examination by Hammond, had to admit that it was only because "it stands to reason" that he held his beliefs about accelerated fires.

The fire marshal was confronted about the possibility that the burn patterns he attributed to accelerants could have been caused by radiation, although his trial testimony repeatedly stated that accelerants were the only possible explanation. At the hearing, he admitted that alternative explanations existed that he had not disclosed to the judge or jury at the trial. He was allowed to "explain" his reason for not disclosing such alternatives. His answer, reflected in the record, was that he did not see his function as requiring him to "discuss every possibility that I ever considered." Among other findings, the judge ruled, "The fire in the Girdler mobile home could reasonably be attributed to noncriminal causes. Specifically, scientific evidence now available supports the finding that the fire may have been a 'flashover' fire ignited by accidental or other noncriminal means and not involving the use of flammable liquids or accelerants."

A new trial was ordered, but the judge called a temporary halt in the proceedings. Judge Sult was thoroughly chastened by hearing this evidence and overturned the first verdict on his own initiative rather than requiring Hammond to appeal to a higher court.[15] Girdler had been Judge Sult's first capital case, and he had been particularly harsh in sentencing. In a scathing rebuke, the judge had told Girdler that he was going to impose consecutive life sentences, making him eligible for parole at age 100. He stated:

> Jennifer Ann Girdler was a baby of the age of two years, a totally innocent and helpless victim, completely dependent upon you not only for the necessities of life but for life itself. Your killing of her constitutes the most vile betrayal of the highest trust which is placed on a human being… .
>
> It is mandated in our law that the Court consider in every sentencing the need for establishing a deterrent to others like you who will inevitably consider the same method you adopted to rid yourself of your problems. Such deterrence can only occur if in this case you are required to spend the maximum amount of time possible imprisoned for your deed.

Larry Hammond had retained David M. Smith to assist in the defense of the second trial, and the state sought assistance from Barker Davie, in addition to using its own "in-house" experts. Presented with dueling experts, the judge ordered the parties to identify a third expert to review the evidence for the benefit of the court. Both Smith and Davie agreed to recommend this author as a neutral, qualified, independent expert and I accepted the court appointment.

I was provided with all the expert testimony from both the trial and penalty phase, as well as some post-conviction hearing transcripts. In addition, all the 8″ × 10″ photographs used in the original trial were provided, as was some of the actual physical evidence, including the crazed glass. As with most of these cases, laboratory analysis failed to identify the presence of any ignitable liquid residue in any of the samples collected from the scene.

My charge from Judge Sult was to help him understand why Girdler had been convicted the first time and whether he would be convicted if tried again. The answer to the first question was not difficult to determine. The testimony of the fire marshal had been devastating, although it was internally contradictory and based on mythology and rank speculation. The alleged pour pattern was nothing more than damage to a vinyl floor caused by radiant heat. Numerous boxes were stored in the hallway, and the edges of the alleged pour pattern were straight lines with right angles in them, shown in Figure 9.9e. Even the irregular pattern on the bedroom floor shown in Figure 9.9f was clearly the result of protection of various unburned areas. There was nothing even vaguely liquid-appearing about most of the alleged pour patterns, other than that they were on the floor and liquids lay on the floor if they are spilled.

Crazed glass results from the application of liquid to hot glass, not rapid heating, as has been repeatedly discussed. Applying the flame from a propane blowtorch to a piece of the glass from the Girdler residence did not result in crazing, just thermal cracking. Applying a few drops of water to the hot glass did result in crazing. That the fire marshal testified that crazing indicated the presence of accelerants near the glass was excusable given the state of the

[15] Because the judge overturned himself, there is no appellate record of the grant of a new trial.

art in 1982, but his further speculation that Girdler had broken the glass in a struggle or a fit of rage was pure fantasy. Nonetheless, it was allowed into evidence without objection from defense counsel and the jury gave this ridiculous testimony the special weight that juries give to expert testimony.

I provided Judge Sult with a 20-page, single-spaced report, outlining all the myths and misconceptions testified to by the fire marshal. My report also stated that if Girdler were to be tried again, Smith would certainly point out all these errors, assuming the judge decided to let the fire marshal put the same rubbish in front of the jury again. The prosecutor agreed to dismiss the case.

Ray Girdler had served more than 8 years in the penitentiary and brought suit for malicious prosecution against the fire marshal, a suit that was eventually settled.

> The fire in the Girdler mobile home could reasonably be attributed to noncriminal causes. Specifically, scientific evidence now available supports the finding that the fire may have been a "flashover" fire ignited by accidental or other noncriminal means and not involving the use of flammable liquids or accelerants.
>
> **—Judge James Sult, January 2, 1991**

9.10.6.1 Error analysis

The first fire marshal was thinking in only two dimensions when he perceived "multiple origins" in this mobile home that no longer had a roof. The supervisor, once he actually looked at the evidence, misinterpreted several irrelevant artifacts, particularly the "pour patterns" on the floor and the crazed glass. He also misinterpreted the thermal injuries to the skulls of Mrs. Girdler and her daughter as resulting from physical trauma. His failure to read the autopsy can be most kindly described as overlooking critical evidence but is more accurately characterized as dereliction of duty.

The medical examiner, in providing support to the unsupportable musings of the fire marshal, misinterpreted critical data by assigning significance to the carboxyhemoglobin readings where none existed.

Even the 1990 review of the fire patterns was flawed when they were again identified as "pour patterns," although Davie, to his credit, published photos of the fire patterns and requested feedback from the fire investigation community on their significance or lack thereof [34].

9.10.6.2 Significance

The *Girdler* case was one of the first in which an arson conviction was reversed based on the new understanding of flashover. Critics of the decision called it "the flashover defense," but reasonable fire investigators began to take the phenomenon seriously. *Newsweek* and the *Los Angeles Times* published Girdler's story and brought the relatively new fire science to the public's attention.

9.10.7 State of Louisiana v. Amanda Gutweiler [35]

Amanda and Flint Gutweiler lived in the village of Dry Prong, Louisiana, near Alexandria, in a two-story house that Flint had built. They were poor, relying on income from Flint's struggling construction business. On January 9, 2001, a check arrived in the mail and Amanda Gutweiler immediately left to cash the check at the bank and to buy some food, for there was none in the house. She left her three children at home alone, with her 10-year-old daughter in charge. She returned home less than 30 min later to find the house on fire with the children trapped inside. She entered the burning structure but was unable to reach the children.

The initial fire investigators were intrigued when they saw spalling on the concrete slab floor and misinterpreted it as evidence of the use of an accelerant. In order to find more evidence of spalling, they pushed the remains of the structure over with a backhoe and collected samples with the aid of a canine, all of which tested negative, with the exception of a sample from the backyard next to an all-terrain vehicle (ATV). In their haste to find more spalling, they failed to properly document the scene before razing the structure. They took a total of seven photographs before the house was pushed over. Based on the spalling, the canine alerts, the positive sample, and the rapid spread of the fire, they declared the fire to be incendiary. Four exterior views of the scene are shown in Figure 9.10a through d. Most of the photographs taken were images of spalling, such as that shown in Figure 9.10e.

Figure 9.10 Four photographs [(a) through (d)] taken immediately before the Gutweiler residence was pushed over to look for more "spalding" *[sic]*.

Figure 9.10 Four photographs [(a) through (d)] taken immediately before the Gutweiler residence was pushed over to look for more "spalding" *[sic]*.

Figure 9.10 (e) One of many photographs of spalling, incorrectly interpreted as evidence of an incendiary fire. (f) Police photograph of Gutweiler's sweatshirt, showing soot just below the collar.

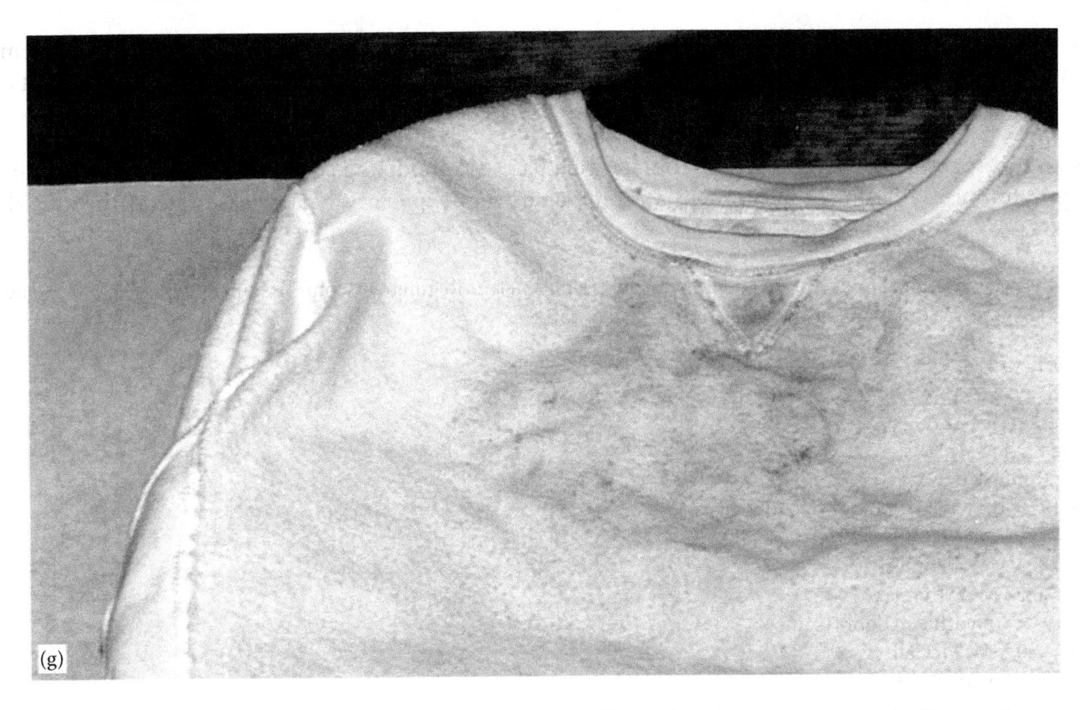

Figure 9.10 (g) Police photograph of the inside of Gutweiler's sweatshirt, which she used as a mask to filter smoke.

The assistant district attorney in charge knew he had a weak case and sought advice from senior investigators at both the state and federal levels. They told him that no matter what caused the fire, the fact that the initial investigators pushed the house over to look for spalling after failing to properly document the scene meant that the case could not be successfully prosecuted. They suggested that he confirm this with a leading authority on arson investigation.

Much to the surprise of the first set of reviewers, this expert was able to "rehabilitate" the case, using timelines of dubious quality, assumptions of even more dubious quality, and his own "expectations" of the fire's ability to spread in the allotted time. The expert's first report stated, "Based upon the estimated fuel load of the rooms involved, the size of the rooms, and the reported rate of fire development, multiple points of origin or the use of an accelerant are very strongly indicated [36]." Presumably, he believed that Gutweiler came home from the bank, where she was videotaped, and set the house on fire. In late 2002, the expert appeared before a grand jury and Gutweiler was indicted on one count of arson and three counts of first-degree murder. The state announced that it intended to seek the death penalty.

Attorneys Michael Small and James Doyle were appointed to defend Gutweiler, and they challenged the reliability of the arson evidence. During the time that challenge was ongoing, they retained a retired ATF agent, George Barnes, who had never accepted a defense case previously. They later retained this author to assist them with some of the scientific issues and later brought in Douglas Carpenter to address the specifics of the modeling issues.

Following an unsuccessful *Daubert* challenge, the expert nevertheless modified his report to remove the reference to the use of accelerants but still stated that the fire "was the result of deliberate ignition of room contents in separate rooms—kitchen and master bedroom." The second report relied heavily on hand calculations using fire dynamics equations, as well as more sophisticated computer modeling (CFAST). The defendant continued to be held without bond.

For the CFAST modeling, the state engaged a second expert, who used a beta version of the model. Mr. Carpenter's analysis of the model's output revealed that the temperature rose as one moved away from the fire, a violation of the laws of thermodynamics. The modeler was so focused on making the model "prove" what he wanted that he failed to adequately scrutinize the output. What with the laws of physics being violated, the CFAST output lost any semblance of credibility. That would not have deterred the prosecution from using the CFAST data, but it was the first expert's "hand" calculations that showed how flawed this approach was.

Here is an excerpt from the expert's 2004 report, which used fire-modeling equations to compare the heat release rate required to cause flashover to occur in a room versus the "expected" heat release rate of the available fuel packages:

> Using the computational tools in Fire Dynamics Tools (U.S. Nuclear Regulatory Commission, 2004), the QFO can be calculated using three different methods (assuming ⅝″ drywall).
>
> Master Bedroom QFO:
> McCaffrey: 1.25 MW
> Babrauskas: 3.4 MW
> Thomas: 2.4 MW
> Living Room/Kitchen:
> 29′ × 19′ × 10′ with 3′ × 10′ + 4′ × 10′ doors:
> QFO = 3.1 MW McCaffrey
> QFO = 8.5 MW Babrauskas
> QFO = 6.25 MW Thomas
> 31′ × 19′ × 10′ with two doors (same):
> QFO = 3.0 MW McCaffrey
> QFO = 8.5 MW Babrauskas
> QFO = 5.8 MW Thomas
> 31′ × 19′ × 10′ with three doors (8′):
> QFO = 3.15 MW McCaffrey
> QFO = 9.6 MW Babrauskas
> QFO = 6.34 MW Thomas
>
> The disparity between calculated values is dependent on different factors in each formula. The numerical average of all three is the best approximation [37].

It is one thing to use a fire model inappropriately, for example, using it to attempt to resolve an issue involving close tolerances. Despite the appearance of three or even four significant figures in some equations, fire models are properly used to provide "order of magnitude" approximations of times, temperatures, species concentrations, and so on. It is quite something else to make a statement such as, "The numerical average of all three is the best approximation." These flashover calculations are rough approximations, and the fire protection engineering profession has clearly stated that they should be used as *bounding* approximations. An arguably appropriate use of the predictions of these three equations would be to state that the *range* of flashover requirements is from 3 to 10 MW. **Nowhere in any peer-reviewed scientific literature does any fire protection engineer suggest averaging the three flashover equations.** The appropriate use of the equations is not to find one "best approximation," but to define a range of possibilities. The expert's approach in this and other fires was completely invalid.

The state's expert also relied on the failure of the passerby to see fire at the residence ten minutes before a large fire was discovered. He opined on the unlikelihood of a six-year-old boy playing with fire. He opined on the description of events given by the mother who entered the house in an attempt to save her children and got soot on her clothes, but not sufficient soot to meet the expert's "expectation" of how much soot she should have received had she been exposed to the fire that this expert "expected." Despite the fact that he had never seen the "lightly sooted" sweatshirt, he stated that the mother's description of events was not possible. The sweatshirt, which is shown in Figures 9.10f and 9.10g, definitely had soot on it, particularly in the area used as a mask. Gutweiler stated that she had pulled the shirt up to cover her nose and mouth.

More than five years after the fire, having spent 4 years in jail, the defendant was granted a bail hearing and presented evidence that the fire was not really a set fire and also that the prosecutor had provided grand jury testimony to his experts in violation of Louisiana law. In the first recorded instance relating to a capital case in Louisiana, the judge granted bond, stating, "[T]he defendant has met her burden of showing both proof is not evident and the presumption is not great that she is guilty." He then dismissed the indictment.

> The defendant has met her burden of showing both proof is not evident and the presumption is not great that she is guilty.
>
> **—Judge Donald Johnson, November 30, 2009**

After two appellate courts upheld the dismissal, a new prosecutor said he wanted to re-indict, and requested a new report from his expert. The third report, issued in December 2008, stated, "Due to the limitations of the evidence and analytical methods, the original conclusion of the undersigned that this fire was ignited in several areas and spread to the extent observed by Mr. Taylor in 10 minutes cannot be defended to a suitable degree of scientific certainty [38]."

The third report referenced the work of Salley, and the study by Rein, Torero, et al. described in Chapter 3 on fire modeling. The expert further stated that despite his previous analysis,[16] child fire-play could not be ruled out, and Gutweiler's description of events might have been inaccurate not because she was lying but because her comments "must be considered in light of the stress and emotion of the situation." The 2008 report also acknowledged, for the first time, "the nearly total lack of documentation of the post-fire scene." The expert stated that he had changed his opinion about the nature of the fire more than a year before his report was requested but that he felt no obligation to report that change until he was requested to prepare the third report.

The prosecutor dismissed the murder charges as a result of the expert's change of opinion, but he filed a "criminal information" charging the defendant with three counts of cruelty to juveniles, carrying a maximum of 40 years per count. After some negotiations, the defendant finally, some 9 years after the fire, agreed to plead guilty to three counts of negligent homicide (for leaving her children home in the charge of her 10-year-old daughter, *not* for setting the fire), provided that her sentence would not exceed time already served. Thus, this travesty of justice was finally brought to an end.

9.10.7.1 Error Analysis

The conclusion by the fire marshal that the fire was incendiary was based on a misinterpretation of irrelevant evidence—the spalling. The reviewing expert's errors were both scientific and ethical. He reached conclusions in the absence of adequate data, he reached conclusions not supported by the limited data available, and he applied methodologies that were not appropriate to the question at hand. Most troubling, however, was his failure to timely report his change of opinion.

9.10.7.2 Significance

This case demonstrates what can go wrong when a fire investigator bent on achieving a certain result misuses science and computer fire models. At one point, because they used the beta version, the state's expert and his modeler actually predicted a fire violating the first law of thermodynamics. The publishing of three reports with three different conclusions did not do the expert's reputation any good. This case became very well known in the fire investigation community and was the subject of an ABC 20/20 episode called "Burned." The case caused not only embarrassment but outrage among some forensic scientists. Gutweiler went on to write a book about her experiences [39] and has been an invited speaker at fire investigation conferences.

9.10.8 David and Linda Herndon v. First Security Insurance [40]

David Herndon returned to his home at approximately 4 p.m. on the afternoon of November 9, 2009, to find it full of smoke. He called 911, and the fire department came and extinguished the fire. The room of origin was the kitchen, and it never became fully involved.

---

[16] His second report stated, "None of the children exhibited any interest in fire setting and appeared to her to be obedient about parental instructions about the wood stove."

Figure 9.11a shows the overall view of the area of origin. A cabinet above the microwave and one mounted to the side fell off the wall, and the countertop, shown in Figure 9.11b, was both burned and broken. There was a receptacle found in the center of the origin area, which is shown in Figure 9.11c.

According to the Herndons, there was an Air Wick air freshener plugged into the receptacle. A burned air freshener was found in the debris adhering to some utensils from the drawer below the receptacle, as shown in Figure 9.11d.

Figure 9.11 (a) Origin area in the Herndon kitchen.

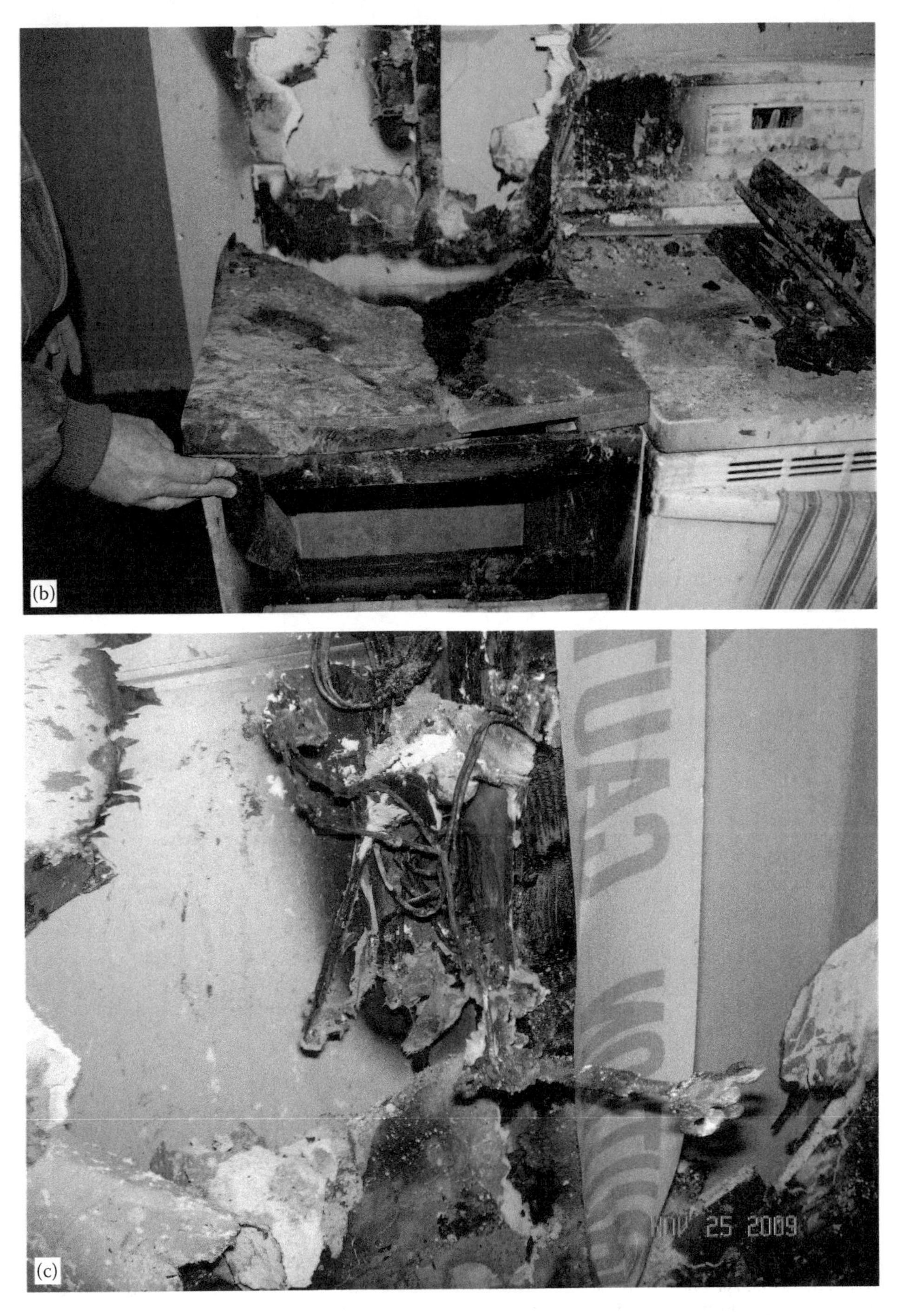

Figure 9.11 (b) Reconstruction of the base cabinet and counter top below the origin. (c) Receptacle at the center of the area of origin. Only the metallic parts remained. The first engineers who examined this receptacle failed to notice small but obvious signs of electrical arcing on both the brass and steel parts.

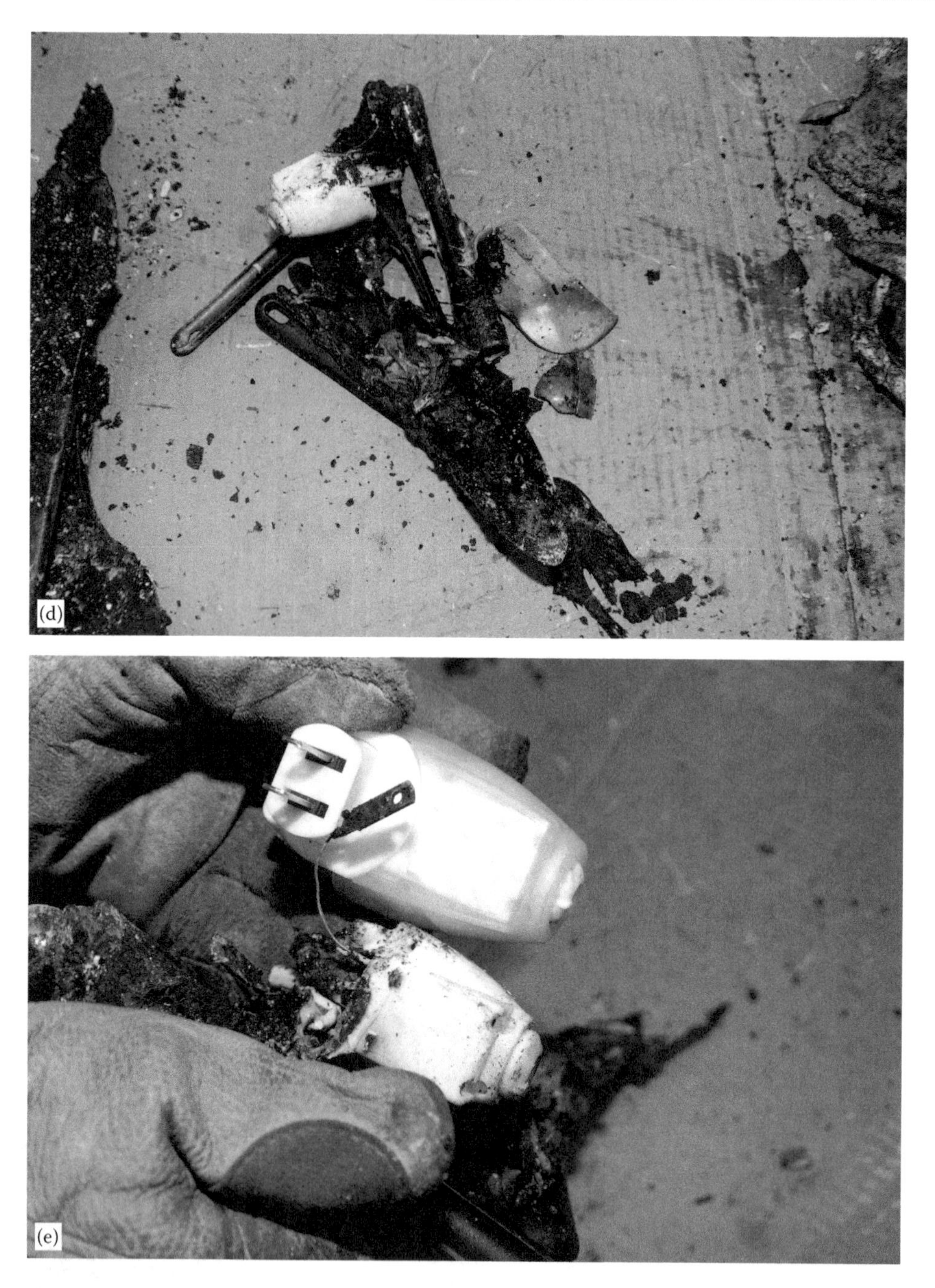

Figure 9.11 (d) Air Wick brand air freshener found adhering to plastic utensils from the kitchen drawer, where it fell during the fire. (e) Comparison of the damaged Air Wick from the origin with an undamaged exemplar. The side of the damaged unit in contact with the receptacle is the side that burned, while the side facing the kitchen was relatively undamaged.

Figure 9.11 (f) Irregular pattern falsely attributed to burning ignitable liquid. The damage was actually caused by burning melted plastic. Ignitable liquid is incapable of causing this kind of damage on a smooth surface. (g) Arcing damage on one of the brass buses in the receptacle at the origin.

Figure 9.11 (h) Arcing damage on the steel grounding strap in the receptacle at the origin. (i) Medium petroleum distillate burning on unfinished plywood. The fire caused practically no damage.

Figure 9.11 (j) (*top*) Medium petroleum distillate burning on a section of the Herndons' floor. (*bottom*) Herndon flooring after exposure to burning MPD.

Figure 9.11 (k) (*top*) Polystyrene foam cube burning on a section of the Herndons' floor. (*bottom*) Herndon flooring after exposure to burning melted plastic foam.

This damaged unit was compared with another Air Wick device found in an undamaged condition. This comparison is shown in Figure 9.11e. The most likely explanation for the finding of the Air Wick in the origin area was that it fell out of the receptacle and landed in the drawer when the counter above the drawer failed. Clearly, the side of the Air Wick device that was plugged in was far more heavily damaged than the side of the device that was facing the kitchen. Hoping to find some subrogation, the insurance carrier put the Air Wick manufacturer on notice of potential liability, and they dispatched an electrical engineer to the scene.

Despite its *condition*, the engineer interpreted the *location* of the Air Wick as evidence that it was not plugged in. He wrote in his report, "Because neither evidence of any failed Airwick device components in the debris nor any Airwick plug prongs were found in the receptacle, the evidence indicates that an Airwick device was not plugged into the receptacle as reported by the Herndons and therefore could not have caused or contributed to the fire." The engineer's examination of the receptacle resulted in the following conclusion: "There were neither plug prongs found in the receptacle jaws nor any electrical activity on the current carrying parts of the receptacle." Later, in his deposition, the engineer stated that the receptacle "showed no evidence of malfunction or electrical activity [41]." As will be shown, this was an error, but the fire investigator for the insurance company believed the engineer, despite the engineer's client's obvious interest in not finding the air freshener to be involved.

At the fire investigator's recommendation, the insurance company sent its own electrical engineer to the scene, and he also opined, "The wall receptacles and air freshener exhibited no evidence of an electrical failure [42]."

The fire investigator was unshakeable in his conviction about the alleged arson. He stated in his deposition, "There was no reasonable hypothesis reached to explain how the fire damage to the cabinets and microwave oven could have created the irregular patterns on the floor [43]." He also reached the erroneous conclusion that there were two points of origin: one on the counter and another on the floor below it. Actually, the counter sustained damage when the liquid from the Air Wick was released on to the counter. The fire investigator had not even bothered to learn whether that liquid was combustible. (According to the easily obtainable MSDS, it was combustible with a flash point in excess of 160°F.)

What caused three people to reach the wrong conclusion (and overlook critical evidence in the process) was the presence of irregular fire patterns on the hardwood floor, shown in Figure 9.11f. Further, four samples from the hardwood floor were analyzed in the laboratory, and medium petroleum distillate (MPD) was detected. The laboratory analyst called the fire investigator and requested a comparison sample, and he complied. The comparison sample also tested positive for MPD, but the investigator was so enthralled by the irregular patterns that he ignored this inconsistent data.

The insurance company therefore denied the claim of the Herndons and they filed suit in order to enforce their contract. The Herndons' attorneys, Ty Tyler and Clark Hamilton, hired Richard Underwood, an electrical engineer, and the author to examine the evidence. Figure 9.11g shows electrical melting on the brass buss from the receptacle, and Figure 9.11h shows an area on the grounding strap that clearly experienced an electrical arc. This evidence could not have been clearer. It is not clear that there was anything wrong with the Air Wick, as this kind of arcing damage can be the result of arc tracking, which in turn can result from contamination on a nonconductive surface, and contamination is certainly likely to occur on receptacles next to kitchen stoves.

The insurance investigators' hypothesis that the irregular patterns were caused by the application of ignitable liquid, specifically MPD, was tested and was disproved. First, as demonstrated by the analysis of the comparison sample, MPD is not an unusual finding on hardwood flooring. It is used as a solvent for the floor coating, and remains trapped indefinitely in the polymer matrix [44]. Second, MPD burning on a smooth surface burns itself out in less than a minute and causes almost no damage. The heavy damage seen on the floor was due to the presence of melted plastic, specifically melted polystyrene foam, that was stored in the cabinets located above the origin of the fire. Mrs. Herndon used the used foam shapes in her craft making, and the shapes fell onto the floor. The insurance company's fire investigator learned about this but did not believe the foam pieces could move four or five feet. Figure 9.11i shows an experiment conducted by the author burning medium petroleum distillate on unfinished plywood, and Figure 9.11j shows the burning on a section of flooring cut from the Herndon residence. Neither of these experiments caused any damage to the flooring beyond a slight blistering of the finish.

Figure 9.11k shows a subsequent experiment in which a polystyrene foam cube was burned on top of the hardwood, and it resulted in exactly the kind of damage to the hardwood as was seen in the residence.

Despite having all this information, both affirmative evidence showing that the fire started at the receptacle and the negative evidence disproving that the damage was the result of burning ignitable liquid, the insurance carrier refused to settle. The case was tried to a jury in November 2012, more than three years after the fire. The jury had no difficulty understanding the evidence and quickly returned a verdict in favor of the Herndons.

9.10.8.1 Error analysis

Serious errors were committed by all three of the investigators who came early to the scene. The fire investigator was misled by the irregular patterns on the floor because he did not keep up with the literature and was unaware of the work that had been published in 2000 by Anthony Putorti showing that damage on smooth surfaces was short-lived and did not cause much damage when the fuel was an ignitable liquid [45]. This caused him to misinterpret critical data. The electrical engineers overlooking the obvious damage to the receptacle, and failing to account for the condition of the Air Wick device, were negligent at best. One of them even said that the device was not plugged in. They overlooked critical data. None of the insurance company's witnesses changed their mind when they were confronted with the melting of the receptacle parts that they saw for the first time when they were deposed. Even after they saw it, they chose to disregard the inconsistent data that was visible on the receptacle parts. The fire investigator chose to ignore the inconsistent data when he learned that his comparison sample contained MPD.

9.10.8.2 Significance

The errors made by two electrical engineers demonstrate that it is easy to overlook small but critical artifacts. Almost all accidental fires start in very small areas with a high energy density, as this one did. There is no ignition scenario that will account for the electrical artifacts other than one with an origin in or on the receptacle. Yet both engineers had an incentive to "rule out" the receptacle. One engineer ruled out his client's product, while the other supported the incorrect interpretation of the irregular patterns favored by the fire investigator who suggested that he be hired. The testing provided additional validation of the work of Mealy, Benfer, and Gottuk on ignitable liquid fires referred to in Chapter 3, and pointed out for the jury the difference between liquids and melted solids. The fire investigator was making the same mistake he had been making for 30 years, and no doubt continues to make to this day.

9.10.9 Tennessee v. Terry Jackson

Terry Jackson lived with his partner, Kandie Brown, in her mobile home in rural Williamson County, Tennessee. They shared the home with six feral cats, and on October 5, 2006, one of the cats apparently knocked over an unattended candle, resulting in serious but not total fire damage. The evidence for the cat lighting the fire was wax found on the tail of the yellow cat [46]. The home after the first fire is shown in Figure 9.12a.

Jackson and Brown boarded up the residence and moved all of the materials they thought they could salvage into the master bedroom. This included three mattresses that were placed vertically against the wall.

On October 20, they were staying in the Best Western Inn while they looked for another place to rent. That morning, they went to the home and set up a generator, two 1,500 W space heaters, and two 500 W halogen work lights in the master bedroom. They also turned on the television. All of these devices were connected using two extension cords that were plugged into a three-way splitter connected to the generator by another extension cord. They had left the door closed in order to keep the cats out of the bedroom. They left the scene for about 45 min, to give the heaters a chance to drive off the morning chill, then returned and began cleaning. Jackson left again to get some soft drinks from his father's house, which was less than a 5-minute drive away. When he returned, the house was on fire and he was unable to reach Brown. She died of smoke inhalation in the master bathroom, with a 51% COHb concentration. The windows of master bedroom area of the home after the October 20 fire are shown in Figure 9.12b.

Beneath the front window, there was a large hole burned through the floor, shown in Figure 9.12c. The fire did not appear to follow the direction of the floor joists, and there is no evidence to indicate it started under the floor, although that was the opinion on which criminal charges were eventually based.

Not understanding the extreme danger involved in stacking mattresses vertically, the original fire investigators believed that the fire had spread "too quickly" to be an electrical fire, despite the fact that there were some seriously overloaded extension cords leading into the bedroom. Figure 9.12d shows a diagram of the residence, and Figure 9.12e shows a hand-drawn sketch of the electrical equipment that was running at the time of the fire. This sketch was attached to the Williamson County report.

It is dangerous to use extension cords on high wattage appliances, and these appliances drew approximately 4,000 W, 3,000 of which were routed through one small extension cord. The larger extension cord attached to the generator was designed to carry no more than 13 amperes (amps); 3,000 W divided by 120 V is 25 amps. There is no question that the cord was overloaded and had been overloaded for more than an hour. One of the heaters was a radiant oil-filled unit, shown in Figure 9.12f, while the second was equipped with a fan and a combustible plastic housing, shown in Figure 9.12g. One of the 500 W halogen lights was found next to one of the mattresses, as shown in Figure 9.12h. The

Figure 9.12 (a) Photograph of the home after the October 5, 2006, fire. (b) Photograph of the fire patterns on the exterior of the master bedroom area after the October 20, 2006, fire.

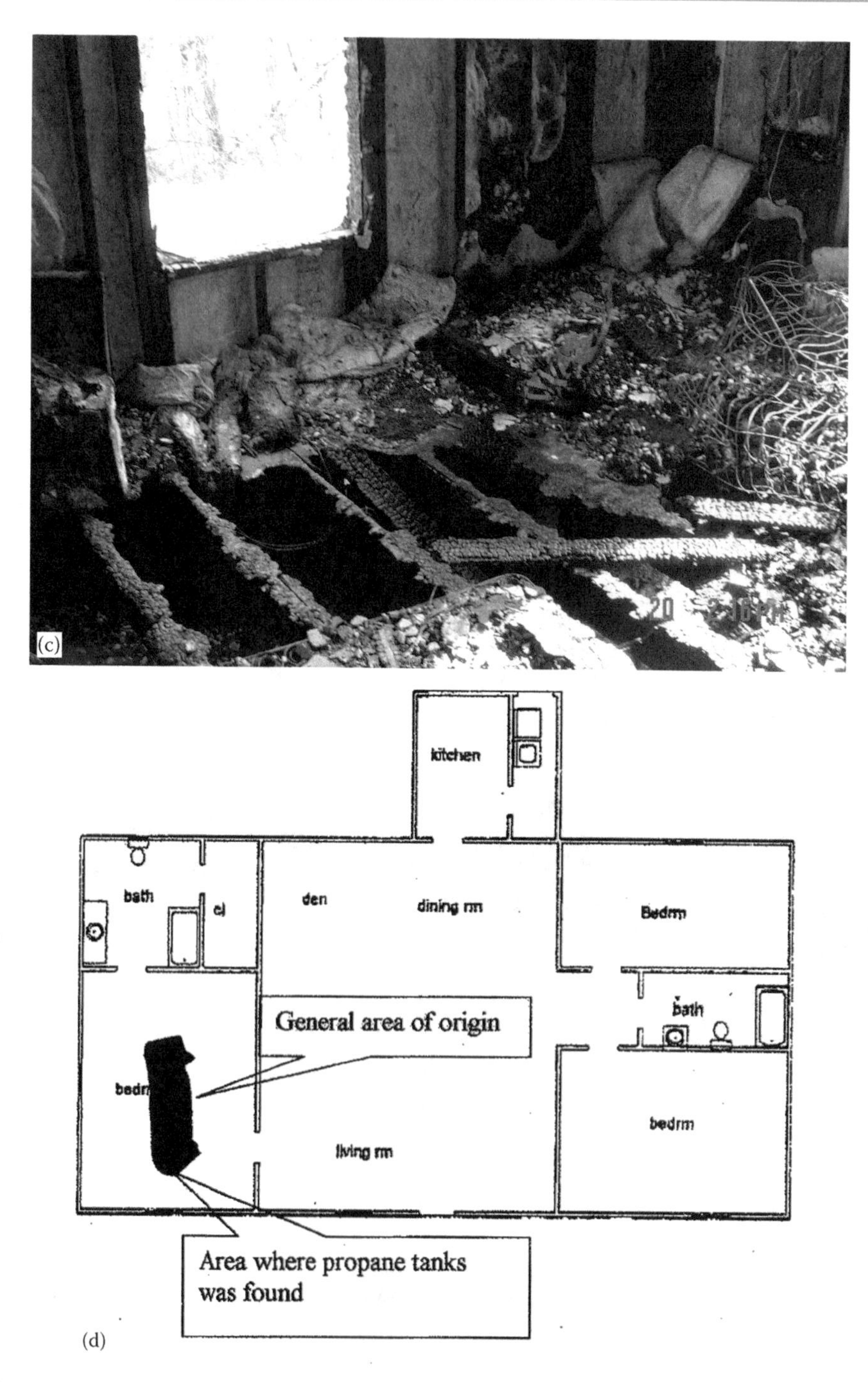

Figure 9.12 (c) Hole burned through the master bedroom floor, inappropriately characterized as the area of origin. (d) Diagram of the Brown/Jackson residence from the Williamson County Fire Report. Despite the fact that the fire burned for at least 13 min in a fully involved condition, investigators believed they could narrow the area of origin to the darkened area shown on this diagram.

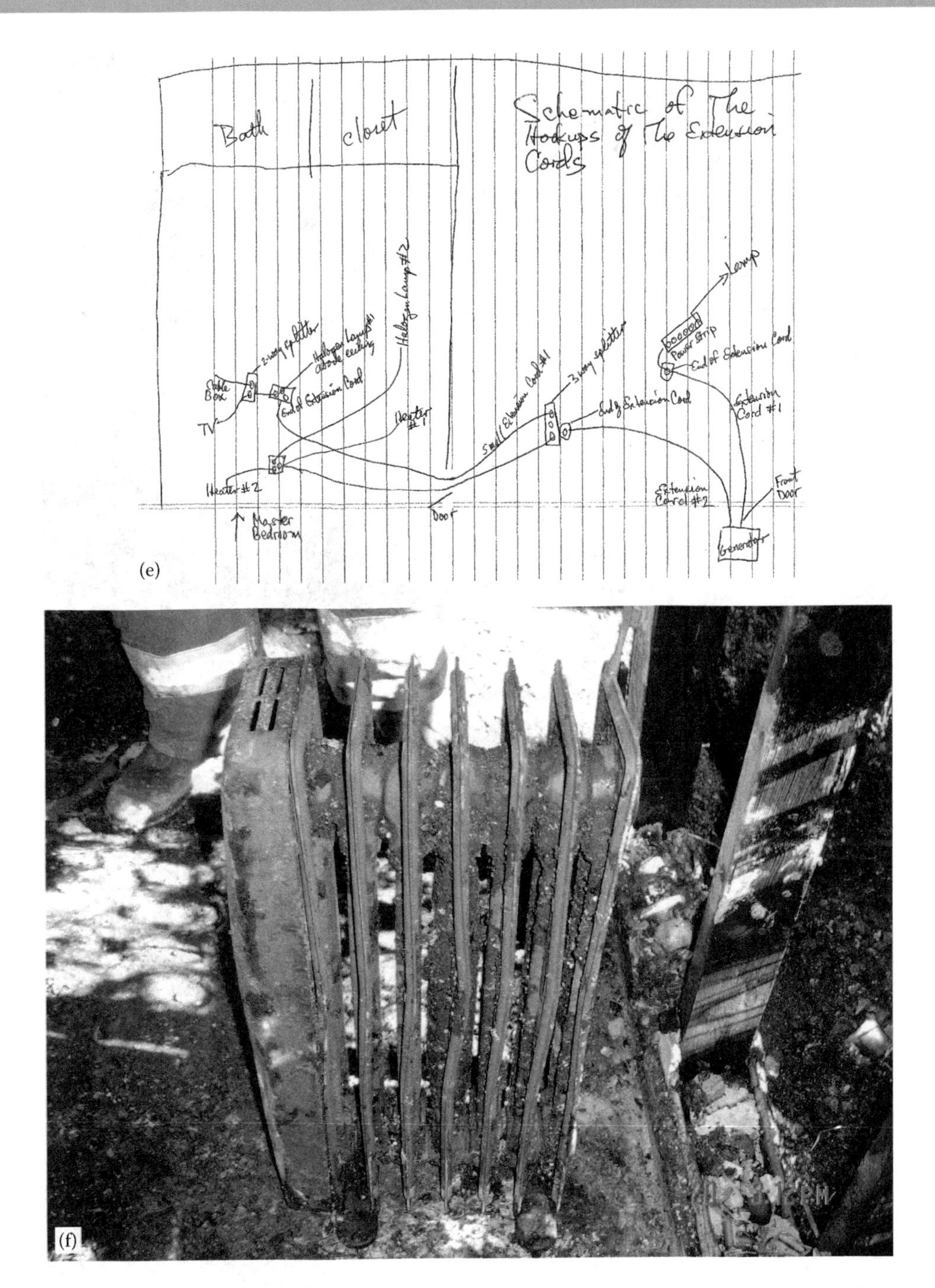

Figure 9.12 (e) Hand-drawn sketch of the generator, extension cords, and electrical devices in use at the time of the fire. (f) Oil-filled radiant heater found beneath the master bedroom.

Figure 9.12 (g) Portable electric space heater. Coil is underneath the fan blade. This unit had a combustible plastic housing. (h) 500 W halogen work light next to a mattress that was in the vertical position.

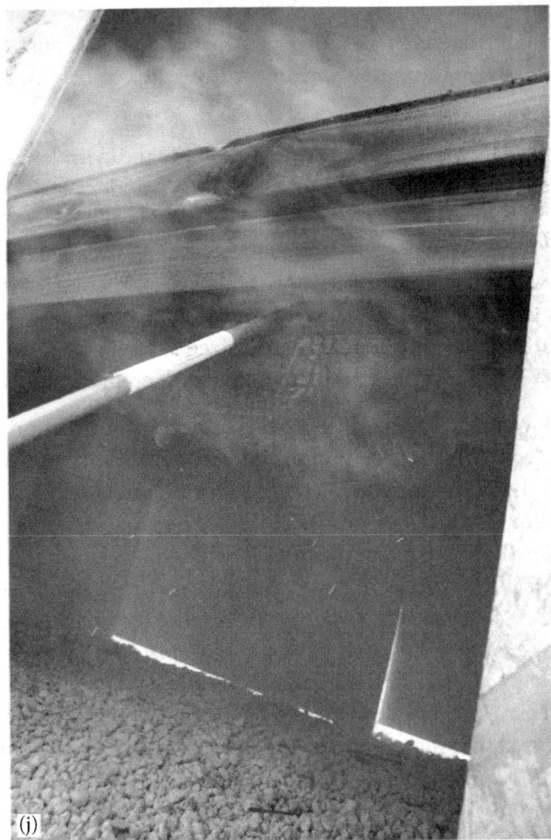

Figure 9.12 (i) 20-pound propane cylinder and weed burner torch retrieved from the crawlspace beneath the master bedroom, several feet from the crawlspace access door. The cylinder valve was wide open, but the torch valve was fully closed. (j) Testing that demonstrated that using a 100,000 Btu/h weed burner to ignite the floor would not result in penetration for at least 12 min.

possibility that one of the lights ignited a mattress or that one of the extension cords shorted out in a shower of sparks both needed to be considered. Once a vertical mattress is ignited, a fire can spread very quickly.

The first water was not put on this fire until at least 13 min after Jackson called 911. Nonetheless, based on the hole in the floor, fire investigators said they were able to narrow the area of origin to that portion of the floor or, in the case of one investigator, *under* the floor in the master bedroom. They also claimed the ability to eliminate electrical causes, despite the fact that not all of the extension cords were recovered. The Williamson County fire report concluded:

> At the conclusion of my fire scene examination and based on the information and facts I have at this time, it is my professional opinion the fire was incendiary and originated underneath the floor area (in the crawl space) of the master bedroom in the center and front area. The incendiary cause was developed as the final hypothesis by determination that the propane tank was found in the "On" position and burn patterns revealed fire had traveled from this area and into the master bedroom. No other sources of ignition were located in this area and other sources of ignition which were found above the crawl space were eliminated [47].

A similar conclusion was reached by one of the private investigators for the homeowner's insurance company:

> Systematic evaluation of this fire scene produced clear and convincing evidence that established the fire originated beneath the center of the bedroom floor located in the southeast section of the house. To establish a fire's cause, sufficient evidence must remain to identify a specific cause while excluding others. The remaining physical evidence and the elimination of natural and accidental causes support the opinion that the introduction of a competent ignition source to combustible fuels initiated this event.
>
> Mr. Terry Jackson reported that he had left the residence and had not been gone for over ten minutes when he returned and found the bedroom totally involved in fire. It is not probable that the fire could have progressed from an ignition from an electrical arc into full room involvement in that period of time [48].

Still another report by the state fire marshal was more circumspect. That report concluded, "Potential sources of accidental electric ignition could not be eliminated within the fire's area of origin, thus the cause of the fire is undetermined in nature [49]."

The county investigator was intrigued by the finding of a 20-pound propane cylinder and 100,000 Btu/h weed burner. He hypothesized, without much basis, that the fire had been intentionally set under the floor using the weed burner. The valve on the propane cylinder, shown in Figure 9.12i, was found in the open position, but *the valve on the torch was fully closed.* So the hypothesis was that, after spending however long it took to light the fire, Jackson turned off the torch, then threw it and the tank under the house. Another investigator said he thought Jackson had to have crawled under the house. That hypothesis can be discarded because doing that would have resulted in Jackson's certain death. It is clear that none of the state's witnesses put much thought into exactly how the hypothesized subfloor ignition was accomplished. (According to Jackson, he had lent the weed burner to his brother, who brought it back empty and just stored it under the house.)

As if the weed burner hypothesis were not sufficiently "novel," another interpretation of the evidence was put forward, again implicating Jackson. The electrical engineer brought to the scene by the insurance company did not find any evidence of electrical activity in the bedroom where the fire was the most intense and where everyone agreed the origin was, but he did find arcing on the wires in the next room upstream electrically of the origin. His new "theory" was that when the fire started, the generator was not running (though it is unclear how Brown would have been able to see given that the windows in the bedroom were boarded up). Sometime after the fire got going, he speculated that Jackson then fired up the generator. There was absolutely zero evidence for this, but the engineer's position was that he "could not rule it out." *And **if** the generator was off at the beginning of the fire, **then** he could rule out an electrical ignition source in the bedroom even if he could not find all of the extension cords* [50]. (An excerpt of an interview with the engineer appears in the next sidebar.) The fire department found the generator running when they arrived. Further, the deputy state fire marshal reported that there were several areas of arcing on the wiring that he saw in the bedroom. The engineer may have failed to find evidence of arcing in the bedroom because he did not come to the scene until at least four days after the fire. This would explain the differences between his observations and the state fire marshal's.

Jackson was indicted on March 10, 2009, more than two years after the fire. He was arrested in Mexico, where he frequently traveled, almost a year later.

Some of the data in "support" of the investigators' opinion was that Jackson reported seeing blue flames while he was in the house trying to gain access to the bedroom. This was a totally specious piece of data, as propane torches do not burn with a blue flame, especially when they are burning uncontrolled. Blue flames only happen when the propane is premixed with air, which could not have happened in the crawlspace under the bedroom.

Ofman: Is your findings that you are telling me now ... are they inconsistent with your findings when you made your report?
Bishop: Uh ... they're not inconsistent. My report is still valid. There is new information, and uh ... or a new theory I should say.
Ofman: And what's the new theory?
Bishop: The new theory is that the generator may have been off when the fire originated and turned on later during the progress of the fire.
Ofman: And, according to what you told me, and you correct me if I'm wrong, the progress of the fire would have already been into the living room at the time it was turned on.
Bishop: Yeah.
Ofman: And, how does that change anything in terms of your original report?
Bishop: Uh ... well had I known that fact I could uh ... in the original report I could *not* have ruled out electrical activity during the course of the fire.
Ofman: Totally.
Bishop: Yeah.
Ofman: Even with a missing piece of wire in the bedroom.
Bishop: Yeah.
Ofman: Can you do that now?
Bishop: If uh ... that's a fact and that theory is true it can be eliminated, yes.

This author was retained by Lee Ofman, Jackson's attorney, and it was immediately apparent that a test of this weed burner hypothesis would be necessary. I built a mock-up of the crawlspace and floor, using the same materials as were present in the mobile home, including "Plytanium" plywood flooring, a Georgia-Pacific product. In three separate tests, I learned that it was very difficult to light the floor on fire because when the torch was too close to the wood, it would go out. Further, holding the torch in place at the crawlspace entrance was almost impossible due to the heat coming out from under the floor. There was no possible way to arrange an ignition in the area shown as the origin in Figure 9.12d. Weed burners are quite loud. They make a sound very much like a jet engine. Brown would have had to be deaf not to have heard the sound of the weed burner if this fantastic scenario had actually taken place. Further, in the three tests, the shortest period of time between ignition of the bottom and penetration of the floor was 12 min. It was necessary to arrange a prop to hold the torch in place because standing at the edge of the crawlspace opening would not have been possible. See Figure 9.12j. This testing was all videotaped and ready to take to trial, but it turned out that no trial would happen.

Ofman was able to avoid Tennessee's usual prohibition on deposing states' witnesses prior to a criminal trial. Jackson and Brown each had $10,000 life insurance policies naming the other as a beneficiary. This was supposedly Jackson's motive for killing his long-time partner. The Garden State Life Insurance Company had denied Jackson's claim, and he filed a civil suit, which allowed Ofman to depose the witnesses that the state planned to use to support their charges. During the deposition of the state's lead expert, it became clear that despite his claim to a bachelor of science degree in fire science from Columbia Southern University (*Magma* [sic] *Cum Laude*), he knew nothing about the behavior of fire. He could not state the energy content of a cubic foot of propane, nor did he know the chemical formula for propane, nor did he know anything else that he should have known to hold himself out as a professional fire investigator. He clearly did not meet the requirements of NFPA 1033. The following are excerpts from his deposition:

Page 10

Q: If you would, state your theory as to the origin and cause of the fire on October 20th, 2006.

A: October 20th, the origin of the fire was in the crawlspace below the master bedroom, which is the left front room. The cause of the fire was initiated from a propane weed burner and direct contact with the building construction of the floor.

Q: And what is your theory as to how the propane weed burner was ignited?

A: Human factor.

Q: And am I correct in saying that your opinion states that this fire was an arson?

A: That's correct, sir.

Page 281

Q: Okay. What's the basic unit of energy called?

A: I'm unsure at this time.

Q: You ever heard of a joule?

A: I have.

Q: What is it?

A: It's a measurement of energy or that's how—it has to do with electricity as well.

Page 282

Q: What are the basic units of power called?

A: AC and DC.

Q: I'm sorry?

A: AC and DC.

Q: Have you ever heard of a watt?

A: Yes, sir.

Q: Would that be the correct answer?

A: More than likely.

Q: What is a watt?

A: I mean I'm unsure. If you want me to look at a manual and give you these answers—

Q: No, I'm asking you—

A: Off the top of my head, I mean you're asking—it's like I'm taking a quiz right now over—

Q: You're taking a deposition.

A: I understand that. But you're asking me questions that are found in a manual.

Q: Do you know what a watt is?

A: No, sir.

Q: Okay. How is the size of a fire measured?

A: I'm unsure at this time.

Q: Okay. What is radiant heat flux?

A: I'm unsure at this time.

Q: What is air entrainment?

A: I'm unsure.

Q: What is a ceiling jet?

A: Unsure.

Page 275

Q: What does pyrolysis mean?

A: Basically the—like, wood will distort, take a different shape. It will—pyrolysis is—the wood is igniting and then the change of the wood is showing like charred and things like that.

Q: Do you know what the chemical formula for hydrogen is?

A: No, sir.

Q: Do you know what the chemical reaction for the combustion of hydrogen is?

A: No, sir.

Q: Do you agree that the combustion of methane is the simplest of all the combustion reactions involving a hydrocarbon?

A: No, sir.

Q: You don't agree with that?

A: Oh, you're asking me if I agree?

Q: Yeah.

A: I'm unsure.

Page 277

Q: Do you know how many BTUs are present in a typical cubic foot of propane?

A: Not at this time.

Q: Do you know what the chemical formula for propane is?

A: I'm unsure at this time.

Q: Can you write down the chemical equation that describes the burning of propane in air?

A: I'm unsure.

Q: How many volumes of oxygen are required to burn a volume of propane?

A: Unsure.

Q: How many volumes of air are required to burn a volume of propane?

A: Unsure.

Q: Have you ever tried to set wood on fire using a propane torch?

A: No, sir.

Q: Do you agree that there's both a liquid phase and a vapor phase inside the propane tank?

A: Yes, sir.

Q: Do you know how much vapor a given volume of liquid produces?

A: No, sir, unsure at this time [51].

Since NFPA 1033 was updated to include the list of required knowledge areas in 2009, this kind of questioning is now becoming common in contested fire cases. Not only is the expert's methodology challenged, his qualifications are questioned as well. The Unified investigator was no more qualified than the county investigator, as he demonstrated in a similar deposition in 2014 [52].

Following the deposition of the county investigator, which the prosecutor attended, Mr. Ofman filed a *Daubert* motion, which read in part:

> Their opinions as to the origin and cause of this fire are not formed on a factual basis supporting their opinions; their opinions are not grounded in the scientific method; their opinions and methodology have not been tested; the evidence they base their opinions on have not been subjected to peer review or publication; there is no known rate of error; and their method of conducting a fire investigation, in this case, is not generally accepted in the scientific community.

The *Daubert* hearing was scheduled for Monday November 3, 2011, but on the Friday before that, the prosecutor dismissed the charges against Jackson. She declined to blame the experts' lack of knowledge for the dismissal. Instead, she stated that Jackson's brother was a credible witness, and he had told investigators from her office that the weed burner was *always* kept in the storage space below the bedroom.

Such a result is common in cases of false accusations. Jackson was unsatisfied with a mere dismissal, however, and Ofman was able to not only get the charges dropped but to get the entire record of this fiasco expunged.

9.10.9.1 Error analysis

The fire investigators failed to consider the obvious possibility that the grossly overloaded extension cords started the fire. They failed to consider the possibility that the 500 W halogen lamps found in close proximity to a vertical mattress represented the ignition source and first fuel ignited. They misinterpreted the critical data (or failed to understand the significance) of the mattresses being vertical. They relied on an electrical engineer who did not find all of the cords but still thought that he could come up with a helpful theory to "eliminate" an electrical fire. They failed to consider the possibility that the weed burner was innocently stored in the storage area under the bedroom. They discarded the inconsistent data that the torch valve was turned off. They discarded the inconsistent data that the state fire marshal saw evidence of arcing in the room of origin and called the fire undetermined. The hypothesis about Jackson setting the fire and then starting the generator was made up of whole cloth. Their limited knowledge of fire and energy demonstrated that they were wholly unqualified to be rendering determinations about the causes of fires. Sadly, all of these actors are still investigating fires for a living.

9.10.9.2 Significance

The prosecution of this case shows how even the most incompetent fire investigators can be believed by the non-scientists who populate prosecutor's and defense attorney's offices. This case generated significant attention in the Tennessee fire community, with everyone but the principal perpetrator eventually backing away from the arson murder hypothesis. Unfortunately for Jackson, he spent almost 2 years awaiting trial on capital murder charges. By the time the charges were dismissed, both he and his family were financially ruined.

9.11 CONCLUSION

This chapter described common errors seen by this investigator during 30 years of reviewing cases. These errors have been categorized and condensed into a checklist to assist in evaluating a fire investigator's determination that a fire was intentionally set.

Numerous examples of mishandled investigations illustrate the way that certain kinds of errors continuously occur and the consequences of those errors. It was not only technical errors but also the legal context of these cases that led to miscarriages (or potential miscarriages) of justice. The actions of these fire investigators, sometimes innocently and other times with at least partial knowledge, resulted in wrongful prosecutions or wrongful denials of insurance claims. The common denominator among these individuals is that they viewed themselves as "arson investigators" rather than "fire investigators," or as "case makers" rather than "truth seekers." They failed to consider alternate hypotheses involving an accidental cause. The attorneys overseeing these cases, who should have known better, failed to serve either their clients or the interests of justice when they soldiered on, even when evidence central to their cases evaporated. Because of the actions, or sometimes the inaction, of these so-called professionals, innocent people were doubly victimized: first by the fire, then by the justice system or their insurance companies.

Review questions

1. When you provide a colleague with your data and report and ask him to it, what kind of review are you requesting?
   a. An administrative review
   b. A peer review
   c. A technical review
   d. A technical peer review
2. What is the importance of motive in an arson investigation?
   a. No criminal conviction can be obtained unless the state can show that the defendant had a motive to commit arson.
   b. A jury is unlikely to believe that the defendant committed arson unless there is a credible motive.
   c. Motive is the same thing as intent.
   d. All of the above.
3. What is the proper role of an accelerant detection canine at a fire scene?
   a. To assist the fire investigator in locating fire debris samples that have a higher likelihood of testing positive in the laboratory

b. To test the investigator's hypothesis that the fire was intentionally set
c. To allow for an immediate determination, as opposed to waiting for the laboratory to complete its analysis
d. To help avoid the necessity of shoveling off the floor in the area of origin

4. What is the significance of a melted aluminum threshold?
   a. Melted aluminum at floor level means that the fire burned at an abnormally high temperature and must have been accelerated.
   b. Melted aluminum indicates locally high temperatures, indicating the presence of accelerants nearby.
   c. Melted aluminum indicates temperatures in excess of 1,200 °F, which is a common temperature at floor level in a fully involved compartment.
   d. Because of thermal inertia, melted aluminum indicates that the fire burned at that point for a long period of time, indicating a point of origin.
5. Which of the following statements regarding a fully involved compartment are true?
   I. Because flashover produces uniform burning, irregular burn patterns are an indicator of the use of an accelerant.
   II. In most situations, after flashover, there will be differences in the amount of damage because of the variation in the amount of ventilation available.
   III. The rules for the interpretation of fire patterns are different from the rules that apply in rooms that have not reached flashover.
   IV. The absence of a fuel source and ignition source at the area of lowest and deepest char indicates that the ignition source was an open flame that has been removed from the scene.
   a. I, II, III and IV
   b. II, III and IV
   c. III and IV only
   d. II and III only

Questions for discussion

1. Provide an example of poor communication resulting in a false conclusion.
2. Presented with a colleague's report, how do you propose to evaluate its accuracy?
3. Although NFPA 921 states, "It is unusual for a hypothesis to be totally consistent with all of the data," why is it a poor practice to ignore inconsistent data?
4. What is the main problem with an incorrect determination that a fire was intentionally set?
5. Why is failure to consider an alternate hypothesis one of the more serious errors a fire investigator can make?

References

1. Doyle, J. (2014) Learning from error in the criminal justice system: Sentinel event reviews, in *Mending Justice: Sentinel Event Reviews*, National Institute of Justice. Available at https://www.ncjrs.gov/pdffiles1/nij/247141.pdf (last visited on November 25, 2017).
2. National Registry of Exonerations website (2017). Available at https://www.law.umich.edu/special/exoneration/Pages/detaillist.aspx?View={FAF6EDDB-5A68-4F8F-8A52-2C61F5BF9EA7}&FilterField1=Group&FilterValue1=A (last visited on January 1, 2018).
3. Lentini, J. (2013) *Scientific Protocols for Fire Investigation*, 2nd ed., Taylor & Francis Group, Boca Raton, FL, p. 565.
4. NFPA (2017) NFPA 921, *Guide for Fire and Explosion Investigations*, National Fire Protection Association, Quincy, MA, §18.7.2, 215 and §19.7.2, 221.

5. Lentini (2013) *supra* at 559.

6. Lentini, J. (1999) "A calculated arson," *The Fire and Arson Investigator*, Vol. 49, No. 3. Available at www.firescientist.com/publications.

7. Lentini (2013) *supra* at 572.

8. NFPA (2017) NFPA 921, *Guide for Fire and Explosion Investigations*, National Fire Protection Association, Quincy, MA, §6.3.7.8, 59.

9. DeHaan, J. (1997) Proposal 921-219 (log # 113), Report on Proposals, NFPA Technical Committee Documentation, Fall 1997, NFPA, Quincy, MA, 410.

10. American Association for the Advancement of Science (2017) *Forensic Science Assessments: A Quality and Gap Analysis-Fire Investigation* (Report prepared by Almirall, J., Arkes, H., Lentini, J., Mowrer, F., and Pawliszyn, J.) doi:10.1126/srhrl.aag2872.

11. *WI v. Joseph Awe*, No. 07-CF-54.

12. *WI v. Joseph Awe*, No. 07-CF-54, Testimony of James Sielehr, December 18, 2007, Trial Transcript at page 192/258.

13. Id. at 119/234.

14. Id. at 182/258.

15. *WI v. Joseph Awe*, No. 07-CF-54, Decision and Order by Hon. Richard Wright, March 21, 2013.

16. *State of Georgia v. Weldon Wayne Carr*, in the Superior Court of Fulton County, Case No. Z-58558-A (1994), see also *State of Georgia v. Weldon Wayne Carr*, 482 S.E. 2d 314 (1997).

17. IAAI Forensic Science Committee (1994) Position paper on accelerant detection canines, *Fire and Arson Investigator* 45(1):22–23.

18. *State of Iowa v. Roy Laverne Buller*, in the Supreme Court of Iowa, No. 146/93-701, 1994.

19. Renaud, T. (2004) The inside story of the *Wayne Carr* case, *Fulton County Daily Report*, Atlanta, GA, October 5, 2004, 1.

20. Avato, S. J., and Cox, A. T. (2009), "Science and circumstance: Key components in fire investigation," *Fire and Arson Investigator* 59(4):47–49.

21. NFPA (2017) NFPA 921, *Guide for Fire and Explosion Investigations*, National Fire Protection Association, Quincy, MA, §24.4.1, 259.

22. *Maynard Clark and Clark Hardware et al. v. Auto-Owners Insurance Company et al.*, in the Circuit Court of the Thirteenth Judicial Circuit, in and for Hillsborough County, Florida, Civil Division, Case No: 92-09683, Division B.

23. Lentini, J., and Waters, L. (1982) "Isolation of accelerant-like residues from roof shingles using headspace concentration," *Arson Analysis Newsletter* 6(3):48.

24. Lentini, J. (1998) "Differentiation of asphalt and smoke condensates from liquid petroleum products using GC/MS," *Journal of Forensic Science* 43(1):97.

25. Eisenstadt, J. (2009) Autopsy Report, GBI DOFS, dated April 4, 2009.

26. Gourley B. (2009) Georgia Insurance and Safety Fire Commissioners Office State Fire Marshals–Arson Unit, undated first report on the Dahlman fire.

27. Gourley B. (2010) Georgia Insurance and Safety Fire Commissioners Office State Fire Marshals—Arson Unit, Supplemental report on the Dahlman fire, prepared on or before March 8, 2010.

28. Sellers, W. (2010) *State of Georgia v. Scott David Dahlman and Valerie Lynn Dahlman*, Case No. 2009R-0533, *Nolle Prosequi* dated May 10.

29. Garner, M., "Investigative error causes murder case against couple to be dropped," *Atlanta Journal and Constitution*, May 16, 2010.

30. http://www.law.umich.edu/newsandinfo/features/Pages/gavitt_exoneration.aspx
31. Plummer, C., and Syed, I. (2012), "Shifted Science" and Post-Conviction Relief, 8 STAN. J. C.R. & C.L. 259.
32. *State of Arizona, Plaintiff, v. Ray Girdler, Jr.,* Defendant, in the Superior Court of the State of Arizona in and for the County of Yavapai, No. 9809.
33. Dale, D. (1981) Office of the State Fire Marshal, Report of Fire, Mobile Home Fire with Two Fatalities, DR #81-05, Phoenix, AZ.
34. Davie, B. (1993) Flashover, *National Fire and Arson Report* 11(1):1.
35. *State of Louisiana, Plaintiff v. Amanda Gutweiler, Defendant.* Ninth Judicial Circuit Court, Rapides Parish, LA, Criminal Docket # 265037.
36. DeHaan, J. (2002) Report re: *State v. Gutweiler* (Tioga Fire), Fire-Ex File Number 01-1101, page 11.
37. DeHaan, J. (2004) Supplemental Report re: *State v. Amanda Gutweiler,* Fire-Ex File Number 01-1101 page 9.
38. DeHaan, J., (2008) Supplemental Report re: *State v. Gutweiler* (Tioga Fire), Fire-Ex File Number 01-1101 page 4.
39. O'Bryan, A. (2011) *Ashes of Innocence*, AuthorHouse, Bloomington, IN.
40. *Donald Scott Herndon and Linda Tranter Herndon, Plaintiffs, v. Security First Insurance Company,* a Florida insurance corporation, Defendant. In the Circuit Court, Fourth Judicial Circuit, in and for Duval County, Florida Case No.: 2010-Ca-007262 Division: Cv-C.
41. *Herndon v. Security First*, August 29, 2011 Deposition of Eric Jackson, P.E., page 20.
42. Martini, H. (2009) Report on Herndon fire, Unified Investigations and Sciences File No. FL010900813.
43. *Herndon v. Security First*, April 21, 2011 Deposition of Herbert Webber, page 94.
44. Lentini, J. (2001) "Persistence of floor coating solvents," *Journal of Forensic Science*, 46(6):1470.
45. Putorti, A. (2000) Flammable and combustible liquid spill/burn patterns, NIJ Report 604-00, U.S. Department of Justice, Office of Justice Programs, National Institute of Justice. Available at http://fire.nist.gov/bfrlpubs/fire01/art023.html
46. Edge, D. (2006) Southern Fire Analysis letter report on October 5 fire loss.
47. Edge, D., and Sanders, J. (2006) Williamson county fire and explosion investigation unit, Origin and Cause Report, page 1.
48. Hooten, J. (2007) Unified Fire and Sciences, Origin and Cause Report on the October 20, 2006 fire.
49. Vaden, D. (2006) Tennessee State Fire Marshal's Bomb and Arson Unit, report on the October 20, 2006 fire.
50. Bishop, A. (2010) Recorded interview with Lee Ofman, page 10.
51. In the Chancery Court for Williamson County, Tennessee, Terry R. Jackson, Plaintiff/Counter Defendant, v. Garden State Life Insurance Company, Defendant, Counter Plaintiff, Case No. 35222, November 11, 2010 Deposition of David L. Edge III.
52. In the US District Court for the Middle District of Tennessee, Chubb National Insurance Company and Travelers Personal Security Insurance Company, Plaintiffs, v. Dale & Maxey, Inc. and Williamson County Heating and Plumbing, Defendants. Case No. 3:13-Cv-0528, July 29, 2014 deposition of Jesse Charles Hooten.

CHAPTER 10

The professional practice of fire investigation

Forensic science is the product of an uneasy and unholy mating of Science, the objective seeker of truth and knowledge, and Forensic, the argumentative persuader of courtroom advocacy. It is not called Justice Science, Law Science, or Truth Science, as many of us would like to imagine. We are a bastard child, an orphan, but still the subject of an intense child custody battle between our estranged parents, the truth seeker and the advocate.

—D. H. Garrison
Bad Science, 1991

LEARNING OBJECTIVES

After reviewing this chapter, the reader should be able to:

- Understand the organization of a quality assurance program
- List the stakeholders with an interest in the fire investigator's work product
- Understand the processes involved in participating in civil and criminal litigation
- Recognize the role of the fire investigator in helping counsel prepare for depositions and hearings
- Be aware of the pitfalls of participating in litigation as a consulting or testifying expert witness

10.1 INTRODUCTION

The previous nine chapters of this volume have been presented with the goal of helping fire investigators and other scientists working in the field of fire investigation become more proficient at interpreting evidence found at fire scenes. The goal of this chapter is to help the proficient investigator be more comfortable in the system that is the ultimate consumer of the investigator's work product.[1] An investigator may have an error rate well below 1%, but no matter the quality of the investigator's work, unless the results of the investigator's work are communicated clearly and, more important, understood and accepted by the investigator's clients, adversaries, and the courts, the investigator's efforts will be wasted.

10.2 IDENTIFYING YOUR STAKEHOLDERS

When designing a set of guidelines for a professional practice, it is useful to identify the stakeholders. What people or groups of people care if the investigator does a good job or not? Whose interests will be affected? What relationships are important in a successful practice?

The first stakeholder in an investigative practice is *the investigator* himself. The investigator has the responsibility to maintain his or her knowledge, skills, and abilities at a level consistent with the state of the art. This requires taking the time to read the trade publications, such as *The Fire and Arson Investigator*, and to keep up with new

[1] Although this is a book about fire investigation, there is nothing in this chapter that does not apply equally well to any forensic science practice.

modules of CFITrainer. A laboratory chemist has a responsibility to keep up with improvements that are reported in *The Journal of Forensic Sciences, Science and Justice*, and other technical journals. All professionals have an obligation to be familiar with the current standards of care. The International Association of Arson Investigators' (IAAI's) *Code of Ethics* states that:

> I will regard it my duty to know my work thoroughly. It is my further duty to avail myself of every opportunity to learn more about my profession.

Keeping up is not an option—it is an obligation. The skills that one learns when first training as a fire investigator or a chemist must be kept sharp and current. The state of the art at the end of an investigator's career will be markedly different than it was at the beginning.

One important way to stay current is to participate in a certification program. Individuals who see the attainment of certification as the capstone of their careers are looking at it the wrong way. Certification is just the *starting point* toward achieving excellence, which is the real goal. Certainly, it is satisfying to be recognized by an independent credentialing body for having achieved certain minimum criteria for practicing in one's chosen field and for having passed the initial examination. The whole point of any meaningful certification program, however, is to improve the quality of work in the field by encouraging professionals to improve their knowledge, skills, and abilities. *Certification is a gateway to professional development.*

The next stakeholder is the investigator's employer, whether that is a public agency, a private company, or a sole proprietorship. Competent, professionally involved investigators reflect well on their agency, and a responsible agency will support the continuous professional development of its employees. This support should be more tangible than moral support. Moral support without tangible benefits is inexpensive, but it is only "lip service." Programs such as tuition reimbursement for technical courses, covering expenses associated with certification, and sending employees to conferences where they can interact with other professionals are the kinds of tangible support that agencies should provide. Despite the costs involved, employees who work for organizations that support professional development are likely to be more loyal and supportive of their employers.

The next group of stakeholders in an investigative practice is the *clients*. Without these people, no practice would exist. In both the public and private sectors, these are the people and organizations that pay our bills or sign our paychecks. They are the very reason for the existence of our organizations. They deserve thorough investigations; understandable, defendable conclusions; honest opinions; and fair business practices. They need to be told what they need to know, not what the investigator thinks they want to hear, unless that also happens to be the truth.

The *courts* represent an important stakeholder in an investigator's practice. As expert witnesses, we are guests in the courtroom and are present solely for the purpose of helping a judge or a jury to understand the facts and evidence. While many investigators only appear in court a handful of times over the course of their career, they should nonetheless be prepared to put every one of their investigations in front of a jury. More important, they should be prepared to sit across a table and answer hard questions from an adversary who does not want to accept their conclusions. This author is frequently asked at depositions what fraction of our investigations is "litigation related." The stock answer to this question is, "I treat all investigations as if they are going to litigation." Only about 10% of the cases in which this author becomes involved actually progress as far as a deposition, but when going into the case, it is never clear which 10% will require sitting across the table from an adversarial attorney, so all investigations need to be treated like they are going to be litigated.

The fire investigation *profession* has a stake in each investigator's individual practice. Cases like the ones discussed in the previous chapter reflect poorly on the entire profession, not just on the incompetent or unskilled part of the profession. (Someone outside the profession often cannot discriminate between a professional and a hack.) Investigators who learn something new have an obligation to share that knowledge with their fellow investigators. Professional participation includes not only attending meetings but also teaching, writing, and taking part in efforts by the profession to improve the overall quality of work through standardization. It is one thing to complain about a passage in NFPA 921; it is something entirely different to formulate a proposal to improve that passage. The vast majority of criticisms heard regarding NFPA 921 come from people who do not participate in the process.

Teaching not only benefits the profession but also benefits the teacher. A seminar speaker might spend up to 8 hours preparing a presentation lasting only an hour. During those hours, the speaker focuses on and thinks critically about the subject matter. This reflection and distillation of ideas cannot help but improve the speaker's understanding of the subject. The same benefits accrue to an author preparing an article for a professional journal.

Perhaps the most valuable learning experience that any fire investigator can have is participation in a test fire. Although expensive and difficult to coordinate, test fires, if done correctly, provide new insights to every participant. Concepts that the investigator has read about in textbooks come alive and take on a clearer meaning. Watching the growth and development of a test fire can be a humbling experience, as the expectations of the participants are shown to be wide of the mark.[2] Such experiences allow for a more accurate "calibration" of an investigator's expectations.

The final stakeholder in a fire investigation practice is *the public*. The goal of every investigator should be to find the true cause of each fire investigated so that future similar incidents can be prevented. Whether this is accomplished by aiding the prosecution of an arsonist or helping to get a dangerous product withdrawn from the marketplace, the goal is the same—to protect the public.

10.3 DOING CONSISTENT WORK

Every major manufacturer of products or provider of services has an interest in supplying consistent quality, and fire investigators are no different. The way that industries and professions attempt to ensure consistent quality work is by following a program outlined in a *quality assurance (QA) manual*. Every agency, public or private, should have a manual that commits to paper the policies and procedures that govern how that agency does business. An organization's QA manual forms the basis for becoming accredited.

The first step in designing a QA manual is to identify the stakeholders in the business. We have already done that. Most QA manuals are organized on four levels, called, appropriately enough, Levels 1 through 4 [1]. Level 1 contains an organization's mission statement, organizational chart, and an outline of the organization's goals and responsibilities to each group of stakeholders. The QA manual is itself a Level 1 document.

Level 2 documents describe the operation of the QA program. This includes procedures for soliciting feedback from clients, procedures for responding to customer complaints, and procedures for issuing corrective or preventive actions, and states the frequency with which internal and external audits of a company or agency's work product will take place. Level 2 of the manual also describes how distribution of the QA manual is controlled and how changes are to be made and documented in the manual. In an audit of a fire investigation practice, the auditor, whether an employee or a contractor from an outside agency retained to perform the audit, typically examines several randomly selected files to make sure that they contain the documents they need to contain and that the appropriate procedures were followed. *The basic rule for audits is that if no documentation of an activity exists, it did not happen.*

Level 3 documents are the procedures followed in conducting any major activity. The written procedures in Chapters 4, 5, and 6 are examples of Level 3 documents. The important thing about Level 3 documents is to say what you do and do what you say. Level 3 documents must be reviewed at least annually and updated as necessary. Going through this exercise at the inception of a QA program seems like so much paperwork, but after a few years of such activity, a culture develops where people actually think critically about what they are doing and whether what they are doing comports with what they have written down.

Level 4 documents are the forms that an agency or company uses to facilitate the collection and preservation of information. These forms are described in the procedures written down in Levels 2 and 3. A typical QA documentation scheme is shown in Figure 10.1.

Accreditation is an increasingly important way by which agencies and companies demonstrate that their work meets certain minimum levels. Accreditation represents one side of the quality triangle in forensic science (Sidebar 10.1). Forensic science laboratories can be accredited by the ANSI-ASQ National Accreditation Board (ANAB) (the successor to American Society of Crime Laboratory Directors Laboratory Accreditation Board ASCLD/LAB), by the American Association for Laboratory Accreditation (A2LA), or by other registered companies that audit the

[2] During the re-creation of the Lime Street fire in Jacksonville, Florida, in 1991, the author and colleagues, who were working for the state, set up a test that we expected would refute the defendant's story. As the room of origin reached flashover in less than 5 minutes, a city fire marshal exclaimed, "That may prove the defendant's case." In fact, it did. The defendant was released from custody within days of the test.

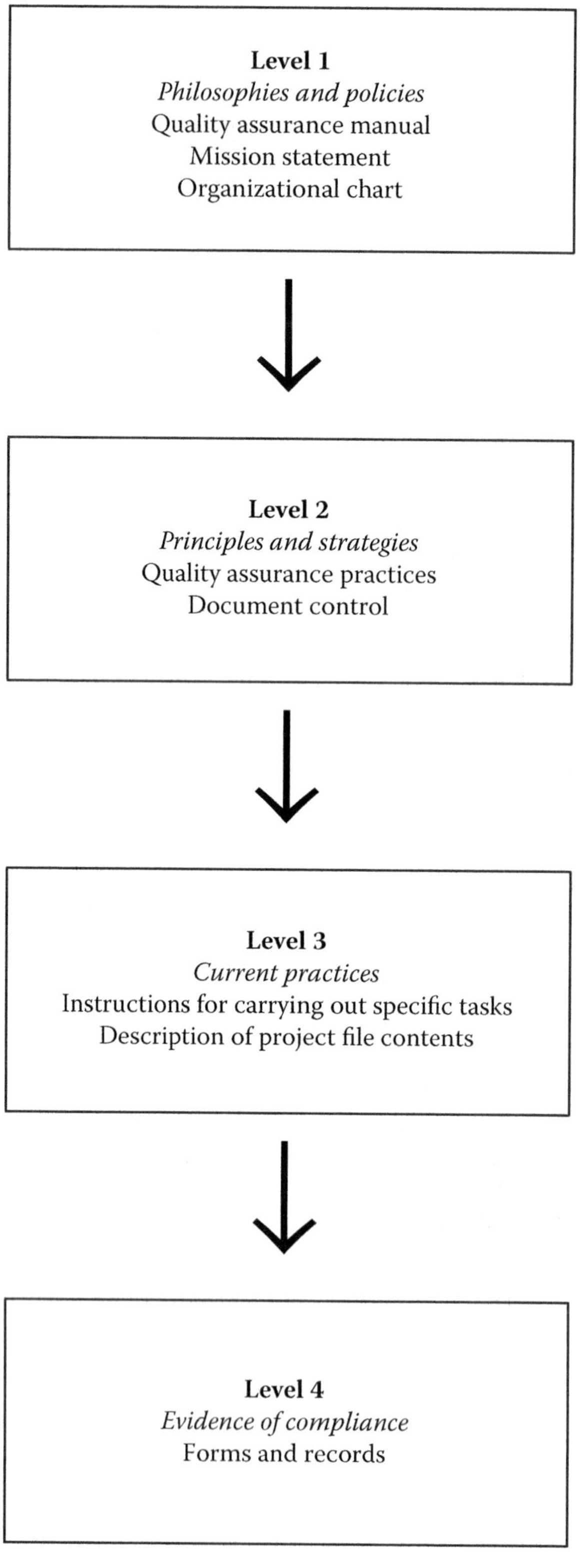

Level 1
Philosophies and policies
Quality assurance manual
Mission statement
Organizational chart

Level 2
Principles and strategies
Quality assurance practices
Document control

Level 3
Current practices
Instructions for carrying out specific tasks
Description of project file contents

Level 4
Evidence of compliance
Forms and records

Figure 10.1 Typical QA documentation scheme.

accredited organization to a particular standard. Testing and calibration laboratories, such as the one where this author worked for 28 years, are generally accredited to ISO 17025, *General Requirements for the Competence of Testing and Calibration Laboratories* [2]. A modified version of ISO 17025 is used by ANAB in its accreditations [3]. Accreditation is also becoming more common for police and fire departments. As of the end of 2017, only two fire investigation agencies, one private and one public, had been accredited. But in early 2018, the NFPA Standards Council began a project to develop a standard for fire investigation units (FIUs).[3] As a result of this effort, accreditation of FIUs will likely become more common in the future. Even if an agency is not currently planning to become accredited, establishing a quality assurance program, including the preparation of a QA manual, is a worthwhile endeavor, even if the audits are done internally. Establishing a QA program of sufficient robustness to achieve accreditation is not a trivial exercise. Murley provides a brief description of the commitment required [4].

SIDEBAR 10.1 The quality triangle in forensic sciences

The concept of the quality triangle in the forensic sciences was first presented in the early 1990s. Lawrence Pressley, a former quality assurance director at the FBI Laboratory, is generally credited with the image shown in Figure 10.2, which ties together accreditation, certification, standardization, and proficiency testing. As interdependent and seamless as these concepts might now seem, they have not come together without some resistance.

Fire investigators embraced certification in the late 1980s, but many fiercely resisted the idea of standardization (only a few still do). Crime laboratory directors embraced the idea of accreditation in the early to mid-1980s but resisted the idea of individual certification, as did some individual practitioners (some still do). The Criminalistics Section of the American Academy of Forensic Sciences voted in the mid-1990s to join with the American Board of Criminalistics (ABC) in supporting certification of individuals, and the American Society of Crime Laboratory Directors (ASCLD) finally decided to support the ABC program in the late 1990s. Some conflicts still exist, such as the debate over who should pay for proficiency testing—the individual who needs it to maintain his or her certification, or the laboratory, which needs proficiency testing to maintain its accreditation. Despite these peripheral disagreements, the forensic quality triangle is the paradigm of the current era.

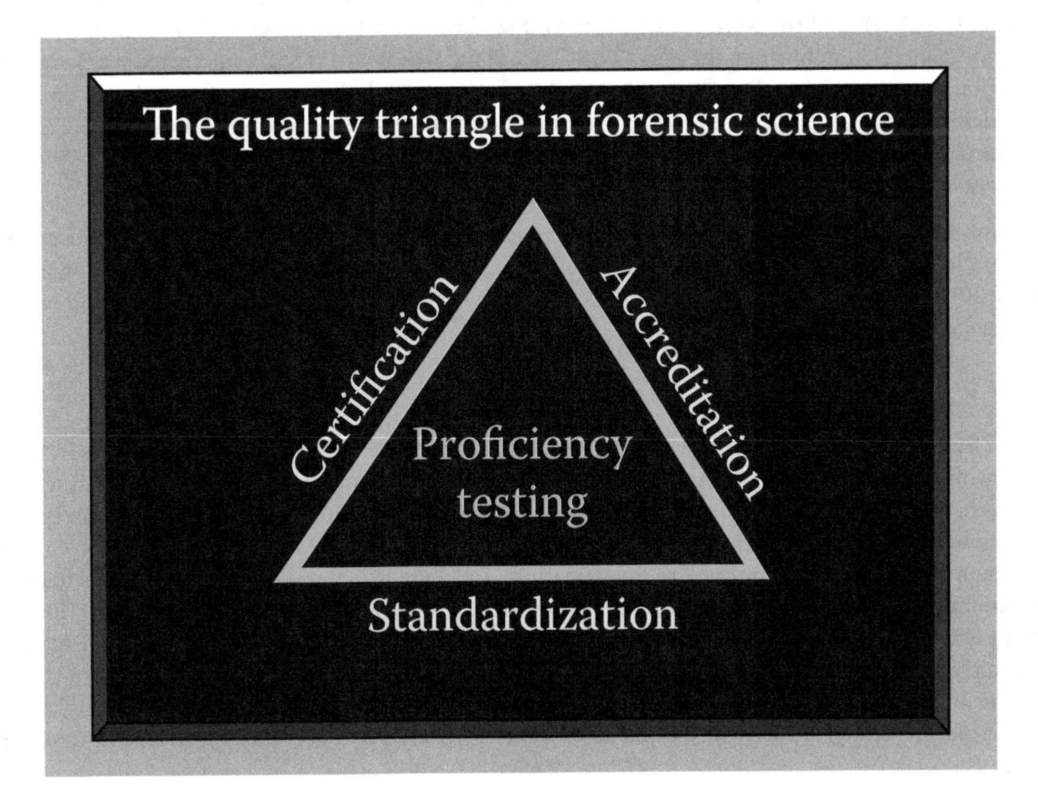

Figure 10.2 The quality triangle in forensic science.

[3] NFPA 1730, *Standard on Organization and Deployment of Fire Prevention Inspection and Code Enforcement, Plan Review, Investigation, and Public Education Operations*, contains a bare-bones description of requirements for FIUs. The proposed standard will be significantly more detailed.

On August 9 and 10, 2004, the American Bar Association House of Delegates, representing the ultimate consumers of the work product of all forensic scientists (including fire investigators), passed the following resolution:

RESOLVED, That the American Bar Association urges federal, state, local and territorial governments to reduce the risk of convicting the innocent, while increasing the likelihood of convicting the guilty, by adopting the following principles:

1. Crime laboratories and medical examiner offices should be accredited, examiners should be certified, and procedures should be standardized and published to ensure the validity, reliability, and timely analysis of forensic evidence.
2. Crime laboratories and medical examiner offices should be adequately funded.
3. The appointment of defense experts for indigent defendants should be required whenever reasonably necessary to the defense.
4. Training in forensic science for attorneys should be made available at minimal cost to ensure adequate representation for both the public and defendants.
5. Counsel should have competence in the relevant area or consult with those who do where forensic evidence is essential in a case.

In February 2009, the National Academy of Sciences, through its Committee on Identifying the Needs of the Forensic Science Community, echoed the ABA resolution when it called for

1. Mandatory accreditation of medical examiners offices and crime laboratories
2. Mandatory certification of all forensic scientists
3. Mandatory implementation of standard methods of analysis and reporting

Consumers are now demanding what the leaders of the forensic science and fire investigation professions have been advocating for more than a decade. These reasonable demands should be accommodated with all deliberate speed.[4]

The National Academy of Sciences, in its 2009 report, *Strengthening Forensic Science in the United States: A Path Forward*, advocates mandatory use of standard methods, mandatory certification for all forensic scientists, and mandatory accreditation for all forensic science laboratories. In response to these recommendations, the National Institute of Justice (NIJ)[5] and the National Institute for Standards and Technology (NIST) signed a memorandum of understanding in 2013 establishing the Forensic Science Standards Board (FSSB), supported by 25 subcommittees from the various forensic science disciplines. These subcommittees, including one for fire and explosion scenes and one for fire debris analysis, comprise the Organization of Scientific Area Committees (OSAC). The OSAC subcommittees replace the various scientific and technical working groups set up in the 1990s. Figure 10.3 shows the structure of this organization [5].

The subcommittees' mandate is to recognize and help develop standards for all of the forensic science disciplines. This is normally accomplished by working with professional organizations and standards development organizations. In late 2017, the Subcommittee on Fire and Explosion Investigations placed both NFPA 921 and NFPA 1033 on the OSAC Registry of Approved Standards.

10.3.1 One state's solution

The State of Texas Fire Marshal's Office (SFMO) acquired a certain infamy in 2010 when the Texas Forensic Science Commission examined two famous arson/homicide cases, known as Willis and Willingham. Willis served 19 years on death row for arson and murder prior to being exonerated in 2004. Willingham spent 12 years on death row before being executed that same year. Following the Forensic Science Commission's inquiry, they issued 17 recommendations to improve fire investigation. The state fire marshal at that time was dismissed, and his replacement endeavored to implement all 17 recommendations, which are shown in Sidebar 10.2. The most far reaching of those recommendations was the establishment of a science advisory workgroup (SAW). The SAW meets quarterly to review all arson convictions involving the SFMO. In addition, the members of the SAW provide training to meeting attendees (the meetings are open and free of charge). Since 2013, working with the Innocence Project of Texas, the SAW has found at least six individuals incarcerated based at least in part on faulty fire science.

[4] Some of these recommendations were implemented by NIJ, and some came to fruition through OSAC. The National Commission on Forensic Science, established in 2013, functioned through early 2017, when its charter was not renewed.

[5] NIJ is the research arm of the US Department of Justice.

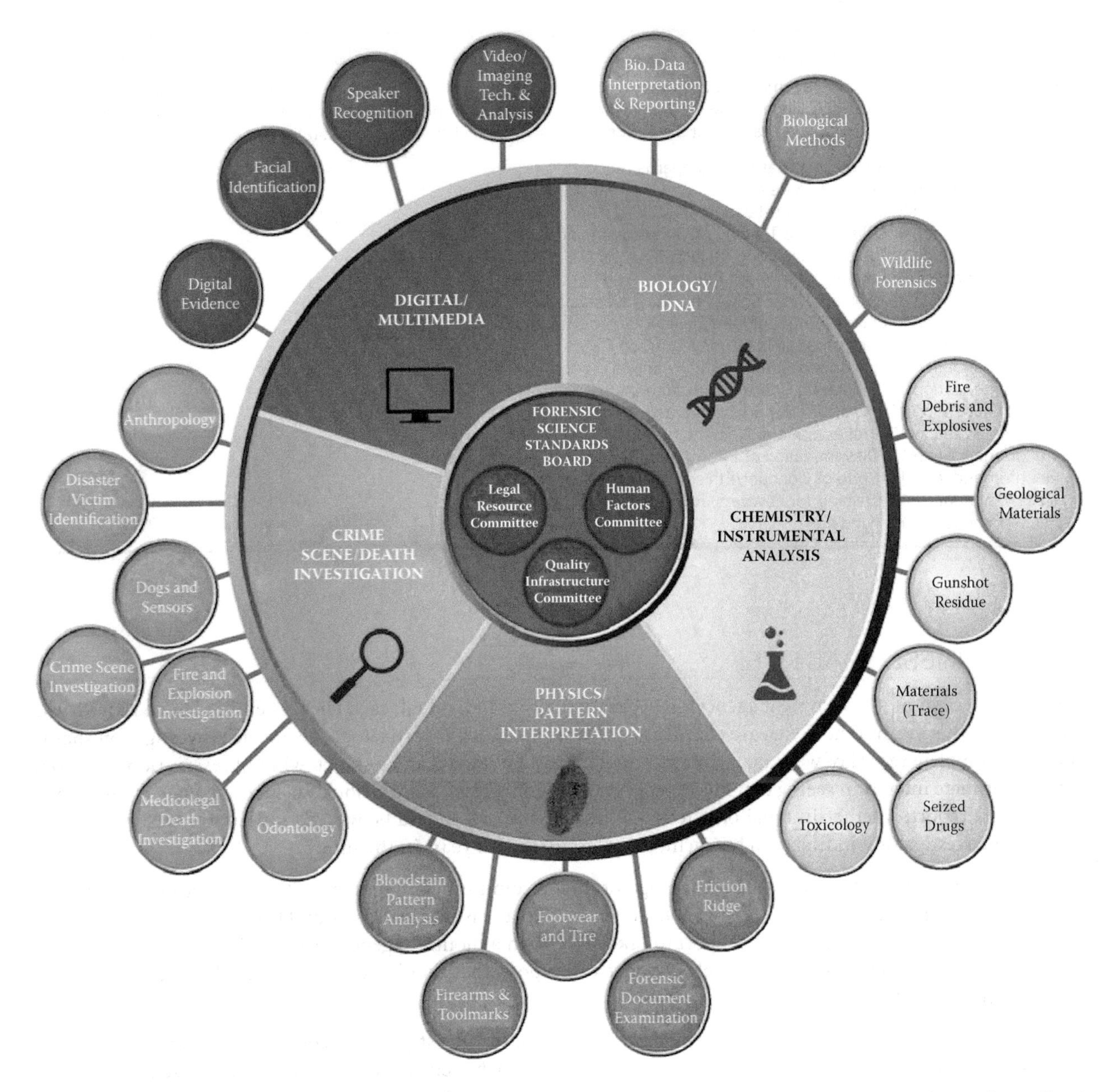

Figure 10.3 OSAC organizational structure. Fire and Explosion Investigation is a subcommittee of the Crime Scene/Death Investigation SAC, and Fire Debris and Explosives is a subcommittee of the Chemistry/Instrumental Analysis SAC.

The SAW does not comment on guilt or innocence, only on the quality of the forensic science used in obtaining the conviction. The IAAI has endorsed the Texas approach and has offered to assist other jurisdictions in setting up similar programs [6].

Texas is leading the forensic sciences in another way. Beginning in 2019, all forensic experts, including fire investigators, will be required to be licensed in order to appear as an expert witness. Details of the requirements for licensing were being worked out as this text went to press. The Forensic Science Commission has been tasked with setting up the program.

SIDEBAR 10.2 Recommendations of the Texas Forensic Science Commission regarding fire investigations

Following are the titles of the 17 recommendations that the Texas Forensic Science Commission made after its three-year investigation into the Willis and Willingham cases. The full text of these recommendations is included in the Commission's 893-page report dated April 15, 2011, and available at the Commission's website: http://www.fsc.state.tx.us/documents/FINAL.pdf

Recommendation 1: Adoption of National Standards
Recommendation 2: Retroactive Review
Recommendation 3: Enhanced Certification
Recommendation 4: Collaborative Training on Incendiary Indicators
Recommendation 5: Tools for Analyzing Ignition Sources
Recommendation 6: Periodic Curriculum Review
Recommendation 7: Involvement of SFMO in Local Investigations
Recommendation 8: Establishment of Peer Review Group/Multidisciplinary Team
Recommendation 9: Standards for Testimony in Arson Cases
Recommendation 10: Enhanced Admissibility Hearings in Arson Cases
Recommendation 11: Evaluating Courtroom Testimony
Recommendation 12: Minimum Report Standards
Recommendation 13: Preservation of Documentation
Recommendation 14: Dissemination of Information Regarding Scientific Advancements
Recommendation 15: Code of Conduct/Ethics
Recommendation 16: Training for Lawyers/Judges
Recommendation 17: Funding.

10.4 BUSINESS PRACTICES

Whether the client is a government agency, a corporation, or an individual, each client is entitled to good faith and fair dealing. This means that clients are entitled to *our best efforts.* An investigator's file for each investigation should contain a complete record of work done and information gathered that can be shown without reservation to anyone with a legitimate interest in seeing that file, especially the client. These records should include adequate photographs, notes, and a report, whether internal or external. These documents should be written so that, on reasonably short notice, the investigator's findings can be intelligently discussed two years in the future.

During the course of an investigation, the investigator should communicate regularly with the client so that the client knows what is happening with the case. Clients are entitled to a reasonable turnaround time. Unless information is still incomplete, no reason exists for a report to take more than a month to generate after all the evidence has been collected. If some factor prevents the timely issuance of a report, that should be communicated to the client so that he or she understands the reason for the delay.

In the private sector, the clients also have a right to know what the investigation is costing them and should receive regular bills. Keeping up with invoicing is an administrative pain, but clients do not like to be surprised at the end of a case with a bill for two years' efforts. When a client asks for an estimate, it is always best to estimate on the high side so that the client will be pleasantly surprised rather than disappointed when he or she gets the bill.

Some investigators and companies have decided that they will seek a competitive edge by competing on price. While this is certainly the American way, an investigator's clientele will be different if he is selected based on the hourly rate. Some companies quote a flat rate for fire scene inspections. This can provide a competitive advantage by allowing the clients to know ahead of time exactly how much money they are going to spend, but flat rate investigations are easier to discredit than those billed on a "time and materials" basis. If an investigator or company charges $795 per investigation, the motivation is to do as many investigations as possible in as short a period of time as possible. Except for laboratory analyses, where the time and materials commitment is usually well known in advance, this author has never seen any utility in flat rate billing.

Some investigators offer a discounted hourly rate for travel. Again, if this provides more work, and more work is needed, nothing is ethically wrong with the practice. It makes little sense from a business standpoint, however, because while one is doing low-value work for one client, it is not possible to do high-value work for another. (Reading while driving is not advisable.) The investigators who get into trouble with their rates are the ones who charge a premium (sometimes double or triple) rate for depositions and court appearances. This rate structure gives the appearance that it is in the investigator's interest to find something to litigate. It is not uncommon to be accused of finding arson because that results in more work, but when the work pays double or triple, who could blame adverse counsel for asking? Complicated rate structures also result in clients having difficulty in understanding invoices. In this investigator's opinion, one should charge for one's time at the same rate for all services.

The extension of credit is an issue that every private-sector company must address. Certainly, if one's clients are large insurance companies, creditworthiness is not an issue as long as one is willing to abide by the company's billing practices and wait the requisite 45, 60, or even 90 days that some companies take to pay their bills. On the other hand, extending credit to a plaintiff's attorney, who will not make a dime on the case unless he prevails, is probably not wise. Many professional and legal organizations have taken the position that it is unethical for an expert witness to be retained on a contingent basis [7]. Further, the common law rule in most jurisdictions is that it is improper to pay an expert witness a contingency fee [8]. Unless the plaintiff's attorney pays in advance, the argument can be made that the investigator's compensation depends on the outcome of the case. Unless an investigator wishes to risk working *pro bono* involuntarily, plaintiffs' attorneys (other than those handling subrogation cases for large insurance companies) and criminal defense attorneys should be required to purchase services in advance via the use of a retainer. These clients expect to be asked for a retainer and are generally not offended when asked to provide one. At times, an investigator's "client" is actually an adverse attorney wishing to take a discovery deposition. Because many cases settle immediately after expert depositions, it is entirely appropriate to request advance payment for these depositions. Extending credit in this situation usually means that the attorney will pass the bill to his or her client, who is the adversary of the client who hired the fire investigator. As such, the investigator's invoice is not likely to reach the top of the payables stack anytime soon. Unless an investigator wants to hire someone to do collections, adverse attorneys should be placed in the same category as plaintiffs' attorneys and criminal defense attorneys; that is, payment in advance should be the rule.

Expenses associated with an investigation and chargeable to a client should be documented with receipts, and those receipts should be held until the case is over. Clients have a right to expect to be fairly charged for business expenses, and many insurance carriers and law firms require the submission of receipts to document those expenses, along with an invoice.

In any business relationship it is important to know what each party expects of the other. A written retainer agreement effectively communicates these expectations. Such written agreements are not usually necessary when doing repeat business with large corporations or long-time clients, but relationships with much smaller clients are well served by contracts that both parties understand. Posting a standard retainer agreement on a website is one way to ensure that every client knows the terms of engagement.

The basic rule in business practices is to treat other people the way you would like to be treated in a business relationship.

10.4.1 *Pro bono* work

Pro bono work is work done without monetary compensation. Fire investigation professionals are sometimes presented with the opportunity to provide their services at no charge. Despite the fact that an investigator's family requires food, shelter, clothing, and tuition, it is a good practice to take the occasional *pro bono* case. Most law firms have policies that commit their associates and partners to 40 to 60 hours per year of *pro bono* work, and such a policy would be useful for fire investigators and for other forensic scientists to adopt as well. This author has frequently had to decline work for someone who needed an expert, simply because they were unable to afford the cost. About once a year, however, a case or a cause comes along where I have felt obligated to either work for free or at a substantially reduced rate. One such case involved a woman falsely accused of killing her husband by setting him on fire. I began the case with a reduced retainer and by the time the trial came along, the defendant had spent all her money on defense costs. It would not have been right to refuse to testify because she could not afford my services.[6] *Pro bono* work can also involve assistance to public agencies such as fire departments or child welfare agencies, or it might involve assuming a position of leadership in a professional organization.

[6] The report accompanying the American Bar Association's recommendations on forensic science, discussed in Sidebar 10.1, goes into great detail on the need for better funding of experts on both sides of criminal cases. The report addresses several court decisions wherein the lack of expert assistance was found to be equivalent to the denial of a citizen's Sixth Amendment right to *effective* assistance of counsel.

While the motives for this investigator's *pro bono* work started out as entirely altruistic, experience has shown that *pro bono* work actually pays. Somehow, in some unpredictable way, work done for the public good almost always results in business coming my way. It may happen in a month or in ten years, but it happens.

10.5 SERVING AS AN EXPERT WITNESS

Excavating, reconstructing, and accurately determining the cause of a fire, particularly a difficult one, is a most rewarding experience. Many times, however, an investigator's determination of the origin, cause, and responsibility for a fire results in legal action being brought against an individual or a corporation, and those individuals or corporations may take a dim view of the investigator's findings. The proffered hypothesis must be thoroughly tested before going public because parties affected by an investigator's determination are going to insist on the opportunity to give that hypothesis the "careful and serious challenge" described in NFPA 921. The affected parties will almost certainly retain counsel, and counsel will, in turn, retain an expert to review the first investigator's findings. At this point the investigator must open up an entirely different toolbox—that of the expert witness.

Rule 702 of the Federal Rules of Evidence (and many similar state rules) reads as follows:

> If scientific, technical, or other specialized knowledge will assist the trier of fact to understand the evidence or to determine a fact in issue, a witness qualified as an expert by knowledge, skill, experience, training, or education may testify thereto in the form of an opinion or otherwise, provided that (1) the testimony is sufficiently based upon reliable facts or data, (2) the testimony is the product of reliable principles and methods, and (3) the witness has applied the principles and methods reliably to the facts of the case.[7]

Although Rule 702 specifically refers to testimony, testifying at trial is actually the *last* function performed by a fire investigator serving as an expert. Trials are becoming battles of experts, and the side that presents its case most convincingly wins. If an expert does the job right, courtroom testimony may not be necessary because the parties often settle a case after learning the details of each other's positions. In arson cases, however, settlements are the exception rather than the rule. Accused persons who are innocent are usually unwilling to state, even in a civil setting, that they did something that they did not do. On the other side, insurance carriers would prefer to spend their money to make an insured, who they believe has committed fraud, spend money rather than use the same money on a settlement. Most other fire-related cases, however, settle out of court.

The first job of an investigator serving as an expert is to help the client understand the facts in evidence. It is then necessary to bring that understanding to the client's lawyer as well. (When the client is a lawyer, it may be necessary to explain the case to the lawyer's client.) It is also the investigator's job to help counsel understand the position of the adverse party and to explain clearly why the expert from the "other" side has missed the mark in his or her determination. When communicating at this stage of litigation, the expert should carefully evaluate the positions of each side. More important, if the client's case has significant weaknesses, it is the expert's job to point those out. Manufacturers do not like to hear that their product malfunctioned and caused a fire. Service providers do not like to hear that their employee was negligent and caused a fire. Insurance companies do not like to hear that the fire was their insured's fault and there is no hope of recovery through subrogation, and criminal defense attorneys do not necessarily like to hear that their client is probably guilty as charged.[8] It is pointless, however, to waste time and money defending a case when the liability is clear or prosecuting a case when it is not. The fire investigator is often the only person involved in the case who is capable of recognizing an incorrect hypothesis.

---

[7] The three numbered provisions added to Rule 702 in 2000 were intended to codify the Supreme Court's interpretation of Rule 702 in the *Daubert* decision and its progeny.

[8] Actually, most criminal defense attorneys suspect that most of their clients are guilty most of the time. Their job is to get the client a fair trial (or a fair plea agreement) regardless. Having an innocent client has been described as a defense attorney's worst nightmare because her or his obligation goes beyond getting the client a fair hearing. If the client is innocent, it becomes the defense attorney's obligation to obtain a dismissal or an acquittal.

It is also important to recognize that the attorney or client may have information developed after the investigator's initial investigation was completed. The investigator should insist on being kept up-to-date with developments in a case, particularly when new data are discovered. The same rule applies to investigators retained later in a case. Tell counsel to show you everything, and then decide what you need to review. It is unpleasant to be ambushed or embarrassed by being presented with important evidence for the first time by adverse counsel.

10.5.1 Advocacy

Some would argue that a fire investigator or a forensic scientist has no business "advocating" in the context of civil or criminal litigation. Rules of conduct for experts have been set down by various professional organizations to which expert witnesses might belong. These codes generally require members to express opinions based on the facts, and some explicitly warn the experts not to be an advocate for one particular side.

"As fire, arson, and explosion and explosion investigators, we shall bear in mind that we are first and foremost truth-seekers, not case makers" comes from the International Association of Arson Investigators' *Code of Ethics*. The much more specific California Association of Criminalists' *Code* states:

> It is not the object of the criminalist's appearance in court to present only that evidence which supports the view of the side which employs him. He has a moral obligation to see to it that the court understands the evidence as it exists and to present it in an impartial manner.

While experts clearly should not be viewed as advocates for one side or the other, they are constantly asked to be an advocate for "the truth." The use of court-appointed experts, who depend on neither side for encouragement or sustenance, is entirely too rare, but that is the subject of another discussion.

The data gathered at a fire scene or in a laboratory do not speak for themselves. Interpretations of the data only come about through the power of sound argument and good science. Experts who are not advocates for their own carefully considered and reasonably held opinions are not likely to make effective witnesses for either side.

While the expert witness does not advocate for one side or the other, he or she must advocate for the evidence and for the truth. If an investigator determines the cause of a fire and sincerely believes that he is correct, and another investigator finds a different cause, at least one of them is wrong. The justice system's function is to provide a forum where both sides have a fair opportunity to present their views of a case so that a judge or jury can try to sort out which investigator is right and which investigator is wrong. The expert's job is to help counsel present the evidence convincingly but without being buffaloed by counsel into rendering "helpful" opinions that cannot be justified. Such opinions are likely to be discredited on cross-examination in any case, and it is the investigator's job to resist counsel's efforts to push him or her farther out on a limb than is comfortable.

Assisting counsel includes crafting one's own direct examination and helping counsel design the cross-examination of the opposing expert. A word of caution is in order here. Investigators should always assume that anything they write down might be discovered by the adverse party. If a cross-examination strategy is discovered before the cross-examination actually takes place, the effectiveness of the cross-examination can suffer. Consult with counsel to make sure that he or she actually wants a written communication regarding cross-examination.

Written communications regarding direct examination are not quite as problematic. This author once had the experience of having sponsoring counsel lose the outline for direct examination, and counsel was required to request a copy from the author, who was sitting on the witness stand at the time. Adverse counsel, on cross-examination, attempted to make some hay out of this exchange, but was politely informed that, of course, I had written the outline for my own direct examination because I am the expert and counsel requested my assistance. It is this author's opinion that if an expert witness takes the stand without having first met with counsel and gone over direct testimony in great detail, both the lawyer and the expert have failed to serve their client.

10.5.2 Discovery

Once the investigator has made certain that counsel understands his or her opinions, it is usually necessary to designate the investigator as an expert, and in that designation, counsel must reveal, at least in outline form, the opinions to be elicited from the expert at the time of trial. This expert disclosure is an important document, and the investigator should insist on the opportunity to review the disclosure *before* it is submitted.

It may also be necessary at this time for the investigator to craft a new report that comports with the format required in a particular jurisdiction. This can be as simple as distilling the opinions to two or three major areas of testimony and attaching a copy of the investigator's original report as an appendix. In other cases, particularly in those jurisdictions where expert depositions are not allowed, the report should be much more detailed to avoid having the testimony excluded for failure to properly disclose opinions before trial. Once the disclosures and reports have been made, the next step is likely that of discovery depositions. Usually, the plaintiffs' or state's experts are deposed first, followed by the defense experts. In most jurisdictions, depositions are not taken in criminal cases[9]; discovery is accomplished by other means, such as preliminary hearings and mandatory disclosures.

In civil cases, the expert discovery depositions are frequently the last step in the process before settlement negotiations. The expert is cautioned, however, that settlement negotiations sometimes fall through, and the deposition is a permanent record that may be used at trial to impeach the witness. Prior to a discovery deposition, the investigator should meet with sponsoring counsel to learn the *purpose* of the deposition. Ostensibly, the purpose of the deposition is to allow the adverse party to learn the opinions of the expert but, in most cases, those opinions are already well known. The adverse party's purpose is to make a record that can be used to attack the credibility of the expert's opinions. As such, the expert needs to think about how the exchange of questions and answers will appear on the written page. Assume that if a statement can be misconstrued, it will be. Counsel for both sides spent at least three years in law school learning to use words to their advantage. Very few expert witnesses possess the requisite skills to engage professional wordsmiths in a battle of words. The important thing to keep in mind, therefore, is to tell the truth, and to tell it so that even someone who wants to misconstrue the words cannot do it. This is best accomplished using simple declarative sentences. Assume that the transcript will be read by someone who has no knowledge of fire investigation. Avoid the use of pretentious words, and if it is necessary to use special terminology, make sure that your definition of that terminology appears in the transcript.

Counsel will frequently elicit the opinions of the expert, and then, in the guise of making sure that he or she understands, will begin the next question with the phrase, "In other words," and then proceed to paraphrase the investigator's opinion in ways that may ever so subtly spin the opinion in a way that might be misleading. My standard response to such a question is, "If I had wanted to use other words, I would have used them."

Testimony in a discovery deposition is usually very different from testimony in a trial. In a trial, lawyers are trained not to ask questions to which they do not know the answers. In a discovery deposition, their job is to ask questions to which they do not know the answers. They will know the answers at the time of trial, and they may or may not ask them then, depending on the answers received in deposition. If questions in a deposition are based on a false premise, or worded so that answering yes or no would be misleading, the expert should stop and either ask counsel to rephrase the question or offer a sufficient explanation so that someone reading the transcript will not be misled. It is also possible to offer explanations to misleading questions from the witness stand, but in a trial, such explanations should be kept to a minimum, and objections to questions containing false premises can only be made by counsel.

The first part of a discovery deposition is usually a detailed examination of the investigator's background, education, training, and experience. This may include irrelevant questions of a personal nature, and the witness should feel free to not answer them. Unless a judge "certifies" a particular question, counsel has no right to know your social security number, home address, home telephone number, or any other personal information that you would not be comfortable having printed in the newspaper. Certainly, all of an investigator's prior experience in investigating fires is fair game, as is all of an investigator's prior testimony or any written articles. A standard question in the background part of almost all expert depositions now is one that asks whether any court has ever excluded any part of the witness' testimony.

Since NFPA 1033 was changed in 2009, many discovery depositions now include a short quiz or even a long quiz to examine an investigator's knowledge of the subjects of fire chemistry, fire dynamics, and the other subjects listed in § 1.3.7. The lawyer is not asking such questions for his or her own edification; the goal is to show that the deponent is unqualified. Thus, "I can look that answer up for you" is not an adequate answer [9]. An investigator who does not know the definitions and the units of measurement for energy, power, and flux is likely to be excluded as unqualified or, even more likely, the case will be dismissed or settled.

[9] Although depositions may be very helpful to both sides in criminal cases, only a few states—Florida, Indiana, Iowa, Missouri, New Hampshire, North Dakota, and Vermont—allow them as of 2018. Some states allow depositions only if a witness will not be available for the trial.

Eventually, the questions will focus on the subject at hand, that is, the investigator's opinions about the origin, cause, and responsibility of the fire in question. The advice typically given to novice witnesses is to answer the question as briefly as possible and to volunteer nothing. This is certainly good advice for the courtroom but may not be the best advice for a discovery deposition. The investigator should meet with sponsoring counsel and learn what goals, if any, counsel may have for this expert witness deposition. Maybe the investigator has possession of information that conclusively proves the case, and maybe adverse counsel has not prepared sufficiently to ask the question that will elicit that information. It may be in the best interests of all concerned for adverse counsel to learn of this information prior to settlement negotiations. The salient question for sponsoring counsel to answer is whether he or she believes that this case will ultimately go to trial or settle. In the end, however, regardless of the purpose of the deposition, it is imperative that all answers be absolutely true.

The expert witness should be alert for questions that indicate that adverse counsel is considering a *Daubert* challenge. Although such challenges are rarely successful, they do require some foundation. Typical questions setting up a *Daubert* challenge include the following:

- Has your interpretation of this data been tested, or can it be tested?
- Have other investigators reached similar conclusions to yours based on similar data and published that data in the literature of your profession?
- With respect to your methodology, is there a known or potential rate of error?
- Are there standards controlling the operation of this methodology?
- Do you believe that the methodology you employed is generally accepted within the relevant scientific or investigative community in which you practice?
- Does the opinion that you propose to testify about grow naturally and directly out of research that you conducted independent of this litigation, or have you developed these opinions expressly for the purposes of testifying?
- Is the field of expertise that you claim one that is known to reach reliable results?

When presented with these questions, the investigator who has followed the general procedures of NFPA 921 will have few concerns. NFPA 921 represents a published standard of care that can be said to be "generally accepted" in the fire investigation community.[10]

Adverse counsel will use the deposition as a trial run of the investigator's cross-examination and may be argumentative and contentious. There is no need to engage in arguments with counsel, nor is there a need to dominate counsel during the deposition. As with using words, cross-examining witnesses is a skill that lawyers have sharpened over the years, and they may pride themselves on their ability to weaken a witness's resolve with a withering glare. The fact is that the transcript of the deposition does not reflect whether the witness is making eye contact with adverse counsel. If counsel wants to engage in a staring contest, it is perfectly permissible to stare at the table. (It is also extremely frustrating for adverse counsel.)

In a deposition, any time the witness wants to take a break, it is entirely appropriate to ask for a break.[11] If adverse counsel gets particularly exercised, as some are known to do, it is perfectly acceptable to ask that the deposition be recessed for a few minutes to allow counsel to regain his or her composure. Anger and intimidation have no place in the discovery process but sometimes creep in. These are times when it is especially appropriate to take a 5- or 10-minute break.

Toward the end of most expert discovery depositions, the following overly broad question is often asked: "Do you have any opinions that we have not yet discussed?" If the investigator really does have important opinions that have not been discussed, they should be brought out at this time. Nuances in previously discussed opinions that have not been deeply explored can probably be left alone.

Another question asked toward the end of many depositions is: "Do you plan to do any further work on this case prior to trial?" This is a place to leave one's options open. Certainly, there will be trial preparations, but it may be necessary

[10] "General acceptance" can only be said to have occurred in 2000, when both the US Department of Justice (DOJ) and the IAAI endorsed the document for the first time.

[11] In a trial, there is a judge and there are 12 innocent civilians whose valuable time is being spent, so it is not possible to take a break simply because it seems convenient. This is not the case in depositions.

to run additional tests if an adverse expert puts forward a novel hypothesis. It is safe enough to answer, "I have not been requested to do any further testing, but may do so if asked."

Once adverse counsel has completed his or her questions, sponsoring counsel has an opportunity to ask questions. Unless counsel is concerned that there has been a serious miscommunication during the deposition, counsel should not take advantage of this opportunity. Such questions alert adverse counsel to sponsoring counsel's perception of weaknesses in the case. These can be discussed privately after the deposition.

The witness will be asked whether he or she chooses to read and sign the deposition or waive signature. It is never a good idea to waive signature at this point, even if one has been deposed in front of the same court reporter a dozen times, and knows that the reporter's skills are impeccable. It is important to read the deposition transcript to look for those passages that could be misconstrued or where communication was less than clear. It may not be possible to change the record, but by reading the transcript, the witness becomes aware of potential pitfalls.

In addition to providing a discovery deposition, the investigator may also be asked by counsel to review the depositions of other witnesses, including other experts in the case. This can be exceedingly tedious work, but it is indoor work that does not involve heavy lifting and it is important. The investigator may have learned about the perceptions of eyewitnesses early in the case, but when those witnesses are asked to describe those same observations two years later, and under oath, the answers may be different. The investigator should be prepared to seek out those differences and understand what they mean, if anything. Similarly, an opposing expert's deposition testimony is likely to shed new light on information previously supplied in the opposing expert's report. Finally, the discovery process may reveal new aspects of the fire scenario that should be explored or even tested.

10.5.3 Courtroom testimony

The ability to make an accurate, intelligible presentation in court is one of the investigator's most important skills. While it is critical that an investigator is able to correctly determine the origin and cause of a fire, or some other aspect of the fire's consequences, those skills are wasted if the investigator's conclusion cannot be understood and believed by a jury. When competing theories exist as to the origin and cause of a fire, the jury's decision might rest entirely on the differences in the quality of presentation between two experts. Alternatively, jurors might ignore the expert testimony altogether and decide the case on other factors. Over a hundred years ago, Judge Learned Hand summed up the problem with competing experts as follows:

> How can the jury judge between two statements each founded upon an expertise confessedly foreign in kind to their own? It is just because the [jurors] are incompetent for such a task that the expert is necessary at all.

According to Hand, testimony from competing experts is not helpful to the jury:

> One thing is certain, they will do no better with the so-called testimony of experts than without, except where it is unanimous. If the jury must decide between such, they are as badly off as if they had none to help [10].

Judge Hand's observations notwithstanding, expert witnesses continue to be found on both sides of many court cases, both civil and criminal. Thirty years ago, it was not unusual for the prosecution's or insurance company's expert to be the only expert called on a particular issue in a case. These days, the credibility of experts has been so shaken by revelations of wrongful convictions, bad science, and outright perjury that they are no longer accorded the respect that they were given in the past. It is now much more common for adverse parties to obtain the services of experts to challenge the opinions of other experts.

Keeping in mind the skepticism that will be directed toward them, investigators should follow some commonsense rules in preparing and presenting their findings to the jury. First, it is important to be prompt. A box full of jurors and alternates, a judge, lawyers and parties for both sides, and the court staff await your testimony. Being late for a court appearance is inexcusable. Being on time usually means waiting for hours or days before being called, but that is an occupational hazard.

Next, show the court the respect that it deserves by dressing appropriately. Even if the jurors are all dressed in blue jeans and T-shirts, an expert witness should dress properly to go to court. Take your lead from the attorneys and dress at least as well as they do. Because it is your ideas and opinions and not your clothing that you want to be the focus of the jury's attention, choose your attire so that it blends in with that of the other professionals in the courtroom.

In planning for your testimony, find out if any issues exist that the court has ruled may not be discussed. For example, the discussion of liability insurance is usually forbidden in a negligence or product liability case, while in a civil arson case, the lack of a criminal prosecution is usually out of bounds. The investigator might be relying on hearsay evidence for part of his or her opinion, and while nothing is inherently wrong in considering such evidence, it usually cannot be presented to the jury. The court may also have granted other motions *in limine*. Awareness of these court orders is essential because if one is violated, even unintentionally, the result may be a mistrial.

Above all, keep in mind the stakes involved for the parties to the litigation. Despite frequent comparisons to the arena, what is about to happen in the courtroom is not a sporting event.[12]

10.5.3.1 Direct examination

You will be introduced to the jury by way of a presentation of your qualifications. While the witness stand is no place for arrogance, do not be overly modest. Let the jury know the full extent of your education, training, and experience. Sometimes, purportedly to save time, but actually to keep the jury from learning the true extent of a witness's accomplishments, adverse counsel will stipulate that the witness is an expert. This situation should be discussed beforehand with sponsoring counsel and a decision made whether to go along with the stipulation or to press on with the description of the witness's qualifications.

Your demeanor in the courtroom will be as important as the words that come out of your mouth. Jurors are skeptical of experts and all have heard stories about experts who may not have been entirely truthful on the witness stand. They will want to look the expert in the eye, and the expert should accommodate that wish. Your role as an expert witness is to help the jury understand the evidence. You are the teacher. It is important to establish eye contact with each individual juror so that all of them can see that you are sincere in your beliefs and opinions.

The role of the expert witness in a courtroom is to *translate* or interpret data that would otherwise not be meaningful to the jury. The data might be a gas chromatogram that is absolutely dispositive on the issue of whether a sample contains ignitable liquid residue, but unless someone familiar with the "language" of gas chromatography interprets it, the data will remain unintelligible to the average juror. The same holds true for other types of evidence commonly encountered in fire investigations, such as electrical artifacts and fire patterns. It is the expert's job to interpret those artifacts so that the jurors understand their meaning.

While this interpretation should be accurate and informative, the expert witness should avoid the use of jargon while at the same time avoiding talking down to the jury or oversimplifying a concept. If it is necessary to use jargon or terms that ordinary citizens are unlikely to understand, be sure to explain the concept. Use simple declarative sentences and avoid the use of pretentious words.[13] The idea that juries are not capable of understanding complex scenarios is simply false. The combined education and experience of the members of the jury is far broader and deeper than many people assume.

It is important to explain concepts of fire behavior to juries in a fashion to which they can relate. For example, it is frequently important for the juries to understand that what they think they know about the behavior of fire may not be true. Almost everyone thinks that they know how a fire behaves, based on their observations of campfires, brush fires, or trash fires, but very few people have actually experienced a compartment fire close up. "Heat rises" is a simplistic and untrue explanation for fully involved fires. Jurors need to understand the basic concepts of fire pattern production, and the best way to convey these concepts is with photographs and diagrams. Most of the observations that a fire investigator makes are visual observations. A fire investigator's testimony, therefore, must also be highly visual if his or her observations are to be adequately conveyed.

One of the best methods currently available for presenting visual and textual information to a jury is with the use of a PowerPoint presentation. This powerful tool is used not only for the presentation of testimony but also for its organization. The handout tool can be used by sponsoring counsel to walk the witness through the direct examination.

---

[12] Clarence Darrow may have expressed this sentiment best. "A courtroom is not a place where truth and innocence inevitably triumph. It is only an arena where contending lawyers fight not for justice, but to win."

[13] You do not "reside" at home—you "live" there. You did not "exit your vehicle"—you "got out of your car." Speak to the jury in the same manner as you speak to your colleagues. The use of pretentious language not only distracts from your message but also leaves the impression that you are pretending to be someone you are not.

When using PowerPoint, it is important to verify that the technology is going to work. Make sure that you have sufficient time prior to your testimony to check the operation of the laptop computer and the liquid-crystal display (LCD) projector. And always have a backup. The backup consists of 8.5″ × 11″ color prints of each PowerPoint slide. Three sets will be necessary: one for adverse counsel, one for sponsoring counsel, and one for the court that will become part of the permanent record of the trial. All of this preparation is expensive, but it is not as expensive as losing the case because a bulb blew out or a computer crashed.

The PowerPoint presentation should start with a black slide so that the system can be left on during the preliminary parts of the witness's presentation. The next slide should be a title slide, with the case name and the words "Exhibits to the Testimony of John Investigator (or Jane Chemist)." If there are principles of fire behavior that require explanation, these should be put in a "primer" section of the presentation so that the jury can understand the fire patterns that are going to be shown later. It is often necessary to put a photograph in the program three times. First, show the unmarked photograph. Next, show the same photograph with circles and arrows and a minimum of explanatory text. Finally, show the photograph again so that the jurors can now look at it and see those salient features without the circles and arrows.

When interpretations are made, particularly interpretations that are the subject of controversy, it is important for the jury to know that the witness is not simply expressing an opinion as to what he thinks a particular piece of evidence means. If there is a section of NFPA 921 or any other learned text that addresses that interpretation, a slide with a quotation is invaluable for showing both the judge and the jury the reliability of the methodology that is being applied to the case.

To the extent possible, make the presentation *interesting.*[14] Make it something that the jury will remember. As an investigator, you would not be doing this work if you did not find it fascinating. Try to communicate some of that fascination and enthusiasm in your presentation.

One drawback to using PowerPoint is that the sequence of slides is pretty much fixed. While it is possible to jump around if necessary, the presentation goes more smoothly if counsel has the discipline to follow the plan laid out in the handout. An alternative to using the PowerPoint program is to use a visual presenter to show the photos or text slides, still using an LCD projector.[15] This provides for more flexibility, at a slight cost in image quality and a major cost in the complexity of the audiovisual setup requirements. Some law firms have sophisticated software that allows more flexibility than PowerPoint.

Direct examination should include not only an expert's observations and opinions but also a gentle "cross-examination" by sponsoring counsel that explores opinions that will be (or have been) testified to by experts called by the adverse party. An expert is not allowed to comment on the credibility of another expert's opinion (that being the "province of the jury"), but it is possible to craft hypothetical questions that let the jurors know that consideration has been given to alternate hypotheses or alternate interpretations of the data.

While direct examination is the place to put forward the strong points of a case, it is a mistake to let the weaknesses come out only on cross-examination. The witness will be correctly perceived as being biased if the jurors only learn of weaknesses in an opinion from adverse counsel's questions. Point out the weaknesses and deal with them in a straightforward manner during your direct testimony. The tone of the questions will be easier to take if they come from sponsoring counsel rather than adverse counsel, and you will be given adequate time to explain why a particular piece of evidence appears the way it does. Jurors will appreciate the honesty and sincerity of the witness who tells "the whole truth," and cross-examination on these issues will likely be less damaging as a result.

10.5.3.2 Cross-examination

Cross-examination has been described as an exercise in humility, and while it is often humbling, it need not be a humiliating experience. Routine questions that come from the cross-examining attorney will include misleading questions such as whether you are being "paid for your testimony."[16] While you are probably being paid (even public servants draw a salary), you are being paid for your *time and your expertise* and not for your testimony, and that

[14] There are few circumstances more humbling than having a juror fall asleep during one's testimony. Rest assured that the sleepers do not become the leaders of the jury.

[15] A visual presenter is a high-resolution video camera mounted on a light box, which allows images of photographs, documents, and even small items of evidence to be sent to an LCD projector.

[16] A great comeback question for redirect, if the issue of payment comes up, is, "Mr. Investigator, would you sell your integrity for $2000?"

should be made clear. The jurors know that all of the professionals in the courtroom on both sides of the case are getting paid (or hoping to get paid), and they understand the concept of making a living. Appearing as a witness is simply part of an investigator's job, and juries understand that. Counsel has a choice of attacking a witness's opinion or integrity. When the attack on the witness's integrity comes, it means that counsel has run out of ways to attack the accuracy of the expert's opinion.

Witnesses have a right to expect a certain amount of protection from abusive cross-examination. If adverse counsel becomes overly provocative, take a breath before answering each question to allow sponsoring counsel an opportunity to object. If questions are phrased so that answering yes or no would mislead the jury, insist on your right to explain but try to keep explanations to the bare minimum. A witness who offers too many long-winded explanations (or speeches) is likely to draw an admonition from the judge. Accept any such admonitions with a humble, "Yes, your Honor."

A word is in order about adverse counsel's use of deposition transcripts at trial. The transcript is only *supposed* to be used when testimony from the witness stand is inconsistent with testimony given at the deposition. By the time the attorney drags out the transcript, however, and asks,"Do you remember when I took your depositions last year?" followed by "Do you also remember that at that time you took an oath and swore to tell the truth, the whole truth and nothing but the truth?" the jurors already think that the witness has contradicted himself, even if there is no difference in the testimony. It is a cheap trick to which sponsoring counsel should object. At times such as this the witness who has studied the deposition transcript can point out to counsel exactly why today's testimony comports with that offered previously.

Certainly, adverse counsel will point out every weakness in the case that he or she knows about and that may not have been discussed in your direct examination. Be prepared to concede those points and do it quickly. If a point needs to be conceded and the witness argues over it for 10 minutes, the jury will focus on that point and forget the rest of the testimony. Cross-examining attorneys enjoy nothing better than a witness who resists conceding a point that he or she has already conceded in deposition or in his or her report.

It is important to have the same demeanor toward the cross-examining attorney as you had toward the sponsoring attorney. That attorney might be a completely different person from the one you met in deposition, even if he or she has the same name. Generally, attorneys behave differently in courtrooms than they do in conference rooms. It is important to continue to answer at least some of the questions in the jury's general direction during cross-examination, although it is usually the goal of the person doing the cross-examining to keep you from doing so. If permitted to walk around the courtroom, sponsoring counsel will typically walk toward the jury box for your direct examination to make it easy for you to look at the jurors when answering. Adverse counsel is likely to walk in the other direction, making it necessary for you to look one way to hear the question and another way to answer the question. If this is uncomfortable, let it go. The jury can tell if you are being artificial. You should just be yourself and tell the truth.

Truth is the accurate and sincere description of reality. If the jurors are unable to determine which expert is more accurate, they will decide based on who they perceive to be more sincere. Sincerity is a difficult state of mind to fake, and most people know that.

The most important questions that you will be asked may not come from the attorneys. If the judge asks you a question, bear in mind that everyone will pay very close attention to the answer. In some courts, jurors are allowed to write questions and the judge, in consultation with the attorneys, decides whether to read them to the witness. In the military, members of the court-martial, who serve as the jury, are allowed to question witnesses. Unlike the attorneys, the judge and jurors have no fear of asking a question to which they do not already know the answer. Questions from the judge or jury help the witness understand which parts of his or her testimony the trier of fact considers important, or which facts or opinions were not clearly communicated.

Numerous courses are available that investigators can take to learn testifying skills. Some of these courses are offered in conjunction with the IAAI Certified Fire Investigator Program. Many forensic science laboratories stage mock trials to prepare scientists to testify. Having an experienced colleague cross-examine a potential witness is one of the more useful ways to prepare for testimony. In the end, however, the most valuable learning experience is gained in the courtroom, as painful as that might be. It is a rare expert witness who has never had a bad day in court. Such days always come to an end. The important thing is to learn from the experience and always tell the truth.

10.6 CONCLUSION

The professional practice of fire investigation requires an understanding of the stakeholders in the practice and the investigator's responsibility to those stakeholders. These responsibilities are best discharged by producing, maintaining, and following a written program committed to paper (or stored on a website) in a QA manual.

The investigator's practice should allow him to earn a living, but it should also serve the public. An investigator's business practices should be structured to be efficient, ethical, transparent, and understandable. Timely and effective communication is an essential part of all business relationships. Working for free is sometimes the best way for an investigator to invest his or her resources.

Participation in the court system as an expert witness is a challenging undertaking that must be approached with a respect for and an understanding of the way the justice system works. While investigators serving as witnesses should avoid bias or even the appearance of bias, they should be prepared to advocate for their accurate observations, carefully tested hypotheses, and well-reasoned opinions.

When testifying, it is important not only to tell the truth but also to leave the correct impression. It is the job of the expert witness to help the judge and jurors understand the evidence. Proficiency at this job, as with all jobs, comes with preparation and practice.

Review questions

1. In the hierarchy of QA documentation, at which level is an organization's mission statement?
   a. Level 1
   b. Level 2
   c. Level 3
   d. Level 4
2. In the Federal Rules of Civil Procedure, which rule governs the admissibility of expert testimony?
   a. Rule 703
   b. Rule 702
   c. Rule 716
   d. Rule 403
3. Which are the most important questions an expert witness is asked?
   a. Questions to which adverse counsel objects
   b. Questions asked on cross-examination
   c. Questions asked by the judge or by the jury
   d. Questions about an expert's qualifications
4. By what means are an expert's opinions disclosed prior to trial?
   a. Sponsoring counsel makes a "disclosure."
   b. The expert prepares a report.
   c. Adverse counsel deposes the expert.
   d. Any or all of the above, depending on the jurisdiction
5. What is the *most* important thing to remember when testifying?
   a. Do not let opposing counsel put words in your mouth.
   b. Do not let sponsoring counsel get you out on a limb.
   c. Try to avoid getting into arguments on cross-examination.
   d. Always tell the truth.

Questions for discussion

1. Who are your stakeholders?
2. What is the best strategy for dealing with an adverse lawyer who seems to be losing his or her composure?
3. What is the difference between telling the truth and leaving a correct impression? Provide an example of telling the truth but leaving an incorrect impression.
4. Why is it important to read transcripts of your testimony?
5. Why should an investigator be deeply involved in the preparation of his or her direct examination?

References

1. Goult, R. (1977) Quality System Documentation, in Peach, R., (Ed.), *The ISO 9000 Handbook*, Irwin, Chicago, IL, p. 315.
2. ISO 17025 (2017) *General Requirements for the Competence of Testing and Calibration Laboratories*, International Organization for Standardization, Geneva, Switzerland, 2017.
3. ANAB (2017) Accreditation Manual for Forensic Service Providers. Available at https://anab.qualtraxcloud.com/ShowDocument.aspx?ID=7183 (last visited January 8, 2018).
4. Murley, C. (2015) Three tips for a smooth ISO 17025 accreditation process, *Quality Magazine*. Available at https://www.qualitymag.com/articles/92810-three-tips-for-a-smooth-iso-17025-accreditation-process (last visited January 8, 2018).
5. Stolorow, M. (2014), Overview of NIST activities in the forensic sciences, Presentation to the American Academy of Forensic Sciences. Available at https://www.nist.gov/sites/default/files/documents/forensics/AAFS-Overview-of-NIST-Activities-in-FS-2014-FINAL.pdf (last visited on January 6, 2018).
6. IAAI (2015), The International Association of Arson Investigators Endorses the use of Multidiscipline Science Review Panels. Available at https://www.firearson.com/Publications-Resources/Fire-Investigation-Resources/Multidiscipline-Science-Review-Panels.aspx (last visited January 8, 2018).
7. California Association of Criminalists, (2015) The Code of Ethics of the California Association of Criminalists, Section IV. B. Available at http://www.cacnews.org/membership/California-Association-of-Criminalists-Code-of-Ethics-09-23-2015.pdf (last visited January 8, 2018).
8. ABA Committee on Ethics and Professional Responsibility (1987) Formal Op. 87-354, Lawyer's Use of Medical–Legal Consulting Firm.
9. Reis, J. (2015), Expert challenges and the revised NFPA 1033, *The Fire and Arson Investigator*, 66(1):30.
10. Hand, L. (1902) Historical and practical considerations regarding expert testimony, *Harvard Law Review* 40(54):15.

Index

Note: Page numbers in italic and bold refer to figures and tables respectively.

9781138037014